Biochemistry of Peptide Antibiotics

Biochemistry of Peptide Antibiotics

Recent Advances in the Biotechnology of β-Lactams and Microbial Bioactive Peptides

Editors
Horst Kleinkauf · Hans von Döhren

Walter de Gruyter Berlin · New York 1990

Editors

Horst Kleinkauf, Professor Dr.
Hans von Döhren, Priv. Doz., Dr.
Institut für Biochemie und Molekulare Biologie
Technische Universität Berlin
Franklinstraße 29
D-1000 Berlin 10
Federal Republic of Germany

Library of Congress Cataloging-in-Publication Data

Biochemistry of peptide antibiotics : recent advances in the biotechnology of β-lactams and microbial bioactive peptides / editors, Horst Kleinkauf, Hans von Döhren.
Bibliography: p.
Includes index.
ISBN 0-89925-551-5 (U.S.)
1. Beta lactam antibiotics — Synthesis. 2. Microbial peptides — Synthesis. 3. Beta lactam antibiotics — Biotechnology. 4. Microbial peptides — Biotechnology. I. Kleinkauf, Horst, 1930—. II. Döhren, Hans von, 1948—
RS431.B48B56 1989 615'.329—dc20 89-12083

Deutsche Bibliothek Cataloging in Publication Data

Biochemistry of peptide antibiotics : recent advances in the biotechnology of β-lactams and microbial bioactive peptides / ed. Horst Kleinkauf; Hans von Döhren. — Berlin; New York : de Gruyter, 1990
ISBN 3-11-011928-5 (Berlin)
ISBN 0-89925-551-5 (New York)
NE: Kleinkauf, Horst [Hrsg.]

Printed on acid free paper.

Printed in Germany.
Typesetting: Arthur Collignon GmbH, Berlin — Printing: Gerike GmbH, Berlin — Binding: Lüderitz & Bauer GmbH, Berlin.

Foreword

Amino acid-derived compounds, as modified or linear peptides, continue to have a high impact in research as well applied sciences, as in pharmacy. This impact can be traced to a number of unique properties. Most important, these compounds contain a linear sequence of amino acids, that permits easily sequence variations in the study of structure-activity relations. The enzymology of peptide formation is well advanced, and applications of enzymes in production processes are emerging. A more detailed knowledge of the cellular targets permits new screening approaches and more extensive chemical variation studies. Such knowledge is intimately connected to the advancement of biochemistry, also made possible with the application of bioactive peptides as research tools.

In this monograph we have assembled major lines of research of the past years, that in each case has reached a certain level of completion. By this we provide an overview of the current directions of the field, that should provide a useful orientation for running projects as well as those to be planned. Current developments in fast moving projects have been included in an introductory chapter. Here we summarize aspects of biosynthesis of peptides formed by ribosomal and nonribosomal pathways, as well as compounds recently detected in new sources or by their unusual effects. Especially important is the progress made in the analysis and cloning of enzymatic pathways, such as β-lactam biosynthesis. This as a rapidly developing field is discussed here together with the advances in cloning of peptide synthetases. The best known multienzyme system, the gramicidin S synthetase, has been treated by Joachim Vater. A more general approach in the multienzymic field is given by Andreas Billich and Rainer Zocher on N-methylated peptide bonds, a feature very common in cyclic structures, and contributed by an integrated methylase function within multienzymic synthetases. A general consideration of enzymatically catalysed peptide bond formation is carried out by Volker Kasche and Günther Michaelis. Their conclusions already have found applications in synthetic chemistry of enzymic acylation reactions. For the genetics of an enzymic pathway Volkmar Braun gives a description of siderophore systems, as models for peptide bond containing metabolites.

Turning then to the more structural aspects, the following five chapters summarize work on the elucidation of structures and effects, the application of screens, and chemical modification studies in structure-function work. Hideo Ono and Setsuo Harada present an overview of the results of their screening for new β-lactams and β-lactam like antibiotics. Giancarlo Lancini and Bruno Cavalleri give an account of the extensive search for D-analyl-D-alanine binding glycopeptides of the vancomycin type. Phytotoxic peptides are reviewed by Jonathan Walton representative for the many highly toxic peptides frequently discovered by their ecological impact. The structural variation of antibiotics, a key problem in compound development, has been over the years most extensively executed with cyclic peptides of the gramicidin S type. This work is considered in perspective by Michinori Waki and

Nobuo Izumiya. A highlight in synthetic peptide chemistry has been the total synthesis of the potent immunomodulator cyclosporin. Now, as the structural work is in a more advanced state, Roland Wenger and Hans Fliri describe the chemical and biochemical implications in detail.

The following two chapters are devoted to Hamao Umezawa's school pioneering in the introduction of peptides as modulators of enzyme activity, with vast implications in disease treatment. Tomohisa Takita and Yasuhiko Muraoka summarize the work on biosynthesis and total synthesis of bleomycin, a highly complex glycopeptide, from which by extensive screenings useful derivatives for cancer chemotherapy have been developed. Takaaki Aoyagi, who has introduced numerous microbial proteinase inhibitors, focuses on the biochemical studies, that provide a basis for possible pharmacological applications.

The final contributions are concerned with the production of peptides. The structural variation of peptides by amino acid exchange upon substrate feeding has been investigated for neoviridogrisein by Yasushi Okumura. Immobilization of biocatalysts, a key step in commercial production processes, has been reviewed by Erick Vandamme.

As an appendix, a table of well characterized compounds has been compiled, arranged according to their structural properties, with some notes on their biochemical actions. This approach of compilation, that we have suggested earlier, should aid the researcher in the selection of comparable compounds, and permit search for structural homologies.

We are indebted to all our colleagues participating in this project, who took the burden of again presenting results and evaluating their fields in perspective. We are sure that this book will be a valuable tool for those working in the peptide field trying to correlate chemistry and biochemistry at various levels of application. We also thank the publishers for their interest and support.

Berlin, January 1990

Horst Kleinkauf
Hans von Döhren

Contents

Chapter 1
Bioactive Peptides – Recent Advances and Trends

Horst Kleinkauf and Hans von Döhren

1. Peptides of Ribosomal Origin 2
   1.1 Protein antibiotics 2
   1.2 Peptide hormones and neuropeptides 2
   1.3 Lantibiotics 4
2. Enzymatically Formed Peptides and Polypeptides 6
   2.1 Single enzymes or multienzyme 7
   2.2 Size of peptide formed 7
   2.3 Structure of multienzymes 8
   2.4 The thiotemplate mechanism 8
   2.5 Enzymic synthesis of peptides 12
3. Genetics of Peptide Synthetases 14
   3.1 Siderophores 14
   3.2 Bacillus peptides: Gramicidin S, tyrocidine, bacitracin 16
   3.3 Genes involved in β-lactam-biosynthesis 17
   3.4 More unified approaches to biosynthetic pathways 18
4. New and Old Sources of Peptides 20
5. Conclusions 23

References 26

Chapter 2
Gramicidin S Synthetase

Joachim Vater

1. Introduction 33
2. Purification of Gramicidin S Synthetase 34
3. Physical and Biochemical Properties of Gramicidin S Synthetase 36
4. Reaction Mechanism of the Biosynthesis of Gramicidin S and Product Patterns of Gramicidin S Synthetase 37
5. Reaction Centers for Substrate Activation 40
6. Inhibitor Studies 43
7. Structural Aspects 46
8. Invesigation of Gramicidin S Negative Mutants 49
9. Prospects 50

References 51

Chapter 3
Formation of *N*-methylated Peptide Bonds in Peptides and Peptidols
Andreas Billich and Rainer Zocher

1. Introduction 57
2. Occurrence of *N*-methylated Peptide Bonds 57
3. Biosynthesis of *N*-methylated Peptides 60
4. The *N*-methyltransferase Function of Enniatin Synthetase 62
4.1 *N*-methylation of enzyme-bound amino acids 63
4.2 Kinetic properties of the *N*-methyltransferase function 64
4.3 Inhibition studies of the *N*-methyltransferase function 64
4.4 Photoaffinity labeling of enniatin synthetase 68
4.5 Monoclonal antibodies to enniatin synthetase 70
5. The *N*-methyltransferase Function of other Peptides Synthetases 72
5.1 Beauvericin synthetase 72
5.2 Cyclosporin synthetase 73
5.3 Actinomycin synthetases 75
6. Conclusions 76
References 77

Chapter 4
Peptide Bond Synthesis by Enzyme-Catalyzed Acyl-Transfer
Volker Kasche and Günther Michaelis

1. Introduction 81
2. Mechanism of the Kinetically Controlled Peptide Synthesis 83
3. Maximum Yields in the Kinetically Controlled Peptide Synthesis 85
4. Selecting the Optimal Enzyme 89
5. Yield Controlling Factors in the Synthesis of a Peptide Bond with One Enzyme 94
5.1 Protection of the P_1' carboxyl group; stereospecifity 94
5.2 pH value 95
5.3 Temperature 96
5.4 Ionic strength 97
5.5 Solvent composition 97
6. Conclusions 98
References 99

Chapter 5
Genetics of Siderophore Biosynthesis and Transport
Volkmar Braun

1. Structure and Function of Siderophores 103
2. F^{3+}-Aerobactin Transport System 104

2.1 Biosynthesis of aerobactin 104
2.2 Fe^{3+}-aerobactin transport 111
3. Enterochelin Biosynthesis and Fe^{3+}-Enterochelin Transport 113
4. Fe^{3+}-Dicitrate Transport 115
4.1 Transport 115
4.2 Regulation 117
5. Regulation of Iron Transport Systems 118
5.1 Regulation of gene expression 118
5.2 Regulation of the Fe^{3+}-siderophore receptor activity 120
6. Antibiotics Containing Siderophore Structures 122
6.1 Naturally occurring sideromycins 122
6.2 Synthetic antibiotics with siderophore structures 123
7. Concluding Remarks 124
Acknowledgements 125
References 125

Chapter 6
Discovery of new β-Lactam and β-Lactam like Antibiotics from Bacteria

Hideo Ono and Setsuo Harada

1. Introduction 131
2. Screening Method and Producing Organisms 134
3. Fermentation 135
4. Isolation Procedure 135
5. Chemical Characterization 140
6. Structure Determination 142
7. Biological Activities of Cephabacin, Formadicin and Lactivicin 147
8. Conclusion 155
References 155

Chapter 7
Glycopeptide Antibiotics of the Vancomycin Group

C. Lancini and B. Cavalleri

1. Introduction 159
2. Producing Organisms 161
3. Chemistry 161
4. Biosynthesis 167
5. Mechanism of Action 169
5.1 Effect on growing cultures 169
5.2 Studies with cell free systems 169
5.3 Action at molecular level 170
6. Relation of Antimicrobial Activity and Mechanism of Action 171
References 172

Chapter 8
Peptide Phytotoxins from Plant Pathogenic Fungi

Jonathan D. Walton

1. Introduction 179
2. Host-Selective Toxins 181
2.1 AM-toxin 181
2.2 Host-specific toxin from Alternaria brassicae 183
2.3 Victorin 184
2.4 HC-toxin 188
2.5 PC-toxin 192
3. Non-Selective Toxins 193
3.1 Tentoxin 194
3.2 Enniatins 196
3.3 Peptidic siderophores 196
References 198

Chapter 9
Chemical Synthesis and Bioactivity of Gramicidin S and Related Peptides

Michinori Waki and Nobuo Izumiya

1. Introduction 205
2. Structures of Gramicidin S and Related Natural Peptides 206
2.1 Gramicidin S and its congeners 206
2.2 Tyrocidines 207
2.3 Gratisin 208
3. Chemical Synthesis of Gramicidin S and Related Peptides 208
3.1 Synthesis of linear precursor peptides 208
3.2 Synthesis of cyclic peptides 209
4. Structure-Activity Relationships of Gramicidin S, Tyrocidines and Gratisin 212
4.1 Gramicidin S 212
4.1.1 Funktion of the amino acid residues at positions 1 and 1′ . . . 216
4.1.2 Funktion of the amino acid residues at positions 2 and 2′ . . . 217
4.1.3 Function of the amino acid residues at positions 1,1′ and 3,3′ . 217
4.1.4 Function of the amino acid residues at positions 3 and 3′ . . . 217
4.1.5 Function of the amino acid residues at positions 4 and 4′ . . . 218
4.1.6 Function of the amino acid residues at positions 5 and 5′ . . . 218
4.1.7 Summary by "sidedness" hypothesis 218
4.2 Tyrocidines 219
4.3 Gratisin 221
5. Conformations of Gramicidin S and Tyrocidines 221
5.1 Optical rotatory dispersion (ORD) and circular dichroism (CD) . . . 222
5.2 Nuclear magnetic resonance (NMR) 223
5.3 Other conformational analyses 224
6. Design of Highly Active Analogs of Gramicidin S 224

7. Active Analogs of Gramicidin S Against Gram-negative Bacteria 228
8. Mechanism of Antimicrobial Action of Gramicidin S 230
8.1 Interactions with model membranes 230
8.2 Interactions with bacterial membranes 230
9. Concluding Remarks 234
References 234

Chapter 10
Cyclosporine: Synthetic Studies, Structure-Activity Relationships, Biosynthesis and Mode of Action

Hans G. Fliri and Roland M. Wegner

1. Introduction 246
2. Cyclosporine 250
2.1 Isolation 250
2.2 Structure determination 250
2.3 Other naturally occurring cyclosporines. Nomenclature 252
3. Total Synthesis 253
3.1 Synthesis of MeBmt 254
3.1.1 The Wenger synthesis 254
3.1.2 The Evans synthesis 256
3.1.3 The Seebach synthesis 258
3.1.4 The syntheses of Rich 258
3.1.5 The Schmidt synthesis 259
3.2 Synthesis of cyclosporine 262
4. Chemical Modification of Cyclosporine 266
5. Structure-Activity Relationships 267
6. Conformation of Cyclosporine in the Crystal, in Aprotic Solvents and in Biological Fluids 270
6.1 Crystal and solution conformation 270
6.2 Cyclosporine conformation in biological fluids 271
6.3 Immunochemical study of the cyclosporine conformation in aqueous medium 271
6.4 Immunological studies of cyclosporine and their interpretation in terms of the conformation of the compound in aqueous media 271
6.5 Calculations using molecular dynamics programs indicate that the crystal structure conformation is in water more stable than any others 273
7. Biosynthesis 274
8. Mechanism of Action of Cyclosporine 276
8.1 Effects on T cells 277
8.2 Effects on B cells 279
8.3 Possible intracellular targets for cyclosporine 280
9. Conclusions and Outlook 280
References 282

Chapter 11
Biosynthesis and Chemical Synthesis of Bleomycin

Tomohisa Takita and Yasuhiko Muraoka

1. Introduction 289
2. Biosynthesis and Semisynthesis of Unnatural Congeners 292
   2.1 Biosynthesis 292
   2.2 Semisynthesis 294
3. Biosynthesis and Chemical Synthesis of Building Blocks of Bleomycin Molecule 295
   3.1 Biosynthesis of unusual amino acids 295
   3.2 Chemical synthesis of unusual amino acids 297
      3.2.1 Synthesis of pyrimidine-containing amino acid (PBA) 297
      3.2.2 Synthesis of *erythro*-β-hydroxy-L-histidine 298
      3.2.3 Synthesis of (2S,3S,4R)-4-amino-3-hydroxy-2-methylpentanoic acid (AHM) 299
      3.2.4 Synthesis of 2′-(2-aminoethyl)-2,4′-bithiazole-4-carboxylic acid (ABC) 300
4. Biosynthetic and Chemical Construction of Bleomycin Molecule 300
   4.1 Biosynthesis of peptide part of bleomycin: deglycobleomycin 300
   4.2 Chemical synthesis of deglycobleomycin 302
   4.3 Chemical synthesis of the disaccharide moiety for total synthesis of bleomycin 303
   4.4 Total synthesis of bleomycin 304
5. Conclusion 306
References 307

Chapter 12
Small Molecular Protease Inhibitors and Their Biological Effects

Takaaki Aoyagi

1. Introduction 311
2. Protease Inhibitors 312
3. Inhibitors Against Endopeptidases 313
4. Enzymatic Activities of Cellular Membrane 317
5. Inhibitors Against Cell Surface Enzymes 320
   5.1 Inhibitors against exopeptidases 320
   5.2 Inhibitors against enzymes on cellular membrane 328
6. Enzymatic Changes in Various Pathologic Conditions 330
   6.1 Roles of catabolic enzymes in muscular dystrophy 330
      6.1.1 Murine muscular dystrophy 330
      6.1.2 Duchenne muscular dystrophy 335
   6.2 Spontaneously hypertensive rats and aminopeptidases 337
   6.3 Immunologic diseases and hydrolytic enzymes 342

7. Modification of Metabolic Homeostasis Caused by Low-Molecular-Weight Enzyme Inhibitors . . . 346
7.1 Enzymatic oscillations caused by bestatin . . . 346
7.2 Influence of angiotensin-converting enzyme inhibitor, foroxymithine 349
7.3 Effects of bestatin on hydrolytic enzymes in progressive muscular dystrophy . . . 352
8. Pharmacologic Applications of Enzyme Inhibitors . . . 356
9. Summary . . . 358
Acknowledgements . . . 359
References . . . 359

Chapter 13
Directed Biosynthesis of Neoviridogriseins

Yasushi Okumura

1. Introduction . . . 365
2. Selective Production of Neoviridogriseins by Directed Biosynthesis . . . 367
2.1 Effect of exogenous amino acid on the production of neoviridogriseins 368
2.2 Specific accumulation of neoviridogrisein II by inhibition of proline hydroxylation reaction . . . 371
2.3 Specific accumulation of neoviridogrisein II by the mutant blocked in *trans*-4-hydroxy-L-proline formation from L-proline . . . 374
3. Summary . . . 377
References . . . 377

Chapter 14
Biochemical Genetical and Biotechnical Aspects of Antibiotic Production via Immobilised Biocatalysis

Erick J. Vandamme

1. Novel Antibiotic Compounds . . . 379
2. Novel Production Ways for Antibiotics . . . 381
3. Total Enzymatic Synthesis of Antibiotics . . . 381
4. Partial Enzymatic Synthesis of Antibiotics . . . 386
5. Antibiotic Fermentation via Immobilised Viable Cells . . . 390
6. β-Lactam-Antibiotic Bioconversions with Immobilised Biocatalysts . . . 392
6.1 Penicillin bioconversions . . . 392
6.2 Cephalosporin bioconversions . . . 397
7. Chiral Side-Chain Production for β-Lactam Antibiotics Synthesis . . . 402
8. Bioconversion of Novel β-Lactam Antibiotics . . . 403
9. Bioconversion of Other Important Antibiotics . . . 404
10. Perspectives . . . 406
References . . . 407

Chapter 15
Compilation of Peptide Structures – A Biogenetic Approach
Hans von Döhren

Introduction . . . 411
General biosynthetic mechanism . . . 411
Compound data . . . 412
Abbreviations and Symbolism . . . 412
Configuration . . . 412
Bonds . . . 412
Abbreviated nomenclature used . . . 412
Cyclic structures . . . 413
Side chain modifications . . . 413
Chain length . . . 413
Combinations of chains . . . 413
Piperidine-type cyclizations . . . 414
Unusual constituents . . . 414
Compilation . . . 415
Index of table . . . 495
List of abbreviations . . . 502
a) compounds . . . 502
b) sources . . . 507

List of Contributors . . . 509
Index . . . 511

Chapter 1
Bioactive Peptides – Recent Advances and Trends

Horst Kleinkauf and Hans von Döhren

1. Peptides of Ribosomal Origin
   1.1 Protein antibiotics
   1.2 Peptide hormones and neuropeptides
   1.3 Lantibiotics
2. Enzymatically Formed Peptides and Polypeptides
   2.1 Single enzymes or multienzyme
   2.2 Size of peptide formed
   2.3 Structure of multienzymes
   2.4 The thiotemplate mechanism
   2.5 Enzymic synthesis of peptides
3. Genetics of Peptide Synthetases
   3.1 Siderophores
   3.2 Bacillus peptides: gramicidin S, tyrocidine, bacitracin
   3.3 Genes involved in β-lactam-biosynthesis
   3.4 More unified approaches to biosynthetic pathways
4. New and Old Sources of Peptides
5. Conclusions

References

Metabolites derived from amino acids, like penicillin, gramicidin S, and bacitracin, represent the classical type of antimicrobial agent, related to their effects on microbial cells. With the development of biochemistry, many effects and actions of these and other peptide antibiotics have been discovered, and research has been directed into more detailed analysis of biosynthesis and mode of action of these compounds. Our rising understanding of the basic life processes owes much to the frequent uses of interference of small compounds with essential cellular operations.

There have been for quite obvious reasons attempts to a more general terminology instead of *peptide antibiotics* to use *bioactive peptides. Bioactive* as a general term includes *antibiotic* as a property, but eventually every compound has some *bioactive* character. For several enzyme effectors antimicrobial properties have not yet been detected, but traditionally such compounds have been grouped in the context of antibiotics. For that reason we have maintained *peptide antibiotics* as a general title for this material.

1. Peptides of Ribosomal Origin

1.1 Protein antibiotics

The structures of compounds treated in the context of peptide antibiotics can be arranged in various ways for presentation. Ribosomally manufactured linear peptides, often with sizes well above 100 amino acid residues, are referred to as protein antibiotics [1]. These polypeptides are composed of the 20 protein amino acids, but may contain highly reactive chromophores [2, 3], as in neocarzinostatine, macromomycin, and actinoxanthin. These facilitate covalent attachment and modification of DNA and lead to antitumor and mutagenic properties. The labile chromophore is specifically bound and protected by its carrier protein, but "free" *second generation* chromophores esperamicin from *Actinomadura verrocosospora,* and calichemicin from *Micromonospora echinospora* have been detected as highly potent antitumor drugs [4, 5]. The action of the neocarzinostatin chromophore (NCS-chrom) has been studied in detail [6, 7, 8]. NCS-Chrom (Figure 1) binds to DNA by intercalation with a preference for the minor groove and AT-sequences [9]. A thiol-activated form of the bicyclo(7,3,0)dodecadiyne system generates a radical, binds dioxgen to generate a peroxyl radical leading to DNA strand breaks [10]. Under anaerobic conditions, a stable covalent chromophore-DNA sugar adduct is formed. As is shown in Figure 2, there are considerable sequence homologies between these three *apo*-proteins [11–13], produced by strains of *Streptomyes.*

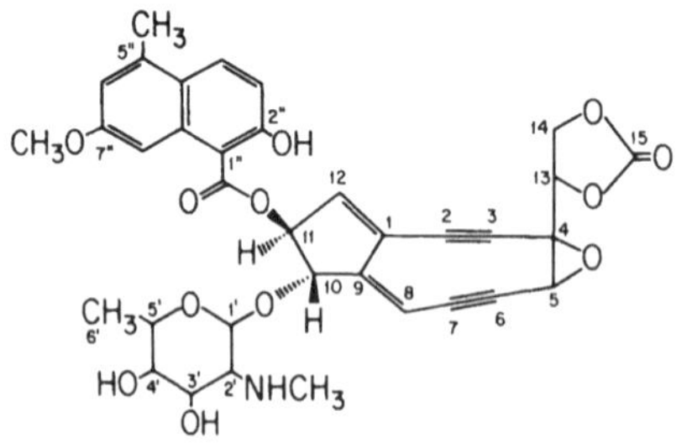

Figure 1 Structure of the neocarzinostatin chromophore NCS-Chrom.

A screening program aimed at protein antibiotics in broths of Actinomyctes by Miyashiro and Udaka [14, 15] has led to several compounds referred to as AN-peptides [16, 17]. Again, for their various actions including antitumor and mutagenic properties the presence of a chromophore constituent appears to be essential.

1.2 Peptide hormones and neuropeptides

Well known are various peptide hormones with frequent N-terminal cyclization to pyroglutamyl-residues, or C-terminal amidation [18–20]. Such an amidation reaction connected to the cleavage of a terminal glycine residue has been characterized

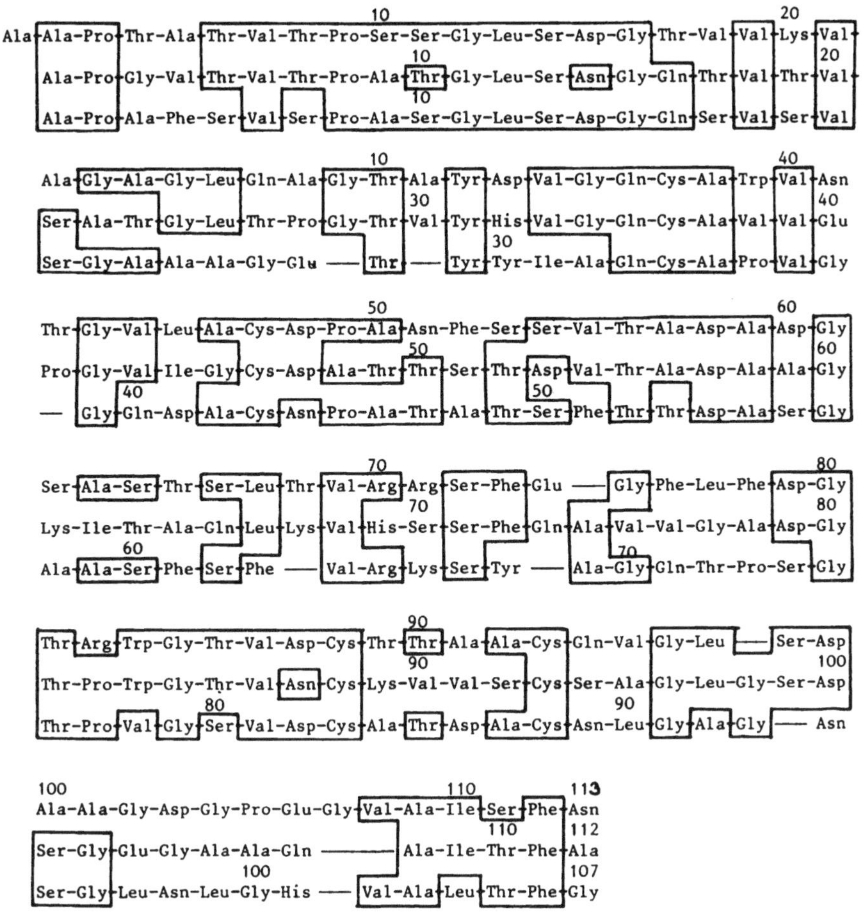

Figure 2 Sequences of the antitumor apo-proteins neocarzinostatin, macromomycin, and actinoxanthin.

as an ascorbate dependent peptidylglycine α-amidating monooxygenase [21–25]. Pyroglutaminyl formation of glutaminyl peptides may occur by nonenzymatic conversion [27], but a plant enzyme has been known [26], and recently a glutaminyl cyclase has been detected in pituitary extracts [28].

Polypeptides with antibacterial properties, like cecropins or attacins, are well known from insects to be inducible upon infection [29]. Vertebrate antimicrobial peptides like defensins, thought to be operating within phagocytic cells [30], and amphibian skin peptides, like magainins [31] and caeruleins [32], show some resemblance to peptide hormones and neurotransmitters [33]. As has been suggested for the 12-peptide bactenecin, which has been isolated from cytoplasmic granules of bovine neutrophils, such antibiotics may originate from larger precursor peptides [152]. Still not understood is the formation of a D-alanine in frog skin dermorphin originating from an all-L chain of the precursor [34].

The treatment of neuropeptides and related regulatory peptides is not in the scope of this review, but it should be noted, that the employed immunodetection procedures, which led to many structures in a variety of organisms, have already found application in the antibiotic field, too.

1.3 Lantibiotics

Modifications of ribosomally derived linear structures beyond peptide bond cleavage and terminal cyclization or amidation have been found in bacterial peptides. A major group containing the thioether amino acids *meso*-lanthionine and 3-methyllanthionine have been termed *lantibiotics* [35]. These peptides contain dehydroamino acids, D-amino acids, and sulfide rings of various sizes, and are formed by proteolytic cleavage of modified precursors (Figure 3). The subtilin precursor contains a

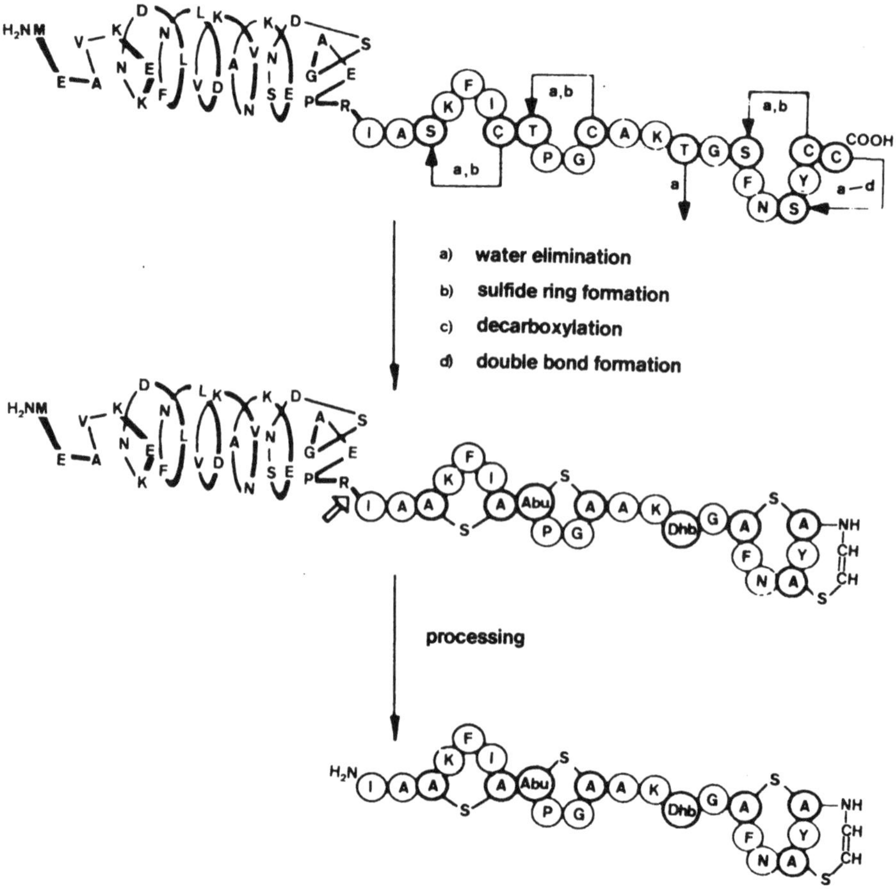

Figure 3 Hypothetical modification and processing of pro-epidermin to the lantibiotic epidermin (Adapted from ref. 35).

24-residue leader region fused to a 32-residue structural region [36]. The epidermin prepeptide contains an N-terminal 30-peptide region with an assumed partially amphiphilic α-helical conformation, to be cleaved at an Arg-Ile bond for secretion of the 21-peptide amide antibiotic [35]. The postulated biosynthetic events are the formation of dehydro-residues from serine and threonine by dehydration followed by addition of Cys-thiols to the respective C-C double bonds with configurational inversion leading to sulfide rings of various sizes (Figure 4) [37].

1

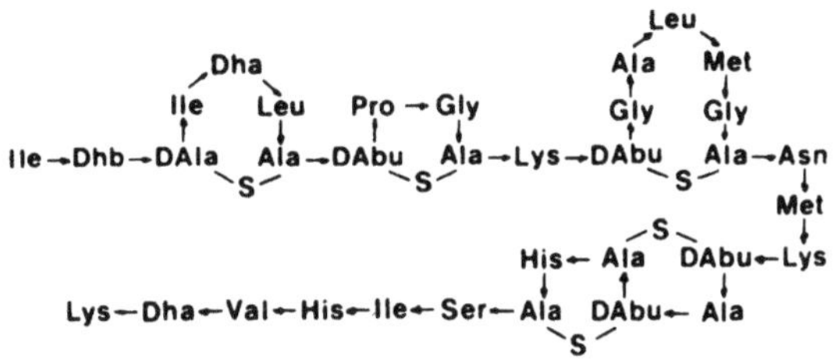

2

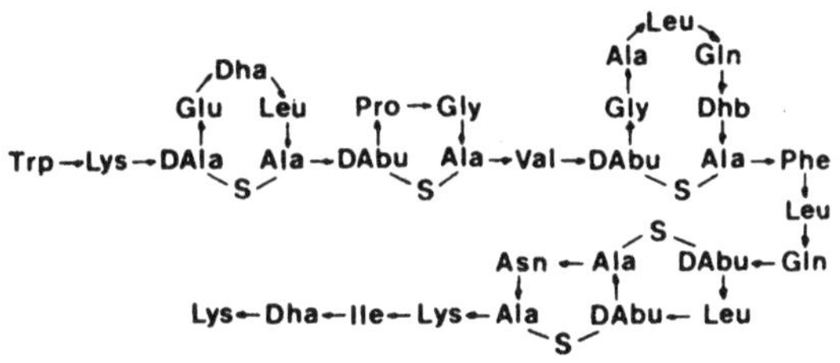

3

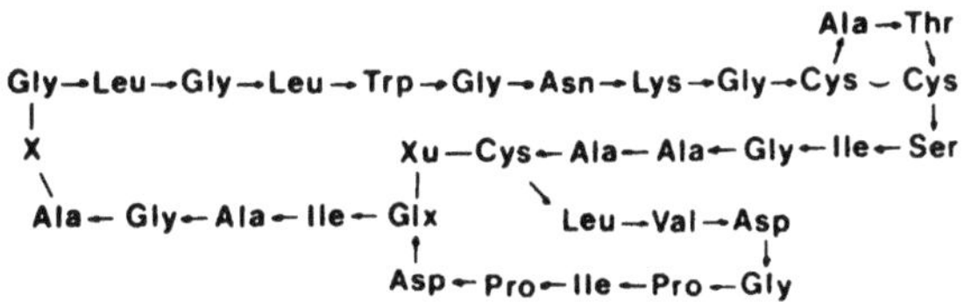

4

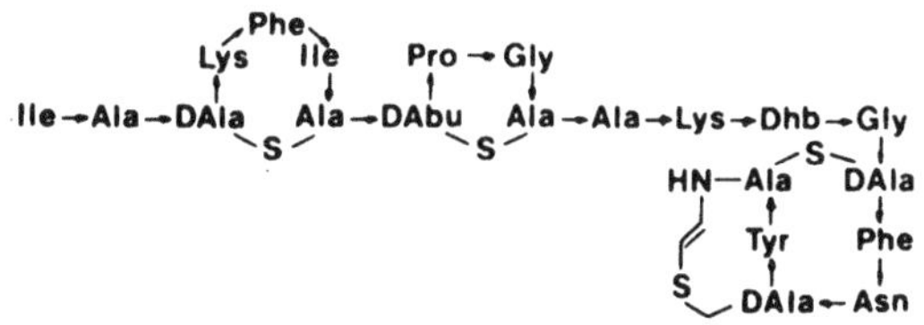

5

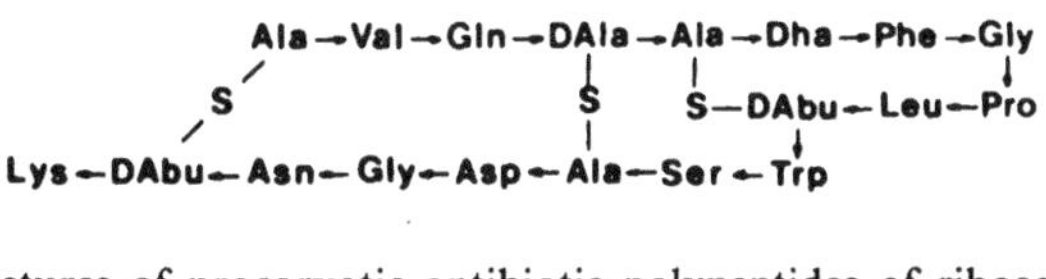

Figure 4 Structures of procaryotic antibiotic polypeptides of ribosomal origin. (1) Nisin, used as a food preservative, produced by *Streptococcus lactis*; (2) subtilin and (3) subtilosin, antibiotics produced by *Bacillus subtilis* (X and X_u represent unknown compounds); (4) epidermin, a peptide found effective against *Propionebacterium acne*, and (5) ancovenin, an inhibitor of angiotensin coverting enzyme.

While the subtilin gene has been isolated from *Bacillus subtilis ATCC 6633* genomic DNA [36], the epidermin sequence has been detected on a 54-kb-plasmid from *Staphylococcus epidermidis* [35]. Evidence for an extrachromosomal location for nisin production had also been obtained for *Streptococcus lactis*, where antibiotic biosynthesis and resistance, together with sucrose fermenting ability, had been transferred in a conjugation like process involving a pSN-plasmid (named for *s*ucrose and *n*isin) [38–40]. The nisin structural gene, as well as those of pep 5 and of gallidermin have all been identified on streptococcal and staphylococcal plasmids [41, 42]. Biosynthetic enzymes catalysing specific cyclization reactions could be of considerable importance in peptide synthesis. Other unusual cyclic structures are contained in ancovenin, an inhibitor of angiotensin converting enzyme found in *Streptomyces* [43], Ro 09-0198 [44a] or lanthiopeptin [44b], immunostimulating and active against *Herpes simplex* virus, from *Streptoverticillium griseoverticillatum* identical to cinnamycin and just differing by a Lys residue from duramycin, both produced by strains of *Streptomyces cinnamoneus*, or have not been elucidated completely in subtilosin A, a cyclic 32-peptide isolated from *Bacillus subtilis* 168 [44]. This latter peptide is formed at the end of vegetative growth and is one of the major antibiotics of at least ten produced by this strain in sporulation medium. Similar compounds have been detected in the subtilin producing strain ATCC 6633 and in a strain of *Bacillus natto*.

2. Enzymatically Formed Peptides and Polypeptides

The enzymatic synthesis of polypeptides may have some implications to violate Crick's central dogma, that *information is lost, once it has passed from nucleic acids into protein structure*. However, the information of construction of the protein template for each polypeptide or peptide is lost indeed, and there is no evidence of any role of a product on the assembly of its template. We would have to trace a case where a protein template, consisting of enzymic units each controlling the addition of a single amino acid are assembled either at the gene or transcriptional level upon informational coupling with the peptide product formed. From the genetic investigations carried out so far, there is no evidence of such kind of interactions. Instead, genes and multigenes code respective enzyme systems, multienzymes, or multienzyme systems; instead of versatility and flexibility, there is more evidence for an interchange of complete biosynthetic enzyme sets at the gene level.

It is not possible from the given structure of a peptide to predict its biosynthetic path. From the past years a number of essential cases has been resolved, and thus permit definite experimental approaches.

2.1 Single enzymes or multienzyme

The first enzymic system resolved, the formation of glutathione, has long been thought to be resembled in the biosynthesis of the β-lactam precursor tripeptide δ-(α-L-aminoadipyl)-L-cysteinyl-D-valine (ACV), from the structural homology to γ-glutamyl-L-cysteinyl-glycine (GSH). So in analogy to γ-Glu-Cys-synthetase and GSH-synthetase two enzymes each catalyzing a single peptide addition where sought for. Instead, the two peptide bonds and the epimerization of valine are catalyzed by a single multienzyme, ACV-synthetase [45–47]. This could hardly have been predicted, except that no enzyme has been known accepting an L-amino acid, and adding the epimer to a peptide. Only this extra function could have been a clue in prediction of a multienzyme.

2.2 Size of peptide formed

The *size* of a peptide is no indication of its biosynthetic origin. The classical case of searching in a wrong direction was the search for a multienyme forming the peptide hormone pyroGlu-His-GlyNH_2 (TRH) in analogy to the just characterized gramicidin S-synthetase. Although evidence for such an enzyme system had been

Table 1 Peptaibols and related peptides.

Name	Producer organisms	Structures[4])
Antiamoebins[1])	*Emericellopsis poonensis* *E. synnematicola* *Streptomyces pimprina*	AcPheAib$_3$IvaGlyLeuAib$_2$HypGlnIva^{13}-HypAibProPhol
Alamethicins	*Trichoderma reesei*	AcAibProAibAlaAib^{6}AlaGlnAibVal-AibGlyLeuAibProValAib$_2$GluGlnPhol
Emericins[2])	*Emericellopsis microspora*	AcPheAib$_3$ValGlyLeuAib$_2$HypGln-IvaHyp^{14}AibPhol
Hypelcin A	*Hypocrea peltata*	AcAibProAibAlaAib$_2$GlnLeuAib-GlyAib$_3$ProValAib$_2$Gln$_2$Leuol
Leucinostatin B[3])	*Paecilomyces lilacinus*	MheMeProAhmodHyLeuAibLeu$_2$-Aib$_2$βAlaDpd
Suzukacillin	*Trichoderma reesei*	Ac(AibAla)$_3$GlnAib$_3$GlyLeuAibPro-ValAibIvaGluGlnPhol
Trichopolyn	*T. polysporum*	MedaAlaAib$_2$IleAlaAib$_2$Tda
Trichotoxins	*T. reesei*	AcAibGlyAibLeuAibGlnAib$_3$AlaAib$_2$-ProLeuAib^{16}Iva^{17}GlnValol
Zervamycins	*Emericellopsis salmosynnemata*	AcTrp^{2}Ile^{3}Gln^{4}Iva^{5}ValThrAib^{8}LeuAib-HypGlnAibHypAibProPhol

[1]) identical with Tü165, [2]) samarosporin or stilbellin, [3]) P168 or antibiotic 1907, [4]) known replacements are ^{2}Ile-Val, ^{3}Glu-Gln, ^{4}Iva-Aib, ^{5}Val-Ile, ^{6}Ala-Aib, ^{8}Leu-Val, ^{13}Hyp-Pro, ^{14}Aib-Ala, ^{16}Iva-Aib, ^{17}Glu-Gln

Abbreviations: Ahmod 2-amino-6-hydroxy-4-methyl-8-oxodecanoic acid, Aib α-aminoisobutyric acid, Dpd (2*S*)-N^1-methylpropane-1,2-diamine, HyLeu hydroxy-Leu, Hyp 4-hydroxy-Pro, Iva isovaline (α-ethylalanine), Meda (*R*)-2-methyldecanoic acid, Mhe (4*S*)-(2*E*)-4-methylhex-2-enoic acid, -ol carboxyl reduced to alcohol, Tda trichodiaminol.

obtained [48], the ribosomal pocessing/modification pathway is now established [18]. In search for the *limit* of size of a peptide formed by an enzymic system, Mohr and Kleinkauf studied the peptaibol alamethicin, a modified linear 20-peptide [49, 50]. An *in vitro* system had been established with a multienzyme fraction, although the number of multienzyme species has not been detected. So far, no peptaibols exceeding 20 amino acids in sequence are known (Table 1). *Bacillus brevis*, the classical source of the first peptide antibiotics tyrocidine and gramicidin S [53], has also been reported to synthesize peptides of 29 and 34 amino acids containing ornithine [52]. We may thus assume, that enzyme systems may well operate beyond 30 definite sequential steps. The limits of the multienzymic systems are quite obvious. They permit the assembly of non-protein constituents, especially cyclic structures. Since each peptide is formed as a linear molecule first, formation of secondary structure may interfere with the addition of amino acids. Each elongation step involves a transport of the growing chain, that may not, as in the ribosomal system, pass an *exit tunnel* before folding. The size of the enzymic machinery involved, as compared to the ribosomal system (with amino acid activating enzymes, tRNA, factors, ribosome, mRNA), is quite similar in the range of 10 to 30 amino acids. Since the enzyme machinery has not been shown to repeatedly use an activation site, but instead has to repeat each enzyme unit in the protein template, soon the ribosomal system turns out to be more effective from an economic point of view.

2.3 Structure of multienzymes

The question of *subunits* of the multienzymes involved in peptide biosynthesis has been clearly resolved in favour of integrated enzyme systems [54]. However, it is not understood why a certain set of sequential reactions is integrated. From proteolytic nicking studies, the integrated structure seems to be essential for the functioning of the sequential elongation mechanism [55]. From the bacterial systems studied, gramicidin S, tyrocidine, gramicidin, bacitracin, or mycobacillin, an apparent limit of six elongation steps in an integrated multienzyme has been predicted (Figure 5) [56]. However, the recent studies of the cyclosporin synthetase from *Beauveria nivea* (the former *Tolypocladium inflatum*) have revealed a single enzyme fraction catalyzing the formation of a cyclic 11-peptide [57]. Still it has not been established if all elonagtion steps are integrated in a single multienzyme structure, but certainly multienzyme structures may turn out much more complex than presently discussed [58].

2.4 The thiotemplate mechanism

This term, introduced by Zimmer and Laland back in 1973 [59], refers to the thiolester attachment mode of the activated amino acids as covalent intermediates. As enzyme studies progressed, Kurahashi extended it to *thiotemplate multienzyme mechanism* [60]. The vast majority of enzymes investigated so far (Table 2) is in

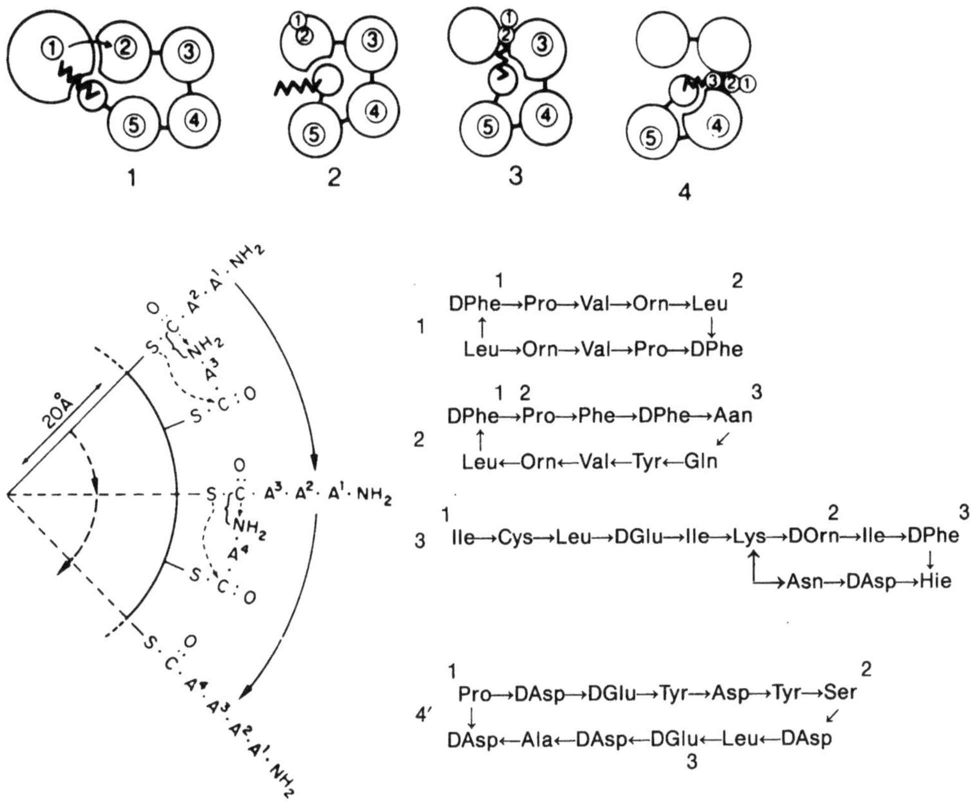

Figure 5 Organization of multienzymic templates forming peptides.
Top: A model of gramicidin S synthetase interpreting the sequential mechanism of polymerization of the amino acids 1 to 5 by conformational changes. (1) shows the transfer of the first amino acid to the multienzyme 2 carrying amino acids 2 through 5. (2) A dipeptide is formed and (3) transported to the site 3. (4) The tripeptide formed is transported to the site 4.
Bottom left: Functioning of the cofactor 4′-phosphopantetheine in the transthiolation and transpeptidation reactions between different activation sites. Bottom right: The organization of multienzyme systems that have been characterized so far. Numbered shaded areas indicate individual multifunctional enzymes.

accordance with this set of requirements: Activation of amino or hydroxy acid constituents as aminoacyl- or acyladenylates, followed by acylation of enzyme-thiols, multienzyme structures containing the cofactor 4′-phosphopantetheine, intermediates are transferred and stored covalently as thiolesters, and only the final product is released [54]. This mechanism has been found for linear peptides (edeine, gramicidin, alamethicine, ACV), cyclopeptides (gramicidin S, tyrocidine, cyclosporin), branched cyclopeptides (bacitracin, polymyxin), depsipeptides (enniatin, beauvericin), lactones (actinomycin), and is now under study for branched lactones (Lysobactin).

While all these compounds contain *a,a*-peptide bonds in their chains (except for the final ring closure reactions in bacitracin or polymyxin) or respective ester bonds,

Table 2 Current state of enzymology of peptides.

Type	Name	Organisms	Structural type[a]	Thio-template	Remarks[b]
Linear					
short	carnosin	rat/chick	L-2	–	p
	sarcophagine	*Sarcophaga peregrina*	L-2	–	pp
	glutathion	diverse	L-3	–	p
	4′-P-pantetheine	diverse	L-3	–	p
	leupeptin	*Streptomyces roseus*	L-3	–	p, me
	ACV	*Aspergillus nidulans*	L-3	+	p
medium	edeine	*Bacillus brevis*	L-5	+	pp, 3 me
long	gramicidin	*Bacillus brevis*	L-15	+	pp, 4 me
	alamethicin	*Trichoderma viride*	L-19	+	pp, me
polymers	multi-Arg-poly (Asp)	*Anabaena cylindrica*	L-(50)	–	pp
	γ-DGlu-capsule	*Bacillus licheniformis*	L-(750..10^4)	–	pp
Cyclic	cyclopeptin	*Penicillium cyclopium*	C-2	+	pp
	enterochelin	*E. coli*	C-3	–	pp
	HC-toxin	*Helminthosporium carbonum*	C-4	(+)	pp
	tentoxin	*Alternaria alternata*	C-4	(+)	pp
	ferrichrome	*Aspergillus quadricinctus*	C-6	+	p, me
	gramicidin S	*Bacillus brevis*	C-10	+	p, 2 me
	tyrocidine	*Bacillus brevis*	C-10	+	p, 3 me
	cyclosporin	*Tolypocladium inflatum*	C-11	+	pp, ? me
	mycobacillin	*Bacillus subtilis*	C-13	–	p, 3 me
Branched cyclic	polymyxin	*Bacillus polymyxa* *Aerobacter aerogenes*	10-C-7	+	pp, ? me
	bacitracin	*Bacillus licheniformis*	12-C-7	+	p, 3 me
lactone	destruxin	*Metarhizium anisopliae*	C-6	(+)	pp
	actinomycin	*Streptomyces chrysomallus*	C-5	+	pp, 2 me
Depsi-peptides	enniatin	*Fusarium oxysporum*	D-6	+	p, me
	beauvericin	*Beauveria bassiana*	D-6	+	pp, me

[a] L: linear, C: cyclic, D: depsipeptide
[b] pp: partially purified, p: purified enzyme fractions, me: number of multienzymes, if known

some structures do contain γ,α-linkages as mycobacillin, β-hydroxy groups in lactone formation as surfactin, or the β-alanyl carboxyl in the chain, as destruxin. Possible differences in the general reaction sequences leading to these peptides are already indicated in the first step, the carboxyl activation. It can be assumed from known one-step enzymes in the formation of murein, glutathione [52], or carnosin [54], that an acyl-phosphate mechanism is likely to be found in the case of β, γ, or δ-carboxyl activation or peptide carboxyl activation (Table 3). No covalent intermediates are formed, but the formed peptides could be obtained in a soluble form.

Table 3 Types of carboxyl activation reactions.

	Type I (AMP)[a]	Type II (NDP)[b]
Acid-CoA ligases	Acetate, medium chains (C_4-C_{12}), intermediate chains (C_3-C_7), long chains ($C_{12}-C_{18}$), benzoate, cholate, luciferin, lysergate, phenylacetate, oxalate, biotin	Succinate, succinate (GDP), medium chains (C_4-C_{12}) (GDP), glutarate
Aminoacyl-tRNA	Ala, Arg, Asn, Asp, Cys, Glu, Gln, Gly, His, Ile, Leu, Lys, Met, Phe, Pro, Ser, Thr, Trp, Tyr, Val	
Peptide synthetases	AcLeu	Pro, DAla, Leu, DLeu, Glu
(1) amino acids α-carboxyl	Aib, Ala, DAla, Asn, Cys, Dab, Dap, Glu, Gln, Gly, His, Ile, DLeu, Leu, Lys, Orn, DPhe, Phe, Pro, Thr, Trp, Tyr, Val, DVal	
(2) amino acids other carboxyls	Aad, Dha, Ise, βTyr	βAla, Gly (γ), DGlu (γ), DAsp (β)
(3) peptides α-carboxyl	AcLeu-Leu	Glu-Cys, UDP-Mur-*N*AcGly (Ala), UDP-Mur*N*Ac-A1-DGlu-A3 (Ay = Lys, Dpm, Dbu, Orn, Hsr)
(4) other carboxyls		UDP-Mur*N*Ac-A1-DGlu (γ)
(5) carboxylic acids	Dhb, pantoate, DHiv	UDP-G*N*Ac-lactate

[a] Carboxyl activation by cleavage of ATP to AMP and PP_i
[b] Carboxyl activation by cleavage of NTP to NDP and P_i (ATP if not indicated differently)

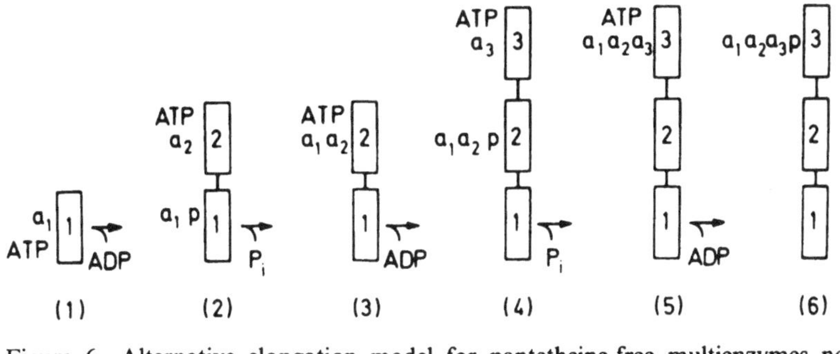

Figure 6 Alternative elongation model for pantetheine-free multienzymes not activating the constituent amino acids, except for the starter. Peptides are formed and transported to individual sites, where carboxyl functions are activated using ATP. ATP has also been postulated to be required for substrate transport.

Studies on surfactin from *Bacillus subtilis* [61] and, more advanced on mycobacillin from *Bacillus subtilis B3* [62–67] led to the proposal of a different mechanism [Figure 6]. In the latter case of the cyclic 13-peptide no activation of amino acids has been detected, except for the starter imino acid proline. No enzyme bound intermediates were detected, and no pantetheine cofactor. Instead, intermediates could be released from three multienzymes by extraction, and apparently intermediate peptides are activated by γ-phosphate cleavge of ATP. Contrary to the thioltemplate mechanism two moles ATP were consumed for each peptide bond formed. It has been suggested, that these extra ATP are consumed to direct a transport of the intermediate peptides [66]. The thioltemplate mechanism as well as glutathion formation seem to be highly effective requiring the energy of a single phosphate bond for each peptide bond formed [68].

2.5 Enzymic synthesis of peptides

The enzymic synthesis of peptides by means of peptide synthetases supplements current methods in use employing reversal of hydrolase reactions in the formation of single peptide bonds [69]. Convincing methods have been developed for single

Gramicidin S: Multienzymic Synthesis

2 L-Phe, 2 L-Pro, 2 L-Val, 2 L-Orn, 2 L-Leu, 10 ATP —GS1, GS2 / Mg^{2+}→

D-Phe → Pro → Val → Orn → Leu
↑ ↓
Leu ← Orn ← Val ← Pro ← D-Phe

10 AMP
10 PP_i

TUB 1978

Enniatin: Synthesis by a single multienzyme

3 D-Hiv, 3 L-Val, 6 ATP, 3 SAM —ESyn / Mg^{2+}→

D-Hiv → NMeVal → D-Hiv
↑ ↓
NMeVal ← D-Hiv ← NMeVal

6 AMP, 6 PP_i, 3 Sahc

ESyn in solution or adsorbed TUB 1984/85

Cyclosporin: Synthesis by a single enzyme fraction

2 L-Ala, 4 L-Leu, 2 L-Val, L-Bmt, L-Abu, Gly, 7 SAM —CySyn* / Mg^{2+}→

D-Ala → NMeLeu → NMeLeu → NMeVal → NMeBmt
↑ ↓
Ala ← NMeLeu ← Val ← NMeLeu ← Sar-Abu

11 AMP, 11 PP_i, 7 Sahc

TUB 1987

δ-(α-Aminoadipyl)-L-cysteinyl-D-valine: Multienzymic Synthesis

Aad, L-Cys, L-Val, 3 ATP? —ACVS / Mg^{2+}→ Aad → Cys → D-Val

3 AMP, 3 PP_i?

TUB 1988

Figure 7 Schemes for single pot enzymatic processes in the production of peptide antibiotics.

step enzymes, as in glutathion (GSH) production. The two enzymes involved, γ-glutamyl-cysteine ligase, and GSH-synthetase, have been cloned from *E. coli* [70, 71], and by combining high enzyme levels obtained by gene dosage, and substrate feeding, a significant improvement of production has been achieved [72]. Similar improvements of enzyme levels are thought to have evolved by extensive mutagenesis of β-lactam producers. This appears to be obvious comparing the level of a key enzyme like isopenicillin N synthase in a low producer like *Flavobacterium* with a high producing strain of *Cephalosporium* [73].

The exploitation of single step enzymes, well known from specificity studies, reached exciting heights with the cloning and large-scale availability of isopenicillin N synthase [74, 75]. A large number of tripeptide precursors have been transformed by Baldwin and coworkers to various bicyclic ring systems [75–80]. An understanding of this biocatalysts action will be of great use in an old challenge in synthetic chemistry still not solved.

Table 4 Directed biosynthesis of peptides in vivo.

Compound	Organism	Precursor	Replacement
Arphamenins	*Chromobacterium violaceum*	*Phe*	Tyr
Gliotoxin	*Trichoderma viride*	*Ser*	Ala
Bleomycins	*Streptomyces verticillus*	Various terminal amines	
Gramicidin S	*Bacillus brevis* ATCC 9999	D-Phe	thienyl-Ala, F-Phe
		Orn	Lys
Tyrocidine	*Bacillus brevis* ATCC 8185	D-Phe	thienyl-Ala
		D-Trp	Phe
		Leu, Val	Ile, alle
Cyclosporine	*Tolypocladium inflatum*	*Abu*	Ala, Thr, Val, Nva
Enniatins	*Fusarium oxysporum*	*Val*	Ile, Leu
Viridogriseins	*S. griseus*	*hyPro*	Pro
Ergot-peptides	*Claviceps purpurea*	*Phe*	F-Phe, Cl-Phe
		Val, Ile, Leu	Nva, Nle, Abu, 5,5,5-F-Leu, hyLeu
		Pro	thioPro
Actinomycins	*S.* various	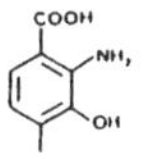	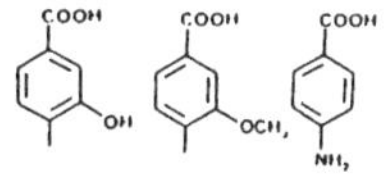
		Pro	Sar, Pip, Aze,
		Val	thioPro, 4-Me/F/Cl/Br-Ile
		DVal	Pro Ile
Echinomycins	*S. echinatus*	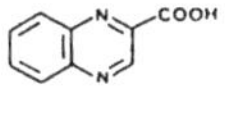	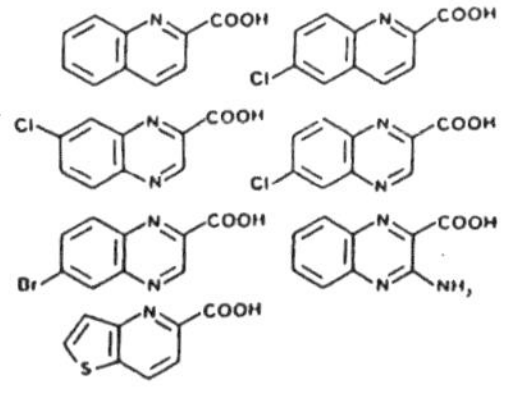
Triostins	*S. triostinicus*		

The application of multistep systems has been useful in the production of gramicidin S and analogs [81–83], enniatins [84, 85], beauvericins [86, 87], and cyclosporins [88] on a small scale (Figure 7). Such *in vitro* systems directly lead to defined structures selectable by precursor addition. The instability of enzyme systems still does not permit a scale up, although some improvements have been made concerning enzyme fermentation [89, 90] and preparation, including stabilization with aqueous polymer solutions [91].

Of more general use are the procedures of directed biosynthesis *in vivo*, feeding substrates to the fermentation broth (Table 4). Several examples are in use in peptide production. Studies on *in vivo* stabilization of multienzymes have been carried out for gramicidin S synthetase [92–97]. Addition of substrates and control of dissolved oxygen prolongs activity and peptide yield.

The production of new types of compounds by interfering with single enzyme activities or by feeding disfunctionalized precursors has been introduced in neoviridogrisein fermentation with the inhibition of hydroxylase by Fe^{2+}-depletion [98], in suppression of peptide acylation of actinomycin by Fe^{2+} and sarcosine addition [99], and in the synthesis of actinomycin halves with analogs of 4-methyl-3-hydroxyanthranilate unable to form the phenoxazinone chromophore [100].

3. Genetics of Peptide Synthetases

The pathway to a metabolite may well involve a set of ten or more genes and enzymes, but the studies conducted so far on siderophores, β-lactam antibiotics, and peptide synthetases have established gene clusters and multigenes coding for multienzymes [101–107]. This is well in accord with studies on nonpeptide antibiotic pathways [108, 109], except that extensive fusions of enzymic functions are established so far only in the peptide field with its most advanced enzymology.

3.1 Siderophores

Most advanced are studies on the siderophores aerobactin and enterochelin (Figure 8), with key functions in pathogenicity of enterobacteria, produced by a set of single-step enzymes. The aerobactin biosynthetic cluster has been detected together with an outer membrane transport system on the 130 kb ColV-plasmid of *E. coli* [110], but also on other plasmids or chromosomal DNA in other bacterial strains. The four enzymes involved catalyze (a) the N^6-hydroxylation of lysine, (b) the acetylation of the N^6-hydroxyl group, (c, d) formation of peptide bonds with the two citrate carboxyl groups. The enzymology and transport system has been reviewed by Braun in this volume. An enterochelin cluster has been mapped on the *E. coli* chromosome [111–114], encoding biosynthesis, membrane receptor and transport functions in a 24 kb-region. Biosynthesis involves conversion of choris-

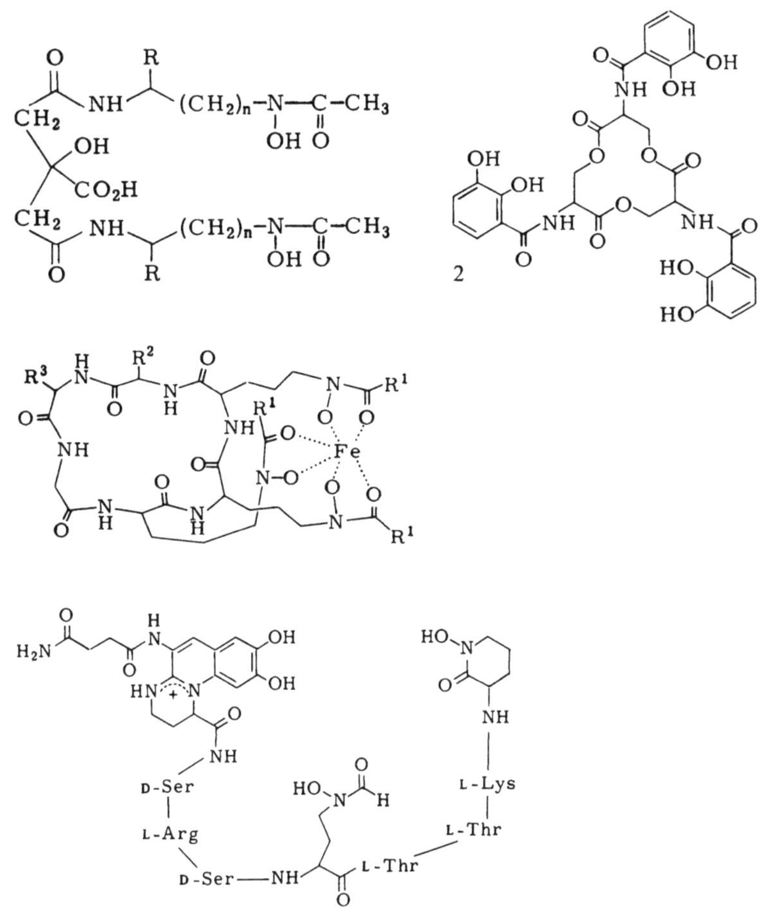

Figure 8 Structures of some siderophores: 1-aerobactin (R = H, n = 4), 2-enterobactin (enterochelin), 3-ferrichrome ($R_1 = CH_3$, $R_2 = R_3 = H$), 4-pyoverdin.

mate to 2,3-dihydroxybenzoate (Dhb), followed by coupling to serine and trimerization. The assembly steps, including activation of Dhb and Ser as adenylates, are assumed to proceed on a multienzyme complex [115, 116]. So far there has been no evidence for other than physical association of these single step functions [117].

More complex peptide siderophores like the fluorescent pyoverdine type from strains of *Pseudomonas* or the fungal ferrichrome type (Figure 8) can be expected to be assembled on multifunctional enzyme templates. Studies on *Pseudomonas syringae* [118], *Pseudomonas putidae* [119], *Pseudomonas* sp. [120], or *Pseudomonas aeruginosa* [121] indicate the presence of 4 to 5 clusters of at least 7 to 12 genes involved in the formation of fluorescent pigment, siderophore, pseudobactin, or pyoverdine, respectively. In case of ferrichrome, a multienzyme or multienzyme complex has been partially characterized from *Aspergillus quadricinctus* [122]. This could indicate again a higher grade of structural integration in eucaryotic systems. The observation of a heptapeptide analog of ferrichrome, tetraglycylferrichrome, from a strain of *Neovossia indica* [123], should lead to highly interesting studies on the origin of templates as well as cyclization mechanisms.

3.2 Bacillus peptides: Gramicidin S, tyrocidine, bacitracin

Non-producer mutants of antibiotic strains have been isolated for decades, with the aim to understand the possible function of antibiotics in nature. In many cases, where nonproducers became available, antibiotics seemed not to be essential, dispensable, *secondary metabolites*. In some cases, it has not been possible to obtain non-producers, such as in *Streptomyces tsusimaensis*, forming valinomycin [124]. A major difficulty is the differentiation of non-producers from low level producers. In the analysis of peptide producers the new quality of multienzyme structure with iuntegrated functional units presents additional problems. If a set of functions is assembled on a multienzyme, a single defect may inactivate this complete set of functions. Still the enzyme structure may be detectable. Single functions may be present, but the overall reaction missing. Complementation of two multienzymes bearing different defects can generally not be expected. Some of these problems are well known from studies of less complex enzymes.

In the work on peptides of *Bacillus* mutants have been isolated from the producers of gramicidin S [125–130], tyrocidine [131], bacitracin [132], polymyxin [133], and most recently, mycobacillin [67]. While only in the gramicidin S (GS) system the enzymic defects have been partially identified, all studies supported the concept of multifunctional enzyme structures. First attempts to isolate the respective multigenes tried to trace an extrachromosomal location. Evidence for this approach came from the high frequency of non-producers obtained by treatment of cells with intercalators known for curing properties [134]. In case of *Bacillus brevis* a membrane bound plasmid [135, 136] and a defective phage [137] have been isolated, with no connection however, to antibiotic production. The production of polyglutamate in *Bacillus natto*, on the other hand, has been linked to a 5.7 kb plasmid [138].

In the meantime, with general cloning procedures aimed at expression from large DNA-fragments (20 kb) [106], or expression of lacZ-fusion fragments of multienzymes [139] with immunogenic detection, some advances have been achieved. In the tyrocidine (TY)-system, consisting of three multienzymes (TY1–3) catalyzing the formation of *cyclo*-D-Phe-Pro-Phe-D-Phe-Asn-Gln-Tyr-Val-Orn-Leu, steps 1, 2–4, and 5–10 are located on proteins of 120, 230 and 450 kDa. As extensive cross reactions of antibodies directed against the multienzyme gramicidin S synthetase 2 (GS2) with these and other peptide synthetases had been established [140], a lac independent expression of TY1 has been achieved in *E. coli* [141]. The corresponding promoter region of *tycA* (TY1 structural gene), constitutively used by *E. coli* RNA polymerase, has been fused to the *lacZ* structural gene and introduced into the *Bacillus subtilis 168*-chromosome with phage SPβ-constructs utilizing Tn917, to study a possibly developmentally regulated expression. Expression markers have been spoO-genes, corresponding to the time of appearance of

TY-multienzymes in *Bacillus brevis ATCC 8185*. Mutants carrying defects in the spoO-loci A, B, and E all fail to utilize the *tycA*-promoter, and at the same time also fail to produce *Bacillus subtilis* antibiotics. However, mutants with defects at the spoO-loci F, H, and J permit transcription. Since it has been known that a spoOA-defect can be overcome with mutations in abrB, abrB suppressor mutants have been found to utilize the *tycA*-promoter even during growth [142]. This implies that the abrB-gene product could act directly as suppressor, which then is inactivated by the spoOA gene product.

These fundamental studies on regulation of expression demonstrate the possible application of genetically relatively well known organisms such as *Bacillus subtilis 168* in the elucidation of regulatory events in genetically unexploited strains. Such applications could have been predicted from the frequent occurence of similar pathways in different organisms. So the expression of TY1 and possibly of TY2 and TY3 could be repressed in *B. brevis* by an abrB-related protein, and regulated by a spoOA-related function. Such kind of analogy would considerably speed up the elucidation of regulatory coordinated events.

3.3 Genes involved in β-lactam biosynthesis

The biochemistry of β-lactams is now in an exciting phase of rapid progress, and several reviews are available on various aspects of the field [143–145]. The wide distribution of β-lactams in pro- and eucaryotic organisms, ascertained through the use of sensitive detection procedures, has been illustrated here with work by Ono and Harada and their colleagues on new bacterial compounds [146]. Genes encoding ACV-synthetase (ACVS), isopenicillin N synthase (IPNS), isopenicillin N epimerase (IPNI), deacetoxy cephalosporin synthase (DAOCS), and deacetyl cephalosporin synthase (DACS) have been or will be isolated from various bacterial and fungal sources, and their expression and organization will be compared. It can be reflected if these biosynthetic pathways could have evolved from a common ancestor, or if they rather could have been transferred horizontally. Results obtained so far indicate that ACVS is a single chain multienzyme in *Streptomyces, Aspergillus, Cephalosporium,* and *Penicillium* [147]. The key enzyme IPNS has been cloned and sequenced from strains of the Gram-negative rod *Flavobacterium* [155], the Gram-positive filamentous *Streptomyces* [148, 149], and the fungi *Aspergillus nidulans* [150], *Penicillium chrysogenum* [151], and *Cephalosporium acremonium* [74] (Figure 9). A high degree of similarity has been detected, and a transfer of a pathway from *Streptomyces* to *Cephalosporium* has been suggested based on the similarity of G/C-content [150]. However, this is not the case with the structural genes of *Aspergillus* and *Penicillium.* Hypothetically derived rates of evolution of DNA sequences suggest the divergence of pro- and eucaryotic biosynthesis genes much later than the divergence of bacteria and fungi (Figure 10) [152]. So the analysis of pathways to various β-lactams may provide insights into possible transfers of gene clusters within microorganisms. At the same time the isolation of genes will permit the reconstruction of the events of the most intensive random mutational strain im-

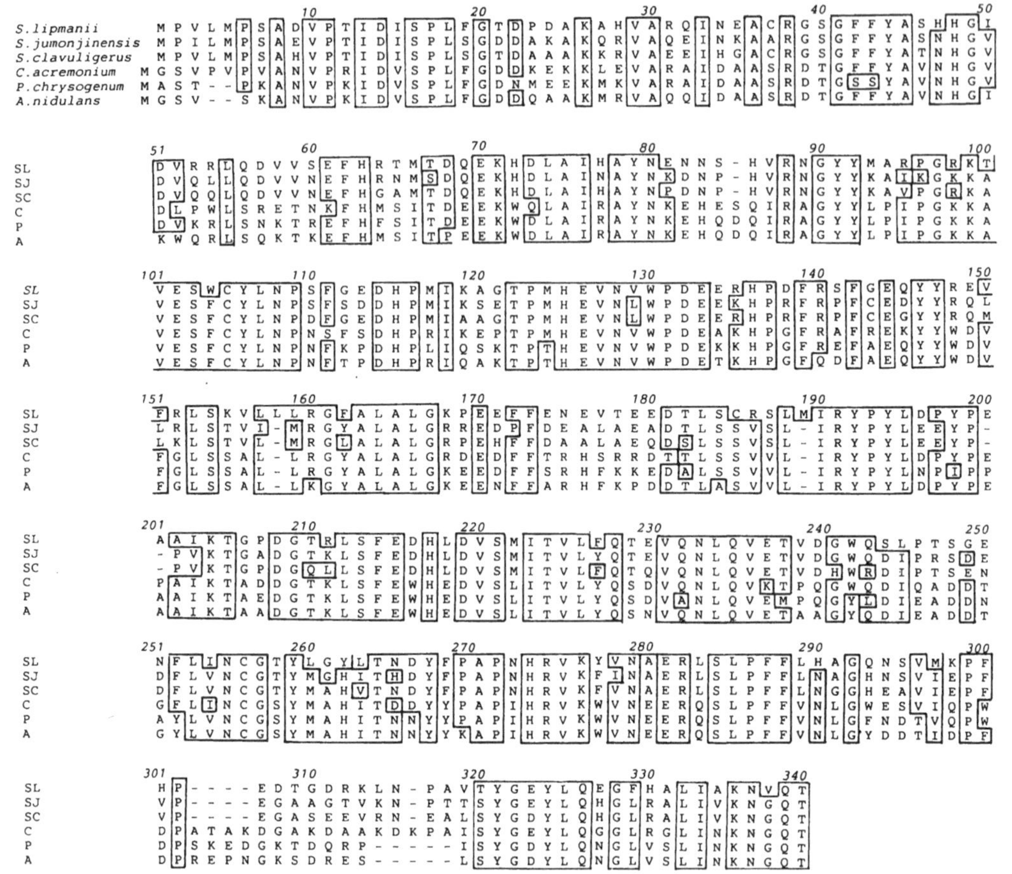

Figure 9 Comparison of sequences of isopenicillin N synthases isolated from the procaryotes *Streptomyces clavuligerus, S.lipmanii, S.jumojenensis,* and the eucaryotes *Aspergillus nidulans, Penicillium chrysogenum,* and *Acremonium cephalosporium.*

provements that have been carried out in penicillin production, and permit more rational approaches. Finally more detailed studies on the enzymology of the involved reactions should find applications in new synthetic approaches.

3.4 More unified approaches to biosynthetic pathways

Although little is known about the structure and functional properties of polyketide synthases when compared to polypeptide synthetases, extensive homologies are expected, due to the similarity of catalytic functions. Indeed, in *Streptomyces*, certain DNA-homologies have been detected and ascribed to gene clusters of polyketide forming systems [153]. In addition, the application of orthogenal-field-alternation gel electrophoresis has helped to resolve chromosomes and giant linear plasmids.

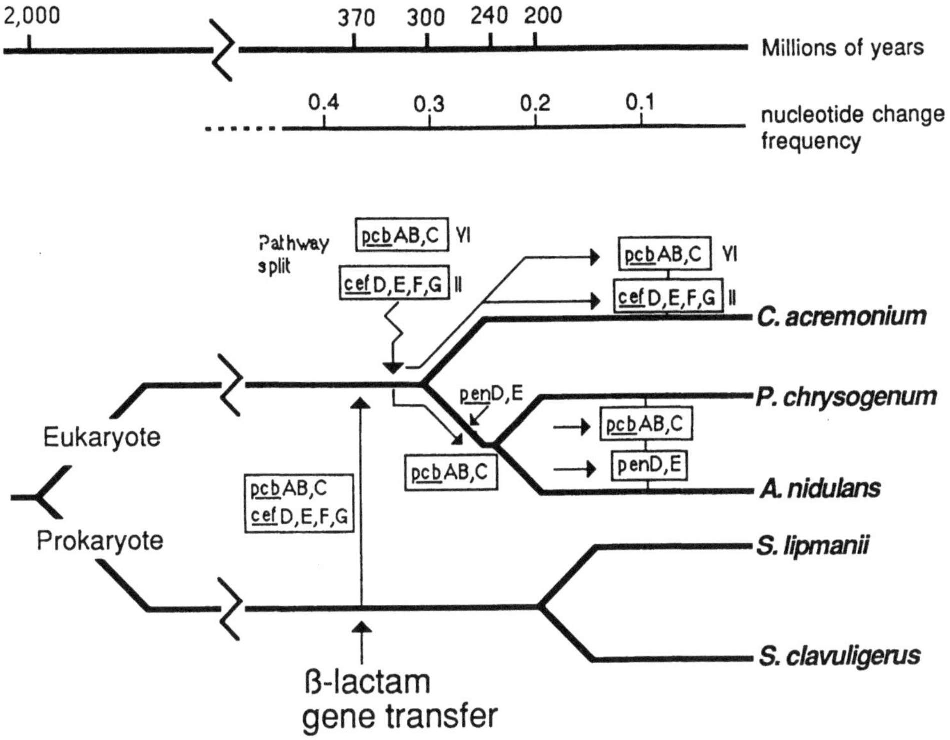

Figure 10 Phylogenetic tree of isopenicillin N (IPN) synthases derived from gene sequences by Ingolia et al. [148], and extended by Skatrud (personal communication). The time scale indicates the divergence of pro- and eucaryotes two billion years ago, based on 5 S RNA sequences, while fungi diverged several hundred million years ago [148]. Gene and gene product identities of IPN synthases are with 53 to 57% much higher than those of other highly conserved proteins known in bacteria and fungi, ranging between 42 and 52%. A possible interpretation is a horizontal transfer of the set of genes (*pcb* AB, C and *cef* D, E, F, G) from bacteria to fungi, where the pathway is split to cephalosporin producers (*Cephalosporium acremonium*) and penicillin producers (e. g. *Aspergillus nidulans* and *Penicillium chrysogenum*). Note that in *Cephalosporium* the two sets of genes reside on chromosomes II and VI (Skatrud and Queener, *Gene*, in press), while in the penicillin pathway the modification genes *pcb* D and *pcb* E participate. The notation *pcb* AB indicates the fusion of the two peptide bond forming reactions into a single multifunctional gene/gene product in ACV synthetase [45, 46].

The latter type of DNA may be involved in carrying clustered biosynthetic information [154]. The successful application of gene probes to identify respective heterologous genes has been demonstrated for the isopenicillin N-synthase from the fungi *Cephalosporium acremonium, Penicillium chrysogenum* [151], and *Aspergillus nidulans* [150], as well as for the procaryotes *Streptomyces* [148, 149] and *Flavobacterium* [155]. This approach will enable us to compare the genetic organization of a number of pathways. Protein homology on the other hand has been used successfully in the isolation of multigenes of multienzymes forming the peptides gramicidin S and tyrocidine. Such homologies are readily detected by the cross reactions of antibodies. Cross reactions have been useful in the detection and identification of multienzymes of unknown multienzyme systems. In the near future we can expect to obtain information on the organization and assembly of enzyme systems catalyzing reaction sequences. Since metabolites are frequently found that combine structural features of polyketides and polypeptides, or originate from similar sequential reactions, new fields of synthetic biological chemistry are emerging.

4. New and Old Sources of Peptides

There are different approaches leading to new or already known structures. A promising way has always been the investigation of unknown, unexploited types of organisms. This approach has been used by Reichenbach and coworkers in the study of Myxobacteria [156]. From 1100 newly isolated strains of these gliding bacteria with remarkable life cycles involving macroscopic fruiting bodies a number of antibiotics have been isolated, including the peptides myxothiazol, althiomycin, and myxovalargin (Figure 11) [157, 158]. While althiomycin has been known as a *Streptomyces* metabolite [159], the others represent new compounds with remarkable properties. Myxothiazol has been found to inhibit mitochondrial respiration binding to the bc_1 complex, blocking the reduction of cytochrome c_1, and inhibiting the reduction of the iron-sulfur protein by ubiquinol [160]. Antibacterial action is observed only in strains like *Paracoccus denitrificans*, containing a similar ubiquinol: cytochrome c_2 oxidoreductase [199]. An obvious problem is the sensitivity of detection, clearly related to the level of production of certain metabolites. From the related Cytophaga group, new compounds have been detected employing various

1 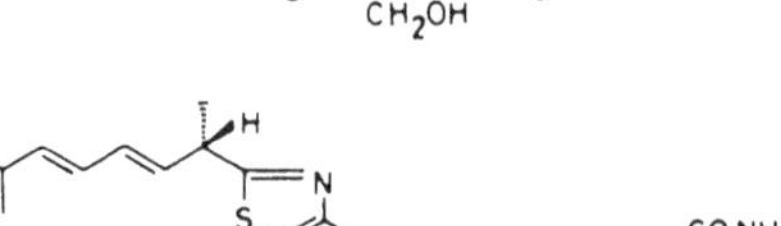

2 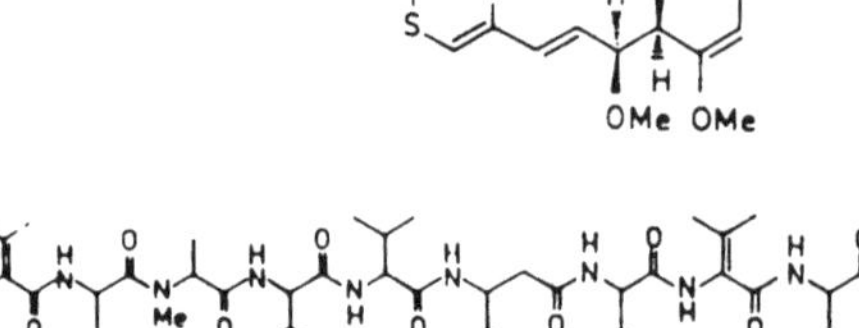

3 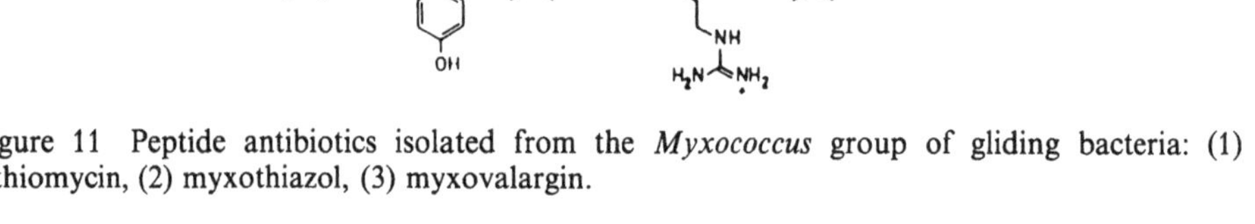

Figure 11 Peptide antibiotics isolated from the *Myxococcus* group of gliding bacteria: (1) althiomycin, (2) myxothiazol, (3) myxovalargin.

selective screening procedures (Figure 12). High sensitivity β-lactam detection assays led to monobactams in *Flexibacter* [161, 162], and to new cephalosporin derivatives in *Lysobacter* [163–166]. The latter have independently been characterized also from strains of *Flavobacterium* [167, 168] and *Xanthomonas* [163–166, 169, 170]. Some of these data have been discussed by Ono and Harada in their chapter on β-lactam directed screenings [146]. Screens aiming at the antibacterial spectrum of the vancomycin/ristocetin group of glycopeptides led to the peptide lactones lysobactin from *Lysobacter* [171] and katanosin from *Cytophaga* [172, 173]. So it appears, that certain compounds are restricted to specific groups of organisms, while others are widely distributed. This seems evident from the wide distribution of β-lactams in pro- and eucaryotes [174], as compared to the occurence of the vancomycin type of glycopeptides, occuring only in Actinomycetes/Streptomycetes. The latter, complexing the D-Ala-D-Ala terminal of the murein-pentapeptide thus interfering with the crosslinking reaction, have been screened by competition tests employing D-Ala-D-Ala analogs [175], or by solid phase binding assays [176]. One out of 45 strains analyzed contained detectable amounts of glycopeptide.

Such an approach differs radically from attempts to trace a group or even all metabolites from a given strain. So in genetic work carried out e. g. with *Bacillus subtilis* until recently the property *antibiotic production* has not been correlated with any metabolite. Even for the most commonly used strain 168 no studies have been available on the antibiotic spectrum produced until recently [44]. At least ten different antibiotics have been detected, and the complex structure of subtilosin (Figure 4) has almost been resolved. In addition to bacilysin (Figure 12), which is produced in a variety of *B.subtilis* strains, two antifungal peptides, fengymycin [177] and rhizocticins [178], have been identified recently from the strains F-29-3 and ATCC 6633, respectively, the latter also known to produce the lantibiotic subtilin. From the work of Bushell on secondary metabolite formation in *Streptomyces cattleya* [179], it is well known that highly specific conditions may lead to the expression and formation of a metabolite. Such conditions are related to the extracellular situation, leading to concentration changes of certain key compounds like glucose, phosphate, or metal ions, or less documented, as response to other competing organisms. So the genetic information for the production of a certain metabolite could be present, but not expressed under the conditions of screen. An additional problem could be the genetic instability of such information, being subject to change by yet unknown rearrangenments.

New sources yet in the beginning of exploration are marine organisms [180–183]. It has been pointed out by Ireland in a recent review on marine peptide metabolites, that the paucity of nitrogenous metabolites in earlier reports, often attributed to the fact that nitrogen is a limiting nutrient in the ocean, is highly dependent on the choice of organism [183]. More than 40% of so called secondary metabolites that have been characterized in the last decade from sponges, marine microbes, bryozoans, tunicates, and blue-green algae, contained nitrogen, in contrast to less than 10% of those identified from red and brown algae, coelenterates, molluscs and echinoderms. As this heterogenous field is not in a state of directed screens, toxin identifications and the search for anti-cancer drugs have been the leading activities.

1

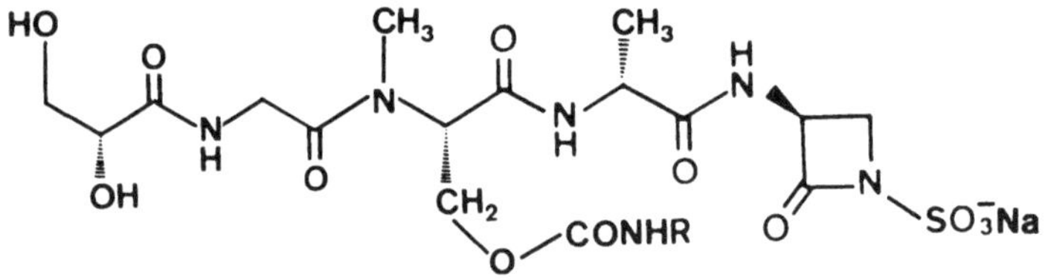

2

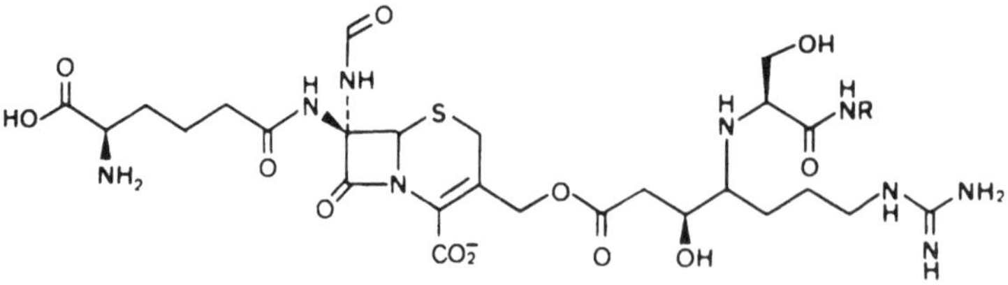

3

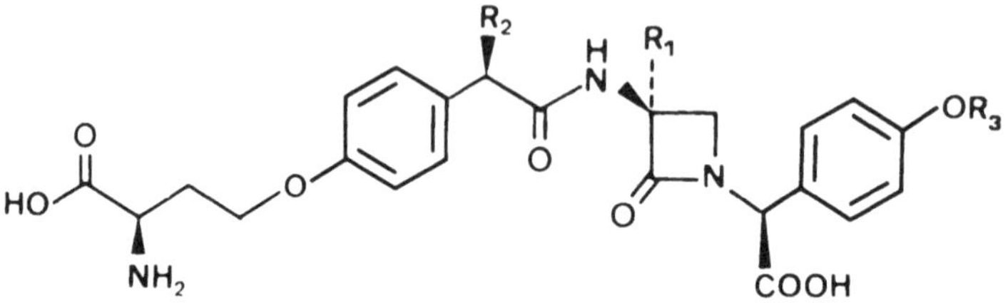

4

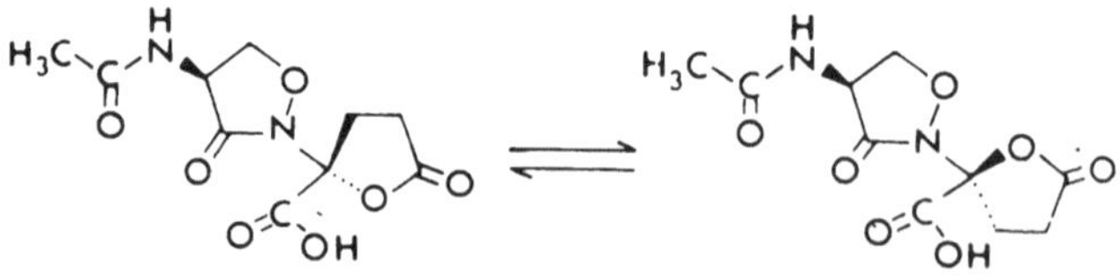

5 DLeu → Leu → PhSer → HyLeu → Leu → Arg → Ile
＼ ↓
Ser ← HyAsn ← Gly ← Thr

6 DLeu → Leu → PhSer → HyLeu → Leu → DArg → Val/Ile
＼ ↓
Ser ← HyAsn ← Gly ← aThr

Figure 12 Recently isolated β-lactams and peptides from the *Cytophaga* group of gliding bacteria. (1) SQ 28,332 (*Flexibacter sp.*), (2) cephabacin (*Lysobacter lactamgenus*, also *Flavobacterium sp.*, *Flavobacterium chitinovorum*, *Xanthomonas lactamgena*), (3) formadicins (*Flexibacter alginoliquefaciens*), and TAN-667 (*Cytophaga phytophila*), the latter contains tetrasaccharides and an unknown amino acid in position R_3, (4) lactivicin (*Lysobacter albus* and *Empedobacter lactamgenus*), (5) Katanosin (*Cytophaga sp.*), and (6) Lysobactin (*Lysobacter sp.*).

Cyanoginosins, complex cyclo-heptapeptides, have been identified as causes of fresh water toxification by the blue green alga *Microcystis aeruginosa* (Figure 13) [184, 185, 200]. A marine cyanophyte, *Lyngbya majuscula*, produces Lyngbyatoxin A, a potent irritant causing a form of contact dermatitis ("swimmer's itch"). The prenylated cyclodipeptide is identical to one of the teleocidins produced by *Streptomyces mediocidicus* [186, 187]. The compounds are tumor promoters and toxic to fish [186, 188] suggesting a possible function as chemical defense agent.

The question of unambiguous origin has again to be adressed to the antibacterial sponge isolate discodermin, containing t-Leu known from Streptomcetes peptides [189]. *Discodermia* is known to contain algal and bacterial symbionts.

A number of cyclic peptides with unusual structures and significant biological activities have been isolated from tunicates (sea squirts). A group of peptides from *Lissoclinum patella*, containing thiazole and oxazoline residues, are highly effective against viral infections and possess remarkable anti-tumor properties [190–195] (Figure 13). The structures of patellamides have finally been identified by chemical synthesis [194, 195]. Didemnin B, a cyclic depsipeptide also isolated from a tunicate [196, 197] is now in advanced clinical trials with prospective applications as antiviral, antitumor, and immunosuppressive drug [198].

From marine molluscs a number of hormon-like peptides and toxins have been identified, and have been shown or are likely to be of ribosomal origin. An exception are the dolastatins, isolated from the sea hare *Dolabella auricularia*, with interesting cytostatic properties [201]. Following the proposed structure, all 16 possible sequence and stereo isomers have been synthesized, none of which however corresponded to the natural isolate [202–206].

5. Conclusions

Peptide bonds containing compounds of both enzymatic and ribosomal origin have been and will continue to be useful in our basic understanding of biochemistry, and at the same time find applications, mainly in medical areas. A considerable gap exists generally in our understanding of the intrinsic role of these compounds in the producing organisms. It is evident that such an understanding would be of considerable advantage in the evaluation of properties and functional requirements, and thus lead to a more biochemical approach of adressing the question *how a certain function, property etc. is controlled or regulated.*

Peptides, from their construction as linear polymers of structurally variable building blocks, offer unique options of modification of structural themes. The studies available so far indicate that minor changes can be essential in drug-receptor interactions as well as in high dose toxicity effects. In addition the vast number of specific peptidases permit in many cases an efficient control of activities.

1

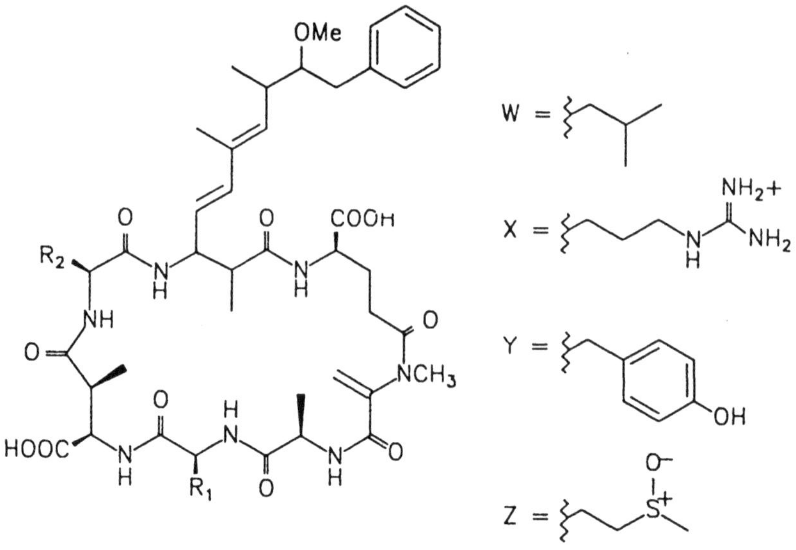

7 R_1 = W, R_2 = Me
8 R_1 = W, R_2 = X
9 R_1 = Y, R_2 = X
10 R_1 = Y, R_2 = Me
11 R_1 = Y, R_2 = Z

2

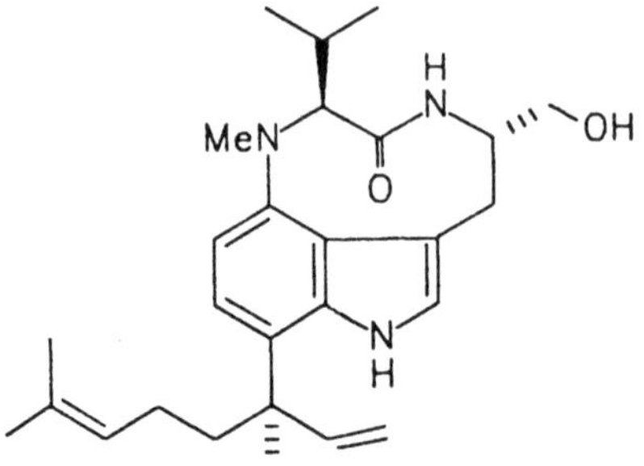

12

3 HCO-DAla-Phe-Phe-Pro-(DVal / DtLeu)-(Val / tLeu)-DTrp-Arg-DCya-Thr-

MeGln-DLeu-Asn-Thr-Sar

4

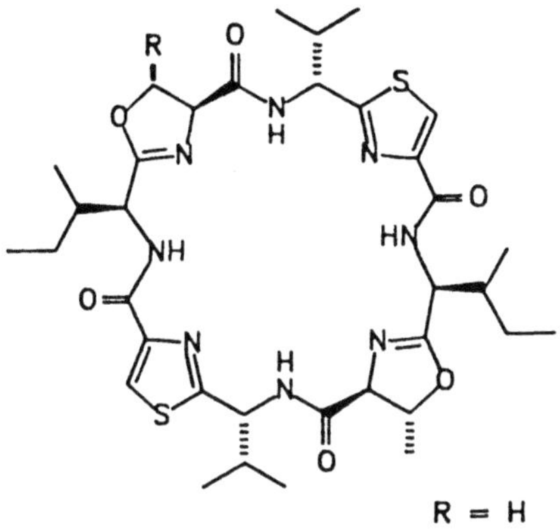

5

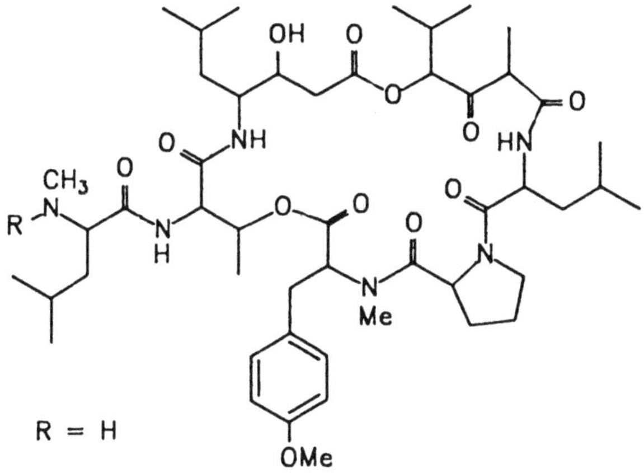

6

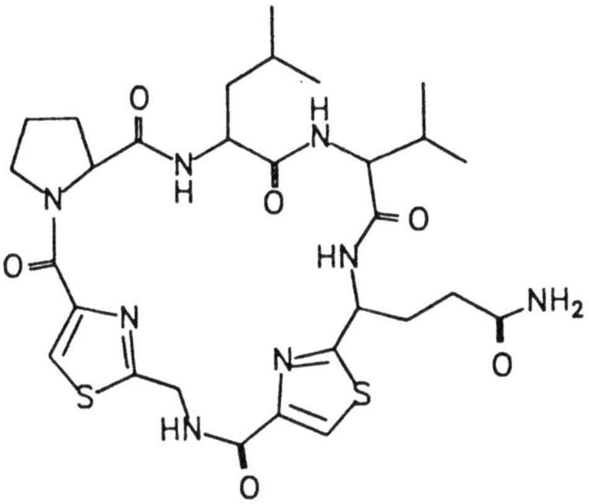

Figure 13 Peptides identified from marine sources. (1) cyanoginosin (*Microcystis aeroginosa*), (2) lyngbyatoxin (*Lyngbya majuscula*), (3) discodermin (*Discodermia*), (4) patellamide (*Lissoclinum patella*), (5) didemnin (*Didemnidae*), and (6) Dolastatin (*Dolabella auricularia*).

References

1. Bérdy, J. (1980) *Handbook of Antibiotic Compounds*. Vol. IV, Pt. 1, 2, Boca Raton, Florida, CRC Press.
2. Hensens, O., Dewey, R., Liesch, J., Napier, M., Reamer, R., Smith, J., Albers-Schönberg, G., and Goldberg, I. H. (1983) *Biochem. Biophys. Res. Comm. 113*, 538.
3. Edo, K., Mizugaki, M., Koide, Y., Seto, H., Furihata, K., Otake, N., and Ishida, N. (1985) *Tetrahedron Lett. 265*, 331.
4. Lee, M. D., Dunne, T. A., Chang, C. C., Ellestad, G. A., Siegel, M. M., Morton, G. O., McGahren, W. J., and Borders, D. B. (1987) *J. Am. Chem. Soc. 109*, 3466.
5. Golik, J., Dubay, G., Groenewold, G., Kawaguchi, H., Konishi, M., Drishnan, B., Onkuma, H., Saitoh, K., and Doyle, T. W. (1987) *J. Am. Chem. Soc. 109*, 3462.
6. Goldberg, I. H. (1988) in *The Roots of Modern Biochemistry* (H. Kleinkauf, H. v. Döhren and R. Jaenicke, eds.), p. 389–396, de Gruyter, Berlin, New York.
7. Goldberg, I. H. (1986) in *Mechanisms of DNA Damage and Repair* (M. Simic, L. Grossmann, and A. C. Upton, eds.) p. 231–244, Plenum Press.
8. Goldberg, I. H. (1987) in *Molecular Mechanisms of Carcinogenic and Antitumor activity* (C. Chagas and B. Pullman, eds.), p. 425–462, Pontifical Academiae Scientiarum Scripta Varia 70.
9. Chen, K.-X., Gresh, N., and Pullman, B. (1987) *Nucl. Acids Res. 15*, 2175.
10. Goldberg, I. H. (1987) *Free Radical Biol. Med. 3*, 41.
11. Gibson, B. W., Herlihy, W. C., Samy, T. S. A., Hahm, K.-S., Maedo, H., et al. (1984) *J. Biol. Chem. 259*, 10 801.
12. Samy, T. S. A., Hahm, K.-S., Modest, E. J., Lampmann, G. W., Keutmann, H. T., et al. (1983) *J. Biol. Chem. 258*, 183.
13. Khoklov, A. S., Reshetov, P. D., Chupova, L. A., Cherchez, B. Z., Zhigis, L. S., and Stoyachenko, I. A. (1976) *J. Antibiot. 29*, 1026.
14. Miyashiro, S., and Udaka, S. (1982) *J. Antibiot. 35*, 1319.
15. Miyashiro, S., and Udaka, S. (1983) *J. Antibiot. 36*, 1136.
16. Miyashiro, S., and Udaka, S. (1984) *J. Antibiot. 37*, 20.
17. Miyashiro, S., and Udaka, S. (1984) *J. Antibiot. 37*, 27.
18. Krieger, D. T. (1983) *Science 222*, 975.
19. Lechan, R. M., Wu, P., Jackson, I. M. D., Wolfe, H., Cooperman, S., Mandel, G., and Goodman, R. H. (1986) *Science 231*, 159.
20. Lynch, D. R., and Snyder, S. H. (1986) *Annu. Rev. Biochem. 55*, 773.
21. Bradbury, A. F., Finnie, M. D. A., and Smyth, D. G. (1982) *Nature 298*, 686.
22. Husain, I. and Tate, S. S. (1983) *FEBS Lett. 152*, 277.
23. Glembotski, C. C. (1984) *J. Biol. Chem. 259*, 13 041.
24. Kizer, J. S., Busby, W. H., Cottle, C., and Youngblood, W. W. (1984) *Proc. Natl. Acad. Sci. 81*, 3228.
25. von Zastrow, M., Tritton, T. R., and Castle, J. D. (1986) *Proc. Natl. Acad. Sci. 83*, 3297.
26. Messer, M., and Ottesen, M. (1964) *Biochim. Biophys. Acta 92*, 409.
27. Blombäck, B. (1968) *Methods. Enzymol.* 11, 398.
28. Fischer, W. H., and Spiess, J. (1987) *Proc. Natl. Acad. Sci. 84*, 3628.
29. Boman, H. G., and Hultmark, D. (1987) *Annu. Rev. Microbiol. 41*, 103.
30. Ganz, T., Selsted, M. E., Szklarek, D., Harwig, S. S. L., Daher, K., Bainton, D. F., and Lehrer, R. I. (1985) *J. Clin. Invest. 76*, 1427.
31. Zasloff, M. (1987) *Proc. Natl. Acad. Sci. 84*, 5449.
32. Richter, K., Egger, R., and Kreil, G. (1986) *J. Biol. Chem. 261*, 3676.
33. Erspamer, V., and Melchiorri, C. (1980) *Trends Pharmacol. Sci. 1*, 391.
34. Richter, K., Egger, R., and Kreil, G. (1987) *Science 238*, 200.
35. Schnell, N., Entian, D.-D., Schneider, U., Götz, F., Zähner, H., Kellner, R., and Jung, G. (1988) Nature 333, 276.
36. Banerjee, S., and Hansen, J. N. (1988) *J. Biol. Chem. 263*, 9508.
37. Rayman, K., Hurst, A. (1984) in *Biotechnology of Industrial Antibiotics* (E. J. Vandamme, ed.) p. 607–28, Dekker, New York.
38. Gasson, M. J. (1984) *FEMS Microbiol. Lett. 21*, 7.

39. Gonzalez, C. F., and Kunka, B. S. (1985) *Appl. Environm. Microbiol. 49*, 627.
40. Steele, J. L., and McKay, L. L. (1986) *Appl. Environm. Microbiol. 51*, 57.
41. Schnell, N., Entian, K.-D., Götz, F., Hörner, T., Kellner, R., and Jung, G. (1989) *FEMS Microbiol. Lett. 58*, 263.
42. Kaletta, C., Entian, K.-D., Kellner, R., Jung, G., Reis, M., and Sahl, H.-G. (1989) *Arch. Microbiol. 152*, 16.
43. Wakamiya, T., Veki, Y., Shiba, T., Kido, Y., Motoki, Y. (1985) *Tetrahedron Lett. 26*, 665.
44a. Kessler, H., Steuernagel, S., Gillesen, D., and Kamiyama, T. (1987) *Helv. Chim. Acta 70*, 726.
44b. Wakamiya, T., Fukase, K., Naruse, N., Konishi, M., and Shiba, T. (1988) *Tetrahedron Lett. 29*, 4771.
44. Babasaki, K., Takao, T., Shimonishi, Y., and Kurahashi, K. (1985) *J. Biochem. Tokyo 98*, 585.
45. van Liempt, H., Palissa, H., and von Döhren, H. (1988) *ASM-Meeting Bloomington, Abstract.*
46. van Liempt, H., von Döhren, H., and Kleinkauf, H. (1989) *J. Biol. Chem.*
47. van Liempt, H. (1988) *Thesis, TU Berlin.*
48. Mitnick, M. A., and Reichlin, S. (1971) *Science 172*, 1241.
49. Mohr, H. (1977) *Thesis TU Berlin.*
50. Mohr, H., and Kleinkauf, H. (1978) *Biochim. Biophys. Acta 526*, 375.
51. Kleinkauf, H., and von Döhren, H. (1986) in *Biotechnology* (H.-J. Rehm and G. Reed, eds.), Vol. 4 *Microbial Products* (H. Pape and H.-J. Rehm, eds.), pp. 283–308, VCH Weinheim.
52. Kleinkauf, H., and Koischwitz, H. (1978) *Progr. Mol. Subcell. Biol. 6*, 59.
53. von Döhren, H., and Kleinkauf, H. (1988) in *The Roots of Modern Biochemistry* (H. Kleinkauf, H. von Döhren and R. Jaenicke, eds.) pp. 355–368, de Gruyter, Berlin, New York.
54. Kleinkauf, H., and von Döhren, H. (1987) *Annu. Rev. Microbiol. 41*, 259.
55. Altmann, M., von Döhren, H., El-Samaraie, A., Kittelberger, R., Pore, M. S., and Kleinkauf, H. (1982) in *Peptide Antibiotics-Biosynthesis and functions* (H. Kleinkauf and H. von Döhren, eds.) pp. 243–252. de Gruyter, Berlin, New York.
56. Kleinkauf, H., and von Döhren, H. (1983) in *Biochemistry and Genetic Regulation of Commercially Important Antibiotics.* (L. C. Vining, ed.) pp. 95–145, Addison-Wesley, Reading.
57. Billich, A., and Zocher, R. (1986) *J. Biol. Chem. 262*, 17258.
58. Kleinkauf, H., and von Döhren, H. (1990). Progr. Drug Res. 34, in press.
59. Zimmer, T.-L., and Laland, S. G. (1973) *Essays Biochem. 9*, 31.
60. Kurahashi, K. (1981) in *Antibiotics*, Vol. 4, *Biosynthesis* (W. Corcoran, ed.), pp. 325–352, Springer, Berlin, New York.
61. Vater, J., unpublished data.
62. Ghosh, S. K., Mukhopadyay, N. K., Majumder, S., and Bose, S. K. (1983) *Biochem. J. 215*, 539.
63. Ghosh, S. K., Mukhopadyay, N. K., Majumder, S., and Bose, S. K. (1985) *Biochem. J. 230*, 785.
64. Majumder, S., Ghosh, S. K., Mukhopadyay, N. K., and Bose, S. K. (1985) *J. Gen Microbiol. 131*, 119.
65. Mukhopadyay, N. K., Ghosh, S. K., Majumder, S., and Bose, S. K. (1985) *Biochem. J. 225*, 639.
66. Mukhopadyay, N. K., Majumder, S., Ghosh, S. K., and Bose, S. K. (1986) *Biochem. J. 240*, 265.
67. Majumder, S., Mukhopadyay, N. K., Ghosh, S. K., and Bose, S. K. (1988) *J. Gen. Microbiol. 134*, 1147.
68. Fujikawa, K., Suzuki, T., and Kurahashi, K. (1966) *Biochim. Biophys. Acta 161*, 232.
69. Kasche, V., and Michaelis, G. (1988) this volume, pp. 81–101.
70. Murata, K., and Kimura, A. (1982) *Appl. Environm. Microbiol. 44*, 1444.
71. Gushima, H., Miya, T., Murata, K., and Kimura, A. (1983) *Agric. Biol. Chem. 47*, 1927.
72. Kimura, A., Murata, K., Miya, T., and Gushima, H. (1984) *Eur. Patent 107, 400*; (1982) *Jap. Patent 82/170, 727.*
73. Palissa, H. (1989) *Thesis TU Berlin.*

74. Samson, S. M., Belagaje, R., Blankenship, D. T., Chapman, J. L., Perry, D., Skatrud, P. L., Van Frank, R. M., Abraham, E. P., Baldwin, J. E., Queener, S. W., and Ingolia, T. D. (1985) *Nature 318*, 191.
75. Baldwin, J. E. (1988) *Nat. Prod. Rep. 5*, 101.
76. Baldwin, J. E., Abraham, E. P., Adlington, R. M., Murphy, J. A., Green, N. B., Ting, H.-H., and Usher, J. J. (1983) *J. Chem. Soc. Chem. Comm.*, 1319.
77. Baldwin, J. E., Abraham, E. P., Adlington, R. M., Chakravarti, B., Derome, A. E., Murphy, J. A., Field, L. D., Green, N. B., Ting, H.-H., and Usher, J. J. (1983) *J. Chem. Soc. Chem. Comm.*, 1317.
78. Baldwin, J. E., Adlington, R. M., Basak, A., and Ting, H.-H. (1986) *J. Chem. Soc. Chem. Comm.*, 1280.
79. Baldwin, J. E., Adlington, R. M., Basak, A., Flitsch, S. L., Forrest, A. K., and Ting, H. H. (1986) *J. Chem. Soc. Chem. Comm.*, 273.
80. Baldwin, J. E., Adlington, R. M., Derome, A. E., Ting, H.-H., and Turner, N. J. (1984) *J. Chem. Soc. Chem. Comm.*, 1211.
81. Hamilton, B. K., Montgomery, J. P., and Wang, D. I. C. (1974) in *Enzyme Engineering*, Vol. 2 (Pye, E. K., and Wingard, L. B., Jr., eds.) pp. 153–159, Plenum Press New York.
82. Wang, D. I. C., Stramondo, J. G., and Fleischaker, R. (1977) in *Biotechnological Applications of Proteins and Enzymes* (Sharon, N., and Bohak, Z., eds.) pp. 183–201, Academic Press New York London.
83. Stramondo, J. G., Solomon, B. A., Colton, C. K., and Wang, D. I. C. (1978) *AICHE Symp. Ser. 74 (172)*, 1.
84. Madry, N., Zocher, R., Grodzki, K., and Kleinkauf, H. (1984) *Appl. Microbiol. Biotechnol. 20*, 83.
85. Siegbahn, N., Mosbach, K., Grodzki, K., Zocher, R., Madry, N., and Kleinkauf, H. (1985) *Biotechnol. Lett. 7*, 297.
86. Peeters, H., Zocher, R., Madry, N., Oelrichs, P. B., Kleinkauf, H., and Kraepelin, G. (1983) *J. Antibiot. 36*, 1762.
87. Peeters, H. (1987) *Thesis, TU Berlin.*
88. Billich, A., and Zocher, R. (1987) *J. Biol. Chem. 262*, 17258.
89. von Döhren (1982) in *Peptide Antibiotics-Biosynthesis and Functions* (H. Kleinkauf and H. von Döhren, eds.) pp. 169, de Gruyter, Berlin, New York.
90. Kleinkauf, H., and von Döhren, H. (1981) in *Advances in Biotechnology*, Vol. 3 (C. Vezina and K. Singh, eds.) p. 69, Pergamon Press London.
91. Kirchner, A., Simonis, M., and von Döhren, H. (1987) *J. Chromatogr. 396*, 199.
92. Agathos, S. N., and Demain, A. L. (1983) *Ann. N.Y. Acad. Sci. 434*, 44.
93. Agathos, S. N., and Demain, A. L. (1983) *ACS Symp. Ser. No. 207*, 53.
94. Demain, A. L., and Agathos, S. N. (1985) *Can. J. Microbiol. 32*, 208.
95. Agathos, S. N., and Demain, A. L. (1986) *Enzyme Microb. Technol. 8*, 465.
96. Agathos, S. N., and Demain, A. L. (1986) *Appl. Microbiol. Biotechnol. 24*, 319.
97. Agathos, S. N., and Demain, A. L. (1987) in *Horizons of Biochemical* Engineering (S. Aiba, ed.) pp. 147–161, University of Tokyo Press.
98. Okumura, Y. (1988) this volume pp. 365–378.
99. Fujimoto, Y., Gotoo, M., Abe, K., Suzuki, K., Kido, Y., Uyeda, M., Kinjo, J., Nohara, T., and Shibata, M. (1987) *Agric. Biol. Chem. 51*, 803.
100. Keller, U. (1984) *J. Biol. Chem. 259*, 8226.
101. Braun, V. (1988) this volume, pp. 103–129.
102. Bagg, A., and Neilands, J. B. (1987) *Microbiol. Rev. 51*, 509.
103. Burnham, M. K. R., Hodgson, J. E., and Normansell, I. D. (1987) *Eur. Patent Appl. o-233-715-A2.*
104. Chen, C. W., Lin, H.-F., Kuo, C. L., Tsai, H.-L., and Tsai, J. F.-Y. (1986) *5th Int. Symp. Genetics of Industrial Microorganisms*, Split, p. 164, abstract W7–P14.
105. Ingolia, T. D., and Queener, S. W. (1989) *Medicinal Chem. Rev.* in print.
106. Mittenhuber, G. (1988) *Thesis TU Berlin.*
107. Kleinkauf, H., Marahiel, M. A., von Döhren, H., Czekay, G., Frese, D., Krause, M., and Mittenhuber, G. (1987) *18th FEBS Meet.* Ljubljana, p. 10.

108. Malik, V. S. (1986) in *Biotechnology* (H.-J. Rehm and G. Reed, eds.) Vol. 4: *Microbial Products II* (H. Pape and H.-J. Rehm, eds.) pp. 39–68, VCH Verlagsgesellschaft Weinheim.
109. Hopwood, D. A. (1986) in *Biological, Biochemical and Biomedical Aspects of Actinomycetes* (G. Szabo, S. Biro and M. Goodfellow, eds.) pp. 3–14, Akademia Kiado, Budapest.
110. Braun, V. (1988) this volume pp. 103–129.
111. Laird, A. J., Ribbons, D. W., Woodrow, G. C., and Young, I. G. (1980) *Gene 11*, 347.
112. Laird, A. J., and Young, I. G. (1980) *Gene 11*, 359.
113. Pickett, C. L., Hayes, L., and Earhart, C. F. (1984) *FEMS Microbiol. Lett. 24*, 77.
114. Pettis, G. S., and McIntosh, M. A. (1987) *J. Bacteriol. 169*, 4154.
115. Greenwood, K. T., and Luke, R. K. J. (1976) *Biochim. Biophys. Acta 454*, 285.
116. Woodrow, G. C., Young, I. G., and Gibson, F. (1979) Biochim. *Biophys. Acta 582*, 145.
117. Nahlik, M. S., Fleming, T. P., and McIntosh, M. A. (1987) *J. Bacteriol. 169*, 4163.
118. Loper, J. E., Orser, C. S., Panopoulos, N. J., and Schroth, M. N. (1984) *J. Gen. Microbiol. 130*, 1507.
119. Marugg, J. D., van Spanje, M., Hoehstra, W. P. M., Schippers, B., and Weisbeek, P. J. (1985) *J. Bacteriol. 164*, 563.
120. Moores, J. C., Magazin, M., Ditta, G. S., and Leong, J. (1984) *J. Bacteriol. 157*, 53.
121. Hohnadel, D., Haas, D., and Meyer, J.-M. (1986) *FEMS Microbiol. Lett. 36*, 195.
122. Hummel, W., and Diekmann, H. (1981) *Biochim. Biophys. Acta 617*, 313.
123. Deml, G., Voges, K., Jung, G., and Winkelmann, G. (1984) *FEBS Lett.* 173, 53.
124. Keller, U., unpublished results.
125. Kambe, M., Imae, Y., and Kurahashi, K. (1974) *J. Biochem. Tokyo 75*, 481.
126. Kanda, M., Hori, K., Kurotsu, T., Miura, S., Nozoe, A., and Saito, Y. (1978) *J. Biochem. Tokyo 84*, 435.
127. Hori, K., Kanda, M., Kurotsu, T., Miura, S., Yamada, Y., and Saito, Y. (1981) *J. Biochem. Tokyo 90*, 439.
128. Saito, Y. (1982) in *Peptide Antibiotics-Biosynthesis and Functions* (H. Kleinkauf and H. von Döhren, eds.) pp. 195–208, de Gruyter, Berlin, New York.
129. Hori, K., Kurotsu, T., Kanda, M., Miura, S., Yamada, Y., and Saito, Y. (1982) *J. Biochem. Tokyo 91*, 369.
130. Hori, K., Kanda, M., Miura, S., Yamada, Y., and Saito, Y. (1983) *J. Biochem. Tokyo 93*, 177.
131. Symons, D. C., and Hodgson, B. (1982) *J. Bacteriol. 151*, 580.
132. Haavik, H. I., and Froyshov, O. (1975) *Nature 254*, 79.
133. Vasantha, N., Balakrishnan, R., Kaur, S., and Jayaraman, K. (1980) *Arch. Biochem. Biophys. 200*, 40.
134. Marahiel, M. A., Danders, W., Krause, M., and Kleinkauf, H. (1979) *Eur. J. Biochem. 99*, 49.
135. Dobritsa, A. T., Dobritsa, S. V., and Tanyashin, V. I. (1978) *Mol. Gen. Genet. 164*, 195.
136. Marahiel, M. A., Lurz, R., and Kleinkauf, H. (1980) *J. Antibiot. 34*, 323.
137. Stärko, K. (1982) *Diplomarbeit TU Berlin.*
138. Hara, T., Aumayr, A., Fujio, Y., and Ueda, S. (1982) *Appl. Environm. Microbiol. 44*, 1456.
139. Krause, M., Marahiel, M. A., von Döhren, H., and Kleinkauf, H. (1985) *J. Bacteriol. 162*, 1120.
140. Bothe, D. (1986) *Thesis TU Berlin.*
141. Marahiel, M. A., Krause, M., and Skarpeid, H.-J. (1985) *Mol. Gen. Genet. 201*, 231.
142. Marahiel, M. A., Zuber, P., Czekay, G., and Losick, R. (1987) *J. Bacteriol. 169*, 2215.
143. Nüesch, J., Heim, J., and Treichler, H.-J. (1987) *Annu. Rev. Microbiol. 41*, 51.
144. Jensen, S. E. (1985) *Crit. Rev. Biotechnol. 3*, 277.
145. Demain, A. L. (1983) in *Antibiotics containing the β-lactam structure*, Part I (A. L. Demain and N. A. Solomon, eds.), pp. 198–228, Springer, Berlin, New York.
146. Ono, H., and Harada, S. (1988) this volume pp. 131–158.
147. von Döhren et al., unpublished.
148. Weigel, B. J., Burgett, S. G., Chen, V. J., Skatrud, P. L., Frolik, C. A., Queener, S. W., and Ingolia, T. D. (1988) *J. Bacteriol. 170*, 3817.
149. Leskiw, B. K., Aharonowitz, Y., Mevarech, M., Wolfe, S., Vining, L. C., Westlake, D. W. S., and Jensen, S. E. (1989) *Mol. Gen. Genetics*, in press.

150. Ramon, D., Carramolino, L., Patino, C., Sanchez, F., and Penalva, M. A. (1987) *Gene 57*, 171.
151. Carr, L. G., Skatrud, P. L., Scheetz, M. E., Queener, S. W., and Ingolia, T. D. (1986) *Gene 48*, 257.
152. Romeo, D., Skerlavaj, B., Bolognesi, M., and Gennaro, R. (1988) *J. Biol. Chem. 263*, 9573.
153. Malpartida, F., Hallam, S. E., Kieser, H. M., Motamedi, H., Hutchinson, C. R., Butler, M. J., Sugden, D. A., Warren, M., McKillop, C., Bailey, C. R., Humphreys, G. O., and Hopwood, D. A. (1987) *Nature 325*, 818.
154. Kinashi, H., Shimaji, M., and Sakai, A. (1987) *Nature 328*, 454.
155. Aharonowitz, Y. et al., unpublished results.
156. Reichenbach, H., Gerth, K., Irschik, H., Kunze, B., and Höfle, G. (1988) *Trends Biotechnol. 6*, 115.
157. Irschik, H., Gerth, K., Kemmer, T., Steinmetz, H., and Reichenbach, H. (1983) *J. Antibiot. 35*, 1454.
158. Irschik, H., and Reichenbach, H. (1985) *J. Antibiot. 38*, 1237.
159. Sakakibiara, H., Naganawa, H., Ohno, M., Maeda, K., and Umezawa, H. (1974), *J. Antibiot. 27*, 879.
160. von Jagow, G., Ljungdahl, P. O., Graf, P., Ohnshi, T., and Trumpower, B. L. (1984) *J. Biol. Chem. 259*, 6318.
161. Cooper, R., Bush, K., Principe, P. A., Trejo, W. H., Wells, S. J., and Sykes, R. B. (1983) *J. Antibiot. 36*, 1252.
162. Singh, P. D., Johnson, J. H., Ward, P. C., Wells, J. S., Trejo, W. H., and Sykes, R. B. (1983) 1245.
163. Tsubotani, S., Hida, T., Kasahara, F., Wada, Y., and Harada, S. (1984) *J. Antibiot. 37*, 1546.
164. Ono, H., Nozaki, Y., Katayama, N., and Okazaki, H. (1984) *J. Antibiot. 37*, 1528.
165. Harada, S., Tsubotani, S., Ono, H., and Okazaki, H. (1984) *J. Antibiot. 37*, 1536.
166. Nozaki, H., Okonogi, K., Katayama, N., Ono, H., Harada, S., Kondo, M., and Okazaki, H. (1984) *J. Antibiot. 37*, 1555.
167. Singh, P. D., Young, M. G., Johnson, J. H., Cimarusti, C. M., and Sykes, R. B. (1984) *J. Antibiot. 37*, 773.
168. Shoji, J., Kato, T., Sakazaki, R., Nagata, W., Terui, Y., Nakagawa, Y., Shiro, M., Matsumoto, K., Hattori, T., Yoshida, T., and Kondo, E. (1984) *J. Antibiot. 37*, 1486.
169. Nozaki, Y., Katayama, N., Tsubotani, S., Ono, H., and Okazaki, H. (1985) *J. Antibiot. 38*, 1141.
170. Tsubotani, S., Hida, T., Ono, H., and Harada, S. (1985) *J. Antibiot. 38*, 1152.
171. Tymiak, A. A., Kirsch, D. R., O'Sullivan, J., and McCullough, J. E. (1986) *Jpn. Kokai 227*, 788.
172. Shoji, J., Hinoo, H., Matsumoto, K., Hattori, T., Yoshida, T., Matsuura, S., and Kondo, E. (1988) *J. Antibiot. 41*, 713.
173. Kato, T., Hinoo, H., Terui, Y., Kikuchi, J., and Shoji, J. (1988) *J. Antibiot. 41*, 719.
174. Elander, R. P. (1983) in *Antibiotics containing the Beta-Lactam Structure* (A. L. Demain and N. A. Solomon, eds.), p. 97, Springer, Berlin, Heidelberg, New York, Tokyo.
175. Rake, J. B., Gerber, R., Mehta, R. J., Newman, D. J., Oh, Y. K., Phelen, C., Shearer, M. C., Sitrin, R. D., and Nisbet, L. J. (1986) *J. Antibiot. 39*, 58.
176. Corti, A., and Cassani, G. (1985) *Appl. Biochem. Biotechnol. 11*, 101.
177. Vanittanakom, N., Loeffler, W., Koch, U., and Jung, G. (1986) *J. Antibiot. 39*, 888.
178. Rapp, C., Jung, G., Kugler, M., and Loeffler, W. (1988) *Liebigs Ann. Chem.*, 655.
179. Bushell, M. E., and Fryday, A. (1983) *J. Gen. Microbiol. 129*, 1733.
180. Faulkner, D. J. (1984) *Nat. Prod. Rep. 1*, 251.
181. Faulkner, D. J. (1985) *Nat. Prod. Rep. 2*, 552.
182. Faulkner, D. J. (1986) *Nat. Prod. Rep. 3*, 1.
183. Ireland, C. M., Molinski, T. F., Roll, D. M., Zabriskie, T. M., McKee, T. C., and Swersey, J. C. (1989) in press.
184. Botes, D. P., Tuiman, A. A., Wessels, P. L., Viljoen, C. C., Kruger, H., Williams, D. H., Santikarn, S., Smith, R. J., and Hammond, S. J. (1984) *J. Chem. Soc. Perkin Trans. I*, 2311.
185. Botes, D. P., Wessels, P. L., Kruger, H., Runngar, M. T. C., Santikarn, S., Smith, R. J., Barna, J. C. J., and Williams, D. H. (1985) *J. Chem. Soc. Perkin Trans. I*, 2747.

186. Sakai, S., Hitotsuyanagi, Y., Aimi, N., Fujiki, H., Suganuma, M., Sugimura, T., Endo, Y., and Shudo, K. (1986) *Tetrahedron. Lett. 27*, 5219.
187. Fujiki, H., and Sugimura, T. (1983) *Cancer Surveys 2*, 539.
188. Cardellina II, J. H., Marner, F.-J., and Moore, R. E. (1979) *Science 204*, 193.
189. Matsunaga, S., Fusetani, N., and Konosu, S. (1985) *J. Nat. Prod. 48*, 236.
190. Matsunaga, S., Fusetani, N., and Konosu, S. (1984) *Tetrahedron Lett. 25*, 5165.
191. Matsunaga, S., Fusetani, N., and Konosu, S. (1985) *Tetrahedron Lett. 26*, 855.
192. Yasumasa, H., Kato, S., and Shioiri, T. (1985) *Tetrahedron Lett. 26*, 3223.
193. Schmidt, U., and Weller, D. (1986) *Tetrahedron Lett. 27*, 3495.
194. Schmidt, U., and Griesser, H. (1986) *Tetrahedron Lett. 27*, 163.
195. Hamada, Y., Shibata, M., and Shioiri, T. (1985) *Tetrahedron Lett. 26*, 6501.
196. Rinehart, K. L., Jr., Gloer, J. B., Hughes, R. J., Jr., Renis, H. E., McGovren, J. P., Swynenberg, E. B., Stringfellow, D. A., Kuentzel, S. L., and Li, L. H. (1981) *Science 212*, 933.
197. Rineheart, K. L., Kishore, V., Nagarajan, S., Lake, R. J., Gloer, J. B., Bozich, F. A., Li, K.-M., Maleczka, R. E., Todsen, W. L., Munro, M. H. G., Sullins, D. W., and Sakai, R. (1987) *J. Am. Chem. Soc. 109*, 6846.
198. Chun, H. G., Davies, B., Hoth, D., Suffness, M., Plowman, J., Flora, K., Grieshaber, C., and Leyland-Jones, B. (1986) *Invest. New Drugs 4*, 279.
199. Thierbach, G., and Reichenbach, H. (1983) *Arch. Microbiol. 134*, 104.
200. Painuly, P., Perez, R., Fukai, T., and Shimizu, Y. (1988) *Tetrahedron Lett. 29*, 11.
201. Pettit, G. R., Kamano, Y., Brown, P., Gust, D., Inoue, M., and Herald, C. L. (1982) *J. Am. Chem. Soc. 104*, 905.
202. Pettit, G. R., and Holzapfel, C. W. (1986) *J. Org. Chem. 51*, 4580.
203. Schmidt, U., Utz, R., Lieberknecht, A., Griesser, H., Potzolli, B., Bahr, J., Wagner, K., and Fischer, P. (1987) *J. Org. Chem. 233.*
204. Schmidt, U., Utz, R., Lieberknecht, A., Griesser, H., Potzolli, B., Bahr, J., Wagner, K., and Fischer, P. (1987) *J. Org. Chem. 236.*
205. Schmidt, U. (1986) *Pure Appl. Chem. 58*, 295.

Chapter 2
Gramicidin S Synthetase

Joachim Vater

1. Introduction
2. Purification of Gramicidin S Synthetase
3. Physical and Biochemical Properties of Gramicidin S Synthetase
4. Reaction Mechanism of the Biosynthesis of Gramicidin S and Product Patterns of Gramicidin S Synthetase
5. Reaction Centers for Substrate Activation
6. Inhibitor Studies
7. Structural Aspects
8. Investigation of Gramicidin S Negative Mutants
9. Prospects
References

Abbreviations:
GS: gramicidin S synthetase (complete enzyme), GS1: gramicidin S synthetase 1 (light enzyme, phenylalanine racemase), GS2: gramicidin S synthetase 2 (heavy enzyme), CmK: chloromethylketone, NEM: *N*-ethylmaleimide, DTNB: 3,3′-dithiobis(6-nitrobenzoic acid), DTE: dithioerythritol

1. Introduction

In the decade between 1960 and 1970 it was demonstrated in several laboratories that numerous bioactive peptides, especially those of microbial origin, were synthesized by their producer organisms nonribosomally [1–4]. This feature was mainly based on the observation that the biosynthesis of such compounds was insensitive against the action of inhibitors for protein biosynthesis, as, for example, chloramphenicol, streptomycin etc. In the late 1960's it was demonstrated mainly in the laboratories of Lipmann [5], Laland [6] and Kurahashi [7] that polyenzymes were used for the production of certain peptide antibiotics, as the gramicidins, tyrocidines or bacitracins. A few peptide synthetases were isolated and partially purified. One of the most prominent systems among such multienzymes is gramicidin S synthetase. The investigation of this multienzyme contributed basic knowledge on the mechanism of nonribosomal peptide synthesis. Gramicidin S is a cyclic decapeptide antibiotic produced by certain *B. brevis* strains

$$\begin{array}{ccccccccc} \text{D-Phe} & \rightarrow & \text{Pro} & \rightarrow & \text{Val} & \rightarrow & \text{Orn} & \rightarrow & \text{Leu} \\ \uparrow & & & & & & & & \downarrow \\ \text{Leu} & \leftarrow & \text{Orn} & \leftarrow & \text{Val} & \leftarrow & \text{Pro} & \leftarrow & \text{D-Phe} \end{array}$$

Two pathways have been characterized for nonribosomal biosyntheses of bioactive peptides. Most of the peptide synthetases characterized so far, as, for example, gramicidin S synthetase, use the thiotemplate mechanism [5–8]. Such multienzymes activate their substrate amino acids in two steps. The first stage involves aminoacyl adenylates of the constituent amino acids which are subsequently transferred to a reactive sulfhydryl group at the specific thiotemplates. It is postulated [9] that the substrates which are continuously provided at the reaction centers are linked together in a series of transpeptidation and transthiolation processes by interaction of an internal central phosphopantetheine carrier with the peripheral thiol groups at the reaction centers.

Representatives of the second pathway used in peptide biosynthesis are the enzyme systems which catalyze the formation of glutathione [10] or mycobacillin [11, 12], a tridecapeptide from *B. subtilis* B3. In this case a one-step phosphate activation mechanism is used. Here elongation proceeds by addition of nonactivated acceptor amino acids to carboxyl activated intermediate peptides attached noncovalently to the reaction centers as aminoacyl phosphates.

Gramicidin S synthetase has been intensively reviewed by several authors [5–8, 13–25]. Though the reaction sequence of gramicidin S formation has been studied in detail, knowledge is still lacking concerning the architecture of the multienzyme, in particular. Recent research in this field is concentrated on the proteinchemical, physicochemical, genetic and immunological analysis of gramicidin S synthetase. This article is mainly focussed on the progress obtained during the last decade.

2. Purification of Gramicidin S Synthetase

Gramicidin S synthetase has been purified in several laboratories. Generally combinations of gel filtration, ion exchange – and adsorption chromatography as well as gradient centrifugation techniques were used [26–33]. The progress in purification of this multienzyme reached until 1975 was reviewed by Zimmer and Laland [34]. Using such procedures GS2 could be purified to a nearly homogeneous state. GS1, however, usually appeared only in partially purified form. In recent years advanced purification techniques were developed to obtain enzyme preparations of good quality and high specific activities.

Pass et al. [35, 36] reported on affinity chromatography of GS1 and GS2 using 3,3′-diaminodipropylamine substituted Sepharose to which amino acid components of gramicidin S were coupled via their carboxyl group. The authors claimed that binding of both enzymes to such columns is determined by their affinity to the substrate ligands attached to the Sepharose matrix. A more careful analysis of these effects [37, 38] revealed that binding of GS1 and GS2 occurs mainly non-biospecifically and has rather to be ascribed to adsorptive effects of the spacer molecules than to specific, active site directed interactions with the ligands. Nevertheless, such columns allow efficient purification of both components of gramicidin S synthetase. For example, if proline is attached to Sepharose, a rapid and convenient separation of GS1 and GS2 in yields between 80–100% could be obtained. If GS1 labeled with ^{14}C-D-Phe in thioester bound form was loaded onto the column, the tracer appeared transferred to GS2 upon elution and separation of both enzymes. No transfer occurred when valine, ornithine and leucine were used as the ligands at the Sepharose matrix. Obviously, charging of proline to the heavy enzyme is required for the transfer of D-Phe to take place. The label appeared incorporated into gramicidin S, if GS2 was incubated with GS1, ATP, Mg^{2+} and the other substrate amino acids. This indicates that in the biosynthesis of gramicidin S D-Phe is transferred to an acceptor site on GS2 prior to chain elongation. If L-Val, L-Orn and L-Leu were used as the ligands for preparation of the Sepharose columns both enzymes were retained. Using a salt step GS1 was always eluted in front of GS2.

A similar technique was applied by Hori et al. [39] and Kanda et al. [40] for efficient purification of gramicidin S synthetase in combination with gel filtration and ion exchange procedures. Ornithine-Sepharose 4B was used for chromatography of GS2. GS1 was obtained in highly purified form by combination of chromatography on free-amino and free-carboxyl-D-Phe-Sepharose 4B. Also in this case the results suggest that the interaction between enzymes and gel might be mainly due to ion exchange effects.

Recently efficient techniques for the purification of GS1 were introduced. Schröter-Kermani et al. [41] showed that this enzyme is inhibited by certain disulfides. The most potent compound tested was cystamine which interacts with GS1 forming a mixed disulfide between cysteamine and a sulfhydryl group of the enzyme. Cystamine coupled to epoxy-activated Sepharose 6B has been used for covalent chromatography of GS1. The immobilized enzyme can be cleaved from the matrix after disulfide reduction with thiol reagents, as dithioerythritol (DTE), either in the presence of the substrates Phe, ATP and Mg^{2+} or by a salt step.

The preparation of both enzymes of gramicidin S synthetase was efficiently improved by fast protein liquid chromatography (FPLC) using a Pharmacia Mono Q HR 5/5 anion exchange column after prepurification of GS1 and GS2 by gel filtration on AcA34 and ion exchange chromatography on DE-52 [42]. This procedure is distinguished by a high resolution and allows chromatography of a few milligrams of enzyme material in a time as short as 20–60 min. Using the FPLC technique GS1 and GS2 preparations of high purity were obtained. Important features of the FPLC elution profiles are their excellent reproducibility and the high recovery of the enzyme activity. Highly purified gramicidin S synthetase obtained using the procedures summarized above shows specific activities in the range of 20–35 nmol gramicidin S produced per min and mg protein.

3. Physical and Biochemical Properties of Gramicidin S Synthetase

Sedimentation coefficients $s_{20,w}$ of 6.1 and 11.6 S have been determined for GS1 and GS2, resp. [27, 34]. The molecular weight of GS2 has been extensively investigated by Koischwitz and Kleinkauf [43]. They obtained 280 kDa as an average of their data. For GS1 100 kDa was reported by Kleinkauf et al. [26], as estimated from sucrose gradient density centrifugation. Recent studies using SDS-polyacrylamide gel electrophoresis and gel filtration techniques show an appreciably higher molecular weight for GS1 [44]. From a diagram log MW versus the electrophoretic mobility of GS1 and various standard proteins a molecular weight of 118 kDa was derived. Gel filtration of GS1 on a Sephacryl S-200 superfine column calibrated with standard proteins showed an even higher value of 132 kDa which implies that the shape of GS1 may deviate from the globular conformation to some extent which is in agreement with electron micrographs of GS1.

The amino acid composition of GS1 and GS2 has been determined [44]. Both proteins are distinguished by a high number of acidic amino acid residues in agreement with their isoelectric points which range between 4.5 and 5 [31]. Their cysteine content was estimated measuring the number of cysteic acids after hydrolysis of the proteins oxidized with performic acid [44]. 6–7 and 20–23 sulfhydryl groups were obtained for GS1 and GS2, resp. These results are in good agreement with previous thiol titration experiments using *N*-ethylmaleimide (NEM) and 3,3′-dithiobis(6-nitrobenzoic acid) (DTNB). Under saturating inhibitor concentrations Schlumbohm et al. [45, 46] found an uptake of NEM by GS1 and GS2 equivalent to a modification of 6 and 24 thiols, resp. These data are significantly higher than those reported previously by Kanda et al. [47] who obtained a labeling of 3 and 20 mol sulfhydryl groups per mol GS1 and GS2, resp. The data published by both groups [45, 48] for GS1 are in good agreement using DTNB. In this case 6 thiols were found. There is again a discrepancy in their data concerning GS2 using this modifier. Schlumbohm et al. [46] detected 16 sulfhydryl groups, while Kanda et al. [47] found an appreciably higher value of 27 thiols.

Substrate amino acid induced protection of GS1 and GS2 against deactivation by NEM indicates that in addition to the specific thiols at the thiotemplates of gramicidin S synthetase reactive SH-groups are also involved in the primary aminoacyl adenylate activation reactions [45, 46]. At low NEM concentrations in the range of 20 μM the catalytically active thiols of the thioester binding sites are preferentially modified. Additional substrate induced protection effects have been observed at NEM concentration between 20 and 200 μM at which the ATP-PP_i-exchange reactions of both enzymes are affected. Obviously, one sulfhydryl group is involved in the primary activation of Phe at GS1, while 2 cysteines seem to be essential for the adenylation of each L-Pro, L-Val and L-Leu. Even 3 thiols were detected for the ornithine activation.

4. Reaction Mechanism of the Biosynthesis of Gramicidin S and Product Patterns of Gramicidin S Synthetase

The mechanism of the biosynthesis of gramicidin S has been explored in detail. The reaction sequence is well documented by the research performed in several laboratories [5–8, 13–25]. Gramicidin S formation comprises at least 18 individual steps which are demonstrated in Scheme 1.

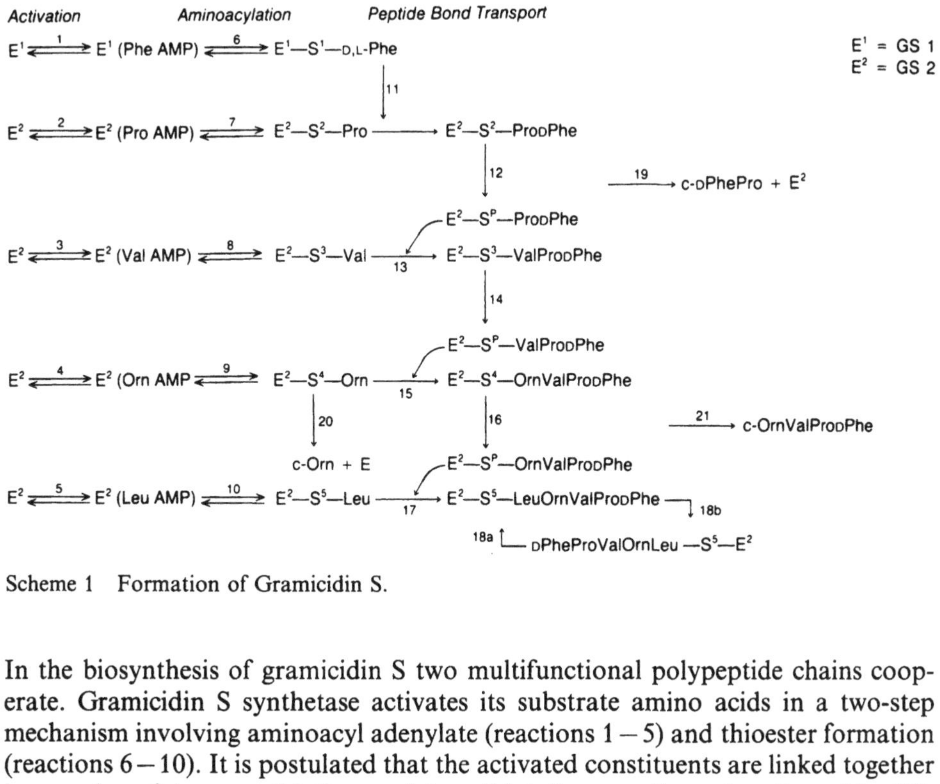

Scheme 1 Formation of Gramicidin S.

In the biosynthesis of gramicidin S two multifunctional polypeptide chains cooperate. Gramicidin S synthetase activates its substrate amino acids in a two-step mechanism involving aminoacyl adenylate (reactions 1–5) and thioester formation (reactions 6–10). It is postulated that the activated constituents are linked together in a series of transpeptidation and transthiolation reactions (reactions 11–17). Phenylalanine racemase [gramicidin S synthetase 1 (GS1), EC 5.1.1.11; 118 kDa] activates and racemizes phenylalanine. The elongation process starts with the transfer of activated D-Phe to a condensing, phosphopantetheine-containing multienzyme [gramicidin S synthetase 2 (GS2); 280 kDa] which activates the other substrate amino acids. Two pentapeptides are finally assembled head to tail forming the cyclodecapeptide antibiotic (reactions 18a and b). The mechanism of the termination reaction still remains to be clarified. Intramolecular [49] and intermolecular [50] cyclization of two pentapeptide intermediates has been discussed.

Gramicidin S synthetase shows a complex product pattern. In addition to gramicidin S several other compounds can be formed by early termination of the growing peptide chain as well as by substrate substitution and modification reactions. Already in an early stage of the investigation of the gramicidin S synthetase system D-Phe-Pro-piperazinedione formation was observed [51]. This compound is the main product, if GS is supplied with the first two amino acid components in the reaction sequence of gramicidin S formation (reaction 19). This process has been used by Kleinkauf et al. [24] for the production of several 2,5-piperazinediones. Depending mainly on the nature of the amino acid which determines the tendency for cyclization, the rate of the reaction may compete with gramicidin S formation. 3,4-Dehydroproline, azetidine-2-carboxylic acid and especially γ-thioproline in combination with phenylalanine promote high rates of cyclodipeptide formation.

The thioester complex of GS2 with L-Orn is decomposed converting the activated substrate into 3-amino-2-piperidone (cyclo-L-Orn) which is formed in a rapid internal cyclization process by nucleophilic attack of the δ-amino group at the carbon atom of the thioester bond [52] (reaction 20). In a similar reaction D-Phe-Pro-Val-cyclo-Orn was obtained as a product of the multienzyme by omission of L-Leu from the complete assay mixture [42]. Also this tetrapeptide was formed by cyclization of the C-terminal ornithine to 3-amino-2-piperidone (reaction 21). This compound was identified and characterized by chromatographic and spectroscopic procedures using chemically synthesized reference compounds.

The experimental material so far available implies that both aminoacyl adenylate and thioester formation between GS and its substrate amino acids are reversible processes. This result is mainly based on the detection of a) GS-mediated, specific ATP-AMP exchange reactions reported by several authors [27, 53–55], and b) a decrease in thioester formation in the presence of both AMP and PP_i applied in the millimolar range [56]. Gramicidin S synthetase also catalyzes AMP-ADP and ADP-ATP exchange reactions [57]. AMP and ADP form reactive intermediates which probably represent the mixed anhydrides of these nucleotides and the corresponding amino acid entering the back reaction of substrate activation by interaction with the covalent aminoacyl-thiol-enzyme. These processes are strictly dependent on the presence of ATP and are stimulated by inorganic pyrophosphatase indicating that both exchange reactions are dependent on the initial formation of aminoacyl adenylates and involve the phosphorylation of active intermediates of AMP and ADP by transfer of the γ-phosphate of ATP to yield ADP and ATP, resp.

The thiotemplates of GS can be saturated at substrate amino acid concentrations as low as 10–20 μM [56]. Adenylate formation and ATP-PP_i exchange reactions become apparent at 10–100-fold higher substrate concentrations. Obviously, the dissociation constants K_D for the gramicidin S synthetase – substrate amino acid – thioester complexes are 100–1000-fold lower compared to those of the aminoacyl adenylates and the K_M-data of the corresponding ATP-PP_i exchange processes. The affinity for the substrate amino acids is appreciably higher at the thiotemplate sites than at the aminoacyl adenylate reaction centers. Therefore the activation equilibria are quantitatively shifted toward thioester formation.

A set of thermodynamic parameters was derived for the activation processes from the temperature dependence of the K_M and K_D data [56]. Reaction enthalpies were obtained from a van't Hoff analysis of these constants. The thermodynamic data indicate that the enthalpy increments cannot account for the relatively large negative changes of the Gibbs free energy that is observed in the activation of substrate amino acids of GS2. Here ΔG° is predominantly controlled by entropy contributions indicating that amino acid recognition and selection in the biosynthesis of gramicidin S is mainly dependent on hydrophobic and van der Waals interactions between their specific side chains and complementary regions of the active sites of the enzyme. In contrast, the overall activation and concomitant racemization of phenylalanine by GS1 is an exothermic process which is mainly determined by the enthalpy increment of ΔG°.

The intermediate steps in the biosynthesis of gramicidin S shown in Scheme 1 have been studied as a function of temperature between 0 and 40°C [58]. Substrate amino acid activation is observed all over this temperature range. The aminoacyl adenylation reactions show a characteristic temperature dependence. With exception of the ornithine-activation they are endothermic processes. Anomalous Arrhenius plots have been obtained for the substrate amino acid dependent ATP-PP_i exchange reactions catalyzed by GS. These results indicate multiple reaction states of the multienzyme. On the other hand, the thioester formation reactions are almost independent of temperature showing low activation energies.

Gramicidin S formation stops below 10°C, while D-Phe-Pro-piperazinedione is still formed at an appreciable rate. D-Phe-Pro-Val-tripeptide formation has been detected as the temperature sensitive step. The production of this intermediate is diminished in parallel with gramicidin S formation under these conditions. These effects imply that either the mobility of the phosphopantetheine arm may be strongly reduced below 10°C or that the elongation cycle may be blocked, for example, by a conformational change of the multienzyme which would be induced between 8 and 12°C.

Table 1 Half-lives $t_{1/2}$ for the thioester formation of GS2 (heavy enzyme) with substrate amino acids.

Amino acid	Temperature (°C)	$t_{1/2}$ (s)
L-Pro	1	135
L-Val	1	570
	10	300
	20	25
	30	12
L-Orn	1	20
L-Leu	1	15
	25	2.6

Reaction conditions: a) 9.3 μM L-Pro; b) 18.6 μM L-Val; c) 10.3 μM L-Orn or d) 8.8 μM L-Leu; 2.5 mM ATP; 10 mM Mg^{2+}; 0.25 mM EDTA; 1 mM DTE in 20 mM phosphate buffer, pH = 7.2. Enzyme concentration: 0.27 μM.

Studies of the temperature dependence of the biosynthesis of gramicidin S show that the valine activation and essentially the D-Phe-Pro-Val-tripeptide formation are the rate limiting steps in this process. This is corroborated by kinetic measurements. The kinetics of the thioester formation reactions of GS2 have been measured under substrate saturation conditions. The half-life times for these processes which are summarized in Table 1 were found in the range of seconds to a few minutes. The detection of the kinetic parameters at temperatures higher than 20°C needs the application of the quenched flow technique, as demonstrated for the leucine activation in Table 1.

Thioester formation of GS2 with L-Orn and L-Leu is appreciably faster than with L-Pro and L-Val. The direct measurement of the activation kinetics demonstrates again that the valine activation is one of the rate limiting steps in the biosynthesis of gramicidin S. A characteristic feature of the valine activation is its exceptional temperature sensitivity indicating a relatively high activation energy.

5. Reaction Centers for Substrate Activation

Kinetics of the substrate amino acid dependent ATP-PP_i exchange reactions and inhibition studies showed a random addition of substrates to the aminoacyl adenylate formation sites in the primary activation steps of the biosynthesis of gramicidin S [59]. From experiments with substrate analogous amines, carboxylic acids and *N*-acylated derivatives of Phe, Pro, Val, Orn and Leu it has been inferred that similar to tRNA synthetases the α-amino group of the substrate amino acids is essential for their binding to the aminoacyl adenylate activation centers of gramicidin S synthetase, while the carboxyl group is of minor importance in this respect [59, 60].

A collection of substrate analogues has been tested concerning replacement of the natural substrates in the activation processes and product formation [14, 24, 62]. Generally activation of an amino acid or one of its analogues coincides well with its incorporation into the peptide product. A few of such compounds, as, for example, the Phe analogues threo-phenylserine or β-2-thionylserine as well as some D-isomers (D-Leu or D-Ile) are readily activated and accepted at the reaction centers of the multienzyme, but not incorporated into the peptide product. D-Phe required for the biosynthesis of gramicidin S is provided by the racemase activity of GS1. A specific sulfhydryl group at the active site of this protein is aminoacylated by phenylalanine. After alkaline hydrolysis both enantiomers of Phe appear in a ratio D:L = 7:3 [7, 62]. In the subsequent transfer reaction to GS2 the D-form is exclusively involved demonstrating the stereo-specificity of the initiation reaction.

The activation site of GS2 for L-ornithine is distinguished by a relatively high substrate variability [60, 62]. Studies with analogues of L-Orn specifically modified at its δ-amino group, like L-arginine, *N*-δ-*t*-butyloxycarboxyl-L-ornithine or *N*-δ-tri-fluoroacetyl-L-ornithine show that this reaction center is not rigidly closed and that a hydrophobic region extends from the site normally occupied by the δ-amino group of L-Orn.

Several structural analogues of ATP have been studied assuming their effects in aminoacyl adenylate formation catalyzed by GS. ATP can be replaced by 2-chloro-ATP as well as 2′- and 3′-dATP, while analogues substituted in position 8 are rejected at the reaction centers. 2′-O-methyl and 3′-O-methyl-ATP inhibit the aminoacyl adenylation processes. Obviously, the activation sites of gramicidin S synthetase do not resemble those of known aminoacyl tRNA ligase families, but exhibit a specific class of ATP-binding sites.

Most of the studies reported so far present qualitative or semi-quantitative information. K_M, K_i and K_D data are needed for analysis of a) the intermolecular forces which determine substrate recognition and binding and b) the topography of the reaction centers of GS. We have performed such measurements of the incorporation of substrate analogues into the thioester binding sites of the multienzyme. In Table 2 the dissociation constants of some proline and ornithine analogues are shown. L-Pro can be efficiently replaced by thioproline and azetidine carboxylic acid, as previously demonstrated for aminoacyl adenylation, piperazinedione and cyclodecapeptide formation [14, 24, 63]. The dissociation constants for the GS2-substrate analogue-thioester complexes are 20–60-fold higher than the K_D for thioester bound proline. Also L-Orn can be replaced both in the activation reactions and product formation by L-Lys and L-Arg.

Table 2 The specificity of thioester activation sites of GS.

Activation site for	Substrate or substrate analogue	K_D (μM)
L-proline	L-proline	0.44
	L-thioproline	10
	L-azetidine carboxylic acid	31
L-ornithine	L-ornithine	0.3
	L-lysine	430
	L-arginine	13,700

Reaction conditions: Variable concentrations of substrate amino acids or substrate analogues; 2.5 mM ATP; 10 mM Mg^{2+}; 0.25 mM EDTA; 1 mM DTE in 20 mM phosphate buffer, pH = 7.2; temperature: 37°C.

Table 3 demonstrates that L-Val and L-Leu which show a close structural relationship can be bound to the thiotemplate of either substrate. L-Val binds more tightly to the thioester binding site of L-Leu than vice versa L-Leu to the activation center of L-Val. Scatchard plots for thioester formation of GS2 with these substrates and a collection of structurally related analogues, like L-Ile, L-*allo*-Ile, L-Norval, L-Norleu or L-Thr, are biphasic. This pattern is consistent with binding of these components to both reaction centers. Substrate analogues with branched side chains, as L-Ile and especially L-*allo*-Ile bind much more tightly than the linear derivatives L-Norval and N-Norleu. Analogues, like L-Thr, which are related to the natural

Table 3 The specificity of thioester activation sites of GS.

Activation site for	Substrate or substrate analogue	Substrate concentration for 50% thioester formation (μM)
a) L-valine	L-valine	a) 0.18 b) 12
b) L-leucine	L-leucine	a) 100 b) 0.11
	L-norvaline	a) 20 b) 700
	L-norleucine	a) 1500 b) 12
	L-isoleucine	a) 5 b) 20
	L-allo-isoleucine	a) 0.8 b) 1.4
	L-threonine	a) 175 b) 385

Reaction conditions as indicated in Table 2.

substrate by replacement of a hydrophobic structural element by a polar group are still accepted at the thioester binding sites of L-Val and L-Leu. As has been shown previously by other authors [14, 24, 63] all these compounds mediate efficient ATP-PP_i exchange and cyclodecapeptide formation. Testing the fidelity of GS Laland et al. [64] demonstrated that cyclohexylalanine functions as an analogue for phenylalanine at GS1 and for leucine at GS2. In a similar manner phenylalanine can replace L-Leu. D- and L-Phe are activated and thioesterified by the heavy enzyme to a significant degree. The manifold substrate substitutions recognized so far show that peptide producing enzyme systems, like gramicidin S synthetase, usually are less specific concerning substrate binding than tRNA synthetases in the ribosomal system.

Thioester are stable in acidic media and are efficiently hydrolyzed in alkaline solution. This has been confirmed for the reactive thioester complexes involved in the biosynthesis of gramicidin S [54]. Gadow et al. [52] showed that these intermediates are slowly hydrolyzed at neutral pH under mild conditions. These processes are accelerated 10–100-fold in the presence of thiol compounds, as dithioerythritol, cysteine, cysteamine or β-mercaptoethanol, while disulfides as pantetheine or DTNB do not significantly promote these reactions. At 3°C in the presence of 1 mM dithioerythritol, for example, the half-life times for the decomposition of most of these thioester complexes are in the range of a few hours. The high stability of the GS2-D-Phe-Pro-Val-thioester complex with a half-life $t_{1/2} = 90$ h is remarkable. This effect is possibly due to steric hindrance of hydrolysis by the attached hydrophobic moiety of the tripeptide. The rate of the decomposition processes increases appreciably with temperature. The half-lives of the thioester intermediates at 37°C are in the range of 10–20 min. The catalytic effects of thiol compounds presumably reflect acyl transfer processes from the reactive sulfhydryl groups at the thiotemplates of the multienzyme to the accelerator, followed by hydrolysis of the *S*-acyl-accelerator derivatives. Roskoski et al. [50] applied thioesters of thiophenol and *N*-acetyl cysteine with phenylalanine as active-site directed agents for transfer of this substrate to the thioester binding sites of GS1. The rapid hydrolysis of the reactive thioester complexes of gramicidin S synthetase in neutral buffer solutions in the presence of thiol reagents represents an effective technique for hydrolysis of these intermediates under mild conditions. Fully active synthetase is recovered. Therefore, this procedure is superior to the usually applied cleavage with alkali, performic acid or hydroxylamine which cause a complete deactivation of the multienzyme.

Schlumbohm [46] has developed two techniques for specific labeling of the thiotemplate sites of gramicidin S synthetase. In the first protocol the specific substrate amino acid is attached directly to the reaction center in radioactive form. The other procedure involves a selective labeling of the active sites by ^{14}C-*N*-ethyl-maleimide or other SH-modifiers. This can be achieved by a combination of the following steps: a) substrate protection of a thiotemplate by thioester formation, b) alkylation of the residual sulfhydryl groups, c) isolation and hydrolysis of the reactive thioester complexes and d) subsequent selective labeling of the recovered reaction center. These techniques form the basis for the isolation of active site peptides of the thioester binding sites of gramicidin S synthetase. Using the direct thioester labeling technique the amino acid sequence around the reactive cysteine at the thioester binding site of L-Val was determined. Labeled enzyme was digested by BrCN. The peptide fragment carrying ^{14}C-Val was purified by gel filtration using Sephadex G-75 superfine and reversed phase HPLC methods and split into smaller fragments by *S. aureus* protease. The active site peptide was isolated by a combination of reversed phase HPLC and preparative thin layer chromatography. The following nonapeptide was obtained by gas phase sequencing which represents the first sequence information observed for GS:

Leu-Gly-Gly-His-Cys-Leu-Arg-Ala-HSer

The indirect labeling procedure using thiol-modification reagents, as [^{14}C]-NEM, was applied for labeling of the thioester binding site of L-Pro. The purification and sequencing of the corresponding active site peptide is in progress.

6. Inhibitor Studies

Studies with specific inhibitors in enzyme catalysis are of high importance

a) for the determination of the number and properties of the functional groups which determine substrate recognition and binding at the reaction centers,
b) for investigation of active site topology,
c) for selective blocking of intermediate reactions or
d) for affinity labeling of the active sites.

Under such aspects of collection of substrate analogous inhibitors for GS has been tested by several authors.

The amines, amides and aminoalcohols derived from the substrate amino acids represent specific, reversible, competitive inhibitors of the activation reactions of GS [60, 61]. In particular, the amides of L-Pro and L-Leu inhibit specifically the ATP-PP_i exchange reactions mediated by these substrates without affecting the other activation processes. Their inhibition constants are approx. ten-fold lower than those of the amides of valine and ornithine as well as the K_M-data for L-Pro, L-Val and L-Leu. The K_i-values of the substrate related amines are of similar order of magnitude as the Michaelis constants for the substrate specific ATP-PP_i exchange reactions. *N*-acyl derivatives of the amino acid components of gramicidin S function either as inhibitors of substrate activation or are not accepted by the multienzyme. AMP and adenosine inhibit these processes competitively to ATP and noncompetitively to the amino acid substrates. AMP is a more potent inhibitor than adenosine.

Specific reagents combining a substrate-like design and a chemically reactive group are of interest as active site directed inhibitors. Their substrate analogous moiety is important to localize such agents at the active sites of an enzyme, while its chemically reactive part is designed to form a covalent bond leading to an irreversible attachment of the inhibitor to the reaction centers. Halogenomethylketones, for example, are representatives of such compounds which are qualified as specific alkylating modifiers for the thioester binding sites of GS. The first inhibition experiments using chloromethylketones of substrate amino acids were reported by Leung and Baxter [65]. These authors showed that the chloromethylketone derived from L-Pro (ProCmK) efficiently blocked the L-Pro dependent ATP-PP_i exchange, while the exchange reactions mediated by the other substrates remained essentially unchanged. ProCmK exhibited a non-competitive type of inhibition, as expected of an irreversible inhibitor with a 1 : 1 stoichiometry between enzyme and inhibitor. The active site nature of the inactivation of GS2 by ProCmK was indicated by the protective effect shown by the fully competitive inhibitor L-2-hydroxymethylpyrrolidine. Similar results have been obtained by Nguyen Huu et al. [66] for GS1 using L-Phe-chloro- or bromomethylketone which function as irreversible inhibitors of the phenylalanine dependent ATP-PP_i exchange catalyzed by phenylalanine racemase. The substrates L-Phe and ATP protected the enzyme efficiently against inactivation. Kittelberger et al. [33] reported that L-PheCmK in addition to its inhibitory effect on GS1 also blocks the thioester formation activities of GS2. These effects may be related to the binding and activation of Phe by GS2 [67]. Incubation of GS2 with its substrate amino acids, ATP and Mg^{2+} show a complete protection of the aminoacylation reactions at the 4 reactive sulfhydryl groups of GS2 and the biosynthesis of gramicidin S. Recent results [68] show that the valine chloromethylketone inhibits specifically the thioester incorporation of L-Val. At higher concentrations (>1 mM) also the leucine incorporation is impaired. These effects are reversible indicating a noncovalent attachment of ValCmK to the thioester binding sites. The ATP-PP_i exchange reactions are not affected under these conditions. Lower specificities have been observed for ProCmK and LeuCmK which preferentially block the thioester formation of L-Pro and L-Leu, resp. Under the same conditions, however, also the other thiolation reactions are inhibited to some extent. In the presence of the ProCmK a transfer of phenylalanine from GS1 to GS2 was observed which supports the view that binding of proline is necessary in thioester bound form at GS2 for the transfer of Phe to take place.

3-Amino-2-piperidone (cyclo-Orn) selectively blocks the ornithine activation center of GS2 at the thioester binding site within a similar range of concentration as observed for the inhibition of gramicidin S formation. All other activation reactions of GS2 remain unaffected by this compound. Obviously, the thioester binding site of L-Orn has a preference for a cyclic conformation of this substrate, whereas the aminoacyl adenylate site recognizes only the extended form.

Inhibitor studies are informative for the detection of the functional groups at a reaction center which participate in catalysis. Such experiments have been performed also for GS. Kanda et al. [47, 48] and Schlumbohm et al. [45, 46] have demonstrated that the reactive sulfhydryl groups at the thiotemplates as well as at the aminoacyl adenylation sites can be discriminated using sulfhydryl reagents as NEM, DTNB or iodoacetamide. Such compounds inhibit the thiolation of the substrate amino acids of GS and the whole process of decapeptide formation more severely than the aminoacyl adenylate activation reactions which are affected at 10–100-fold higher modifier concentrations. These processes can be studied separately, if the thioester formation sites are blocked at low concentrations of sulfhydryl inhibitors. At a concentration of 10 μM NEM, for example, both gramicidin S formation and the thiolation reactions decline almost completely. Under such conditions 4–6 sulfhydryl groups of GS are labeled. On the other hand the L-Val-specific ATP-PP_i exchange reaction in the first activation step is blocked at 250 μM NEM, when 18–22 thiols of GS2 are modified. Similar results have been obtained for the activation of the other substrates. The inhibition effects of DTNB are comparable with those observed in the presence of NEM. If iodoacetamide is used as inhibitor, 10–100-fold higher concentrations are needed to obtain a similar inhibition of GS, as shown by the other sulfhydryl reagents.

GS1 is inhibited by several disulfides [41], while the corresponding sulfhydryl compounds are either ineffective or less effective. The most potent disulfide tested was cystamine. Its binding to the enzyme does not affect the Phe-dependent ATP-PP_i exchange reaction. Mg^{2+} alone or in combination with the substrates ATP and phenylalanine causes significant protection against inactivation. The cystamine inactivated enzyme can be reactivated by treatment with DTE or other sulfhydryl reagents. Cystamine and related compounds obviously form a mixed disulfide connecting cysteamine with sulfhydryl groups of the enzyme. These effects have been used for covalent chromatography of GS on cystamine-Sepharose.

Kanda et al. [69, 70] reported that the aminoacyl adenylate activation reactions of GS are inhibited by phenylglyoxal competitively to ATP. Both ATP and the amino acid substrates prevent the inactivation of GS1 and GS2 by phenylglyoxal. In the presence of ATP one arginine per mol of GS1 and 4 such residues per mol of GS2 are protected from modification by this agent as determined by amino acid analysis and the incorporation of [7-^{14}C]phenylglyoxal into the protein. It has been demonstrated that 2 mol of phenylglyoxal react with 1 mol of arginine. Similar effects have been obtained also for the isoleucyl tRNA synthetase from *E. coli*. These results support the participation of one arginine residue in ATP binding to GS1 and GS2 and indicate that such functions are essential for the formation of aminoacyl adenylates both in the nonribosomal and ribosomal system.

7. Structural Aspects

Though the reaction sequence of gramicidin S formation has been studied in detail, knowledge is still lacking concerning the architecture of gramicidin S synthetase. GS1 and GS2 cannot be split into subunits by treatment with 2–3% SDS and high concentrations of reducing agents. The molecular weight of GS2 remained also unchanged upon denaturation using the following conditions [19, 31, 32, 43]:

a) Gel filtration in the presence of 6 M urea;
b) sulfonation of thiols in 8 M urea with subsequent SDS-polyacrylamide gel electrophoresis or
c) modification of carboxymethylated protein with citraconic acid anhydride in 5 M guanidine hydrochloride and gel electrophoresis in 6 M urea.

From such studies it was concluded that both GS1 and GS2 represent multifunctional polypeptide chains. A domain structure has been proposed [19, 71] for GS2 arranging the covalently linked amino acid activating domains around the 4′-phosphopantetheine containing carrier unit.

Specific cleavage of the native multienzyme structure using limited proteolysis techniques is an efficient tool for the investigation of structure-function relationships. Lee and Lipmann [72, 73] demonstrated a subunit structure for tyrocidine synthetase 2 and 3. They isolated a proteolytic activity from *B. brevis* ATCC 8185 solubilized by Triton X-100 which induced an irreversible dissociation of both enzymes into amino acid activating domains of similar size (65–70 kDa) and a smaller, pantetheine-containing fragment of 17–20 kDa which obviously represents the central peptidyl carrier unit. Similar attempts to dissociate GS2 by detergents or denaturing agents have not been successful so far. Limited proteolysis studies have been performed by several authors [19, 21, 74, 75]. Various proteases show specific cleavage and inhibition patterns. Altmann et al. [74] tested 14 proteinases concerning their effects on the functions of gramicidin S synthetase. Experiments with papain showed a rapid loss of activity at the ornithine activation site. Cleavage products were separated by DEAE-cellulose chromatography and gel filtration. Fragments activating proline, valine and leucine were detected. Analysis of the fragmentation pattern by SDS-polyacrylamide gel electrophoresis showed 3 fragments of similar size with molecular weights in the range of 180–190 kD. Also, if trypsin was used as protease, the ornithin activation site was affected most severely. Substrate amino acid protection demonstrated the active site directed nature of the cleavage process. In the presence of valine and ornithine a complete protection against trypsin degradation was obtained. The effect of trypsin on the structure and function of GS has also been studied by Aarstadt and Froyshov [75]. At a protease concentration of 2 μg/ml a 50% decrease in the activation activities of

GS2 for all of its substrate amino acids was observed which led to a complete loss of product formation. In contrast the phenylalanine activation of GS1 was more stable against trypsin treatment. AcA34-gel filtration experiments show a cleavage of both enzymes into smaller fragments. For GS2 4 peaks corresponding to molecular weights of approx. 210, 125, 100 and 65 kDa were resolved. The last fragment was assumed as the smallest cleavage product of GS2 showing activation of individual amino acids.

Limited proteolysis experiments were reviewed by Kleinkauf et al. [19, 21]. With chymotrypsin 10 fragments of sizes between 100 to 220 kD were observed, while using the proteolytic activity of an extract of *B. brevis* 12 fragments in the range between 35 to 220 kDa were produced, as analyzed by SDS-polyacrylamide gel electrophoresis. Chymotryptic fragmentation does not affect the ornithine dependent ATP-PP_i exchange reaction, whereas the other activities were lost (Val) or reduced (Pro, Leu). Obviously, chymotrypsin attacks GS2 primarily in the valine activation domain. Another type of cleavage was found for the extract of *B. brevis*. Here the proline activation was lost completely, while the activation of Val and Leu was only moderately reduced. A summary of the limited proteolysis experiments on GS2 was presented in [21].

Electron micrographs of GS2 were first obtained by Wecke et al. [in ref. 16] showing ring-shaped particles with an outer and inner diameter of 12 and 6 nm, resp. Vater et al. [44] investigated the quaternary structure of both enzymes of gramicidin S

b

a

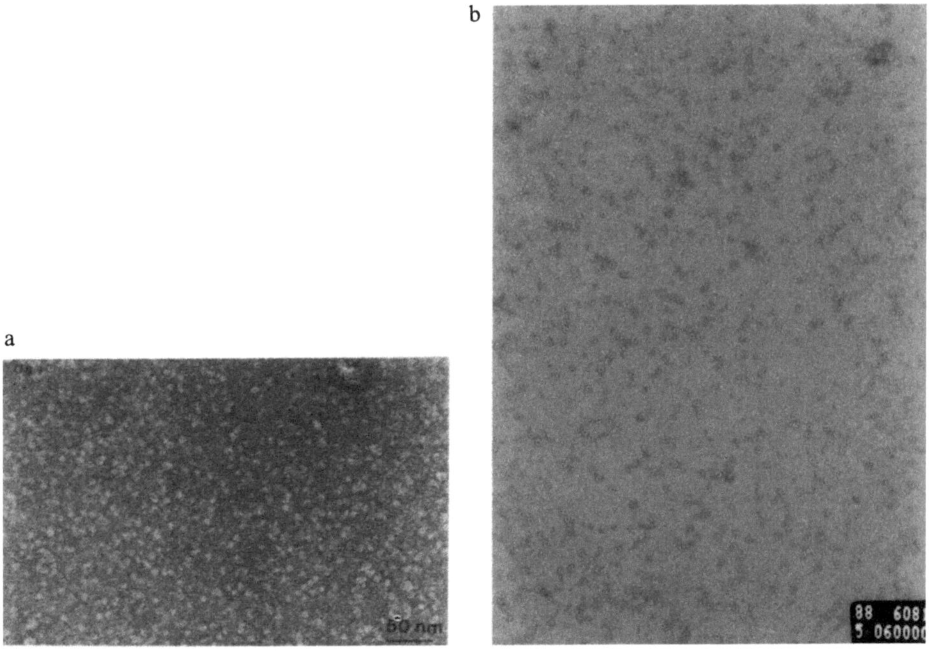

Figure 1 Electron micrographs of a) GS1 (× 500 000; scale bar = 50 nm) and b) GS2 (× 180 000) negatively stained with 2% uranylacetate.

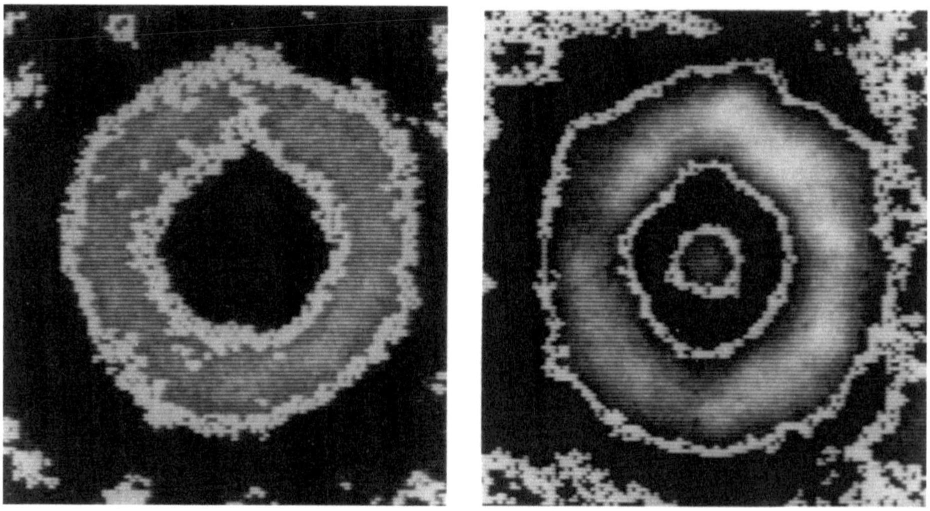

a b

Figure 2 Statistical computer analysis of GS2 particles. Halftone images of averaged particles with contour lines added. Particles a) without and b) with a central domain.

synthetase by electron microscopical techniques using chemical as well as physical fixation and staining procedures. Electron micrographs of GS1 show particles of oblate ellipsoidal conformation with a diameter of approx. 7.5 nm, as demonstrated in Figure 1. For GS2 also a ring structure was obtained with similar dimensions as reported by Wecke et al. [in ref. 16]. The height of the GS2-particles was determined from freeze dried preparations shadowed with tungsten either unidirectionally or by rotational evaporation. From their shadow cast a height of 6 nm was estimated. From a statistical computer analysis of such micrographs the GS2 particles could be divided into 2 classes of A) with, and B) without a central structural domain (A : B = 45 : 55). This central part of the multifunctional protein has a diameter of 2.5 nm and may be related to the 4′-phosphopantetheine-containing peptidyl carrier domain. A statistical computer analysis of GS2 particles negatively stained with 2% uranylacetate A) without and B) with a central domain is shown in Figure 2A and B.

Information is still lacking concerning the primary, secondary and tertiary structure of gramicidin S synthetase. Studies pointing to these topics are in progress. Current activities in this field are concentrated on the isolation and sequencing of active site peptides [46] and the determination of the nucleotide sequences of the genes coding for both enzymes of gramicidin S synthetase [76].

A secondary structure analysis of GS1 and GS2 was performed on the basis of circular dichroism spectra which were measured in the range between 192 and 240 nm [44]. The data obtained show that both proteins are rich in β-sheet structure (45% for both GS1 and GS2), while α-helical regions occur with a lower frequency in the architecture of these multifunctional polypeptides (13% for GS1 and 17% for GS2).

8. Investigation of Gramicidin S Negative Mutants

Valuable information on the structure and function of gramicidin S synthetase was obtained from studies of gramicidin S negative mutants of *Bacillus brevis*. Such research was essentially performed in the laboratories of Kurahashi [22, 78] and Saito [39, 40, 79–83]. Numerous mutations of this kind were induced in *B. brevis Nagano* and ATCC 9999 by treatment with mutagenic agents.

Kambe et al. [78] classified their collection of mutants into 3 groups lacking activities of a) GS1, b) GS1 and GS2 and c) GS2. Among these, two produced intact GS1, but defective multienzyme GS2 which were designated hh and n – 7. Both mutant enzymes showed lower sedimentation constants than the wild type proteins (11.3 and 6.3 S compared to 12.2 of native GS2) corresponding to molecular weights of 250 and 100 kDa. hh did not activate L-Leu, n – 7 activated L-Pro only. Their structural relation to the wild type GS2 has been demonstrated by immunological techniques [84]. Queener et al. [85] suggested that n – 7 is a proline activating subunit and that hh represents a GS2-complex missing the leucine subunit. These mutations can be ascribed to mutations in the structural genes of GS2. D-Phe-Pro-diketopiperazine formation was not observed for the mutant enzyme of hh, but was detected for n – 7. Obviously, the L-Pro activating subunit of GS2 alone can interact with the racemase, while an intact leucine activating domain is important for the interaction between both components of gramicidin S synthetase.

Saito's group obtained about 20 mutant strains of *B. brevis Nagano* which were classified into five groups. The first group lacks GS1. The second possesses GS1, but does not show a GS2 complex activating its 4 substrate amino acids. The third group lacks both GS1 and GS2. All these mutants could form neither gramicidin S nor D-Phe-L-Pro dipeptide. The mutants of group 4 are of high relevance, because they have retained Phe activating activity, but show an incomplete GS2 multienzyme from which one specific amino acid activating unit among L-Pro, L-Val and L-Leu is deficient. The mutants of group 5 possess all five amino acid activating activities as the wild type strain, but are not able to synthesize gramicidin S.

GS2 from mutant enzymes which are deficient in the activation of one specific substrate amino acid (for example, BII-3, BI-3 or BI-9 lacking activation of L-Pro, L-Val and L-Leu, resp.) had the same $s_{20,w}$ = 12.2 S as the wild type enzyme [39]. For all species the radioactivity of bound substrates were associated with a protein band on SDS-polyacrylamide gel electrophoresis showing a molecular weight of 280 kDa. These results confirm the conclusion of other authors [19, 32, 43] that GS2 represents a multifunctional polypeptide chain. The failure of specific amino acid activation reactions presumably are due to modifications of the active sites rather than to a complete absence of the amino acid activating domains.

Saito et al. [80, 82] characterized a class of mutants (E-1; E-2; BI-4 or C-3, for example) which could not form even D-Phe-L-Pro-diketopiperazine despite of the presence of the phenylalanine activating activity in GS1 and the proline activating activity in GS2. Both mutant enzymes had the same molecular weight as the wild type proteins. The mutants showed double defects.

1. The GS1 component can activate phenylalanine as adenylate, but not as thioester. Also the racemization of Phe and the Phe dependent ATP-[^{14}C]-AMP exchange activity are lost. Obviously, the reactive cysteine at the reaction center is still present, as demonstrated by substrate protection against inhibition by thiol inhibitors, like DTNB, but is not catalytically competent, probably due to a modified tertiary structure of the thiotemplate.
2. GS2 of these mutants lacked 4′-phosphopantetheine and was not able to accept the activated Phe from normal GS1.

In contrast, the light enzyme of diketopiperazine (+) mutants which belong to group IV (BI-3, BI-6, BI-9, BIII-1 and E-4) had both Phe activating and racemizing activities.

Complementation studies showed that 4′-phosphopantetheine deficient GS2 would not synthesize D-Phe-L-Pro-diketopiperazine and gramicidin S in the presence of the wild type GS1, even though these mutant enzymes activated all constituent amino acids of gramicidin S and had racemase activity in GS1. Pantothenic acid containing GS2 can form diketopiperazine, if the complementary GS1 and the proline activation domain of GS2 are intact. Under these conditions ^{14}C-Phe was associated with a peptide containing 4′-phosphopantetheine in thin layer- and Sephadex G-50 column chromatography of pepsin digests of GS2. These results suggest that 4′-phosphopantetheine participates in the transfer of phenylalanine from GS1 to GS2 and in D-Phe-Pro-dipeptide formation. On the other hand Roskoski et al. [50] and Kleinkauf et al. [9] reported that D-Phe is not transferred to GS2 in the absence of L-Pro. Pass et al. [36] confirmed that proline binding to GS2 was required for the transfer of D-Phe and assumed a thiol group as acceptor site located near the proline activating domain.

It is interesting that 3 of 4 mutants of group IV deficient in the leucine activation site of GS2 could not synthesize D-Phe-L-Pro-diketopiperazine. This pattern is in agreement with the observation of Kambe et al. [78] that their hh mutant enzyme which does not catalyze leucine activation shows a strongly reduced rate of dipeptide formation. From the experimental material so far available it follows that a) an intact proline site is essential for the initiation of gramicidin S synthesis and b) also the leucine activation domain seems to be important for this process.

9. Prospects

The efforts of current research are concentrated on the proteinchemical, genetic, immunological and physicochemical characterization of gramicidin S synthetase. In particular, these approaches yield details of the structure of this multienzyme system. Targets of proteinchemical analysis of GS are the labeling and analysis of the reactive domains for substrate activation and peptide elongation [46]. This sequence information is the prerequisite for the localization of the active sites in the framework of the amino acid sequence derived from nucleotide sequencing of the genes coding for GS1 and GS2. Of special interest is the central peptidyl carrier domain of GS2 which has to be labeled and isolated by proteolytic degradation of the multienzyme. Sequencing this part of GS2 is of high relevance for detection of evolutionary relationships in comparison with the acyl carrier protein of fatty acid synthetases. The genetic analysis [76, 77] is concentrated a) on the chromosomal location of the genes involved in the biosynthesis of gramicidin S, b) on the analysis of the regulatory elements which control the expression of the genes coding for peptide antibiotic synthesis in *B. brevis*, and c) on the sequence determination of the structural genes. Deciphering of the DNA-structures will contribute to the knowledge of the structure of peptide forming proteins and to the investigation of their common evolutionary origin.

A collection of poly- and monoclonal antibodies has been raised against GS1 (Ommerborn et al., unpublished results) and GS2 [86] which can be used for identification of the antigenic determinants of both enzymes. Monospecific antibodies are efficient tools for the localization of functional domains of these proteins by immunoelectron microscopy or immunoreaction with fragments of GS1 or GS2 obtained from expression cloning of gene segments.

Physicochemical techniques are focussed on the three-dimensional structure by crystallization and x-ray analysis of both enzymes of gramicidin S synthetase as well as on the geometry and topography of the reaction centers mainly by NMR and EPR-techniques. From this research insights into the thiotemplate mechanism on the molecular level as well as into the architecture of gramicidin S synthetase and related multienzymes are to be expected.

References

1. Mach, B., Reich, E., and Tatum, E. L., Separation of the biosynthesis of the antibiotic polypeptide tyrocidine from protein biosynthesis, *Proc. Natl. Acad. Sci. USA 50*, 175, 1963.
2. Paulus, H., and Gray, E., The biosynthesis of polymyxin B by growing cultures of *Bacillus polymyxa, J. Biol. Chem. 239*, 865, 1964.
3. Eikhom, T. S., Jonsen, J., Laland, S., and Refsvik, T., Studies on the biosynthesis of gramicidin S in whole cells of *Bacillus brevis, Biochim. Biophys. Acta 80*, 648, 1964.
4. Daniels, M. J., Studies of the biosynthesis of polymyxin B, *Biochim. Biophys. Acta 156*, 119, 1968.
5. Lipmann, F., Grevers, W., Kleinkauf, H., and Roskoski, jr., R., Polypeptide synthesis of protein templates: The enzymatic synthesis of gramicidin S and tyrocidine, *Adv. Enzymol. 35*, 1, 1971.
6. Laland, S. G., and Zimmer, T.-L., The protein thiotemplate mechanism of synthesis for the peptide antibiotics produced by *Bacillus brevis, Essays Biochem. 9*, 31, 1973.
7. Kurahashi, K., Biosynthesis of small peptides, *Ann. Rev. Biochem. 43*, 445, 1974.
8. Lipmann, F., Attempts to map a process evolution of peptide biosynthesis, *Science 173*, 875, 1971.
9. Kleinkauf, H., Roskoski, jr., R., and Lipmann, F., Pantetheine-linked peptide intermediates in gramicidin S and tyrocidine biosynthesis, *Proc. Natl. Acad. Sci. USA 68*, 2069, 1971.
10. Meister, A., and Tate, S., Glutathione and related γ-glutamyl compounds: Biosynthesis and utilization, *Ann. Rev. Biochem. 45*, 559, 1976.

11. Sengupta, S., and Bose, S. K., Peptides from a mycobacillin-synthesizing cell-free system, *Biochem. J. 128*, 47, 1972.
12. Ghosh, S. K., Majumder, S., Mukhopadhyay, N. K., Bose, S. K., Functional characterization of constituent enzyme fractions of mycobacillin synthetase, *Biochem. J. 230*, 785, 1985.
13. Perlman, D., and Bodanszky, M., Biosynthesis of peptide antibiotics, *Ann. Rev. Biochem. 40*, 449, 1971.
14. Katz, E., and Demain, A. L., The peptide antibiotics of Bacillus: Chemistry, biogenesis and possible functions, *Bacteriol. Rev. 41*, 449, 1977.
15. Frøyshov, Ø., Zimmer, T.-L., and Laland, S. G., Biosynthesis of microbial peptides by the thiotemplate mechanism, in *Internat. Rev. Biochem., Amino Acid and Protein Biosynthesis II*, Vol. 18, 49, 1978, H. R. V. Arnstein, ed., Univ. Park Press, Baltimore.
16. Kleinkauf, H., and Koischwitz, H., Peptide bond formation in non-ribosomal systems, *Progr. Mol. Subcell. Biol.* 6, 59, 1978.
17. Kleinkauf, H., Antibiotic polypeptides – biosynthesis on multifunctional protein templates, *Planta Medica 35*, 1, 1979.
18. Zimmer, T.-L., Frøyshov, Ø., Laland, S. G., Peptide antibiotics, in *Economic Microbiology* Vol. III, 123, 1979, A. H. Rose, ed., Academic Press, New York.
19. Kleinkauf, H., and Koischwitz, H., Gramicidin S-Synthetase, in H. Bisswanger and E. Schmincke-Ott, eds., *Multifunctional Proteins*, p. 217, Wiley, New York, 1980.
20. Kleinkauf, H., and Koischwitz, H., Gramicidin S-Synthetase: On the structure of a polyenzyme template in polypeptide synthesis, *Mol. Biol. Biochem. Biophys. 32*, 205, 1980.
21. Kleinkauf, H., and von Döhren, H., Nucleic acid independent synthesis of peptides, *Curr. Top. Microbiol. Immunol. 91*, 129, 1981.
22. Kurahashi, K., Biosynthesis of peptide antibiotics, *Antibiotics (N.Y.) 4*, 325, 1981.
23. Kleinkauf, H., and von Döhren, H., eds. *Peptide Antibiotics – Biosynthesis and Functions*, de Gruyter, Berlin, 1982.
24. Kleinkauf, H., and von Döhren, H., A survey of enzymatic biosynthesis of peptide antibiotics. In H. Umezawa, A. L. Demain, T. Hata, C. R. Hutchinson, eds., *Trends in Antibiotic Research*, p. 220, Japan. Antibiot. Res. Assoc., Tokyo. 1982.
25. Kleinkauf, H., and von Döhren, H., Biosynthesis of peptide antibiotics, *Ann. Rev. Microbiol. 41*, 259, 1987.
26. Kleinkauf, H., Gevers, W., and Lipmann, F., Interrelation between activation and polymerization in gramicidin S biosynthesis, *Proc. Natl. Acad. Sci. USA 62*, 226, 1969.
27. Yamada, M., and Kurahashi, K., Further purification and properties of adenosine triphosphate-dependent phenylalanine racemase of *Bacillus brevis Nagano, J. Biochem. (Tokyo) 66*, 529, 1969.
28. Otani, S., Yamanoi, T., and Saito, Y., Biosynthesis of gramicidin S; ornithine activating enzyme, *J. Biochem. (Tokyo) 66*, 445, 1969.
29. Otani, S., jr., Yamanoi, T., and Saito, Y., Fractionation of the enzyme system responsible for gramicidin S biosynthesis, *Biochim. Biophys. Acta 208*, 496, 1970.
30. Koischwitz, H., and Kleinkauf, H., Gramicidin S-Synthetase – preparation of the multienzymic complex with a high specific activity, *Biochim. Biophys. Acta, 429*, 1041, 1976.
31. Vater, J., and Kleinkauf, H., A further characterization of phenylalanine racemase, the light enzyme of gramicidin S-synthetase, *Biochim. Biophys. Acta 429*, 1062, 1976.
32. Christiansen, C., Aarstadt, K., Zimmer, T.-L., and Laland, S. G., A rapid method for the preparation of pure heavy enzyme of gramicidin S synthetase, *FEBS Lett. 81*, 121, 1977.
33. Kittelberger, R., Palacz, Z., von Döhren, H., Salnikow, J., and Kleinkauf, H., Inhibition of gramicidin S-synthetase 2 by L-phenylalanine chloromethylketone, *FEBS Lett. 151*, 248, 1983.
34. Zimmer T.-L., and Laland, S. G., Gramicidin S Synthetase, *Meth. Enzymol. 43*, 567, 1975.
35. Pass, L., Zimmer, T.-L., and Laland, S. G., The use of affinity chromatography in determining the sites of protein-protein interaction relative to the binding sites of substrates in gramicidin S synthetase, *Eur. J. Biochem. 40*, 43, 1973.
36. Pass, L., Zimmer, T.-L., and Laland, S. G., On the use of affinity chromatography in demonstrating the transfer of thioester-bound D-phenylalanine from the light enzyme of gramicidin S synthetase to an acceptor site for this amino acid on the heavy enzyme, *Eur. J. Biochem. 47*, 607, 1974.

37. Schröter, C., Rönspeck, W., Altmann, M., von Döhren, H., and Kleinkauf, H., Use of affinity chromatography in purification of gramicidin S synthetase in [*23*] 259, 1982.
38. Christiansen, C., Nordvi, B., Zimmer, T.-L., and Laland, S. G., A survey on the use of affinity chromatography for studying the mechanism of gramicidin S formation in [*23*], 265, 1982.
39. Hori, K., Kurotsu, T., Kanda, M., Miura, S., Nozoe, A., and Saito, Y., Studies on gramicidin S synthetase – purification of the heavy enzyme obtained from some mutants of *Bacillus brevis*, *J. Biochem. (Tokyo) 84*, 425, 1978.
40. Kanda, M., Hori, K., Kurotsu, T., Miura, S., Nozoe, A., and Saito, Y., Studies on gramicidin S synthetase – purification and properties of the light enzyme obtained from some mutants of *Bacillus brevis*, *J. Biochem. (Tokyo) 84*, 435, 1978.
41. Schröter-Kermani, C., von Döhren, H., and Kleinkauf, H., Active thiol-directed binding and adsorption of gramicidin S-synthetase 1 to disulfide-containing matrices, *Biochim. Biophys. Acta 883*, 345, 1986.
42. Vater, J., Schlumbohm, W., Palacz, Z., Salnikow, J., Gadow, A., and Kleinkauf, H., Formation of D-Phe-Pro-Val-cyclo-Orn by gramicidin S synthetase in the absence of L-leucine, *Eur. J. Biochem. 163*, 297, 1987.
43. Koischwitz, H., and Kleinkauf, H., Gramicidin S-synthetase – electrophoretic characterization of the multienzyme, *Biochim. Biophys. Acta 429*, 1052, 1976.
44. Vater, J., Schmiady, H., Tesche, B., Schlumbohm, W., Salnikow, J., Zepmeusel, R., and Kleinkauf, H., Structural features of gramicidin S synthetase, manuscript in preparation.
45. Schlumbohm, W., Vater, J., and Kleinkauf, H., Reactive sulfhydryl groups involved in the aminoacyl adenylate activation reactions of the gramicidin S synthetase 2, *Biol. Chem. Hoppe-Seyler 366*, 925, 1985.
46. Schlumbohm, W., Gramicidin S-synthetase: Characterization of the sulfhydryl groups involved in substrate amino acid activation reactions and isolation of active site peptides of the thioester binding centers, *Thesis, Technical University of Berlin*, 1987.
47. Kanda, M., Hori, K., Kurotsu, T., Miura, S., and Saito, Y., A comparative study of sulfhydryl groups required for the catalytic activity of gramicidin S synthetase and isoleucyl tRNA synthetase, *J. Biochem. (Tokyo) 96*, 701, 1984.
48. Kanda, M., Hori, K., Kurotsu, T., Miura, S., Yamada, Y., and Saito, Y., Sulfhydryl groups related to the catalytic activity of gramicidin S synthetase 1 of *Bacillus brevis*, *J. Biochem. (Tokyo) 90*, 765, 1981.
49. Stoll, E., Frøyshov, Ø., Holm, H., Zimmer, T.-L., and Laland, S. G., On the mechanism of gramicidin S formation from intermediate peptides, *FEBS Letters 11*, 348, 1970.
50. Roskoski, R., jr., Ryan, G., Kleinkauf, H., Gevers, W., and Lipmann, F., Polypeptide biosynthesis from thioesters of amino acids, *Arch. Biochem. Biophys. 143*, 485, 1971.
51. Otani, S., Yamanoi, T., Saito, Y., and Otani, S., Fractionation of an enzyme system responsible for gramicidin S biosynthesis, *Biochem. Biophys. Res. Commun. 25*, 590, 1966.
52. Gadow, A., Vater, J., Schlumbohm, W., Palacz, Z., Salnikow, J., and Kleinkauf, H., Gramicidin S Synthetase – Stability of reactive thioester intermediates and formation of 3-amino-2-piperidone, *Eur. J. Biochem. 132*, 229, 1983.
53. Gevers, W., Kleinkauf, H., and Lipmann, F., The activation of amino acids for biosynthesis of gramicidin S, *Proc. Natl. Acad. Sci. USA 60*, 269, 1968.
54. Gevers, W., Kleinkauf, H., and Lipmann, F., Peptidyl transfers in gramicidin S biosynthesis from enzyme bound thioester intermediates, *Proc. Natl. Acad. Sci USA 63*, 1335, 1969.
55. Saxholm, H., Zimmer, T.-L., and Laland, S. G., The mechanism of the inhibition of gramicidin S synthesis by D-leucine, *Eur. J. Biochem. 30*, 138, 1972.
56. Vater, J., Mallow, N., Gerhardt, S., Gadow, A., and Kleinkauf, H., Gramicidin S synthetase. Temperature dependence and thermodynamic parameters of substrate amino acid activation reactions, *Biochemistry 24*, 2022, 1985.
57. Rapaport, E., Remy, P., Kleinkauf, H., Vater, J., and Zamecnik, P. C., Aminoacyl – tRNA synthetases catalyze AMP-ADP-ATP exchange reactions, indicating labile covalent enzyme-amino acid intermediates, *Proc. Natl. Acad. Sci. USA 84*, 7891, 1987.
58. Vater, J., Mallow, N., Gerhardt, S., and Kleinkauf, H., The temperature dependence of the partial processes involved in the biosynthesis of gramicidin S, in [*23*], 219, 1982.
59. Kittelberger, R., Altmann, M., and von Döhren, H., Kinetics of amino acid activation of gramicidin S synthesis, in [*23*], 209, 1982.

60. Kleinkauf, H., Koischwitz, H., Vater, J., Zocher, R., Keller, U., Mahmutoglu, J., Bauer, K., Altmann, M., Kittelberger, R., Marahiel, M., and Salnikow, J., Nonribosomal biosynthesis of biologically active peptides, in M. Luckner and K. Schreiber, eds., *Regulation of Secondary Product and Plant Hormone Metabolism*, p. 37, Pergamon Press, Oxford/New York, 1978.
61. Yamada, M., and Kurahashi, K., Adenosine triphosphate and pyrophosphate dependent phenylalanine racemase of *Bacillus brevis Nagano, J. Biochem (Tokyo) 63*, 59, 1968.
62. Vater, J., and Kleinkauf, H., Substrate specificity of the aminoacyl adenylate activation sites of gramicidin S-synthetase, *Acta Microbiol. Acad. Sci. Hung. 22*, 419, 1975.
63. Kleinkauf, H., and von Döhren, H., Cell-free biosynthesis of peptide antibiotics, *Adv. Biotechnol. 3*, 83, 1981.
64. Laland, S. G., Aarstadt, K., and Zimmer, T.-L., The fidelity of gramicidin S-synthetase with particular reference to the amino acids cyclohexylalanine and phenylalanine in [*23*], 185, 1982.
65. Leung, D. C., and Baxter, R. M., Substrate derived reversible and irreversible inhibitors of the multienzyme I of gramicidin S biosynthesis, *Biochim. Biophys. Acta 279*, 34, 1972.
66. Nguyen Huu, M. C., von Dungen, A., and Kleinkauf, H., Irreversible inhibition of the light enzyme of gramicidin S synthetase by halogenomethylketones of phenylalanine, *FEBS Lett. 67*, 75, 1976.
67. Aarstadt, K., Zimmer, T.-L., and Laland, S. G., The fidelity of gramicidin S synthetase, *Eur. J. Biochem. 112*, 335, 1980.
68. Kittelberger, R., Gramicidin S-synthetase: Kinetic and proteinchemical studies of the structure and function of the multienzyme system, *Thesis, Technical University of Berlin*, 1983.
69. Kanda, M., Hori, K., Kurotsu, T., Yamada, Y., Miura, S., and Saito, Y., Essential arginine residue in gramicidin S synthetase 1 of Bacillus brevis, *J. Biochem. (Tokyo) 91*, 939, 1982.
70. Kanda, M., Hori, K., Miura, S., Yamada, Y., and Saito, Y., A comparative study of essential arginine residues in gramicidin S synthetase 2 and isoleucyl tRNA synthetase, *J. Biochem. (Tokyo) 92*, 1951, 1982.
71. Kleinkauf, H., Die Biosynthese von Antibiotika-Peptiden, *Chemie in unserer Zeit 14*, 105, 1980.
72. Lee, S. G., and Lipmann, F., Isolation of a peptidyl-pantetheine-protein from tyrocidine-synthesizing polyenzymes, *Proc. Natl. Acad. Sci. USA 71*, 607, 1974.
73. Lee, S. G., and Lipmann, F., Isolation of amino acid activating subunit-pantetheine protein complexes: Their role in chain elongation in tyrocidine synthesis, *Proc. Natl. Acad. Sci. USA 74*, 2343, 1977.
74. Altmann, M., v. Döhren, H., El-Samaraie, A., Kittelberger, R., Pore, M. S., and Kleinkauf, H., Limited proteolysis: Studies on the multienzyme GS2 of gramicidin S-synthetase, in [*23*], 243, 1982.
75. Aarstadt, K., and Frøyshov, Ø., Tryptic cleavage of the heavy enzyme of gramicidin S synthetase, in [*23*], 253, 1982.
76. Krause, M., Marahiel, M. A., von Döhren, H., and Kleinkauf, H., Molecular cloning of an ornithine-activating fragment of the gramicidin S synthetase 2 gene from *Bacillus brevis* and its expression in *Escherichia coli, J. Bacteriol. 162*, 1120, 1985.
77. Marahiel, M. A., Krause, M., and Skarpeid, H.-J., Cloning of the tyrocidine synthetase 1 gene from *Bacillus brevis* and its expression in *Escherichia coli, Mol. Gen. Genet. 201*, 231, 1985.
78. Kambe, M., Imae, Y., and Kurahashi, K., Biochemical studies on gramicidin S non-producing mutants of *Bacillus brevis* ATCC 9999, *J. Biochem. (Tokyo) 75*, 481, 1974.
79. Shimura, K., Iwaki, M., Kanda, M., Hori, K., Kaje, E., Hasegawa, S., and Saito, Y., On the enzyme system obtained from some mutants of *Bacillus brevis* deficient in gramicidin S formation, *Biochim. Biophys. Acta 338*, 557, 1974.
80. Hori, K., Kanda, M., Kurotsu, T., Miura, S., Yamada, Y., and Saito, Y., Absence of pantothenic acid in gramicidin S synthetase 2 obtained from some mutants of *Bacillus brevis, J. Biochem. (Tokyo) 90*, 439, 1981.
81. Hori, K., Kurotsu, T., Kanda, M., Miura, S., Yamada, Y., and Saito, Y., Evidence for a single multifunctional polypeptide chain on gramicidin S synthetase 2 obtained from a wild strain and mutant strain of *Bacillus brevis, J. Biochem. (Tokyo) 91*, 369, 1982.
82. Saito, Y., Some characteristics of gramicidin S-synthetase obtained from mutants of *Bacillus brevis* which could not form D-phenylalanyl-L-prolyl diketopiperazine, in [*23*], 195, 1982.

83. Hori, K., Kanda, M., Miura, S., Yamada, Y., and Saito, Y., Transfer of D-phenylalanine from GS1 to GS2 in gramicidin S synthesis, *J. Biochem. (Tokyo) 93*, 177, 1983.
84. Bothe, D., von Döhren, H., Zschiedrich, H., El-Samaraie, A., Krause, M., and Kleinkauf, H., Further characterization of multienzyme fragments of gramicidin S synthetase obtained from gramicidin S nonproducer mutants, in [*23*], 233, 1982.
85. Queener, S. W., Sebek, O. K., and Vézina, C., Mutants blocked in antibiotic synthesis, *Ann. Rev. Microbiol. 32*, 593, 1978.
86. Bothe, D., Immunological investigation of multifunctional enzymes catalyzing peptide and depsipeptide biosynthesis in *Bacillus brevis* and *Fusarium oxysporum, Thesis, Technical University of Berlin*, 1986.

Chapter 3
Formation of *N*-methylated Peptide Bonds in Peptides and Peptidols

Andreas Billich and Rainer Zocher

1. Introduction
2. Occurrence of *N*-methylated Peptide Bonds
3. Biosynthesis of *N*-methylated Peptides
4. The *N*-methyltransferase Function of Enniatin Synthetase
   4.1 *N*-methylation of enzyme-bound amino acids
   4.2 Kinetic properties of the *N*-methyltransferase function
   4.3 Inhibition studies of the *N*-methyltransferase function
   4.4 Photoaffinity labeling of enniatin synthetase
   4.5 Monoclonal antibodies to enniatin synthetase
5. The *N*-methyltransferase Function of other Peptides Synthetases
   5.1 Beauvericin synthetase
   5.2 Cyclosporin synthetase
   5.3 Actinomycin synthetases
6. Conclusions
References

1. Introduction

Peptides and peptolides from the secondary metabolism of microorganisms are synthezised enzymatically by non-ribosomal pathways [1]. Besides common L-amino acids these substances may contain D-amino acids and unusual amino or hydroxy acids not present in proteins. A further peculiar property is the occurrence of *N*-methylated peptide bonds in such compounds. So far this structural feature of many peptides has received little attention. In this review we therefore want to summarize the knowledge concerning the occurrence and the biosynthetic origin of *N*-methylated amide groups in microbial peptides and peptidols.

2. Occurrence of *N*-methylated Peptide Bonds

A synopsis of the occurrence of *N*-methylated peptides and peptolides in nature is given in Table 1. The compounds usually exhibit antibiotic activity, some of them also immunomodulating, cytostatic or antiviral properties.

Table 1 *N*-methylated peptides and peptolides.

		Compound	Producing organism	Methylated amino acid	Reference
Actinomycetes	Actinomycins	various strains of *Streptomyces; Micromonospora floridensis*	Gly and Val (or *allo*-isoleucine or Ala)	2	
	Peptolides	Virginiamycin group B-I antibiotics[a]	various *Streptomycetes*	Phe or p-dimethylamino-phenylalanine	3
	Peptolides	Virginiamycin group B-II antibiotics[b]	various *Streptomycetes*	Gly and 3-methylleucine	3
	Peptolides	Quinoxaline antibiotics[c]	various *Streptomycetes*	Cys, *S*MeCys[d] and Val (or *allo*-isoleucine or γ-methyl-*allo*-isoleucine)	3
	Peptolides	Luzopeptins	*Actinomadura luzonensis*	L-3-hydroxyvaline and Gly	4
	Peptolides	Globomycin	*S. hygroscopicus*	Leu	3
	Peptolides	Monamycins	*S. jamaicensis*	D-Leu	5
	Peptolides	Stendomycin	*S. endus, S. species*	Thr	5
	Peptolides	Grisellimycin	*S. griseus, S. coelicus*	Val, Thr, and Leu	5
	Peptolides	Cycloheptamycin	*S. species*	5-methoxy-tryptophan, Ala	5
		Mycoplanecins	*Actinoplanes awajiensis*	Thr, Val and Leu or 2-aminoheptanoic acid	6,7
		Neopeptins	*S. species*	Phe and Asn	8,9
	Depsipeptide	Azinothricin	*S. species*	D-Ala	10
	Peptide	Stenothricin	*S. species*	Gly	3
	Cyclopeptide	Ilamycins	*S. islandicus, S. insubtus*	Leu, Ala, and 2-amino-4-methyl-5-oxo-pentanoic acid	5
	Diketopiperazine	(3Z, 6E)-1-*N*-Methylalbonoursin	*S. albus*	4-methyl-penten-(2)-oic acid	11
		Tryptophan-dehydrobutyrine diketopiperazine	*S. spectabilis*	Trp	12

Table 1 contined

		Compound	Producing organism	Methylated amino acid	Reference
Cyanobacteria	Peptolide	Majusculamides	*Lyngbya majuscula*	Val, Ile, O-methyl-tyrosine	13
Fungi		Destruxins	*Aspergillus ochraceus, Metarrhizium anisopliae*	Val and Ala	5
	Depsipeptides	Enniatins	several strains of *Fusarium*	Val or Leu or Ile	5
		Beauvericin	*Beauveria bassiana, Paecilomyces fumosoroseus, Polyporus sulphureus*	Phe	5
		Bassianolide	*Beauveria bassiana, Verticillium lecanii*	Leu	3
		Sporidesmolides	*Pithomyces chartarum*	Leu	5
		Aspochracin	*Aspergillus ochraceus*	Val and Ala	5
		Isotentoxin	*Alternaria mali*	Ala and 2,3-dehydrophenylalanine	5
	Cyclopeptides	Cyclosporins	*Tolypocladium inflatum, Cylindrocarpon lucidum*	Leu, Val, Gly, and Bmt (or deoxyBmt or 2-aminooctanoic acid)	5,14
		Cycloaspeptides	*Aspergillus sp.*	Tyr, Phe	15
	Diketopiperazine	*epi*-Oligothiadiketopiperazine antibiotics[f]	*Chaetomium uniporum;* various *Hyphomycetes*	Ala or Ser	5
Sponges	Peptolide	Theonellamine B	*Theonella sp.*[g]	Val, D-Leu, Ile, and *allo*-isoleucine	16
	Peptide	Discodermins	*Discoderma kiiensis*[g]	Gln	17
Tunicates	Peptolide	Didemnins	*Trididemnum sp.*[g]	Leu	18

[a]Including virginiamycins, vernamycins, ostreogrycin, and patricins. [b]Including viridogriseins and neoviridogriseins. [c]Including quinomycins and triostins. [d]In quinomycins. [e]Bmt = (4*R*)-4-(E)-2-Butenyl-4-methyl-L-threonine. [f]Including gliotoxin, chaetocin, verticillins, melinacidins, chaetomin, sporidesmins, hyalodendrins, and sirodesmins, [g]It cannot be excluded that these compounds are synthesized by symbiotic or associated microorganisms.

The overwhelming number of these metabolites are produced by mycelium-forming bacteria, the *Actinomycetes*, especially by the genus *Streptomyces*. There is only one example for a *N*-methylated peptolide (majusculamide) from a procaryote not belonging to this group, a blue-green alga (*Lyngbya*). By contrast *N*-methylation of peptide bonds is never observed in structurally similar metabolites from lower, unicellular procaryotes. Especially the numerous peptide metabolites from the bacilli, e. g. gramicidins, tyrocidins, surfactin, polymyxins, are all unmethylated.

On the other hand, fungi produce methylated peptides, too. Almost all of them are *Fungi imperfecti* belonging to the *Hyphomycetes*; exceptions are *Chaetomium uniporum*, an ascomycete forming chaetomin, and *Polyporus sulphureus*, a basidiomycete forming beauvericin. *N*-methylated peptides have also been isolated from marine sponges and some tunicates. It is an open question, whether the metabolites are synthezised by these higher developed eucaryotes themselves or by some symbiotic or associated microorganisms.

The antibiotics listed in Table 1 include linear peptides (e. g. stenothricin), cyclic peptides (e. g. ilamycins, cyclosporins), diketopiperazines (e. g. gliotoxin), true depsipeptides (e. g. enniatins), and, as the most numerous group, peptolides.

With the exception of glycine, *N*-methylation usually is restricted to hydrophobic amino acids, i. e. alanine, valine, leucine, isoleucine, phenylalanine, tyrosine, and tryptophan, and some non-proteinogenic amino acids, e. g. *allo*-isoleucine, 3-methylleucine, and 5-methoxytryptophan. In addition, some amino acids with polar side groups may be methylated: a threonine residue in griselli- and stendomycins, asparagine in neopeptins, glutamine in discodermins, and cysteine in quinoxaline antibiotics. Besides L- also D-amino acids, e. g. D-leucine in the monamycins and D-alanine in azinothricin, can be methylated.

Some antibiotics contain only one *N*-methylated peptide bond, others from two (e. g. cycloheptamycin, two out of six peptide bonds methylated) up to seven (e. g. cyclosporin, seven out of eleven). The only example of antibiotics, in which all peptide bonds are methylated is provided by the depsipeptides of the enniatin type, including enniatins, beauvericin and bassianolide.

3. Biosynthesis of *N*-methylated Peptides

Peptide and depsipeptide antibiotics are synthesized by large multifunctional enzymes following a thiotemplate mechanism. The precursor amino or hydroxy acids are activated as thioesters via their adenylates and then linked together to form peptide or ester bonds. Sulfhydryl groups both from cysteine residues and from enzyme-bound 4-phosphopantetheine are involved in this mechanism [1].

In the case of methylated peptides, the *N*-methyl amino acids have been shown to be no precursors, neither *in vivo* [19–23] nor in studies with isolated enzyme systems [19, 24–29]. *In vivo* studies showed that the methyl moiety origins from the methyl group of methionine, indicating that *S*-adenosyl-L-methionine would be the methylating coenzyme [19–22, 30].

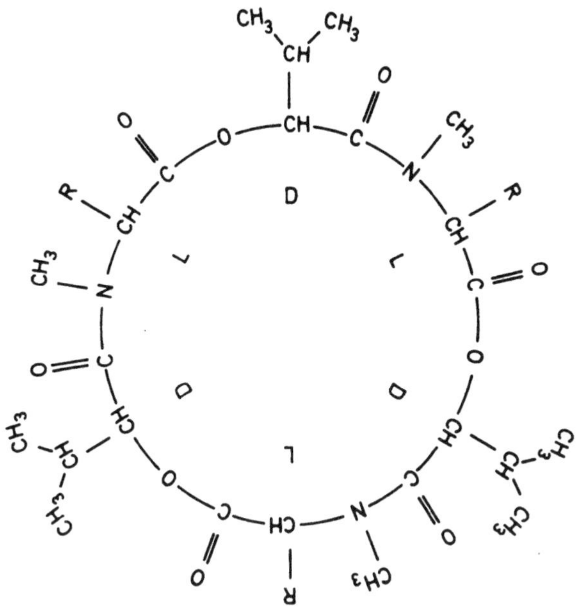

Figure 1 Enniatin A, R = $-Ch(CH_3)CH_2CH_3$; enniatin B, R = $-CH(CH_3)_2$; enniatin C, R = $-CH_2CH(CH_3)_2$; beauvericin, R = $-CH_2C_6H_5$.

The question arose, at what stage of the biosynthetic process the *N*-methylation would take place. In the case of enniatin (see Figure 1), Audhya and Russell [31] suggested that this depsipeptide is methylated after ring closure, because they found trace amounts of unmethylated product. In addition, Tamura and Suzuki [32] postulated a similar mechanism in the case of destruxins, i. e. formation of destruxin by successive methylation of the unmethylated product (protodestruxin).

By contrast, in our laboratory it was found by *in vitro* studies with an isolated enzyme that *N*-methylation of the valyl residues in enniatin B does not occur following the synthesis of the depsipeptide ring, but on the stage of the thioesterified amino acid [24]. Thus part of the synthesizing enzyme must be a *N*-methyltransferase, capable of methylating an enzyme-bound intermediate.

It has recently been demonstrated that the biosynthesis of other *N*-methylated peptides from *Actinomycetes* and from fungi proceeds via similar mechanisms (see below). Therefore, this pathway seems to be of general importance in the synthesis of peptides and depsipeptides.

The methyltransferase function of the multifunctional enzyme enniatin synthetase has recently been studied in detail. Due to the relative simplicity of the enzyme and its reaction product compared to other systems, it should provide a convenient model for other synthetases involved in the biosynthesis of *N*-methylated peptides.

4. The *N*-methyltransferase Function of Enniatin Synthetase

Enniatins are cyclic hexadepsipeptide antibiotics with ionophoretic properties produced by various strains of *Fusarium* [33, 34]. They are composed of alternating residues of D-2-hydroxyisovaleric acid (D-Hiv) and of a *N*-methyl-L-amino acid, linked by amide and ester bonds (see Figure 1). Enniatin synthetase, the multifunctional enzyme responsible for the biosynthesis of these compounds from their primary precursors, has been isolated from *Fusarium oxysporum* and studied by Zocher and co-workers [19, 24–26]. The protein consists of a single polypeptide chain with a M_r of about 250 kDa. The reaction sequence leading to the cyclic enniatin molecule can be divided into five partial reactions [19, 24, 25] (see Figure 2): 1. activation of D-Hiv and a branched chain amino acid as adenylates; 2. transfer of adenylates to specific thiol groups to form thioesters; 3. *N*-methylation of the covalently bound amino acid residue with *S*-adenosyl-L-methionine (AdoMet) as the donor of the methyl group; 4. peptide bond formation; 5. condensation of enzyme bound dipeptidols and cyclisation (enniatin formation). In the

Substrate activation

1) ATP + D-Hiv + E ⇄ (D-Hiv ~ AMP)E + PP_i

2) ATP + L-Val + E ⇄ (L-Val ~ AMP)E + PP_i

Thioester formation

3) (D-Hiv ~ AMP)E → D-Hiv ~ S-E + AMP

4) (L-Val ~ AMP)E → L-Val ~ S-E + AMP

N-*Methylation*

5) L-Val ~ S-E —AdoMet→ L-MeVal ~ S-E

Peptide-bond formation

6) (D-Hiv ~ S)(L-MeVal ~ S)>E → D-Hiv-L-MeVal ~ S(HS)>E

Ester-bond formation

7) D-Hiv-L-MeVal ~ S-E —Cyclisation→

cyclo-[D-Hiv-L-MeVal]$_3$ + E

Enniatin B

Figure 2 Scheme of the partial reactions of enniatin synthesis.

absence of AdoMet from the reaction mixture, unmethylated enniatins are synthesized, but at a reduced reaction rate [25].

4.1 *N*-methylation of enzyme-bound amino acids

The occurrence of the *N*-methylation on the stage of the enzyme-bound amino acid could be demonstrated after incubation of the enzyme with ATP, [^{14}C]-valine, and AdoMet and subsequent release of *N*-methyl-L-valine by performic acid treatment [24] (see lane 5 in Figure 3). In the absence of AdoMet, only valine could be split off the enzyme (lane 2).

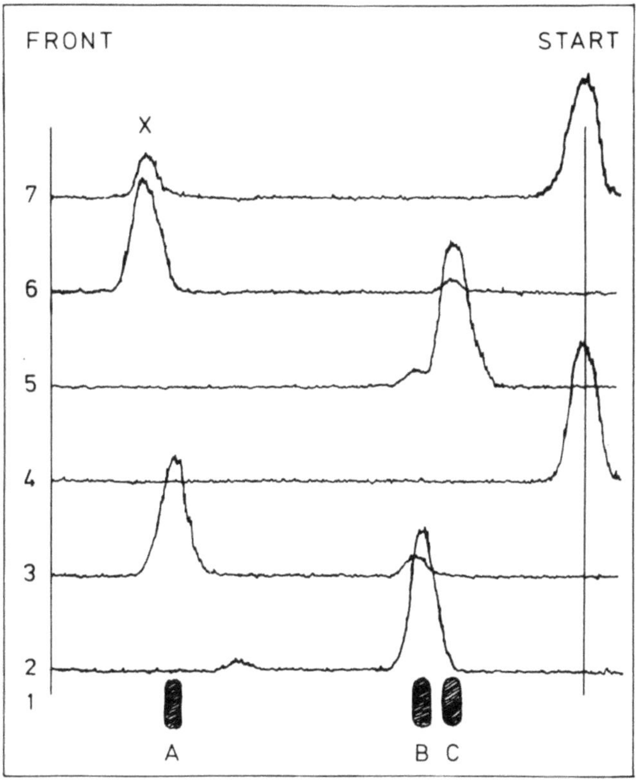

Figure 3 Influence of isovaleric acid on the formation of covalently bound intermediates (separation by tlc).
First the enzyme was loaded with [^{14}C]-valine. In parallel experiments the same conditions were used with the additional presence of AdoMet. After 5 min, to two of the incubations isovaleric acid was added and after a further 5 min the reaction was stopped by addition of 7% trichloroacetic acid. Lane 1: Marker substances: A = isovalerylvaline, B = valine, C = *N*-Methyl-L-valine. Lane 2: valine-enzyme-complex + performic acid. Lane 3: valine-enzyme-complex + isovaleric acid + performic acid. Lane 4: valine-enzyme-complex + isovaleric acid + formic acid. Lane 5: *N*-Methyl-valine-enzyme-complex + performic acid. Lane 6: *N*-Methylvaline-enzyme-complex + isovaleric acid + performic acid. Lane 7: *N*-Methylvaline-enzyme-complex + isovaleric acid + formic acid (Compound X is obviously isovaleryl-*N*-methyl-L-valine).

To clarify the reaction sequence of enniatin synthesis following the *N*-methylation step, the enzyme was loaded with either [^{14}C]-L-valine or [^{14}C]-D-Hiv and then isovaleric acid was added. The latter compound is able to occupy both the hydroxy and the amino acid binding sites of the enzyme. From the possible reaction products (ester or amide) only the formation of isovalerylvaline was observed after treatment with performic acid, indicating that the next step after *N*-methylation is peptide bond formation. In the presence of AdoMet, isovaleryl-*N*-methylvaline was formed.

4.2 Kinetic properties of the *N*-methyltransferase function

The K_m-value for AdoMet was found to be 10 μM [35] and lies in the range of K_m-values reported for other *N*-methyltransferases [36]. A double-reciprocal plot of kinetic measurements with AdoMet as the varied substrate is linear over the whole range of concentrations used; thus there is no sign of a positive cooperativity of AdoMet binding as observed in the case of glycine *N*-methyltransferase from rat liver [37].

The K_m-values for the other substrates of the enzyme are identical for both enniatin and demethyl-enniatin synthesis [35]. Obviously, AdoMet does not affect the affinity of the enzyme to ATP, valine, or D-Hiv. However, the maximal velocity of the reaction is about 6 times higher for enniatin synthesis than for the formation of its unmethylated congener. The *N*-methylated enzyme-bound amino acid residue seems to be more reactive than the unmethylated one, leading to an increase of the rate of peptide bond formation; this might be due to the higher basicity or to a favourable orientation of the secondary amino group.

4.3 Inhibition studies of the *N*-methyltransferase function

A characteristic property of all *N*-methyltransferases that have been studied so far is their sensitivity to inhibition by the reaction product *S*-adenosylhomocysteine (AdoHcy); in addition, sinefungin, another compound structurally related to AdoMet (see Figure 4), is known to act as a competitive inhibitor of a variety of *N*-methylating enzymes [36]. Therefore, the effect of AdoHcy and sinefungin on the product formation by enniatin synthetase was tested [35]. As can be seen in Figure 5, sinefungin inhibited *N*-methylation, but its presence even in excessive amounts still allowed synthesis of demethyl-enniatin (lane D). At lower concentrations of sinefungin, besides enniatin B and demethyl-enniatin B further two bands appeared (2 and 3, lane C). These bands were identified as mono- and dimethyl-enniatin B, respectively. In contrast to the findings with sinefungin, AdoHcy not only blocked formation of enniatin B, but also that of the unmethylated product (see Figure 5, lane E, F). A 50% inhibition of demethyl-enniatin B synthesis was observed at a concentration of about 30 μM of AdoHcy.

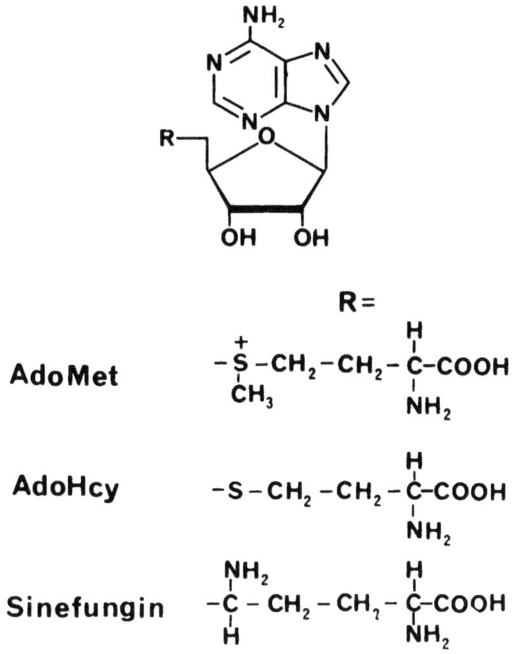

Figure 4 Structural formulas of *S*-adenosyl-L-methionine (AdoMet), *S*-adenosyl-L-homocysteine (AdoHcy), and sinefungin.

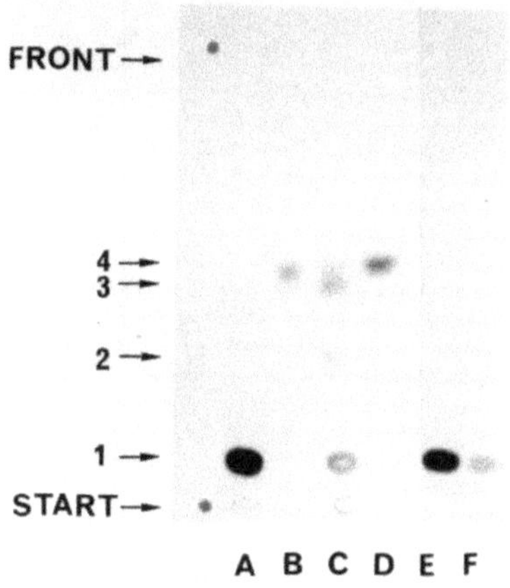

Figure 5 Influence of AdoHcy and sinefungin on product formation of enniatin synthetase. The enzyme was incubated with all necessary substrates for enniatin synthesis with [^{14}C]-valine as the radiolabel. Products formed in the presence and absence of AdoHcy and sinefungin were separated by reversed phase tlc on RP18. Lane A: enniatin B (marker), B: demethyl-enniatin B (marker), C: +510 μM sinefungin, D: +6.7 mM sinefungin, E: +16 M AdoHcy, F: +160 μM AdoHcy.

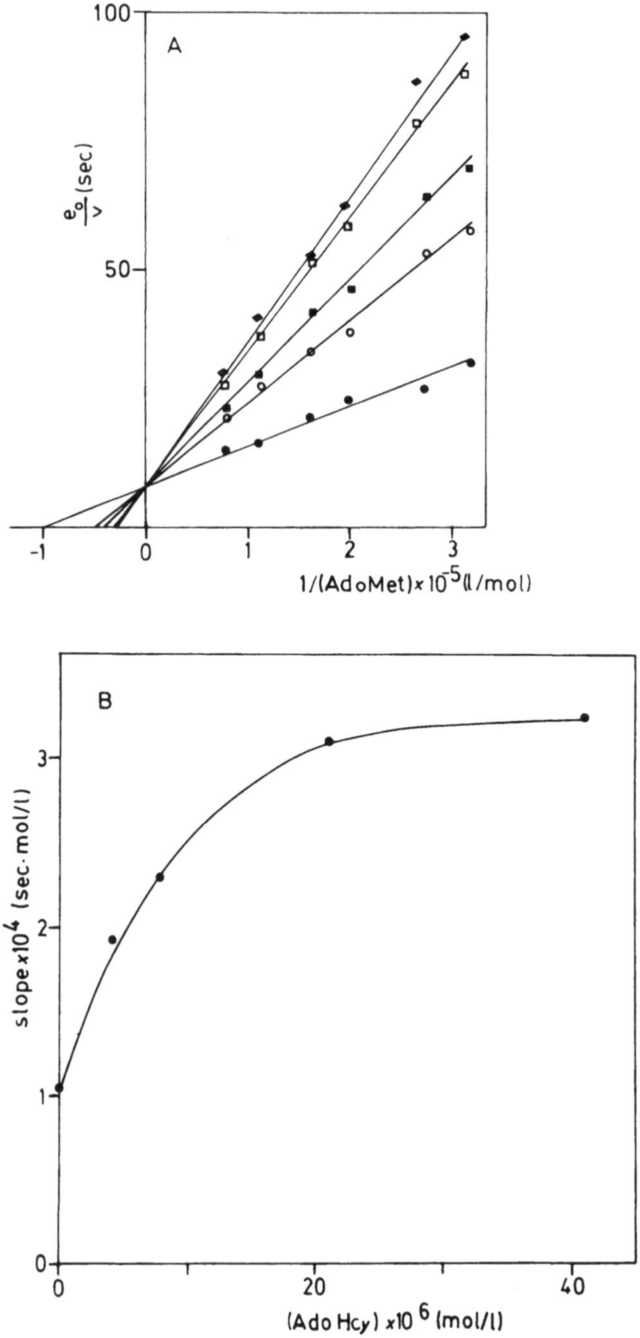

Figure 6 (A) Effect of AdoHcy on the synthesis of enniatin B by enniatin synthetase. AdoMet was used as radiolabel and was varied from 3 to 12.5 μM. The concentrations of AdoHcy used were: 0 (●), 4.2 (○), 8.3 (■), 20.8 (□), and 41.6 (◆) μM.
(B) Secondary plot of the slopes of Figure 6 A versus the concentration of AdoHcy.

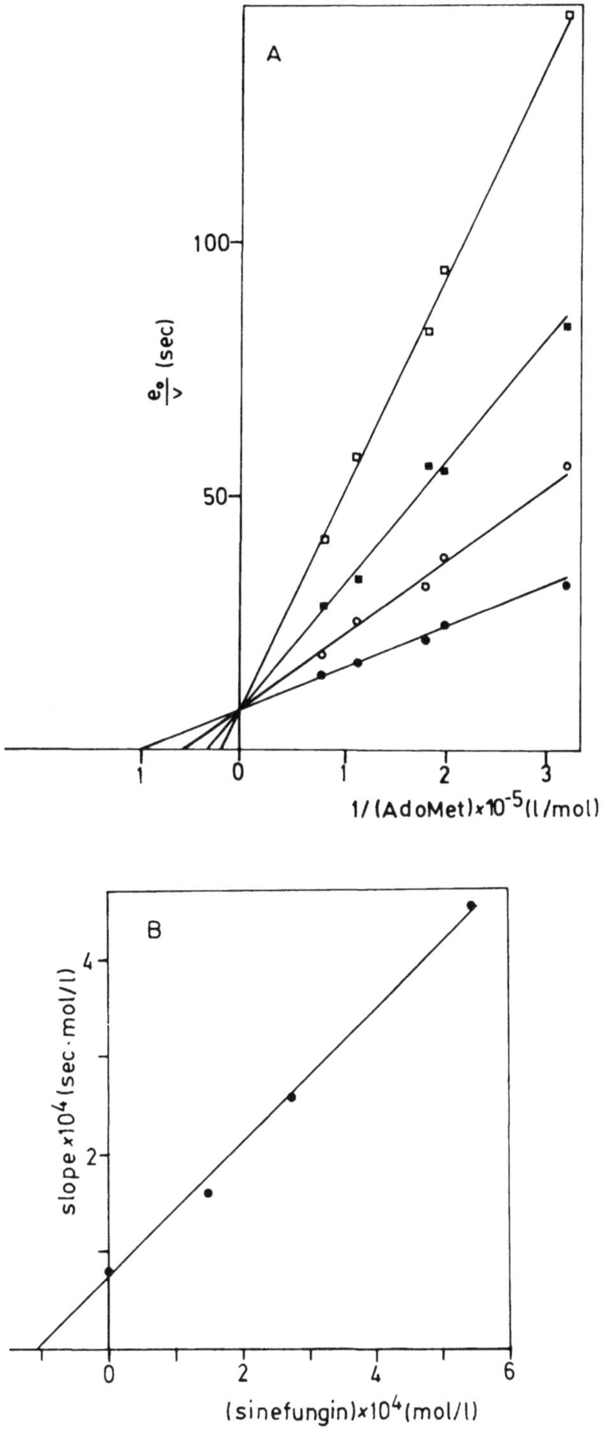

Figure 7 (A) Effect of sinefungin on the synthesis of enniatin B by enniatin synthetase. AdoMet was used as radiolabel and was varied from 3 to 12.5 μM. The concentrations of sinefungin used were: 0 (●), 135 (○), 271 (■), and 542 (□) μM.
(B) Secondary plot of the slopes in Figure 7A versus the concentration of sinefungin.

Kinetic measurements of demethyl-enniatin B synthesis revealed, that AdoHcy acts as a noncompetitive inhibitor with respect to all of the necessary substrates, i. e. ATP, valine, and D-Hiv [35]. However, neither the rates of valyl or D-hydroxy-isovaleryl adenylate synthesis as studied by means of the valine- or D-Hiv-dependent ATP-pyrophosphate exchange reaction, were reduced, nor was the formation of valine or D-Hiv thioester inhibited. Therefore, AdoHcy must exhibit an influence on the elongation or cyclisation reactions of depsipeptide synthesis.

When the synthesis of the methylated depsipeptide, enniatin B, was studied, AdoHcy also acted as a noncompetitive inhibitor with respect to ATP, valine, and D-Hiv. However, kinetic measurements with AdoMet as the varied substrate reveal another pattern of inhibition; the double-reciprocal plots in the presence of different concentrations of AdoHcy show a family of lines intersecting at a common ($1/v_{max}$)-value (see Figure 6 A). The secondary plot of the slopes of these lines versus the inhibitor concentration (Figure 6 B) yields an hyperbola, which is indicative of a so-called partial competitive inhibition [38]. This pattern can only be explained, if one assumes that there is a discrete inhibitory site to which the reaction product AdoHcy is able to bind. When the site is occupied, the affinity of the enzyme to AdoMet is lowered, while the v_{max} of the reaction is not diminished.

In contrast to AdoHcy, sinefungin acts as a pure competitive inhibitor preferably interacting with the AdoMet binding site (Figure 7). This is not surprising, since at pH 7.2 sinefungin carries a positive charge at its α-NH_2-group, which is situated at about the same location as the charge of the sulphur in the AdoMet molecule; AdoHcy does not possess a positive charge in this position. This may be the discriminating factor for the binding to the two sites.

In addition to AdoHcy and sinefungin, some other compounds known to inhibit specific transmethylation reactions were tested; 5′-deoxy-5′-(isobutylthio)adenosine, 5′-deoxy-5′-(methylthio)adenosine, and 5-methyl-tetrahydropteroylpentaglutamate did not have any effect on the formation of the unmethylated or the methylated depsipeptide.

4.4 Photoaffinity labeling of enniatin synthetase

A method for a site specific affinity labeling of methyltransferases has been introduced by Yu [39] and by Hurst et al. [40]. Enniatin synthetase, too, was radiolabeled when irradiated in the presence of AdoMet labeled at the methyl group [35]. In order to determine the number of methyl groups that can be transferred from AdoMet to enniatin synthetase, the protein was irradiated in presence of different concentrations of its methyl-labeled substrate and determined covalently bound radioactivity after acid precipitation of the protein (Figure 9). As can be seen in Figure 8, the enzyme was labeled with the [^{14}C]-methyl group in a molar ratio of about 1 : 1.

With sinefungin a dose-dependent reduction of labeling was observed when added to the incubation mixture. This is expected for a competitive inhibitor; the concen-

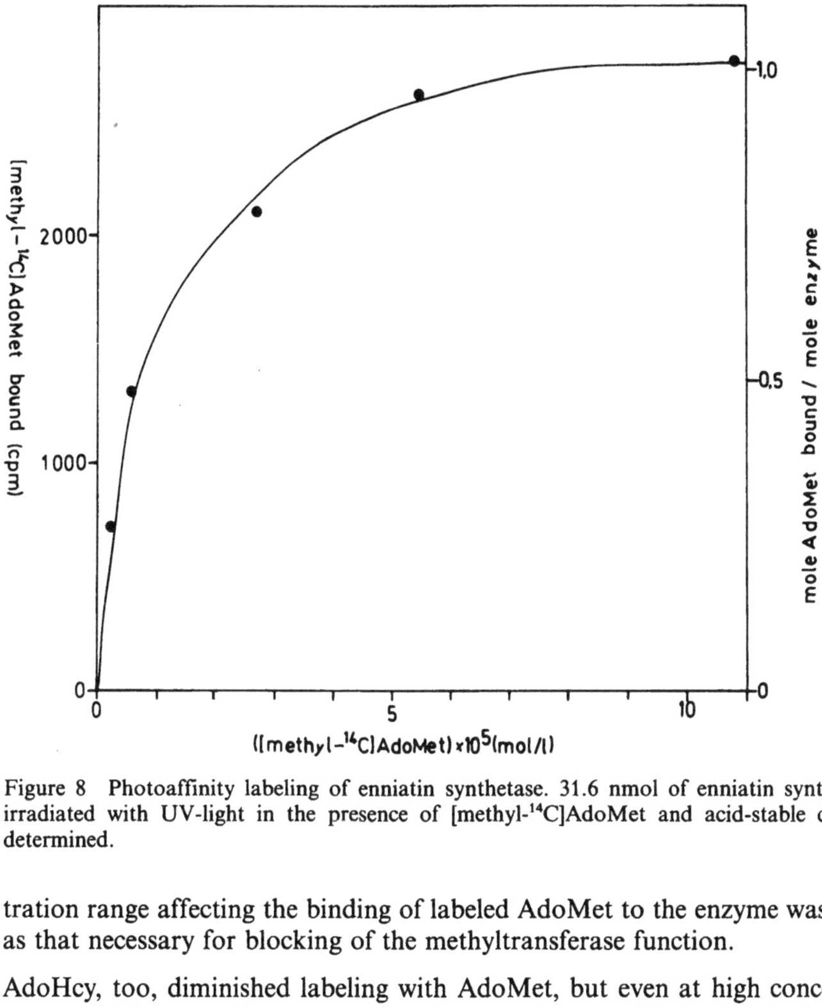

Figure 8 Photoaffinity labeling of enniatin synthetase. 31.6 nmol of enniatin synthetase were irradiated with UV-light in the presence of [methyl-^{14}C]AdoMet and acid-stable counts were determined.

tration range affecting the binding of labeled AdoMet to the enzyme was the same as that necessary for blocking of the methyltransferase function.

AdoHcy, too, diminished labeling with AdoMet, but even at high concentrations was not able to totally prevent the photoreaction, as did sinefungin; in contrast, the enzyme-bound radioactivity reached a reduced but constant level. This behaviour is expected for a partial competitive inhibitor, which reduces the affinity of the substrate to the enzyme, but even at infinite high inhibitor concentrations allows formation of an enzyme-substrate-inhibitor complex.

Limited proteolytic digestion of the enzyme irradiated in the presence of [methyl-^{3}H]AdoMet yielded one radioactively labeled fragment, which was relatively stable against further degradation (see Figure 9). It has to be concluded, that this fragment represents all or at least part of a catalytic domaine of the methyltransferase of enniatin synthetase. The size of the 25 kDa-fragment is in agreement with reported molecular weights for *N*-methyltransferases (or their monomers), which lie around 30 kDa [37, 41, 42].

The question arose, whether this fragment contains the binding site for the other substrate of the methyltransferase, i. e. valine, too. Attempts to isolate a radioac-

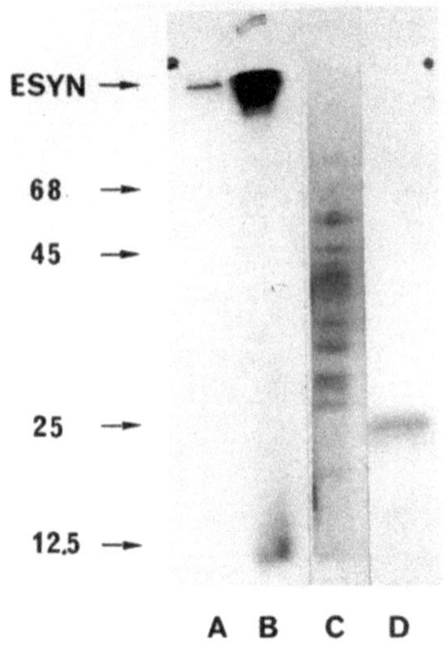

Figure 9 Polyacrylamide-sodium dodecyl sulfate-gel electrophoresis of labeled enniatin synthetase.
The enzyme was labeled with: lane A, [^{14}C]-iodoacetamide; lane B, [methyl-^{3}H]AdoMet; lane C, [^{14}C]-iodoacetamide, subsequent digestion with protease V8; lane D, [methyl-^{3}H]AdoMet, subsequent digestion with protease V8. Numbers at the side indicate molecular mass in kDa.

tively labeled fragment after covalent loading of the enzyme with [^{14}C]-valine were unsuccessful due to the instability of the thioester under the alkaline conditions required during electrophoresis. As an alternative, iodoacetamide was used, which has been shown to act as a selective inhibitor of the valine binding site of enniatin synthetase by its reaction with a catalytic thiol group [25]. Figure 9, lane F, shows that a variety of fragments of the V8 digest were labeled, when [^{14}C]-iodoacetamide was used. However, in the position of the AdoMet labeled fragment no radioactive band appeared, indicating the absence of the catalytic thiol group from the 25 kDa fragment.

4.5 Monoclonal antibodies to enniatin synthetase

Five monoclonal antibodies designated 1.56, 21.1, 25.91, 28.7, and 28.34 have been prepared against enniatin synthetase [43]. All antibodies inhibited enniatin formation; their ability to inhibit different partial reactions of the multifunctional enzyme was studied (see Table 2). Antibodies 21.1 and 25.91 inhibited valyl thioester formation, while 1.56, 28.7, and 28.34 additionally inhibit D-2-hydroxyisovaleric acid thioesterification. None of the antibodies affected the formation of L-valyl or D-hydroxyisovaleryl adenylate by the enzyme.

Since all five antibodies inhibited the formation of a thioester between enzyme and valine, their effect on the *N*-methylation site of the multienzyme could not be

Table 2 Properties of monoclonal antibodies to enniatin synthetase.

Monoclonal antibody	Subclass	Binding to denatured antigen	K_a x 10^{-7} M^{-1}	Influence on partial reactions Step[a] 1	2	3	4	5
1.56	IgG	+	n.d.	−	−	+	+	−
21.1	IgG	+	n.d.	−	−	−	+	−
25.91	IgG	+	n.d.	−	−	−	+	−
28.7	IgM	−	1.2	−	−	+	+	+
28.34	IgM	−	1.9	−	−	+	+	+

[a] The numbers of the reaction steps refer to Figure 2. Steps 6 and 7 are left out, since all antibodies studied inhibit earlier reaction steps.

Table 3 Influence of monoclonal antibodies on photoaffinity labeling of enniatin synthetase with [^{3}H-methyl]AdoMet.

Preincubation with antibody	Radioactive labeling of the enzyme (cpm)	(%)
−	116495	100
1.56	118390	102
21.1	127560	109
25.91	108067	93
28.7	7480	6
28.34	23299	20

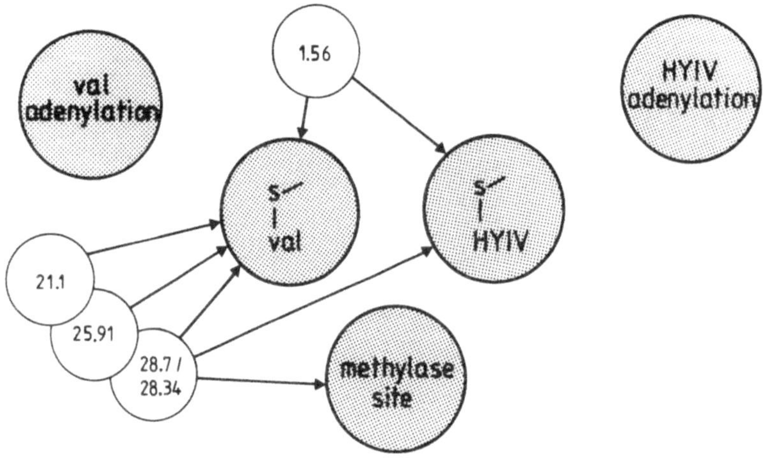

Figure 10 Scheme of the interactions of monoclonal antibodies with the catalytic sites of enniatin synthetase.

studied directly. Therefore, the influence of the antibodies on the binding of [methyl-^{3}H]AdoMet after UV-irradiation (see above) was measured. The data shown in Table 3 reveal that antibodies 1.56, 21.1, and 25.91 were not capable to prevent photoaffinity labeling of the enzyme, whereas 28.7 and 28.34 considerably reduced binding of radioactivity to the enzyme. The results indicate that there must be distinct thioester activation sites for valine and D-hydroxyisovalerate close to each other and in the neighbourhood of the methyltransferase site. These deductions are illustrated in Figure 10.

It is of interest that antibodies 28.7 and 28.34, which inhibit the photoaffinity labeling with AdoMet, also bound to an enzyme from *Streptomyces chrysomallus*, which is involved in the synthesis of the peptidolactone moiety of actinomycins and its *N*-methylation (see below) [29].

5. The *N*-methyltransferase Function of other Peptide Synthetases

5.1 Beauvericin synthetase

In the cyclodepsipeptide beauvericin the branched chain amino acid present in the enniatins is replaced by L-phenylalanine (see Figure 1). An enzyme capable of its total synthesis from the amino and hydroxy acid precursors with the consumption of ATP and AdoMet has been isolated from *Beauveria bassiana* [27]. This beauvericin synthetase bears a great deal of similarity to enniatin synthetase; however,

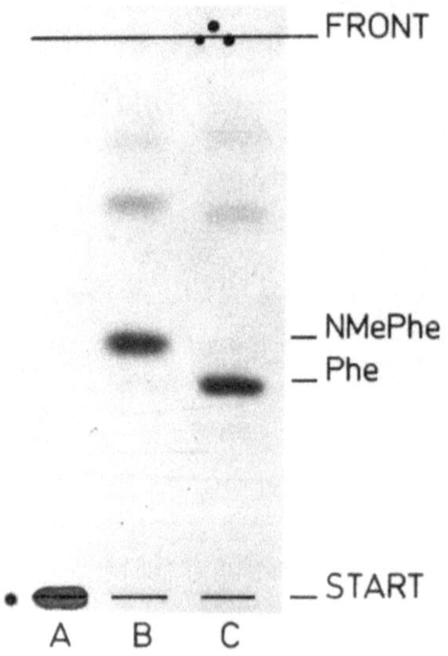

Figure 11 Liberation of covalently bound [^{14}C]-phenylalanine and [^{14}C]-*N*-methylphenylalanine by treatment with performic acid after labeling of the enzyme with [^{14}C]-phenylalanine (lane C) or with [^{14}C]-phenylalanine in the additional presence of AdoMet. Lane A represents a control experiment with formic acid instead of performic acid using the [^{14}C]-phenylalanine labeled enzyme.

it exhibits another substrate specificity at the amino acid binding site, which renders it incapable of enniatin B synthesis. It has been shown that in this system, too, *N*-methylation occurs on the stage of the enzyme bound amino acid, i. e. phenylalanine (see Figure 11) [44].

5.2 Cyclosporin synthetase

Cyclosporin is a cyclic undecapeptide with immunosuppressive properties produced by the fungus *Tolypocladium inflatum* [45]. Its structure is shown in Figure 12. Besides the unusual amino acids 2-aminobutyric acid, D-alanine, and 4-(E)-butenyl-4-methyl-L-threonine, it also contains a number of *N*-methylated peptide bonds.

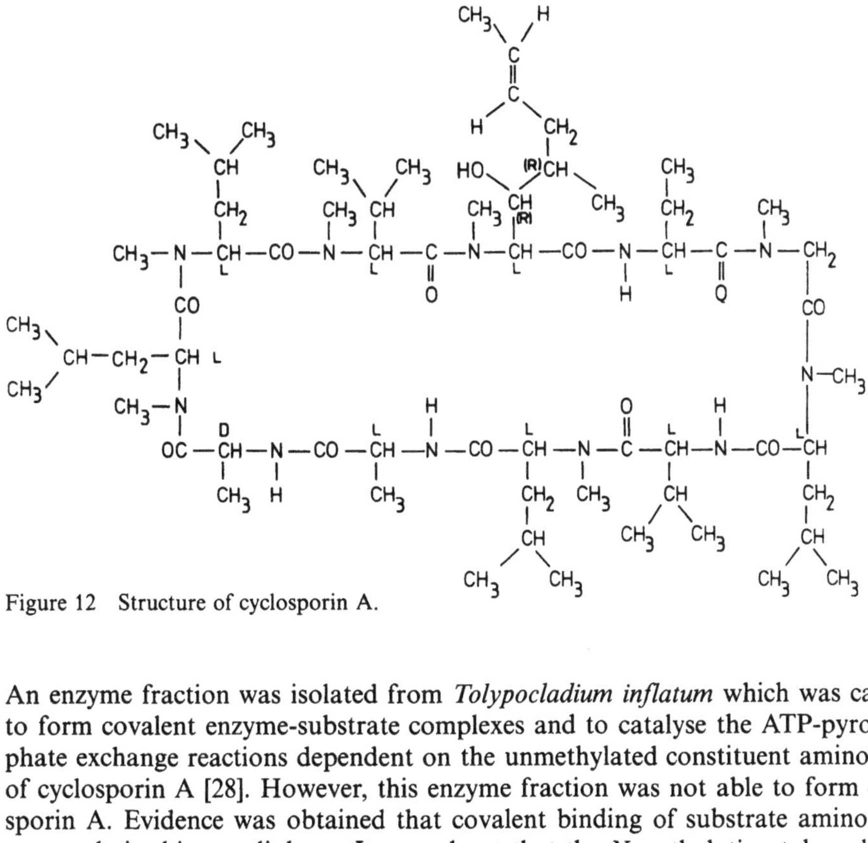

Figure 12 Structure of cyclosporin A.

An enzyme fraction was isolated from *Tolypocladium inflatum* which was capable to form covalent enzyme-substrate complexes and to catalyse the ATP-pyrophosphate exchange reactions dependent on the unmethylated constituent amino acids of cyclosporin A [28]. However, this enzyme fraction was not able to form cyclosporin A. Evidence was obtained that covalent binding of substrate amino acids occurred via thioester linkage. It turned out that the *N*-methylation takes place at the stage of the thioesterified amino acid, because the *N*-methyl amino acid can be split off the protein by performic acid treatment, when the enzyme had been incubated in the additional presence of AdoMet. In the case of leucine most of the radioactivity was localized in a band corresponding to *N*-methylleucine. This fits with the exclusive presence of *N*-methylleucine residues in cyclosporin A. On the other hand, the same experiment performed with radioactive valine revealed an almost equal distribution of the radioactivity between valine and *N*-methylvaline as shown in Figure 13 (lane 2). This latter finding is in agreement with the presence of one residue of each valine and *N*-methylvaline in the peptide chain of cyclosporin A.

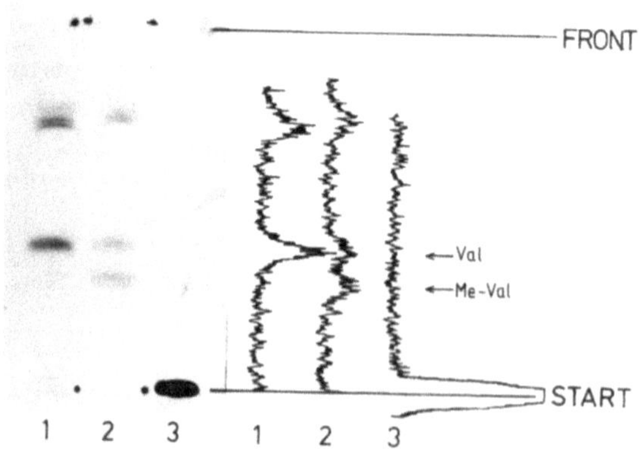

Figure 13 Liberation of covalently bound [^{14}C]-valine and [^{14}C]-methylvaline by treatment with performic acid after labeling of the enzyme with [^{14}C]-valine (lane 1) or with [^{14}C]-valine in the additional presence of AdoMet (lane 2). Lane 3 represents a control experiment with formic acid instead of performic acid using the [^{14}C]-valine-labeled enzyme.

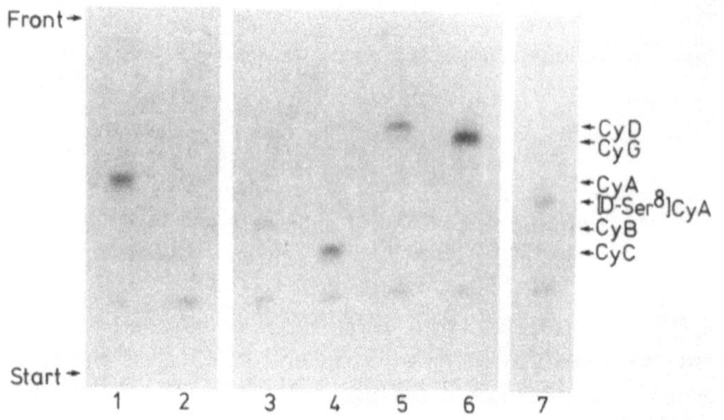

Figure 14 Autoradiograph of separation of enzymatically synthesized cyclosporins by thin-layer chromatography.
Lane 1: reaction product of the incubation of enzyme with the constituent amino acids of cyclosporin A, ATP/Mg^{2+} and [^{14}C-methyl]AdoMet; lane 2: as 1, but omission of ATP; lane 3: as 1, but replacement of L-2-aminobutyric acid by L-alanine; lane 4, replacement by L-threonine; lane 5: replacement by L-valine; lane 6: replacement by L-norvaline; lane 7: as 1, but replacement of D-alanine by D-serine. Positions of reference compounds (cyclosporins A, B, C, D, G and [D-serine^8]cyclosporine) are indicated.

Recently, the isolation of an enzyme complex from *Tolypocladium inflatum*, strain 7939/F has been achieved, which was capable to synthesize the whole cyclosporin A molecule as well as some of its homologues [46] (see Figure 14).

5.3 Actinomycin synthetases

The actinomycins are bicyclic pentapeptide lactone antibiotics in which two pentapeptide lactone rings are attached via amide bonds to the carboxylic group of a phenoxazinone dicarboxylic acid (Figure 15a). It was shown that actinomycins are most likely derived from the non-enzymatic, oxidative condensation of two monocyclic 4-methyl-3-hydroxyanthranilic acid pentapeptide lactones (Figure 15b) [47].

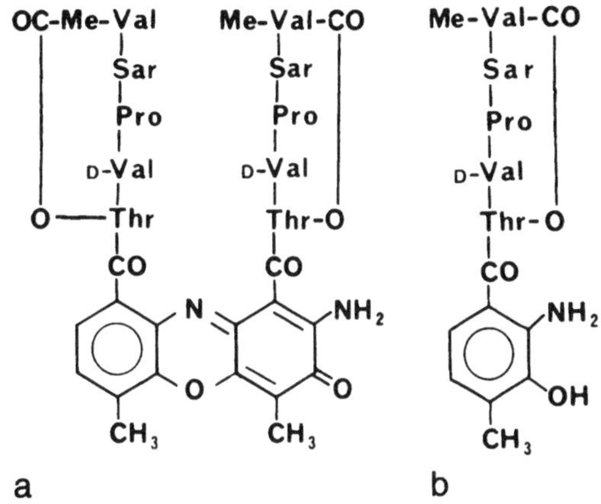

Figure 15 Structure of actinomycin D and its immediate precursor 4-methyl-3-hydroxyanthranilate pentapeptide lactone.

Keller has isolated three enzymes from *Streptomyces chrysomallus* involved in the biosynthesis of actinomycins [29]. The enzyme designated actinomycin synthetase I (M_r about 55,000 Da) activates the precursor of the chromophore moiety, 4-methyl-3-hydroxyanthranilic acid, by formation of the corresponding adenylate [47]. Actinomycin synthetase II activates the first two amino acids of the peptide chains, threonine and valine, as thioesters via their adenylates; it is a single polypeptide chain of M_r 225,000 Da. Similarly, actinomycin synthetase III activates proline, glycine, and valine (the remaining three amino acids in the antibiotic) as thioesters and is a single polypeptide chain of about M_r 280,000 Da.

Actinomycin D contains two *N*-methylated amino acids in its peptide chains. These are in the 4-position sarcosine and in the 5-position *N*-methyl-L-valine. A methyltransferase activity responsible for *N*-methylation of thioesterified glycine and valine is harbored in the actinomycin synthetase III. Figure 16 illustrates how fluorography of electropherograms with ^{14}C-aminoacylated synthetases II and III

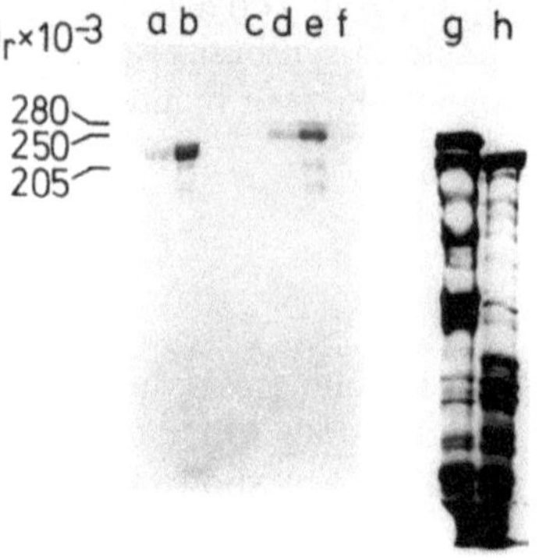

Figure 16 Fluorogram of an electrophoretic separation of [^{14}C]-amino-acylated actinomycin synthetases. Actinomycin synthetase II was aminoacylated with L-[^{14}C]-threonine (lane a) or L-[^{14}C]-valine (lane b). After incubation, samples were loaded on a 5% polyacrylamide gel. Similarly actinomycin synthetase III was aminoacylated with [^{14}C]-proline (lane c), [methyl-^{14}C]AdoMet in the presence of glycine (lane d) or valine (lane e), and without amino acid (lane f). Lanes g and h show the corresponding electropherograms stained with Coomassie Blue with actinomycin synthetase II and III, respectively.

helped to identify the multifunctional polypeptides. The first fraction contained one band with 225 kDa, which could be labeled with valine and threonine (Figure 16, lanes a and b). In the second fraction labeling of two distinct bands with [methyl-^{14}C]AdoMet in the presence of each glycine or valine is visible (lanes d and e); these bands have molecular masses of about 280 and 255 kDa. [^{14}C]-Proline labeled only the 280-kDa band, as revealed by the faintly visible band in lane c. The 255-kDa band obviously was a product of proteolysis from the 280-kDa band during enzyme purification.

Attempts to obtain the total synthesis of the pentapeptide lactones (see Figure 15 b) by incubating the three synthetases with all necessary substrates were as yet unsuccessful. However, the observations with this system provide first evidence that the principle of *N*-methylation occuring on the stage of the enzyme-bound amino acid is followed in actinomycetes, too. Furthermore, it is remarkable that although actinomycins are synthesized by a complex of enzymes, *N*-methylation occurs by the same enzyme which also activates the amino acids to be methylated and not by a separate methylase.

6. Conclusions

A variety of peptides and peptolides containing *N*-methylated peptide bonds are known from procaryotic and eucaryotic origin. Although these compounds may strongly differ in their structure it seems that they are synthesized by a common mechanism: 1. Activation of amino acids as adenylates, 2. transfer of amino acids to specific thio groups, 3. methylation of covalently bound amino acid residues, 4. peptide bond formation, 5. peptide chain elongation and in most cases cyclisation. In all cases studied so far the methyltransferase functions involved in this process are integral parts of the multifunctional antibiotic synthetases; in the case of enniatin and beauvericin synthetases all functions, including the methylases, are located on one polypeptide chain, while in the case of the actinomycin synthetases the methyltransferase is fused with an amino acid activating enzyme belonging to the multienzyme complex. From the occurrence of partially methylated peptides it has to be concluded that each amino acid residue to be methylated must have its own methyltransferase unit, e. g. it is proposed that the enzyme synthesizing the undecapeptide cyclosporin contains eleven activation sites, four of which are vicinal to unique methyltransferase sites.

Studies with monoclonal antibodies to enniatin synthetase lead to the conclusion that the methyltransferase of the multienzymes must be immunologically related, irrespective of their procaryotic or eucaryotic origin. It will be of interest to expand these studies using molecular genetics to reveal similarities of sequence and structure between the different methyltransferases.

References

1. Kleinkauf, H., and von Döhren, H., Nucleic acid independent synthesis of peptides, *Curr. Top. Microbiol. Immunol. 91*, 129, 1981.
2. Mauger, A. B., and Katz, E., Actinomycins, in *Antibiotics: Isolation, Separation and Purification*, Weinstein, M. J., and Wagman, G. H., eds., Elsevier, Amsterdam, 1978, 1.
3. Okumura, Y., Peptidolactones, in *Biochemistry and genetic regulation of commercially important antibiotics*, Vining, L. C., ed., Addison-Wesley Publishing Company, Massachusetts, 1983, chap. 6.
4. Konishi, M., Ohkuma, H., Sakai, F., Tsuno, T., Koshiyama, H., Naito, T., and Kawaguchi, H., BBM-928, a new antitumor antibiotic complex; III: structure determination of BBM-928 A, B, and C, *J. Antibiot. 34*, 148, 1981.
5. Berdy, J., *Handbook of antibiotic compounds*, 1980, CRC Press, Boca Raton, Florida, Vol. IV, 1980.
6. Arai, M., Haneishi, T., and Nakajima, M., *Ger. Offen. 3,041,130*, 1981; *Japan Appl. 79/141,650*, 1979.
7. Arai, M., Haneishi, T., Nakajima, M., Torikata, A., and Enokita, R., *Eur. Pat. Appl. EP37,736*, 1981; *Japan Appl. 80/45,402*, 1980.
8. Ubukata, M., Uramoto, M., and Isono, K., The structure of neopeptins, inhibitors of fungal cell wall biosynthesis, *Tetrahedron Lett. 25*, 423, 1984.
9. Satomi, T., Kusakabe, H., Nakamura, G., Nishio, T., Uramoto, M., and Isono, K., Neopeptins A and B, new antifungal antibiotics, *Agric. Biol. Chem. 46*, 2621, 1982.
10. Maehr, H., Liu, C.-M., Palleroni, N. J., Smallheer, J., Todaro, L., Williams, T. H., and Blount, J. F., Microbial Products V: azinothricin, a novel hexadepsipeptide antibiotic, *J. Antibiot. 39*, 17, 1985.
11. Freer, A. A., Gardner, D., and Poyser, J. P., X-Ray structure of (3Z,6E)-1-N-methylalbonoursin (antibiotic M145725) from a Streptomyces sp., *J. Chem. Res. 283*, 1984.
12. Kondo, S., Shiba, T., Suzuki, A., Takita, T., Maeda, K., and Kimura, Y., Appendix tables, in *Bioactive peptides produced by microorganisms*, Umezawa, H., Takita, T., and Shiba, T., eds., Halsted Press, New York, 1978, 183.
13. Carter, D. C., Moore, R. E., Mynderse, J. S., Niemczura, W. P., and Todd, J. S., Structure of majusculamide C, a cyclic depsipeptide from *Lyngbya majuscula*, *J. Org. Chem. 49*, 236, 1984.

14. von Wartburg, A., and Traber, R., Chemistry of the natural cyclosporin metabolites, in *Ciclosporin*, Borel, J. F., Karger, Basel, 1986, 28.
15. Kobayashi, R., Samejima, Y., Nakajima, S., Kawai, K. I., and Udagawa, S. I., Studies on fungal products: XI. Isolation and structures of novel cyclic peptides, *Japan. Chem. Pharm. Bull. 35*, 1347, 1987.
16. Nakamura, H., Kobayashi, J., Nakamura, Y., and Ohizumi, Y., Theonellamine B, a novel peptidal Na,K-ATPase inhibitor from an okinawan marine sponge of the genus *Theonella, Tetrahedron Lett. 27*, 4319, 1986.
17. Matsunaga, S., Fusetani, N., and Konosu, S., Bioactive marine metabolites VII: structures of discodermins B, C, and D, antimicrobial peptides from the marine sponge *Discodermia kiiensis, Tetrahedon Lett. 26*, 855, 1985.
18. Rinehart, K. L., Gloer, J. B., Cook, J. C., Mizsak, S. A., and Scahill, T. A., Structures of the didemnins, antiviral and cytotoxic depsipeptides from a carribbean tunicate, *J. Am. Chem. Soc. 103*, 1857, 1981.
19. Zocher, R., Salnikow, J., and Kleinkauf, H., Biosynthesis of enniatin B, *FEBS Lett. 71*, 13, 1976.
20. Peeters, H., Zocher, R., Madry, N., and Kleinkauf, H., Incorporation of radioactive precursors into beauvericin produced by *Paecilomyces fumosoroseus, Phytochemistry 22*, 1719, 1983.
21. Mauger, A. B., and Katz, E., Two distinct site-dependent biosynthetic pathways for the incorporation of sarcosine into actinomycins, *Arch. Biochem. Biophys. 176*, 181, 1976.
22. Ciferri, O., Albertini, A., and Cassani, G., Origin of the sarcosine molecules of actinomycins, *Biochem. J. 96*, 853, 1965.
23. Arroyo, V., Hall, M. J., Hasall, C. H., and Yamasaki, K., Incorporation of amino acids into the cyclohexadepsipeptide monamycin, *J. Chem. Soc., Chem. Commun. 21*, 845, 1976.
24. Zocher, R., and Kleinkauf, H., Biosynthesis of enniatin B: partial purification and characterization of the synthesizing enzyme and studies of the biosynthesis, *Biochem. Biophys. Res. Commun. 81*, 1162, 1978.
25. Zocher, R., Keller, U., and Kleinkauf, H., Enniatin synthetase, a novel type of multifunctional enzyme catalyzing depsipeptide synthesis in *Fusarium oxysporum, Biochemistry 21*, 43, 1982.
26. Zocher, R., Keller, U., and Kleinkauf, H., Mechanism of depsipeptide formation catalysed by enniatin synthetase, *Biochem. Biophys. Res. Commun. 110*, 292, 1983.
27. Peeters, H., Zocher, R., Madry, N., Oelrichs, P. B., Kleinkauf, H., and Kraepelin, G., Cell-free synthesis of the depsipeptide beauvericin, *J. Antibiot. 12*, 1762, 1983.
28. Zocher, R., Nihira, T., Paul, E., Madry, N., Peeters, H., Kleinkauf, H., and Keller, U., Biosynthesis of cyclosporin A: partial purification and properties of a multifunctional enzyme from *Tolypocladium inflatum, Biochemistry 25*, 550, 1986.
29. Keller, U., Actinomycin synthetases, *J. Biol. Chem. 262*, 5852, 1987.
30. Zocher, R., Madry, N., Peeters, H., and Kleinkauf, H., Biosynthesis of cyclosporin A, *Phytochemistry 23*, 549, 1984.
31. Audhya, T. K., and Russell, D. W., Natural enniatin A, a mixture of optical isomers containing both *erythro-* and *threo-N*-methyl-L-isoleucine, *J. Chem. Soc. Perkin Transact. 1*, 743, 1974.
32. Tamura, S., and Suzuki, A., Microbial peptides and amino acid derivatives biologically active to insects or higher plants, in *Bioactive peptides produced by microorganisms*, Umezawa, H., Takita, T., and Shiba, T., eds., Wiley, chap. 6.
33. Plattner, P. A., Nager, U., and Boller, A., Über die Isolierung neuartiger Antibiotika aus Fusarien, *Helv. Chim. Acta 31*, 594, 1948.
34. Wipf, H. K., Pioda, L. A. R., Stefanac, Z., and Simon, W., Complexes of enniatins and other antibiotics with alkyl metal ions, *Helv. Chim. Acta 51*, 377, 1968.
35. Billich, A., and Zocher, R., The *N*-methyltransferase function of the multifunctional enzyme enniatin synthetase, *Biochemistry*, 1987, 26, 8417.
36. Cantoni, G. L., Richards, H. H., and Chiang, P. K., Inhibitors of *S*-adenosylhomocysteine hydrolase and their role in the regulation of biological methylation, in *Transmethylation*, Usdin, E., Borchardt, R. T., and Creveling, C. R., eds., North Holland, Amsterdam, 1979, 155.
37. Ogawa, H., and Fujioka, M., Purification and properties of glycine *N*-methyltransferase from rat liver, *J. Biol. Chem. 257*, 3447, 1982.
38. Segel, I. H., *Enzyme kinetics*, Wiley, New York, 1975, pp. 161–166 and pp. 182–185.

39. Yu, P. H., Specific photoactivated covalent binding of *S*-adenosylmethionine to phenylethanolamine N-methyltransferase, *Biochim. Biophys. Acta 742*, 517, 1983.
40. Hurst, J. H., Billingsley, M. L., and Lovenberg, W., Photoaffinity labelling of methyltransferase enzymes with *S*-adenosylmethionine: effects of methyl acceptor substrates, *Biochem. Biophys. Res. Commun. 122*, 499, 1984.
41. Im, Y. S., Chiang, P. K., and Cantoni, G. L., Guanidinoacetate methyltransferase, *J. Biol. Chem. 254*, 11047, 1979.
42. Irace, G., Colonna, G., Camardella, M., Pietra, G. D., and Porta, R., Purification and molecular properties of rabbit lung indolamine *N*-methyltransferase, *Biochemistry 21*, 1464, 1982.
43. Billich, A., Zocher, R., Kleinkauf, H., Braun, D. G., Lavanchy, D., and Hochkeppel, H., Monoclonal antibodies to the multienzyme enniatin synthetase, *Biol. Chem. Hoppe-Seyler 368*, 521, 1987.
44. Peeters, H., Zocher, R., and Kleinkauf, H., Synthesis of beauvericin by a multifunctional enzyme, *J. Antibiotics 41*, 352 1988.
45. Borel, J. R., *Ciclosporin*, Karger, Basel, 1986.
46. Billich, A., and Zocher, R., Enzymatic synthesis of cyclosporin A, *J. Biol. Chem. 262*, 17258, 1987.
47. Keller, U., Kleinkauf, H., and Zocher, R., 4-Methyl-3-hydroxyanthranilic acid activating enzyme from actinomycin-producing *Streptomyces chrysomallus, Biochemistry 23*, 1479, 1984.
48. Keller, U., Acylpentapeptide lactone synthesis in actinomycin-producing *Streptomycetes* by feeding with structural analogs of 4-methyl-3-hydroxyanthranilic acid, *J. Biol. Chem. 259*, 8226, 1984.

Chapter 4
Peptide Bond Synthesis by Enzyme-Catalyzed Acyl-Transfer

Volker Kasche and Günther Michaelis

1. Introduction
2. Mechanism of the Kinetically Controlled Peptide Synthesis
3. Maximum Yields in the Kinetically Controlled Peptide Synthesis
4. Selecting the Optimal Enzyme
5. Yield Controlling Factors in the Synthesis of a Peptide Bond with One Enzyme
   5.1 Protection of the P_1' carboxyl group; stereospecifity
   5.2 pH value
   5.3 Temperature
   5.4 Ionic strength
   5.5 Solvent composition
6. Conclusions
References

1. Introduction

The biosynthesis of biologically active peptides or peptide derivatives can be subdivided in ribosomal or non-ribosomal peptide synthesis and subsequent processing of the initial products.

Ribosomal peptide synthesis. The peptide is synthesized as part of a polypeptide sequence in the normal translation process involving peptidyltransfer reactions on ribosomes. The polypeptide chain is then posttranslationally processed by sequence specific proteinases yielding the desired peptide [1]. In cases where the carboxyl terminal is amidated as in some peptide hormones a multifunctional enzyme that catalyzes the amidation step also participates in this processing [2]. In this process only L-amino acids can be used for the synthesis. Important products are peptide hormones, neuropeptides and antimicrobial peptides as the magainins from *X. leavis* [3].

Non-ribosomal peptide synthesis. In this process both L- and D-amino acids are involved. They are activated using other groups for the activation than in the normal translation procedure. The peptides are formed by ordered peptidyltransfer mechanism catalyzed by multifunctional enzymes [4]. In some cases the formed peptide alone has a low biological activity. The biological activity can be increased by acyl- or peptidyltransfers to the peptide in vivo or in vitro. For β-lactam antibiotics acyl-transfer mechanisms are important in vivo in the formation of penicillin G and V [5]. Some important semisynthetic β-lactam antibiotics that are peptidyl-transfer products, as ampicillin or cephalexin, can only be formed in high yields in vitro [6]. Important products here are peptide antibiotics and other enzyme inhibitors [7].

These enzyme catalyzed reactions for the synthesis of peptides in vivo and in vitro can be described by the general mechanism given in Figure 1. In vivo these reactions

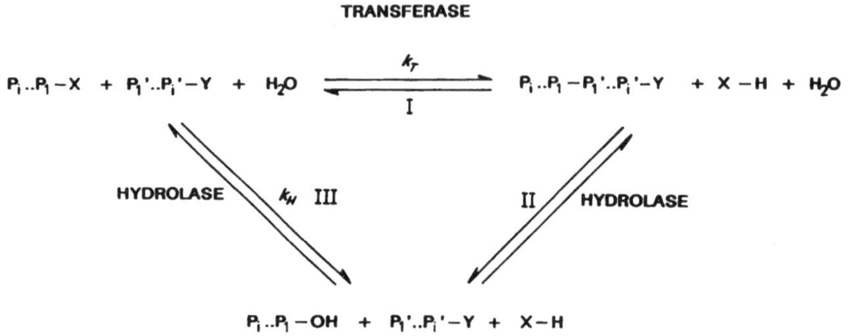

Figure 1 Enzyme catalyzed kinetically (I + II + III) and equilibrium (II) controlled synthesis of peptides $P_i \cdots P_1 - P_1' \cdots P_i'$. The enzyme catalyzing these reactions must have a transferase (II) and a hydrolase (II, III) function.

k_T, k_H = apparent second order transferase and hydrolase rate of the used enzyme.
P_i and P_i' = amino acid residues.
$-X$ = group used to activate the carboxyl end of peptide $P_i \cdots P_1^-$
$-Y$ = group used to protect the carboxyl end of peptide (amino acid) $P_1' \cdots P_i'$ (P_1').

are catalyzed by transferases or ligases with transferase/hydrolase functions. In this *kinetically controlled reaction* (I + II + II in Figure 1) the maximum yield of the condensation product $\cdots P_1 - P_1' \cdots$* depends on the ratio of the transferase to hydrolase rate of the used transferase. This justifies the term kinetically controlled reaction. When this ratio is large the product $\cdot\cdot P_1 - P_1' \cdot\cdot$ can be formed in high yields even when the thermodynamically favourable products are $\cdot\cdot P_1 - OH$ and $P_1' \cdot\cdot$ i. e. the equilibrium lies far to the hydrolysis side. This formation of non-

* The interaction between a peptide and the active site on an enzyme is described by the following scheme [8] the bond that can be hydrolyzed or formed in the enzyme catalyzed reaction

peptide $H_3 + N P_i \cdots P_1 \;\; \downarrow \;\; P_1' \cdots P_i' \cdots COO^-$
$\updownarrow \;\; \updownarrow \qquad \updownarrow_1 \;\; \updownarrow$
enzyme active site $S_i \cdots S_1 \quad C \quad S_1' \cdots S_i'$

where P_i are amino acids, S_i (S_i') the binding site for P_i (P_i') on the enzyme and C the groups directly involved in the catalytic reactions.

equilibrium concentration of peptides (and other condensation products) in vivo is only possible when activated substrates are used. The transferases (ligases with a transferase function as amino acid-tRNA ligases) that are used in vivo may also catalyze the hydrolysis reactions (II or III in Figure 1). Examples here are hydrolytic steps in proofreading mechanisms (amino acid activation or DNA-synthesis) [9, 10]. The ratio of the apparent transferase to hydrolase rate for these transferases can be estimated to be $>10^6$ from the few experimental data published here [11]. Such high ratios are necessary to minimize the energy waste involved in the acyl- or peptidyl transfer to H_2O (≈ 55 M) at the low concentrations ($<10^{-3}$ M) of peptides in vivo.

The enzymes involved in the biosynthesis of peptides and their derivatives can be used as catalysts also for the in vitro synthesis. The ligases and transferases used in ribosomal protein synthesis will probably not be of practical importance for the biotechnical synthesis of peptides due to the complex system required here and the non-availability of these enzymes in large quantities [12]. For other condensation products as oligosaccharides or oligonucleotides the transferases that are used for the in vivo synthesis can also be applied for the biotechnical synthesis of these compounds [13, 14]. Transferases involved in the non-ribosomal peptide synthesis of some peptide antibiotics (gramicidin S) have been used for the peptide synthesis in vitro [4]. Generally, however, the complex aminoacid or acyl-activations involved in the biosynthesis, and the limited availability of the transferases and ligases used for peptide synthesis in vivo necessitates the development of other systems for the enzyme catalyzed synthesis or modifications or peptides in vitro.

The kinetically controlled peptide synthesis reactions I–III given in Figure 1 can also be catalyzed by hydrolases that can act as transferases. Such hydrolases are those with covalent acyl-enzyme intermediates as serine or cysteine proteinases [11]. The use of these proteinases for peptide synthesis has been extensively reviewed recently [11, 15–19]. This reflects the current interest to evaluate the potential of proteinases as catalysts for peptide synthesis as an alternative or complement to chemical peptide synthesis. In this review some newer results on the quantitative analysis of the yield controlling factors and the P_1' sequence- and stereospecifity in the kinetically controlled synthesis not covered in existing reviews will be analyzed.

2. Mechanism of the Kinetically Controlled Peptide Synthesis

Studies on the kinetically controlled synthesis of condensation products as oligosaccharides [20], peptides [21–24], and β-lactam antibiotics [25] using hydrolases as catalysts have presented kinetic evidence that the nucleophile NH must be bound to the acyl-(glycosyl-)enzyme before it can deacylate this intermediate. The general mechanism that has been confirmed in these and other studies is given in Figure 2. As it applies also for other condensation products a more general formulation

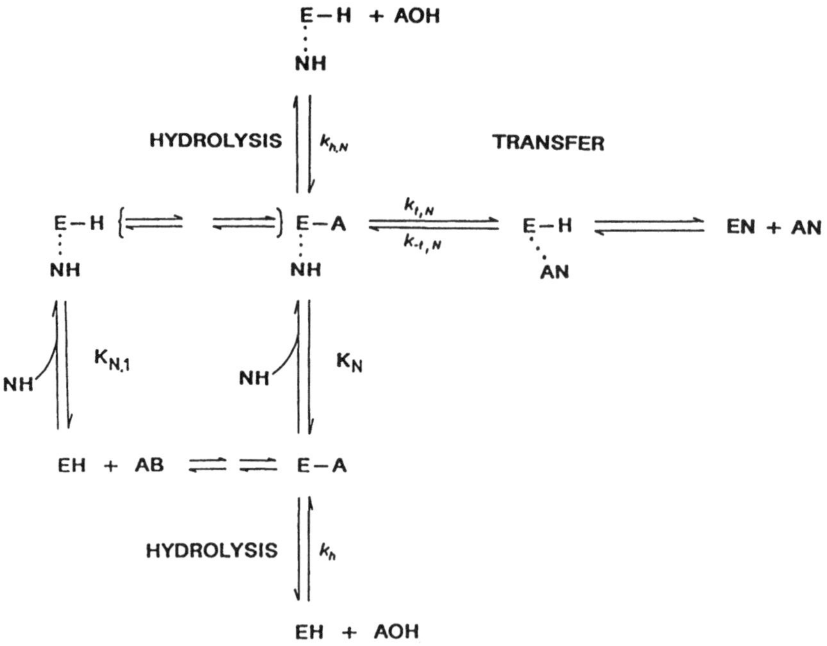

Figure 2 Mechanism for the kinetically controlled synthesis of a peptide (AN) or other condensation products. The nucleophile NH must be bound to the acyl-enzyme before the nucleophilic deacylation reaction. The reactions in brackets are not possible when the nucleophile is a competitive inhibitor in the hydrolysis of P_1-X. Dots indicate noncovalent binding. For simplicity only the acyl enzyme intermediates have been given.

for the activated substrate (AB) and the nucleophile (NH), than in Figure 1 is used here.

Depending on whether the activated substrate AB and the nucleophile NH have overlapping or non-overlapping binding sites on the enzyme the nucleophile is a competitive or mixed inhibitor for the substrate. In the former case the acyl-enzyme cannot be formed with an enzyme that has bound the nucleophile NH.

For the transferases (ligases) that catalyze the biosynthesis of these condensation products in vivo, non-overlapping binding sites for the activated substrate AB and the nucleophile NH can be postulated. For hydrolases that have not been evolutionary selected based on their transferase function, both cases are possible. The P_1'-specificity depends on the strength in the binding of the nucleophile NH to the acyl-enzyme prior to the deacylation reaction.

This demonstrates qualitatively that the mechanism and nucleophile binding may influence the yield of condensation product (peptide) that may be obtained in the kinetically controlled synthesis with hydrolases as catalysts. Thus contrary to the *equilibrium controlled synthesis* (II and III in Figure 1), where the enzyme cannot influence the yield, the enzyme properties influence the yield in the *kinetically controlled synthesis*.

3. Maximum Yields in the Kinetically Controlled Peptide Synthesis

The peptide formed in the kinetically controlled synthesis is a substrate for the hydrolase. The hydrolysis of this product can only influence the maximum yield when the hydrolysis rate is of the same order of magnitude or larger than the rate of the synthesis of the condensation product. Thus two cases must be considered:

I. Hydrolysis rate $\geq$ synthesis rate,
II. Hydrolysis rate $\ll$ synthesis rate.

I. Hydrolysis rate of condensation product not negligible compared with the synthesis rate. In this case the product concentration has a maximum when the rate of hydrolysis equals to the rate of synthesis. The maximum concentration of the condensation product formed in the kinetically controlled synthesis is given by the relation [11] (Figure 2):

$$[\mathrm{AN}]_{\max} = \frac{(k_\mathrm{T}/k_\mathrm{H})_\mathrm{app} \cdot [\mathrm{NH}]}{[\mathrm{H_2O}] + (k_\mathrm{T}/k_\mathrm{H})_\mathrm{app} \cdot [\mathrm{NH}]} \cdot \frac{\left(\dfrac{k_\mathrm{cat}}{K_\mathrm{m}}\right)_{\mathrm{AB,\,NH}}}{\left(\dfrac{k_\mathrm{cat}}{K_\mathrm{m}}\right)_{\mathrm{AN,\,NH}}} \cdot [\mathrm{AB}] \qquad (1)$$

where the $(k_\mathrm{T}/k_\mathrm{H})_\mathrm{app}$- and $(k_\mathrm{cat}/K_\mathrm{m})$-values (for the substrates given in the subscript) applies for the concentration of the nucleophile at this maximum. In other studies a partition constant p has been used as a measure for the apparent hydrolase to transferase rate of the proteinase [24]. The relation between p and $(k_\mathrm{T}/k_\mathrm{H})_\mathrm{app}$ is

$$(k_\mathrm{T}/k_\mathrm{H})_\mathrm{app} = [\mathrm{H_2O}]/p \qquad (2)$$

Both quantities can be determined from the ratio of initial rates for the formation of the transferase v_AN and hydrolase v_AOH product

$$\frac{v_\mathrm{AN}}{v_\mathrm{AOH}} = \left(\frac{k_\mathrm{T}}{k_\mathrm{H}}\right)_\mathrm{app} \cdot \frac{[\mathrm{NH}]}{[\mathrm{H_2O}]} \qquad (3)$$

A relation between $(k_\mathrm{T}/k_\mathrm{H})_\mathrm{app}$ and the 'microscopic' constants in Figure 2

$$\left(\frac{k_\mathrm{T}}{k_\mathrm{H}}\right)_\mathrm{app} = \frac{k_\mathrm{t,\,N}}{k_\mathrm{h,\,N} \cdot K_\mathrm{N} + k_\mathrm{h,\,N} \cdot [\mathrm{NH}]} \qquad (4)$$

can be derived assuming that the nucleophile binding to the acyl-enzyme can be described by equilibrium conditions [22, 25]. Equation (4) shows that $(k_\mathrm{T}/k_\mathrm{H})_\mathrm{app}$ depends on the nucleophile content. This has been observed in experimental studies [20–25]. Such results have given kinetic evidence for the mechanism in Figure 2 with nucleophile binding prior to deacylation. Direct evidence for this by equilibrium binding studies has only been given for penicillin amidase [26].

Relations for $[AN]_{max}$ for this case are given in Table 1. They can generally be written as

$$[AN]_{max} = \alpha \cdot \beta \cdot [AB]\,[NH] \tag{5}$$

where α depends only on deacylation rates and the strength of nucleophile binding, and β on the substrate properties of the activated substrate and the condensation product in the presence of the nucleophile.

II. *The hydrolysis rate is negligible compared with the synthesis rate.* The differential equation for the ratio of activated substrate consumption to condensation product formation [24].

$$\frac{-d[AB]}{d[AN]} = \left(\frac{k_H}{k_T}\right)_{app} \cdot \frac{[H_2O]}{[NH]} + 1 \tag{6}$$

can be integrated after variable separation. This gives the following integral equation

$$\int_{[AB]_0}^{0} -d[AB] = \int_{0}^{[AN]_{max}} \left\{\left(\frac{k_H}{k_T}\right)_{app} \cdot \frac{[H_2O]}{[NH]_0 - [AN]} + 1\right\} d[AN] \tag{7}$$

where the subscript 0 denotes conditions at the start of the kinetically controlled synthesis. For $[NH]_0 \gg [AN]_{max}$ the nucleophile content and $(k_H/k_T)_{app}$ is practically constant during the synthesis. Then the integration of Equation (7) gives

$$[AN]_{max} = \frac{[AB]_0 \cdot [NH]_0}{(k_H/k_T)_{app} \cdot [H_2O] + [NH]_0} \tag{8}$$

When the assumption ($[NH]_o \gg [AN]_{max}$) does not apply, $(k_H/k_T)_{app}$ depends on [NH]. Then the integration of Equation (7) yields a more complex expression for $[AN]_{max}$ given in Table 1. In contrast to case I the maximum yields in this case does not depend on the ratio of the substrate and condensation product properties (β in Equation (5)). In both cases the maximum does not depend on the enzyme content. The latter only influences the time required to reach the maximum. This has also been experimentally verified [27]. The relations in Table 1 have been derived assuming that the ratio (E—A)/[E—A··NH] can be derived from the equilibrium condition

$$\frac{[E-A] \cdot [NH]}{[E-A \cdot\cdot NH]} = K_N \tag{9}$$

Such equilibrium in binding is a general assumption used to derive expressions for the influence of activators and inhibitors in enzyme kinetics [28, 29]. Equation (4) can be inverted yielding

$$\left(\frac{k_H}{k_T}\right)_{app} = \frac{k_h \cdot K_N}{k_{t,N}} + \frac{k_{h,N} \cdot [NH]}{k_{t,N}} \tag{10}$$

Table 1 Expressions for the maximum yield of the condensation product AN in the kinetically controlled synthesis. The constants are defined in Figure 2 and Equation (3); Subscript 0 denotes initial concentrations.

	Maximum concentration of condensation product $[AN]_{max}$	Has been shown to apply for the kinetically controlled synthesis of
Case I Rate of product hydrolysis ≈ rate of product synthesis	$\frac{k_{t,N} \cdot [NH]}{k_h \cdot [H_2O] \cdot (K_N)_{app} + (k_{h,N} \cdot [H_2O] + k_{t,N}) \cdot [NH]} \cdot \frac{\left(\frac{k_{cat}}{K_m}\right)_{AB,NH}}{\left(\frac{k_{cat}}{K_m}\right)_{AN,NH}} \cdot [AB]$ $\alpha = \frac{(k_T/k_H)_{app}}{(k_T/k_H)_{app} \cdot [NH] + [H_2O]} \cdot [NH] \cdot \beta \cdot [AB]$	-β-lactam antibiotics (penicillins, cephalosporins) -oligosaccharides
Case II Rate of product hydrolysis ≪ rate of product synthesis	$\frac{k_{t,N} \cdot [AB]_0}{k_{t,N} + k_{h,N} \cdot [H_2O]} + \frac{k_h \cdot [H_2O] \cdot (K_N)_{app}}{k_t + k_{h,N} \cdot [H_2O]} \ln\left(1 - \frac{[AN]_{max}}{[NH]_0}\right)$ for $[NH]_0 \gg [AB]_0, [-AN]_{max}$ $\frac{k_{t,N} \cdot [AB]_0}{k_{t,N} + k_{h,N} \cdot [H_2O]} = \frac{[AB]_0 [NH]_0}{(k_H/k_T)_{app} \cdot [H_2O] + [NH]_0}$	-peptides with proteinases whose esterase activity ≫ amidase activity

a relation, linear in [NH], whose slope for the same nucleophile and acyl should be independent on the substrate content, and on the group used for the acyl activation. This has been investigated in the synthesis of penicillins. For activated substrates that were competitively inhibited by the nucleophile 6-APA (6-amino-penicillanic acid) as D-phenylglycine amid or n-butylester the slopes of $(k_H/k_T)_{app}$ as a function of [NH] (Equation (10)) differed from the ones observed for substrates for which 6-APA was found to be a mixed inhibitor [25, 30]. From this follows that the equilibrium relation (Equation (9)) cannot be used to determine the ratio [E–A]/[E–A··N] in the kinetically controlled synthesis. This especially applies for the case where NH is a competitive inhibitor for AB. A steady state concentration in E–A··NH can, however, always be assumed. In thise case

$$\frac{d[E-A\cdot\cdot NH]}{dt} = k_{h,N}\cdot[E-A]\cdot[NH] - k_{N,-1}\cdot[E-A\cdot\cdot NH] - \\ - k_{h,N} + k_{t,N}\cdot[E-A\cdot\cdot NH] = 0 \qquad (11)$$

with equilibrium concentrations in [E–A], [NH] and [E–A··NH] the first two terms in Equation (11) are zero, and the rate of product formation from [E–A··NH] must be zero. The latter is not observed. From this and other observations follows that the nucleophile binding in the mechanism given in Figure 2 generally cannot be described by equilibrium relations. Steady state relations are required to describe this [11, 31]. In this context this implies that the quantities determined from the slopes and intercepts of graphs of Equation (10) are not true molecular properties. The equations in Table 1 and the experimentally determined $(k_T/k_H)_{app}$-values can, however, still be used for a rational analysis of the factors that influence the maximum yield. The dissociation constants for the nucleophile binding to the acyl-enzyme are then apparent constants $(K_N)_{app}$ that are no molecular properties as they are concentration dependent. It can be shown that

$$(K_N)_{app} > K_N \qquad (12)$$

where K_N is the equilibrium constant given by Equation (9).

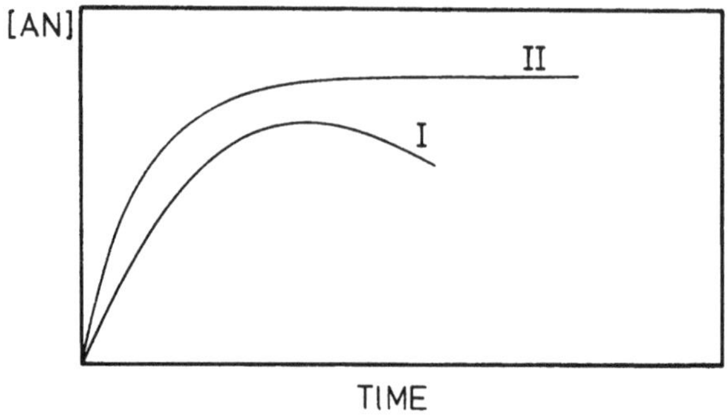

Figure 3 Time dependence in the kinetically controlled synthesis of condensation products AN.
Case I: Synthesis rate of the same order of magnitude as the rate of the hydrolysis of the product.
Case II: Synthesis rate ≫ hydrolysis rate.

Whether a specific synthesis can be described by the case I or II in Table 1 cannot be generally stated. It has to be determined from the time course of [AN] in the kinetically controlled synthesis (Figure 3). Generally case II applies for peptide synthesis using activated esters and proteinases with large ratios in the esterase to amidase hydrolysis rate. This ratio has been shown to decrease in the following order α-chymotrypsin > trypsin ≫ papain > penicillinamidase [26].

The relations for $[AN]_{max}$ given in Table 1 can then be used for a rational

- selection of the optimal enzyme for the kinetically controlled synthesis of a given peptide bond;
- optimization of the maximal yield for a given peptide using one enzyme.

4. Selecting the Optimal Enzyme

For the synthesis of a peptide bond P_1-P_1' it is necessary to use a proteinase that forms a covalent acyl-enzyme intermediate with a high P_1-specificity for the desired P_1 amino acid side chain. Data on the $P_i \cdot\cdot P_1$-specificities of proteinases, based on extensive studies on hydrolysis of different substrates have been compiled in several reviews [15, 18, 33]. For substrates P_1-P_1' where P_1 is constant and P_1' varied (k_{cat}/K_m) should, based on the principle of microscopic reversibility, increase with $(k_T/k_H)_{app}$ for P_1' as a nucleophile in the kinetically controlled synthesis of P_1-P_1' [22]. Thus hydrolysis data could be used to determine the P_1'-specificity in the synthesis reaction. The substrates used in the studies on hydrolase P_1'-specificity were mainly esters ($-O-X$ in P_1') or amides (with NH_2 or chromogenic amines as p-nitroanilides in P_1'). Consequently they are of limited use for the determination of the specificities of amino acids or peptides in the P_1' ($\cdot\cdot P_i'$) positions. For this aim data on $(k_T/k_H)_{app}$ and maximal yields in kinetically controlled synthesis of different peptides can be used.

For case II systems (product hydrolysis rate ≪ product synthesis rate), the yield increases with $(k_T/k_H)_{app}$ (Equation (8) and Table 1). For a given peptide bond the enzyme with the largest $(k_T/k_H)_{app}$-value should give the highest yield. Data for different proteinases are summarized in Figures 4–7 [34–45].

From these data follows that these proteinases have broad P_1' specificities. For the endoproteinases only P_1' amino acids with protected-COO^- groups can be used. These enzymes have a negative charge near the S_1'-binding site that repels-COO^- groups and limits their use as exoproteinases. The cysteine proteinases have a preference for apolar P_1'-amino acids and have at similar nucleophile concentrations $(k_T/k_H)_{app}$ values that are an order of magnitude larger than for the best serine proteinases. Compared with the cysteine proteinases, α-chymotrypsin and trypsin have a preference for basic P_1'-amino acids. The exoproteinase carboxypeptidase Y has the broadest nucleophile specificity. For this enzyme the $(k_T/k_H)_{app}$ values are generally lower than for the other enzymes. Of the enzymes discussed here only trypsin has been used for kinetically controlled synthesis of peptides with Asp or Glu in the P_1'-position. Only Clostripain could transfer acyl or peptidyl groups to Pro and CPD-Y was able to transfer these groups to Cys [37].

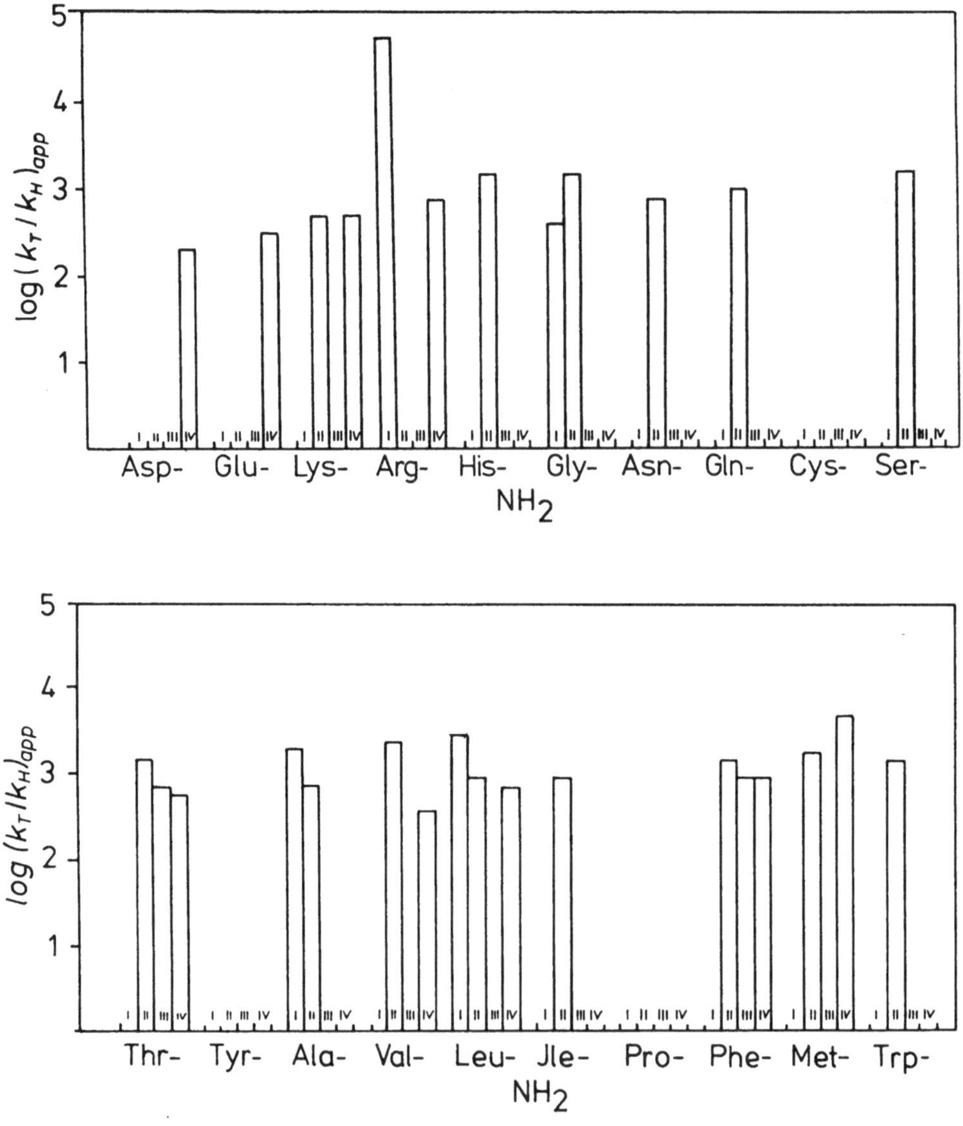

Figure 4 P_1'-specificity of serine proteinases expressed as log $(k_T/k_H)_{app}$ for different amino acid amides used as nucleophiles.

Conditions.

I. Bovine α-chymotrypsin (EC 3.4.21.1); P_1 = Tyr; 25°C; pH 8.0 (50 mM Tris-HCl, total ionic strength 0.55 M, 1.5% MeOH); $k_{t,N}/(k_h \cdot K_N)$ was determined from Equation (10) [24].

II. Yeast carboxypeptidase Y (EC 3.4.16.1); P_1 = Ala; 25°C; pH 9.5 (pH-Stat. total ionic strength 0.3–0.6 M; 10% MeOH); nucleophile concentration 0.2–0.5 M [36].

III. Human leucocyte elastase (EC, 3.4.21,37); P_1 = Ala; 25°C; pH 9.0 (0.1 M 2-(cyclo-hexyl-amino)ethane sulfonic acid, total ionic strength 0.5 M); $k_{t,N}/(k_h \cdot K_N)$ was determined from Equation (10) [40].

IV. Bovine trypsin (EC 3.4.21.4); P_1 = Arg; 25°C; pH 8.0 (50 mM Tris-HCl, total ionic strength 0.55 M, 1.5% MeOH); $k_{t,N}/(k_h \cdot K_N)$ was determined from Equation (10) [39].

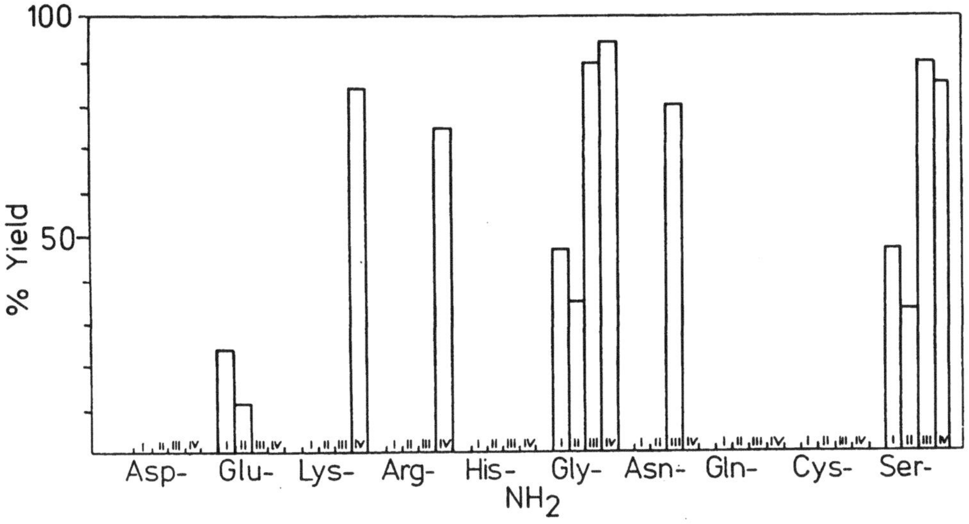

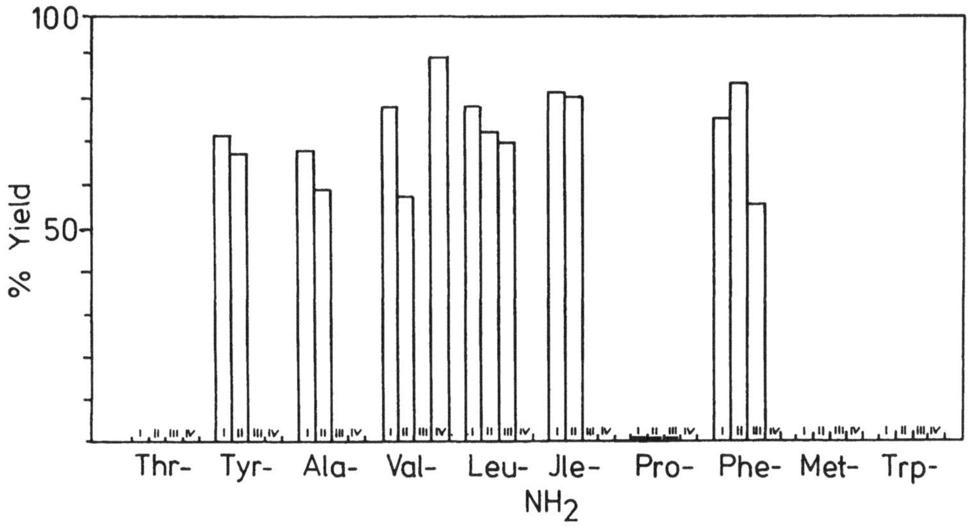

Figure 5 P′₁-specificity of serine proteinases expressed as maximum peptide yield (in % of activated substrate) for different amino acid amides, used as nucleophiles.

Conditions:

I. Bovin α-chymotrypsin; P_1 = Phe; 25°C; pH 10 (0.2 M $Na_2CO_3/NaHCO_3$, 10% dimethylformamide); nucleophile concentration 0.1 M [41].

II. Bovin trypsin; P_1 = Arg; 25°C; pH 10.4 (0.5 M carbonate buffer); nucleophile concentration 0.2 M [35].

III. Yeast carboxypeptidase Y; P_1 = Ala; 25°C; pH 9.5 (pH-Stat. total ionic strength 0.3–0.6 M, 10% MeOH); nucleophile concentration 0.2–0.5 M [36].

IV. Malt carboxypeptidase II (EC 3.4.16.1); P_1 = Arg; 25°C, pH 8.5 (pH-Stat.) [38].

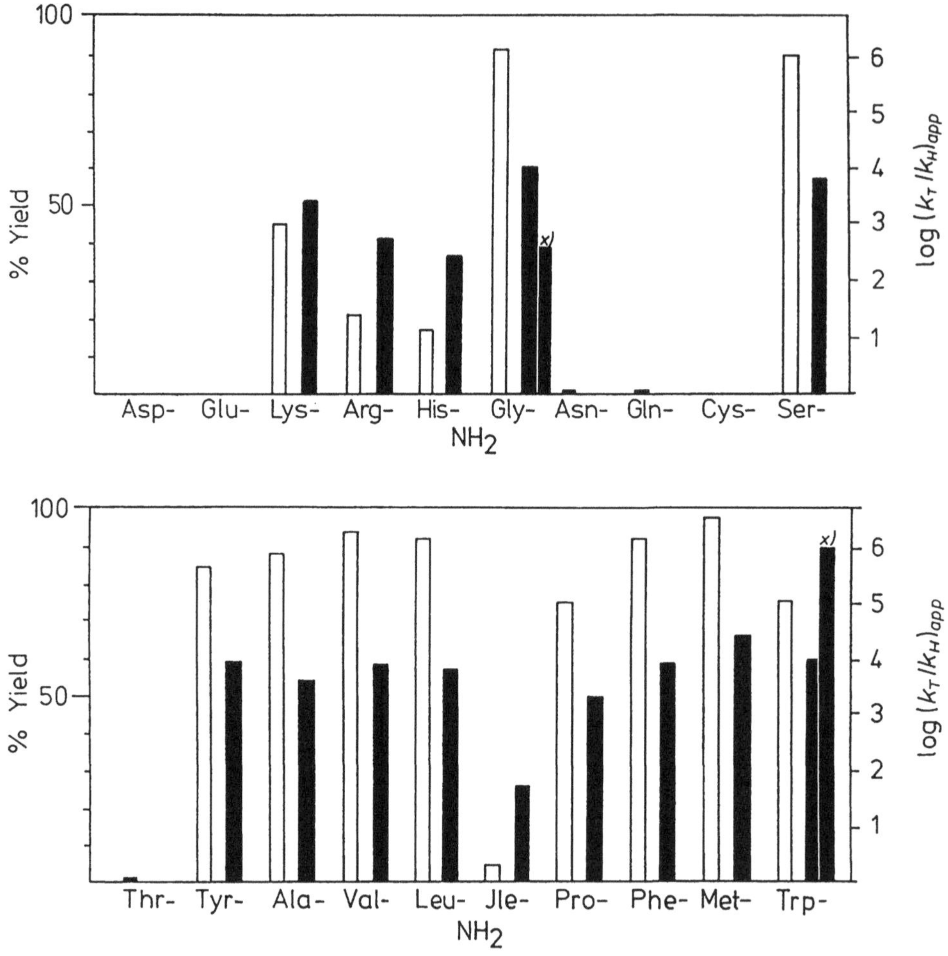

Figure 6 P_1'-specificity of cysteine proteinases expressed as log $(k_T/k_H)_{app}$ (shaded bars), and maximum peptide yield in % of activated substrate (open bars) for different amino acid amines as nucleophiles.
Conditions:
I. Clostripain (EC 3.4.22.8); P_1 = Arg; 25°C; pH 7.3 (0.2 M Tris-HCl); nucleophile concentration 0.1 M [45].
II. Papain (EC 3.4.22.2) (X-labeled bars); P_1 = *trans*-cinnamoylimidazole, 25°C; pH 9.0 (borate buffer, total ionic strength 0.1 M); $k_{t,N}/(k_h \cdot K_N)$ was determined from Equation (10) [42].

Direct kinetic evidence that the kinetically controlled synthesis using these enzymes and ester as the activated substrate can be described as a case II system has only been given for chymotrypsin [41], trypsin [27] and carboxypeptidase Y [50]. Then the peptide yield should increase with $(k_T/k_H)_{app}$ (Equation (8) and Table 1). For the experimental studies where $(k_T/k_H)_{app}$ and maximal yields have been determined at the same conditions the observed yields equals the values calculated from Equation (8) within the experimental error.

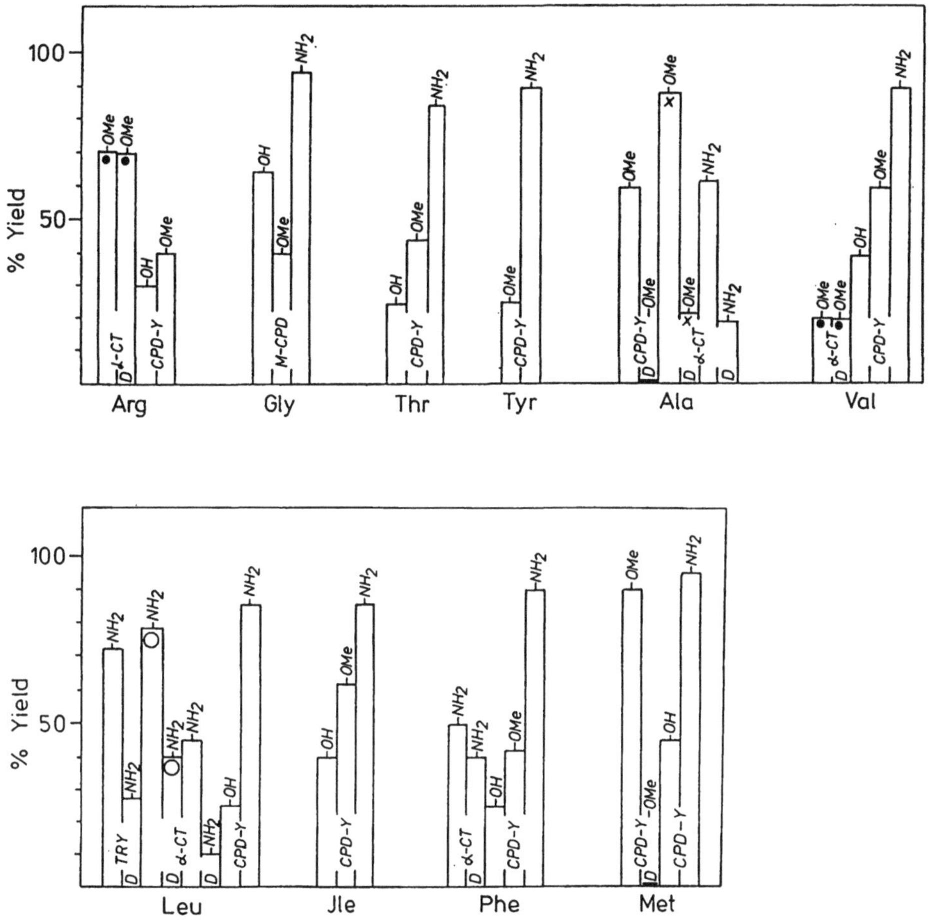

Figure 7 P_1'-specificity of serine proteinases expressed as maximum peptide yield in % of activated substrate. Stereospecificity and influence on the group (P_2') used to protect the carboxyl group given above the bars, of the P_1' amino acid (L- or D-form indicated on the bar).
Enzymes and conditions:
CPD-Y (yeast carboxypeptidase Y): P_1 = Ala; 25°C, pH 9.5 (pH Stat. total ionic strength 0.3–0.6 M), nucleophile concentration 0.2–0.5 M [36, 37].
M-CPD (Malt carboxypeptidase): P_1 = Arg; 25°C; pH 8.5 (pH-Stat.) [38].
TRY (bovine trypsin): P_1 = Arg; 25°C; pH 10.4 (0.5 M carbonate buffer nucleophile concentration 0.2 M [35].
α-CT (bovine α-chymotrypsin): P_1 = Tyr; 25°C; pH 9.0 (carbonate buffer, total ionic strength adjusted to 0.2 M with NaCl); nucleophile concentration 80 mM [43].
X: P_1 = Phe; 25°C; pH 10 (0.2 M carbonate buffer, 50% dimethylformamide); nucleophile concentration 0.2 M [44].
o: P_1 = Phe; 25°C; pH 10 (0.2 M carbonate buffer, 10% dimethylformamide); nucleophile concentration 0.1 M) [41].
•: P_1 = Tyr; 25°C; pH 9.1 (0.2 M carbonate buffer, 50% dimethylformamide); nucleophile concentration 0.8 M) [34].

For kinetically controlled synthesis where the product hydrolysis rate ≈ product synthesis rate (Case I) more enzyme properties than for case II influence the maximal product yield (Equation (5)). The term $a \cdot [NH]$ increases with $(k_T/k_H)_{app}$ (Equation (1)). The quantity β may increase or decrease as a function of [NH]. The latter case is possible when the activated substrate is a more specific substrate than the product, and both are competitively inhibited by the nucleophile. This shows that the properties of the substrates and the nature of the nucleophile as an inhibitor may influence the maximum product yield markedly. To obtain maximal yield of a desired product as a function of enzyme properties, substrate activation and nucleophile content it is, therefore contrary to case II, not sufficient to use the activated substrate and enzyme with the highest value of $a \cdot (k_T/k_H)$. For this aim the maximum value of $(a \cdot \beta \cdot [NH])$ as function of enzyme properties, substrate activation and [NH] has to be determined. The separate determination of a and β as function of these variables is in this case unpractical as the number of measurements required here are much larger than for case II systems. This is demonstrated by the following experimental observations.

In the semisynthesis of ampicillin and cephalexin higher yields were obtained with D-phenylglycine-O-Me than with D-phenylglycine-O-Et, $-NH_2$ even when the latter are more specific substrates and have larger a-values for the used enzyme (*E. coli* penicillin amidase) [11, 46]. In the semisynthesis of human insulin, where a large excess of nucleophile is used there is a maximum in the yield as a function of nucleophile concentration. Under these conditions the equilibrium controlled synthesis may even be favourable i. e. be more rapid than the kinetically controlled synthesis [47, 48].

Once the enzyme has been selected $(k_T/k_H)_{app}$ can be further optimized by changes in: the group used to protect the P_1'-carboxyl end, pH, temperature, solvent composition and ionic strength.

5. Yield Controlling Factors in the Synthesis of a Peptide Bond with One Enzyme

5.1 Protection of the P_1' carboxyl group; stereospecificity

For some proteinases $(k_T/k_H)_{app}$ has been shown to be much larger for the amino acid amide than for the corresponding ester (Figure 7). This may be due to the better binding of $-NH_2$ than $-O-X$ ($= P_2'$) in the S_2'-binding site. Proteinases that are more specific for L-amino acids in the S_1'-binding site can also be used for the kinetically controlled synthesis of L-D peptides. This is illustrated in Figure 8 where it is shown that the relative importance of a productive binding, when it can deacylate the acyl-enzyme, depends on the relative magnitude of the $S_1' \cdot\cdot P_1'$ and $S_2' \cdot\cdot P_2'$ interaction. The stereospecificity should increase with the strength of these interactions. When they are of equal magnitude, as for amides, productive binding of D-nucleophiles to L-specific enzymes is less likely. When the $S_2' \cdot\cdot P_2'$

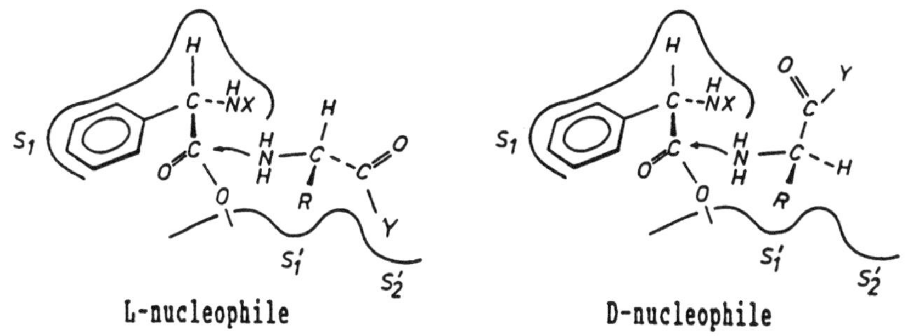

Figure 8 Stereospecificity in the transferase function of proteinases specific for L-amino acids in the P_1'-position. Diffent productive binding modes, with correct orientation of the amino group relative to the carbonyl carbon in the acyl-enzyme, of L- and D-amino acids with or without protected carboxyl as nucleophiles group. The most specific binding mode with occupied S_1'-binding sites is productive for the L- and non-productive for D-form.

interaction is weaker than the $S_1' \cdot\cdot P_1'$ interaction, as for esters, the probability for productive binding of D-nucleophiles should be larger than for the amides. This is also observed in the stereospecificity with α-chymotrypsin. For amides a larger stereospecificity is observed then for esters of the amino acids (Ala, Leu, Phe) where comparative data are available (Figure 7). Thus the L-specific endoproteinases α-chymotrypsin and trypsin can be used for the synthesis of L-D peptides [43, 44, 49]. This has not been observed for the exoproteinase carboxypeptidase Y.

5.2 pH value

Case II. A pH-optimum above the pK value for the P_1' aminogroup (7−8 for amino acid amides or esters; 9−10 for amino acids) is generally observed [34, 35, 41, 50]. Below the pH-optimum the increase in $(k_T/k_H)_{app}$ can be explained by the increase in the concentration of the uncharged P_1'-amino group that is required for the nucleophilic deacylation. Above the pH-optimum the decrease is probably due to the increase in the concentration of OH^- as a nucleophile, although this has not been studied in detail. This will increase k_H and reduce $(k_T/k_H)_{app}$. In some cases where Tris (Tris(hydroxymethyl)-aminomethase) is used as a buffer component this can reduce the pH-optimum and the yield [27, 39, 45]. Tris acts as a nucleophile and more Tris is required to obtain larger pH-values [51]. This will reduce the yield of the desired peptide. Consequently here as in other enzyme kinetic studies Tris must not be used as a buffering component.

Some amidases, as penicillin amidase [52] and subtilisin [53] lack S – S bridges. This reduces their pH stability above pH 9.0 and limits their use for the kinetically controlled synthesis of peptides in this pH range. Due to its specificity for phenacetyl or D-amino acids in the P_1-position *E. coli* penicillin amidase that has been shown to have a proteinase function, could have applications as an amino terminal deblocking enzyme, in peptide synthesis [54, 55]. It has also been used to deblock protected cysteine [56].

Case I. When the activated substrate and the product have the same charge the ratio of the (k_{cat}/K_m)-values should be practically pH-independent. Then the pH-optimum that is also observed for this case is mainly due to the pH-dependence of $(k_T/k_H)_{app}$ as discussed for case I.

The pH-dependence in this case has been extensively studied for the kinetically controlled synthesis of β-lactam antibiotics with penicillin amidases from different sources [11, 57, 58]. In this case peptides with low p*K*-values (≈ 5) of the NH_2-group as 6-aminopenicillanic acid and 7-amino-3-deacetoxycephalosporanic acid have been used as nucleophiles. The low pH-optimum (≈ 7) cannot be explained due to reactions with OH^- as a nucleophile. In these cases the pH-dependence of $(K_N)_{app}$ seems to cause the reduction in $(k_T/k_H)_{app}$ at pH values above the optimum. For the *E. coli* enzyme two positive charges in or near the S_1' and S_2' binding sites on the enzyme have been shown to influence the binding of the negatively charged nucleophile [11, 25]. With increasing pH the number of positively charged group on the enzyme is reduced. This causes an increase in $(K_N)_{app}$ and a decrease in $(k_T/k_H)_{app}$ and acylation yield. The decrease in yield below the pH-optimum can also be influenced by charges on the activated substrate. For the important semisynthetic penicillins, ampicillin, amoxicillin and cephalexin the side chain (D-phenylglycin) amino group has a p*K*-value 6.9. At pH lower than 6.9 the activated substrate (ester, amid) has a positive charge and is repelled by positive groups in the active site of penicillin amidase from *E. coli*. This results in an increase in K_m for the substrate. The corresponding decrease in β (Equations 1 and 5) causes the observed decrease in the yield below pH 7.0 [11]. For uncharged side chains the pH optimum is lower [59]. For the kinetically controlled synthesis of ampicillin, amoxicillin and cephalexin amidases with less positive charges, or more negative charges, in or near the S_1, binding site than the *E. coli* penicillin amidase could decrease the pH optimum. This and product yield increases have been observed for an enzyme from *X. citi* [59]. In this case, however, substrate and nucleophile concentrations of biotechnical relevance (> 100 mM) increase the ionic strength. This causes a reduction in $(k_{cat}/K_m)_{AB, NH}$ in Equation (1) and the product yield when D-phenylglycin ester or amid is used as an activated substrate (see also Section 5.4 below) [57, 59].

5.3 Temperature

Only few studies have been published on the influence of the temperature [11, 44]. In these, for both case I and II systems, temperature optima (< 20°C) have been observed. For case II systems this implies that $(k_T/k_H)_{app}$ decreases with temperature above >20°C [60]. This is, however, not generally valid as temperature maxima have been observed for case I systems where this ratio increases with temperature [11].

5.4 Ionic strength

For both case I and II systems the ionic strength influences the binding of the nucleophile to a charged S_1' binding site. When opposite charges are involved, as for the synthesis of β-lactam antibiotics catalyzed by *E. coli* penicillin amidase, an increase in the ionic strength causes a decrease in the binding of the nucleophile and the product yield [11]. In this case where both the activated substrate and the nucleophile contribute to the ionic strength the amount of ions to buffer the reaction medium should be minimized. When equal charges are involved as in peptide synthesis with amino acids as nucleophiles catalyzed by endoproteinases an increase in the yield with increasing ionic strength is expected. This has been confirmed by experimental data [61].

5.5 Solvent composition

From Table 1 follows that the product yield can be increased when $[H_2O]$ is reduced by the addition of organic solvents that influence the properties of the used enzyme marginally. Detailed studies on the latter have not yet been performed. For peptide synthesis (case II) increase in yields with reduced water content have been observed with dimethylformamide [34, 41, 44], dichlormethane and dichlorethane as organic solvents [61]. The available data cannot yet be used to analyze whether the organic solvents influence the P_1'-specificity of these enzymes selectively. In kinetically controlled synthesis of peptides using substrates and nucleophiles with low solubilities in aqueous systems the solubility can be increased by addition of *inert* organic solvents. Alcohols are not suitable with proteinases as catalysts as transesterification products may be formed [51].

For the synthesis of semisynthetic β-lactams addition of organic solvents have been shown to have a marginal influence on the yields [62]. In this case the reactants and/or products are ions and their solubility decreases in organic solvents. Methanol has, however, been shown to increase the product yield due to recycling of the activated substrate [63].

For a rational selection of suitable *non-toxic* solvents to increase the product yields, more studies must be performed. Such investigations must also include investigations on the enzyme stability in these solvents.

6. Conclusions

The results summarized in Figures 4–7 show that serine and cysteine proteinases can be used for the kinetically controlled synthesis of peptide bonds by the acyl-(peptidyl-)transfer mechanism shown in Figures 1 and 2. To obtain more than 50% acyl transfer to P_1' at nucleophile concentrations ≤ 1 M, case II conditions (Figure 3) and nucleophiles with $(k_T/k_H) > 100$ must be used (Equation (8) and Table 1). From Figures 4–7 follows that for the used enzymes this can be achieved with all amino acids except for Cys as P_1'. The acylation yields generally increase in the following order $P_1'-OH < P_1'-O-X < P_1'-NH_2$. The enzymes differ in their P_1'-specificity, and the data in Figures 4–7 can be used to select the optimal enzyme for the acyl-(peptidyl-)transfer to a given P_1' amino acid.

From Figure 7 follows that the L-specific (for P_1') serine endoproteinases α-chymotrypsin and trypsin also can be used for acyl-(peptidyl-)transfers to D-P_1' amino acids. In this case the yields generally increase in the order D-$P_1'-O-X <$ D-$P_1'-NH_2$. This may be due to the relative strengths of the $P_1' \cdot\cdot S_1'$ and $P_2' \cdot\cdot S_2'$ interactions (Figure 8). Thus the kinetically controlled synthesis can also be used for the synthesis of L-D peptide bonds. From Figure 7 follows that the serine exoproteinase carboxypeptidase cannot be used for that purpose.

Once the optimal enzyme has been selected the acyl-(peptidyl-)transfer yields can be increased using the relations in Table 1 and a rational choice of the yield influencing factors analyzed in Section 5. For this aim, however, more studies especially on the influence of solvent composition are required. These should include their effects on the P_1'-specificity and enzyme stability [64].

In this review the factors influencing the formation of a single peptide bond have been analyzed. The conclusions derived here can also be used to design experimental conditions for the enzyme catalyzed stepwise synthesis of peptides or peptide modifications (amidation of carboxyl end groups in peptides by transpeptidation) in vitro. For this aim either a combination of chemical, equilibrium and kinetically controlled synthesis of the peptide bonds, or the stepwise addition of $P_1'-O-X$ groups to a peptide by kinetically controlled synthesis have been used [65, 66]. In the latter case the P_1' specificity is lower than when using P_1'-$-NH_2$ as nucleophiles. The product is, however, a better substrate than an amide for proteinases whose esterase activity is larger than their amidase activity. With these, the more favourable conditions for kinetically controlled synthesis (case II in Figure 3), can generally be achieved with esters as activated substrates but not with amides [67].

The results from studies on the kinetically controlled peptide bond synthesis are here mainly used to analyze the factors that influence the product yields. They may, however, also be used to analyze more basic enzymological problems. Based on the kinetic data on the P_1' specificity reviewed here, systems for direct studies on the $P_1' \cdot\cdot S_1'$ interaction by X-ray crystallography can be selected. Such investigations can be used for a comparative analysis of the sequence (P_1') specificity of proteinases using natural systems or enzymes designed by site directed mutagenesis. The results could provide basic information on interactions between substrates and the active site on enzymes that influence their properties as catalysts.

References

1. Wold, F., In vivo chemical modification of proteins (post-translation), *Ann. Rev. Biochem. 50*, 783, 1981.
2. Glembotski, Chr. C., Further characterization of the peptidyl α-amidating enzyme in rat anterior pituitary secretory granules, *Arch. Biochem. Biophys. 241*, 673, 1985.
3. Cannon, M., Antimicrobial peptides a family of wound healers, *Nature 328*, 478, 1987.
4. Kleinkauf, H., and von Döhren, H., Nucleic acid independent synthesis of peptides, *Curr. Topics in Microbiol. Immun. 91*, 129, 1981.
5. Queener, St. W., and Neuss, N., The biosynthesis of β-lactam antibiotics, *Chemistry and Biology 3*, 2, 1982.
6. Demain, A. L., Wolfe, S., Jensen, S. E., and Westlake, D. W. S., Enzymatic approach to synthesis of unnatural beta-lactams, *Science 226*, 1386, 1984.
7. Schindler, P., Enzyme inhibitors of microbial origin, *Phil. Trans. R. Soc.* Lond. *290*, 291, 1981.
8. Berger, A., and Schechter, I., Mapping the active site of papain with the aid of peptide substrates and inhibitors, *Phil. Trans. Roy. Soc.* Lond. *257*, 249, 1970.
9. Fersht, A. R., Enzymic editing mechanisms in protein synthesis and DNA replication, *Trends Biochem. Sci. 5*, 262, 1980.
10. Fersht, A. R., Shi, J.-P., and Tsui, W.-C., Kinetics of base misinsertion by DNA polymerase I of escherichia coli, *J. Mol. Biol. 165*, 655, 1983.
11. Kasche, V., Mechanism and yields in enzyme catalysed equilibrium and kinetically controlled synthesis of β-lactam antibiotics, peptides and other condensation products, *Enzyme Microb. Technol. 8*, 5, 1986.
12. Imahori, K., Iwasaki, T., Yamamoto, K., Nakajima, H., and Tomioka, I., Process for synthesizing peptides or peptide derivaties, *European Patent Application*, 83300362.7, 1983.
13. Ditson, S. L., Sung, St. M., and Mayer, R. M., Dynamic reactivities of dextransucrase, *Arch. Biochem. Biophys. 249*, 53, 1986.
14. Trip, E. M., and Smith, M., Enzymatic synthesis of oligo-deoxyribonucleotides of defined sequence. Polynucleotide phosphorylase catalyzed addition of deoxyribonucleotides to primers which are good or poor acceptors, *Nucl. Acids Res. 5*, 1529, 1978.
15. Fruton, J. S., Proteinase-catalyzed synthesis of peptide bonds, *Adv. Enzymol. Relat. Areas. Mol. Biol. 53*, 239, 1982.
16. Chaiken, I. M., Komoriya, A., Ohno, M., and Widmer, F., Use of enzymes in peptide synthesis, *Appl. Biochem. Biotechn. 7*, 385, 1982.
17. Jakubke, H.-D., Kuhl, P., and Könnecke, A., Grundprinzipien der proteasekatalysierten Knüpfung der Peptidbindung, *Angew. Chemie 97*, 79, 1985.
18. Morihara, K., Using proteases in peptide synthesis, *Trends Biotechn. 5*, 164, 1987.
19. Breddam, K., Serine carboxy peptidases. A review, *Carlsberg Res. Commun. 51*, 83, 1986.
20. Wallenfels, K., and Malhotra, O. P., Galactosidases, *Advance Carbohydr. Chem. 16*, 239, 1961.
21. Fink, A. L., and Bender, M. L., Binding sites for substrate leaving groups and added nucleophiles in papain-catalyzed hydrolases, *Biochemistry 8*, 5109, 1969.
22. Riechmann, L., and Kasche, V., Kinetic studies on the mechanism and the specificity of peptide semisynthesis catalyzed by the serine proteases α-chymotrypsin and β-trypsin, *Biochem. Biophys. Res. Commun. 120*, 686, 1984.
23. Breddam, K., and Ottesen, M., Malt carboxypeptidase catalyzed aminolysis reactions, *Carlsberg Res. Commun. 49*, 473, 1984.
24. Schellenberger, V., and Jakubke, H.-D., A spectrophotometric assay for the characterization of the S′ subsite specificity of α-chymotrypsin, *Biochem. Biophys. Acta 869*, 54, 1986.
25. Kasche, V., Haufler, U., and Zöllner, R., Kinetic studies on the mechanism of the penicillin amidase-catalysed synthesis of ampicillin and benzylpenicillin, *Hoppe-Seyler's Z. Physiol. Chem. 365*, 1435, 1984.
26. Haufler, U., Penicillinamidase aus Escherichia coli: Untersuchungen zu Struktur, Stabilität und Katalysemechanismus der Hydrolyse- und Transferreaktion, *Dissertation Universität Bremen*, 1987.
27. Riechmann, L., and Kasche, V., Peptide synthesis catalyzed by the serin proteinases chymotrypsin and trypsin, *Biochim. Biophys. Acta 830*, 164, 1985.

28. Cornish-Bowden, A., *Fundamentals of enzyme kinetics*, Butterworth, London, 4–6, 1979.
29. Wang, Z. X., and Tsou, C. L., Kinetics of substrate reaction during irreversible modification of enzyme activity for enzymes involving two substrates, *J. Theor. Biol. 127*, 253, 1987.
30. Markowsky, D., unpublished data, 1986.
31. Ingraham, L. L., and Makower, B., Variation of the michaelis constant with the concentrations of the reactants in an enzyme-catalyzed system, *J. Physical Chem. 58*, 266, 1954.
32. Riechmann, L., Die Synthese von Peptiden mit Hilfe der Proteasen Chymotrypsin, Trypsin und Thermolysin, *Dissertation Universität Bremen*, 1986.
33. *Enzyme nomenclature*, Academic Press, Inc., London, 1984.
34. West, J. B., and Wong, C.-H., Enzyme-catalyzed irreversible formation of peptides containing D-amino acids., *J. Org. Chem. 51*, 2728, 1986.
35. Oka, T., and Morihara, K., Trypsin as a catalyst for peptide synthesis, *J. Biochem. 82*, 1055, 1977.
36. Widmer, F., Breddam, K., and Johansen, J. T., Influence of the structure of amine components on carboxypeptidase Y catalyzed amide bond formation, *Carlsberg Res. Commun. 46*, 97, 1981.
37. Widmer, F., Breddam, K., and Johansen, J. T., Carboxypeptidase Y catalyzed peptide synthesis using amino acid alkyl esters as amine components, *Carlsberg Res. Commun. 45*, 453, 1980.
38. Breddam, K., Enzymatic properties of malt carboxypeptidase II in hydrolysis and aminolysis reactions, *Carlsberg Res. Commun. 50*, 309, 1985.
39. Hanisch, U.-K., Könnecke, A., Schellenberger, V., and Jakubke, H.-D., Characterization of the S′ subsite specificity of trypsin, *Biocatalysis 1*, 129, 1987.
40. Stein, R. L., Strimpler, A. M., Catalysis by human leukocyte elastase. aminolysis of acyl-enzymes by amino acid amides and peptides, *Biochemistry 26*, 2238, 1987.
41. Morihara, K., and Oka, T., Chymotrypsin as the catalyst for peptide synthesis, *Biochem. J. 163*, 531, 1977.
42. Brubacher, L.-J., and Bender, M. L., The preparation and properties of trans-cinnamoyl-papain, *J. Am. Chem. Soc. 88*, 5871, 1966.
43. Kasche, V., and Michaelis, G., unpublished data, 1987.
44. Nilsson, K., and Mosbach, K., Peptide synthesis in aqueous-organic solvent mixtures with α-chymotrypsin immobilized to tresyl chloride-activated agarose, *Biotechnology and Bioengineering 26*, 1146, 1984.
45. Fortier, G., and MacKenzie, S. L., Peptide bond synthesis by clostridiopeptidase B., *Biotechn. Letters 8*, 777, 1986.
46. Kasche, V., Mechanism and maximum yield in kinetically controlled peptide synthesis, *Adv. in the Biosciences 65*, 151, 1987.
47. Jonczyk, A., and Gattner, H.-G., Eine neue Semisynthese des Humaninsulins Tryptisch-katalysierte Transpeptidierung von Schweineinsulin mit L-Threonin-Tert-Butylester, *Hoppe-Seyler's Z. Physiol. Chem. 362*, 1591, 1981.
48. Morihara, K., Ueno, Y., and Sakina, K., Influence of temperature on the enzymatic semisynthesis of human insulin by coupling and transpeptidation methods, *Biochem. J. 240*, 803, 1986.
49. Kato, K., Kawahara, K., Takahashi, T., and Igarasi, S., Enzymatic synthesis of amoxicillin by the cell-bound α-amino acid ester hydrolase of xanthomonas citri, *Agric. Biol. Chem. 44*, 821, 1980.
50. Widmer, F., and Johansen, J. T., Enzymatic peptide synthesis. Carboxypeptidase Y catalyzed formation of peptide bonds, *Carlsberg Res. Commun. 44*, 37, 1979.
51. Kasche, V., and Zöllner, R., Tris(hydroxymethyl)methylamine is acylated when it reacts with acyl-chymotrypsin, *Hoppe-Seyler's Z. Physiol. Chem. 363*, 531, 1982.
52. Schumacher, G., Sizmann, D., Haug, H., Buckel, P., and Böck, A., Penicillin acylase from E. coli: unique gene-protein relation, *Nucleic Acid Res. 14*, 5713, 1986.
53. Handbook of Biochemistry and Molecular Biology: Proteins Vol. III (G. D. Fasman, ed.) 3rd. ed., CRC Press, Cleveland 1976.
54. Wang, Q.-C., Fei, J., Cui, D.-F., Zhu, S.-G., and Xu, L.-G., Application of an immobilized penicillin acylase to the deprotection of N-phenylacetyl insulin, *Biopolymers 25*, 114, 1986.
55. Fuganti, C., and Grasselli, P., Immobilized Penicillinamidase: application to the synthesis of the dipeptide aspartame, *Tetrahedron Letters 27*, 3391, 1986.

56. Hermann, P., personal communication, 1986.
57. Kato, K., Kawahara, K., Takahashi, T., and Igarasi, S., Enzymatic synthesis of amoxicillin by the cell-bound α-amino acid ester hydrolase of xanthomonas citri, *Agric. Biol. Chem. 44*, 821, 1980.
58. Kasche, V., and Galunsky, B., Ionic strenght and pH effects in the kinetically controlled synthesis of benzylpenicillin by nucleophilic deacylation of free and immobilized phenyl-acetyl-penicillin amidase with 6-aminopenicillanic acid, *Biochem. Biophys. Res. Commun. 104*, 1215, 1982.
59. Haß, W., unpublished observation, 1987.
60. Riechmann, L., and Kasche, V., Peptide synthesis catalyzed by the serine proteinases chymotrypsin and trypsin, *Biochem. Biophys. Acta 830*, 164, 1985.
61. Kuhl, P., Walpuski, J., and Jakubke, D.-D., Untersuchungen zum Einfluß der Reaktionsbedingungen auf die α-chymotrypsinkatalysierte Peptidsynthese im wäßrig-organischen Zweiphasensystem, *Pharmazie 37*, 766, 1982.
62. Kasche, V., Haufler, U., and Riechmann, L., Equilibrium and kinetically controlled synthesis with enzymes: semisynthesis of penicillins and peptides, *Meth. in Enzymol. 136*, 280, 1987.
63. Kasche, V., Ampicillin- and cephalexin-synthesis catalyzed by E. coli penicillin amidase. Yield increase due to substrate recycling, *Biotechnol. Lett. 7*, 877, 1985.
64. Klibanov, A. M., Substrate specificity of enzymes in organic solvents vs. water is reversed, *J. Am. Chem. Soc. 108*, 2767, 1986.
65. Kullmann, W., Protease-catalyzed peptide bond formation: Application to synthesis of the COOH-terminal octapeptide of cholecystokinin, *Biochemistry 79*, 2840, 1982.
66. Breddam, K., Widmer, F., and Johansen, J. T., Amino acid methyl esters as amine components in CPD-Y catalyzed peptide synthesis: control of side reactions, *Carlsberg Res. Commun. 48*, 231, 1983.
67. Breddam, K., Johansen, J. T., and Ottesen, M., Carboxypeptidase Y catalyzed transpeptidation and condensation reactions, *Carlsberg Res. Commun. 49*, 457, 1984.

Chapter 5
Genetics of Siderophore Biosynthesis and Transport

Volkmar Braun

1. Structure and Function of Siderophores
2. Fe^{3+}-Aerobactin Transport System
   2.1 Biosynthesis of aerobactin
   2.2 Fe^{3+}-aerobactin transport
3. Enterochelin Biosynthesis and Fe^{3+}-Enterochelin Transport
4. Fe^{3+}-Dicitrate Transport
   4.1 Transport
   4.2 Regulation
5. Regulation of Iron Transport Systems
   5.1 Regulation of gene expression
   5.2 Regulation of the Fe^{3+}-siderophore receptor activity
6. Antibiotics Containing Siderophore Structures
   6.1 Naturally occurring sideromycins
   6.2 Synthetic antibiotics with siderophore structures
7. Concluding Remarks
Acknowledgements
References

1. Structure and Function of Siderophores

Bacteria and fungi synthesize structurally rather diverse compounds of low molecular weight which have in common a very high affinity for Fe^{3+} ions. These compounds are designated siderophores [1] to account for their ability to bind Fe^{3+} and to transport it into cells. Formerly, the iron compounds were termed sideramines [2] because they promoted growth of microorganisms, in contrast to the sideromycins [3] which inhibit growth. Another name in use was siderochromes [4], which described the redish color developed when the complex between the colorless iron-binding compounds and Fe^{3+} is formed. Usually siderophores are released into the culture medium where they scavenge Fe^{3+}. They are taken up into the microbes as Fe^{3+}-siderophore complex.

Most siderophores contain hydroxamate or phenolate groups which bind Fe^{3+} in an octahedral array. In this report, a few examples of such compounds will be presented. A list of presently known siderophores can be found in recent reviews [5–8].

The need for siderophores comes from the insolubility of Fe^{3+} at physiological pH values. Microbes are either faced with insoluble iron hydroxide polymers, or with Fe^{3+} bound to carrier and storage proteins (such as transferrin, lactoferrin and ferritin), by which higher organisms solved the problem of iron supply. The solution to this problem found by microbes are the siderophores, which either solubilize the iron from insoluble sources or withdraw it from the iron proteins. A clear indication that most microbes which grow in Fe^{3+} environments use siderophores as iron suppliers can be deduced from the fact that siderophores are usually synthesized and secreted as a response to iron starvation. In anaerobic ambients iron may occur as Fe^{2+} which is readily soluble and does not require the employment of special iron-complexing agents.

The function of siderophores as iron carriers has most clearly been shown in *Escherichia coli*. With the combination of selective growth conditions and genetics, five Fe^{3+} transport systems and one Fe^{2+} transport system have been identified in *E. coli* K-12. In the following, concepts and methods employed will be described because they are exemplary for the elucidation of the iron supply systems of other bacteria and fungi. In addition, recent developments will be outlined which show that the iron transport systems may be used to bring antibiotics into cells which as a result reduces their minimal inhibitory concentration more than a hundredfold.

2. Fe^{3+}-Aerobactin Transport System

2.1 Biosynthesis of aerobactin

In 1969 aerobactin was detected in the culture fluid of *Aerobacter* (now *Enterobacter*) *aerogenes* 62-1 [11]. The way it was found shows how the different iron transport systems can be separated from each other. The first siderophore characterized in enteric bacteria, to which *Enterobacter* belongs, was enterochelin of *E. coli* and the identical enterobactin of *Salmonella typhimurium* [12]. Since *E. coli* K-12, the classical strain of *E. coli* genetics, forms only enterochelin as the sole siderophore, a mutant in enterochelin biosynthesis depended entirely on the addition of 2,3-dihydroxy-benzoate, the precursor of enterochelin (see Section 3 for further details) to grow in media of low iron content. Such media are formed by passing a minimal salts medium through a Chelex-100 column, or by removing iron by extraction with 8-hydroxyquinoline, or by addition of 0.2 mM 2,2′ dipyridyl or 0.3 μM ovotransferrin. The latter two strongly bind iron and fail to deliver it to *E. coli* cells. In contrast to *E. coli* K-12 mutants in enterochelin synthesis, an equivalent *Enterobacter* mutant still grew on a low iron medium. This suggested the existence of an additional siderophore which provided the necessary iron. It took ten further years until the Fe^{3+}-aerobactin transport system gained a broader interest. By then it was recognized that pathogenic strains of *E. coli* carrying ColV plasmids determine a high-affinity iron transport system which was different to enterochelin [13] and instead secreted a siderophore with hydroxamate groups [14]. It was then shown that aerobactin was the iron carrier [15–17]. Because of the relationship to the virulence of pathogenic *E. coli*, numerous studies followed aimed at establishing the role of the aerobactin iron supply system in bacterial pathogenicity. The ability to use aerobactin as siderophore was frequently found to be encoded on ColV plasmids, but also on other plasmids and on the chromosome. In addition to *Enterobacter* and *E. coli*, aerobactin synthesis and Fe^{3+}-aerobactin uptake was determined in *Shigella* [20], *Salmonella* [21] and *Yersinia* [22].

Aerobactin consists of two residues of 6-(*N*-acetyl-*N*-hydroxy) lysine linked to citrate (Figure 1). It is the first hydroxamate-siderophore whose biosynthesis was elucidated. The use of modern genetics was instrumental to determine the pathway, the enzymes involved, the structural genes which encode the enzymes, the arrangement of the genes on the DNA, their transcriptional polarity and the regulation of transcription.

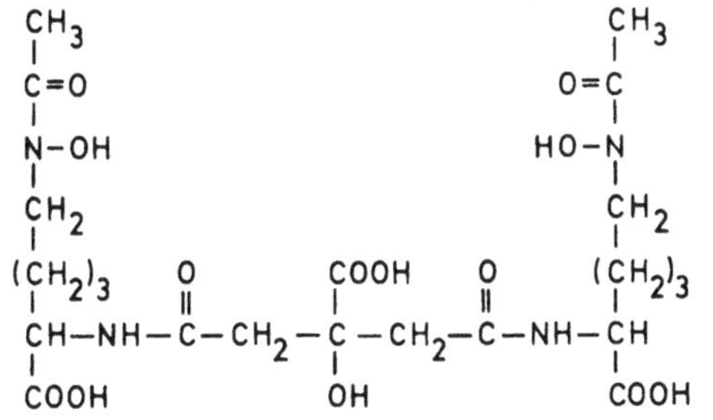

Figure 1 Structure of aerobactin. Fe^{3+} is thought to be bound by the two hydroxamate residues and the central hydroxyl and carboxyl group of the citrate moiety.

We used a straightforward approach which is generally applicable for the determination of biosynthetic pathways. Mutants in aerobactin synthesis were isolated using *N*-methyl-*N*′-nitro-N-nitroso guanidine (MNNG) as a chemical mutagen, and transposons as biological mutagens. Both procedures give rise to more or less statistically distributed mutations. Treatment with MNNG or other chemical mutagens results in point mutations and deletions whereas the insertion of transposons interrupts a gene. With chemical mutagens more than one mutation is frequently created in contrast to transposons which usually insert only once. Restriction analysis determines the approximate site of the transposon insertion in a given gene. Determination of the number and the kind of chemically induced mutations require sequencing of the DNA. Transposon insertions usually exert polar effects on genes located downstream (reduction of transcription) which can also occur in chemically induced mutants. Genes with transposon insertions are usually inactivated, unless the insertions are close to the 3′ end, which may result in truncated polypeptides from which the direction of transcription can be deduced. Point mutants and in frame deletions synthesize polypeptides which can still show partial activity. Obviously, each method has its advantages and disadvantages, and they complement each other.

Defects in aerobactin synthesis render cells unable to grow in iron-deficient media, for example on nutrient agar plates on which iron has been bound to dipyridyl making it unusuable by *E. coli*. However, aerobactin only serves as sole siderophore after synthesis of enterochelin has been interrupted. We used an *aroB* mutant with a defect in the synthesis of chorismate from which enterochelin is derived (see Section 3 for further details). In genetics positive selection is preferred because it allows one to isolate in one step the desired mutants among the large pool of wild-type cells, and cells with mutations in other genes. We introduced the inhibitor streptonigrin for selection of iron supply mutants [23] since it has been shown that iron enhances the bactericidal action of streptonigrin [24]. At iron-limited growth conditions enterochelin- and aerobactin-negative cells grow very slowly and are not killed by streptonigrin in contrast to aerobactin-producing cells with a proficient iron supply system.

The various mutants obtained could be divided in functional groups by cross-feeding tests. This method depends on the secretion of precursors formed by the mutants and the ability of other mutants, which require them for completion of the biosynthetic pathway, to take up the precursors. We developed a test system in which cells of a presumed donor strain were on a small paper disk placed on an iron-deficient nutrient agar plate seeded with the presumed recipient strain. The donor strain was pregrown under rich iron conditions so that it continues to grow on the disk and gradually comes under iron stress. This induces the formation of aerobactin precursors which are secreted and in turn feed the recipient strain. As a result, a growth zone is observed around the filter paper disk. Since 10 filter papers can be placed on one plate this method is fast and reliable for assaying a large number of mutants.

The structure of aerobactin suggested a biosynthesis starting from lysine which is modified in two steps and then linked to citrate. In fact *N*-oxidation of lysine was achieved by a soluble and particulate fraction of *Aerobacter aerogenes* 62-1 [25]. Therefore, it was likely that *N*-oxidation preceded *N*-acetylation. *N*-hydroxy groups can be determined by oxidation with iodine. The resulting azo compound is used to form an azo dye from sulfanilic acid and naphthyl 1-amine. The same reaction can be used for determining the hydroxamate group but the acetyl residue has first to be cleaved off by acid hydrolysis. These two reactions permitted the determination of the presumptive precursors *N*-hydroxylysine and *N*-hydroxy-*N*-acetyl-lysine in culture supernatants of mutants with enzymatic blocks in subsequent steps. *N*-hydroxy-*N*-acetyl-lysine was also quantitatively determined using an amino acid analyzer.

An additional efficient method to complement mutants is the transformation with DNA fragments cloned on plasmids. In the case of the aerobactin system, cloning of the aerobactin biosynthetic genes was rather easy since i) it turned out that the genes are closely linked, and ii) the genes are "concentrated" on the ColV plasmid (130 kb in contrast to the 4700 kb of the *E. coli* chromosome). It was also fortunate that *E. coli* K-12 is devoid of aerobactin biosynthesis so that for cloning of the entire biosynthetic gene cluster a suitable strain was available. Otherwise one is never sure with mutants devoid of a biosynthetic pathway that all the genes are missing.

After cloning of the complete biosynthetic operon, insertion mutants with the transposon Tn*1000* were constructed and mapped on the cloned DNA fragment [23, 26]. In addition, the cleavage sites of restriction endonucleases were determined and used to construct defined deletions by enzymatically excising DNA fragments [26]. They were used to complement mutants in the ColV plasmid which had been isolated by insertion of the transposons Tn*5* and Mud1. Furthermore, different fragments of the aerobactin region were cloned on the compatible vectors pBR322 and pACYC184, which can be maintained together in one cell, and which complemented each other to the complete synthesis of aerobactin or to a precursor [26].

Expression of cloned genes on multi-copy plasmids or in expression vectors is performed in so-called minicells which contain the small plasmid DNA but are devoid of the large chromosomal DNA. They are formed by certain *E. coli* strains and are separated from normal cells by gradient centrifugation. Another method uses so-called maxicells in which the chromosomal DNA is preferentially damaged by UV treatment leaving plasmids largely intact. They replicate so that transcription occurs mainly from plasmid DNA. A third method of preferential expression of cloned DNA employs coupled transcription/translation in a cell-free system programmed by cloned DNA. By these methods it was possible to assign polypeptides to genes of aerobactin biosynthesis [26].

The result obtained [26] is depicted in Figure 2. Aerobactin is synthesized by 4 genes termed *aerA* to *aerD* by us and *iucA* to *iucD* by others [27]. The molecular weights of the polypeptides and the enzymatic activities assigned to the gene products are also shown. The same number of genes was found by a third group which, however, did not determine their specific biosynthetic functions [28]. The 4 genes are transcribed from left to right and they form an operon. Aerobactin is the first hydroxamate-siderophore for which the biosynthesis was elucidated. Aerobactin has been chemically synthesized [29]. Lysine *N*(6)-oxidase [30] and *N*(6)-hydroxylysine : acetyl coenzyme A *N*(6)-acetyltransferase [30, 31] has been isolated from cells carrying multi-copy recombinant plasmids. Such cells express the genes 10 times more highly than cells carrying the genes on large single-copy plasmids such as pColV, or on the chromosome. Gel filtration of the enzymatically active *N*-oxidase revealed a molecular weight of 52 000 [30] which agrees with the molecular weight determined by gel-electrophoresis of the radioactively labeled polypeptide obtained in mini- or maxicells [26, 28]. For the active *N*-acetylase a molecular weight twice [30], and three to four times [31], the size of the single polypeptide was determined suggesting that it consists of a dimer, trimer or tetramer. The enzyme acetylates *N*-hydroxylysine and *N*-hydroxyornithine, but not lysine or ornithine, supporting the biosynthetic scheme deduced from the combined genetic and biochemical studies summarized in Figure 2.

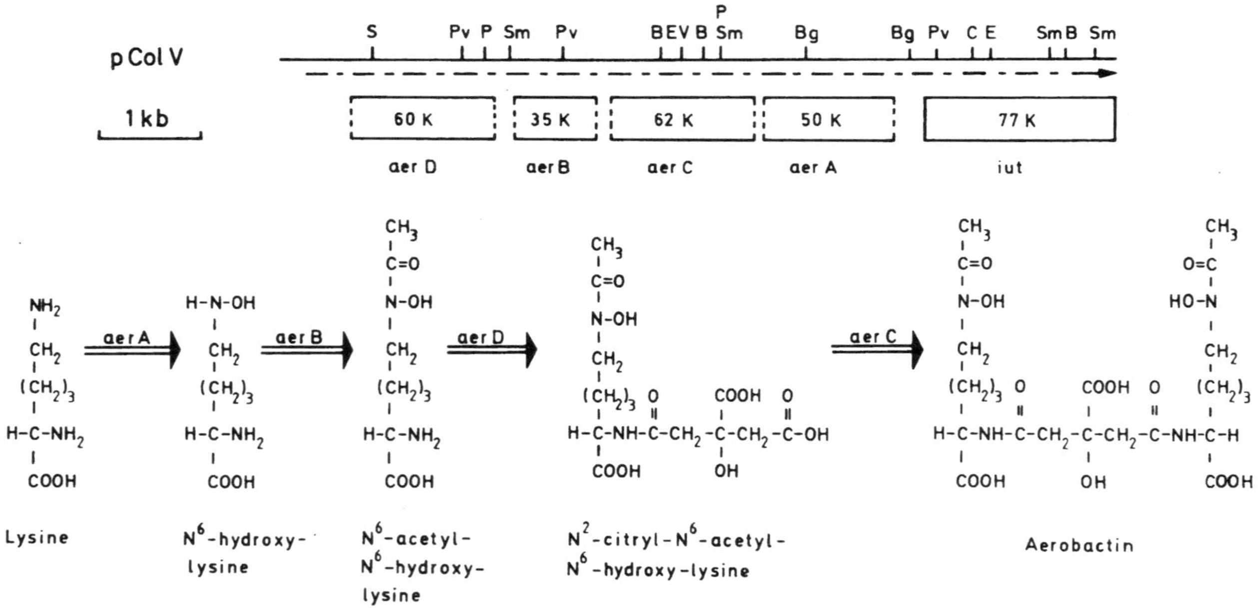

Figure 2 Arrangement of the genes and scheme of aerobactin biosynthesis. The genes *aerA*, *aerB*, *aerC* and *aerD* [26], also designated *iucD*, *iucB*, *iucC* and *iucA* [27], encode the 4 polypeptides which catalyze oxidation of lysine to *N*(6)-hydroxylysine, acetylation of *N*(6)-hydroxylysine to N(6)-hydroxy-*N*(6)-acetyllysine and the stepwise attachment of two lysine derivatives to citrate. The precise location and the exact size of the genes is only known for *iut* [71], the structural gene of the outer membrane receptor transport protein, and the *iucD* gene which encodes the *N*-oxidase [46] of which both have been sequenced. 60 k and 62 k denote the apparent molecular weights of the polypeptides as determined by polyacrylamide gel electrophoresis. The arrow indicates the transcriptional polarity. Cleavage sites of restriction endonucleases are marked B, *Bam* HI; Bg, *Bgl* II; C, *Cla* I; E. *Eco* RI; EV, *Eco* RV, P, *Pst* I; Pv, *Pvu* I; S, *Sal* I; Sm, *Sma* I.

Identification of the polypeptides by radioactive labeling depended on the presence of multiple copies of the structural genes in the cells. As an example, the *N*-oxidase of lysine will be presented. The enzymatic activity is difficult to detect in crude cell extracts and purification of the enzyme seemed to be impossible. After cloning in the vector pBR322, of which about 20 copies are present per cell, the *N*-oxidase was readily detectable and showed a specific activity 14 times higher than observed in the original *Aerobacter* strain [30]. The enzyme could also be purified by column chromatography (Heydel and Diekmann, personal communication). W. Köster (this institute) cloned the *N*-oxidase gene *aerA* behind the strong promoter of gene 10 of phage T7. The T7 RNA polymerase selectively recognizes T7 promoters and is resistant to rifampicin. If the *E. coli* polymerase is inhibited by this antibiotic, cells exclusively synthesize m-RNA from genes under T7 promoter control. Since the *E. coli* m-RNAs have half-lives of about 2 min, proteins synthesized after some time of T7 directed transcription will only be translated from the cloned *aerA* gene. The results obtained are shown in Figure 3. The amount of *N*-oxidase increases to an extent that the enzyme can readily be detected among all the proteins of a cell extract by staining, after separation of the proteins by electrophoresis on a polyacrylamide gel (Figure 3, lanes a to e). The weak band at the same position as the *N*-oxidase (Figure 3, lane a) is an unknown protein. The exclusive synthesis of the *N*-oxidase is demonstrated by radioactive labeling with [^{35}S]methionine for 1.5 min after the T7 polymerase had transcribed the *aerA* gene for 45 min. The experiment shown in Figure 3, lanes f to i, also demonstrates the stability of the *N*-oxidase in contrast to the truncated Iut outer membrane receptor protein (see Fe^{3+}-aerobactin transport below) of which half of the structural gene *iut* had been cloned together with *aerA*. The lanes, g, h, i of Figure 3 show the chase period, in the presence of a large surplus of unlabeled methionine. In the same experiment, the labeled cells were disrupted and the homogenate separated into the soluble and membrane fractions. Nearly all of the *N*-oxidase present in the homogenate (Figure 3, lane k) was recovered in the membrane fraction (Figure 3, lane m) and virtually none in the soluble fraction (Figure 3, lane l).

Additional, indirect evidence for a membrane association of the *N*-oxidase has been gained by studying Tn*phoA* and *lac* fusions to the *iucD* (= *aerA*) gene under *lac* control of the multicopy plasmid pUC9. Furthermore, the nucleotide sequence of *iucD* of pColV-K30 was determined. An *N*-terminal hydrophobic region with features of a signal peptide directed some of the alkaline phosphatase (PhoA) to the periplasm indicating that this segment may also fix the *N*-oxidase to the cytoplasmic membrane [46a].

Under the assumption that oxidation of aliphatic *a*-amino groups occurs in the biosynthesis of most hydroxamate siderophores, the *N*-oxidase is considered to be an interesting enzyme for searching for inhibitors of aerobactin synthesis. The inhibition of such synthesis would disrupt the iron supply and thus inhibit growth of bacteria and fungi. A quantitative chemical assay of *N*-oxidase activity, which can easily be automated for screening large numbers of samples, was developed by us based on a strain which overproduces *N*-oxidase but is lacking the enzymes which catalyse the subsequent steps of aerobactin synthesis. In this strain the *N*(6)-hydroxylysine formed is not metabolized further [46].

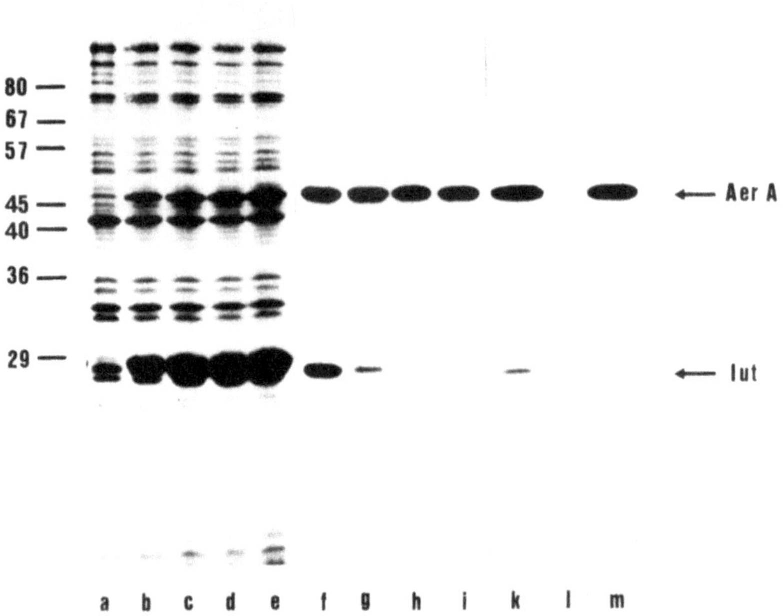

Figure 3 Overproduction of the lysine *N*(6)-oxidase. The *aerA* gene was cloned behind the strong promoter of the phage T7 Ø10 gene present on the plasmid pT7-5. The structural gene of the T7 RNA polymerase is contained on another plasmid, pGP1-2, under the control of the temperature-inducible λP_L promoter. The *E. coli* RNA polymerase can be inhibited by adding rifampicin which leaves the T7 polymerase unaffected. The T7 polymerase has a high selectivity for T7 promoters [72]. The *Bam*HI, *Cla*I fragment of Figure 2 was cloned into pT7-5 resulting in a clone with a complete *aerA* gene and a *iut* gene truncated at the 3′ end. Lanes *a* to *e* show protein bands after electrophoresis on polyacrylamide slab gels in the presence of 0.1% sodium dodecyl sulfate after staining with Coomassie brilliant blue, lanes *f* to *m* are autoradiographs of [^{35}S]methionine labeled proteins. Cells of *E. coli* WM1576 containing plasmid pWK349 were grown at 27°C until the absorbance of the culture at 578 nm reached 1.0. Then the temperature was rised to 42°C for 30 min. Rifampicin (0.1 mg/ml) was added and the culture incubated further at 37°C. Lanes *a* to *e* contain total cellular protein. To lane *a*, a sample was applied which was taken at the time, the temperature was rised from 27°C to 42°C, *b* after 30 min of temperature shift, *c* to *d* after 60, 90 and 180 min. Note the increase in the concentration of the AerA protein which on this gel showed an electrophoretic mobility which was closer to 48,000 than 50,000 (50 k). The truncated Iut′ protein was very strongly expressed and showed an apparent molecular weight of 28,000. The numbers on the left indicate the positions of standard proteins with known molecular weights. The samples in lanes *f* to *i* show a pulse-chase experiment to see whether the proteins formed remain stable or are degraded again, *f* after 1.5 min pulse labeling with [^{35}S]methionine, *g* to *i* after 5, 15 and 60 min, respectively, chase with a large surplus of non-radioactive methionine. Note that the AerA protein remains stable but the truncated Iut′ protein is rapidly degraded. Lane *k* contains the total cell protein content, *l* the soluble fraction and *m* the membrane fraction. Note that virtually the entire amount of the overexpressed oxidase is found in the membrane fraction (Data from W. Köster).

2.2 Fe^{3+}-aerobactin transport

In addition to the genes encoding proteins for aerobactin biosynthesis the ColV plasmids studied in most detail, K30 and K311, encode the structural gene, termed *iut*, for a transport protein which is located in the outer membrane [32]. Receptor proteins in the outer membrane are typical for Fe^{3+}-siderophore transport systems of Gram-negative bacteria. They are essential transport constituents and apparently serve for binding the Fe^{3+}-siderophores at the cell surface from where they are transported across the outer membrane into the periplasm, and through the cytoplasmic membrane into the cytoplasm. It is not clear where Fe^{3+} and the siderophores are dissociated. It is generally assumed that release of iron requires reduction to Fe^{2+} whose stability constant is orders of magnitude below that of Fe^{3+} [8]. Thus, Fe^{3+} has to cross 4 compartments to enter cells. It is not transported as a metal ion but as a siderophore complex. Moreover, the permeability barrier of the outer membrane is not overcome by diffusion, which holds true for almost all substrates. Instead, uptake of Fe^{3+}-siderophores through the outer membrane requires energy and the TonB protein. Through the action of both, the outer membrane receptor proteins assume different conformations [7, 9, 10]. It is assumed that in one conformation the receptor proteins bind the Fe^{3+}-siderophores at the cell surface and in the other conformation the Fe^{3+}-siderophores are released, presumably in a vectorial manner, into the periplasm. After unloading, aerobactin is secreted and can serve for the transport of additional Fe^{3+} ions [33]. Recycling of aerobactin stands in contrast to enterochelin, which is hydrolysed and the products cannot be reused for iron transport nor are they precursors of enterochelin biosynthesis [12]. Ferrichrome, another siderophore used by *E. coli* and related Enterobacteria, is inactivated by acetylation of one of the iron binding sites [34].

Elucidation of transport mechanisms are hampered in several ways. Firstly, transport is a vectorial process and requires a membrane. Therefore, isolation of individual constituents of transport systems may provide clues for binding properties of substrates, for energy providing mechanisms, but no information about the molecular events taking place during the translocation of a substrate from one side of the membrane to the other. Secondly, several proteins are frequently involved which probably interact with each other in a proper way only within the membrane. Therefore, reconstitution studies have to reestablish the complex situation in the membrane. Thirdly, the concentration of transport proteins in bacterial cytoplasmic membranes is very low, and they are difficult to isolate due to their hydrophobic properties. A (partial) solution to these biochemical problems is provided by genetic methods. As has been described before, the employment of classical genetics and of recombinant DNA technology allows the identification of the genes, their products, and frequently their functions. The amount of a protein can be increased such that it becomes a prominent band on polyacrylamide gel electropherograms. Thus the protein can clearly be identified among hundreds of other membrane proteins. The production of a protein can be scaled up to a degree that is lethal to the cells. If controlable expression vectors are used, cells are grown and maintained under conditions whereby the membrane proteins are not synthesized. Synthesis is only induced for isolation and characterization of the protein. Interaction of

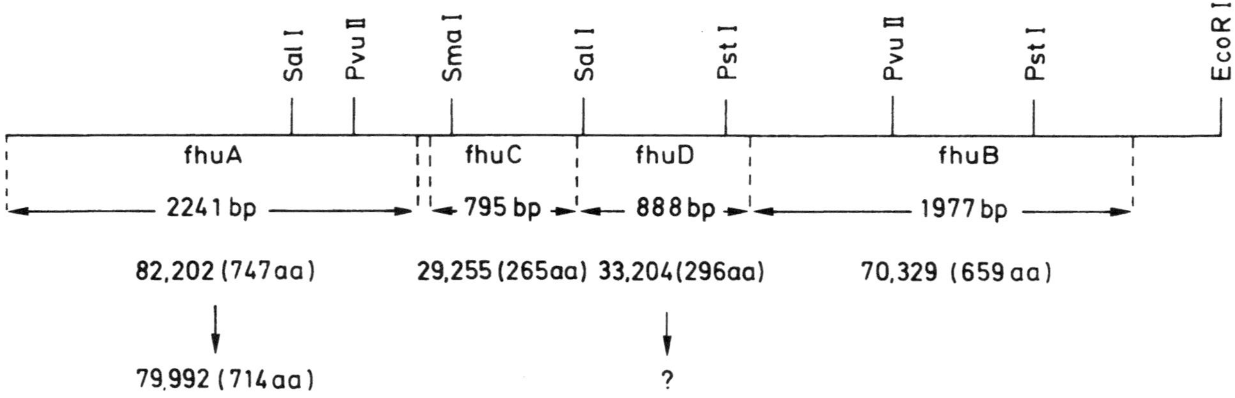

Figure 4 *fhu* operon of *E. coli* K-12 consisting of the structural gene *fhuA* of the outer membrane receptor protein and of the *fhuB*, *fhuC* and *fhuD* genes which determine transport proteins localized in the cytoplasmic membrane. The entire region has been sequenced. The *fhuA* gene consists of 2244 base pairs [bp] including the stop codon, and is separated by 50 base pairs from the *fhuC* gene comprising 798 base pairs. The translation initiation site of *fhuD* is uncertain. *fhuD* overlaps by 1 base pair with *fhuB*. Transcription proceeds from left to right.

different proteins can be established by isolating mutants which suppress mutations in the reaction partner. The activity of a defective protein can (partially) be restored by physical interaction with the reaction partner which has been altered in such a way that a functional binding occurs again. The question whether an active protein forms a monomer or a multimer can be approached by negative complementation in that 2 gene copies, one encoding an active and one an inactive polypeptide, are assembled in a single cell. If a multimer is formed both kinds of polypeptides assemble giving rise to a (partially) inactive protein. Mutations are also of great value to identify the binding sites for substrates. Since genes are now isolated and sequenced faster than proteins, primary structures can be determined and compared with other proteins which often provides a clue which function a protein may fullfill. Moreover, site directed mutagenesis and chemical synthesis of entire genes opens the way of almost unlimited modification of genes and their products in order to study their function and mode of action.

Employment of some of the above techniques provided the following results regarding Fe^{3+}-aerobactin transport. The only aerobactin-specific transport component is the above mentioned outer membrane protein encoded by the *iut* gene which is closely linked to the 4 aerobactin biosynthetic genes. Three additional genes, designated *fhuB, fhuC, fhuD* [35], which map at min 4 of the *E. coli* chromosome, encode proteins which are constituents of the Fe^{3+}-aerobactin transport system [23] (Figure 4). The genes have been sequenced and the FhuB, C proteins were localized in the cytoplasmic membrane [35–37]. The FhuB protein is an extremely hydrophobic protein. The FhuC protein contains sequences which are also found in F1-ATP'ases and in certain other transport proteins and which are considered to be ATP binding sites. It is assumed that FhuC energizes Fe^{3+}-aerobactin transport by ATP hydrolysis. Interestingly, the FhuB, C, D proteins are involved in the uptake of all Fe^{3+}-hydroxamate siderophores. In *E. coli* these are aerobactin, ferrichrome, coprogen and rhodotorulic acid. Ferrichrome binds to the FhuA outer membrane receptor protein, Fe^{3+}-coprogen and Fe^{3+}-rhodotorulic acid to the FhuE receptor [10, 38]. Transport of the Fe^{3+}-siderophores of the hydroxamate type is unique in that uptake across the outer membrane requires specific receptor proteins, and uptake across the cytoplasmic membrane is catalysed by three common proteins.

3. Enterochelin Biosynthesis and Fe^{3+}-Enterochelin Transport

The occurrence of enterochelin is apparently confined to the Enterobacteriaceae. It has been found in *E. coli*, *Salmonella typhimurium*, in *Klebsiella pneumoniae*, *Aerobacter aerogenes* and in *Shigella sonnei* and in a few strains of *Shigella flexneri* [20]. Enterochelin is composed of a cyclic triester of 2,3-dihydroxybenzoylserine (Figure 5). The latter is formed by condensation of L-serine and 2,3-dihydroxybenzoic acid. The pathway of the latter branches from chorismate and requires the

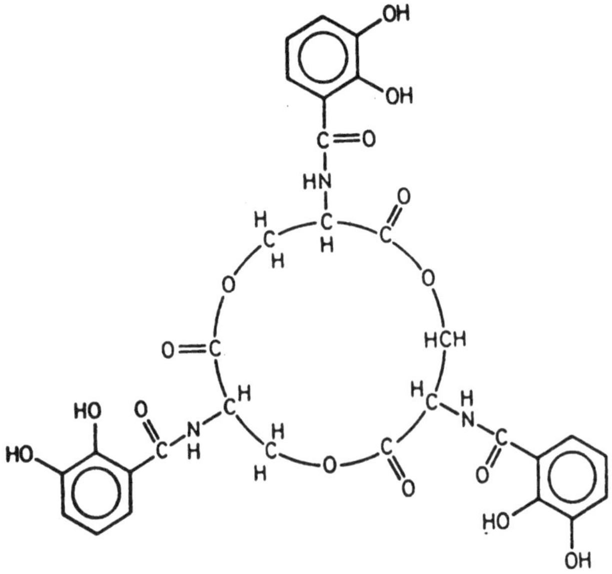

Figure 5 Structure of enterochelin (enterobactin). Fe^{3+} is bound by the 6 aromatic hydroxyl groups.

activity of three structural genes designated *entA, entB, entC* coding for three enzymes. For the synthesis of dihydroxybenzoylserine four genes, *entD, E, F, G,* were identified. The enzymology, using cell-free extracts and early complementation tests, were summarized [12] and has not been extended much further [39]. However, application of recombinant DNA technology confirmed the number of biosynthetic genes and established their order at minute 13.5 of the 100 min *E. coli* linkage map [40, 41].

Transport of Fe^{3+}-enterochelin requires an outer membrane receptor protein whose structural gene *fepA* has been sequenced [42]. In addition, a transport protein encoded by the *fepB* gene was localized in the periplasm [43, 44]. Two further genes termed *fepC* [43] and *fepD* [44] seem to belong to the transport system and may facilitate transport across the cytoplasmic membrane.

An esterase encoded by the *fes* gene hydrolyses enterochelin to the trimer, dimer and monomer of dihydroxybenzoylserine. It is still a matter of debate whether the esterase is involved in the release of iron and thus an essential component of iron uptake. Mössbauer spectroscopy revealed rapid uptake of Fe^{3+} enterochelin which, however, could be released by osmotic shock treatment, suggesting a periplasmic location of the initially transported Fe^{3+}-enterochelin [45]. The radioactively labeled complex could also be exchanged by addition of an excess of nonradioactive complex. Under the conditions employed it took more than 30 min until a major fraction of Fe^{3+} was released from enterochelin and converted to Fe^{2+}. Apparently, Fe^{3+}-enterochelin is rapidly taken up into the periplasm and slowly released by reduction to Fe^{2+} and built into proteins and other components.

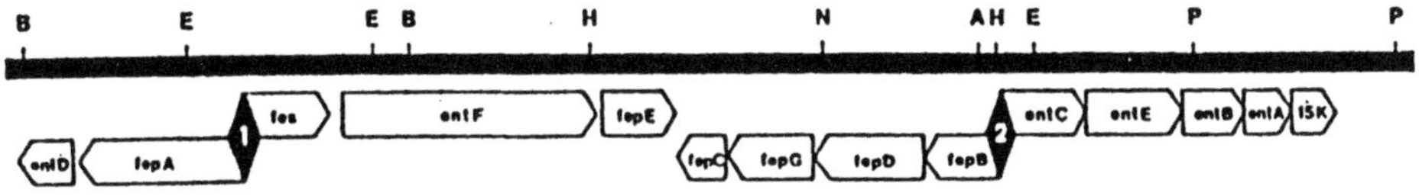

Figure 6 Organization of enterobactin biosynthesis and transport genes. The physical location and transcriptional orientation of the siderophore genes (as well as an uncharacterized cistron encoding a 15 kDA protein) is depicted. Direction of transcription is indicated by arrowheads for each cistron. Two iron-controlled regulatory regions are located between *fepA* and *fes* (region 1) and between *fepB* and *entC* (region 2). Abbreviations: A, *AvaI*; B, *Bam* HI; E, *Eco* RI; H, *Hin* dIII; N, *Nru* I; and P, *Pvu* II. Figure provided by Mark A. McInstosh, University of Missouri, Columbia. The actual number of genes in the *fepD* region and the transcriptional linkage of the central *fep* gene cluster are unresolved questions (Charles F. Earhart, University of Texas, Austin).

The 13 genes identified for enterochelin synthesis and Fe^{3+}-enterochelin transport are scattered over 22 kb on the *E. coli* chromosome (Figure 6). Neither the biosynthetic nor the transport genes are organized together in a single operon. Instead, they are mixed and transcribed in different directions.

4. Fe^{3+}-Dicitrate Transport

4.1 Transport

The first indication that *E. coli* K-12 expresses an iron uptake system mediated by citrate came from the growth promotion by citrate of Fe^{3+}-enterochelin transport (*fep*) and enterochelin synthesis (*ent*) mutants [12]. The apparent K_M of the transport system was determined to be 0.2 μM and the V_{Max} was 66 atoms of Fe^{3+} per mg cell dry weight. Mutants, termed *fec*, which are unable to respond to the addition of citrate in iron-limited media fell into two classes. One type of mutation was cotransducible by phage P1 with the nearby *argF* gene, the other was not [47]. With a Tn*10* transposon inserted close to the *fec* region, both types of *fec* mutants could be cotransduced suggesting the same map position [48]. It was then shown that the mutation close to *argF* affected an outer membrane protein to which Fe^{3+}-dicitrate could be bound and which was essential for the proper function of the transport of iron provided as a citrate complex [49].

Attempts to clone this transport system from *E. coli* K-12 succeeded in yielding a DNA fragment containing most, but not all, of the genes required specifically for the citrate-mediated iron uptake. However, from *E. coli* B a larger DNA fragment could be cloned which complemented all of the *fec* mutants. The comparison of both fragments revealed a piece of DNA at the 5′ end of the coding strand derived from the K-12 strand which differed from the equivalent region of the *E. coli* B derived strand (Figure 7). The DNA fragment containing a *fec* unrelated piece of DNA was already present in the cosmid collection used for cloning the *fec* region.

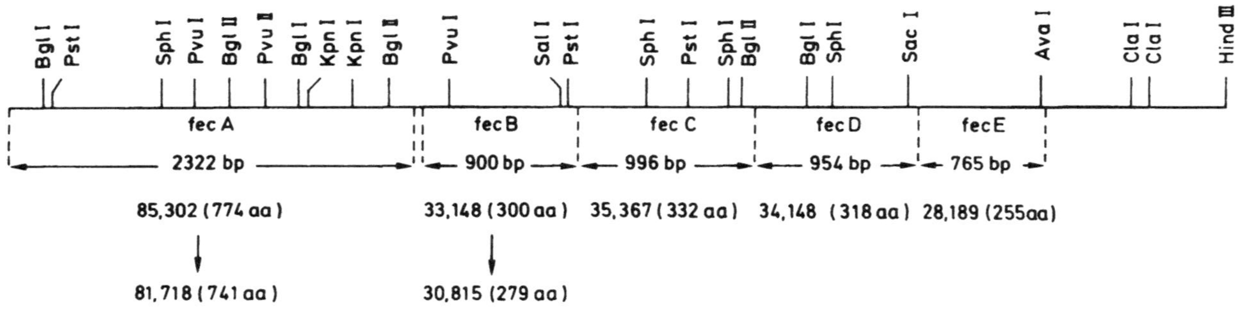

Figure 7 Cloned DNA fragment of the *E. coli* chromosome which complements all mutants in the Fe^{3+}-dicitrate transport system of *E. coli* K-12. Gene *fecA* encodes the outer membrane receptor protein, *fecB* a protein found in the periplasm [49a]. The genes *fecC* and *fecD*, *E* were cloned and sequenced by Z. Yaraghi and B. Van Hove, respectively.

Further attempts to isolate the entire *fec* region of *E. coli* K-12 failed up to now for unknown reasons. There are other peculiarities associated with the genetics of this region. The exclusive cotransduction of *fecA* with *argF* cannot be understood since *fecB* immediately follows *fecA*. All genes have been sequenced. The known accumulation of IS elements and inversions in this region may cause some of these peculiarities.

Gene *fecA* encodes the outer membrane receptor protein, and *fecB* a protein containing a typical signal sequence. The FecB protein has been found in the perplasm [49a]. Proteins secredted into the periplasm are synthesized as precursors from which the signal peptides are cleaved off during export across the cytoplasmic membrane. The location of FecB in the periplasm is of particular interest. It suggests that the Fe^{3+}-dicitrate transport system falls into the category of the so-called binding protein-dependent transport systems which include amino acid and sugar transport systems [50]. They contain soluble proteins in the periplasm which bind the substrates and deliver them to the transport components in the cytoplasmic membrane. The Fe^{3+}-dicitrate transport mechanism could thus be similar to the Fe^{3+}-enterochelin transport mechanism and both may have common features with the binding protein-dependent transport systems. The *fhuD* gene of the Fe^{3+}-hydroxamate operon determines a protein with a typical signal sequence, if the largest open reading frame is in fact transcribed and translated. Recently, the FhuD protein was localized in the periplasm and thus could function like a binding protein (V. Braun, unpublished results).

Aside from the periplasmic binding proteins, these transport systems exhibit another common characteristic. They contain at the cytoplasmic face of the cytoplasmic membrane a rather hydrophilic protein with amino acid sequences typical for ATP-binding proteins. For some of them, ATP binding has been demonstrated. Indeed, recent sequencing determination of the entire *fec* locus revealed a gene, termed *fecE*, which contains the typical conserved sequences for ATP-binding and hydrolysing proteins (B. Van hove, personal communication). The FhuC protein of the Fe^{3+}-hydroxamate transport system also exhibits these structural features [37]. The *fec* operon also encodes the two very hydrophobic proteins FecC and FecD which resemble most each other but also show strong homology to the FhuB protein of the Fe^{3+}-hydroxamate transport system. These structural features which are found in all periplasmic binding protein-dependent transport systems strongly suggest the same transport mechanism for all Fe^{3+}-siderophore compounds across the cytoplasmic membrane.

4.2 Regulation

A particular feature of the Fe^{3+}-dicitrate transport system is its regulation. It is induced by citrate [12] and by iron [48] but the amount of iron available in the medium must be low. The system can be induced in mutants unable to transport iron supplied as citrate complex [48, 49, 51]. The mutations have to be outside regions which prevent expression of the outer membrane receptor protein since

FecA minus mutants can no longer be induced. In addition, the TonB protein has to be active. Fluoro- and phosphocitrate are strong inducers but transport iron very poorly. From these data the concept of an exogenous induction has been postulated. Fe^{3+}-dicitrate has to be transported across the outer membrane. This transport step requires the activity of the outer membrane receptor and the TonB protein. Once in the periplasm Fe^{3+}-dicitrate does not have to be taken up across the cytoplasmic membrane into the cytoplasm. Instead, it binds to a transmembrane protein which when occupied with the inducer, positively regulates transcription of the *fec* genes. Citrate as such is inactive. The amount of citrate in the cytoplasm can exceed 10 times the concentration of Fe^{3+}-dicitrate (0.1 mM) required for induction of the system from without.

5. Regulation of Iron Transport Systems

5.1 Regulation of gene expression

It has long been known that siderophores are usually only synthesized under iron-limiting growth conditions [1–4, 12]. It remained to be studied whether iron regulates enzyme synthesis, or enzyme activity, or both. Later it was found that the activity of transport systems was only high when the cells had been grown in low-iron media. With the detection of outer membrane receptor proteins as essential transport constituents, which could be identified by staining on polyacrylamide electropherograms, a semi-quantitative method was available to investigate the concentration of a transport protein in cells grown at different iron supply conditions. Pulse-chase experiments using radioactive amino acids permitted the differentiation between iron dependence of protein synthesis and stability. A direct way to quantitatively examine expression of iron-regulated genes was provided by the operon fusion technique developed by Casadaban and Cohen [52]. A derivative of phage Mu, designated Mud1 (Ap, *lac*), inserts randomly like a transposon into DNA, and mutants are screened which express β-galactosidase, the product of the *lacZ* gene, under the control of iron. The β-lactamase gene contained in Mud1 selects for Mu insertions which render cells ampicillin resistant (Ap^R). Using this method expression of most of the known iron-regulated proteins has been studied and shown that iron regulates mainly at the level of transcription (summarized in [10, 53]). Different regulatory units (operons) respond to a different extent and to different iron concentrations.

The Mud1 (Ap, *lac*) operon fusion in iron-regulated genes made in possible to score mutants in which regulations was omitted. A single mutant [54] expressed all the genes constitutively suggesting a single repressor protein, which inhibited transcription, was inactive in the mutant [56]. The mutation was mapped at min 15.5 of the *E. coli* linkage map, the gene, designated *fur*, was cloned and sequenced [57]. Additional point mutants in *fur*, which formed an altered inactive protein, were isolated. When such a protein was expressed in a fur^+ wild-type strain, negative complementation was obtained, suggesting that altered Fur* polypeptides did bind to active Fur^+ polypeptides and the multimers formed were more or less inactive. This shows that several identical polypeptides associate with each other to form an active Fur regulatory protein (K. Hantke, S. Schäffer, this institute). It was proposed that the Fur protein is an aporepressor which is converted to an active repressor by binding Fe^{2+} as corepressor [53–57]. This concept has recently been supported by binding of the isolated Fur repressor protein to the upstream region of the *aer* operon shown in Figure 8. In the presence of Mn^{2+}, which has coordination properties similar to Fe^{2+} but is less subject to oxidation in vitro, the Fur protein protected primarily an area 31 base pairs in length which covers 7 base pairs upstream of the "–35" promoter site down to the first base pair of the "–10" site [58]. This sequence exhibits a dyad symmetry typical for binding sites of transcription regulatory proteins. Similar sequences were found upstream of the genes *fhuA*

iucA	GATAATGAGAATCATTATT
fhuA	CTTTATAATAATCATTCTC
fepA	TATTATGATAACTATTTGC
fur	TATAATGATACGCATTATC
fhuE (1)	TACAAACAAAATTATTCGC
fhuE (2)	GCGTATATTTCTCATTTGC
tonB	GAATATGATTGCTATTTGC
exbB	GAGAACGACTATCATTACG
sltA	GAATATGATTATCATTTTC
fecA	GAAAATAATTCTTATTTCG
consensus (3)	GATAATGATAATCATTATC
	gAt(a/t)ATgAt(t/a)at(c/t)ATTttc

Figure 8 Possible binding sites of the Fur repressor protein at presumed operators of iron-regulated genes of known sequence. Binding of Mn^{2+}-Fur has actually been shown to the upstream region of *iucA* = *aerD* by DNaseI protection experiments [58]. In addition a deletion analysis of the promoter of *sltA* confined the iron-regulated region to the sequence shown in the Figure [61]. For the other genes the assignment is tentative. For example, in the *fhuE* promoter region there are two sequences, (1) fits best the consensus sequence (3), and (2) covers the "–10" promoter region. The *exbB* sequence was determined by K. Eick-Helmerich. Note that only a few sequences exhibit the dyad repeat. In addition to the consensus sequence derived from *iucA*, *sltA*, *fur*, *fhuA* and *fepA* [58, 61] we added a sequence in which the strongly homologous sites are indicated by capital letters to take into account the less "conserved" sequences.

[59], *fhuE* [60], *fepA* [42] and *fecA* (H. Staudenmaier, this institute). In addition, transcription of the *fur* repressor gene seems to be autoregulatory since the Fur binding region is also present upstream of the structural gene [57]. Moreover, it has recently been shown that expression of the shiga-like toxin in *E. coli* is mediated by Fur and that the same consensus sequence apparently serves as a repressor binding site [61]. It can be expected that the genes of other iron-regulated toxins, most notably diphtheria toxin, are controlled by an iron-regulatory box of similar structure. The sequence homology between the promoters of known iron regulated genes is shown in Figure 8.

5.2 Regulation of the Fe^{3+}-siderophore receptor activity

The receptor proteins are constituents of the outer membrane. It has tacitly been assumed that their activity is not regulated because there are no regulatory metabolites in the outer membrane. Metabolism takes place in the cytoplasm and in the cytoplasmic membrane and regulatory devices are expected to occur only in these two compartments. However, binding of phage T1 and Ø80 to the FhuA outer membrane receptor protein, in a way that the phage DNA was released from the phase head and taken up into the cytoplasm of *E. coli*, required cellular energy and the TonB function [62]. T1 host range mutants could infect *tonB* mutants showing that DNA uptake once released from the phage head was not *tonB*-dependent [63]. In contrast, phage T5 which also binds to the FhuA receptor infects cells independent of *tonB*. The DNA is released from the phage T5 by adsorption to the isolated outer membrane, or the isolated FhuA protein [64].

Ferrichrome inhibits binding of T5 to the FhuA protein in a state where it is disconnected from the TonB protein and cellular energy, for example in energy-starved cells, in *tonB* mutants, in isolated outer membranes, or after the FhuA proteins has been extracted from cells [64]. In energized *tonB*$^+$ cells ferrichrome fails to inhibit T5 binding but inhibits T1 and Ø80 binding showing that it is present at the receptor also under this condition. We interpreted these results such that the FhuA protein can assume two different conformations, one in which ferrichrome inhibits T5 binding, and one where ferrichrome cannot interfere with T5 adsorption [63]. Phage T1 and Ø80 binding differentiates also between these two conformational states in that only the energized conformation of *tonB*$^+$ cells is active as receptor. We further proposed that the TonB protein serves as a coupling device between the cytoplasmic and the outer membrane in that it mediates the energy state of the cytoplasmic membrane to the receptor proteins in the outer membrane. In this concept the TonB protein acts like an allosteric regulatory protein for outer membrane receptor proteins. In this model the TonB protein would have to be localized in the cytoplasmic membrane where it actually has been found [65]. It would also have to interact directly with outer membrane proteins which are large enough to extend through the periplasm to the cytoplasmic membrane. The sequence of the TonB protein [66] suggests a rather hydrophilic protein so that a large portion of TonB could be located in the periplasm. A genetic means to obtain evidence for the interaction of two proteins are mutations in one protein which functionally suppress mutations in another protein. Such suppressor mutants were isolated by Knut Heller, this institute. A point mutation in the receptor protein of the vitamin B_{12} transport system [67] was suppressed by mutations in the *tonB* gene. Eight independently isolated mutants had single nucleotide replacements in the same triplett, three of which led to the exchange of Gln to Lys and five from Gln to Leu. This finding strongly supports the notion of a physical binding of the TonB protein in the regulation of outer membrane receptor proteins. The results obtained with the vitamin B_{12} transport system are certainly representative for the Fe^{3+}-siderophore transport systems because they apparently follow very similar mechanisms [73].

If the TonB protein binds to the outer membrane receptor proteins, one would expect binding domains in the receptors which can be recognized by a similarity in the amino acid sequence. Such a "TonB box" has indeed been found. Although the receptor proteins have very different amino acid sequences they contain two consensus sequences, one close to the amino-terminal and one close to the carboxy-terminal end. The consensus sequence at the amino-terminus (Figure 9) is exactly at the site where the point mutation in the vitamin B_{12} receptor occurs. This consensus sequence is also present in colicins which bind to outer membrane

FecA	23	Phe	<u>Thr</u>	Leu	Ser	<u>Val</u>
FhuE	6	Glu	<u>Thr</u>	Val	Ile	<u>Val</u>
FhuA	7	Asp	<u>Thr</u>	Ile	Thr	<u>Val</u>
BtuB	6	Asp	<u>Thr</u>	Leu	Val	<u>Val</u>
Iut	6	Glu	<u>Thr</u>	Phe	Val	<u>Val</u>
FepA	12	Asp	<u>Thr</u>	Ile	Val	<u>Val</u>
Colicin B	16	Asp	<u>Thr</u>	Met	Val	<u>Val</u>
Colicin M	2	Glu	<u>Thr</u>	Leu	Thr	<u>Val</u>
Colicin I	23	Glu	Ile	Met	Ala	<u>Val</u>

Figure 9 "TonB box" of outer membrane receptor proteins and of colicins which are constituents of TonB-dependent uptake processes. The numbers indicate the position of the amino acid residue in the sequence of the mature protein. The most conserved sites are underlined. The BtuB and colicin sequences are described in references 67, 69 to 71.

receptors and require the TonB protein for their uptake (Figure 9). There is no evidence for the involvement of the carboxy-terminal consensus sequence of the receptor proteins in the TonB-dependent transport of Fe^{3+}-siderophores or vitamin B_{12} across the outer membrane. This sequence is also absent in the colicins so it may serve another function common to the receptor proteins.

6. Antibiotics Containing Siderophore Structures

6.1 Naturally occurring sideromycins

Siderophores were detected as antibiotics (sideromycins). In 1947 grisein and in 1951 albomycin was described [74]. Both compounds are probably identical and are composed of a group of structurally related substances. A recently described sideromycin (antibiotic No. 216) apparently belongs to the same class of substances [75]. In 1957 it was shown that the action of albomycin was antagonized by ferrichrome which was the first siderophore whose structure was elucidated [76]. The mechanism by which the siderophore (former names sideramine, siderochrome) inhibited the action of the sideromycin was unknown until it was shown that the uptake of danomycin (antibiotic A-22,765) and ferrimycin A1 was inhibited by ferrioxamine B [74]. This finding explained the original observation of an antagonistic effect between ferrioxamine B and ferrimycin A1 when *Staphylococcus aureus* was exposed to both substances at the same time [77]. These observations paved the way for a long lasting interest in siderophores. Research was stimulated by attempts to find a useful antibiotic, to understand the mechanism of action of the ferrimycins, to elucidate the function of the siderophores for the producing strain, to understand the antagonism between siderophores and sideromycins in molecular terms, to use siderophores as iron donors in clinical cases, and to treat iron overload diseases with the iron-free form of siderophores. From the therapeutic concepts only the latter found wide application. Iron chelation therapy is practiced with desferrioxamine B (trade name Desferal®) which is the best compound to remove excessive iron resulting from genetic defects (primary hemochromatosis, anemias), or repeated blood transfusions.

Regarding the mode of action it has been shown for danomycin (antibiotic A-22,765) that it inhibits polyU-directed polyphenylalanine synthesis in an in vitro system of *S. aureus* [74]. In the same system ferrimycin A1 and albomycin were not inhibitory. Further studies with albomycin (E. Fischer, personal communication) suggest multiple target sites which is also supported by the failure to isolate among albomycin-resistant mutants derivatives with altered target sites.

It took 24 years from the detection to the determination of the structure of albomycin [78]. In Figure 10 the albomycins are presented such that their structural analogy to ferrichrome becomes obvious. Albomycin is taken up into cells of *Escherichia coli* by the same transport system as ferrichrome (see Figure 4). To study transport albomycin was labeled with tritium at the iron ligand moiety and with radioactive sulfur at the antibiotically active thionucleoside [79]. Albomycin

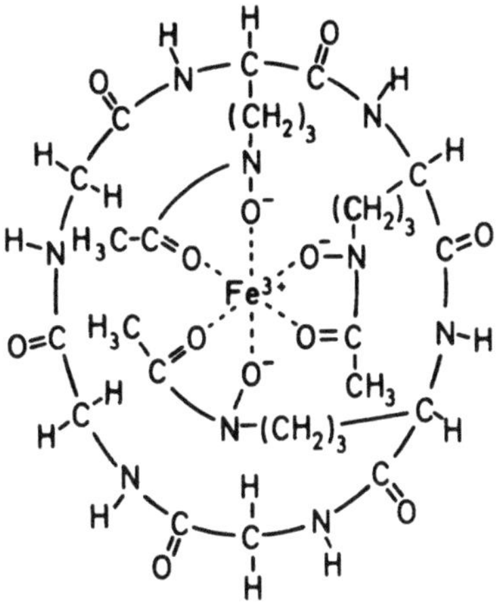

Ferrichrome

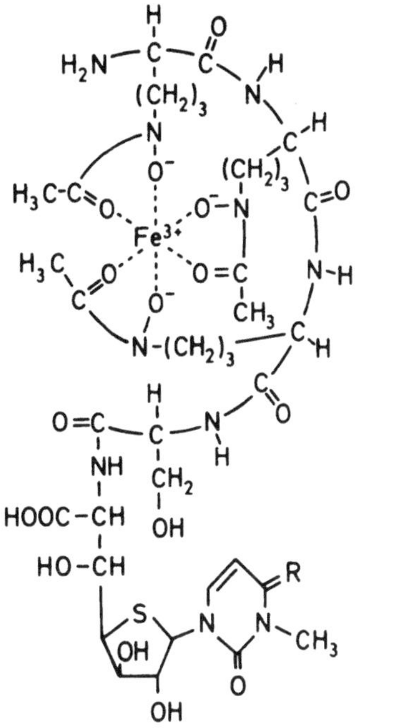

Albomycin δ1 R= O

Albomycin δ2 R= N-C(=O)-NH_2

Albomycin ε R= NH

Figure 10 Structure of ferrichrome and of albomycin.

was cleaved in the cells, the iron ligand (tritium label) released into the culture medium whereas the thionucleoside (sulfur label) was 500 fold concentrated within the cell [79]. Albomycin was cleaved by peptidase N. In mutants devoid of peptidase N albomycin supplied iron and functioned as a siderophore. Apparently, the iron ligand moiety serves as a carrier for the antibiotically active thionucleoside [80].

6.2 Synthetic antibiotics with siderophore structures

The enrichment of sideromycins in target cells via active transport systems decreases their minimal inhibitory concentration by two orders of magnitude. Therefore, they were taken as leads to synthesize compounds composed of siderophore moieties as carriers to which inhibitors were covalently linked. For example, sulfanilamidonicotinic acid was coupled to ferricrocin, a naturally occurring ferrichrome derivative with one glycine residue replaced by serine. The resulting ester inhibited *Staphylococcus aureus* at a concentration of 0.05 μM in contrast to the free sulfonamide of which 50 μM were required [81]. Apparently, the antibiotic was carried into the cells by the siderophore. Nothing is known about the iron transport system(s) of *S. aureus*.

A synthetic cephalosporin (E-0702) was substituted with a dihydroxybenzoyl moiety [82]. The susceptibility of *E. coli* was increased 250 fold compard with the same compound in which one of the two vicinal hydroxy groups was replaced by a methoxy group. Iron was bound by the hydroxy derivative. Mutants in the transport of all known iron compounds altered in the *tonB* gene were 100–1000 fold more resistant than *tonB* $^{+}$ cells. Since the TonB protein is only active in iron and vitamin B_{12} transport, and the vicinal hydroxy groups on the benzene ring resemble enterochelin, it is fairly convincingly shown that this cephalosporin compound enters *E. coli* cells via the outer membrane transport system for iron (see Section 3 and 5.2). This notion is supported by the enhanced susceptibility of cells grown at iron-limiting conditions at which synthesis of the iron transport systems is derepressed (see Section 5.1). E-0702 at 0.1 µg per ml killed within 2 h 99.9% of iron-starved *E. coli* but left iron-repleted cells unaffected. It was not reported which outer membrane receptor protein is used. Very recently two such receptors were identified for compounds of the same class (C. Curtis, personal communication). Interestingly, it was shown earlier [83] that synthetic enterochelin derivatives do not use exclusively the enterochelin receptor (FepA). Instead, only multiple mutants, in *fepA fecA* and *fepA fhuA cir*, showed a reduced growth response on some of these compounds as sole iron source. These studies demonstrate that siderophores can be used as very effective carriers of antibiotics. For *β*-lactams uptake has only to occur into the periplasm so that only the transport route across the outer membrane of gram-negative bacteria has to be operative. The only known transport system in the outer membrane, in contrast to the mere diffusion through the water-filled channels formed by the porins, are those for iron and vitamin B_{12}. Mutants in these transport systems, such as in the *tonB* gene, which is essential for all of them, are less susceptible to siderophore antibiotics but such mutants would also be depleted from the essential iron and consequently would not properly grow in the infected host.

In addition to the compounds described above a rifamicin derivative which is at least 200 times more active than rifampicin was shown to be taken up into *E. coli* via the FhuA receptor and the TonB protein [84]. Surprisingly, no structural similarity to ferrichrome or to other Fe^{3+}-hydroxamates is recognizable. It also remained unclear how this compound, with an unchanged inhibition of the DNA-dependent RNA polymerase, is taken up across the cytoplasmic membrane. Apparently, it is sufficient to overcome the permeability barrier of the outer membrane and to be concentrated in the periplasm in order to enter the cytoplasm at a much faster rate.

7. Concluding Remarks

Through the employment of molecular genetics elucidation of the iron transport systems has become possible. The genes of siderophore biosynthesis and of Fe^{3+}-siderophore transport were identified, the gene products were determined and localized in the different cell compartments, nucleotide sequences were analysed and results were obtained which supported existing models and suggested new concepts. The latter can now be tested by applying site directed mutagenesis, gene fusion techniques and gene synthesis. The results obtained so far are of general significance. Iron serves as an environmental signal which through the action of the Fur repressor adapts the cell's physiology to the nearly ubiquitous iron starvation conditions encountered in most environments. *E. coli* regulates about 30 genes through the action of a single repressor protein. Especially extraintestinal pathogens may profit from the iron regulatory device as can be deduced from the synthesis of a number of virulence factors such as toxins and extracellular hydrolases in response to iron stress. The extreme low amounts of iron available to microorganisms led to the evolution of particular iron supply systems which involve siderophores, outer membrane receptor proteins, regulated uptake and not mere diffusion through the outer membrane, transport into the cytoplasm, and release of iron from the siderophores which are either destroyed, or released and reused. After having defined the transport proteins and their location in the cell, the transport mechanism in molecular terms will have to be unraveled. For the transport of iron it will be of special interest where the iron is released from the siderophores. Preliminary experiments in *E. coli* indicate membrane-bound and soluble reductase activities which can release Fe^{3+} through reduction to Fe^{2+} and oxidation of NADH to NAD^+ (E. Fischer, this institute). It will also be important to determine whether uptake across the outer membrane is unidirectional. The type of energy used for transport through the outer and the cytoplasmic membranes is another problem pertinent for understanding the transport mechanism in molecular terms. Finally, knowledge of the functional interaction of the transport proteins will provide the ultimate model for desribing the translocation steps across both membranes.

Acknowledgements

The author's work was supported by the Deutsche Forschungsgemeinschaft (SFB 76, SFB 323, Br 330/8-1) and the Fonds der Chemischen Industrie. I like to thank Robin Harkness for his comments on the manuscript.

References

1. *Microbial Iron Metabolism* (Neilands, J. B., Ed.) Academic Press, New York, London, 1974.
2. Keller-Schierlein, W., Prelog, V., and Zähner, H., Siderochrome, natürliche Eisen(III)-trihydroxamat-Komplexe, *Fortschr. Chem. Org. Naturst. 22*, 279, 1964.
3. Knüsel, F., and Zimmermann, W., Sideromycins, in *Antibiotics*, Vol. 3, Corcoran, J. W., Hahn, F. E., Ed., Springer Verlag, Heidelberg, 1974, p. 653.
4. Neilands, J. B., Microbial iron compounds. *Annu. Rev. Biochem. 50*, 715, 1981.
5. Diekmann, H., Siderochromes, iron(III)-trihydroxamates, in *Handbook of Microbiology*, 2. edition, Vol. 5, Laskin, A. J., Lechevalier, H. A., Ed., CRC Press, Boca Raton, Florida, 1984, p. 783.

6. Winkelmann, G., Iron complex products (siderophores), in *Biotechnology*, Vol. 4, Rehm, H.-J., Reed, G., Ed., VCH Verlagsgesellschaft, Weinheim, 1986, p. 216.
7. Braun, V., and Winkelmann, G., Microbial iron transport. Structure and function of siderophores, in *Progr. Clinic. Biochem. Med. Vol. 5*, 67, 1987.
8. Raymond, K. N., Müller, G., and Matzanke, B. F., Complexation of iron by siderophores, a review of their solution and structural chemistry and biological functions, *Topics in Current Chemistry 123*, 49, 1982.
9. Braun, V., The unusual features of the iron transport system of *Escherichia coli, Trends in Biochem. Sci. 10*, 75, 1985.
10. Braun, V., The iron transport systems of *Escherichia coli*, in *The enzymes of biological membranes*, Vol. 3, Martonosi, A. N., Ed., Plenum Publish. Corp., p. 617, 1985.
11. Gibson, F., and Magrath, D. J., The isolation and characterization of a hydroxamic acid (aerobactin) formed by *Aerobacter aerogenes* 62-1, *Biochim. Biophys. Acta 192*, 175, 1969.
12. Rosenberg, H., and Young, I. G., Iron transport in the enteric bacteria, in *Microbial Iron Metabolism* (Neilands, J. B., Ed.) Academic Press, New York, London 1974, p. 67.
13. Williams, P., Novel iron uptake system specified by ColV plasmids: an important component in the virulence of invasive strains of *Escherichia coli, Infect. Immun. 26*, 925, 1979.
14. Stuart, S. T., Greenwood, K. T., and Luke, R. K., Hydroxamate-mediated transport of iron controlled by ColV plasmids, *J. Bacteriol. 143*, 35, 1980.
15. Braun, V., *Escherichia coli* containing the plasmid ColV produce the iron ionophore aerobactin, *FEMS Microbiol. Lett. 11*, 225, 1981.
16. Warner, P. J., Williams, P. H., Bindereif, A., and Neilands, J. B., ColV plasmid-specified aerobactin synthesis by invasive strains of *Escherichia coli, Infect. Immun. 33*, 540, 1981.
17. van Tiel-Menkveld, G. J., Mentjox-Vervuurt, J. M., Oudega, G., and de Graaf, F., Siderophore production by *Enterobacter cloacae* and a common receptor protein for the uptake of aerobactin and cloacin DF13, *J. Bacteriol. 150*, 490, 1982.
18. Valvano, M. A., and Crosa, J. H., Aerobactin iron transport genes commonly encoded by certain ColV plasmids occur in the chromosome of a human invasive strain of *Escherichia coli* K1, *Infect. Immun. 46*, 159, 1984.
19. Bindereif, A., and Neilands, J. B., Aerobactin genes in clinical isolates of *Escherichia coli, J. Bacteriol. 161*, 727, 1985.
20. Payne, S. M., Niesel, D. W., Peixotto, S. S., and Lawlor, K. M., Expression of hydroxamate and phenolate siderophores by *Shigella flexneri, J. Bacteriol. 155*, 949, 1983.
21. Colonna, B., Nicoletti, M., Visca, P., Casalino, M., Valenti, P., and Maimome, F., Composite IS elements encoding hydroxamate-mediated iron uptake in FI*me* plasmids from epidemic *Salmonella spp., J. Bacteriol. 162*, 307, 1985.
22. Stuart, S., Prpic, J. K., and Robins-Browne, R. M., Production of aerobactin by some species of the genus *Yersinia, J. Bacteriol. 166*, 1131, 1986.
23. Braun, V., Gross, R., Köster, W., and Zimmermann, L., Plasmid and chromosomal mutants in the iron(III)-aerobactin transport system of *Escherichia coli.* Use of streptonigrin for selection, *Mol. Gen. Genet. 192*, 131, 1983.
24. Yeowell, H. N., and White, J. R., Iron requirement in the bactericidal mechanism of streptonigrin, *Antimicrob. Ag. Chemother. 22*, 961, 1982.
25. Parniak, M. A., Jackson, G. E. D., Murray, G. J., and Viswanatha, T., Studies on the formation of N^6-hydroxylysine in cell free extracts of *Aerobacter aerogenes* 62-1, *Biochim. Biophys. Acta 569*, 99, 1979.
26. Gross, R., Engelbrecht, F., and Braun, V., Identification of the genes and their polypeptide products responsible for aerobactin synthesis by pColV plasmids, *Mol. Gen. Genet. 201*, 204, 1985.
27. de Lorenzo, V., and Neilands, J. B., Characterization of *iucA* and *iucC* genes of the aerobactin system of plasmid ColV-K30 in *Escherichia coli, J. Bacteriol. 167*, 350, 1986.
28. Carbonetti, N. H., and Williams, P., A cluster of five genes specifying the aerobactin iron uptake system of plasmid ColV-K30, *Infect. Immun. 46*, 7, 1984.
29. Maurer, P. J., and Miller, M. J., Microbial ïron chelators: total synthesis of aerobactin and its constituent amino acid, N^6-acetyl-N^6-hydroxylysine, *J. Am. Chem. Soc. 104*, 3096, 1982.
30. Heydel, P., Plattner, H., and Diekmann, H., Lysine N-hydroxylase and N-acetyltransferase of the aerobactin system of pColV plasmids in *Escherichia coli, FEMS Microbiol. Lett. 40*, 305, 1987.

31. Coy, M., Paw, B. H., Bindereif, A., and Neilands, J. B., Isolation and properties of N^{ε}-hydroxylysine: acetyl coenzyme A N^{ε}-transacetylase from *Escherichia coli* pABN11, *Biochemistry 25*, 2485, 1986.
32. Bindereif, A., Braun, V., and Hantke, K., The cloacin receptor of ColV-bearing *Escherichia coli* is part of the Fe^{3+}-aerobactin transport system, *J. Bacteriol. 150*, 1472, 1982.
33. Schneider, R., Hartmann, A., and Braun, V., Transport of the iron ionophore ferrichrome in *Escherichia coli* and *Salmonella typhimurium* LT2. *FEMS Microbiol. Lett. 11*, 115, 1981.
34. Hartmann, A., and Braun, V., Iron transport in *Escherichia coli*: uptake and modification of ferrichrome, *J. Bacteriol. 143*, 246, 1980.
35. Fecker, L., and Braun, V., Cloning and expression of the *fhu* genes involved in the iron(III)-hydroxamate uptake into *Escherichia coli*, *J. Bacteriol. 156*, 1301, 1983.
36. Köster, W., and Braun, V., Iron hydroxamate transport of *Escherichia coli*: nucleotide sequence of the *fhuB* gene and identification of the protein, *Mol. Gen. Genet. 204*, 435, 1986.
37. Burkhardt, R., and Braun, V., Nucleotide sequence of the *fhuC* and *fhuD* genes involved in iron(III)-hydroxamate transport: domains in FhuC homologous to ATP-binding proteins, *Mol. Gen. Genet. 209*, 49, 1987.
38. Hantke, K., Identification of an iron uptake system specific for coprogen and rhodotorulic acid in *Escherichia coli*, *Mol. Gen. Genet. 191*, 301, 1983.
39. Greenwood, K. T., and Luke, R. K. J., Studies on the enzymatic synthesis of enterochelin in *Escherichia coli* K-12. Four polypeptides involved in the conversion of 2,3-dihydroxybenzoate to enterochelin, *Biochim. Biophys. Acta 454*, 285, 1976.
40. Pickett, C. L., Hayes, L. D., and Earhart, C. F., Molecular cloning of the *Escherichia coli* K-12 *entACGBE* genes, *FEMS Microbiol. Lett. 24*, 77, 1984.
41. Fleming, T. P., Nahlik, M. S., Neilands, J. B., and McIntosh, M. A., Physical and genetic characterization of cloned enterobactin genomic sequences from *Escherichia coli* K-12, *Gene 34*, 47, 1985.
42. Lundrigan, M. D., and Kadner, R. J., Nucleotide sequence of the gene for the ferrienterochelin receptor FepA in *Escherichia coli*, *J. Biol. Chem. 261*, 10797, 1986.
43. Pierce, J. R., and Earhart, C. F., *Escherichia coli* K-12 envelope proteins specifically required for ferrienterobactin uptake, *J. Bacteriol. 166*, 390, 1986.
44. Ozenberger, B. A., Nahlik, M. S., and McIntosh, M. A., Genetic organization of multiple *fep* genes encoding ferric enterobactin transport functions in *Escherichia coli*, *J. Bacteriol. 169*, 3638, 1987.
45. Matzanke, B. F., Ecker, D. J., Yang, T. S., Huyuh, G. H., Müller, G., and Raymond, K. N., *Escherichia coli* iron enterobactin uptake monitored by Mössbauer spectroscopy, *J. Bacteriol. 167*, 674, 1986.
46. Engelbrecht, F., and Braun, V., Inhibition of microbial growth by interference with siderophore biosynthesis. Oxidation of primary amino groups in aerobactin synthesis by *Escherichia coli*, *FEMS Microbiol. Lett.* 33, 223, 1986.
46a. Herrero, M., de Lorenzo, V., and Neilands, J. B., Nucleotide sequence of the *iucD* gene of the pColV-K30 aerobactin operon and topology of its product studied with *phoA* and *lacZ* gene fusions, *J. Bacteriol. 170*, 56, 1988.
47. Woodrow, G. C., Langman, L., Young, I. G., and Gibson, F., Mutations affecting the citrate-dependent iron uptake system in *Escherichia coli*, *J. Bacteriol. 133*, 1524, 1978.
48. Hussein, S., Hantke, K., and Braun, V., Citrate-dependent iron transport system in *Escherichia coli*, *Eur. J. Biochem. 117*, 431, 1981.
49. Wagegg, W., and Braun, V., Ferric citrate transport in *Escherichia coli* requires outer membrane receptor protein FecA, *J. Bacteriol. 145*, 156, 1981.
49a. Pressler, U., Staudenmaier, H., Zimmermann, L., and Braun, V., Genetics of the iron dicitrate transport system of *Escherichia coli*, *J. Bacteriol. 170*, 2716, 1988.
50. Hengge, R., and Boos, W., Maltose and lactose transport in *Escherichia coli*: examples of two differents types of concentrative transport systems, *Biochim. Biophys. Acta 737*, 443, 1983.
51. Zimmermann, L., Hantke, K., and Braun, V., Exogenous induction of the iron-dicitrate transport system of *Escherichia coli* K-12, *J. Bacteriol. 159*, 271, 1984.
52. Casadaban, M. J., and Cohen, S. N., Lactose genes fused to exogenous promoters in one step using a Mu-*lac* bacteriophage: in vitro probe for transcriptional control sequences, *Proc. Natl. Acad. Sci. USA 76*, 4530, 1979.

53. Braun, W., Regulation of iron uptake in *Escherichia coli*, in *Microbiology*, 1983, Schlessinger, D., Ed., American Society for Microbiology, Washington, D.C., 1984, p. 301.
54. Hantke, K., Regulation of ferric iron transport in *Escherichia coli* K-12: isolation of a constitutive mutant, *Mol. Gen. Genet. 182*, 288, 1981.
55. Braun, V., and Burkhardt, R., Regulation of the ColV plasmid-determined iron(III)-aerobactin transport system in *Escherichia coli, J. Bacteriol. 152*, 223, 1982.
56. Hantke, K., Negative control of iron uptake systems in *Escherichia coli, FEMS Microbiol. Lett. 15*, 83, 1982.
57. Schäffer, S., Hantke, K., and Braun, V., Nucleotide sequence of the iron regulatory gene *fur*, *Mol. Gen. Genet. 200*, 110, 1985.
58. de Lorenzo, V., Wee, S., Herrero, M., and Neilands, J. B., Operator sequences of the aerobactin operon of plasmid ColV-K30 binding the ferric uptake regulation (*fur*) repressor, *J. Bacteriol. 169*, 2624, 1987.
59. Coulton, J. W., Mason, P., Cameron, D. R., Carmel, G., Jean, R., and Rode, H. N., Protein fusions of β-galactosidase to the ferrichrome-iron receptor of *Escherichia coli* K-12, *J. Bacteriol. 165*, 181, 1986.
60. Sauer, M., Hantke, K., and Braun, V., Ferric-coprogen receptor FhuE of *Escherichia coli*: processing and sequence common to all TonB-dependent outer membrane receptor proteins, *J. Bacteriol. 169*, 2044, 1987.
61. Calderwood, S. B., and Mekalanos, J. J., Iron regulation of shiga-like toxin expression in *Escherichia coli* is mediated by the *fur* locus, *J. Bacteriol. 169*, 4759, 1987.
62. Hancock, R. E. W., and Braun, V., Nature of the energy requirement for the irreversible adsorption of bacteriophages T1 and Ø80 to *Escherichia coli* K-12, *J. Bacteriol. 125*, 409, 1976.
63. Hantke, K., and Braun, V., Functional interaction of the *tonA/tonB* receptor system in *Escherichia coli, J. Bacteriol. 135*, 190, 1978.
64. Hoffmann, H., Fischer, E., Kraut, H., and Braun, V., Preparation of the FhuA (TonA) receptor protein form cell envelopes of an overproducing strain of *Escherichia coli* K-12, *J. Bacteriol. 166*, 404, 1986.
65. Plastow, G. S., and Holland, J. B., Identification of an *Escherichia coli* inner membrane polypeptide specified by a λ*tonB* transducing bacteriophage, *Biochem. Biophys. Res. Commun. 90*, 1007, 1979.
66. Postle, K., and Good, R. F., DNA sequence of the *Escherichia coli tonB* gene, *Proc. Natl. Acad. Sci. USA, 80*, 5235, 1983.
67. Heller, K., and Kadner, R. J., Nucleotide sequence of the gene for the vitamin B_{12} receptor protein in the outer membrane of *Escherichia coli, J. Bacteriol. 161*, 904, 1985.
68. Friedrich, M. J., de Veaux, L. C., and Kadner, R. J., Nucleotide sequence of the *btuCED* genes involved in vitamin B_{12} transport in *Escherichia coli* and homology with components of periplasmic-binding-protein-dependent transport systems, *J. Bacteriol. 167*, 928, 1986.
69. Schramm, E., Mende, J., Braun, V., and Kamp, R. M., Nucleotide sequence of the colicin B activity gene *cba*: consensus pentapeptide among TonB-dependent colicins and receptors, *J. Bacteriol. 169*, 3350, 1987.
70. Köck, J., Ölschläger, T., Kamp, R. M., and Braun, V., Primary structure of colicin M, an inhibitor of murein biosynthesis, *J. Bacteriol. 169*, 3358, 1987.
71. Krone, W. J. A., Stegehuis, F., Koningstein, G., van Doorn, C., Roosendaal, B., de Graaf, F. K., and Oudega, B., Characterization of the pColV-K30 encoded cloacin DF13/aerobactin outer membrane receptor protein of *Escherichia coli*; isolation and purification of the protein and analysis of its nucleotide sequence and primary structure. *FEMS Microbiol. Lett. 26*, 153, 1985.
72. Tabor, S., and Richardson, C. C., A bacteriophage T7 RNA polymerase/promoter system for controlled exclusive expression specific genes. *Proc. Natl. Acad. Sci, USA, 82*, 1074, 1985.
73. Heller, K. J., Kadner, R. J., and Günther, K., Suppression of the *btuB451* mutation by mutations in the *tonB* gene suggests a direct interaction between TonB and TonB-dependent receptor proteins in the outer membrane of *Escherichia coli, Gene 64*, 147, 1988.
74. Nüesch, J., and Knüsel, F., "Sideromycins", in *Antibiotics*, Vol. 1 (D. Gottlieb and P. D. Shaw, eds.) pp. 499 – 541, Springer Verlag, Heidelberg, 1967.

75. Miyakawa, S., Noto, T., and Okazaki, H., Mechanisms of resistance in *Escherichia coli* to the sideromycin antibiotic No. 216: isolation and characterization of the resistant mutants. *Microbiol. Immunol. 26*, 885, 1982.
76. Neilands, J. B., Some aspects of microbial iron metabolism, *Bacteriol. Rev. 21*, 101, 1957.
77. Bickel, J., Gäumann, E., Keller-Schierlein, W., Prelog, V., Vischer, E., Wettstein, A., and Zähner, H., Über eisenhaltige Wachstumsfaktoren, die Sideramine, und ihre Antagonisten, die eisenhaltigen Antibiotika Sideromycine. *Experientia 16*, 128, 1960.
78. Benz, G., Schröder, T., Kurz, J., Wünsche, C., Karl, W., Steffens, G., Pfitzner, J., and Schmidt, D., Konstitution der Desferriform der Albomycine δ_1, δ_2 und ε. *Angew. Chem. 94*, 552, 1982.
79. Hartmann, A., Fiedler, H.-P., and Braun, V., Uptake and conversion of the antibiotic albomycin by *Escherichia coli. Eur. J. Biochem. 99*, 517, 1979.
80. Braun, V., Günter, K., Hantke, K., and Zimmermann, L., Intracellular activation of albomycin in *Escherichia coli* and *Salmonella typhimurium. J. Bacteriol. 156*, 308, 1983.
81. Zähner, H., Diddens, H., Keller-Schierlein, W., and Naegeli, H. U., Some experiments with semisynthetic sideromycins. *J. Antibiotics 30 (Suppl) 201*, 1977.
82. Watanabe, N.-A., Nagasu, T., Katsu, K., and Kitoh, K., E-0702, a new cephalosporin, is incorporated in *Escherichia coli* cells via the *tonB*-dependent iron transport system. *Antimicrob. Agents. Chemother. 31*, 497, 1987.
83. Heidinger, S., Braun, V., Pecoraro, V. L., and Raymond, K. N., Iron supply to *Escherichia coli* by synthetic analogs of enterochelin. *J. Bacteriol. 153*, 109, 1983.
84. Pugsley, A. P., Zimmermann, W., and Wehrli, W., Highly efficient uptake of a rifamycin derivative via the FhuA-TonB-dependent uptake route in *Escherichia coli. J. Gen. Microbiol. 133*, 3505, 1987.

Chapter 6
Discovery of new β-Lactam and β-Lactam like Antibiotics from Bacteria

Hideo Ono and Setsuo Harada

1. Introduction
2. Screening Method and Producing Organisms
3. Fermentation
4. Isolation Procedure
5. Chemical Characterization
6. Structure Determination
7. Biological Activities of Cephabacin, Formadicin and Lactivicin
8. Conclusion
References

1. Introduction

Our screening program to search for new inhibitors of cell wall synthesis started in 1978. Of such inhibitors, it is well known that β-lactam antibiotics possess potent activity against Gram-positive and negative bacteria, low toxicities towards mammals, and that improvement of their biological activity is accomplished relatively easily by chemical modification [1]. Naturally-occurring β-lactam antibiotics with a new nucleus or a new partial structure having new biological activities have long been sought in this field in the hope of obtaining better chemotherapeutics. At that time when most research in this field was centered on fungi and actinomycete [1], we considered that these compounds would be produced not only by these organisms but also by bacteria. Our screening method incorporates use of strains that are hypersensitive to β-lactam antibiotics [2], susceptibility to β-lactamases, and observation of morphological changes in test organisms [3].

In 1981, Imada and Asai et al. found the new monocyclic β-lactam antibiotics, sulfazecin and isosulfazecin from *Pseudomonas* species [4–5]. These antibiotics have an *N*-sulfonated function in a β-lactam ring. At almost the same time, R. B. Sykes et al. isolated many *N*-sulfo monocyclic β-lactams (monobactams) from various types of bacteria [5–6].

Thereafter, we made efforts to establish a rapid screening system to check the biological activities of producing bacteria isolated from soils, plants, and water, and we also developed systematic procedures using HPLC for the isolation of

Table 1 Naturally occurring β-lactam and β-lactam like antibiotics from bacteria.

Compound	Producing organism	Organization	Reference
Sulfazecin type			
Sulfazecin	*Pseudomonas acidophila* G-6302	Takeda	4, 49
SQ 26445	*Gluconobacter* sp. ATCC 31581 *Acetobacter pasteurianus* subsp. *pasteurianus* ATCC 6033	Squibb	6
Isosulfazecin	*Pseudomonas acidophila* SB-72310	Takeda	4, 50
SQ 26,180	*Chromobacterium violaceum* SC11378 (ATCC 31532)	Squibb	6, 51
SQ 26700, 26823, 26875, 26970, 26812	*Agrobacterium radiobacter* SC11742 (ATCC 31700)	Squibb	6, 52
SQ 28332	*Flexibacter* sp. SC12681 (ATCC 35208)	Squibb	53
PB-5582-A	*Cytophaga johnsonae* PB-5582	Shionogi	54
TAN-542	*Flexibacter* sp. YK-114	Takeda	55
SQ 28502 & 28503	*Flexibacter* sp. SC11479 (ATCC 35103)	Squibb	56
TAN-643	*Cytophaga aurantiaca* YK-261	Takeda	57
PB-5266 A, B & C	*Cytophaga johnsonae* PB-5266	Shionogi	58
TAN-850	*Pseudomonas* sp. MK-41	Takeda	59
Carbapenem type			
SQ 27,860	*Serratia* sp. SC 11482 (ATCC 39006) *Erwinia carotovora* SC12637 *Erwinia herbicola* SC12638	Squibb	60
TAN-458	*Azospirillum* sp. UK-1521	Takeda	61
Cephabacin type			
SQ 28,516 & 28,517	*Flavobacterium* sp. SC12154	Squibb	62
Chitinovorin A, B & C	*Flavobacterium chitinovorum* PB-5016	Shionogi	63

Table 1 Naturally occurring β-lactam and β-lactam like antibiotics from bacteria.

Compound	Producing organism	Organization	Reference
Cephabacin F_1, F_2, F_3, F_4, F_5 & F_6	*Lysobacter lactamgenus* YK-90 *Xanthomonas lactamgena* YK-280	Takeda	9, 39, 64, 65
Cephabacin H_1, H_2 & H_3	*Lysobacter lactamgenus* YK-90	Takeda	9, 39, 64, 65
Cephabacin H_4, H_5 & H_6	*Xanthomonas lactamgena* YK-280	Takeda	9, 39, 64, 65
Cephabacin F_7, F_8 & F_9	*Xanthomonas lactamgena* YK-278	Takeda	9, 39, 64, 65
Chitinovorin D	*Flavobacterium* sp. PB-5246	Shionogi	66
Cephabacin M_1, M_2, M_3, M_4, M_5 & M_6	*Xanthomonas lactamgena* YK-431	Takeda	40, 67
Nocardicin type			
Formadicin A, B, C & D	*Flexibacter alginolique-faciens* YK-49	Takeda	10, 42
TAN-667	*Cytophaga phytophila* YK-390	Takeda	68
Lactivicin type			
Lactivicin	*Empedobacter lactamgenus* YK-258 *Lysobacter albus* YK-422	Takeda	11, 43, 69, 70

* Relations of the antibiotics: Sulfazecin = SQ 26,445, SQ 28,332 = PB-5582-A = TAN-542, SQ 27,860 = TAN-458 p-nitro-benzylester, SQ 28,516 = N,N'-Diacetyl cephabacin F3, Chitinovorin A, B & C = Cephabacin F1, F2 & 7-Formylamino-deacetylcephalosporin C, respectively.

unstable β-lactam antibiotics from aqueous solutions [7–8]. Using these methods, we found some previous unknown types of β-lactams and a related antibiotic from bacteria as shown in Table 1. The important discoveries that can be attributed to these studies are *N*-sulfo monocyclic β-lactams (sulfazecins, monobactams), formylamino β-lactams at the 3-position of β-lactam carbonyls (cephabacins F [9], formadicin A and C [10]) and an isolated dicyclic dipeptide (lactivicin [11]) having the similar biological activities to β-lactam antibiotics.

In this chapter, we will focus on the detection, isolation, chemical characterization, structure determination, and biological activities of cephabacins, formadicins, and lactivicin.

2. Screening Method and Producing Organisms

It is desirable that a screening method for use with microbial cultures is very sensitive and specific for the target antibiotic and uncomplicated for possibile automation.

For the screening of β-lactams, methods based on the following parameters have been developed.

(1) differential antibacterial activity against β-lactam insensitive Gram-negative bacteria (*P. aeruginosa* and *E. coli*] and their β-lactam hypersensitive mutants [2, 12–13]

(2) susceptibility to β-lactamase [2, 13–15]

(3) inhibitory activity towards β-lactamase [16–20]

(4) induction of morphological changes in test organism [21–23]

(5) no activity against *Mycoplasma* or bacterium which lacks cell wall [24–25]

(6) reactivity with specific chemical reagents [26]

(7) inhibitory activity towards D-alanine carboxypeptidase [27–29]

(8) induction of β-lactamase [5]

(9) affinity to penicillin binding proteins [30]

Our screening system consists of (1), (2), (3) and (4) and in addition data concerning antibacterial spectra and TLC bioautographies are used as a counter screening to eliminate known β-lactams at an early stage.

Bacteria which were found to produce β-lactams are listed in Table 1. All of β-lactam producing bacteria are Gram-negative and can be classified into the four groups according to the Bergey's Manual of Determinative Bacteriology (8th Edition, 1974), namely, the gliding bacteria (Part 2 in the manual), spiral and curved bacteria (Part 6), aerobic rods (Part 7), and facultative anaerobic rods (Part 8). As shown in Table 1, 3-methoxylated monobactams are produced by aerobic rods (*Pseudomonas, Gluconobacter, Acetobacter*, and *Agrobacterium*) and facultative anaerobic rods (*Chromobacterium*); almost all 3-non-methoxylated monobactams and formadicins are found to be produced by the gliding bacteria (*Cytophaga, Flexibacter*); simple carbapenem antibiotics are produced by facultative anaerobic rods (*Serratia* and *Erwinia*) and spiral bacteria (*Azospirillum*); cephalosporins are produced by *Flavobacterium, Xanthomonas, Pseudomonas*, and *Lysobacter*; and lactivicin is the product of *Empedobacter* (high GC *Flavobacterium*) and *Lysobacter*. The gliding bacteria such as *Cytophaga, Flexibacter*, and *Lysobacter* are frequently found to be β-lactam producers. These bacteria are characterized by gliding mobility and an ability to decompose polysaccharides. It remains to be elucidated whether the ability to degrade polysaccharides is related to the ability of these bacteria to produce β-lactam antibiotics or not.

3. Fermentation

At first, β-lactam producing bacteria are usually incubated at 20 to 24°C for 24 to 48 h using a seed culture medium consisting of 2% glucose, 3% soluble starch, 1% soybean flour, 0.3% corn-steep liquor, 0.5% Polypepton and 0.5% NaCl (pH 7.0). Subsequently, the organisms are transferred to the main fermentation medium consisting of e. g. 3% dextrin, 3% soybean flour and 0.5% $CaCO_3$ (pH not adjusted) for cephabacins F and H; 3% dextrin, 1.5% soybean flour, 1.5% corn-gluten meal, 0.2% Polypepton, 0.1% $Na_2S_2O_3 \cdot 5\,H_2O$ (pH not adjusted) for cephabacin M; and 3% glycerol, 2% soybean flour, 1% corn-gluten meal, 0.2% Polypepton, and 0.5% $CaCO_3$ for formadicin production. The main fermentation is carried out on the scale of 1200 liters of medium in a 2000 liter fermentor. The cultivation is performed at 20 to 24°C for 48 to 66 h with aeration (1200 l/min) and agitation (120 rpm) using 0.05% Actcol as an antifoaming agent.

4. Isolation Procedure

The β-lactam and β-lactam like antibiotics shown in Table 1 can be separated into two groups, i. e. acidic (monocyclic β-lactams, lactivicin) and basic (cephabacins) substances. Furthermore, acidic substances are separable into two sub-groups depending on whether their acidities are based on carboxylic acid (formadicins, lactivicin) or sulfonic acid (sulfazecin, monobactams). Thus, three types of procedure are used for isolation of these antibiotics based on their physico-chemical properties. The typical isolation procedures for sulfazecins, formadicins, and cephabacins M are shown in Figures 1, 2, and 3. They are purified mainly by column chromatography using ion-exchange resins, adsorptive resins, and ion-exchange Sephadex.

Ion-paired extraction using alkylammonium halides or hydroxides is used for separation of acidic compounds (sulfazecin-type) from desalted aqueous solutions. Reversed phase preparative HPLC is valuable in the final stages of the isolation of each component. The isolation of antibiotics that are unstable in aqueous solution and whose half-lives are usually less than 1 hour at 60°C, is carried out in a cold room at below 7°C. It is also always necessary to adjust the pH of the solutions and eluates within the stable range for the compounds.

For the detection of antibiotics, the antibacterial activity against mutant strains which are hypersensitive to β-lactam antibiotics, for example, *Escherichia coli* PG 8 and *Pseudomonas aeruginosa* C 141, obtained by an agar diffusion method is useful in the early separation stages. The analytical HPLC is monitored by the antimicrobial activity measured by the disc method, we call this a "peak survey". After obtaining clear relationships between the antimicrobial activities and peak heights of HPLC, the samples in the later separation stages were rapidly analyzed by HPLC before decomposition occurred.

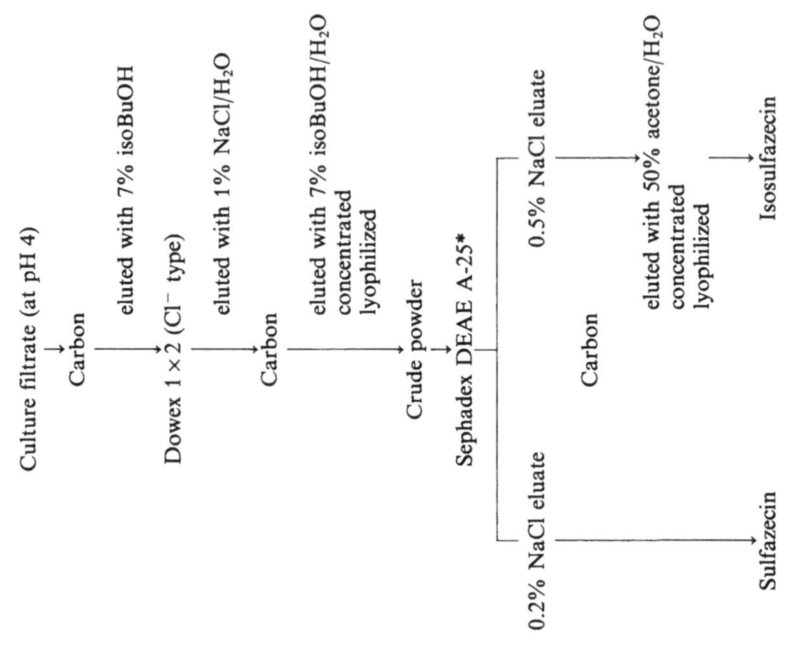

Figure 1 Isolation procedure of sulfazecin and isosulfazecin.

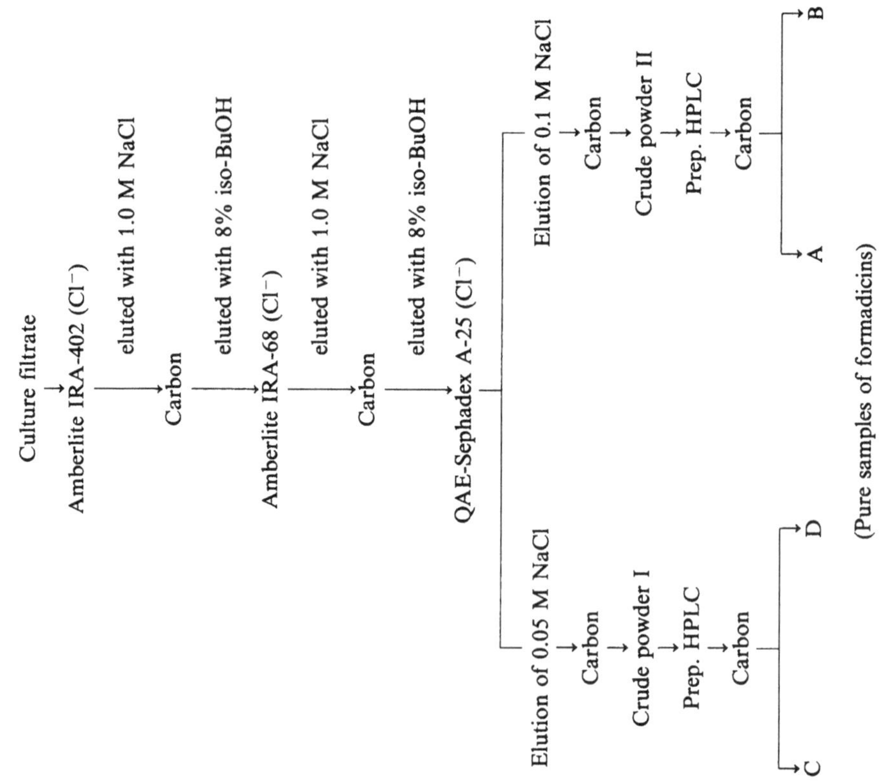

Figure 2 Isolation procedure of formadicins A, B, C, and D.

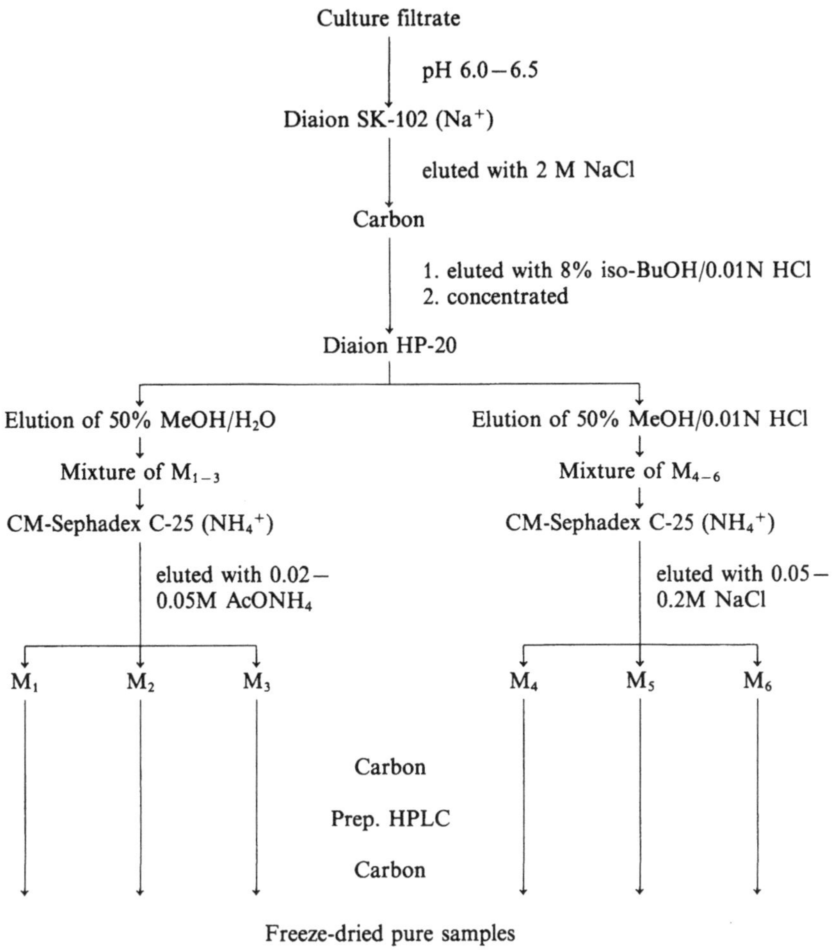

Figure 3 Isolation procedure of cephabacin M_{1-6}.

The conditions for analytical HPLC are shown in Table 2. The best medium for analysis of β-lactam antibiotics is octadecyl silane (ODS). A suitable mobile phase can be obtained by changing the pH of the buffer or the concentrations of organic solvents. For sulfazecin-type antibiotics, alkylammonium halides are added in the mobile phase to delay retention times by blocking sulfonic acid groups. Preparative HPLC can be classified into three typical methods. The conditions are described below. Model 64 (Knauer) and JASCO UVIDEC-100-II were used for a solvent supplier and a variable UV detector.

(1) TAN-850; Column: ODS (YMC-GEL SH-343, 10 micron, 20 mmID, 25 cm, Yamamura Chem. Lab.), Mobile phase: 23% Acetonitrile/0.05 M Tetrabutylammonium hydroxide/0.01 M phosphate buffer-pH 6.3 (pH 7.4), Detection: 220 nm, Flow rate: 10 ml/min.

Table 2 HPLC data of β-lactam antibiotics and lactivicin*.

(1) Sulfazecin-type

Mobile phase	Retention time (min)		
	Sulfazecin	Isosulfazecin	TAN-850
16% Acetonitrile/0.005M Tetrabutylammonium hydrogen sulfate/0.05M Ammonium sulfate (pH 3.0)	2.20	2.25	7.3
16% Acetonitrile/0.005M Tetrabutylammonium hydroxide/0.01M phosphate buffer (pH 7.4)	2.8	3.1	16.8

Equipment: Model 6000A/660/440 (Waters Assoc.)
Detection: UV absorbance at 214 nm

(2) SQ 28,332-type

Mobile phase	Retention time (min)	
	SQ 28,332	TAN-643
2% MeOH/0.01M phosphate buffer (pH 6.3)	8.5	4.3

Equipment: Liquid chromatograph 638-50 (Hitachi)
Detection: UV absorbance at 214 nm

(3) Formadicin-type

Mobile phase	Retention time (min)				
	Formadicin				TAN-667
	A	B	C	D	
10% MeOH/0.01M phosphoric acid (pH 3.0)	2.5	3.1	4.9	5.9	3.3

Equipment: Liquid chromatograph 638-50 (Hitachi)
Detection: UV absorbance at 220 nm

(4) Cephabacin M-type

Mobile phase	Retention time (min)					
	M_1	M_2	M_3	M_4	M_5	M_6
5% MeOH/0.01M phosphoric acid (pH 3.0)	5.1	5.4	6.7	–	–	–
10% MeOH/0.01 M phosphoric acid (pH 3.0)	2.2	2.2	2.4	5.2	6.4	7.8

Equipment: Model 6000A/660/440 (Waters Assoc.)
Detection: UV absorbance at 254 nm

* Conditions: Column; ODS, YMC-Pack A-312 (Yamamura Chem. Lab.), Flow rate: 2 ml/min.

(5) Lactivicin-type

Mobile phase	Retention time (min)			
	Lactivicin		Lactivicin methylester	
	A	B	A	B
4% MeOH/0.01M phosphate buffer (pH 6.3)	4.8	4.3		
20% MeOH/0.01M phosphate buffer (pH 6.3)			4.9	5.7

Equipment: Model 6000A/660/440 (Waters Assoc.)
Detection: UV absorbance at 214 nm

Table 3 Stabilities of β-lactam antibiotics and lactivicin in phosphate buffer (60 °C, 100 μg/ml).

Compound	Half-life (h)*					
	pH 3	pH 5	pH 6	pH 7	pH 8	pH 9
Epithienamycin B		0.43	1.1	2.2	4.8	
Epithienamycin A		0.5	1.8	5.1	8.5	
Clavulanic acid		0.7		1.4		0.4
Cephabacin F_1	1.4	0.73		0.87		0.5
Cephabacin H_1	0.85	0.85		0.85		0.58
C-19393 H_2		1.0	4.8	4.7	1.4	
C-19393 E_5		1.5	2.4	1.5	0.5	
Cephalosporin C	1.6	1.5		1.5		0.90
Lactivicin	0.39	1.63		0.45		0.35
Cephabacin M_1	1.8	1.8		1.5		0.37
Penicillin N		2.7		8.0		2.0
7-Formylamino-DCPC	1.6	19.7		19.7		3.6
DCPC	1.9	27.4		29.0		6.3
Formadicin A	4.5	>40		>40		3.0
SQ 28332	>8.0	> 8.0		> 8.0		1.5

* The samples prepared in our laboratories were used. The residual amounts were measured by HPLC. Under lines express stable pH ranges. DCPC: Deacetylcephalosporin C.
Underline indicates the stable pH.

(2) Formadicins A–D; Column: ODS (YMC-Pack S-363, 15 micron, 30 mmID, 25 cm, Yamamura Chem Lab.), Mobile phase: 3% MeOH/phosphate buffer (pH 3.0), Detection: 220 nm, Flow rate: 20 ml/min.
(3) Cephabacin H_{1-3}; Column: ODS (TSK-Gel LS-410KG, 15 micron, 21.5 mmID 30 cm, Toyo Soda), Mobile phase: 0.01 M phosphate buffer (pH 3.5), Detection: 260 nm, Flow rate: 20 ml/min.

The stabilities of β-lactam and β-lactam like antibiotics in aqueous solutions are shown in Table 3. These are written in the order of their stability at pH 5, which is the pH of the eluate from ion-exchange resins or Sephadex when the eluant is a dilute NaCl solution. Cephalosporin C, cephabacins M, penicillins, and lactivicin were relatively stable. Carbapenems, clavulanic acids and cephabacins F and H were more unstable, whereas sulfazecins, formadicins, and deacetylcephalosporins were more stable than cephalosporin C.

5. Chemical Characterization

Once a bioactive substance has been isolated and purified the UV and CD spectra are measured. Table 4 shows these data for β-lactam antibiotics isolated in our laboratories. By comparison with these data the chromophores and classification of newly isolated antibiotics can be rapidly elucidated. Cephalosporin or nocardicin type antibiotics have characteristic UV absorption maxima at 260 nm or 224 and at 268 – 269 nm, respectively. Cephabacins F and cephabacin H (cephalosporins) had similar UV and CD spectra to those of cephabacins M (cephamycins). One of carbapenem antibiotics, TAN-458, showed analogous spectral patterns with those

Table 4 UV, CD and IR (β-lactam) spectra in sodium salts of β-lactams and lactivicin.

Compound	UV: λ_{max}nm (ε)	CD: $[\theta]_{nm}$	IR: λ_{max}(cm^{-1})
Penicillin N	End abs.	232 (+18900)	1765
Clavulanic acid	End abs.	215 (−39800), 238 (+27900)	1765
Sulfazecin	End abs.	236 (+11000)	1770
Isosulfazecin	End abs.	228 (+11500)	1770
TAN-850	End abs.	236 (+14000)	1765
SQ 28332	End abs.	221 (−31000)	1765
TAN-643	End abs.	221 (−35900)	1765
TAN-458	253 (3850)	218 (−12100), 255 (+7510)	1750
Cephabacin H_1	260 (9000)	226 (−50700), 258 (+21000)	1765
Cephabacin M_1	264 (9000)	234 (−58800), 260 (+49100)	1780
Cephabacin F_1	260 (9520)	228 (−30900), 260 (+29500)	1775
Formadicin A	224 (24500), 269 (2000)	230 (−113000)	1760
Formadicin B	224 (23300), 268 (2100)	229 (−194000)	1740
TAN-667	222 (20700), 262 (8130)	230 (−105000)	1760
Lactivicin	216 (4050, sh)	212 (+18400), 238 (−22200)	1730[a]

* The samples were prepared in our laboratories and measured at 23 – 25 °C in water by Hitachi Spectrophotometer 320 (UV) and JASCO Automatic Recording Spectropolarimeter J-20A (CD). The IR spectra were measured in KBr.

[a] γ-lactam.

Table 5 ^{13}C NMR spectra of TAN-667 (100 MHz, in D_2O, JEOL GX-400).

Assignment	TAN-667	Formadicin A
C-10	177.67 s	178.09 s
C-11	177.52 s	177.59 s
Glu	177.45 s	177.60 s
C-21	176.76 s	176.88 s
C-2	166.92 s	166.96 s
NHCHO	166.73 d	166.75 d
C-16	161.01 s	161.01 s
C-9	159.44 s	159.47 s
C-13	134.02 s	133.95 s
C-6	132.63 s	132.43 s
C-7	132.51 d*	132.54 d*
C-14	131.31 d*	131.33 d*
C-8	119.94 d*	119.89 d*
C-15	117.89 d*	117.85 d*
Glu	103.08 d	103.11 d
Glu	79.00 d	78.99 d
Glu	78.31 d	78.17 d
C-12	75.94 d	75.95 d
Glu	75.46 d	75.59 d
C-3	74.55 s	74.53 s
Glu	73.33 d	74.53 d
C-18	68.14 t	68.07 t
C-5	63.56 d	63.57 d
C-20	56.20 d	56.22 d
C-4	56.11 t	56.05 t
C-19	32.66 t	32.64 t
CO	168.86 s	
-C=N-	154.54 s	
-CH=	144.45 d	
Pyr	106.18 d	
-CH=	105.42 d	
Pyr	103.38 d	
Fur	91.54 d	
–CHO–	86.02 d	
–CHO–	85.96 d	
–CHO–	84.02 d	
–CHO–	83.96 d	
–CHO–	79.22 d	
–CHO–	77.72 d	
–CHO–	76.45 d	
–CHO–	72.84 d	
–CHO–	72.49 d	
–CHO–	71.78 d	
–CH20-	67.69 t	
–CH20-	67.63 t	
–CH20-	63.71 t	
–CH20-	63.52 t	
–CH(NH–)	57.12 d	

* Duplicate.

of the cephalosporins as had been anticipated from the similarity of their chromophores. The antibiotics having no UV absorption maxima showed distinctive Cotton effects in their CD spectra. Clavulanic acid has two contrasting Cotton effects. Sulfazecins which have a methoxy group in the 3-position of the β-lactam carbonyl showed a positive Cotton effect similar to those of the penicillins, but non-methoxylated sulfazecins (SQ 28,332 or TAN-643) showed a negative Cotton effects at 221 nm. Thus, these UV and CD spectral patterns are can be used to identify the chromophores of β-lactam antibiotics. The UV and CD spectral data for lactivicin showed clearly different patterns to those for known β-lactam antibiotics. The absorptions maxima in the IR spectra that correspond to the β-lactam carbonyl groups are very important for identification of the β-lactam ring (Table 4). Absorption at around 1770 cm^{-1}, except for 1750 cm^{-1} in TAN-458, indicated slight shifts arising from differences in the substituents of the β-lactam rings. Lactivicin has an absorption maximum at 1780 cm^{-1} as is seen in known β-lactams, however, this was assigned to a γ-lactone carbonyl after structure determination. The absorption corresponding to a γ-lactam carbonyl was assigned to 1730 cm^{-1}.

The molecular formulae of new antibiotics are determined by elementary analysis and secondary ionization mass spectrometry (SI – MS). The parent ion peak of the molecule is usually detected as $(M + H)^+$ containing sodium atom when the antibiotics are isolated as sodium salts. In addition, the peaks $(M + Na)^+$ and $(M + 2\,Na)^+$ are also detected. On the samples examined in our laboratories, only TAN-667 did not show clear mass spectral patterns. ^{13}C NMR spectroscopy gives us important information about the carbon number of the molecule as shown in Table 5. The multiplicity of the signals measured by the OFF-RESONANCE or DEPT methods are useful to assign the functional group and to characterize the antibiotics. For example, the ^{13}C NMR spectral data for TAN-667 suggests that it consists of four sugar moieties and an amino acid in addition to a formadicin A moiety.

6. Structure Determination

Figure 4 shows the chemical structures of β-lactam and β-lactam like antibiotics isolated from bacteria. The structure of sulfazecin was eventually determined by X-ray crystallographic analysis [31] because it was difficult to elucidate the binding position of the sulfonyloxy group chemically. The absolute configuration of the methoxy group in isosulfazecin was determined to be *a* by the CD spectral analysis [32]. The most important technique in the structure determination studies of SQ 28,332 was the mass-analyzed ion kinetic energy (MIKE) method in FAB mass spectrometry for determination of the peptide chain sequence. The structures of TAN-850 and TAN-643 were elucidated by comparing their ^{13}C NMR spectral data with those of sulfazecin and SQ 28,332, as shown in Table 6.

The structures of cephabacins and formadicins were determined by comparison of data from degradation studies and spectroscopic analyses with those for cepha-

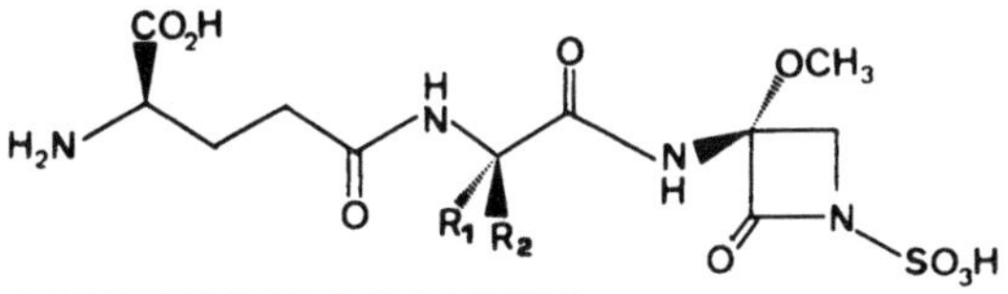

	R_1	R_2
Sulfazecin	H	CH_3
Isosulfazecin	CH_3	H

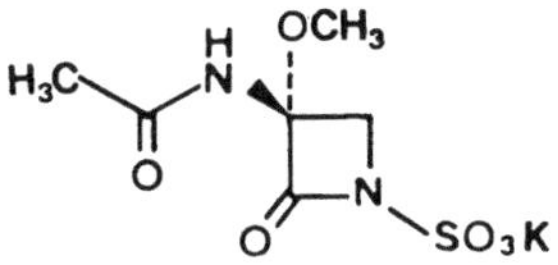

SQ 26,180

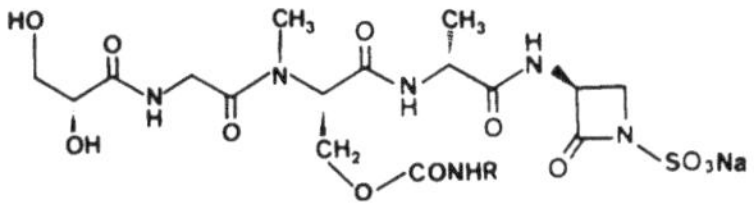

	R
SQ 28,332	–H
TAN-643	–CH–COONa | $CONH_2$

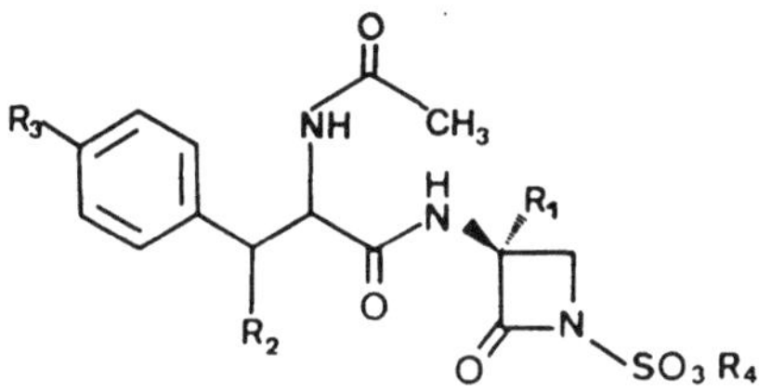

	R_1	R_2	R_3	R_4
SQ 26,823	OCH_3	H	H	Na
SQ 26,875	OCH_3	H	OH	K
SQ 26,700	H	H	OH	K
SQ 26,970	OCH_3	OH	OSO_3Na	Na
SQ 26,812	OCH_3	OSO_3Na	OSO_3Na	Na

Figure 4 Structures of β-lactam antibiotics and lactivicin isolated from bacteria.

(Figure 4, continued)

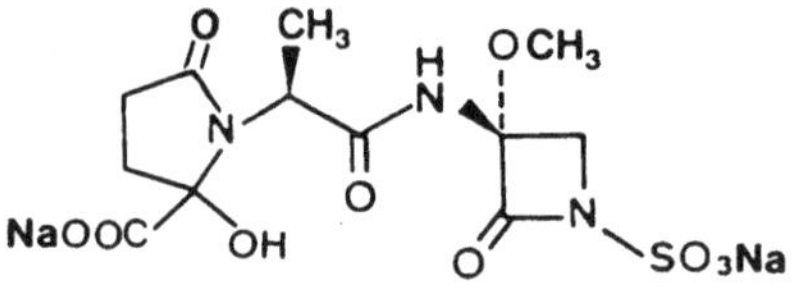

TAN-850

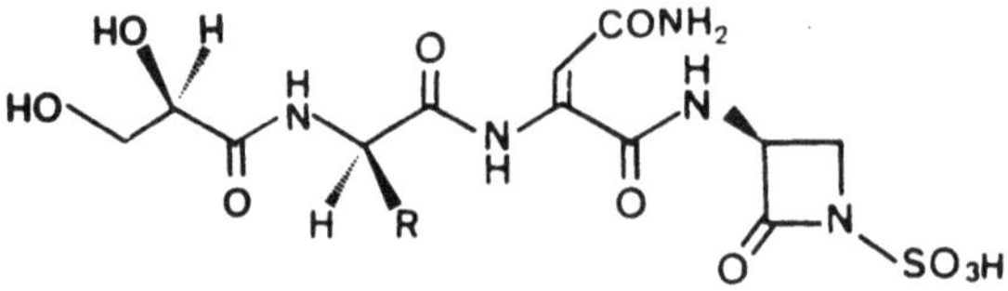

PB-5266	R
A	$-CH_3$
B	$-CH_2OH$
C	$-H$

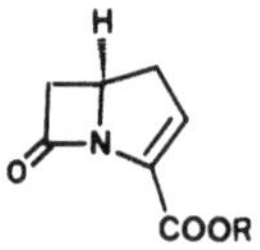

	R
SQ 27,860	PNB
TAN-458	Na

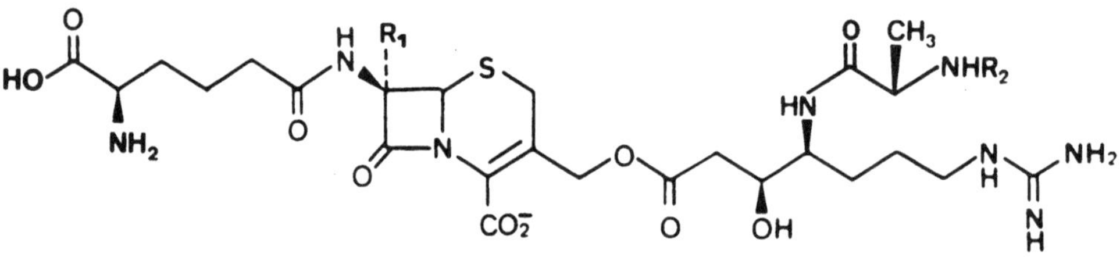

Cephabacin	R_1	R_2
F_1	$-NHC(=O)H$	$-H$
F_2	$-NHC(=O)H$	← L-Ala
F_3	$-NHC(=O)H$	← L-Ala ← L-Ala

Cephabacin	R_1	R_2
H_1	$-H$	$-H$
H_2	$-H$	← L-Ala
H_3	$-H$	← L-Ala ← L-Ala

(Figure 4, continued)

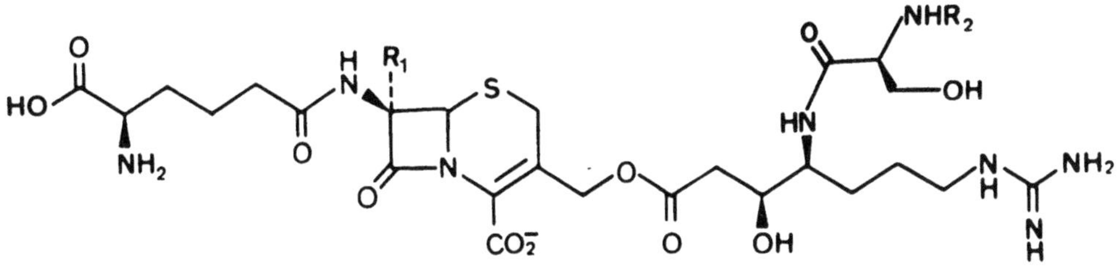

Cephabacin	R_1	R_2
F_4	$-NHC(=O)H$	$-H$
F_5	$-NHC(=O)H$	← L-Ser
F_6	$-NHC(=O)H$	← L-Ser ← L-Ser

Cephabacin	R_1	R_2
H_4	$-H$	$-H$
H_5	$-H$	← L-Ser
H_6	$-H$	← L-Ser ← L-Ala

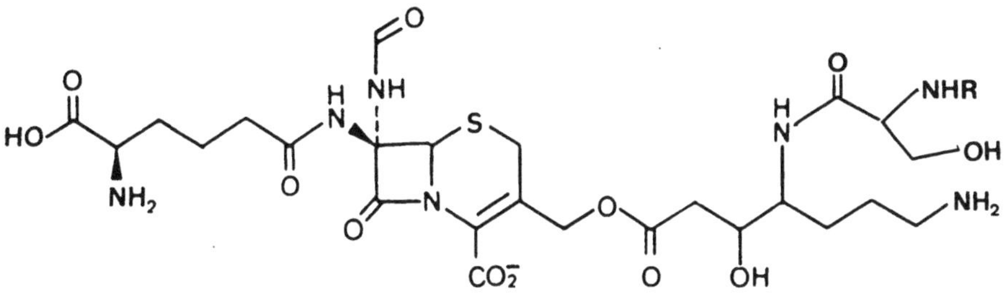

Cephabacin	R
F_7	$-H$
F_8	← L-Ser
F_9	← L-Ser ← L-Ala

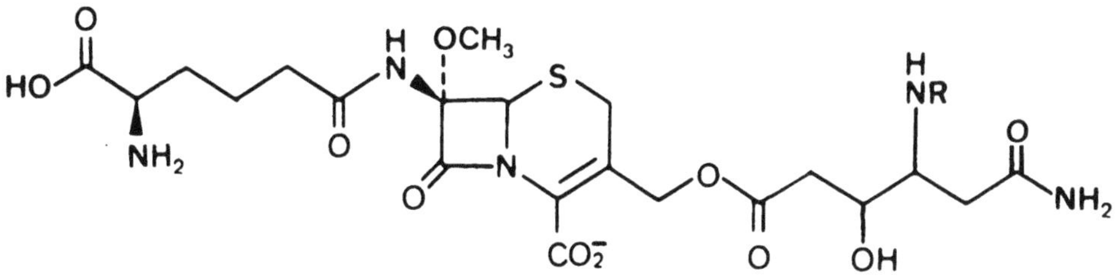

Cephabacin	R
M_1	← L-Val ← L-Orn
M_2	← L-Val ← L-Orn ← L-Ser
M_3	← L-Val ← L-Orn ← L-Ser ← L-Ala
M_4	← L-Val ← L-Orn ← L-Val ← L-Orn
M_5	← L-Val ← L-Orn ← L-Val ← L-Orn ← L-Ser
M_6	← L-Val ← L-Orn ← L-Val ← L-Orn ← L-Ser ← L-Ala

(Figure 4, continued)

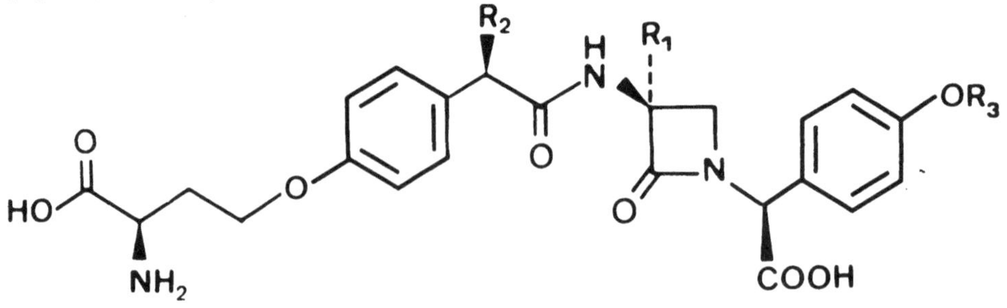

Formadicin	R_1	R_2	R_3
A	–NHCHO	–OH	D-Glucuronic acid
B	–H	–NHCHO	D-Glucuronic acid
C	–NHCHO	–OH	–H
D	–H	–NHCHO	–H
TAN-667	–NHCHO	–OH	Tetrasaccharides and an unknown amino acid

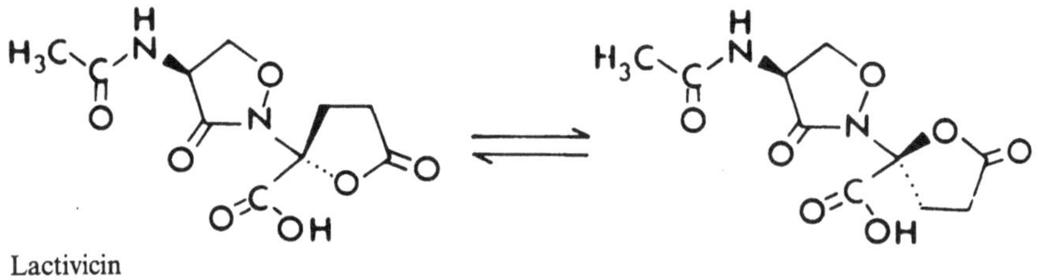

Lactivicin

mycins [33] and nocardicins [34]. On mild alkaline hydrolysis, cephabacins F, M, and H afford 7-formylamino deacetylcephalosporin C (DCPC), 7-methoxy DCPC, and DCPC, respectively, plus oligopeptides. After acidic hydrolysis of each oligopeptide, the known amino acids were identified by amino acid autoanalyzer and TLC, and their absolute configurations were determined by the modified HPLC method [35]. The structures of the three new amino acids were elucidated by ^1H and ^{13}C NMR spectroscopy, the former including proton spin-decoupling studies; degradation studies; and X-ray analysis. *N*-terminal amino acids and the amino acid sequences in cephabacins M were determined by the DNP method and Edman degradation, respectively. When formadicin A was hydrolyzed by 2 N hydrochloric acid, D-glucuronic acid and two aromatic amino acids were obtained. Glucuronidase digestion of formadicin A gave D-glucuronic acid plus formadicin C and enzymatic degradation using *T. variabilis* [36] gave a dehomoseryl compound. All signals in the ^{13}C NMR spectrum of formadicin A were assigned on the basis of the spectral data obtained from these degradation products.

The absolute configuration of the formylamino group in cephabacins F was determined to be *a* by comparing the CD spectral data (Table 4) with those of other cephalosporins [37], but could not be determined by this method for the formadicins

Table 6 ^{13}C NMR spectra of TAN-850 and TAN-643 (100 MHz, in D_2O, JEOL GX-400).

Carbon	TAN-850	Sulfazecin*	Carbon	TAN-643	SQ 28,332
CO	180.16 s	178.76 s	CO	177.90 s	177.80 s
CO	180.09 s	176.82 s	CO	176.65 s	177.64 s
CO	176.71 s	174.37 s	CO	174.37 s	174.38 s
2-CO	165.24 s	164.75 s	CO	173.89 s	
C-12	96.06 s	52.73 d	CO	172.65 s	172.60 s
C-3	93.79 s	93.63 s	C-9	172.65 s	
C-4	57.15 t	57.10 t	C-2	168.96 s	168.85 s
C-7	55.84 d	55.06 d	C-20	161.43 s	161.36 s
OCH_3	54.40 q	55.67 q	C-16	75.15 d	75.18 d
C-10	35.65 t	33.43 t	C-17	66.10 t	66.17 t
C-11	31.80 t	28.21 t	C-19	64.07 t	64.12 t
7-CH_3	16.54 q	18.94 q	C-22	61.74 d	
			C-10	60.38 d	60.65 d
			C-3	58.03 d	57.94 d
			C-7	52.43 d	52.40 d
			C-4	50.51 t	50.56 t
			C-13	44.78 t	43.99 t
			C-25	35.35 q	35.41 q
			C-18	19.06 q	19.10 q

* 75 MHz, Bruker AC-300.

because of interference by the other chromophores. Therefore, we applied the NOESY method in the ^{1}H NMR spectrum. The signals at 3.99 ppm (d, $J = 6.5$ Hz) and 3.52 (d, $J = 6.5$ Hz) in the ^{1}H NMR spectrum in formadicin C were assigned to $H_{4\alpha}$ and $H_{4\beta}$, respectively, from the chemical shifts at 3.73 ppm (t, $J = 5.7$ Hz, $H_{4\beta}$) and 3.05 (dd, $J = 5.7$, 2.5 Hz, H_4) in that of formadicin D. The obvious NOE was observed between the proton signal of the formylamino group and the a proton signal at the 4-position. Thus CD spectroscopy and NOE experiment in the ^{1}H NMR spectroscopy are available to determine the absolute configuration of the substituent at the α-position of β-lactam carbonyl.

Lactivicin exists in aqueous solutions as an equilibrium mixture. Their structures are determined by acidic and basic degradation and spectral analyses to be an isolated dicyclic dipeptide forming epimers at the binding position of γ-lactone and γ-lactam. Various data on its stereochemistry obtained by X-ray analysis [38] was useful to account for its unique biological activities regarding β-lactamases and penicillin-binding proteins.

7. Biological Activities of Cephabacin, Formadicin and Lactivicin

The antibacterial activities of cephabacins F_{1-9}, M_{1-6}, and H_{1-6} compared with those of related cephalosporins are shown in Tables 7, 8, and 9. Cephabacins F_{1-9} possess inhibitory activity against a broad range of Gram-positive and negative

Table 7 Antibacterial activity of cephabacin F and related compound.

Organism	MIC (μg/ml) at 10^6 cfu/ml									
	Cephabacin									7-FA-DCPC
	F_1	F_2	F_3	F_4	F_5	F_6	F_7	F_8	F_9	
Escherichia coli NIHJ JC-2	12.5	25	25	25	25	50	25	25	50	100
Salmonella typhimurium IFO 12529	12.5	25	25	25	25	50	25	25	50	50
Citrobacter freundii IFO 1268	25	50	100	50	50	100	50	50	100	100
Klebsiella pneumoniae IFO 3317	25	50	50	50	50	50	50	50	50	50
Enterobacter cloacae IFO 12937	25	25	100	25	50	100	50	50	100	>100
Serratia marcescens IFO 12648	25	50	50	25	25	50	25	50	50	50
Proteus mirabilis ATCC 21100	12.5	50	25	25	25	25	12.5	12.5	50	25
P. vulgaris IFO 3988	12.5	50	25	25	25	25	25	25	50	25
P. morganii IFO 3168	25	25	50	25	25	50	50	50	50	100
Pseudomonas aeruginosa IFO 3080	25	100	>100	50	100	>100	100	>100	>100	>100
Alcaligenes faecalis IFO 13111	1.56	6.25	12.5	12.5	6.25	12.5	6.25	12.5	50	25
Acinetobacter calcoaceticus IFO12552	0.78	1.56	3.13	1.56	3.13	3.13	1.56	3.13	3.13	12.5
Staphylococcus aureus FDA 209P	50	>100	>100	>100	>100	>100	>100	>100	>100	>100
Bacillus subtilis PCI 219	3.13	12.5	12.5	6.25	12.5	12.5	12.5	25	25	>100

7-FA-DCPC: 7-formylaminodeacetylcephalosporin C.
Data from reference 65.

Table 8 Antibacterial activity of cephabacin M and related compounds.

Organism	MIC (μg/ml) at 10^6 cfu/ml						
	Cephabacin						7-M-DCPC
	M_1	M_2	M_3	M_4	M_5	M_6	
Escherichia coli NIHJ JC-2	12.5	50	100	100	>100	>100	>100
Salmonella typhimurium IFO 12529	6.25	12.5	12.5	12.5	12.5	25	100
Citrobacter freundii IFO 12681	>100	>100	>100	>100	>100	>100	>100
Klebsiella pneumoniae IFO 3317	12.5	12.5	12.5	12.5	12.5	12.5	>100
Enterobacter cloacae IFO 12937	>100	>100	>100	>100	>100	>100	>100
Serratia marcescens IFO 12648	>100	>100	>100	>100	>100	>100	>100
Proteus mirabilis ATCC 21100	6.25	6.25	6.25	25	25	50	25
P. vulgaris IFO 3988	12.5	25	25	100	100	>100	50
P. morganii IFO 3168	12.5	25	12.5	25	25	25	100
Pseudomonas aeruginosa IFO 3080	>100	>100	>100	>100	>100	>100	>100
Alcaligenes faecalis IFO 13111	3.13	6.25	3.13	12.5	12.5	12.5	12.5
Acinetobacter calcoaceticus IFO 12552	50	100	100	50	100	100	>100
Staphylococcus aureus FDA 209P	50	50	50	25	25	25	>100
Bacillus subtilis PCI 219	3.13	6.25	6.25	3.13	3.13	3.13	>100

7-M-DCPC: 7-Methoxydeacetylcephalosporin C.
Data from reference 40.

Table 9 Antibacterial activity of cephabacin H and related compound.

Organism	MIC (μg/ml) at 10^6 cfu/ml						
	Cephabacin						DCPC
	H_1	H_2	H_3	H_4	H_5	H_6	
Escherichia coli NIHJ JC-2	50	50	>100	50	50	>50	100
Salmonella typhimurium IFO 12529	50	50	100	50	100	>50	50
Citrobacter freundii IFO 12681	>100	>100	>100	>100	>100	>50	>100
Klebsiella pneumoniae IFO 3317	100	100	100	100	100	>50	100
Enterobacter cloacae IFO 12937	>100	>100	>100	>100	>100	>50	>100
Serratia marcescens IFO 12648	>100	>100	>100	>100	>100	>50	>100
Proteus mirabilis ATCC 21100	25	25	25	25	25	25	12.5
P. vulgaris IFO 3988	50	50	100	>100	>100	50	12.5
P. morganii IFO 3168	>100	>100	>100	>100	>100	>50	>100
Pseudomonas aeruginosa IFO 3080	>100	>100	>100	>100	>100	>50	100
Alcaligenes faecalis IFO 13111	12.5	12.5	12.5	6.25	6.25	12.5	12.5
Acinetobacter calcoaceticus IFO 12552	50	100	100	25	100	>50	>100
Staphylococcus aureus FDA 209P	25	25	12.5	25	50	50	50
Bacillus subtilis PCI 219	0.78	1.56	1.56	1.56	1.56	1.56	3.13

DCPC: deacetylcephalosporin C.
Data from reference 65.

bacteria. Among these compounds, cephabacin F_1 is the most potent. Cephabacins M_{1-6} are moderately active against Gram-positive and negative bacteria. Among cephabacins M, cephabacin M_1 shows the most potent activity. Cephabacins H_{1-6} are active against Gram-positive bacteria but are less active against Gram-negative bacteria; cephabacin H_1 is most potent. The antibacterial activity of cephabacins M is intermediate between those of cephabacins F and cephabacins H. The C-3 side chains of cephabacins F, M, and H contribute to their antibacterial activities as cephabacins F, M, and H are more active than corresponding compounds without C-3 side chains, namely, 7-formylamino-DCPC, 7-methoxy-DCPC and DCPC, respectively. Introduction of the 7-formylamino group increased the activity against Gram-negative bacteria and decreased the activity against Gram-positive bacteria [65].

Cephabacins F are highly resistant to hydrolysis by various types of β-lactamases. Cephabacins M are moderately hydrolyzed by *Enterobacter cloacae* cephalosporinase and cephabacins H are susceptible to several kinds of cephalosporinase (Table 10). The stability of cephabacins F, M, and H against β-lactamase seems to be determined mainly by the C-7 substituent as DCPC with a 7-formylamino or 7-methoxy substituent group has very high or moderate stability, respectively.

Cephabacins F, M, and H show weak β-lactamase inhibitory activity in a test system using cephalosporinase of *Proteus vulgaris* GN 4413 [39–40].

Cephabacins F_1, M_1, and H_1 have high affinity for penicillin binding proteins (PBPs). As shown in Table 11, among the lethal targets in *E. coli* (PBPs 1, 2, and 3) cephabacin F_1, M_1, and H_1 show the highest affinity for PBPs 1 A and 1 B. In *B. subtilis*, cephabacins F_1 and M_1 show the high affinity for PBP 4 among those thought to be the lethal targets (PBP 1, 2, and 4) in this organism. Cephabacin M_1 also shows high affinity to PBP 4, 5, and 6 of *E. coli* and PBP 5 of *B. subtilis*, as commonly observed in 7-methoxy cephalosporin antibiotics [41]. The affinity to PBP 1 of cephabacin F_1, M_1, and H_1 in *E. coli* is consistent with their ability to induce morphological changes (spheroplasting) in this test organism [39–40].

Formadicins A, B, C, and D show antibacterial activity against several Gram-negative bacteria (Table 12). The activity of formadicins C and D is more potent than that of formadicins A and B. The D-glucuronide moiety decreases antibacterial activity. Formadicins A and C having the 3-formylamino substituents are more active than formadicins B and D which have the same substituent at the C-12 position [41]. The antibacterial spectra of the formadicins are similar to that of nocardicin A [42]. Formadicins are very stable to hydrolysis by β-lactamases (Table 13) and formadicins A and C are more stable than formadicins B and D, and nocardicin A. The 3-formylamino group directly bound to the β-lactam ring gives extreme stability against β-lactamases, as observed in cephabacins F. Formadicins show no inhibitory activity against β-lactamases [42]. Formadicins A and C show high affinity to PBP 1 B and 1 A and formadicin B and nocardicin A show high affinity to PBP 1 B in *P. aeruginosa* (Table 11).

Lactivicin is especially active against Gram-positive bacteria, but is effective against Gram-positive and negative bacteria. However, it has no activity against fungi and

Table 10 Stability of cephabacins and related compounds to hydrolysis by β-lactamases.

Source of enzyme	Cephabacin			7-FA-DCPC	Cephabacin M_1	7-M-DCPC	Cephabacin			DCPC
	F_1	F_2	F_3				H_1	H_2	H_3	
Penicillinase										
S. aureus 1840	<0.01	<0.01	<0.01	<0.01	<0.01	<0.01	0.05	0.04	0.04	0.02
E. coli TN713 (TEM-1)	<0.01	<0.01	<0.01	<0.01	<0.01	<0.01	0.41	0.50	0.55	0.12
K. oxytoca TN1719	<0.01	<0.01	<0.01	<0.01	–	–	0.84	0.74	0.72	0.51
Cephalosporinase										
E. cloacae TN1282	<0.01	<0.01	<0.01	0.02	20.7	74.1	43.1	39.1	37.3	94.1
P. aeruginosa U31	<0.01	<0.01	<0.01	0.02	–	–	29.3	22.0	20.0	132
P. vulgaris GN4413	<0.01	<0.01	<0.01	<0.01	<0.01	<0.01	21.9	14.4	11.7	24.9

The rate of hydrolysis by β-lactamases was determined spectrophotometrically or microbiologically. Relative rate of hydrolysis is expressed by taking the rate for benzylpenicillin (penicillinase), or cephaloridine (cephalosporinase) as 100.
(–: not tested).

Mycoplasma. It is more active against β-lactam hypersensitive mutants of *P. aeruginosa* and *E. coli* than their parent strains (Table 14). Lactivicin inhibits the selective incorporation of radioactive diaminopimelic acid into the bacterial peptidoglycan. It is susceptible to several β-lactamases, especially cephalosporinases from *E. cloacae* and *P. aeruginosa*. Lactivicin induces β-lactamase from *B. liqueni-*

Table 11 Penicillin binding protein binding activity of cephabacins, formadicins, lactivicin and related compounds.

Antibiotic	*E. coli* LD-2	*P. aeruginosa* IFO 3080	*B. subtilis* PCI 219
Cephabacin F_1	1B, 1A	–	4
7-FA-DCPC	1B, 1A	–	–
Cephabacin M_1	1B, 1A, 4, 5, 6	–	4, 5
7-M-DCPC	(1B, 1A), 4, 5, 6	–	–
Cephabacin H_1	1B, 1A	–	–
DCPC	1B, 1A	–	–
Formadicin A	–	1B, 1A	–
Formadicin B	–	1B	–
Formadicin C	–	1B, 1A	–
Nocardicin A	–	1B	–
Lactivicin	1A, 1B	–	1, 2, 4

–: not tested.

Table 12 Antibacterial activity of formadicins and nocardicin A.

Organism	MIC (μg/ml) at 10^6 cfu/ml				
	Formadicin				Nocardicin A
	A	B	C	D	
Escherichia coli NIHJ JC-2	>100	>100	>100	>100	>100
Salmonella typhimurium IFO 12529	>100	>100	100	100	>100
Citrobacter freundii IFO 12681	>100	>100	>100	>100	>100
Klebsiella pneumoniae IFO 3317	>100	>100	>100	>100	>100
Enterobacter cloacae IFO 12937	>100	>100	>100	>100	>100
Serratia marcescens IFO 12648	>100	>100	>100	>100	>100
Proteus mirabilis ATCC 21100	50	100	50	50	12.5
P. vulgaris IFO 3988	25	50	25	50	3.13
P. morganii IFO 3168	>100	>100	>100	>100	>100
Pseudomonas aeruginosa IFO 3080	100	>100	25	>100	50
Alcaligenes faecalis IFO 13111	12.5	25	6.25	12.5	1.56
Acinetobacter calcoaceticus IFO 12552	100	>100	25	>100	100
Staphylococcus aureus FDA 209P	>100	>100	>100	>100	100
Bacillus subtilis PCI 219	>100	>100	50	50	50

Data from reference 42.

formis and it also weakly inhibits β-lactamase. It shows high affinity to PBP 1 A and 1 B of *E. coli* and PBPs 1, 2 and 4 of *B. subtilis* [43].

Lactivicin is the first compound without a β-lactam ring in its structure that shows biological activities similar to those of β-lactam antibiotics [69–70].

Table 13 Stability of formadicins and related compound to hydrolysis by β-lactamases.

Source of enzyme	Relative rate of hydrolysis				
	Formadicin				Nocardicin A
	A	B	C	D	
Penicillinase					
S. aureus 1840	<0.01	0.33	<0.01	0.69	<0.01
E. coli TN713 (TEM-1)	<0.01	1.11	<0.01	1.56	0.11
K. oxytoca TN1719	<0.01	0.28	<0.01	0.41	3.52
Cephalosporinase					
E. cloacae TN1282	<0.01	0.01	<0.01	0.08	0.01
P. aeruginosa U31	<0.01	0.65	<0.01	1.45	0.01
P. vulgaris GN4413	<0.01	6.33	<0.01	6.73	6.55

Data from reference 42.

Table 14 Antibacterial activity of lactivicin.

Organism	MIC (μg/ml) at 10 cfu/ml
Escherichia coli NIHJ JC-2	50
E. coli LD-2	50
E. coli PG-8*	0.78
Salmonella typhimurium IFO 12529	50
Citrobacter freundii IFO 12681	100
Klebsiella pneumoniae IFO 3317	100
Enterobacter cloacae IFO 12648	>100
Serratia marcescens IFO 12648	50
Proteus mirabilis ATCC 21100	25
P. vulgaris IFO 3988	25
P. morganii IFO 3168	100
Pseudomonas aeruginosa IFO 3080	>100
P. aeruginosa C-141*	0.2
Alcaligenes faecalis IFO 13111	50
Acinetobacter calcoaceticus IFO 12552	50
Staphylococcus aureus FDA 209P	3.13
Micrococcus luteus IFO 12708	0.39
Bacillus subtilis PCI 219	3.13
Bacteroides fragilis ATCC 2509	12.5
Fusobacterium necrophorum VPI 2891	0.78
Clostridium perfringens PB6K	6.25
Peptostreptococcus anaerobius B-40	3.13

* β-lactam hypersensitive mutant.
Data from reference 43.

8. Conclusion

In the past 50 years since the discovery of penicillin, several key β-lactam antibiotics have appeared, penicillin and cephalosporin from fungi (1925–45), cephamycin, clavulanic acid, carbapenem, and nocardicin from actinomycetes (1972–76), and sulfazecin from bacteria (1981) each having great impact on and giving direction to antibiotic researchers. Through screening targeting at β-lactam antibiotics produced by bacteria, formylamino-cephem (cephabacins F, formadicins A and C) and methoxy-cephem as well as cephem, formylamino-monocyclic β-lactam, and β-lactam like antibiotic (lactivicin) have been found in addition to monobactam. The research and development of synthetic analogues of these compounds is in progress. The discovery of N-sulfonated monocyclic β-lactams already led the launch of aztreonam [44] and carumonam [45]. The formylamino-substituted cephems are in clinical trials [46–47]. Improvement of the biological activity by chemical modification is also in progress for lactivicin, a new group of antibiotics which has high affinity to PBP [48].

The discovery of naturally-occurring compounds sometimes give us unique, unknown structures. As we have indicated in this chapter, if the screening system possesses both high specificity and sensitivity for the target compounds, and isolation and purification is carried out carefully, fruitful results can be expected. We believe that the more precise the request, the clearer the reply.

References

1. Morin, R. B., and Gorman, M., *Chemistry and Biology of β-Lactam Antibiotics*, 1–3, Academic Press, New York, 1982.
2. Kitano, K., Kintaka, K., Suzuki, S., Katamoto, K., Nara, K., and Nakao, Y., Screening of microorganisms capable of producing β-lactam antibiotics, *J. Ferment. Technol. 53*, 327, 1975.
3. Imada, A., Recent progress in β-lactam antibiotics, *J. Takeda Res. Lab. 41*, 194, 1982.
4. Imada, A., Kitano, K., Kintaka, K., Muroi, M., and Asai, M., Sulfazecin, and isosulfazecin, novel β-lactam antibiotics of bacterial origin, *Nature 289*, 590, 1981.
5. Sykes, R. B., and Wells, J. S., Screening for β-lactam antibiotics in nature, *J. Antibiotics 38*, 119, 1985.
6. Sykes, R. B., Cimarusti, C. M., Bonner, D. P., Bush, K., Floyd, D. M., Georgopapadakou, N. H., Koster, W. H., Liu, W. C., Parker, W. L., Principe, P. A., Rathnum, M. L., Slusarchyk, W. A., Trejo, W. H., and Wells, J. S., Monocyclic β-lactam antibiotics produced by bacteria, *Nature 291*, 489, 1981.
7. Harada, S., Screening of antibacterial antibiotics, *J. Takeda Res. Lab. 45*, 99, 1986.
8. Harada, S., in G. H. Wagman and R. Cooper, Eds., *natural products isolation*, J. Chromat. Lib. Vol. 43, New cephalosporins, Elsevier, Amsterdam, 1989.
9. Tsubotani, S., Hida, T., Kasahara, F., Wada, Y., and Harada, S., Cephabacins, new cephem antibiotics of bacterial origin. III. Structural determination, *J. Antibiotics 37*, 1546, 1984.
10. Hida, T., Tsubotani, S., Katayama, N., Okazaki, H., and Harada, S., Formadicins, new monocyclic β-lactam antibiotics of bacterial origin. II. Isolation, characterization and structures, *J. Antibiotics 38*, 1128, 1985.
11. Harada, S., Tsubotani, S., Hida, T., Ono, H., and Okazaki, H., Structure of lactivicin, an antibiotic having a new nucleus and similar biological activities to β-lactam antibiotics, *Tetrahedron Letters 27*, 6229, 1986.

12. Aoki, H., Kunugita, H., Hosoda, J., and Imanaka, H., Screening of new and novel β-lactam antibiotics, *Jap. J. Antibiotics 30*, Suppl., S 207, 1977.
13. Okamura, K., Koki, A., Sakamoto, M., Kubo, K., Mutoh, Y., Fukagawa, Y., Kouno, K., Shimauchi, Y., Ishikawa, T., and Lein, J., Microorganisms producing a new β-lactam antibiotic, *J. Ferment. Technol. 57*, 265, 1979.
14. Stapley, E. O., Jackson, M., Hernandez, S., Zimmerman, S. B., Currie, S. A., Mochales, S., Mata, J. M., Woodruff, H. B., and Hendlin, D., Cephamycins, a new family of new β-lactam antibiotics, *Antimicrob. Agents Chemother. 2*, 122, 1972.
15. Inouye, S., Kojima, M., Shomura, T., Iwamatsu, K., Niwa, T., Kondo, Y., Niida, T., Ogawa, Y., and Kusama, K., Discovery, isolation and structure of novel cephamycins of *Streptomyces chartreusis, J. Antibiotics 36*, 115, 1983.
16. Umezawa, H., Mitsuhashi, S., Hamada, M., Iyobe, S., Takahashi, S., Utahara, R., Osato, Y., Yamazaki, S., Ogawara, H., and Maeda, K., Two β-lactamase inhibitors produced by a *Streptomyces, J. Antibiotics 26*, 51, 1973.
17. Brown, A. G., Butterworth, D., Cole, M., Hanscomb, G., Hood, J. D., and Reading, C., Naturally-occurring β-lactamase inhibitors with antibacterial activity, *J. Antibiotics 29*, 668, 1976.
18. Iwami, M., Komori, T., Kohsaka, M., and Imanaka, H., FR-900318, a novel penicillin with β-lactamase inhibitory activity, *J. Antibiotics 36*, 1774, 1983.
19. Butterworth, D., Cole, M., Hanscomb, G., and Rolinson, G. N., Olivanic acids, a family of β-lactam antibiotics with β-lactamase inhibitory properties produced by *Streptomyces* species, *J. Antibiotics 32*, 287, 1979.
20. Uri, J. V., Actor, P., and Weisbach, J. A., Rapid and simple method for detection of β-lactamase inhibitors, *J. Antibiotics 31*, 789, 1978.
21. Spratt, B. G., Distinct penicillin binding proteins involved in the division, elongation, and shape of *Escherichia coli* K12, *Proc. Natl. Acad. Sci. USA 72*, 2999, 1975.
22. Nagarajan, R., Boeck, L. D., Gorman, M., Hamill, R. L., Higgens, C. E., Hoehn, M. M., Stark, W. M., and Whitney, J. G., β-lactam antibiotics from *Streptomyces, J. Am. Chem. Soc. 93*, 2308, 1971.
23. Inukai, M., Enokita, R., Torikita, A., Nakahara, M., Iwado, S., and Arai, M., Globomycin, a new peptide antibiotic with spheroplast-forming activity. I. Taxonomy of producing organisms and fermentation, *J. Antibiotics 31*, 410, 1978.
24. Gilpin, R. W., Young, F. E., and Chatterjee, A. N., Characterization of a stable L-form of *Bacillus subtilis* 168, *J. Bacteriol. 113*, 486, 1973.
25. Omura, S., Tanaka, H., Oiwa, R., Nagai, T., Koyama, Y., and Takahashi, Y., Studies of bacterial cell wall inhibitors IV. Screening method for the specific inhibitors of peptideglycan synthesis, *J. Antibiotics 32*, 978, 1979.
26. Wanning, M., Zaehner, H., Krone, B., and Zeeck, A., Ein neues antifungisches β-Lactam-Antibioticum der Clavam-Reihe, *Tetrahedron Lett. 22*, 2539, 1981.
27. Fleming, I. D., Nisbet, L. J., and Brewer, S. J., Target directed antimicrobial screens in *Bioactive Microbial Products: Search and Discovery*, Bu'Lock, J., Nisbet, L., and Winstanley, D., Eds., Academic Press, 1982, p. 107.
28. Leyh-Bouille, M., Nguyen-Dische, M., Pirlot, S., Veithen, A., Bourguignon, C., and Ghuysen, J., *Streptomyces* K15 DD-peptidase-catalysed reactions with suicide β-lactam carbonyl donors, *Biochem. J. 235*, 177, 1986.
29. Schindler, P. W., and Koenig, W., Chatterjee, S., and Ganguli, B. N., Improved screening for β-lactam antibiotics. A sensitive, high-throughput assay using DD-carboxypeptidase and a novel chromophore-labeled substrate, *J. Antibiotics 39*, 53, 1986.
30. Waji, M., Ishino, N., Ito, Y., Ueda, K., Matsuhashi, M., and Strominger, J. L., Crystalization of high molecular weight PBP, a target enzyme of β-lactam antibiotics in *Abst. Ann. Meeting Jap. Agr. Chem. Soc.*, 1987, 369.
31. Kamiya, K., Takamoto, M., Wada, Y., and Asai, M., Structure of sulfazecin-Methanol (1/1), *Acta Cryst. B37*, 1626, 1981.
32. Shinagawa, S., Kasahara, F., Wada, Y., Harada, S., and Asai, M., Structures of bulgecins, bacterial metabolites with bulge-inducing activity, *Tetrahedron 40*, 3465, 1984.
33. Nagarajan, R., Boeck, L. D., Gorman, M., Hamill, R. L., Higgens, C. E., Hoehn, M. M., Stark, W. M., and Whitney, J. G., β-lactam antibiotics from *Streptomyces. J. Am. Chem. Soc. 93*, 2308, 1971.

34. Hashimoto, M., Komori, T., and Kamiya, T., Nocardicin A and B, novel monocyclic β-lactam antibiotics from a *Nocardia* species, *J. Am. Chem. Soc. 98*, 3023, 1976.
35. Weinstein, S., Engel, M. H., and Hare, P. E., The enantiomeric analysis of a mixture of all common protein amino acids by high-performance liquid chromatography using a new chiral mobile phase, *Anal. Biochem. 121*, 370, 1982.
36. Serizawa, N., Nakagawa, K., Haneishi, T., Kamimura, S., and Naito, A., Enzymatic conversion of cephamycin C by D-amino acid oxidase from *Trigonopsis variabilis*. *J. Antibiotics 33*, 585, 1980.
37. Nagarajan, R., and Spry, D. O., The 3-cephem chromophore, *J. Am. Chem. Soc. 93*, 2310, 1971.
38. Wada, Y., Takamoto, M., Tsubotani, S., and Kamiya, K., Structure of 4-amino-lactivinic acid, *Acta Cryst. C43*, 1786, 1987.
39. Ono, H., Nozaki, Y., Katayama, N., and Okazaki, H., Cephabacins, new cephem antibiotics of bacterial origin. I. Discovery and taxonomy of the producing organisms and fermentation, *J. Antibiotics 37*, 1528, 1984.
40. Nozaki, Y., Katayama, N., Tsubotani, S., Ono, H., and Okazaki, H., Cephabacin M_{1-6}, new 7-methoxy cephem antibiotics of bacterial origin I. A producing organism, fermentation, biological activities and mode of action, *J. Antibiotics 38*, 1141, 1985.
41. Curtis, N. A. C., Ross, G. W., and Boulton, M., Effect of 7-*a* methoxy substitution of cephalosporins upon their affinity for the penicillin-binding proteins of *E. coli* K12, *J. Antimicrob. Chemother. 5*, 391, 1979.
42. Katayama, N., Nozaki, Y., Okonogi, K., Ono, H., Harada, S., and Okazaki, H., Formadicins, new monocyclic β-lactam antibiotics of bacterial origin I. A producing organism, fermentation and biological activities, *J. Antibiotics 38*, 1117, 1985.
43. Nozaki, Y., Katayama, N., Ono, H., Tsubotani, S., Harada, S., Okazaki, H., and Nakao, Y., Binding of a non-β-lactam antibiotic to penicillin-binding proteins, *Nature 325*, 179, 1987.
44. Cimarusti, C. M., Bonner, D. P., Breur, H., Chang, H. W., Fritz, A. W., Floyd, D. M., Kissick, T. P., Koster, W. H., Kronenthal, D., Massa, F., Mueller, R. H., Pluscec, J., Slusarchyk, W. A., Sykes, R. B., Taylor, M., and Weaver, E. R., 4-Alkylated monobactams, Chiral synthesis and antibacterial activity, *Tetrahedron 39*, 2577, 1983.
45. Sendai, M., Hashiguchi, S., Tomimoto, M., Kishimoto, S., Matsuo, T., Kondo, M., and Ochiai, M., Chemical modification of sulfazecin. Synthesis of 4-(substituted methyl)-2-azetidinone-1-sulfonic acid derivatives, *J. Antibiotics 38*, 346, 1985.
46. Milner, P. H., Guest, A. W., Harrington, F. P., Ponsford, R. J., Smale, T. C., and Stachulski, A. V., 6(7)-Formamido penicillins and cephalosporins. *J. Chem. Soc., Chem. Commun.*, 1335, 1984.
47. Murakami, K., Doi, M., and Yoshida, T., 7*a*-Formylamino substituent confers β-lactamase-inactivating potency on 1-oxacephalosporins, *Antimicrob. Agents Chemother. 30*, 447, 1986.
48. Natsugari, H., Kawano, Y., Morimoto, A., Yoshioka, K., and Ochiai, M., Synthesis of lactivicin and its derivatives, *J. Chem. Soc. Chem. Commun.*, 62, 1987.
49. Asai, M., Haibara, K., Muroi, M., Kintaka, K., and Kishi, T., Sulfazecin, a novel β-lactam antibiotic of bacterial origin; Isolation and chemical characterization, *J. Antibiotics 34*, 621, 1981.
50. Kintaka, K., Haibara, K., Asai, M., and Imada, A., Isosulfazecin, a new β-lactam antibiotic, produced by an acidophilic pseudomonad; Fermentation, isolation and characterization, *J. Antibiotics 34*, 1081, 1982.
51. Parker, W. L., Koster, W. H., Cimarusti, C. M., Floyd, D. M., Liu, W., and Rathnum, M. L., SQ 26, 180, a novel monobactam. II Isolation, structure determination and synthesis, *J. Antibiotics 35*, 189, 1982.
52. Parker, W. L., and Rathnum, M. L., EM 5400, a family of monobactam antibiotics produced by *Agrobacterium radiobacter* II. Isolation and structure determination, *J. Antibiotics 35*, 300, 1982.
53. Singh, P. D., Johnson, J. H., Ward, P. C., Wells, J. S., Trejo, W. H., and Sykes, R. B., SQ 28, 332, a new monobactam produced by a *Flexibacter* sp. Taxonomy, fermentation, structure determination and biological properties, *J. Antibiotics 36*, 1245, 1983.
54. Tanaka, K., Kondo, E., Matsumoto, K., Hattori, M., and Shoji, M., β-lactam antibiotic PB-5582-A and its production, *Japanese Patent Application*, S 59 – 151 895 (Feb. 17, 1983).

55. Harada, S., Nozaki, Y., and Ono, H., Antibiotic TAN-542 and its production, *Japanese Patent Application*, S 60 – 41 490 (July 12, 1984).
56. Cooper, R., Bush, K., Principe, P. A., Trejo, W. H., Wells, J. S., Two new monobactam antibiotics produced by a *Flexibacter* sp. I. Taxonomy, fermentation, isolation and biological properties, *J. Antibiotics 36*, 1252, 1983.
57. Ono, H., Nozaki, Y., and Harada, S., Antibiotic TAN-643, its production and a new species of *Cytophaga, Japanese Patent Application*, S 60 – 210 990 (Apr. 3, 1984).
58. Kato, T., Hinoo, H., Terui, Y., Nishikawa, J., Nakagawa, Y., Ikenishi, Y., and Shoji, J., PB-5266 A, B and C, new monobactams II. Physicochemical properties and chemical structures, *J. Antibiotics 40*, 139, 1987.
59. Katayama, N., Koyama, K., Nozaki, Y., Harada, S., and Ono, H., A new sulfazecin-type β-lactam antibiotic, TAN-850, *J. Takeda Res. Lab. 46*, 62, 1987.
60. Parker, W. L., Rathnum, M. L., Wells, J. S., Trejo, W. H., Principe, P. A., and Sykes, R. B., SQ 27, 860, a simple carbapenem produced by species of *Serratia* and *Erwinia, J. Antibiotics 35*, 653, 1982.
61. Kintaka, K., Harada, S., Ono, H., and Okazaki, H., and Sakane, T., Production of a carbapenem antibiotic by a spiral bacterium, *Azospirillum* sp., *J. Takeda Res. Lab. 44*, 17, 1985.
62. Singh, P. D., Young, M. G., Johnson, J. H., Cimarusti, C. M., and Sykes, R. B., Bacterial production of 7-formamidocephalosporins. Isolation and structure determination, *J. Antibiotics 37*, 773, 1984.
63. Shoji, J., Kato, T., Sakazaki, R., Nagata, W., Terui, Y., Nakagawa, Y., Shiro, M., Matsumoto, K., Hattori, T., Yoshida, T., and Kondo, E., Chitinovorins A, B and C, novel β-lactam antibiotics of bacterial origin, *J. Antibiotics 37*, 1486, 1984.
64. Harada, S., Tsubotani, S., Ono, H., and Okazaki, H., Cephabacins, new cephem antibiotics of bacterial origin. II. Isolation and characterization, *J. Antibiotics 37*, 1536, 1984.
65. Nozaki, Y., Okonogi, K., Katayama, N., Ono, H., Harada, S., Kondo, M., and Okazaki, H., Cephabacins, new cephem antibiotics of bacterial origin. IV. Antibacterial activity, stability to β-lactamases and mode of action, *J. Antibiotics 37*, 1555, 1984.
66. Shoji, J., Sakazaki, R., Kato, T., Terui, Y., Matsumoto, K., Tanimoto, T., Hattori, T., Hirooka, K., and Kondo, E., Isolation of chitinovorin D, *J. Antibiotics 38*, 538, 1985.
67. Tsubotani, S., Hida, T., Ono, H., and Harada, S., Cephabacin M_{1-6}, new 7-methoxycephem antibiotics of bacterial origin. II. Isolation, characterization and structural determination, *J. Antibiotics 38*, 1152, 1985.
68. Hida, T., Maki, M., Saga, O., Nozaki, Y., Ono, H., and Harada, S., A new formadicin-type monocyclic β-lactam antibiotic, TAN-667, *J. Takeda Res. Lab. 46*, 55, 1987.
69. Harada, S., Tsubotani, S. Hida, T., Koyama, K., Kondo, M., and Ono, H., Chemistry of a new antibiotic: lactivicin, Tetrahedron *44*, 6589, 1988.
70. Nozaki, Y., Katayama, N., Harada, S., Ono, H., and Okazaki, H., Lactivicin, a naturally occurring non-β-lactam antibiotic having β-lactam-like action: biological activities and mode of action, J. Antibiotics *42*, 84, 1989.

Chapter 7
Glycopeptide Antibiotics of the Vancomycin Group

G. C. Lancini and B. Cavalleri

1. Introduction
2. Producing Organisms
3. Chemistry
4. Biosynthesis
5. Mechanism of Action
   5.1 Effect on growing cultures
   5.2 Studies with cell free systems
   5.3 Action at molecular level
6. Relation of Antimicrobial Activity and Mechanism of Action
References

1. Introduction

Vancomycin [1], an antibiotic effective in the treatment of severe Gram-positive infections, was isolated in 1956 from cultures of *Streptomyces orientalis* [2]. Although its chemical structure was completely elucidated only several years later [3, 4], it was soon recognized as the first member of a new family of antibiotics known at present as "glycopeptide antibiotics" or more precisely "glycopeptide antibiotics of the vancomycin group".

The common characteristics of these products are easily defined on the basis of chemical structure and biological properties. From the chemical point of view they are linear heptapeptides in which at least five of the amino acid residues are benzenic rings linked each other to form a triphenylether moiety and a diphenyl group. The aromatic rings may carry various substituents as hydroxy or methyl groups, sugars, chlorine atoms.

From the biological point of view the most striking feature is their mechanism of action. In fact, they inhibit the growth of sensitive bacteria by binding to the D-alanyl-D-alanine terminus of one or more intermediates of the cell wall synthesis, thus determining the cell disruption [5, 6].

Although initially scanty used vancomycin became more and more clinically important because of its activity on staphylococci resistant to the β-lactams and other commonly used antibiotics [1]. This spurred the research of new members of this

Table 1 Glycopeptide antibiotics.

Name	Producing organism		Ref.	Year*	Company
Ristocetin A, B	*Nocardia lurida*	NRRL 2430	7, 8	1953	Abbott
Vancomycin	*Nocardia orientalis*	NRRL 2450	2, 8	1955	Lilly
Actinoidin A, B	*Proactinomyces actinoides*		8	1956	Res. Inst. For The Discovery Of New Antibiotics, Moscow
K-288 (Actinoidin)	*Streptomyces haranomachiensis* n. sp.		15	1961	Tohoku University, Sendai
Ristomycin A, B	*Proactinomyces fructiferi* var. *ristomycini*		7, 8	1962	Res. Inst. For The Discovery Of New Antibiotics, Moscow
Avoparcin (LL-AV290 complex)	*Streptomyces candidus*	NRRL 3218	8, 16	1966	American Cyanamid
AM374	*Streptomyces eburosporeus*	NRRL 3582	17	1969	American Cyanamid
A477	*Actinoplanes* sp.	NRRL 3884	18	1971	Lilly
Actaplanin (A4696 complex)	*Actinoplanes missouriensis*	ATCC 23342	19, 20	1971	Lilly
AB-65	*Saccharomonospora viride*	T-80	21	1973	Dainippon Pharmaceuticals
Teicoplanin (Teichomycin complex)	*Actinoplanes teichomyceticus*	ATCC 31121	22–24	1975	Lepetit
A35512 complex	*Streptomyces candidus*	NRRL 8156	8, 25, 26	1976	Lilly
OA-7653	*Streptomyces hygroscopicus* subsp. *hiwasaensis*	ATCC 31613	27	1978	Otsuka Pharmaceuticals
A51568A (N-demethylvancomycin), B	*Nocardia orientalis*	NRRL 15232	28, 29	1982	Lilly
A41030 complex	*Streptomyces virginiae*	NRRL 15156	30, 31	1982	Lilly
A47934	*Streptomyces toyocaensis*	NRRL 15009	32	1982	Lilly
Ardacin (AAD-216 complex, aridicin)	*Kibdelosporangium aridum*	ATCC 39323	33, 34	1983	Smith, Kline and French
Chloropolysporin A, B, C	*Faenia interjecta* sp. nov.	BP-583	35, 36	1983	Sankyo
A40926 complex	*Actinomadura* sp.	ATCC 39726	37	1984	Lepetit
M43 complex	*Nocardia orientalis*	NRRL 2450	38	1984	Lilly
Kibdelin (AAD-609 complex)	*Kibdelosporangium aridum* subsp. *largum*	ATCC 39922	39, 40	1985	Smith, Kline and French
Izupeptin A, B	*Nocardia* sp.	FERM P-8656	41	1986	Kitasato Institute, Tokyo
Parvodicin (AAJ-271)	*Actinomadura parvosata*	ATCC 53463	42	1986	Smith, Kline and French
Synmonicin A, B, C (CWI-785)	*Synnemomyces mamnoorii*	ATCC 53296	43	1986	Eskayef Ltd., Smith, Kline and French
Actinoidin A_2	*Nocardia* sp.	SKF-AAJ-193	44	1987	Smith, Kline and French
Orienticin (PA-42867-A, B, C, D)	*Nocardia orientalis*	PA-42867	45	1987	Shionogi
Eremomycin	Actinomyces	INA-238	46	1987	Institute Of New Antibiotics, Academy Of Medical Sciences Of The USSR, Moscow
A42867	*Nocardia* sp.	ATCC 53492	47	1987	Lepetit
A82846 A, B, C	*Nocardia orientalis*	NRRL 18098, 18099, 18100	48	1987	Lilly

* of patent or first publication.

family with potential improved activity in respect to the older ones, among which the best known are ristocetin [7] and avoparcin [8]. The first, used for some time in clinical practice, was later withdrawn because it causes plateled aggregation, a property now exploited for diagnostic purposes. Avoparcin is used as animal feed additive [9].

The results of this research, that it is still ongoing, are among others actaplanin, used as animal growth promoter [10], and teicoplanin, a clinically effective agent endowed with a higher activity of vancomycin on several pathogens and easier to be administered, due to its favourable pharmacokinetics [11, 12].

2. Producing Organisms

At present about thirty microbial strains producing glycopeptide antibiotics have been reported in the literature. These are the result of screening programs initially generically aimed at antimicrobial activities but later designed to select products active on cell wall synthesis [13] and lately very specifically to isolate substances which bind to D-alanyl-D-alanine peptides [14]. The strains and their products are listed in Table 1. In most of the cases complexes of several closely related substances are produced by a single strain as often is the case with antibiotic producers.

It may be noted that all the producing microorganisms belong to the order of Actinomycetales. A high proportion is represented by non-Streptomyces species, the so-called "rare Actinos". This may reflect the tendency, prevalent in the last decade, to screen for these unusual microorganisms in order to improve the chances of isolating novel products.

3. Chemistry

Vancomycin structure (see Table 2) was elucidated in an effort lasting 25 years using several techniques: chemical degradation, NMR spectroscopy and X-ray analysis [8]. Similar chemical and spectroscopic studies were performed to establish the structure of ristocetin [7, 8].

One of the difficulties encountered was the low degree of purity of the compounds. At present, high purity samples can be easily obtained by the use of analytical and preparative reversed-phase HPLC (see for example Refs. 24, 31, 34, 36, 40, 42, 49) and affinity chromatography [40, 50, 51]. The determination of complete structures, including the configuration and spatial arrangements, is now accomplished in a reasonable time by computer assisted high field NMR (Refs. 52 – 54 and references cited in Ref. 42) and FAB-MS [55, 56]. Thus, most of the structures of glycopeptide antibiotics are known and reported in Tables 2 – 5.

Table 2 Vancomycin-type glycopeptides.

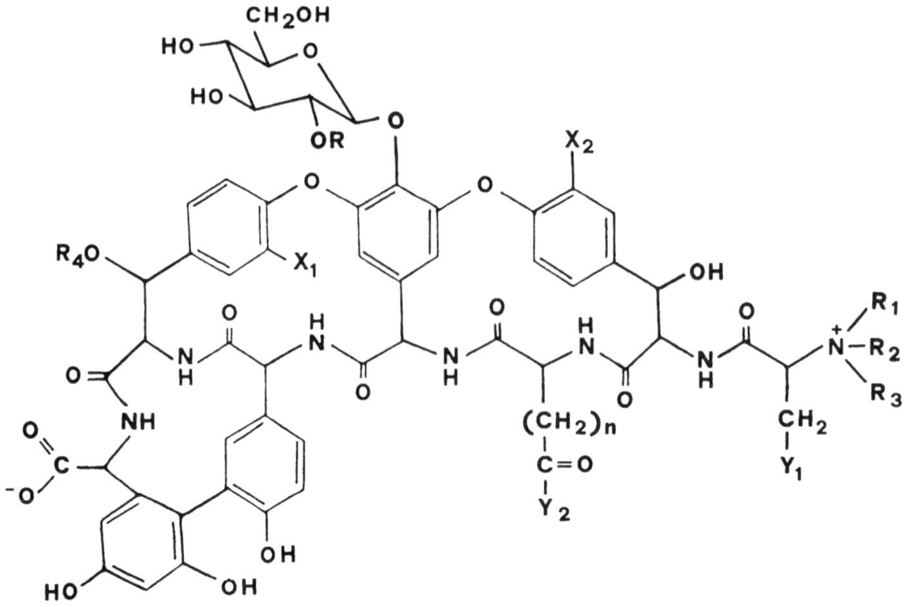

Antibiotic	n	R	R_1	R_2	R_3	R_4	X_1	X_2	Y_1	Y_2	Ref.
Vancomycin	1	L-van	H	H	CH_3	H	Cl	Cl	iC_3H_7	NH_2	3, 4, 8
OA-7653	2	*	H	CH_3	CH_3	glu	Cl	Cl	H	NH_2	57
A51568A	1	L-van	H	H	H	H	Cl	Cl	iC_3H_7	NH_2	
A51568B	2	L-van	H	H	H	H	Cl	Cl	iC_3H_7	NH_2	
M43A	1	L-van	CH_3	CH_3	CH_3	H	Cl	Cl	iC_3H_7	NH_2	38
M43B	1	L-van	CH_3	CH_3	CH_3	H	Cl	Cl	iC_3H_7	NH_2	
M43C	1	H	CH_3	CH_3	CH_3	H	Cl	Cl	iC_3H_7	NH_2	
M43D	1	L-van	H	CH_3	CH_3	H	Cl	Cl	iC_3H_7	NH_2	
Orienticin A	1	4-epi-van	H	H	CH_3	4-epi-van	Cl	H	iC_3H_7	NH_2	58
Orienticin B	1	L-oli	H	H	CH_3	4-epi-van	Cl	H	iC_3H_7	NH_2	
Orienticin C	1	4-epi-van	H	H	CH_3	4-epi-van	H	H	iC_3H_7	NH_2	
Orienticin D	1	4-epi-van	H	CH_3	CH_3	4-epi-van	Cl	H	iC_3H_7	NH_2	
Eremomycin	1	ere	H	H	CH_3	ere	H	Cl	iC_3H_7	NH_2	59
A82846A	1	4-epi-van	H	H	CH_3	4-epi-van	H	Cl	iC_3H_7	NH_2	48
A82846B	1	4-epi-van	H	H	CH_3	4-epi-van	Cl	Cl	iC_3H_7	NH_2	
A82846C	1	4-epi-van	H	H	CH_3	4-epi-van	H	H	iC_3H_7	NH_2	

van = vancosamine
glu = glucose
oli = olivose
ere = eremosamine (4-epi-van)
* = the hydroxyl of the triphenyl ether does not carry any sugar.

Table 3 Actinoidin-type glycopeptides.

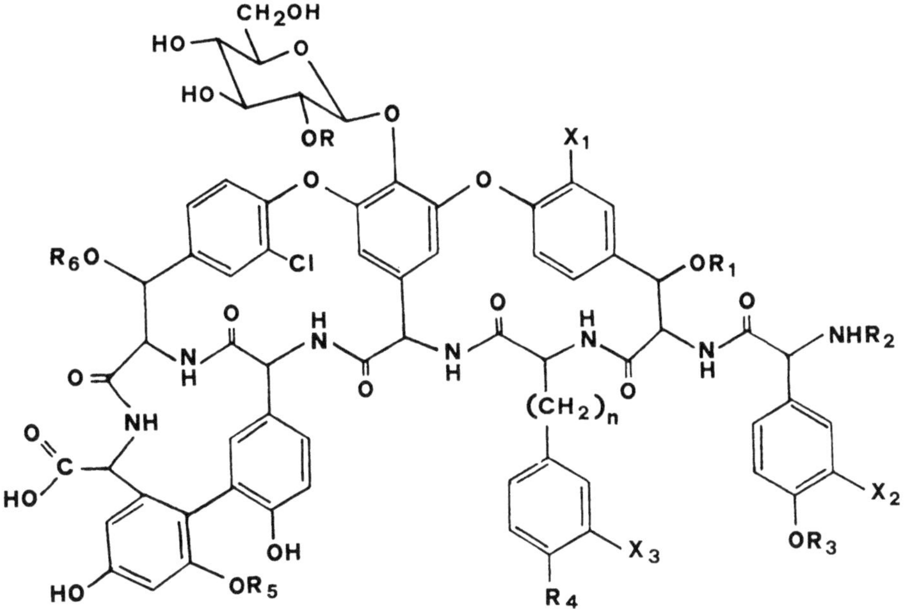

Antibiotic	n	R	R_1	R_2	R_3	R_4	R_5	R_6	X_1	X_2	X_3	Ref.
Actinoidin A	1	L-aco	H	H	H	H	D-man	L-aca	H	H	H	8, 53, 54
Actinoidin B	1	L-aco	H	H	H	H	D-man	L-aca	H	Cl	H	
Avoparcin-*α*	0	L-ria	D-man	CH_3	L-rha	OH	H	L-ria	H	H	H	8, 60, 61
Avoparcin-*β*	0	L-ria	D-man	CH_3	L-rha	OH	H	L-ria	H	H	Cl	
Avoparcin-*ε*	0	L-ria	H	CH_3	L-rha	OH	H	L-ria	H	H	Cl	
Chlorpoly-sporin B	0	H	man	CH_3	rha	OH	H	ria	Cl	H	Cl	62
Chloropoly-sporin C	0	H	man	CH_3	H	OH	H	ria	Cl	H	Cl	
Actinoidin A_2	1	L-rha	H	H	H	H	D-man	L-aca	H	H	H	54

aco = acosamine
aca = actinosamine
ria = ristosamine
man = mannose
rha = rhamnose

The chemistry and the physico-chemical characteristics of several of them have been reviewed [7, 8]. As mentioned before, five amino acids are common to all glycopeptide antibiotics: the first (carboxy terminus) is a dihydroxyphenylglycine. The second and the sixth are substituted tyrosines. The third and the fourth are p-hydroxyphenylglycines.

Table 4 Ristocetin-type glycopeptides.

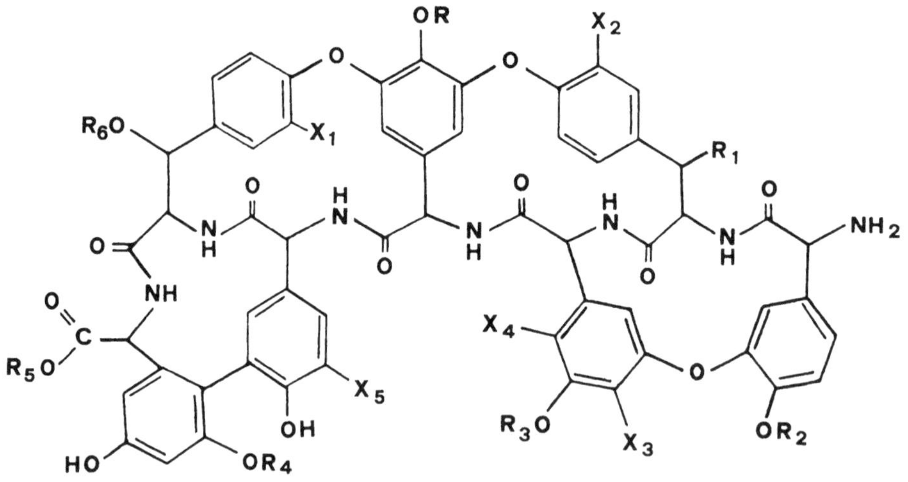

Antibiotic	R	R_1	R_2	R_3	R_4	R_5	R_6	X_1	X_2	X_3	X_4	X_5	Ref.
Ristocetin													
A	L-rha→D-glu ←D-man←D-ara	OH	H	H	D-man	CH_3	L-ria	H	H	CH_3	H	H	7, 8
B	L-rha→D-glu	OH	H	H	D-man	CH_3	L-ria	H	H	CH_3	H	H	
Actaplanin													
A	glu←man	H	H	man	man	CH_3	ria	Cl	H	CH_3	H	H	63
B_1	glu←rha	H	H	man	man	CH_3	ria	Cl	H	CH_3	H	H	
B_2	glu	H	H	man	man	CH_3	ria	Cl	H	CH_3	H	H	
B_3	glu←man	H	H	H	man	CH_3	ria	Cl	H	CH_3	H	H	
C_1	glu←rha	H	H	H	man	CH_3	ria	Cl	H	CH_3	H	H	
C_3	glu	H	H	man	H	CH_3	ria	Cl	H	CH_3	H	H	
G	glu	H	H	H	man	CH_3	ria	Cl	H	CH_3	H	H	
A35512B	glu←rha	OH	H	fuc	man	CH_3	3-epi- L-van	H	H	H	Cl	H	8, 64
A41030													
A	H	H	H	H	H	H	H	Cl	Cl	H	H	Cl	31
B	H	H	H	H	H	H	H	Cl	Cl	H	H	H	
C	H	H	gal	H	H	H	H	Cl	Cl	H	H	Cl	
E	H	H	H	H	H	H	H	Cl	H	H	H	H	
F	H	H	gal→gal	H	H	H	H	Cl	Cl	H	H	Cl	
A47934	H	H	SO_3H	H	H	H	H	Cl	Cl	H	H	Cl	32

ria = ristosamine
van = vancosamine
man = mannose
rha = rhamnose
glu = glucose
gal = galactose
ara = arabinose
fuc = fucose

Table 5 Teicoplanin-group glycopeptides.

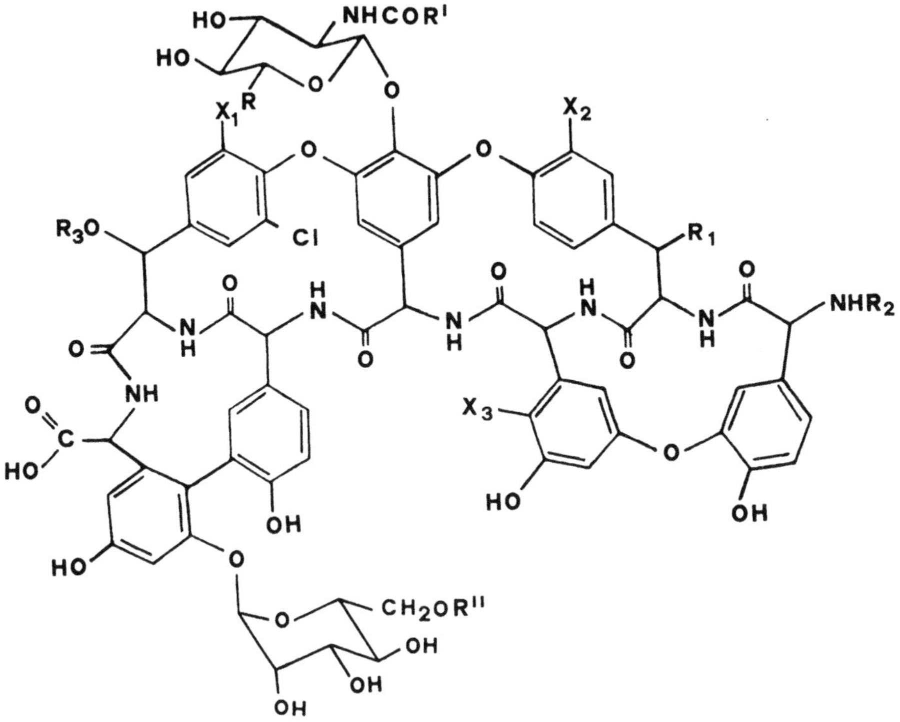

Antibiotic	R	R′	R″	R_1	R_2	R_3	X_1	X_2	X_3	Ref.
Teicoplanin										
T-A2-1	CH_2OH	$(CH_2)_2CH{=}CH(CH_2)_4CH_3$	H	H	H	Ac-N-glu	H	Cl	H	13,65–67
T-A2-2	CH_2OH	$(CH_2)_6CH(CH_3)_2$	H	H	H	Ac-N-glu	H	Cl	H	
T-A2-3	CH_2OH	$(CH_2)_8CH_3$	H	H	H	Ac-N-glu	H	Cl	H	
T-A2-4	CH_2OH	$(CH_2)_6CH(CH_3)CH_2CH_3$	H	H	H	Ac-N-glu	H	Cl	H	
T-A2-5	CH_2OH	$(CH_2)_7CH(CH_3)_2$	H	H	H	Ac-N-glu	H	Cl	H	
Ardacin										
A	COOH	$(CH_2)_8CH_3$	H	OH	CH_3	H	Cl	Cl	Cl	52,55
B	COOH	$(CH_2)_7CH(CH_3)_2$	H	OH	CH_3	H	Cl	Cl	Cl	
C	COOH	$(CH_2)_8CH(CH_3)_2$	H	OH	CH_3	H	Cl	Cl	Cl	
A40926										
A	COOH	$(CH_2)_9CH_3$	H	H	CH_3	H	H	H	Cl	68
B	COOH	$(CH_2)_8CH(CH_3)_2$	H	H	CH_3	H	H	H	Cl	
Kibdelin										
A	CH_2OH	$(CH_2)_8CH_3$	H	OH	CH_3	H	Cl	Cl	Cl	40
B	CH_2OH	$(CH_2)_7CH(CH_3)_2$	H	OH	CH_3	H	Cl	Cl	Cl	
C_1	CH_2OH	$(CH_2)_8CH(CH_3)_2$	H	OH	CH_3	H	Cl	Cl	Cl	
C_2	CH_2OH	$(CH_2)_{10}CH_3$	H	OH	CH_3	H	Cl	Cl	Cl	
D	CH_2OH	$(CH_2)_2CH{=}CH(CH_2)_4CH_3$	H	OH	CH_3	H	Cl	Cl	Cl	
Parvodicin										
A	COOH	$(CH_2)_8CH_3$	H	H	CH_3	H	H	H	Cl	42
B_1	COOH	$(CH_2)_7CH(CH_3)_2$	H	H	CH_3	H	H	H	Cl	
B_2	COOH	$(CH_2)_9CH_3$	H	H	CH_3	H	H	H	Cl	
C_1	COOH	$(CH_2)_8CH(CH_3)_2$	H	H	CH_3	H	H	H	Cl	
C_2	COOH	$(CH_2)_{10}CH_3$	H	H	CH_3	H	H	H	Cl	
C_3	COOH	$(CH_2)_8CH(CH_3)_2$	$COCH_3$	H	CH_3	H	H	H	Cl	
C_4	COOH	$(CH_2)_{10}CH_3$	$COCH_3$	H	CH_3	H	H	H	Cl	

N-glu = glucosamine

Based on their structural feature the glycopeptide antibiotics may be divided into three types, according to the heptapeptide composition. A fourth group is characterized by the presence of fatty acid residues as substituents of an amino sugar; thus, the latter can be defined as lipoglycopeptide antibiotics.

The first type (Table 2) is characterized by asparagine and leucine as the fifth and seventh amino acid, respectively. Aspartic acid and glutamine may replace for asparagine, and in several compounds the terminal amino group of leucine may be methylated to a different extent.

Sugars are glucose and vancosamine or epi-vancosamine, with few exceptions (olivose and eremosamine). In most of the compounds the triphenylether moiety carries two chlorine atoms. This type may be called *Vancomycin-type* from the most known component.

The second type, or *Actinoidin-type* (Table 3), has phenylalanine or p-hydroxyphenylglycine as the fifth amino acid and p-hydroxyphenylglycine as the seventh one. A variety of sugars and amino sugars are present.

The third type, or *Ristocetin-type* (Table 4), has a m-hydroxyphenylglycine as the fifth amino acid and p-hydroxyphenylglycine as the seventh one. The aromatic rings of these two amino acids are linked by an ether bond. The members of this type present a higher variability in the pattern of sugar substituents, including a tetrasaccharide moiety in ristocetin. This compound is the only glycopeptide lacking chlorine substituents.

The *Teicoplanin group* (Table 5) has the same heptapeptide skeleton of the ristocetin-type. Characteristic of this group is the presence of fatty acid residues of 10 to 12 carbon atoms which may be linear or branched (iso or anteiso branching) only exceptionally unsaturated. This acyl chains form an amide bond with the 2-amino group of either glucosamine or aminoglucuronic acid linked to the p-hydroxyphenyl residue of the fourth amino acid.

As noted previously, X-ray data on a degradation derivative of vancomycin led to confirm the absolute configuration of this compound [3, 4, 8]. Although crystalline forms suitable for X-ray studies have not been obtained for other glycopeptides, NMR comparison with vancomycin and NMR experiments have provided good evidence, for all the cases examined, that the absolute configuration of the chiral centers of the seven amino acids are 1S, 2S, 3R, 4R, 5S, 6R, 7R. The stereostructures are similar. That of the aglycone of teicoplanin, obtained from a Dreiding model [67], is shown in Figure 1.

From the structures of glycopeptides reported in Tables 2–5 is evident that the high and diverse degree of substitution confers a wide variation of physico-chemical characteristics.

The isoelectric points vary from 3.2 for A47934 to 8.5 for actaplanin (data from the Author's laboratory). Another important property is the lipophilic character, from very hydrophilic compounds, such as vancomycin and ristocetin to rather lipophilic ones such as teicoplanin and ardacin (formerly called aridicin). These two parameters, charge and lipophilicity, strongly influence the biological activity,

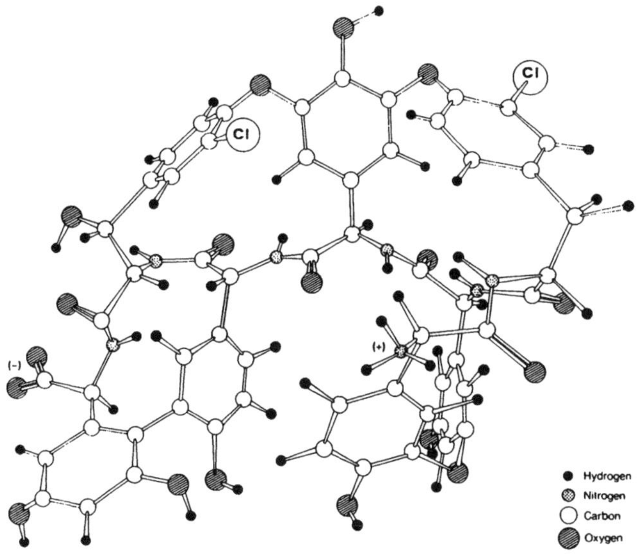

Figure 1 Stereo-structure of teicoplanin aglycone obtained from a Dreiding model.

i.e., the *in vitro* activity against selected strains and the pharmacokinetic (half-lives, serum protein and tissue binding) [69, 70]. The knowledge of the structures, the common molecular basis of action and the correlation between pharmacokinetic and physico-chemical properties have offered ground for selective hydrolysis, oriented semisynthetic modifications and biotransformations.

4. Biosynthesis

Examination of the structure of glycopeptide antibiotics suggests a biosynthetic scheme including the following steps:

a) synthesis of the single uncommon amino acids.
b) assembly of the amino acids probably through the multienzyme thiotemplate mechanism.
c) formation of the ether and C-C bonds.
d) glycosylation.

Although the scheme appears straightforward very little experimental evidence is available to support it. Several papers report data on the origin of the constituent amino acids. Experiments with ^{13}C and ^{2}H enriched precursors and NMR spectroscopy have shown that in vancomycin D-tyrosine is a precursor for the synthesis of D-p-hydroxyphenylglycine and β-hydroxy-m-chlorotyrosine. β-Hydroxylation occurs with retention of configuration at the β-carbon atom. The two m-dihydroxyphenylglycine units are labelled when ^{13}C acetate is added to cultures and the labelling pattern is consistent with formation of the aromatic ring by cyclization of a tetraketide [71]. Similar results using either ^{13}C or radioactively labelled material have been reported for avoparcin [61], ristocetin [72], ardacin [73] and A47934 [74]. The sequence of reactions leading from tyrosine to p-hydroxyphenylglycine has been the subject of some speculation. Addition of potential precursor to fermentations of ardacin [73] or *N*-demethylvancomycin [28] suggests that p-hydroxyphenylglyoxilic acid and p-hydroxymandelic acid (but not p-hydroxyphenylacetic acid) are intermediates of the biosynthesis. The results are consistent with the sequence: tyrosine → β-hydroxytyrosine → p-hydroxy-β-hydroxyphenylpyruvic acid → p-hydroxymandelic acid → p-hydroxyphenylglyoxylic acid → p-hydroxyphenylglycine. The same sequence has been proposed for the biosynthesis of p-hydroxyphenylglycine in nocardicin production [75]. No study has been reported on the assembly of the amino acids into the heptapeptide nor on the reactions by which the aromatic rings are linked to form the triphenylether or the diphenyl moiety.

A noteworthy difficulty encountered in the study of biosynthesis is that the glycopeptide antibiotics, and most probably their aglycones, for reasons discussed in Section 6 do not penetrate readily into bacterial cell. Thus, attempts to elucidate the sequence of last biosynthetic steps by using blocked mutants either in cosynthesis experiments or with addition of intermediates resulted unsuccessful, and apparently the use of protoplast gave also poor results [76].

The final biosynthetic steps have been examined in some detail in the case of ardacins. Characteristic of these antibiotics produced by *Kibdelosporangium aridum* is the presence of a *N*-acyl-2-deoxy-2-aminoglucuronic unit, beside a mannose, as a sugar substituent. A related family of antibiotics, kibdelins, differs from the aridicins for the presence of *N*-acylglucosamine instead of the acylglucuronic acid. Kibdelins are readily converted into the corresponding ardacins by cultures of *K. aridum* indicating that the oxidation of carbon-6 of glucosamine to a carboxy group is the last biosynthetic step [73]. Either ^{14}C labelled ardacin aglycone or mannosyl aglycone give origin to labelled ardacins when added to cultures of *K. aridum.* However, because low yields were observed after a long incubation time further confirmation is required to prove that indeed the aglycone is the precursor of the final molecule. In fact, it is possible to conceive a different sequence of reactions in which the acylglucosamine unit is added before the completion of the aglycone structure.

In this respect, studies recently reported on antibiotic A47934 are of interest. This antibiotic, produced by *Streptomyces toyocaensis*, has a structure very similar to the ardacin aglycone: however it lacks a sugar substituent but does possess a sulfate ester on a phenolic group. Extensive experiments with labelled substrates, blocked mutants and biochemical inhibitors indicate that the sulfate is added prior to the formation of intermediates that possess antimicrobial activity (note that the glycopeptide aglycones are about as active as the whole molecules) [74]. In contrast, the addition of the ardacin sugar, mannose, appears to be a rather unspecific reaction. It proceeds either on the ardacin aglycone or on the aglycone carrying the acylaminoglucuronate, and it is performed by several glycopeptide-producing actinomycetales belonging to different genera [73]. A similar lack to substrate specificity has been observed in the Author's laboratory in the mannosylation of teicoplanin aglycone and pseudoaglycones by *Actinoplanes teichomyceticus* (unpublished results).

The origin of the acyl moieties typical of ardacins, teicoplanin and A40926 is under investigation in the Author's laboratory. It has been found that the yields of given components of an antibiotic complex can be enhanced by addition to fermentation broths of precursors of the fatty acid characteristic of the components. These precursors are branched amino acids in the case of branched fatty acids or long chain unsaturated acids in the case of linear acyl moieties [77, 78].

5. Mechanism of Action

5.1 Effect on growing cultures

Several evidences indicate that members of the glycopeptide group of antibiotics interfere with cell wall assembly and more particularly with peptidoglycan synthesis of sensitive bacteria:

a) Vancomycin and ristocetin inhibit the synthesis of peptidoglycan, as demonstrated by the block of uptake of peptidoglycan precursors well before that of other macromolecular synthesis is affected [79-81]. This was also clearly shown with teicoplanin [82].
b) The inhibition is accompanied by accumulation of peptidoglycan precursors, mainly UDP-MurNAc-pentapeptide. This indicate that the antibiotic does not interfere with the initial cytoplasmatic reactions but with a later step of the byosynthetic process [79, 81, 82].
c) Bacterial protoplasts are not significantly affected by these antibiotics [83]. Similarly, growth of L-forms derived from a sensitive *S. aureus* strain is not inhibited by high concentration of vancomycin, ristocetin or teicoplanin (unpublished results from the Author's laboratory).
d) Pre-incubation of bacterial cells with these antibiotics prevents lysis by lysozime or other lytic enzymes indicating a specific binding to cell wall peptidoglycan [82, 84].

5.2 Studies with cell free systems

Bacterial membrane preparations which synthesize peptidoglycan from UDP-MurNAc-pentapeptide were used by several authors to identify the biosynthetic reaction inhibited by glycopeptide antibiotics [85-87]. It was found that peptidoglycan synthesis was blocked and at the same time the formation of the phospholipid intermediate (*N*-acetylglucosamine-*N*-acetylmuramylpentapeptidephospholipid) was stimulated. This result indicates that the polymerisation step of peptidoglycan synthesis, also called the transglycosylation reaction, is a target of the antibiotic action. However, this does not exclude that also later events as transpeptidation are affected.

The inhibitory action is reversed by the presence of cell wall fragments indicating again the propensity of the antibiotic to bind to cell wall components [82, 88].

5.3 Action at molecular level

The observation by Chatterjee and Perkins [89] that vancomycin complexes with UDP-MurNAc-pentapeptide provided on important clue on the molecular aspects of this antibiotic action [90, 91].

A series of detailed studies done by Nieto and Perkins [90, 91], mainly by UV differential spectroscopy, established that vancomycin and ristocetin bind with high affinity to peptides ending with a D-alanyl-D-alanine terminus. *N,N'*-Diacetyl-L-lysyl-D-alanyl-D-alanine showing the highest affinity for vancomycin was used as the standard model. By measuring the effect on complex formation of altering the tripeptide composition the following observations were made:

a) A free carboxyl group of the terminal amino acid is an absolute requirement for binding. A D configuration (or glycine) for this amino acid is also necessary. Affinitiy for vancomycin decreases when bulkier amino acids replace alanine whereas affinity for ristocetin is substantially unchanged.
b) The presence of an L substituent at residue 2 again prevents complex formation and a methyl side chain gives the highest affinity. An isobutyl side chain decreases slightly the affinity for vancomycin and somewhat more that for ristocetin.
c) An acyl-L-amino acid at residue 3 confers a higher affinity than an acyl-D-amino acid or acetic acid. A free amino group impairs considerably the complex formation.

An association constant with *N,N'*-diacetyl-L-Lys-D-Ala-D-Ala of 2.6×10^6 $lmol^{-1}$ was calculated spectrophotometrically for teicoplanin, as compared with values of 1.5×10^6 and 5.9×10^5 for vancomycin and ristocetin, respectively [82].

Subsequently, using NMR methods more detailed information was obtained on the structure of the complex formed by the antibiotics with the model peptides. On the basis of the N-H temperature coefficients, ^1H chemical shifts and the intermolecular Nuclear Overhauser Effect a model was proposed for the interaction of vancomycin [92-95], ristocetin [95-97], and avoparcin [98] with either acetyl-D-Ala-D-Ala or diacetyl-L-Lys-D-Ala-D-Ala. According to this model the complex is stabilized by several H-bonds, including those between the COO^- terminus of the di- or tri-peptide and a pocket of NH protons near the terminal NH_2 of the antibiotic. Other H-bonds are formed between one NH and one CO of the antibiotic and one CO and one NH, respectively, of the peptide.

Differences have been noted in the complex formation with ristocetin and vancomycin [99]. The first has a rigid structure in which the model peptide fits considerably well and which presents lipophylic pockets accomodating the alanine methyl groups.

With vancomycin, which is more flexible, the fitting is obtained by a large conformational change which stabilizes the complex [100]. The terminal NH_2 group of the antibiotic seems most important for the initial interaction between the antibiotic and the peptide, and thus for the rate of the complex formation which is higher with vancomycin [99].

6. Relation of Antimicrobial Activity and Mechanism of Action

Most aerobic and anaerobic Gram-positive bacteria are sensitive to the action of glycopeptide antibiotics. Minimal inhibition concentration values of representative strains are shown in Table 6 (Ref. 101 and B. Goldstein, personal communication). Gram-negative bacteria are generally insensitive with the partial exception of *Neisseria gonorrhoeae* [1, 101], and the weak activity (from 8 to 64 μg/ml) shown by some teicoplanin semisynthetic derivatives on *Escherichia coli*, *Proteus vulgaris* and *Pseudomonas aeruginosa* [65, 102].

Table 6 Comparison of antibacterial activity (MIC, μg/ml) of various glycopeptides.

Antiobiotic	*Staphylococcus aureus* TOUR	*Staphylococcus epidermidis* ATCC 12228	*Streptococcus pyogenes* C 203	*Streptococcus faecalis* ATCC 7080	*Clostridium perfringens* ISS 30543	*Neisseria gonorrhoeae* ISM 68/126
Ristocetin	4	2	0.25	1	2	64
Vancomycin	0.25	0.5	0.13	0.5	0.13	32
Avoparcin	2	2	0.25	0.25	0.5	128
Actaplanin	1	2	0.13	0.5	1	128
Teicoplanin	0.13	0.13	0.06	0.13	0.003	32
A35512B	1	0.5	0.13	0.5	2	64
A41030	0.03	0.008	0.13	0.13	–	64
A47934	0.06	0.03	0.13	0.13	0.25	8
Ardacin	1	4	0.13	2	0.06	64
A40926	0.06	0.06	0.06	0.06	0.004	2

Since peptidoglycan biosynthesis is similar in Gram-positive and Gram-negative bacteria the insensitivity of the latter ones must be attributed to the lack of penetration of the antibiotic through the outer cell membrane typical of these organisms. In fact, glycopeptides appear too large to enter through the porin channels and on the other hand are too polar to cross the lipid layer. For the same reason most probably they cannot penetrate into the cell through the cytoplasmic membrane. The lack of penetration was shown experimentally with iodovancomycin [103]. Thus, their action is limited to the events of the biosynthetic process taking place at the surface of the cytoplasmic membrane, that is the transglycosylation or the transpeptidation reactions. It was suggested that the growing nascent peptidoglycan chain is the site of interaction [104].

The effect of glycopeptides on growing bacterial cell is mainly bactericidal [1, 105, 106]. Upon addition of the antibiotic to cultures growth inhibition is followed by lysis of the cells. It is probable that, as with β-lactam antibiotcs, lysis is not directly induced by the antibiotic but is due to autolysines liberation.

Of clinical importance is the observation that the emergence of resistant strains is a very rare event for glycopeptide antibiotics [1, 105–108]. This can be again interpreted on the basis of their peculiar mechanism of action: in fact, they constitute the only known example of antibiotics whose target is the substrate of an enzymatic reaction rather than an enzyme or other macromolecule. However, recently an Enterococcus with plasmid-mediated resistance to vancomycin and teicoplanin was described [109]. The biochemical mechanism of resistance has not yet been elucidated.

a) The genus Nocardia has been recently revised and two new genera have been proposed (110): Amycolata and Amycolatopsis. Accordingly the strain producing vancomycin is at present denoted *Amycolatopsis orientalis* and that producing ristocetin (ristomycin) *Amycolatopsis orientalis* subs. *lurida*. Similarly the producers of orienticin and chloroorienticin should be classified as *Amycolatopsis orientalis*.

b) The structure of two glycopeptides previously reported have been now published. A42867 belongs to the cancomycin type and should be added to Table 2 (111). Synmomicin A represents a new type of glycopeptide where the terminal aminoacid is a N-methyl-p-hydroxy phenylglycine carrying a neutrla sugar and the fifth aminoacid is methionine (112).

c) A large number of semisynthetic derivatives have been reported but only a few new natural metabolites: UK-68,597 of the ristocetin type in which the terminal C-NH_2 is replaced by C=O group (113). Helvecardin A, B are of the actinoidin type having as sugar substituents 2-O-methylrhamnose and ristosamine (114). Chloro-orienticins belonging to the vancomycin type (115). A80407 A, B are produced by *Kibdelosporangium aridum*: their structure is still unknown (116).

References

1. Cheung, R. P. F., and DiPiro, J. T., Vancomycin: an update, *Pharmacotherapy 6*, 153, 1986.
2. McCormick, M. H., Stark, W. M., Pittenger, G. E., Pittenger, R. C., and McGuire, J. M., Vancomycin, a new antibiotic. I. Chemical and biological properties, in *Antibiotics Annual*, 1955–1956, Welch, H., and Marti-Ibanez F., Eds., Medical Encyclopedia, Inc., New York, 1956, p. 606.
3. Pfeiffer, R. R., Structural features of vancomycin, *Reviews of infectious diseases 3*, 205, 1981.
4. Harris, C. M., Kopecka, H., and Harris, T. M., Vancomycin: Structure and transformation to CDP-I, *J. Am. Chem. Soc. 105*, 6915, 1983.
5. Barna, J. C. J., and Williams, D. H., The structure and mode of action of glycopeptide antibiotics of the vancomycin group, *Ann. Rev. Microbiol. 38*, 339, 1984.
6. Reynolds, P. E., Inhibitors of bacterial cell wall synthesis, in *The Scientific Basis of Antimicrobial Chemotherapy*, Grennwood, D., and O'Grady, F., Eds., Cambridge University Press, Cambridge, 1985, p. 13.

7. Katrukha, G. S., and Silaev, A. B., The Chemistry of Glycopeptide Antibiotics of the Vancomycin group, in *Chemistry of Peptides and Proteins*, Vol. 3, Voelter, W., Bayer, E., Ovchinnikov, Y. A., and Ivanov, V. T., Eds., de Gruyter, Berlin – New York, 1986, p. 289.
8. Sztaricskai, F., and Bognár, R., The Chemistry of the Vancomycin Group fo Antibiotics, in *Recent Developments in the Chemistry of Natural Carbon Compounds*, Vol. 10, Bognár, R., and Szántay, Cs., Eds., Akadémiai Kiadó Ed., Budapest, 1984, p. 91.
9. Stewart, C. S., and Duncan, S. H., The effect of avoparcin on cellulolytic bacteria of the ovine rumen, *J. General Microbiol. 131*, 427, 1985.
10. Scheifinger, C. C., Anderson, D. B., McGuffey, R. K., Pendlum, L. C., Richardson, L. F., and Wellenreiter, R. H., *The use of actaplanin in ruminant and non-ruminant nutrition*, Abstract N° 439, Program Abstract 23th Intersc. Conf. Antimicrobial Agents Chemother., Las Vegas, 1983.
11. Williams, A. H., and Grüneberg, R. N., Teicoplanin, *J. Antimicrob. Chemotherapy 14*, 441, 1984.
12. Speller, D., and Greenwood, D., Eds., Teicoplanin, clinical use and laboratory correlation. J. Antimicrob. Chemotherapy *21*, Suppl. A, 1, 1988.
13. Coronelli, C., Gallo, G. G., and Cavalleri, B., Teicoplanin: chemical, physico-chemical and biological aspects, *Il Farmaco, Ed. Sci. 42*, 767, 1987.
14. Rake, J. B., Gerber, R., Mehta, R. J., Newman, D. J., Oh, Y. K., Phelen, C., Shearer, M. C., Sitrin, R. D., and Nisbet, L. J., Glycopeptide antibiotics: a mechanism-based screen employing a bacterial cell wall receptor mimetic, *J. Antibiotics 39*, 58, 1986.
15. Matsumoto, K., A vancomycin-related antibiotic from *Streptomyces* sp. K – 288, *J. Antibiotics Ser. A, 14*, 141, 1961.
16. Kuntsmann, M. P., Mitscher, L. A., Porter, J. N., Shay, A. J., and Darken, M. A., LL – AV290, a new antibiotic. I. Fermentation, isolation, and characterization, *Antimicrob. Agents and Chemotherapy 242*, 1968.
17. Kunstmann, M. P., and Porter, J. N., Animal feed composition containing antibiotic AM374 as a growth promoter, *U.S. Patent 3,803,306*, 1974.
18. Hamill, R. L., Haney, M. K., and Stark, W.So, Antibiotic A477 and process for preparation thereof, *U.S. Patent 3,780,164*, 1973.
19. Debono, M., Merkel, K. E, Weeks, R. E., and Cole, H. J., Antibiotic A-4696 Factors B_1, B_2, B_3, C_1, C_3 and E_1, *U.S. Patent 4,322,406*, 1982.
20. Debono, M., Merkel, K. E., Molloy, R. M., Barnhart, M., Presti, E., Hunt, A. H., and Hamill, R. L., Actaplanin, new glycopeptide antibiotics produced by *Actinoplanes missouriensis*. The isolation and preliminary chemical characterization of actaplanin, *J. Antibiotics 37*, 85, 1984.
21. Tamura, A., and Takeda, I. Antibiotic AB-65, a new antibiotic from *Saccharomonospora viride, J. Antibiotics 28*, 395, 1975.
22. Parenti, F., Beretta, G., Berti, M., and Arioli, V., Teichomycins, new antibiotics from *Actinoplanes teichomyceticus*, nov. sp. I. Description of the producer strain, fermentation studies and biological properties, *J. Antibiotics 31*, 276, 1978.
23. Bardone, M. R., Paternoster, M., and Coronelli, C., Teichomycins, new antibiotics from *Actinoplanes teichomyceticus* nov. sp. II. Extraction and chemical characterization, *J. Antibiotics 31*, 170, 1978.
24. Borghi, A., Coronelli, C., Faniuolo, L., Allievi, G., Pallanza, R., and Gallo, G. G., Teichomycins, new antibiotics from *Actinoplanes teichomyceticus* nov. sp. IV. Separation and characterization of the components of teichomycin (teicoplanin), *J. Antibiotics 37*, 615, 1984.
25. Michel, K. H., and Higgens, C. E., Antibiotic A35512 and process for production thereof, *U.S. Patent 4,122,168*, 1978.
26. Michel, K. H., Shah, R. M., and Hamill, R. L., A35512, a complex of new antibacterial antibiotics produced by *Streptomyces candidus*. I. Isolation and characterization, *J. Antibiotics 33*, 1397, 1980.
27. Kamogashira, T., Nishida, T., and Sugawara, M., A new glycopeptide antibiotic, OA-7653, produced by *Streptomyces hygroscopicus* subsp. *hiwasaensis*, *Agric. Biol. Chem. 47*, 499, 1983.
28. Boeck, L. D., Mertz, F. P., Wolter, R. K., and Higgens, C. E., *N*-Demethylvancomycin, a novel antibiotic produced by a strain of *Nocardia orientalis*. Taxonomy and fermentation, *J. Antibiotics 37*, 446, 1984.

29. Hunt, A. H., Marconi, G. G., Elzey, T. K., and Hoehn, M. M., A51568A: *N*-Demethylvancomycin, *J. Antibiotics 37*, 917, 1984.
30. Boeck, L. D., Mertz, F. P., and Clem, S. M., A41030, a complex of novel glycopeptide antibiotics produced by a strain of *Streptomyces virginiae*. Taxonomy and fermentation studies, *J. Antibiotics 38*, 1, 1985.
31. Eggert, J. H., and Michel, K. H., isolation and characterization of A41030, a complex of novel glycopeptide antibiotics. Application of the Michel-Miller High Performance Low Pressure Liquid Chromatography System, *J. Antibiotics 39*, 792, 1986.
32. Boeck, L. D., Mertz, F. P., A47934, a novel glycopeptide-aglycone antibiotic produced by a strain of *Streptomyces toyocaensis*. Taxonomy and fermentation studies, *J. Antibiotics 39*, 1533, 1986.
33. Shearer, M. C., Actor, P., Bowie, B. A., Grappel, S. F., Nash, C. H., Newman, D. J., Oh, Y. K., Pan, C. H., and Nisbet, L. J., Aridicins, novel glycopeptide antibiotics. I. Taxonomy, production and biological activity, *J. Antibiotics 38*, 555, 1985.
34. Sitrin, R. D., Chan, G. W., Dingerdissen, J. J., Holl, W., Hoover, J. R. E., Valenta, J. R., Webb, L., and Snader, K. M., Aridicins, novel glycopeptide antibiotics. II. Isolation and characterization, *J. Antibiotics 38*, 561, 1985.
35. Okazaki, T., Enokita, R., Miyaoka, H., Takatsu, T., and Torikata, A., Chloropolysporins A, B, and C, novel glycopeptide antibiotics from *Faenia interjecta* sp. nov. I. Taxonomy of producing organism, *J. Antibiotics 40*, 917, 1987.
36. Takatsu, T., Nakajima, M., Oyajima, S., Itoh, Y., Sakaida, Y., Takahashi, S., and Haneishi, T., Chloropolysporins A, B, and C, novel glycopeptide antibiotics from *Faenia interjecta* sp. nov. II. Fermentation, isolation and physico-chemical characterization, *J. Antibiotics 40*, 924, 1987.
37. Selva, E., Gastaldo, L., Beretta, G., Borghi, A., Goldstein, B. P., Arioli, V., Cassani, G., and Parenti, F., Antibiotic A40926 complex and its pure factors PA, PB, A, B and BO, *Eur. Patent Appl. 177,882*, 1986.
38. Higgins, H. M., Jr., Mc Cormick, M. H., Merkel, K. E., and Michel, K. H., Improvements in and relating to antibiotics M43A, M43B, M43C and M43D, *Eur. Patent Appl. 159*, 180, 1985.
39. Shearer, M. C., Giovenella, A. J., Grappel, S. F., Hedde, R. D., Mehta, R. J., Oh, Y. K., Pan, C. H., Pitkin, D. H., and Nisbet, L. J., Kibdelins, novel glycopeptide antibiotics. I. Discovery, production, and biological evaluation, *J. Antibiotics 39*, 1386, 1986.
40. Folena-Wasserman, G., Poehland, L. B., Yeung, E. W-K., Straiger, D., Killmer, L. B., Snader, K., Dingerdissen, J. J., and Jeffs, P. W., Kibdelins (AAD-609), novel glycopeptide antibiotics. II. Isolation, purification, and structure, *J. Antibiotics 39*, 1395, 1986.
41. Spiri-Nakagawa, P., Fukushi, Y., Maebashi, K., Imamura, N., Takahashi, Y., Tanaka, Y., Tanaka, H., and Omura, S., Izupeptins A and B, new glycopeptide antibiotics produced by an Actinomycete, *J. Antibiotics 39*, 1719, 1986.
42. Christensen, S. B., Allaudeen, H. S., Burke, M. R., Carr, S. A., Chung, S. K., DePhillips, P., Dingerdissen, J. J., DiPaolo, M., Giovenella, A. J., Heald, S. L., Killmer, L. B., Mico, B. A., Mueller, L., Pan, C. H., Poehland, B. L., Rake, J. B., Roberts, G. D., Shearer, M. C., Sitrin, R. D., Nisbet, L. J., and Jeffs, P. W., Parvodicin, a novel glycopeptide from a new species, *Actinomadura parvosata*: discovery, taxonomy, activity and structure elucidation, *J. Antibiotics 40*, 970, 1987.
43. Arjuna Rao, V., Ravishankar, D., Sadhukhan, A. K., Ahmed, S. M., Goel, A. K., Prabhu, N. S., Verma, A. K., Venkateswarlu, A., Allaudeen, H. S., Hedde, R. H., and Nisbet, L. J., *Synmonicins: a novel antibiotic complex produced by Synnemomyces marmoorii gen. et. sp. nov. I. Taxonomy of the producing organism, fermentation and biological properties*, Abstracts No. 939–942, Program Abstract 26th Intersc. Conf. Antimicrob. Agents Chemother., New Orleans, 1986.
44. Dingerdissen, J. J., Sitrin, R. D., DePhillips, P. A., Giovenella, A. J., Grappel, S. F., Mehta, R. J., Oh, Y. K., Pan, C. H., Roberts, G. D., Shearer, M. C., and Nisbet, L. J., Actinoidin A_2 a novel glycopeptide: production, preparative HPLC separation and characterization, *J. Antibiotics 40*, 165, 1987.
45. Tsuji, N., Matsumoto, K., Kawamura, Y., Tadashi, Y., Matsuura, S., and Kamigauchi, T., Glycopeptide antibiotics, their preparation and use and microorganisms for producing them, *Eur. Patent Appl. 231,111*, 1987.

46. Gauze, G. F., Brazhnikova, M. G., Laiko, A. V., Sveshnikova, M. A., Preobrazhenskaya, T. P., Fedorova, G. B., Borisova, V. N., Tolstykh, I. V., Yurina, M. S., Pokras, L. S., Goldberg, L. E., Malkova, I. V., and Stepanova, E. S., Eremomycin, a novel cyclic glycopeptide antibiotic, *Antibiot. Med. Bioteknol. 32*, 571, 1987.
47. Riva, E., Selva, E., Denaro, M., Cassani, G., and Parenti, F., New antibiotic A42867 produced by new Nocardia sp. useful as antibacterial agent and animal feed additive, *Eur. Patent Appl. 254999*, 1987.
48. Hamill, R. L., Mabe, J. A., Mahoney, D. F., Nakatsukasa, W. M., and Yao, R. C-F., Glycopeptide antibiotics, *Eur. Patent Appl. 265071*, 1987.
49. Pan, C. H., Chan, G., Nisbet, L. J., Polansky, M. J., Vogt-Speth, S., and Sitrin, R. D., Analytical HPlC of the aridicin glycopeptide complex and its application to fermentation development, *J. Indust. Microbiol. 1*, 303, 1987.
50. Corti, A., and Cassani, G., Synthesis and characterization of D-alanyl-D-alanine-Agarose: a new bioselective adsorbent for affinity chromatography of glycopeptide antibiotics, *Appl. Biochem. Biotechnol. 11*, 101, 1985.
51. Folena-Wasserman, G., Sitrin, R. D., Chapin, F., and Snader, K. M., Affinity chromatography of glycopeptide antibiotics, *J. Chromatography 392*, 225, 1987.
52. Jeffs, P. W., Mueller, L., DeBrosse, C., Heald, S. L., and Fisher, R., Structure of aridicin A. An integrated approach employing 2D NMR, energy minimization, and distance constraints, *J. Am. Chem. Soc. 108*, 3063, 1986.
53. Batta, G., Sztaricskai, F., Csanadi, J., Kamaromi, I., and Bognar, R., ^{13}C NMR study of actinoidins: carbohydrate moieties and their glycosidic linkages, *J. Antibiotics 39*, 910, 1986.
54. Heald, S. L., Mueller, L., and Jeffs, P. W., Actinoidins A and A_2: structure determination using 2D NMR methods., *J. Antibiotics 40*, 630, 1987.
55. Roberts, G. D., Carr, S. A., Rottschaefer, S., and Jeffs, P. M., Structural characterization of glycopeptide antibiotics related to vancomycin by Fast Atom Bombardment Mass Spectrometry, *J. Antibiotics 38*, 713, 1985.
56. Naylor, S., Skelton, N. J., and Williams, D. H., Analysis of glycopeptides by Fast Atom Bombardment Mass Spectrometry in relation to their hydrophobicity, *J. Chem. Soc. Commun.*, 1619, 1986.
57. Jeffs, P. W., Yellin, B., Mueller, L., and Heald, S. L., Structure of the antibiotic OA-7653, J. Org. Chem. 53, 471, 1988.
58. Tsuji, N., Kobayashi, M., Kamigauchi, T., Yoshimura, Y., and Terui, Y., New glycopeptide antibiotics. I. The structure of orienticins, *J. Antibiotics 41*, 819, 1988.
59. Brazhnikova, M. G., Properties of eremomycin, a new glycopeptide antibiotic. *Proc. 2nd Intern. Symp. "New Bioactive Metabolites from Microorganisms"*, May, 2–7, 1988, Gera (GDR), Poster no. 40.
60. McGahren, W. J., Leese, R. A., Barbatschi, F., Morton, G. O., Kuck, N. A., and Ellested, G. A., Components and degradation compounds of the avoparcin complex, *J. Antibiotics 36*, 1671, 1983.
61. McGahren, W. J., Martin, J. H., Morton, G. O., Hargreaves, R. T., Leese, R. A., Lovell, F. M., Ellestad, G. A., O'Brien, E., and Holker, J. S. E., Structure of avoparcin components, *J. Am. Chem. Soc. 102*, 1671, 1980.
62. Takatsu, T., Takahashi, S., Nakajima, M., and Haneishi, T., Chloropolysphorins A, B and C, novel glycopeptide antibiotics from *Faenia interjecta* sp. nov. III. Structure elucidation of chloropolysporins, *J. Antibiotics 40*, 933, 1987.
63. Hunt, A. H., Elzey, T. K., Merkel, K. E., and Debono, M., Structures of the actaplanins, *J. Org. Chem. 49*, 641, 1984.
64. Debono, M., Molloy, R. M., Bernhart, M., and Dorman, D. E., A35512, a complex of new antibacterial antibiotics produced by *Streptomyces candidus*. II. Chemical studies on A35512B. *J. Antibiotics 33*, 1407, 1980.
65. Malabarba, A., Strazzolini, P., Depaoli, A., Landi, M., Berti, M., and Cavalleri, B., Teicoplanin, antibiotics from *Actinoplanes teichomyceticus* nov. sp. VI. Chemical degradation: physico-chemical and biological properties of acid hydrolysis products, *J. Antibiotics 37*, 988, 1984.
66. Barna, J. C. J., Williams, D. H., Stone, D. J. M., Leung, T.-W. C., and Doddrell, D. M., Structure elucidation of the teicoplanin antibiotics, *J. Am. Chem. Soc. 106*, 4895, 1984.

67. Malabarba, A., Ferrari, P., Gallo, G. G., Kettenring, J., and Cavalleri, B., Teicoplanin, antibiotics from *Actinoplanes teichomyceticus* nov. sp. VII. Preparation and NMR characteristic of the aglycone of teicoplanin, *J. Antibiotics 39*, 1430, 1986.
68. Waltho, J. P., Williams, D. H., Selva, E., and Ferrari, P., Structure elucidation of the glycopeptide complex A40926, *J. Chem. Soc., Perkin Trans. I*, 2103, 1987.
69. Pitkin, D. H., Mico, B. A., Sitrin, R. D., and Nisbet, L. J., Charge and lipophilicity govern the pharmacokinetics of glycopeptide antibiotics, *Antimicrob. Agents Chemother. 29*, 440, 1986.
70. Wittendorf, R. W., Swagzdis, J. E., Gifford, R., and Mico, B. A., Protein binding of glycopeptide antibiotics with diverse physical-chemical properties in mouse, rat, and human serum, *J. Pharmacokinet. Biopharm. 15*, 5, 1987.
71. Hammond, S. J., Williamson, M. P., Williams, D. H., Boeck, L. D., and Marconi, G. G., On the biosynthesis of the antibiotic vancomycin, *J. Chem. Soc., Chem. Commun.* 344, 1982.
72. Hammond, S. J., Williams, D. H., and Nielsen, R. V., The biosynthesis of ristocetin, *J. Chem. Soc., Chem. Commun.* 116, 1983.
73. Chung, S. K., Taylor, P., Oh, Y. K., DeBrosse, C., and Jeffs, P. W., Biosynthetic studies on aridicin antibiotics. I. Labeling patterns and overall pathways, *J. Antibiotics 39*, 642, 1986.
74. Zmijewski, Jr., M. J., Briggs, B., Logan, R., and Boeck, L. D., Biosynthetic studies on antibiotic A47934, *Antimicrob. Agents Chemother. 31*, 1497, 1987.
75. Towsend, C. A., and Brown, A. M., Nocardicin A: biosynthetic experiments with amino acid precursors, *J. Am. Chem. Soc. 105*, 913, 1983.
76. Chung, S. K., Oh., Y. K., Taylor, P., Gerber, R., and Nisbet, L. J., Biosynthetic studies of aridicin antibiotics. II. Microbial transformations and glycosylations by protoplasts, *J. Antibiotics 39*, 652, 1986.
77. Assi, F., Lancini, G. C., and Gianantonio, A., Selective increase in ratio of single T-A2 major component of teicoplanin A2 complex, *Eur. Patent. Appl. 204179*, 1988.
78. Selva, E., Gastaldo, L., Denaro, M., Cassani, G., and Parenti, F., Production of antibiotic complex A40926 enriched in particular factors by fermenting Actinomadura ATCC 39727 in presence of appropriate precursor, e. g. valine or isoleucine, *Eur. Patent Appl. 259781*, 1986.
79. Reynolds, P. E., Studies on the mode of action of vancomycin, *Biochim. Biophys. Acta. 52*, 403, 1961.
80. Jordan, D. C., Effect of vancomycin on the synthesis of the cell wall mucopeptide of *Staphylococcus aureus, Biochem. Biophys. Res. Commun. 6*, 167, 1961.
81. Wallas, C. H., and Strominger, J. L., Ristocetins, inhibitors of cell wall synthesis in *Staphylococcus aureus, J. Biol. Chem. 238*, 2264, 1963.
82. Somma, S., Gastaldo, L., and Corti, A., Teicoplanin, a new antibiotic from *Actinoplanes teichomyceticus* nov. sp., *Antibicrob. Agents Chemother. 26*, 917, 1984.
83. Reynolds, P. E., Peptidoglycan synthesis in bacilli. II. Characteristics of protoplast membrane preparations, *Biochim. Biophys. Acta, 237*, 255, 1971.
84. Jordan, D. C., and Mallory, H. D. C., Site of action of vancomycin on *Staphylococcus aureus, Antimicrob. Agents Chemother. 1964*, 489, 1964.
85. Anderson, J. S., Matsuhashi, M., Haskin, M. A., and Strominger, J. L., Lipid-phosphoacetylmuramyl-pentapeptide and lipid-phosphodisaccharide-pentapeptide: presumed membrane transport intermediates in cell wall synthesis, *Proc. Natl. Acad. Sci. U.S. 53*, 881, 1965.
86. Anderson, J. S., Matsuhashi, M. Haskin, M. A., and Strominger, J. L., Biosynthesis of the peptidoglycan of bacterial cell walls. II. Phospholipid carriers in the reaction sequence. *J. Biol. Chem. 242*, 3180, 1967.
87. Matsuhashi, M., Dietrich, C. P., and Strominger, J. L., Biosynthesis of the peptidoglycan of bacterial cell walls. III. The role of soluble ribonucleic acid and of lipid intermediates in glycine incorporation in *Staphylococcus aureus, J. Biol. Chem. 242*, 3191, 1967.
88. Sinha, R. K., and Neuhaus, F. C., Reversal of the vancomycin inhibition of peptidoglycan synthesis by cell walls, *J. Bacteriol. 96*, 374, 1968.
89. Chatterjee, A. N., and Perkins, H. R., Compounds formed between nucleoides related to the biosynthesis of bacterial cell wall and vancomycin, *Biochem. Biophys. Res. Commun. 24*, 489, 1966.
90. Nieto, M., and Perkins, H. R., Modification of the acyl-D-alanyl-D-alanine terminus affecting complex-formation with vancomycin, *Biochem. J. 123*, 789, 1971b.

91. Nieto, M., and Perkins, H. R., The specificity of combination between ristocetins and peptides related to bacterial cell wall mucopeptide precursors, *Biochem. J. 124*, 845, 1971c.
92. Williams, D. H., and Kalman, J. R., Structural and mode of action studies on the antibiotic vancomycin. Evidence from 270-MHz Proton Magnetic Resonance, *J. Am. Chem. Soc. 99*, 2768, 1977.
93. Convert, O., Bongini, A., and Feeney, J., A ^{1}H Nuclear Magnetic Resonance Study of the interactions of vancomycin with N-acetyl-D-alanyl-D-alanine and related peptides, *J. Chem. Soc., Perkin Trans. II*, 1262, 1980.
94. Williams, D. H., and Butcher, D. W., Binding site of the antibiotic vancomycin for a cell-wall peptide analogue, *J. Am. Chem. Soc. 103*, 5697, 1981.
95. Williams, D. H., Williamson, M. P., Butcher, D. W., and Hammond, S. J., Detailed binding sites of the antibiotics vancomycins and ristocetin A: determination of intermolecular distances in antibiotic/substrate complexes by use of the time-dependent NOE, *J. Am. Chem. Soc. 105*, 1332, 1983.
96. Kalman, J. R., and Williams, D. H., An NMR study of the interaction between the antibiotic ristocetin A and a cell wall peptide analogue. Negative Nuclear Overhauser Effects in the investigation of drug binding sites, *J. Am. Chem. Soc. 102*, 906, 1980.
97. Williamson, M. P., and Williams, D. H., ^{1}H N.M.R. studies of the structure of ristocetin A and of its complexes with bacterial cell wall analogues in aqueous solution, *J. Chem. Soc., Perkin Trans. I*, 949, 1985.
98. Fesik, S. W., Armitage, I. M., Ellestad, G. A., and McGahren, W. J., Nuclear Magnetic Resonance studies on the interaction of avoparcin with model receptors of bacterial cell walls, *Molecul. Pharmacol. 25*, 281, 1984.
99. Williamson, M. P., Williams, D. H., and Hammond, S. J., Interactions of vancomycin and ristocetin with peptides as a model for protein binding, *Tetrahedron 40*, 569, 1984.
100. Harris, C. M., Kopecka, H., and Harris, T. M., The stabilization of vancomycin by peptidoglycan analogs, *J. Antibiotics 38*, 51, 1985.
101. Goldstein, B. P., Selva, E., Gastaldo, L., Berti, M., Pallanza, R., Ripamonti, F., Ferrari, P., Denaro, M., Arioli, V., and Cassani, G., A40926, a new glycopeptide antibiotic with anti-*Neisseria* activity, *Antimicrob. Agents Chemother. 31*, 1961, 1987.
102. Malabarba, A., Trani, A., Ferrari, P., Pallanza, R., and Cavalleri, B., Synthesis and biological activity of some esters of the *N*-acetylglucosaminyl aglycone and of the aglycone of teicoplanin, *J. Antibiotics 40*, 1572, 1987.
103. Perkins, H. R., and Nieto, M., The preparation of iodinated vancomycin and its distribution in bacteria treated with the antibiotic, *Biochem. J. 116*, 83, 1970.
104. Perkins, H. R., Vancomycin and related antibiotics, *Pharmacol. Ther. 16*, 181, 1982.
105. Pallanza, R., Berti, M., Goldstein, B. P., Mapelli, E., Randisi, E., Scotti, R., and Arioli, V., Teichomycin: *in vitro* and *in vivo* evaluation in comparison with other antibiotics, *J. Antimicrob. Chemother. 11*, 419, 1983.
106. Watanakunakorn, C., Mode of action and *in vitro* activity of vancomycin, *J. Antim. Chemother. 14*, Suppl. D, 7, 1984.
107. Kirby, W. M. M., Vacomycin therapy of Staphyloccocal infections, *Antibiotica et Chemotherapia 11*, 84, 1963.
108. Weidemann, B., and Kresken, M. The incidence and development of resistance in *Staphylococcus aureus* from three European countries, *J. Antim. Chemother. 14*, Suppl. D, 27, 1984.
109. Leclercq, R., Derlot, E., Duval, J. D., and Courvalin, P., Plasmid-mediated resistance to vancomycin and teicoplanin in *Enterococcus faecium, N. Engl. J. Med. 316*, 927, 1987.
110. Lechevalier, M. P., Prauser, H., Labeda, D. P., and Ruan, J.-S., Two new genera of nocardioform actinomycetes: *Amycolata* gen. nov. and *Amycolatopsis* gen. nov., Int. J. Syst. Bacterol. *36*, 29, 1986.
111. Riva, E., Gastaldo, L., Beretta, M. G., Ferrari, P., Zerilli, L. F., Cassani, G., Goldstein, B. P., Berti, M., Parenti, F., and Denaro, M., A42867, a novel glycopeptide antibiotic, J. Antibiotics *42*, 497, 1989.
112. Sitrin, D. R., and Wasserman, G. F., Affinity and HPLC purification of glycopeptide antibiotics, in *Natural Products Isolation,* Wagman, G. H., and Cooper, R., Eds., Elsevier, Amsterdam, 1989, p. 111.

113. Holdom, K. S., Maeda, H., Ruddock, J. C., and Tone, J., New glycopeptide antibiotic. eur. Patent Appl. 265, 143, 1988.
114. Sankio, K. K., Helvercardin A and B – Used for treatment and prophylaxis of human and animal infections. Japan Pat. 63152-397-A, 1987.
115. Tsuji, N., Kamigauchi, T., Kobayashi, M., and Terui, Y., New glycopeptide antibiotics: II. the isolation and structures of chloro-orienticins, J. Antibiotics *41*, 1506, 1988.
116. Doolin, L. E., Gale, R. M., Godfrey, O. W. Jr., Hamill, R. L., Mahoney, D. F., and Yao, R. C.-F., Glycopeptide antibiotics. Eur. Patent Appl. 0219707, 1989.

Chapter 8
Peptide Phytotoxins from Plant Pathogenic Fungi

Jonathan D. Walton

1. Introduction
2. Host-Selective Toxins
   2.1 AM-Toxin
   2.2 Host-specific toxin from Alternaria brassicae
   2.3 Victorin
   2.4 HC-toxin
   2.5 PC-toxin
3. Non-Selective Toxins
   3.1 Tentoxin
   3.2 Enniatins
   3.3 Peptidic siderophores
References

1. Introduction

Researchers interested in the molecular factors involved in plant pathogenesis have over the years discovered in fungi and bacteria a multitude of compounds which are toxic to plants. Chemists have elucidated the structures of many of these compounds, and plant pathologists have studied their possible involvement in the diseases caused by the organisms that produce them. In most cases, the chemistry has outpaced the biology: we know the structures of many phytotoxins, but relatively few have been definitively evaluated for a role in pathogenesis.

This review focuses on those phytotoxins that (1) are at least partially peptidic in structure, and (2) are produced by fungi that (3) are plant pathogens. The first restriction is justified on the grounds that compounds of similar structure might be expected to have similar modes of action and especially similar biosynthetic origins. The second restriction is to emphasize that this review does not cover peptidic phytotoxins made by bacteria. The two best-studied peptidic phytotoxins from plant pathogenic bacteria, phaseolotoxin and tabtoxin, are modified di- and tripeptides, respectively. Biosynthetically they probably arise by a conventional single enzyme/single modification sequence similar to glutathione. Their chemistry and mode of action are reviewed in Daly and Deverall (1983); for the recent cloning of the phaseolotoxin biosynthetic gene(s) from *Pseudomonas syringae* pv. *phaseo-*

licola, see Peet et al. (1986) and Kamdar et al. (1988). The third restriction is in order to emphasize those aspects of toxins which are relevant to the field of plant pathology, that is, to an understanding of the molecular interactions between microorganisms and higher plants. This is meant to redress in part the tendency to neglect research on the ecological roles which cyclic peptide "antibiotics" might play in favor of research on their medical applications.

The phytotoxic peptides and the antibiotic peptides have been studied for very different reasons, and this has dictated what questions concerning them have been deemed important. Interest in the antibiotic peptides stems from their medical significance. Determination of their structures is important as the first step towards synthesizing more efficacious analogs, and studies on their biosynthesis have practical importance for improving production. The physiological modes of action of many cyclic peptide antibiotics are not known, although many *in vivo* and *in vitro* effects have been described. More striking has been our inability to assign a function to most peptide antibiotics in the organisms that produce them, that is, to rationalize what selective advantage they confer.

The study of phytotoxic peptides, on the other hand, begins with the hypothesis that they have an ecological function, namely, to enhance the ability of the producing organisms to cause disease and hence increase their fitness. The major conceptual contribution of studies on toxins to plant pathology has been that low molecular weight compounds *can* be determinants of disease. This point originally was demonstrated with unpurified preparations; efforts to purify and characterize the active chemical species came much later. The chemistry has been justified as necessary to rationally explore the sites of action of the toxins in the plants, with the hope (as yet unrealized) that this will lead to the identification and hence manipulation of genes controlling disease resistance in crops. Toxins from plant pathogens have also been studied as potential herbicides.

A major difference between the peptide antibiotics made by *Bacillus* and *Streptomyces* and the phytotoxins made by plant pathogens is that biosynthesis of the latter is, with one known exception, constitutive. [The exception is HS-toxin, a galactosyl-terpenoid host-selective toxin produced by *Cochliobolus (Helminthosporium) sacchari*, a pathogen of sugarcane (Larkin and Scowcroft, 1981)]. In culture, phytotoxins are produced even by germinating spores, and production continues as long as the fungus is growing. Constitutive production, especially in the earliest stages of spore germination, can be understood by considering the importance of the toxins to initial colonization and continued pathogenicity as the organisms that produce them ramify through the plant. It has been shown that HC-toxin and victorin, two cyclic peptide phytotoxins produced by *Cochliobolus carbonum* race 1 and *Cochliobolus victoriae*, respectively, are present in germinating conidia (infectious spores), and that non-toxin producing isolates of the fungi can successfully colonize plant tissues if the appropriate toxins are applied exogenously at the site of penetration (Yoder and Scheffer, 1969; Comstock and Scheffer, 1973).

Many antibiotics, including the toxins produced by plant pathogens, are not produced under all *in vitro* cultural conditions. Is this an indication of nutritional or environmental regulation of toxin production? For example, lysine in the medium inhibits penicillin synthesis by *Penicillium*, because lysine and penicillin biosynthesis share a common pathway which is negatively regulated by the cytoplasmic concentration of lysine (Drew and Demain, 1977). However, apparent phytotoxin production *in vitro* might depend on the medium only to the extent that the medium influences the stability of the toxin, due, for example, to an unfavorable pH, excessive aeration, or to degradation by extracellular enzymes produced by the fungi. To reach valid conclusions, research on the possible regulation of phytotoxin biosynthesis will have to be done by measuring changes in levels of the biosynthetic enzymes or of the mRNA for those enzymes, rather than on the toxin titers in culture. Only recently have plant pathologists begun to study the enzymes of phytotoxin biosynthesis.

From the study of phytotoxic compounds isolated from plant pathogens, what generalizations can be made? First, not surprisingly, all types of molecules are represented – polyketides, terpenoids, carbohydrates, amino acids, peptides, etc. Second, clearly not all phytotoxic compounds produced by plant pathogens have a role in disease. There are several cases in which it has been shown, most clearly by genetic manipulation of the producing organism, that production of a particular phytotoxin by a pathogen does not correlate positively with pathogenicity (defined as the qualitative ability to cause any amount of disease), virulence (defined as the quantitative ability to cause more or less disease), or host range (Yoder, 1980).

Plant pathologists separate phytotoxins into two major groups: those which are host-selective (or host-specific), and those which are non-selective. This review will cover them in turn.

2. Host-Selective Toxins

The host-selective toxins have attracted a great deal of attention from plant pathologists because they can account at the molecular level for the specificity (that is, host range) of the producing pathogens. All host-selective toxins known are produced by fungi, predominantly species in the genera of *Cochliobolus* (also known by its imperfect stage as *Helminthosporium* or *Bipolaris*) and *Alternaria*. The toxins affect only those genotypes or cultivars of host plants which the producing fungi are able to infect in the field, and naturally occurring isolates or mutants of the fungi which cannot make the toxins are not (or only weakly) pathogenic on those cultivars. For recent reviews see Yoder (1980), Kono et al. (1981), and Daly and Deverall (1983).

2.1 AM-toxin

Species of *Alternaria* produce many selective and non-selective phytotoxins. The host-specific toxins produced by *A. kikuchiana*, *A. alternata* f.sp. *lycopersici*, *A.*

citri, *A. longipes*, and *A. fragariae* are not peptidic and will not be considered here (Nishimura and Kohmoto, 1983).

Alternaria alternata p. v. *mali* (*Alternaria mali*) causes a leaf blotch on certain cultivars of apple (Sawamura, 1962). Filtrates of the culture medium in which the fungus had been grown cause symptoms similar to the fungus when applied to leaves of susceptible cultivars. Among thirty-eight cultivars of apple there is a perfect correlation between reaction to the fungus and reaction to cell-free culture filtrates and to purified AM-toxin I (Sawamura, 1966; Okuno et al., 1974a; Ueno et al., 1975a; Kohmoto et al., 1977; Maeno et al., 1984).

As would be necessary for a role in pathogenesis, AM-toxin is constitutively produced, even by germinating spores (Kohmoto et al., 1976). To date three active components have been isolated and characterized from *A. alternata* f.sp. *mali*. All three are closely related cyclic tetradepsipeptides, known as AM-toxins I, II, and III (Figure 1). AM-toxin I is also known as alternariolide (Okuno et al., 1974). The

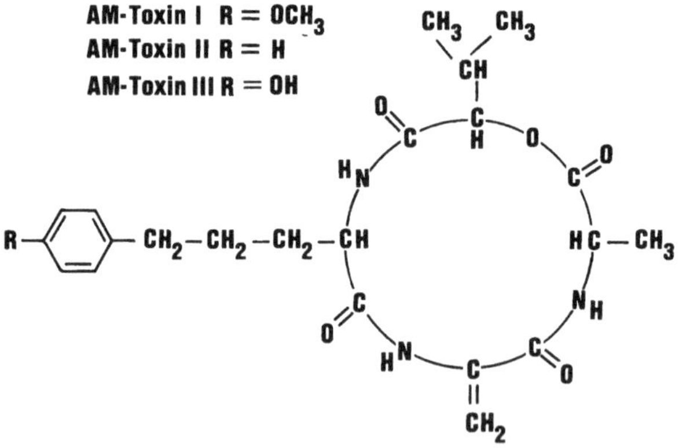

Figure 1 Structures of the AM-toxins made by *Alternaria alternata* pv. *mali*.

constituents of AM-toxin I are L-2-amino-5-(p-methoxyphenyl)pentanoic acid (alternamic acid), dehydroalanine (aminoacrylic acid Dha), L-alanine, and L-2-hydroxy-3-methylbutanoic acid. In AM-toxin II, the methoxy group of alternamic acid is lacking, giving L-2-amino-5-phenylpentanoic acid. In AM-toxin III the methoxy group is replaced by a hydroxyl group, giving L-2-amino-5-(p-hydroxyphenyl)pentanoic acid (Okuno et al. 1974b; Okuno et al., 1975; Ueno et al., 1975b; Ueno et al. 1975c). In addition, several minor host-selective toxins, as yet unidentified, have been detected in culture filtrates of *A. alternata* f.sp. *mali* (Kohmoto et al., 1976).

AM-toxin I is active against susceptible apple varieties (e.g., Indo) at 0.1 to 2 ng/ml, depending on the bioassay (Ueno et al., 1975a). Resistant apple varieties (e.g., Jonathan) are affected only at a concentration 10^4 to 10^5 higher. AM-toxin II and III are ten times less active and as active, respectively, as AM-toxin I (Shimohigashi and Izumiya, 1978; Ueno et al., 1975c).

The AM-toxins have been synthesized (Lee et al., 1976; Shimohigashi et al., 1977a; Shimohigashi and Izumiya, 1978; Kanmera et al., 1981). During the course of these studies a number of derivatives of the AM-toxins were made and tested for biological activity (Shimohigashi et al., 1977a, b; Shimohigashi and Izumiya, 1978; Kanmera et al., 1981; Noda et al., 1980). The dimer of AM-toxin I, a cyclic octadepsipeptide, has been synthesized and has activity against susceptible apple varieties only at 20 to 40 μg/ml (Lee et al., 1976). Whereas AM-toxin containing L-alanine in place of dehydroalanine has very low activity, substitution with D-alanine gives a compound with significant biological activity (Shimohigashi et al., 1978).

Despite several thorough studies on the effects of the AM-toxins on plants, relatively little evidence indicates a particular site of action. Like many other toxins, both selective and non-selective, the AM-toxins cause electrolyte leakage from sensitive cells within a few minutes, indicating membrane damage (Kohmoto et al., 1976; Park et al., 1977; Park et al., 1981). Ultrastructural studies confirm that AM-toxin affects the plasma membrane, but also the chloroplast, morphological changes being seen within one hour after treatment (Park et al., 1977). However, electrolyte leakage is hardly more than an indicator of cell death, and ultrastructural studies could not be expected to elucidate the primary biochemical events that mediate toxicity.

The AM-toxins also induce electrolyte leakage from the Japanese pear cultivar Niijisseiki but not from the pear cultivar Chojuro (Kohmoto et al., 1976). Since pears are not a host of *A. mali*, this demonstrates that production of a host-selective toxin is not by itself sufficient for pathogenicity, even if the producing organism causes diseases on other plants. A similar situation is found in the case of Cyl-2 and chlamydocin, which are structural analogs of the maize host-selective toxin HC-toxin and show to some extent the same type of host selectivity. The former is made by a pathogen of woody plants, the latter by a saprophyte; neither fungus can infect maize (Walton et al., 1985).

2.2 Host-specific toxin from *Alternaria brassicae*

This fungal pathogen causes black spot disease of rapeseed (*Brassica campestris* and *B. napus*), an important source of edible oil. Recently it has been reported that *A. brassicae* produces a host-specific phytotoxin *in vitro* (Bains and Tewari, 1987). Although there are usually only two or three levels of response of plants to host-selective toxins, five levels of differential reaction to *A. brassicae* and its toxin were identified (Bains and Tewari, 1987). At all five levels, susceptibility to the fungus and sensitivity to the toxin were positively correlated without exception. The host-specific toxin produced by *A. brassicae* was identified as destruxin B (Bains and Tewari, 1987), which had been previously identified as a member of a family of metabolites from the fungus *Metarrhizium anisopliae*, a pathogen of insects (Tamura et al., 1964). Structurally, destruxin B is a cyclic peptido-lactone containing L-proline, L-isoleucine, L-N-methylvaline, L-N-methylalanine, L-beta-alanine, and D-alpha-hydroxymethylvaleric acid (Figure 2). The related compounds made by

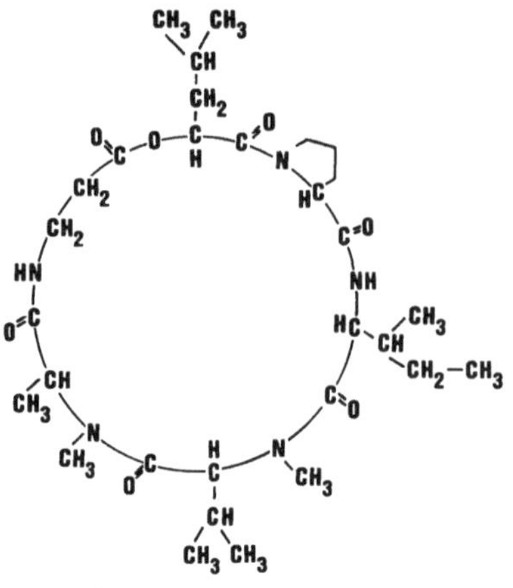

DESTRUXIN B

Figure 2 Structure of destruxin B made by *Metarrhizium anisopliae* and *Alternaria brassicae.*

M. anisopliae (but apparently not by *A. brassicae*) have different side chains on the hydroxy acid, and lack N-methyl residues on the valine and/or alanine (Suzuki and Tamura, 1972). Structurally, the destruxins are similar to the AM-toxins made by another species of *Alternaria*. Destruxin B and a desmethyl derivative, protodestruxin, have been synthesized (Kuyama and Tamura, 1965; Lee et al., 1975).

The destruxins are toxic to mice and insects but not microorganisms. Prior to the report of Bains and Tewari (1987) they had apparently not been tested for phytotoxicity. Nothing is known about their mode of action against plants.

It is not known how the destruxins are biosynthesized. It has been proposed that compounds similar to destruxin B but lacking N-methylation on either the valine residue or both the valine and alanine residues ("protodestruxin") are precursors of destruxin B (Suzuki et al., 1970; Suzuki and Tamura, 1972). In light of the subsequent discovery that N-methylation of other cyclic peptido-lactones occurs on multifunctional enzymes after amino acid binding but before peptide bond formation (Zocher et al., 1982; Zocher et al., 1986; Keller, 1987), these compounds are probably not precursors in the normal sense, but rather are "parallel" products made by the same enzyme or enzymes. It might be relevant that protodestruxin was isolated from a methionine auxotroph of *M. anisopliae* (Suzuki and Tamura, 1972), and that the *N*-methyl groups in enniatin, cyclosporin, and actinomycin are derived from S-adenosylmethionine. If the original report of Bains and Tewari (1987) is confirmed, this would make a very nice system with which to study the biosynthesis of a host-selective phytotoxin.

2.3 Victorin

The most extensively studied host-selective toxin is victorin, also known as HV-toxin, produced by the oat pathogen *Cochliobolus (Helminthosporium) victoriae.*

The appearance of this destructive pathogen in the 1940's in the U.S.A. was preceded by the introduction of a gene for resistance to the agent of the crown rust disease, *Puccinia coronata avenae*, into the major oat varieties of that time. Unfortunately, this same "resistance" gene causes susceptibility to *C. victoriae*, and after a few years of severe losses, the "improved" varieties were abandoned. Shortly after the first description of the disease (Meehan and Murphy, 1946), it was discovered that cell-free culture filtrates of the fungus were selectively toxic to the same oat varieties that were susceptible to *C. victoriae* (Meehan and Murphy, 1947). From this came the first convincing demonstration that low molecular weight fungal metabolites could account for the host-range of plant pathogenic fungi, and sparked a great deal of effort not only to understand the chemistry and biology of victorin, but also to discover other host-selective toxins.

The structure of victorin has finally been elucidated. It is a family of unusual partially cyclic pentapeptides containing novel amino acids and chlorine. The major victorin species, victorin C, contains 5,5-dichloroleucine, *erythro-β*-hydroxyleucine, *threo-β*-hydroxylysine, *α*-amino-*β*-chloroacrylic acid, and a cyclic *α*-amino acid called victalanine (Figure 3). Victorin is glyoxylated at the N-terminus (Macko et

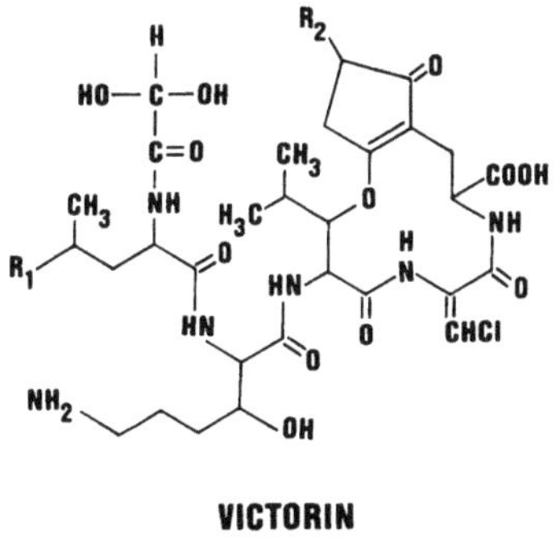

		R_1	R_2
Victorin	B	CH_2Cl	OH
	C	$CHCl_2$	OH
	D	$CHCl_2$	H
	E	CCl_3	OH

Figure 3 Structures of the victorins made by *Cochliobolus victoriae*.

al., 1985; Wolpert et al., 1985). The elemental formula of victorin C and the structures of three of the amino acids have been independently confirmed (Gloer et al., 1985). The structures of the minor species of victorin are similar except for the substitution of chloroleucine or 5,5,5-trichloroleucine for dichloroleucine (victorins B and E, respectively), a hydrogen atom instead of a hydroxyl group on victalanine (victorin D), and 2,5-DOPA (hydroxytyrosine) in place of victalanine (called victoricine). In victoricine, the oxygen bridge to hydroxyleucine is via the hydroxyl group at the 5 position in DOPA (Wolpert et al., 1986).

Victorins B and D are approximately as biologically active as victorin C, but victorin E is more active and victoricine is somewhat less active (Wolpert et al., 1988). The glyoxylic acid residue is absolutely essential for biological activity. Methylation of the carboxyl group or of the *ε*-amino group of hydroxylysine reduces activity by 10 and 100 times, respectively, but the derivatives are still host-selective (Wolpert et al., 1988). Biological activities of the minor toxins and derivatives were evaluated with a dark CO_2 fixation bioassay originally developed to assay HMT-toxin from *Cochliobolus heterostrophus* (Bhullar et al., 1975). This assay might not be the most appropriate for victorin bioassays. The biochemical basis of the inhibition of dark CO_2 fixation by HMT-toxin is most likely depletion of cytoplasmic pools of phosphoenolpyruvate (PEP), since maize contains large amounts of PEP (maize is a C4 plant and hence uses PEP carboxylase as the primary step in photosynthetic carbon assimilation). Regeneration of one mole of PEP from pyruvate requires two moles of ATP, and HMT-toxin disrupts mitochondrial ATP synthesis in sensitive maize cells (Walton et al., 1979). Since oats are a C3 plant, this pathway is much less active than in maize. This is probably reflected in the fact that the dark CO_2 fixation assay is insensitive (EC_{50} 37 nM – Wolpert et al., 1988) compared to other bioassays such as root growth or protoplast survival (EC_{50} 120 pM – Walton and Earle, 1984; Macko et al., 1985; Mayama et al., 1986). An additional problem with the standard dark CO_2 fixation assay is the significant diffusion barrier presented by the cuticle and tissue of leaf disks. Although closely related compounds such as the native victorins and minor derivatives probably enter tissues at approximately the same rate, this is not necessarily true for victorin conjugated to biotin or Bolton-Hunter reagent (Wolpert et al., 1988). Comparative toxicity studies using complex tissues are further confounded by the fact that a major feature of victorin's phytotoxicity is disruption of the symplast/apoplast barrier (Keck and Hodges, 1973). Therefore, the more toxic victorin derivatives would penetrate tissues more effectively, resulting in a further increase in their apparent toxicity.

To accompany its unusual chemistry, victorin has an intriguing biology. It is arguably the most phytotoxic compound known, inhibiting growth of sensitive plants half-maximally at 100 pg/ml (120 pM) yet having no perceptible effects on resistant oats even at a concentration 500,000 times higher (Walton and Earle, 1984). Sensitivity is determined by a single dominant gene, called *Vb* or *Pc*, which the best available evidence indicates also determines resistance (i. e., the opposite disease phenotype) to crown rust (Rines and Luke, 1985).

Despite a large number of studies on the mode of action of victorin, one can conclude little more beyond that victorin has strong and rapid effects on many cellular processes. Probably, although the evidence is far from definitive, the initial lesion induced by victorin is at the plasma membrane and/or at other cellular membranes. This lesion causes electrolyte leakage, inhibition of macromolecule synthesis, and cell death (Samaddar and Scheffer, 1967; Keck and Hodges, 1973; Novacky and Hanchey, 1974; Walton and Earle, 1985a). Pretreatment with inhibitors of protein synthesis can reversibly protect sensitive tissues against victorin (Gardner and Scheffer, 1973; Rancillac et al., 1976; Walton and Earle, 1985a). This finding in combination with the genetic evidence has led to the proposal that the plasma membrane of sensitive oats contains a toxin binding site, or receptor, which has a relatively high turnover rate. According to this hypothesis, insensitive oats and sensitive plants which have been treated with inhibitors of protein synthesis lack this receptor (Gardner and Scheffer, 1973). To date, no direct evidence for this hypothesis has been obtained, but on the basis of our current understanding of the chemistry of victorin it is now feasible to prepare radiolabelled derivatives and undertake *in vitro* binding experiments to identify a biochemical site of action. Wolpert and Macko (1989) have reported that radioiodinated victorin binds host-selectively to oat membranes.

Little is known about the biosynthesis of victorin. L-[4,5-^{3}H] leucine is incorporated into victorin *in vivo* (Gloer et al., 1985). Although it has been suggested that the dichloroleucine in victorin might be biosynthetically derived from dichloropyruvate (Macko et al., 1985), the efficiency of incorporation of tritiated leucine is more consistent with leucine being incorporated into victorin directly and not by degradation to acetyl-CoA and then back into leucine by the leucine biosynthetic pathway. That is to say, chlorine might instead be incorporated at some step after synthesis of the leucine, not before.

Cycloheximide inhibits victorin production (Gloer et al., 1985), which could be taken as *prima facie* evidence that, unlike other cyclic peptides, victorin is synthesized on ribosomes. However, in such long-term experiments it is difficult to separate an inhibition of victorin production due to inhibition of the ribosomal synthesis of the enzymes necessary to make victorin from a direct effect on victorin biosynthesis. Most likely victorin is biosynthesized by a nucleic acid-independent "thioltemplate" mechanism like other cyclic peptides (other chapters in this volume). A logical approach to discovering the enzyme or enzymes involved in biosynthesis of victorin would be to test cell-free extracts of *C. victoriae* for ATP/PP$_i$ exchange dependent on the amino acids present in victorin. β-hydroxyleucine is commercially available; the others would have to be synthesized. However, since amino acid modifications frequently are catalyzed by cyclic peptide synthetases, leucine and lysine might be the appropriate substrates to assay a victorin synthetase by ATP/PP$_i$ exchange. A variety of non-toxin producing natural isolates and mutants of this fungus are available to confirm that any identified enzymatic activities are in fact involved in the biosynthesis of victorin.

Of relevance to attempts to find a "victorin synthetase" is the genetic analysis of toxin production. Although no detailed studies of toxin production by *C. victoriae* alone have been done, in crosses between it and *C. carbonum* race 1, which makes the cyclic tetrapeptide HC-toxin (see below), the progeny segregate 1:1:1:1 for production of victorin, HC-toxin, both toxins, or neither toxin (Scheffer et al., 1967). This striking result is not explained by assuming that these "secondary metabolites" require many biosynthetic steps catalyzed by many independent enzymes encoded by many independent genes. Nor can it be the case that victorin and HC-toxin are a single enzymatic step apart, considering their very different structures. One possibility is that the isolates chosen for this experiment fortuitously had all but one of the structural genes (i. e., genes coding for enzymes) necessary to make the other toxin. A variation of this possibility is that the isolates fortuitously had all the structural genes necessary to synthesize the other toxin, but differed only at a genetic locus regulating the expression of the genes for those enzymes. Another possibility is that victorin and HC-toxin are synthesized by single multifunctional polypeptides encoded by single genetic loci. This must be considered seriously because at least one other cyclic peptide, enniatin B, is known to be synthesized in this manner (Zocher et al., 1982). A third possibility is that the genetic loci in *C. victoriae* and *C. carbonum* race 1 which control toxin production are complex. This will be considered in more detail below in the discussion of HC-toxin.

2.4 HC-toxin

Cochliobolus carbonum (*Helminthosporium carbonum* or *Bipolaris zeicola*) race 1 parasitizes only maize cultivars which are homozygous recessive at the nuclear *Hm* locus. *C. carbonum* produces a compound, HC-toxin, which inhibits root growth of *hmhm* maize at 0.5 µg/ml but inhibits growth of *HmHm* or *Hmhm* maize only at a hundred-fold higher concentration (Scheffer and Ullstrup, 1965; Pringle and Scheffer, 1967). HC-toxin is produced by germinating conidia, and addition of exogenous HC-toxin allows colonization of susceptible maize by *C. victoriae*, a pathogen of oats (Comstock and Scheffer, 1973).

Early work on the structure of HC-toxin indicated that it was a cyclic (or N-terminus blocked) peptide containing alanine and proline in a ratio of 2 to 1 (Pringle, 1971). Many years later three different groups published nearly-identical

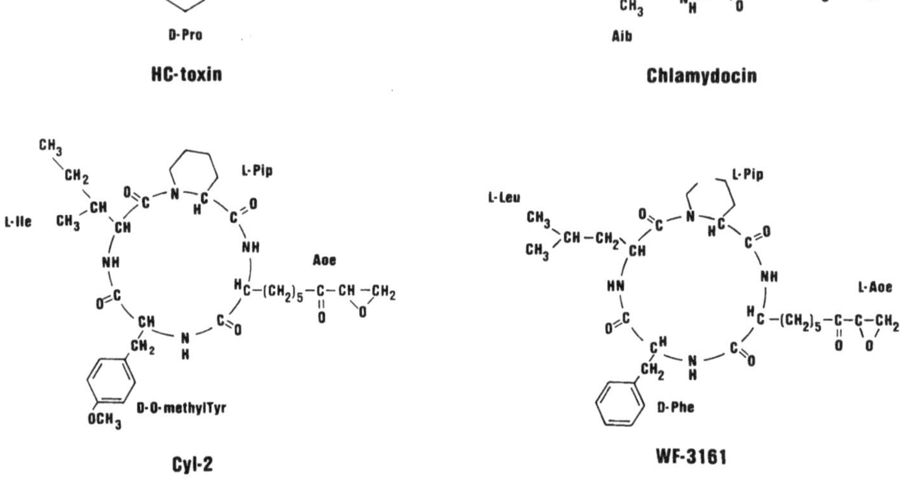

Figure 4 Structures of HC-toxin, chlamydocin, Cyl-2, and WF-3161 made by *Cochliobolus carbonum* race 1, *Diheterospora chlamydosporia, Cylindrocladium scoparium*, and *Petriella guttulata*, respectively.

structures within a short time. All of them concurred that HC-toxin is a cyclic tetrapeptide containing 2 moles of alanine, 1 mole of proline, and 1 mole of 2-amino-8-oxo-9,10-epoxidecanoic acid, referred to hereafter as AOE (Figure 4). AOE had been previously described from other cyclic tetrapeptides. Leisch et al. (1985) were the first to publish that HC-toxin was a tetrapeptide containing AOE. Walton et al. (1982a, b) proposed a different sequence of amino acids, and also chiralities for the alpha carbons of the four amino acids. This revised sequence was confirmed by Gross et al. (1982) and Pope et al. (1983). The stereochemistry of the alanine residues, both of which Walton et al. (1982b) originally had proposed to have the L configuration, was reexamined in light of NMR studies done in the laboratory of Daniel Rich at the University of Wisconsin. Analysis of the amino acids in HC-toxin by HPLC separation of D- and L-alanine with a chiral mobile phase, and also by digestion with D- and L-amino acid oxidases followed by automated amino acid analysis, established that HC-toxin contained one mole of D-alanine and one mole of L-alanine (Kawai et al., 1983; Walton and Earle, 1985b). The difficult job of determining which alanine was D and which L was settled by NMR (Kawai et al., 1983).

With data obtained by nuclear Overhauser effect (NOE) NMR, Kawai et al. (1983) showed that in chloroform HC-toxin adopts a *trans,trans,trans,trans* conformation, with a bis-gamma-turn. The same data were obtained by Mascagni et al. (1983), who, lacking experimental data on the configurations of the alpha carbons of the four amino acids, only were able to suggest a "most-likely" model for the conformation of HC-toxin. The structure of HC-toxin has been confirmed by total synthesis (Kawai and Rich, 1983).

To date, two (and maybe three) minor species of HC-toxin have been found in culture filtrates of *C. carbonum* race 1. In one, HC-toxin II, glycine replaces D-alanine (Kim et al., 1985; Tanis et al., 1986). In another, HC-toxin III, trans-3-hydroxyproline (configuration of the alpha carbon not reported) replaces D-proline (Tanis et al., 1986). All three have the amino acid sequence NH_2-AOE-Pro-COOH. The three HC-toxins exist in culture filtrates in a ratio of 200:2:1 (Tanis et al., 1986). Kim et al. (1985) reported that HC-toxin II is 30 to 40 times less active than HC-toxin I, but Rasmussen and Scheffer (1988) reported that HC-toxin II is half as active as HC-toxin I. Both groups used the same root growth bioassay. HC-toxin III and a new, as yet unidentified compound, HC-toxin IV, are 10 times and 100 times less active than HC-toxin I, respectively (Rasmussen and Scheffer, 1988).

As was found for the related cyclic tetrapeptide chlamydocin (Figure 4) (Closse et al., 1974; and see below), an intact epoxide is required for the biological activity of HC-toxin. Compounds lacking it cannot protect sensitive maize against native HC-toxin (Walton and Earle, 1983; Ciuffetti et al., 1983). Reduction of the vicinal ketone also abolishes activity, but it has not been reported if the 8-hydroxy derivative gives protection against native HC-toxin (Kim et al., 1987).

HC-toxin is a member of a family of four cyclic tetrapeptides all containing AOE and an imino acid, either proline or pipecolic acid (Figure 4). Cyl-2 and chlamydocin have some host-selective activity, but are more active than HC-toxin against HC-toxin-insensitive maize (Walton et al., 1985). All four compounds are active as antimitogens against cultured mammalian cancer cells. On several grounds Shute et al. (1987) have proposed that the biologically active conformation of all four compounds is *cis,trans,trans,trans*. Considering their structural resemblance and biological activities as cytostatic but not cytotoxic agents against both maize and mammalian cells, it is possible that all four cyclic peptides shown in Figure 4 have a similar mode of action (Walton et al., 1985; Shute et al., 1987; but see Kim et al., 1987).

The mode of action of HC-toxin is unknown. Unlike other host-selective toxins, such as victorin, it does not cause rapid cell disruption and death. In fact, the survival of isolated leaf protoplasts from sensitive maize is improved by incubation in HC-toxin (Earle and Gracen, 1982; Wolf, 1988). HC-toxin does not inhibit protein, RNA, or DNA synthesis in maize (Walton and Earle, unpublished observations), and these processes are not primary points of attack of chlamydocin or HC-toxin in mammalian cells (Walton et al., 1985). HC-toxin stimulates uptake of nitrate ions but not other ions or neutral solutes (Yoder and Scheffer, 1973a, b). Interestingly, several synthetic peptide alcohols (e.g., N-carbobenzoxy-L-prolyl-L-valinol) also stimulate nitrate uptake in maize (Lin and Kauer, 1985). It is not known if HC-toxin and the peptide alcohols act similarly.

Like victorin, production of HC-toxin by *C. carbonum* race 1 is controlled by a single Mendelian genetic locus (Scheffer et al., 1967; Yoder 1980; Panopoulus et al., 1984). In fact, some years prior to the discovery of HC-toxin it had been demonstrated that enhanced virulence on *hmhm* maize segregated as a single gene in crosses between races 1 and 2 of *C. carbonum* (Nelson and Ullstrup, 1961). Partly in order to be able to eventually understand how a single genetic locus controls biosynthesis of a pathogenicity factor which happens to be a complex secondary metabolite, experiments have been initiated to identify and purify enzymes involved in the biosynthesis of HC-toxin. Incorporation of [^{14}C]amino acids into HC-toxin *in vivo* is not inhibited by cycloheximide, indicating a non-ribosomal biosynthetic mechanism (Wessel et al., 1987). Cell free extracts of *C. carbonum* race 1 can activate D-alanine, L-alanine, and L-proline, but not D-proline, by ATP/PP_i exchange (Walton, 1987; Wessel et al., 1988a). These activities are clearly involved in the biosynthesis of HC-toxin because they are lacking in race 2 and race 3 isolates of *C. carbonum*, which do not make HC-toxin, and because the presence of these activities segregates with the ability to produce HC-toxin in sexual crosses (Walton, 1987). *In vitro* biosynthesis of the complete HC-toxin has not been attempted, mainly because AOE is difficult to synthesize and its biosynthetic route is unknown.

The ATP/PP_i exchange activities that activate alanine and proline are co-eluted from a gel filtration column with an apparent molecular weight of 300,000 (Walton, 1987). However, they are separated from each other by ion exchange or hydroxyapatite chromatography. The enzyme which activates L-proline has been called HTS-1, and the enzyme which activates D-alanine and L-alanine, HTS-2 (Walton and Holden, 1988). After separation by ion exchange, if the two enzymes are then rerun on gel filtration, they both individually have apparent molecular weights of 300,000. However, by SDS-PAGE the two activities have molecular weights of

220,000 and 170,000, respectively. Wessel et al. (1987, 1988a) have also found enzymes in *C. carbonum* race 1 which can activate D-alanine, L-alanine, and L-proline. Some discrepancies in results from the two groups could result from aggregation of the two enzymes, from differences in the age of the fungus, or to differences in purification protocols (W. Wessel, personal communication).

Although HC-toxin contains D-proline, only L-proline is recognized by "HC-toxin synthetase" (Wessel et al. 1988a; Walton, 1987). In this respect it is different than gramicidin S synthetase (Vater and Kleinkauf, 1976), but similar to actinomycin D and bacitracin synthetases, which recognize only the L form of certain amino acids even though the products contain the D form. Like other cyclic peptide synthetases, HTS-1 binds [^{14}C]L-proline as the aminoacyl thioester (Walton and Holden, 1988). Thus, HTS-1 uses L-proline as substrate and converts it to D-proline, prior to incorporation into HC-toxin. This experiment also conclusively proves that the proline-activating enzyme is not a proline aminoacyl-tRNA synthetase.

HC-toxin contains two moles of alanine, one each of the D and L configurations (Figure 4). Both are activated by a single enzyme. This raises interesting questions about the relation of the two alanines in regard to activation and epimerization. When [^{14}C]D-alanine is fed to the fungus, it is incorporated only into the D-alanine moiety of HC-toxin (unpublished data). Consistent with this experiment, after radiolabelled D-alanine is incubated with the partially purified alanine enzyme under conditions to promote D-alanyl thioester formation, only D-alanine can be released by subsequent treatment with performic acid. In contrast, when the same experiment is done with radiolabelled L-alanine, performic acid releases both L-alanine and D-alanine (Walton and Holden, 1988). Therefore, L-alanine is a biosynthetic precursor of both the L- and D-alanine residues in HC-toxin, but D-alanine is not the precursor of the L-alanine residue. We cannot exclude the possibility that *C. carbonum* has a significant cytoplasmic pool of D-alanine, synthesized by another route, and that the D-alanine in HC-toxin comes both from this pool and from L-alanine.

We have estimated the affinities (K_m values) and capacities (V_{max} values) of the alanine and proline enzymes of HC-toxin synthetase for D-alanine, L-alanine, and L-proline. The affinities are surprisingly low. HTS-1 has a K_m for L-proline of 18 mM, and HTS-2 has K_m values for L-alanine and D-alanine of 101 mM and 3 mM, respectively (Walton and Holden, 1988). ATP/PP_i exchange dependent on D-alanine and L-alanine are strictly additive, indicating that the activation sites for these two amino acids are independent. Our thioester experiments (see above) are consistent with there also being two independent thioester sites, one for L-alanine and one for D-alanine. This might reflect two initial ("peripheral") thioester sites, or might reflect binding of the amino acids to a qualitatively different site, one "peripheral" and the other "central", following the nomenclature of Lynen for fatty acid synthetase (Lee and Lipmann, 1977; Lipmann, 1982). This is being addressed by more detailed studies of D- and L-alanine activation, and competition, using an assay based on thioester formation rather than ATP/PP_i exchange, and also by identification and analysis of the peptide or peptides which bind D- and L-alanine.

The synthesis, activation, and incorporation of AOE into HC-toxin are still poorly understood. Wessel et al. (1988b) have provided the first clues as to its biosynthetic origins by showing that radiolabelled acetate is incorporated into the AOE residue of HC-toxin. Incorporation of acetate into fatty acids and into HC-toxin are differentially sensitive to cerulenin, an inhibitor of fatty acid synthetase. This suggests that AOE is not synthesized by fatty acid synthetase but rather by some type of polyketide synthetase, which may or may not be a part of HTS-1 and HTS-2 (Wessel et al., 1988b). *In vivo* feeding experiments with [^{13}C]acetate and analysis of the resultant HC-toxin by [^{13}C]NMR might provide additional information on the biosynthetic origins of AOE.

Several attempts to detect enzyme-linked pantetheine associated with either the alanine or the proline activating enzyme from *C. carbonum* race 1 have failed (unpublished data). Using a microbiological assay, we can detect pantothenic acid in proteins where it is known to occur, and in extracts of *C. carbonum* race 1, but in the latter the enzyme-bound pantothenic acid is not co-eluted from a gel filtration column with either HTS-1 or HTS-2.

How do the biochemical studies of HC-toxin biosynthesis relate to the genetic results? The simplest rationalization is that the "tox" locus in *C. carbonum* (called *TOX2* by analogy with the *TOX1* gene in *C. heterostrophus* which controls HMT-toxin synthesis) is a gene cluster, in which the two or more protein coding regions are contiguous on the chromosome and hence segregate as a single gene in crosses. Other possible explanations include the existence of additional undiscovered genes, or alternative splicing of a single coding region.

2.5 PC-toxin

Periconia circinata is a fungal pathogen of grain sorghum (*Sorghum bicolor*), causing a root and crown rot. Susceptibility is incompletely dominant, i.e. cultivars with genotype *PcPc* are completely susceptible, those with genotype *Pcpc* are moderately susceptible, and those with genotype *pcpc* are resistant (Schertz and Tai, 1969). In culture, *P. circinata* produces a compound or compounds, called collectively Pc-toxin, which inhibit growth of sorghum cultivars susceptible to the fungus. Demonstration of an absolute correlation between sensitivity to the toxin and susceptibility to the fungus has established PC-toxin as a host-selective toxin (Scheffer and Pringle, 1961).

A preparation of PC-toxin active against susceptible plants at 0.1 μg/ml (with "no toxicity" to resistant plants), reacted positively with ninhydrin and gave five ninhydrin-reactive spots after acid hydrolysis and separation by TLC (Pringle and Scheffer, 1963). These were subsequently identified as aspartic acid, alanine, glutamic acid and serine by automated amino acid analysis (Pringle and Scheffer, 1966). Wolpert and Dunkle (1980) subsequently produced preparations of PC-toxin which were active at 1 ng/ml, using a somewhat different bioassay than that used by Pringle and Scheffer (1963). They identified two active compounds by TLC and HPLC separation, both being of low molecular weight, acidic, and containing aspartic acid (identified by automated amino acid analysis) and an unidentified polyamine. No additional amino acids were found. Work is in progress on the structures of these purified Pc-toxins (L. Dunkle, personal communication).

Concerning the mode of action of this peptidic toxin, little is known. Gardner et al. (1972), Dunkle and Wolpert (1981), and Arias et al. (1983) concluded on the basis of electrolyte leakage and ultrastructural studies that the initial site of action was probably not the plasma membrane. Pretreatment with the protein synthesis inhibitor cycloheximide protects sensitive sorghum against PC-toxin, a situation similar to that found with the oat-selective victorin (see above). Wolpert and Dunkle (1983) tested the possibility that this protection was due to a requirement for synthesis of a new protein or proteins in the chain of events leading to cell death after exposure to PC-toxin. They found that PC-toxin selectively increases levels of four polypeptides (all of MW 16,000 but with pI's between 5.8 and 6.2) in root tips from sensitive plants. These polypeptides were also present in resistant roots but their levels were not affected by PC-toxin. This effect is due to increased transcription of the gene or genes for the 16,000 MW polypeptides as shown by *in vitro* translation of mRNA isolated from PC-toxin-treated tissue. Is this protein involved in the toxic response to PC-toxin? As Wolpert and Dunkle (1983) state in their discussion, this question will be difficult to answer without knowledge of what the 16,000 MW peptide is.

One puzzling experimental finding is that cycloheximide must be applied at least two hours before PC-toxin to protect against the toxin. However, one would expect that if *de novo* protein synthesis were an essential link in the chain of events leading to cell death, that treatment with cycloheximide even simultaneously with PC-toxin would be sufficient to give protection. The necessity for pre-exposure to cycloheximide argues for the involvement of a constitutive protein with a rapid turnover in the reaction to PC-toxin (see the discussion of victorin above), and not *de novo* gene expression. One difficulty in interpreting experiments such as this is that cycloheximide and Pc-toxin might penetrate the tissue at different rates. The poor penetration of cycloheximide into intact plant tissues has confounded attempts to establish the relationship between protein synthesis and action of the plant hormone auxin (Ray, 1985).

Mercuric chloride at a concentration of 40 µM also induces synthesis of these polypeptides in sorghum, but is non-selective in its effects (Traylor et al., 1987). Heat shock or treatment with cadmium chloride induces a different pattern of new protein synthesis. These results demonstrate that although the response originally described by Wolpert and Dunkle (1983) is not specific to PC-toxin, neither can it be discounted as only a general stress response.

3. Non-Selective Toxins

A non-selective toxin is any compound produced by a plant pathogen that shows any degree of phytotoxicity. There are many more non-selective than host-selective toxins. Although one would expect at least some non-selective toxins to be involved in pathogenesis, this is difficult to prove. The simple and definitive genetic analyses, especially from the point of view of the plant, that can be done with the host-selective toxins cannot be done with the non-selective ones. A few non-selective toxins have been demonstrated not to be involved in pathogenesis, usually by genetic analyses (e. g., Holenstein and Defago, 1983). Nonetheless, non-selective toxins are interesting for several reasons. First, although they may not be required for diseases to occur, they can contribute to causing the characteristic symptoms of the disease, e. g., tentoxin (see below). Second, since most plant pathogens, bacterial and fungal, live part of the time in the soil, non-selective toxins might have a role in coping with competition from other microorganisms. The ability to compete outside the plant could indirectly influence the economic importance of a pathogen *qua* pathogen. Third, many non-selective toxins are interesting for their unusual chemistries or effects on plants.

3.1 Tentoxin

Tentoxin was first isolated from *A. tenuis* (now considered a form of *A. alternata*), a recently described pathogen of cotton and other crops (Fulton et al., 1965). Tentoxin is also made by *A. alternata* pv. *mali*, the pathogen of apple which makes the AM-toxins (see above). In the case of tentoxin, the distinction between a host-selective and a non-selective toxin is not clear-cut, because tentoxin affects many but not all plant species throughout the plant kingdom (Templeton, 1972). All plants evolutionarily less advanced than the ferns are sensitive. The situation among dicotyledonous plant species is complex. In some cases all the species in a family are sensitive, in other cases they are all insensitive, and in a few cases a family or even genus contains both sensitive and insensitive species (Durbin and Uchytil, 1977). Most monocotyledonous plants are resistant to tentoxin, exceptions being sorghum (*Sorghum vulgare*) and crabgrass (*Digitaria* spp.). In crosses between interfertile species of *Nicotiana* which differ in their response to tentoxin, sensitivity to tentoxin is maternally inherited, indicating that the gene or genes controlling response to tentoxin are located in a plastid, presumably the chloroplast (Durbin and Steele, 1979). Since relatively few plants which are sensitive to tentoxin are susceptible to the fungi that make it, and although tentoxin contributes to symptom development, it cannot be considered a determinant of disease.

Tentoxin is one of the best understood phytotoxins both chemically and biologically. Structurally tentoxin is a cyclic tetrapeptide, cyclo(L-MeAla-L-Leu-MePhe(Z)-Gly) (Figure 5) (Meyer et al., 1974; Koncewicz et al., 1973). It has been synthesized (Rich and Mathiaparanam, 1974). The conformation of tentoxin has been studied by NMR using synthetic derivatives such as substitution of proline for L-MeAla and D-MeAla for L-MeAla. Substitution of proline constrains the conformation, which allowed a detailed analysis of its conformation by NMR; biological activity is retained (Rich and Bhatnagar, 1978). When D-MeAla replaces L-MeAla, multiple conformers result which are stable enough to be separated into two fractions by TLC. Both fractions are significantly less active than native tentoxin, but also differ

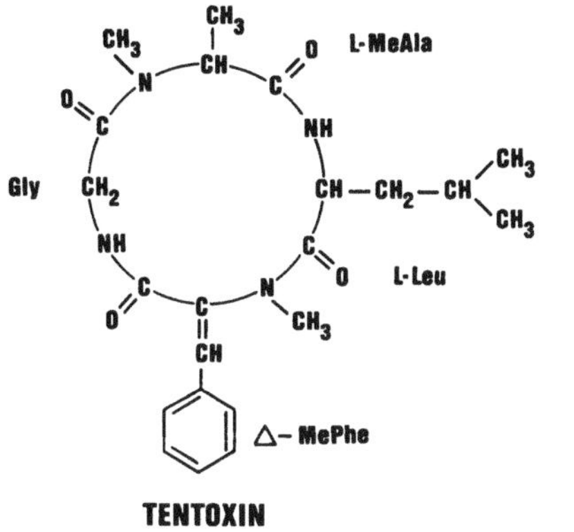

Figure 5 Structure of tentoxin made by *Alternaria alternata* pv. *tenuis* and *A. alternata* pv. *mali.*

significantly from each other. According to the authors, this is the first demonstration that the conformation of a cyclic peptide influences its biological activity (Rich et al., 1978). The conformer of [D-MeAla]tentoxin with the higher activity more closely resembles the conformation deduced for [L-Pro]tentoxin, consistent with this conformer being the one with biological activity. When sarcosine is substituted for L-MeAla, a complex mixture of conformers results and biological activity is almost as high as for native tentoxin; included among the conformers present is the putative "active" one (Rich et al., 1979). Additionally, the double bond in the dehydrophenylalanine residue is essential for biological activity (Koncewicz et al., 1973; Rich et al., 1975).

Because tentoxin causes chlorosis, that is, loss of chlorophyll from treated leaves, studies on tentoxin have focused on the chloroplast as a site of action. Ultrastructural alterations induced by tentoxin are restricted to the chloroplast (Halloin et al., 1970). In due course it was discovered that tentoxin inhibits photophosphorylation (Arntzen, 1972); consistent with this, that tentoxin inhibits chloroplast protein and RNA synthesis dependent on light but not exogenous ATP (Bennett, 1976); and that radiolabelled tentoxin, synthesized by Bhatnagar and Rich, binds to chloroplast ATPase (CF_1) from tentoxin-sensitive but not insensitive plants (Steele et al., 1976, 1977, 1978). For a more detailed historical review of the mode of action of tentoxin, see Gilchrist (1983). Thus the site of action of tentoxin has been well established.

Since tentoxin is a simple cyclic tetrapeptide containing several commercially obtainable amino acids, it should be possible to identify and purify the enzyme or enzymes which catalyze its synthesis. It may be harder to find a "tentoxin synthetase" than it was to find the "HC-toxin synthetase", because, insofar as toxin yields reflect biosynthetic activity, *C. carbonum* race 1 makes three to five times more HC-toxin than *A. alternata* makes tentoxin (Saad et al., 1970; Walton, unpublished data).

3.2 Enniatins

The enniatins are a group of cyclic hexidepsipeptides produced by several species of *Fusarium*, a genus of filamentous fungi in the Ascomycotina which contains many important plant pathogens. The chemically and biochemically closely related beauvericins are produced by fungi in both the Deuteromycotina and the Basidiomycotina (Deol et al., 1978). The biochemistry of enniatin biosynthesis is covered in another chapter in the present volume.

Many of the species of *Fusarium* which produce enniatins are plant pathogens (Lacey, 1950; Booth, 1971). One of these, *F. lateritium*, is an effective biological control agent of a fungal pathogen of apricot, due, perhaps, to its ability to make enniatins (Bishop and Ilsley, 1978). A role for a cyclic peptide antibiotic in biological control is similar to that proposed for the siderophores produced by bacteria in the rhizosphere (see below).

Nothing is currently known about the possible role of enniatins in plant disease. If one considers however, that: (1) enniatins might be important in plant pathogenesis; (2) enniatin B in the plant pathogen *Fusarium oxysporum* is synthesized by a single, large polypeptide (Zocher et al., 1982); (3) the enzyme has been purified and characterized, including with monoclonal antibodies (Billich et al., 1987); (4) the molecular biology of filamentous fungi is developed; and (5) *F. oxysporum* can be transformed (H. C. Kistler, University of Florida, personal communication), it now would be possible to clone the gene coding for the enniatin synthetase and, by appropriate insertion/deletion experiments, definitively evaluate the enniatins for a role in pathogenesis. Two major challenges to such research would be the high degree of variability and host specialization within *F. oxysporum*, and that enniatins might play a role in only one or a few of the many possible interactions between *F. oxysporum* and its host plants.

3.3 Peptidic siderophores

Siderophores are low-molecular weight compounds which bind iron [Fe(III)] strongly and specifically. They are found in many microorganisms, including most if not all fungi, and are important to the organisms that produce them for "scavenging" biologically essential iron in iron-deprived environments (Neilands and Leong, 1986). Structurally there are a variety of siderophores.

Siderophores are not phytotoxins in the same sense as the other compounds discussed in this review. They are dealt with here because they have been studied by plant pathologists for their possible role in either contributing to disease, when produced by a pathogen, or in preventing disease, when they are produced by a microorganism that is antagonistic to a pathogen. The most common situation in the latter case is siderophore production by bacteria that inhibit spore germination or mycelial growth of a proximate fungal pathogen. Thus siderophores may be important factors in what is known as "biological control" of plant pathogens (Schroth and Hancock, 1982).

Of the known siderophores, only ferrichrome and its derivatives are clearly peptidic, although rhodoturulic acid is a related diketopiperazine, and amino acids are found in other siderophores (Nielands and Leong, 1986).

Ferrichromes are cyclic hexapeptides containing three residues of d-N-acyl-δ-N-hydroxyornithine (AHO), and a tripeptide containing serine, glycine, and/or alanine. In ferrichrome itself (Figure 6) the acyl group is an acetyl moiety, and the amino acids are all glycine.

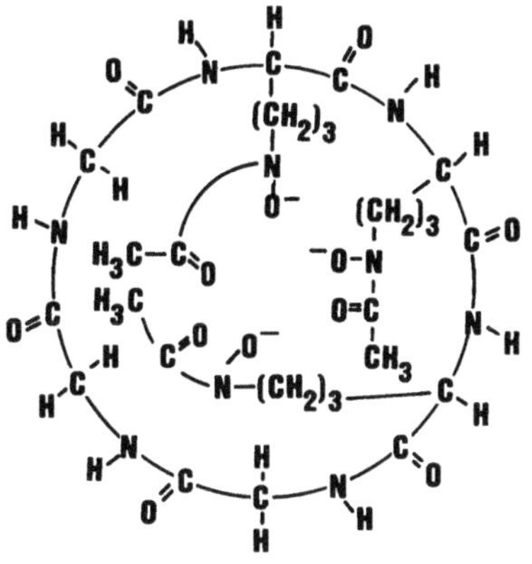

FERRICHROME

Figure 6 Structure of ferrichrome made by many fungi, including *Ustilago maydis*. Fe^{3+} is bound by the oxygen atoms in the center of ferrichrome as drawn.

On the basis that siderophores have been implicated in diseases of animals; that ferrichrome is produced by plant pathogenic fungi; and that one of these fungi, *Ustilago maydis*, is genetically well-understood, Leong and co-workers have begun an intensive investigation into the possible role that ferrichrome-mediated iron uptake might play in pathogenesis (Leong et al., 1987). To date, mutants have been found which are altered in the regulation of ferrichrome biosynthesis or which are unable to synthesize δ-*N*-hydroxyornithine, a precursor of AHO (Leong et al., 1987).

Some work has been done on the biosynthesis of siderophores. From experiments on the biosynthesis of ferrichrome and rhodoturulic acid from *Aspergillus, Fusarium, Rhodotorula,* and *Penicillium*, it appears that ornithine is hydroxylated by one enzyme but that most of the biosynthesis occurs on large multienzymes. ATP/PP_i-dependent exchange driven by AHO and the other amino acid constituents of ferrichrome (glycine, etc.) has been observed (Akers and Neilands, 1978; Hummel and Diekmann, 1981), indicating that at least part of the biosynthesis of ferrichrome is like that of other cyclic peptides. Hummel and Diekmann (1981) found that the large enzyme (1.1 × 10^6 MW) from *Aspergillus quadricinctus* which activates AOH and glycine also has hydroxyornithine acetylase activity. A hydroxyornithine acetylase has been partially purified from *Ustilago sphaerogena* by ammonium sulfate precipitation (Ong and Emery, 1972), but it is not known if that activity is an independent enzyme or associated with other activities on a multienzymic complex. Apparently no search has been made for enzyme-linked pantetheine in these enzymes.

References

Akers, H. A., and Neilands, J. B., Biosynthesis of rhodotorulic acid and other hydroxamate type siderophores, in *Biological Oxidation of Nitrogen*, Gorrod, J. W., Ed., Elsevier/North-Holland Biomedical Press, p. 429, 1978.

Arias, J. A., Dunkle, L. D., and Bracker, C. E., Ultrastructural and developmental alterations induced by *Periconia circinata* toxin in the root tip of sorghum, *Can. J. Bot. 61*, 1491, 1983.

Arntzen, C. J., Inhibition of photophosphorylation by tentoxin, a cyclic tetrapeptide, *Biochim. Biophys. Acta 283*, 539, 1972.

Bains, P. S., and Tewari, J. P., Purification, chemical characterization and host-specificity of the toxin produced by *Alternaria brassicae, Physiol. Mol. Plant Path. 30*, 259, 1987.

Bennett, J., Inhibition of chloroplast development by tentoxin, *Phytochemistry 15*, 263, 1976.

Bhullar, B. S., Daly, J. M., and Rehfeld, D. W., Inhibition of dark CO_2 fixation and photosynthesis in leaf discs of corn susceptible to the host-specific toxin produced by *Helminthosporium maydis*, race T, *Plant Physiol. 56*, 1, 1975.

Billich, A., Zocher, R., Kleinkauf, H., Braun, D. G., Lavanchy, D., and Hochkeppel, H.-K., Monoclonal antibodies to the multienzyme enniatin synthetase, *Biol. Chem. Hoppe-Seyler 368*, 521, 1987.

Bishop, G. C., and Ilsley, A. H., Production of enniatin as a criterion for confirming the identity of *Fusarium lateritium* isolates, *Aust. J. Biol. Sci. 31*, 93, 1978.

Booth, C., *The genus Fusarium*, Commonwealth Mycological Institute, Kew, England, 1971.

Ciufetti, L. M., Pope, M. R., Dunkle, L. D., Daly, J. M., and Knoche, H. W., Isolation and structure of an inactive product derived from the host-specific toxin produced by *Helminthosporium carbonum, Biochemistry 22*, 3507, 1983.

Closse, A., and Huguenin, R., Isolierung und Strukturaufklärung von Chlamydocin, *Helv. Chim. Acta 57*, 533, 1974.

Comstock, J. C., and Scheffer, R. P., Role of host-selective toxin in colonization of corn leaves by *Helminthosporium carbonum, Phytopathology 63*, 24, 1973.

Daly, J. M., and Deverall, B. J., Eds. *Toxins and plant pathogenesis*, Academic Press, Sydney, 1983.

Deol, B. S., Ridley, D. D., and Singh, P., Isolation of cyclodepsipeptides from plant pathogenic fungi, *Aust. J. Chem. 31*, 1397, 1978.

Drew, S. W., and Demain, A. L., Effect of primary metabolites on secondary metabolism, *Ann. Rev. Microbiol. 31*, 343, 1977.

Dunkle, L. D., and Wolpert, T. J., Independence of milo disease symptoms and electrolyte leakage induced by the host-specific toxin from *Periconia circinata, Physiol. Plant Path. 18*, 315, 1981.

Durbin, R. D., and Uchytil, T. F., A survey of plant insensitivity to tentoxin, *Phytopathology 43*, 229, 1977.

Durbin, R. D., and Steele, J. A., What art thou, O specificity? in *Recognition and Specificity in Plant Host-Parasite Interactions*, Daly, J. M., and Uritani, I., Eds., Jap. Sci. Soc. Press, Tokyo, p. 115, 1979.

Earle, E. D., and Gracen, V. E., Effects of *Helminthosporium* phytotoxins on cereal leaf protoplasts, *Plant Tissue Culture 1982: Proc. Fifth Intl. Cong. Plant Tissue and Cell Culture*, p. 663, 1982.

Fulton, N. D., Bollenbacker, K., and Templeton, G. E., *Phytopathology 55*, 49, 1965.

Gardner, J. M., and Scheffer, R. P., Effects of cycloheximide and sulfhydryl-binding compounds on sensitivity of oat tissues to *Helminthosporium victoriae* toxin, *Physiol. Plant Path. 3*, 147, 1973.

Gardner, J. M., Mansour, I. S., and Scheffer, R. P., Effects of the host-specific toxin of *Periconia circinata* on some properties of sorghum plasma membranes, *Physiol. Plant Path. 2*, 197, 1972.

Gilchrist, D. G., Molecular modes of action, in *Toxins and Plant Pathogenesis*, Daly, J. M., and Deverall, B. J., Eds., Academic Press, Sydney, p. 81, 1983.

Gloer, J. B., Meinwald, J., Walton, J. D., and Earle, E. D., Studies on the fungal phytotoxin victorin: structures of three novel amino acids from the acid hydrolyzate, *Experientia 41*, 1370, 1985.

Gross, M. L., McCrery, D., Crow, F., and Tomer, K. B., The structure of the toxin from *Helminthosporium carbonum, Tetrahedron 23*, 5381, 1982.

Halloin, J. M., De Zoeten, G. A., Gaard, G., and Walker, J. C., The effects of tentoxin on chlorophyll synthesis and plastid structure in cucumber and cabbage, *Plant Physiol. 45*, 310, 1970.

Holenstein, J., and Defago, G., Inheritance of naphthazarin production and pathogenicity to pea in *Nectria haematococca, J. Exptl. Bot. 34*, 927, 1983.

Hummel, W., and Diekmann, H., Preliminary characterization of ferrichrome synthetase from *Aspergillus quadricinctus, Biochim. Biophys. Acta 657*, 313, 1981.

Kamdar, H., Clements, D., and Patil, S., Cloning and characterization of genes encoding phaseolotoxin, in Graniti, A., Ballio, A., and Durbin, R., eds., *Phytotoxins and plant pathology*, Elsevier, Rome, in press, 1989.

Kanmera, T., Aoyagi, H., Waki, M., Kato, T., and Izumiya, N., Syntheses of AM-toxin III and its analogs using the Hofmann degradation, *Tet. Lett.*, p. 3625, 1981.

Kawai, M., and Rich, D. H., Total synthesis of the cyclic tetrapeptide, HC-toxin, *Tet. Lett. 24*, 5309, 1983.

Kawai, M., Rich, D. H., and Walton, J. D., The structure and conformation of HC-toxin, *Biochem. Biophys. Res. Comm. 111*, 398, 1983.

Keck, R. W., and Hodges, T. K., Membrane permeability in plants: changes induced by host-specific pathotoxins, *Phytopathology 63*, 226, 1973.

Keller, U., Actinomycin synthetases, *J. Biol. Chem. 262*, 5852, 1987.

Kim, S., Knoche, H. W., and Dunkle, L. D., Essentiality of the ketone function for toxicity of the host-selective toxin produced by *Helminthosporium carbonum, Physiol. Molec. Plant Path. 30*, 433, 1987.

Kim, S., Knoche, H. W., Dunkle, L. D., McCrery, D. A., and Tomer, K. B., Structure of an amino acid analog of the host-specific toxin from *Helminthosporium carbonum, Tet. Lett. 26*, 969, 1985.

Kohmoto, K., Khan, I. D., Renbutsu, Y., Taniguchi, T., and Nishimura, S., Multiple host-specific toxins of *Alternaria mali* and their effect on the permeability of host cells, *Physiol. Plant Path. 8*, 141, 1976.

Kohmoto, K., Taniguchi, T., and Nishimura, S., Correlation between the susceptibility of apple cultivars *Alternaria mali* and their sensitivity to AM-Toxin I, *Ann. Phytopathol. Soc. 43*, 65, 1977.

Koncewicz, M., Mathiaparanam, P., Uchytil, T. F., Sparapano, L., Tam, J., Rich, D. H., Durbin, R. D., The sequence and optical configuration of the amino acids in tentoxin, *Biochem. Biophys. Res. Comm. 53*, 653, 1973.

Kono, Y., Knoche, H. W., and Daly, J. M., Structure: Fungal host-specific, in R. D. Durbin, Ed., *Toxins in Plant Disease*, Academic, New York, 1981.

Kuyama, S., and Tamura, S., Total synthesis of destruxin B, *Agr. Biol. Chem. 29*, 168, 1965.

Lacey, M. S., The antibiotic properties of fifty-two strains of *Fusarium, J. Gen. Microbiol. 4*, 122, 1950.

Larkin, P. J., and Scowcroft, W. R., Eyespot disease of sugarcane. Induction of host-specific toxin and its interaction with leaf cells, *Plant Physiol. 67*, 408, 1981.

Lee, S., and Izumiya, N., Synthesis of a cyclohexidepsipeptide, protodestruxin, *Tet. Lett.* p. 883, 1975.

Lee, S., Aoyagi, H., Shimohigashi, Y., and Izumiya, N., Syntheses of cyclotetradepsipeptides, AM-toxin I and its analogs, *Tet. Lett. 11*, 843, 1976.

Lee, S. G., and Lipmann, F., Isolation of amino acid activating subunit-pantetheine protein complexes: Their role in chain elongation in tyrocidine synthesis, *Proc. Natl. Acad. Sci., U.S.A. 74*, 2343, 1977.

Leisch, J. M., Sweeley, C. C., Staffeld, G. D., Anderson, M. S., Weber, D. J., and Scheffer, R. P., Structure of HC-toxin, a cyclic tetrapeptide from *Helminthosporium carbonum, Tetrahedron 38*, 45, 1985.

Leong, S. A., Wang, J., Budde, A., Holden, D., Kinscherf, T., and Smith, T., Molecular strategies for the analysis of the interaction of *Ustilago maydis* and maize, in *Molecular Strategies for Crop Protection*, Arntzen, C. J., and Ryan, C. A., Eds., Liss, New York, p. 95, 1987.

Lin, W. and Kauer, J. C., Peptide alcohols as promoters of nitrate and ammonium ion uptake in plants, *Plant Physiol. 77*, 403, 1985.

Lipmann, F., On biosynthesis of the cyclic antibiotics gramicidin S and tyrocidine, and of linear gramicidin, in *Peptide Antibiotics, Biosynthesis and Functions*, Kleinkauf, H., and von Döhren, H., Eds., De Gruyter, Berlin, New York, p. 23, 1982.

Macko, V., Wolpert, T. J., Acklin, W., Jaun, B., Seibl, J., Meili, J., and Arigoni, D., Characterization of victorin C, the major host-selective toxin from *Cochliobolus victoriae*: structure of degradation products, *Experientia 41*, 1366, 1985.

Maeno, S., Kohmoto, K., and Otani, H., Different sensitivities among apple and pear cultivars to AM-toxin produced by *Alternaria alternata* apple pathotype, *J. Fac. Agric. Tottori Univ. 19*, 8, 1984.

Mascagni, P., Pope, M., Gibbons, W. A., Ciuffetti, L. M., and Knoche, H. W., The backbone and side chain conformations of the cyclic tetrapeptide HC-toxin, *Biochem. Biophys. Res. Comm. 113*, 10, 1983.

Meehan, F., and Murphy, H. C., A new *Helminthosporium* blight of oats, *Science 104*, 413, 1946.

Meehan, F., and Murphy, H. C., Differential phytotoxicity of metabolic by-products of *Helminthosporium victoriae, Science 106*, 270, 1947.

Meyer, W. L., Kuyper, L. F., Lewis, R. B., Templeton, G. E., and Woodhead, S. H., The amino acid sequence and configuration of tentoxin, *Biochem. Biophys. Res. Comm. 56*, 234, 1974.

Neilands, J. B., and Leong, S. A., Siderophores in relation to plant growth and disease, *Ann. Rev. Plant Physiol. 37*, 187, 1986.

Nishimura, S., and Kohmoto, K., Host-specific toxins and chemical structures from *Alternaria* species, *Ann. Rev. Phytopath. 21*, 87, 1983.

Noda, K., Shibata, Y., Shimohigashi, Y., and Izumiya, N., Syntheses of cyclotetrapeptides, AM-toxin analogs, containing alpha-hydroxyalanine, *Tet. Lett. 21*, 763, 1980.

Novacky, A., and Hanchey, P., Depolarization of membrane potentials in oat roots treated with *victorin, Physiol. Plant Path. 4*, 161, 1974.

Okuno, T., Ishita, Y., Nakayama, S., Sawai, K., Fujita, T., and Sawamura, K., Isolation of a host-specific toxin produced by *Alternaria mali* Roberts, *Ann. Phytopath. Soc. 40*, 375, 1974a.

Okuno, T., Ishita, Y., Sawai, K., and Matsumoto, T., Characterization of alternariolide, a host-specific toxin produced by *Alternaria mali* Roberts, *Chemi. Lett.*, p. 635, 1974b.

Okuno, T., Ishita, Y., Sugawara, A., Mori, Y., Sawai, K., and Matsumoto, T., Structure of the biological active cyclopeptides produced by *Alternaria mali* Roberts, *Tet. Lett.* p. 335, 1975.

Ong, D. E., and Emery, T. F., Ferrichrome biosynthesis: enzyme catalyzed formation of the hydroxamic acid group, *Arch. Biochem. Biophys. 148*, 77, 1972.

Panapoulos, N. J., Walton, J. D., and Willis, D. K., Genetic and biochemical basis of virulence in plant pathogens, in *Genes Involved in Microbe-Plant Interactions*, Verma, D. P. S., and Hohn, T., Eds., Springer, Vienna, p. 339, 1984.

Park, P., Nishimura, S., Keisuke, K., Otani, H., and Tsujimoto, K., Two action sites of AM-toxin I produced by apple pathotype of *Alternaria alternata* in host cells: an ultrastructural study, *Can. J. Bot. 59*, 301, 1981.

Park, P., Tsuda, M., Hayashi, Y., and Ueno, T., Effect of a host-specific toxin (AM-toxin I) produced by *Alternaria mali*, an apple pathogen, on the ultrastructure of plasma membrane of cells in apple and Japanese pear leaves, *Can. J. Bot. 55*, 2383, 1977.

Peet, R. C., Lindgren, P. B., Willis, D. K., and Panopoulos, N. J., Identification and cloning of genes involved in phaseolotoxin production by *Pseudomonas syringae* pv. "*phaseolicola*", *Jour. Bacteriol. 166*, 1096, 1986.

Pope, M. R., Ciuffetti, L. M., Knoche, H. W., McCrery, D., Daly, J. M., and Dunkle, L. D., Structure of the host-specific toxin produced by *Helminthosporium carbonum, Biochemistry 22*, 3502, 1983.

Pringle, R. B., Amino acid composition of the host-specific toxin of *Helminthosporium carbonum, Plant Physiol. 48*, 756, 1971.

Pringle, R. B., and Scheffer, R. P., Amino acid composition of a crystalline host-specific toxin, *Phytopathology 56*, 1149, 1966.

Pringle, R. B., and Scheffer, R. P., Isolation of the host-specific toxin and a related substance with nonspecific toxicity from *Helminthosporium carbonum, Phytopathology 57*, 1169, 1967.
Pringle, R. B., and Scheffer, R. P., Purification of the selective toxin of *Periconia circinata, Phytopathology 53*, 785, 1963.
Rancillac, M., Kaur-Sawhney, R., Staskawicz, B., and Galston, A. W., Effects of cycloheximide and kinetin pretreatments on responses of susceptible and resistant *Avena* leaf protoplasts to the phytotoxin victorin, *Plant Cell Physiol. 17*, 987, 1976.
Rasmussen, J. B., and Scheffer, R. P., Isolation and biological activities of four selective toxins from *Helminthosporium carbonum, Plant Physiol. 86*, 187, 1988.
Ray, P. M., Auxin and fusicoccin enhancement of β-glucan synthase in peas, *Plant Physiol. 78*, 466, 1985.
Rich, D. H., and Bhatnagar, P. K., Conformational studies of tentoxin by nuclear magnetic resonance spectroscopy. Evidence for a new conformation for a cyclic tetrapeptide, *J. Am. Chem. Soc. 100*, 2212, 1978.
Rich, D. H., and Mathiaparanam, P., Synthesis of the cyclic tetrapeptide tentoxin. Effect of an N-methyldehydrophenylalanyl residue on conformation of linear tetrapeptides, *Tet. Lett.*, p. 4073, 1974.
Rich, D. H., Bhatnagar, P. K., Jasensky, R. D., Steele, J. A., Uchytil, T. F., and Durbin, R. D., Two conformations of the cyclic tetrapeptide, [D-MeAla^1]-tentoxin have different biological activities, *Bioorg. Chem. 7*, 207, 1978.
Rich, D. H., Mathiaparanam, P., Grant, J. A., and Bhatnagar, P., Synthesis of cyclic tetrapeptides related to tentoxin, in *Peptides: Chemistry, Structure, Biology*, Walter, R., and Meienhofer, J., Eds., Ann Arbor Science, Ann Arbor, Michigan, p. 943, 1975.
Rines, H. W., and Luke, H. H., Selection and regeneration of toxin-insensitive plants from tissue cultures of oats (*Avena sativa*) susceptible to *Helminthosporium victoriae, Theor. Appl. Genet. 71*, 16, 1985.
Saad, M., Halloin, J. M., and Hagedorn, D. J., Production, purification, and bioassay of tentoxin, *Phytopathology 60*, 415, 1970.
Samaddar, K. R., and Scheffer, R. P., Effect of the specific toxin in *Helminthosporium victoriae* on host cell membranes, *Plant Physiol. 43*, 21, 1968.
Sawamura, K., Studies on spotted diseases of apples. 1. The causal agent of *Alternaria* blotch, *Bull. Tohoku Natl. Agric. Exptl. Stn., Japan 23*, 163, 1962.
Sawamura, K., Studies on spotted diseases of apples. VI. On the host-specific toxin of *Alternaria mali* Roberts, *Bull. Hort. Res. Stn. (Japan), Ser. C, 4*, 43, 1966.
Scheffer, R. P., and Pringle, R. B., A selective toxin produced by *Periconia circinata, Nature, 191*, 912, 1961.
Scheffer, R. P., and Ullstrup, A. J., A host-specific toxic metabolite from *Helminthosporium carbonum, Phytopathology 45*, 1037, 1965.
Scheffer, R. P., Nelson, R. R., Ullstrup, A. J., Inheritance of toxin production and pathogenicity in *Cochliobolus carbonum* and *Cochliobolus victoriae, Phytopathology 57*, 1288, 1967.
Schertz, K. F., and Tai, Y. P., Inheritance of reaction of *Sorghum bicolor* (L.) Moench to toxin produced by *Periconia circinata* (Mang.) Sacc., *Crop Sci. 9*, 621, 1969.
Schroth, M. N., and Hancock, J. G., Disease-suppressive soil and root-colonizing bacteria, *Science 216*, 1376, 1982.
Shimohigashi, Y., and Izumiya, N., Cyclic peptides. V. Synthesis of AM-toxin II and its analog, *Int. J. Peptide Protein Res. 12*, 7, 1978.
Shimohigashi, Y., Lee, S., Aoyagi, H., Kato, T., and Izumiya, N., Cyclic peptides. I. Synthesis of AM-toxin analog containing O-methyl-tyrosine, *Int. J. Peptide Protein Res. 10*, 197, 1977b.
Shimohigashi, Y., Lee, S., Kato, T., and Izumiya, N., Cyclic peptides, IV. Synthesis of diastereomeric dihydro-AM-toxin I and its analogs, *Bull. Chem. Soc. Japan 51*, 584, 1978.
Shimohigashi, Y., Lee, S., Kato, T., and Izumiya, N., Syntheses of cyclotetradepsipeptides, AM-toxin and its analog, *Chemi. Lett.*, p. 1411, 1977a.
Shute, R. E., Kawai, M., and Rich, D. H., Conformationally constrained biologically active peptides: tentative identification of the antimitogenic bioactive conformer of the naturally occurring cyclic tetrapeptides, *Tetrahedron 44*, 685, 1988.
Steele, J. A., Durbin, R. D., Uchytil, T. F., and Rich, D. H., An uncompetitive inhibitor of lettuce chloroplast coupling factor 1, *Biochim. Biophys. Acta 501*, 72, 1978.

Steele, J. A., Uchytil, T. F., and Durbin, R. D., The binding of tentoxin to a tryptic digest of chloroplast coupling factor 1, *Biochem. Biophys. Acta 459*, 347, 1977.

Steele, J. A., Uchytil, T. F., Durbin, R. D., Bhatnagar, P., and Rich, D. H., Chloroplast coupling factor 1: A species-specific receptor for tentoxin, *Proc. Natl. Acad. Sci., U.S.A. 73*, 2245, 1976.

Suzuki, A., and Tamura, S., Isolation and structure of protodestruxin from *Metarrhizium anisopliae, Agr. Biol. Chem. 36*, 896, 1972.

Suzuki, A., Taguchi, H., and Tamura, S., Isolation and structure elucidation of three new insecticidal cyclodepsipeptides, destruxins C and D and desmethyldestruxin B, produced by *Metarrhizium anisopliae, Agr. Biol. Chem. 34*, 813, 1970.

Tamura, S., Kuyama, S., Kodaira, Y., and Higashikawa, S., The structure of destruxin B, a toxic metabolite of *Oospora destructor, Agr. Biol. Chem. 28*, 137, 1964.

Tanis, S. P., Horenstein, B. A., Scheffer, R. P., and Rasmussen, J. B., A new host specific toxin from *Helminthosporium carbonum, Heterocycles 24*, 3423, 1986.

Templeton, G. E., *Alternaria* toxins related to pathogenesis in plants, in *Microbial Toxins*, Vol. VIII, Kadis, S., Ciegler, A., and Ajl, S. J., Eds., Academic Press, New York, p. 169, 1972.

Traylor, E. A., Shore, S. H., Ransom, R. F., and Dunkle, L. D., Pathotoxin effects in sorghum are also produced by mercuric chloride treatment, *Plant Physiol. 84*, 975, 1987.

Ueno, T., Hayashi, Y., Nakashima, T., Fukami, H., Nishimura, S., Kohmoto, K., and Sekiguchi, A., Isolation of AM-toxin I, a new phytotoxic metabolite from *Alternaria mali, Phytopathology 65*, 82, 1975a.

Ueno, T., Nakashima, T., Hayashi, Y., and Fukami, H., Structures of AM-toxin I and II, host specific phytotoxin metabolites produced by *Alternaria mali, Agr. Biol. Chem. 39*, 1115, 1975b.

Ueno, T., Nakashima, T., Hayashi, Y., and Fukami, H., Isolation and structure of AM-toxin III, a host specific phytotoxic metabolite produced by *Alternaria mali, Agr. Biol. Chem. 39*, 2081, 1975c.

Vater, J., and Kleinkauf, H., Gramicidin S-synthetase. A further characterization of phenylalanine racemase, the light enzyme of gramicidin S-synthetase, *Biochim. Biophys. Acta 429*, 1062, 1976.

Walton, J. D., Two enzymes involved in biosynthesis of the host-selective phytotoxin HC-toxin, *Proc. Natl. Acad. Sci, U.S.A. 84*, 8444, 1987.

Walton, J. D., and Earle, E. D., Characterization of the host-specific phytotoxin victorin by high-pressure liquid chromatography, *Plant Sci. Lett. 34*, 231, 1984.

Walton, J. D., and Earle, E. D., Stimulation of extracellular polysaccharide synthesis in oat protoplasts by the host-specific phytotoxin victorin, *Planta 165*, 407, 1985a.

Walton, J. D., and Earle, E. D., Chimica della tossina-HC, una fitotossina gene-specifica, *Phytopath. Meditt. 24*, 319, 1985b.

Walton, J. D., and Earle, E. D., The epoxide in HC-toxin is required for activity against susceptible maize, *Physiol. Plant Pathol. 22*, 371, 1983.

Walton, J. D., and Holden, F. R., Properties of two enzymes involved in the biosynthesis of the fungal pathogenicity factor HC-toxin. *Mol. Plant-Microbe Interactions 1*, 128, 1988.

Walton, J. D., Earle, E. D., Yoder, O. C., and Spanswick, R. M., Reduction of adenosine triphosphate levels in susceptible maize mesophyll protoplasts by *Helminthosporium maydis* race T toxin, *Plant Physiol. 63*, 806, 1979.

Walton, J. D., Earle, E. D., and Gibson, B. W., Purification and structure of the host-specific toxin from *Helminthosporium carbonum* race 1, *Biochem. Biophys. Res. Comm. 107*, 785, 1982.

Walton, J. D., Earle, E. D., Stahelin, H., Grieder, A., Hirota, A., and Suzuki, A., Reciprocal biological activities of the cyclic tetrapeptides chlamydocin and HC-toxin, *Experientia 41*, 348, 1985.

Wessel, W. L., Clare, K. A., and Gibbons, W. A., Enzymic biosynthesis of peptide toxins by plant-specific fungi, *Biochem. Soc. Trans. 15*, 917, 1987.

Wessel, W. L., Clare, K. A., and Gibbons, W. A., Purification and characterization of HC-toxin synthetase from *Helminthosporium carbonum, Biochem. Soc. Trans. 16*, 401, 1988a.

Wessel, W. L., Clare, K. A., and Gibbons, W. A., Biosynthesis of L-AEO, the toxic determinant of the phytotoxin produced by *H. carbonum, Biochem. Soc. Trans. 16*, 402, 1988b.

Wolf, S. J., Selection and characterization of corn (*Zea mays* L.) resistant to *Helminthosporium* spp. using field and tissue culture, *Ph. D. Thesis, Cornell University*, Ithaca, NY, 1988.

Wolpert, T. J., and Dunkle, L. D., Alterations in gene expression in sorghum induced by the host-specific toxin from *Periconia circinata, Proc. Natl. Acad. Sci., U.S.A. 80*, 6576, 1983.

Wolpert, T. J., and Dunkle, L. D., Purification and partial characterization of host-specific toxins produced by *Periconia circinata, Phytopathology 70*, 872, 1980.
Wolpert, T. J., and Macko, V., Specific binding of victorin to a 100-kDa proteins from oats, *Proc. Natl. Acad. Sci., U.S.A. 86*, 4092, 1989.
Wolpert, T. J., Macko, V., Acklin, W., Jaun, B., and Arigoni, D., Structure of minor host-selective toxins from *Cochliobolus victoriae, Experientia 42*, 1296, 1986.
Wolpert, T. J., Macko, V., Acklin, W., Jaun, B., Seibl, J., Meili, J., and Arigoni, D., Structure of victorin C, the major host-specific toxin from *Cochliobolus victoriae, Experientia 41*, 1524, 1985.
Wolpert, T. J., Macko, V., Acklin, W., and Arigoni, D., Molecular features affecting the biological activity of the host-selective toxins from *Cochliobolus victoriae, Plant Physiol. 88*, 37, 1988.
Yoder, O. C., and Scheffer, R. P., Effects of *Helminthosporium carbonum* toxin on nitrate uptake and reduction by corn tissues, *Plant Physiol. 52*, 513, 1973a.
Yoder, O. C., and Scheffer, R. P., Effects of *Helminthosporium carbonum* toxin on absorption of solutes by corn roots, *Plant Physiol. 52*, 518, 1973b.
Yoder, O. C., and Scheffer, R. P., Role of toxin in early interactions of *Helminthosporium victoriae* with susceptible and resistant oat tissue, *Phytopathology 59*, 1954, 1969.
Yoder, O. C., Toxins in pathogenesis, *Ann. Rev. Phytopathol. 18*, 103, 1980.
Zocher, R., Keller, U., and Kleinkauf, H., Enniatin synthetase, a novel type of multifunctional enzyme catalyzing depsipeptide synthesis in *Fusarium oxysporum, Biochemistry 21*, 43, 1982.
Zocher, R., Nihira, T., Paul, E., Madry, N., Peeters, H., Kleinkauf, H., and Keller, U., Biosynthesis of cyclosporin A: partial purification and properties of a multifunctional enzyme from *Tolypocladium inflatum, Biochemistry 25*, 550, 1986.

Chapter 9
Chemical Synthesis and Bioactivity of Gramicidin S and Related Reptides

Michinori Waki and Nobuo Izumiya

1. Introduction
2. Structures of Gramicidin S and Related Natural Peptides
   2.1 Gramicidin S and its congeners
   2.2 Tyrocidines
   2.3 Gratisin
3. Chemical Synthesis of Gramicidin S and Related Peptides
   3.1 Synthesis of linear precursor peptides
   3.2 Synthesis of cyclic peptides
4. Structure-Activity Relationships of Gramicidin S, Tyrocidines and Gratisin
   4.1 Gramicidin S
   4.2 Tyrocidines
   4.3 Gratisin
5. Conformations of Gramicidin S and Tyrocidines
   5.1 Optical rotatory dispersion (ORD) and circular dichroism (CD)
   5.2 Nuclear magnetic resonance (NMR)
   5.3 Other conformational analyses
6. Design of Highly Active Analogs of Gramicidin S
7. Active Analogs of Gramicidin S Against Gram-negative Bacteria
8. Mechanism of Antimicrobial Action of Gramicidin S
   8.1 Interactions with model membranes
   8.2 Interactions with bacterial membranes
9. Concluding Remarks
References

1. Introduction

Gramicidin S (GS) is one of the most popular peptide antibiotics studied extensively by various physico- and organochemical procedures. GS has attracted continuing interest as a model compound for a variety of research objects to chemists, biochemists, and biologists since the discovery in 1944 by Gause and Brazhnikova [1, 2]. It is with the great sadness that we record here the death of one of the discoverers, Professor G. F. Gause [3]. Accumulated studies on chemical synthesis, structure-activity relationship, conformation, mode of action, biosynthesis and function of GS and related peptides have been covered by review articles [4–7], mini-reviews [8–10] and books [11, 12].

In 1979, we published a comprehensive monograph [11] entitled "Synthetic aspects of biologically active cyclic peptides – gramicidin S and tyrocidines". Thus, in this article, progress in the chemistry and biochemistry of GS, tyrocidines and gratisin after 1979 will be mainly described. In particular, we will focus on some newly-coming concepts for designing antibiotic peptides with high activity and also on recent advances in studies on the mode of antimicrobial action of GS. Structure-activity data so far obtained from various analogs and derivatives of GS will be tabulated. New aspects of structure, chemical synthesis and conformation will be briefly described, since detailed descriptions were give on these subjects in the previous monograph [11]. Subjects on linear gramicidins (gramicidin A and congeners) and the biosynthesis of GS and tyrocidines have not been included. The excellent book [12] edited by Kleinkauf and von Döhren and recent literatures [6, 13–21] will help the interested readers to know current research on the biosynthesis and furthermore functions of GS and tyrocidines.

2. Structures of Gramicidin S and Related Natural Peptides

Primary structures of GS, its congeners, tyrocidines and gratisin, will be described in this Section.

2.1 Gramicidin S and its congeners

GS isolated from a strain of *Bacillus brevis* by Gause and Brazhnikova [1, 2] shows strong antimicrobial activity against Gram-positive bacteria and has a cyclic decapeptide structure: cyclo(-Val-Orn-Leu-D-Phe-Pro-)$_2$ (I) (see also Figure 1) [22]. The numbering for amino acid residues in GS shown in Figure 1 will be used throughout this article. The structure was conclusively confirmed by its chemical synthesis by Schwyzer and Sieber in 1957 [23]. This is also the first chemical synthesis of naturally occurring cyclic peptide.

It has often been observed that many natural peptide antibiotics are mixtures of congeners in which one or more amino acid residues are substituted by structurally similar ones. Recently, three minor components were detected in natural GS sample by high performance liquid chromatography (HPLC) analysis [24]. Isolation, sequencing and synthesis of the components by Nozaki and Muramatsu [24, 25] provided proof that two, II and III, are natural congeners of GS containing 2-aminobutyric acid (Abu) in place of one or both Val residues in GS molecule as shown in Figure 1.

GS congeners might be also obtainable by enzymatic synthesis utilizing the broad specificities of GS synthetase [6, 26]. For example, the formation of [Leu$^{1,1'}$]GS

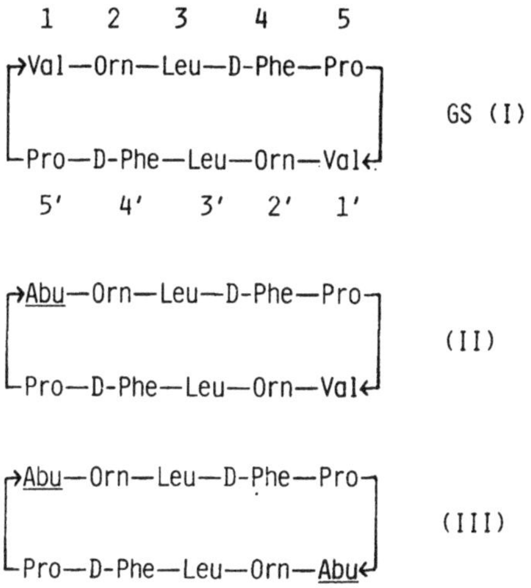

Figure 1 Structures of gramicidin S and its congeners.

[27,28] and various $Phe^{4,4'}$-substituted analogs [29] has been observed in the absence of the constituent amimo acid Val in an incubation mixture and on the addition of homologous amino acid instead of the constituent Phe to an *in vitro* system, respectively.

2.2 Tyrocidines

Dubos's strain of *Bacillus brevis* produces tyrothricin, a mixture of tyrocidines and linear gramicidins [30]. Tyrocidines produced under usual culture conditions are a mixture of congeners of tyrodicine A, B and C [31]. Tyrocidine D [32] and E [33]

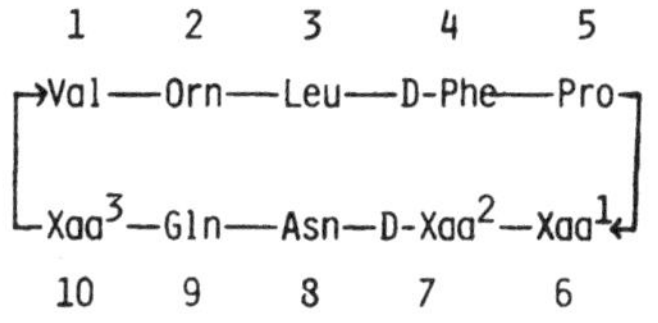

	Xaa^1	Xaa^2	Xaa^3
Tyrocidine A	Phe	Phe	Tyr
Tyrocidine B	Trp	Phe	Tyr
Tyrocidine C	Trp	Trp	Tyr
Tyrocidine D	Trp	Trp	Trp
Tyrocidine E	Phe	Phe	Phe

Figure 2 Structures of tyrocidine congeners.

are another congeners obtained biosynthetically in the Trp-rich growth medium or in the Tyr- and Trp-omitted medium, respectively. Tyrocidines have a cyclic decapeptide structure (Figure 2) and antimicrobial activity against Gram-positive bacteria similar to GS. It is of interest that tyrocidines share the same pentapeptide sequence, -Val-Orn-Leu-D-Phe-Pro-, with GS (see Figures 1 and 2). The structure proposed for tyrocidine A was confirmed by its chemical synthesis [34, 35]. Chemical syntheses of tyrocidine B [36], C [37] and E [38] were also reported.

2.3 Gratisin

A peptide antibiotic, gratisin, which is active against *Bacillus subtilis*, was isolated from *Bacillus brevis* Y-33 by Zharikova et al. [39]. The structure of gratisin was shown to be a cyclic dodecapeptide, cyclo(-Val-Orn-Leu-Phe-Pro-Tyr-)$_2$, but the configurations of the constituent amino acids have not yet been established [40]. Recent synthetic work on gratisin by Tamaki et al. [41, 42] has resulted in the proposal that [D-Phe$^{4,4'}$, D-Tyr$^{6,6'}$] gratisin (IV) is likely to be identical to the natural antibiotic (Figure 3). This will be mentioned in more detail in the Section 4.3.

```
   1     2     3      4      5     6
┌→Val—Orn—Leu—D-Phe—Pro—D-Tyr┐
│                              │        (IV)
└D-Tyr—Pro—D-Phe—Leu—Orn—Val←┘
   6'    5'    4'     3'    2'    1'
```

Figure 3 Possible structure of gratisin.

3. Chemical Synthesis of Gramicidin S and Related Peptides

Cyclic peptides are usually prepared from activated precursor linear peptides by intramolecular cyclization at high dilution [43, 44]. A number of cyclic peptides related to GS have been synthesized following this strategy [11]. Of particular importance for the success of the synthesis are construction step of linear peptides and their cyclization step. Therefore, the two main problems in cyclic peptide synthesis will be treated in this Section.

3.1 Synthesis of linear precursor peptides

At present, synthetic methods for linear oligo- and polypeptides have been well-developed. Thus, any linear oligopeptides related to GS can be built up without difficulty by currently utilized procedures of peptide synthesis: mostly the solution method and sometimes the solid phase method [45, 46] have been used. It should be emphasized, however, that amino acid sequence in linear precursors influences the mode of cyclization and cyclization yield [47–49]. There is no definite way how to select the linear sequence, but selection of Pro or Gly residue at the *C*-terminus is one of the way to avoid racemization in the cyclization reaction.

3.2 Synthesis of cyclic peptides

Schwyzer and Sieber succeeded in the chemical synthesis of GS using the *p*-nitrophenyl ester (ONp) method for the cyclization purpose [23]. Since then many cyclization methods have been investigated intending to produce GS and its analogs in better yields under more mild conditions [11]. Matsuura et al. introduced the *N*-hydroxysuccinimide ester (ONSu) method as a more effective and convenient alternative to the ONp method in GS synthesis [50]. Among various cyclization procedures, the ONSu method keeps the highest popularity [42, 48, 50–60], followed by the azide method [42, 45, 46, 48, 50, 51, 61–63] and scarcely the ONp method [58, 64] in the current synthesis of cyclic peptides related to GS.

Figures 4 and 5 illustrate examples of the synthesis of two analogs of GS (VIII and XI) containing D-2, 3-diaminopropionic acid (D-Dpr) at 4,4' positions and *δ*-hydroxynorvaline (Hnv) at 2,2' positions by the most frequently utilized ONSu and azide methods, respectively [55, 46]. As shown in Figure 4, H-decapeptide active ester (VI) was obtained from Boc-decapeptide acid (V) by the action of *N*-hydroxysuccinimide (HONSu) and 1-ethyl-3-(3-dimethylaminopropyl)carbodiimide (EDC), followed by trifluoroacetic acid (TFA) treatment [55]. The compound VI was subjected to cyclization at high dilution (3 mM) in pyridine to afford a protected cyclic decapeptide (VII) in 41% yield. Selective cleavage of the protecting groups (N^{δ}-For and N^{β}-Z) by respective HCl/MeOH treatment and hydrogenolysis gave the desired analog VIII, [D-Dpr$^{4,4'}$]GS.

Boc-(Leu-D-Dpr(Z)-Pro-Val-Orn(For))$_2$-OH (V)
↓ HONSu-EDC·HCl
↓ TFA
H-(Leu-D-Dpr(Z)-Pro-Val-Orn(For))$_2$-ONSu·TFA (VI·TFA)
↓ pyridine
cyclo(-Leu-D-Dpr(Z)-Pro-Val-Orn(For)-)$_2$ (VII)
↓ HCl/MeOH
cyclo(-Leu-D-Dpr(Z)-Pro-Val-Orn-)$_2$·2HCl
↓ H_2/Pd, HCl
[D-Dpr$^{4,4'}$]GS·4HCl (VIII·4HCl)

Figure 4 Synthesis of [D-Dpr$^{4,4'}$]gramicidin S by the ONSu method.

Boc-(D-Phe-Pro-Val-Hnv(Bzl)-Leu)$_2$-resin

↓ $NH_2NH_2 \cdot H_2O$ in DMF

Boc-(D-Phe-Pro-Val-Hnv(Bzl)-Leu)$_2$-$NHNH_2$ (IX)

↓ HCl/AcOH

H-(D-Phe-Pro-Val-Hnv(Bzl)-Leu)$_2$-$NHNH_2 \cdot 2HCl$

↓ HNO_2

H-(D-Phe-Pro-Val-Hnv(Bzl)-Leu)$_2$-$N_3 \cdot HCl$ (X·HCl)

↓ pyridine

cyclo(-D-Phe-Pro-Val-Hnv(Bzl)-Leu-)$_2$

↓ H_2/Pd

[Hnv$^{2,2'}$]GS (XI)

Figure 5 Synthesis of [Hnv$^{2,2'}$]gramicidin S by the azide method.

On the other hand, as illustrated in Figure 5, Boc-decapeptide hydrazide (IX) was converted to the corresponding hydrazide dihydrochloride with 1 M HCl/AcOH and then transformed into a decapeptide azide (X) with sodium nitrite in acidic medium [46]. The azide X was similarly treated with pyridine at high dilution (1 mM) at 5°C for 3 days. The crude product obtained in 80−90% yield was hydrogenated and then purified by droplet countercurrent chromatography to afford the analog XI, [Hnv$^{2,2'}$]GS, in 28% yield from IX.

Superiority of the ONSu method to the azide method with regard to cyclization yield has been demonstrated in the synthesis of [Pro$^{4,4'}$, Asn$^{5,5'}$]GS (cyclodecapeptide) [51] and gratisin peptides (cyclododecapeptide) [42] by cyclization of the corresponding linear precursor. However, considerable racemization by the active ester method has been observed during cyclization of H-pentapeptide-ONSu to yield diZ-GS [8] and of H-Gly-Ala-D-Val-Leu-Ile-ONp [65]. Such racemization is a serious problem in obtaining optically pure products. In this respect, the azide method is still one of the most useful coupling procedures giving minimal racemization in preparing cyclic peptides under high dilution conditions [8, 48, 65]. One-pot cyclization of free peptides via azide using diphenylphoshoryl azide [66] seems especially promising [47, 67].

Cyclization with the ONSu and azide methods mentioned above necessitates two reaction steps, activation and cyclization-step, however, direct one-pot cyclization of free peptides is also possible with the help of excess coupling reagents. In fact, diZ-GS has been successfully prepared in 80% yield from H-(Val-Orn(Z)-Leu-D-Phe-Pro)$_2$-OH using (5-nitropyridyl)diphenyl phosphinate (5 equiv.) [68]. Reagent, a mixture of EDC·HCl and 1-hydroxybenzotriazole (HOBt) (each, 10 equiv.), has been employed for one-pot cyclization of H-(Abu-Orn(Z)-Leu-D-Phe-Pro-Val-Orn(Z)-Leu-D-Phe-Pro)-OH or H-(Abu-Orn(Z)-Leu-D-Phe-Pro)$_2$-OH to produce a protected GS congener, diZ-[Abu^1]GS or diZ-[Abu$^{1,1'}$]GS, in 75 or 82% yield, respectively [25].

GS is a cyclic decapeptide with the repeated pentapeptide sequences of -Val-Orn-Leu-D-Phe-Pro- as indicated in Figure 1. The symmetrical property of this molecule provides more economical route for the synthesis of GS and its analogs by cyclodimerization of appropriately activated pentapeptides [11, 69]. Waki and Izumiya obtained GS following this strategy [70], then interesting phenomenon was observed during the work that cyclization of a pentapeptide active ester, H-Val-Orn(Z)-Leu-D-Phe-Pro-ONp (XII), gave a mixture composed of a protected GS, cyclic dimer, and a protected semiGS, cyclic monomer, in 7 : 3 ratio by weight.

┌ Val-Orn-Leu-D-Phe-Pro ┐ semiGS (XIII)
└───────────────────────┘

Component analysis of the crude products after cyclization of various activated linear pentapeptides related to GS has shown that the mode of cyclization is greatly dependent on the nature of the terminal amino acids and on the sequence of each pentapeptide [11]. For example, the ratio of cyclic dimer to cyclic monomer decreases with decreasing steric hindrance of the *N*-terminal amino acid residue (Xaa) in the pentapeptide active ester, H-Xaa-Orn(Z)-Leu-D-Phe-Pro-ONp (XIV). Namely, the formation of cyclic dimer decreases in the order of Val, Leu and Ala substitutions at the Xaa, and finally Gly substitution results in only cyclic monomer [11]. Interestingly, exclusive formation of cyclic monomer from H-Leu-D-Dpr(Z)-Pro-Val-Orn(For)-ONSu [55] and H-Aib-Orn(Z)-Leu-D-Phe-Pro-ONSu [58] (Aib, *α*-aminoisobutyric acid) with bulky amino acid at respective *N*-terminus has been reported.

Even *α,β*-dehydrogenation or configurational change of one amino acid residue in the precursor pentapeptide has also influenced the mode of cyclization. For example, only cyclic dimer was produced from H-Val-Orn(For)-Leu-ΔPhe-Pro-ONSu [60] (ΔPhe, *α,β*-dehydrophenylalanine). In other experiment, polymeric products, cyclic dimer and trimer, were formed from H-Val-Orn(Z)-Leu-D-Phe-D-Pro-ONSu [57] which possesses the same sequence as that of XIV (Xaa = Val) except D-Pro. Minematsu et al. [48] examined how variation of the sequence influences the composition of the product by cyclization of pentapeptides by the ONSu or azide method. They prepared five different pentapeptides which are observed in GS. Then, the highest ratio (89 : 11) of cyclic dimer to cyclic monomer was obtained with H-Pro-Val-Orn(Z)-Leu-D-Phe-ONSu (Pro---D-Phe). In the cases of Leu---Orn(Z) and Val---Pro, cyclodimerization predominated also. Preferential monomerization was observed in D-Phe---Leu with D-Phe or Leu as the *N*- or *C*-terminus.

At present, it is still difficult to predict the mode of cyclization of activated pentapeptides based on their primary sequences, although the steric and configurational nature of terminal amino acids and the conformation of the peptide backbone in these precursors may be one of the important factors governing the mode of cyclization. For the better understanding of the mode of cyclization, detailed conformational analysis [58] of precursor peptides and final cyclized peptides in the cyclization medium, e. g., pyridine or *N,N*-dimethylformamide (DMF) should be needed.

4. Structure-Activity Relationships of Gramicidin S, Tyrocidines and Gratisin

Synthetic efforts for a variety of amino acid substituted analogs of bioactive peptides afford valuable information on their structure-activity relationships. In this Section, our attention will be focused on variations of the primary structure of GS, tyrocidines and gratisin toward their bioactivities.

4.1 Gramicidin S

Since the initiating studies on the synthesis of [Lys$^{2,2'}$]GS by Schwyzer and Sieber [69, 71] and of the modified analogs of natural GS by Stepanov et al. [72] both in 1958, large numbers of such GS analogs have been synthesized and their antimicrobial activities have been examined to clarify the relationships between the primary structure and activity of GS. The synthetic analogs with amino acid replacement of GS that have been prepared so far are tabulated in Table 1, together with their antimicrobial activities to allow for easy comparison. Table 2 lists only the Orn-modified analogs derived from natural GS that have recently been prepared, together with some of the previous ones.

It is now well established that GS holds a rigid antiparallel β-sheet structure stabilized by four intramolecular hydrogen bonds between the Val and Leu residues, and by two β-turns (type II′) around the D-Phe-Pro sequences as depicted in Figure 6 [129–132]. For a better understanding of structure-activity relationships of bioactive peptides, conformational considerations are increasingly of importance.

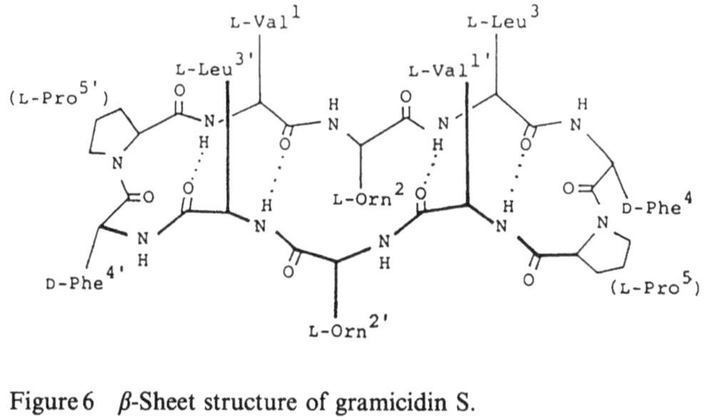

Figure 6 β-Sheet structure of gramicidin S.

Thus, the following discussion will be given, keeping the characteristic β-sheet structure of GS (Figure 6) in mind and judging the summarized data in Tables 1 and 2, on the role of each amino acid residue in GS for manifesting its bioactivity.

Table 1 Antimicrobial activities of amino acid replacement analogs of gramicidin S.

No.	Sequence of cyclic decapeptide[1]					Minimum inhibitory concentration (μg/ml)[2]		Ref.
						B. subtilis	*S. aureus*	
1	Val$^{1,1'}$	Orn$^{2,2'}$	Leu$^{3,3'}$	D-Phe$^{4,4'}$	Pro$^{5,5'}$ (GS)	5	5	70
2	Gly$^{1,1'}$					100	100	46, 73
3	Gly$^{1,1'}$				Gly$^{5,5'}$	>100	>100	74
4	Ala$^{1,1'}$					5	5	73
5	Ala$^{1,1'}$		Ala$^{3,3'}$			>100	>100	75
6	Leu$^{1,1'}$					5	10	76
7	Leu$^{1,1'}$	Lys$^{2,2'}$	Leu$^{3,3'}$	D-Leu$^{4,4'}$	Leu$^{5,5'}$	40	20−40	59
8	Ser$^{1,1'}$		Ser$^{3,3'}$			no activity		77
9	Phe$^{1,1'}$					50		78
10	Phe$^{1,1'}$		Phe$^{3,3'}$			80	>160	79
11	Lys$^{1,1'}$						> 80	80
12	Lys(Z)$^{1,1'}$						20−40	80
13	Lys$^{1,1'}$	Orn(For)$^{2,2'}$					> 80	80
14	D-Val$^{1,1'}$					no activity	no activity	81
15	D-Val$^{1,1'}$			Phe$^{4,4'}$		no activity	no activity	81
16	Abu$^{1,1'}$					5−10	5−20	24, 25
17	Aib$^{1,1'}$					165	>165	58
18	MeVal$^{1,1'}$					160	80	82
19	Hmb$^{1,1'}$					5−20	45−90	83−85
20	Hmb$^{1,1'}$		Hmp$^{3,3'}$			>500	>500	83−85
21	Val^{1}Gly$^{1'}$					5−13	9−25	46
22	Val^{1}Phe$^{1'}$		Phe^{3}Leu$^{3'}$			> 25		78
23	Val^{1}Cys(MBzl)$^{1'}$					10	10	86
24	Val^{1}Abu$^{1'}$					2.5−10	5−20	24, 25

[1] Only the varied residues are shown except in the case of the first compound. All optically active amino acids are of the L-configuration unless otherwise indicated. GS, gramicidin S; Abu, 2-aminobutyric acid; Aib, *α*-aminoisobutyric acid; Hmb, 2-hydroxy-3-methylbutanoic acid; Hmp, 2-hydroxy-4-methylpentanoic acid; MBzl, *p*-methoxybenzyl; Dpr, 2,3-diaminopropionic acid; Dab, 2,4-diaminobutyric acid; Hnv, *δ*-hydroxynorvaline; *δ*-Ava, *δ*-aminovaleric acid; Cha, cyclohexylalanine; ΔAla, *α,β*-dehydroalanine; ΔPhe, *α,β*-dehydrophenylalanine; BTD, bicyclic *β*-turn dipeptide; Pro(4F), 4-fluoroproline.

[2] In certain cases the figures are derived by calculation on the basis of those for GS as a standard.

Table 1 (continued)

No.	Sequence of cyclic decapeptide[1]					Minimum inhibitory concentration (μg/ml)[2]		Ref.
						B. subtilis	*S. aureus*	
25	Val^{1}MeVal$^{1'}$					80	40	82
26		Orn(Boc)$^{2,2'}$		D-Dpr$^{4,4'}$		>160	> 80	52, 55
27		Ser$^{2,2'}$				>310	> 80	87
28		Dab$^{2,2'}$				5	10	88
29		Lys$^{2,2'}$				10	10	71, 88
30		Lys$^{2,2'}$		Gly$^{4,4'}$			>200	89
31		His$^{2,2'}$				40	80	90
32		Hnv$^{2,2'}$				> 85–125	> 85–125	46
33		Orn^{2}His$^{2'}$				5	5	90
34		Orn^{2}Hnv$^{2'}$				9–25	9–25	46
35			Gly$^{3,3'}$			>100	>100	91
36			Ala$^{3,3'}$			20	50	91
37			Val$^{3,3'}$			25	50	92
38			Phe$^{3,3'}$			10–25		78
39			MeLeu$^{3,3'}$			5	5	93
40			Leu^{3}MeLeu$^{3'}$			5	5	93
41			δ-Ava$^{3-4}$			> 80		64
42			δ-Ava$^{3-4,3'-4'}$			> 80		64
43				Gly$^{4,4'}$		50	>100	94
44				D-Ala$^{4,4'}$		40	40	94, 95
45				D-Val$^{4,4'}$		10	10	96
46				D-Leu$^{4,4'}$		5	10	96
47				D-Ser$^{4,4'}$		20	40	53, 97
48				D-Ser(Bzl)$^{4,4'}$		5	2.5	53, 97
49				D-Tyr$^{4,4'}$		9–25	9–25	46
50				D-Trp$^{4,4'}$		5–10	5–10	98
51				D-Pro$^{4,4'}$	Phe$^{5,5'}$	2.5–20	20–40	98
52				D-Pro$^{4,4'}$	Trp$^{5,5'}$	5–10	5–20	98
53				D-Pro$^{4,4'}$	Cha$^{5,5'}$	2.5–10	10–20	98
54				D-Pro$^{4,4'}$	D-Trp$^{5,5'}$	10–40	40–80	98
55				D-Cha$^{4,4'}$		5–10	5–10	98

56			D-Dpr$^{4,4'}$		20	80	52, 55
57			D-Dpr(Z)$^{4,4'}$		5	5	52, 55
58			Ala$^{4,4'}$		>100	>100	94
59			Phe$^{4,4'}$		45–180	300–700	84, 89, 99
60			Pro$^{4,4'}$	Asn$^{5,5'}$	>160	> 80	46, 51
61			Pro$^{4,4'}$	D-Ala$^{5,5'}$	>160	> 80	46, 51
62			Pro$^{4,4'}$	D-Phe$^{5,5'}$	40	155	100
63			D-Phe^{4}Pro$^{4'}$	Pro^{5}D-Phe$^{5'}$	5	10	100
64			Aib$^{4,4'}$		>100	>100	94
65			ΔAla$^{4,4'}$		10	10	52, 55
66			ΔPhe$^{4,4'}$		10	5–10	60, 101
67			δ-Ava$^{4-5}$		5	10	102
68			δ-Ava$^{4-5,4'-5'}$		>100	>100	103
69			BTD$^{4-5,4'-5'}$		5	5	56, 104
70				Gly$^{5,5'}$	10	10	84, 99, 105, 106
71				Leu$^{5,5'}$	5	10	107
72				Ile$^{5,5'}$	10	10	108
73				Phe$^{5,5'}$	10	10	109
74				D-Pro$^{5,5'}$	80	160	57
75				Pro(4F)$^{5,5'}$	2.5		110
76				Aib$^{5,5'}$	>100	>100	111
77				Sar$^{5,5'}$	5	10	112
78				βAla$^{5,5'}$	160	80	113
79				Pro^{5} ^{14}C-Pro$^{5'}$	5	5	45, 46
80				Pro^{5}Gly$^{5'}$	10	10	114
81				δ-Ava$^{5-1'}$	20	40	115
82				δ-Ava$^{5-1',5'-1}$	>100	>100	115
83	[Cys^{1}]GS	[Cys^{1}]GS			>100	>100	86
84	*iso*-GS				>100	>100	116
85	*iso*-[Lys$^{2,2'}$]GS				>100	>100	116
86	*retro*-GS				50	50	84, 99, 117
87	*ent*-GS				5	5	84, 99
88	*retroent-GS*				90–180	300–750	99
89	*retro*-[Gly$^{5,5'}$]GS				60–180	>500	84, 99, 118
90	*ent*-[Gly$^{5,5'}$]GS				45–180	90–180	84, 99
91	*retroent*-[Gly$^{5,5'}$]GS				60–180	>500	99, 118

Table 2 Antimicrobial activities of ornithine-modified analogs of gramicidin S.

No.	Modification at ornithine residues[1]	Minimum inhibitory concentration (μg/ml)[2]		Ref.
		B. subtilis	*S. aureus*	
1	Orn$^{2,2'}$(GS)	5	5	70
2	Orn^{2}Orn(For)$^{2'}$	11	11	119, 120
3	Orn(For)$^{2,2'}$	92	97	119, 120
4	Orn^{2}Orn(Ac)$^{2'}$		9	120
5	Orn(Ac)$^{2,2'}$		63	120–122
6[3]	Orn^{2}Orn(Z)$^{2'}$	40	10	
7	Orn(Z)$^{2,2'}$		(500)[4]	121
8	Orn^{2}Orn(Dns)$^{2'}$		50	123
9	Orn(Dns)$^{2,2'}$		>400	123
10	Orn(Cbm)$^{2,2'}$		56	120
11	Orn(Gly)$^{2,2'}$	20	> 40	121, 124
12	Orn(Ala)$^{2,2'}$	10	20	124
13	Orn(D-Ala)$^{2,2'}$	20	40	124
14	Orn(Leu)$^{2,2'}$	40	> 40	124
15	Orn(D-Leu)$^{2,2'}$	20	20	124
16	Orn(Suc)$^{2,2'}$		50	120, 121
17	Orn(Mal)$^{2,2'}$		33	120
18	Orn^{2}Orn(Cm)$^{2'}$		very low activity	125
19	Orn(Cm)$^{2,2'}$		no activity	121, 125
20	Orn(Me$_2$)$^{2,2'}$	5		121, 122
21	Orn(Me$_3$)$^{2,2'}$	5	5	126
22[5]	MeOrn$^{2,2'}$	5	5	127
23	Arg$^{2,2'}$	5	5	121, 128
24	Hnv$^{2,2'}$		46	120, 121, 128
25	Glu$^{2,2'}$	400		128

[1] Only the varied residues at positions 2 and 2′ in gramicidin S (GS) are shown, and also see the footnote (1) in Table 1. For, formyl; Ac, acetyl; Dns, dansyl; Cbm, carbamyl; Suc, succinyl; Mal, maleyl; Cm, carboxymethyl. [2] See the footnote (2) in Table 1. [3] The compound was synthesized by Minari et al. in our laboratory. [4] The strain name of the microorganisms used is uncertain. [5] The compound contains D-MePhe$^{4,4'}$.

4.1.1 Function of the amino acid residues at positions 1 and 1′

Substitution of Val residues in these positions of GS by an aliphatic amino acid such as Ala, Abu and Leu produces analogs with fully retained activity (Nos. 4, 6 and 16 in Table 1), while that by Phe or Lys(Z) with aromatic character reduces the activity (Nos. 9 and 12 in Table 1). Drastic reduction of the activity is observed on Gly or hydrophilic Lys replacement (Nos. 2 and 11 in Table 1). Introduction of unusual amino acid in these positions results in destabilization of the specific β-sheet structure of GS (Figure 6) to give an inactive compound; D-Val with opposite configuration (No. 14 in Table 1), MeVal without amino hydrogen for hydrogen bonding when incorporated into peptide chain (No. 18 in Table 1) and Aib containing two methyl groups as a side chain (No. 17 in Table 1).

It is interesting to note that a depsipeptide analog containing 2-hydroxy-3-methylbutanoic acid (Hmb) at the 1,1′ positions still retains some activity in spite of the loss of two intramolecular hydrogen bondings (No. 19 in Table 1). One of the Val residues in GS is replaceable by Gly, Abu and S-*p*-methoxybenzyl-cysteine (Cys(MBzl)) without any significant effect on the antimicrobial activity (Nos. 21, 23 and 24 in Table 1), but the substitution by one MeVal residue reduces the activity (No. 25 in Table 1).

4.1.2 Function of the amino acid residues at positions 2 and 2′

The basic character of Orn residues at positions 2 and 2′ is essential for revealing the strong activity of GS. Thus, the change of the basic character to neutral (Nos. 27 and 32 in Table 1 and Nos. 3, 5, 7, 9, 10, 18 and 24 in Table 2) or acidic (Nos. 16, 17, 19 and 25 in Table 2) results in a loss of the activity. Three methylene chains of Orn residues in GS can be replaced by either two methylene chains ([$Dab^{2,2'}$]analog; Dab, 2,4-diaminobutyric acid) or four ([$Lys^{2,2'}$]analog) without loss of the activity (No. 28 or 29 in Table 1). But an aminoacylation of δ-amino groups with α-amino acid reduces the potency (Nos. 11 – 15 in Table 2).

Increase of the basicity by alkylation or guanidination of the δ-amino groups in GS results in a modified analog which retains full activity (No. 20, 21 or 23 in Table 2). However, [$His^{2,2'}$]analog with weakly basic character is only $\frac{1}{8}$ or $\frac{1}{16}$ as active as GS (No. 31 in Table 1). In contrast, single substitution of the Orn residue by His affords a fully active analog (No. 33 in Table 1). Analogs containing one neutral amino acid residue at 2 position still retain some activity (No. 34 in Table 1 and Nos. 2, 4, 6 and 8 in Table 2), showing the necessity of at least one charged group for the activity.

4.1.3 Function of the amino acid residues at positions 1,1′ and 3,3′

The side chains of the four amino acid residues in these positions form hydrophobic side in GS molecule (Figure 6). Simultaneous replacement of these four residues by hydrophilic amino acids such as Ser or smaller aliphatic amino acids results in a complete loss of the activity (No. 8 or 5 in Table 1). A depsipeptide analog containing Hmb at positions 1 and 1′ and 2-hydroxy-4-methylpentanoic acid (Hmp) at positions 3 and 3′ without intramolecular hydrogen bonds is inactive because of conformational instability (No. 20 in Table 1). Replacement of one Val and one Leu residues in GS by aromatic Phe results in a slight loss of activity (No. 22 in Table 1), but simultaneous replacement of the four residues by Phe in a weakly active analog (No. 10 in Table 1).

4.1.4 Function of the amino acid residues at positions 3 and 3′

Loss of the activity by Gly replacement at positions 3 and 3′ and by δ-aminovaleric acid (δ-Ava) replacement at only position 3-4 or both positions 3-4 and 3′-4′ reflects the importance of hydrophobic character at these positions (Nos. 35, 41 and 42 in Table 1). But some optimal bulkiness of the side chains in amino acid residues at positions 3 and 3′ seems to be necessary for the exhibition of strong activity.

Substitution of these residues by aliphatic Ala and Val or by aromatic Phe reduces the activity (Nos. 36–38 in Table 1). Interestingly, analogs containing MeLeu, $[MeLeu^{3}]$GS and $[MeLeu^{3,3'}]$GS, exhibit strong antimicrobial activity unlike $[MeVal^{1}]$- and $[MeLeu^{1,1'}]$GS (Nos. 39 and 40 in Table 1). Three possible conformational models have been proposed for this $[MeLeu^{3,3'}]$GS [11].

4.1.5 Function of the amino acid residues at positions 4 and 4′

D-Amino acid residues at positions 4 and 4′ play a specific role for the maintenance of the stable β-sheet structure of GS by two β-turns around the D-Phe-Pro sequences. Thus, the destabilization of this specific conformation by the introduction of L-amino acids or Aib residues at these positions leads to a complete loss of the activity (Nos. 58, 59 and 64 in Table 1).

Replaced analogs by various bulky aliphatic or aromatic D-amino acids such as D-Leu, D-cyclohexylalanine (D-Cha), D-Trp, D-Ser(Bzl) and D-Dpr(Z) fully retain the activity (Nos. 46, 48, 50, 55 and 57 in Table 1). Decrease of the activity in the order of D-Val, D-Tyr, D-Ala to Gly substitution at 4 and 4′ positions suggests the importance of hydrophobicity at these positions for revealing the activity (Nos. 30, 43–45 and 49 in Table 1). Incorporation of hydrophilic amino acids such as D-Ser or D-Dpr to these positions reduces the potency (Nos. 47 and 56 in Table 1). Single $D\text{-}Phe^{4}\text{-}Pro^{5}$ sequence in the β-turn part of GS can be replaceable with Pro-D-Phe or δ-Ava without affecting the activity (Nos. 63 and 67 in Table 1). However, both substitutions of the sequences by Pro-D-Phe or δ-Ava greatly reduce the activity (Nos. 62 and 68 in Table 1).

4.1.6 Function of the amino acid residues at positions 5 and 5′

Pro residues at 5 and 5′ positions can be replaced by a variety of amino acids (4-fluoroproline (Pro(4F)), Sar, Gly, Leu, Ile and Phe) without loss of the activity (Nos. 70–73, 75 and 77 in Table 1). However, substitution of these positions by unusual amino acids such as D-Pro, Aib, and βAla results in a loss of the activity, reflecting disorder of GS-like conformation (Nos. 74, 76 and 78 in Table 1). Single replacement analog with ^{14}C-Pro at position 5, of course, retains the activity (No. 79 in Table 1). Discussions on the other analogs in Table 1 were already given in the previous monograph [11].

4.1.7 Summary by "sidedness" hypothesis

All the results obtained through the structure-activity relationship studies on GS can be well explained by the "sidedness" hypothesis proposed by Schwyzer [130] and by Kato and Izumiya [133]. That is to say, two hydrophilic side chains are on one side of the GS molecule and four hydrophobic ones are on the other side as revealed in Figure 6. This "sidedness" structure resulted from the characteristic conformation of GS is highly responsible for manifesting the bioactivity of GS.

Thus, an analog that fully satisfies the "sidedness" structure may show strong GS-like activity. In contrast to this, destabilization, disorder, disturbance, destruction and/or loss of this "sidedness" structure may lead to the reduction of the activity.

4.2 Tyrocidines

There are scarce reports on the chemical synthesis of analogs of tyrocidines. The antimicrobial activity of these analogs together with that of natural tyrocidine congeners is summarized in Table 3. Tyrocidines and GS are regarded as intimately related "hybrid" analogs, because tyrocidines share the same pentapeptide sequence with GS (see Figures 1 and 2).

Table 3 Antimicrobial activities of tyrocidines and their analogs.

No.	Cyclic decapeptide[1]	Minimum inhibitory concentration (μg/ml)[2]		Ref.
		B. subtilis	*S. aureus*	
1	Gramicidin S	5	5	70
2	Tyrocidine A (TA)	10	20	35, 134
3	TB	50−100	50−100	134
4	TC	5	5	134
5	TE ([Phe^{10}]TA)	13−25	25	38
6	[Gly^6]TA	50	>50	35
7	[Gly^7]TA	20	50	35
8	[Val^6]TA	10	13−50	15, 135
9	[Orn^7]TA	20	50−100	15, 135
10	[$Leu^{8,9}$]TE ([$Leu^{8,9}$, Phe^{10}]TA)	>80	>40	136
11	[$Phe^{8,9}$]TE ([$Phe^{8,9,10}$]TA)	40	20	136

[1] Optically active amino acids shown in this Table are of the L-configuration. [2] See the footnote (2) in Table 1.

The tyrocidine molecule contains only one basic amino acid residue, Orn, which results in the exhibition of somewhat weaker activity than GS, as in the case of monoacylated GS, e.g., [$Orn(For)^2$]GS (No. 2 in Table 2). Replacement of Phe residue at position 6 in tyrocidine A (TA) by Val residue does not affect the activity (No. 8 in Table 3), but Gly replacement reduces the activity (No. 6 in Table 3). These results indicate the importance of a hydrophobic side chain at position 6 for exhibiting high activity.

Replacement of D-Phe residue at position 7 by Orn residue decreases the activity (No. 9 in Table 3). This result is unexpected because the analog, [Orn^7]TA, has two Orn residues occupying the same positions as in GS. However, the low activity may be explained in terms of molecular topology of TA by Kuo and Gibbons [137]. They have proposed the presence of a highly hydrophobic region consisting of the side chain of Val^1, Leu^3, D-Phe^4, Pro^5, Phe^6, and D-Phe^7 and of a hydrophilic region consisting of Asn^8-Gln^9-Tyr^{10} side chain. The replacement of D-Phe^7 by Orn may

Table 4 Antimicrobial activities of gratisin and its analogs.

No.	Sequence of cyclic dodecapeptide[1]						Minimum inhibitory concentration (μg/ml)[2]		Ref.
							B. subtilis	*S. aureus*	
1	$Val^{1,1'}$	$Orn^{2,2'}$	$Leu^{3,3'}$	$D\text{-}Phe^{4,4'}$	$Pro^{5,5'}$	$D\text{-}Tyr^{6,6'}$ (GR)[3]	5	20	41, 42
2						$Tyr^{6,6'}$	80	160	41, 42, 138
3				$Phe^{4,4'}$		$Tyr^{6,6'}$	> 80	>160	41, 42, 138
4				$Phe^{4,4'}$			>160	160	42
5				$D\text{-}Tyr^{4,4'}$		$D\text{-}Phe^{6,6'}$	40	80	42
6					$D\text{-}Tyr^{5,5'}$	$Pro^{6,6'}$	5	10	42
7						$D\text{-}Phe^{6,6'}$	10	20	42
8				$Pro^{4,4'}$	$D\text{-}Phe^{5,5'}$	$D\text{-}Phe^{6,6'}$	5	40	42
9				$D\text{-}Ala^{4,4'}$			> 80	>160	63
10						$D\text{-}Ala^{6,6'}$	10	40	63
11					$D\text{-}Phe^{5,5'}$	$Pro^{6,6'}$	10	20	139
12			gramicidin S				5	5	70

[1] See the footnote (1) in Table 1. [2] See the footnote (2) in Table 1. [3] Most probable sequence for natural gratisin (GR) (Reference 42).

disturb the topological distribution of hydrophobic and hydrophilic side chains, causing the decrease in the activity. Similarly, introduction of hydrophobic Leu or Phe into the hydrophilic positions 8 and 9 markedly decreases the activity (Nos. 10 and 11 in Table 3).

4.3 Gratisin

In order to determine the structure of natural gratisin, several cyclic dodecapeptides with different configurational sequence have been synthesized [41, 42, 138] on the basis of the structure proposed by Myaskovskaya et al. [40]. Their structures and antibacterial activities examined are summarized in Table 4.

The cyclic dodecapeptide (No. 3 in Table 4) consisting of only L-amino acids showed no activity, but the peptide containing D-Phe$^{4,4'}$ (No. 2 in Table 4) with a structure analogous to GS could inhibit very weakly the growth of Gram-positive bacteria. The peptide (No. 4 in Table 4) with Phe-Pro-D-Tyr sequences was inactive. On the other hand, the cyclic dodecapeptide (XV) (No. 1 in Table 4) having D-Phe-Pro-D-Tyr sequence showed the strongest activity against Gram-positive microorganisms tested. Its activity toward ***B. subtilis*** was the same as that of GS. Thus, it is concluded that the peptide XV is possibly identical to natural gratisin.

Additional analogs have also been prepared for a further investigation of structure-activity relationship of gratisin [42, 63, 139]. The D-Phe$^{4,4'}$ residues have an important role in exhibiting strong activity, whereas the hydroxy group of D-Tyr$^{6,6'}$ residues is not so important (Nos. 5 and 7 in Table 4). The peptides having D-Phe-D-Tyr-Pro (No. 6 in Table 4), D-Phe-D-Phe-Pro (No. 11 in Table 4) and Pro-D-Phe-D-Phe sequences (No. 8 in Table 4) exhibit strong activity, showing that the positioning of Pro residues is not restricted to positions 5 and 5′ for revealing the activity.

5. Conformations of Gramicidin S and Tyrocidines

The rigid β-sheet conformation of GS (see Figure 6) proposed by Hodgkin and Oughton [129] on the basis of X-ray studies and by Schwyzer [130] based on chemical synthesis is now accepted as the most likely structure [131, 132, 140–144]. TA is also known to possess an antiparallel β-sheet conformation with type II′ β-turn at D-Phe^4-Pro^5 and type I β-turn at Gln^9-Tyr^{10} [137, 145–148]. This backbone conformation of TA in solution greatly resembles that of GS, despite the differences in sequence and positioning of D-amino acid residues.

Thus, because of their well established conformations, GS and TA have been often used as model peptides for evaluating newly developed methods for conformational studies or as standard compounds for various analytical methods.

^{1}H-Nuclear magnetic resonance (^{1}H-NMR) studies were carried out using deuterated GS which was prepared biosynthetically [149, 150] and chemically [54, 151]. These experiments were also performed by spin-label relaxation enhancement [152], by decoupling during proton exchange [153], and in the presence of phospholipid membrane [154]. Further, many experiments have been carried out in the conformation field, however in the following, only items and literatures will be given. ^{2}H-NMR [155]; ^{13}C-NMR [101, 156–158]; ^{15}N-NMR [159–163]; NMR studies by nuclear Overhauser effect (NOE) experiment [60, 101, 156, 158, 164–169]; hydrogen exchange kinetics [170]; energy calculation [171–174]; X-ray analysis [141, 175, 176]; Raman spectroscopy as diagnostic analysis for β-turn structure [174, 175, 177]; infrared (IR) spectroscopy [174, 178]; circular dichroism (CD) studies [98, 179] in the presence of phospholipid membrane [59, 154, 180]; chemical transformation, e.g., chlorination [181, 182] and *N*-methylation [94, 126, 127, 161], for conformational analysis; laser desorption [183] and tandem [184] Fourier transform mass spectrometry.

In the previous monograph [11], we presented a detailed review of conformational studies on GS by X-ray, optical rotatory dispersion (ORD) and CD, NMR, IR, and theoretical studies. Consequently, in this Section only major innovation will be mentioned.

5.1 Optical rotatory dispersion (ORD) and circular dichroism (CD)

ORD and CD methods do not provide information on a precise peptide architecture, but can be used as a sensitive indicator for detecting subtle conformational changes. As discussed in Section 4, replacement of the constituent amino acids in GS frequently alters the bioactivity, and such an alternation is related to the change in molecular conformation, which can be observed directly by these optical techniques. Therefore, ORD and CD measurements have been often carried out on many of GS analogs listed in Tables 1 and 2 in connection with their bioactivity.

To fully understand structure and/or conformation-activity relationships of many GS analogs, elucidation of bioactive conformation of them in the environment where they actually act now becomes increasingly important. It is well known that GS first attacks the membrane of sensitive bacteria, altering the permeability of the membrane and releasing intracellular components [11]. Conformational studies of GS and its analogs in the membrane-mimetic environments will provide profound insight on such bioactive conformation.

Recently, Higashijima et al. [154] reported the result on such an approach, analyzing the conformation of GS and its analogs in the presence of phospholipid membrane by CD or ^{1}H-NMR spectroscopy. The analogs employed are [D-Ala$^{4,4'}$]GS (XVI) and [Gly$^{4,4'}$]GS (XVII) with weak activity and inactive [Ala$^{4,4'}$] GS (XVIII) (Nos. 44, 43 and 58 in Table 1). CD profile of the active analog XVI in aqueous solution became essentially the same as that for GS on the addition of lysophosphatidylcholine (lysoPC) vesicles, while the ellipticity was smaller than that for GS by about 10% in the absence of the phospholipid. Remarkable change in CD spectrum of the moderately active analog XVII in aqueous solution was observed; the ellipticity of the CD band (208 nm) was significantly increased and became nearly equal to that of GS by the addition of lysoPC micelles, while the ellipticity was about 50% of that of GS in the absence of the phospholipid.

In contrast, the inactive XVIII showed quite different CD spectrum from that of GS which hardly changed on the addition of lysoPC micelles. These facts indicate that the active analogs XVI and XVII take the GS-like conformation upon binding with the phospholipid membrane, but the inactive analog XVIII does not take the GS-like conformation even in the presence of phospholipids. That is to say, the bioactivity of GS analogs is correlated with the formation of the GS-like conformation upon binding to the phospholipid membrane.

Similarly, Ono et al. [59] examined CD profile of a simple GS analog, cyclo(-Leu-Lys-Leu-D-Leu-Leu-)$_2$ (XIX) (No. 7 in Table 1), with some one-fourth activity of that of GS in the presence of liposomes. Characteristic CD feature (a double minimum of thoughs at 205 and 214 nm) for GS was essentially maintained for XIX in pH 7.4 buffer, though intensity of the troughs largely decreased. In liposomes containing acidic phospholipid, however, the intensity of the double minimum of XIX became much higher than that in the buffer solution, coming considerably close to that of GS. This also suggests that the analog XIX shows the activity in bacterial membranes containing acidic phospholipids by taking the GS-like conformation.

5.2 Nuclear magnetic resonance (NMR)

NMR spectroscopy using ^{1}H, ^{13}C and ^{15}N nuclei is the most powerful tool for studying conformation of peptides in solution. New techniques including NOE and two-dimensional method have widened the applicability of this NMR method. In the following, the topical events on conformational studies on GS by NMR will be highlighted.

Interaction of GS with phospholipid membrane was studied by ^{1}H-NMR (270 MHz) analysis [154]. The N-H proton resonances of GS in aqueous solution were appreciably broadened upon the addition of PC or PC/phosphatidylserine (PS) vesicles, and the broadening was similar for PC and PC/PS vesicles. The broadening effect on the side chain NH_3^+ proton resonance of Orn residues was much smaller than that on the main ring N-H proton resonances by the addition of the phospholipids. This indicates that GS is bound to the vesicles and that the electrostatic interaction between the cationic amino groups of Orn side chains in GS and the polar head groups of phospholipid is not essential for the binding.

Conformation of the Orn side chains in GS in solution was investigated by ^{1}H and ^{15}N-NMR spectroscopy using GS and methylated GS ([Orn(Me$_3$)$^{2,2'}$]GS) in which Orn remains ionized but cannot function as an H-bond donor [161]. The data show the presence of a pair of solvent-labile intramolecular hydrogen bonds between Orn $N^{\delta}H_3^+$ and D-Phe C=O in GS not previously known to exist in solution. In contrast, ^1H-^{13}C selective NOE studies of GS in solution have shown the existence of a single Orn-$N^{\delta}H_2$ to D-Phe-CO hydrogen bond [158], which is compatible with the result obtained by X-ray analysis of the crystalline urea complex of GS [141].

Specific isotopical-labeling has enabled the assignments and interpretations of signals more easy in NMR studies. Tanimura et al. [54, 151] prepared (2*R*, 3*R*)-Phe-2,3-d_2 through asymmetric catalytic reduction of cyclo(-ΔPhe-D-Ala-) under an atmosphere of $^2H_2$ and successive acid hydrolysis. The labeled amino acid was used for the chemical synthesis of [(2*R*,3*R*)-Phe-2,3-$d_2^{4,4'}$]GS. ^{1}H-NMR spectrum of the deuterated GS in dimethyl sulfoxide (DMSO)-d_6 showed a sharp singlet at 2.98 ppm for the (3S)-proton of the residues at 4 and 4′ positions. The result provides the support for earlier proposal [149, 161] that among rotamers of D-Phe aromatic side chain in GS the one with $\kappa_1 = 180°$ is predominant.

5.3 Other conformational analyses

X-ray analysis of crystal of a hydrated GS-urea complex has shown that the backbone conformation of GS is essentially the same as that depicted in Figure 6 [141]. However, the molecular backbone is somewhat twisted and the hydrogen bonds between the two halves of the molecule are no longer parallel. Moreover, there is an intramolecular hydrogen bond between one of the Orn side chain nitrogen atoms and the D-Phe carbonyl oxygen atom as described before. Computation of conformational energy has also reached the same conclusion concerning hydrogen bonding involving the Orn side chain of GS [173].

Simple chemical transformation can be used for the detection of hydrogen-bonded peptide NH's to study the conformation of bioactive peptides in solution. Kondo et al. [181, 182] utilized the replacement reaction of peptide NH in GS using chlorine reagent. They demonstrated that the intramolecularly hydrogen-bonded peptide NH's in GS are much more susceptible to chlorination than the solvent-exposed peptide NH's.

6. Design of Highly Active Analogs of Gramicidin S

GS holds a rigid antiparallel *β*-sheet structure stabilized by four intramolecular hydrogen bonds, and the hydrogen bond stability is provided by two *β*-turns of the D-Phe-Pro sequences as indicated in Figure 6. Consideration of structure and/or conformation-activity relationships of GS based on various GS analogs has now led to following conclusion, as discussed in Section 4.1. The "sidedness" structure, namely, amphiphilic property, induced by the characteristic *β*-sheet and *β*-turn structures in GS molecule is highly responsible for manifesting the strong activity of GS. In other words, the three-dimensional structure as a whole molecule is the essential determinants for the bioactivity of GS. This implies the importance of topochemical consideration, especially in designing active analogs of GS. Such topochemical approach for the design of highly active analogs of GS will be picked up in the following discussions.

Sato and Nagai paid much attention to one of the most important structural elements, the stable β-turn structure formed by the D-Phe-Pro sequence, for GS activity [98, 104]. Firstly, they systematically searched for β-turn preference peptide sequence analyzing CD spectra of *N*-(2,4-dinitrophenyl)tetrapeptide *p*-nitroanilides [51, 57, 185–187]. Some of the conclusions are: (1) configurational sequence of L-D-L-L (or D-L-D-D) prefers β-turn and (2) L-D-D-L (or D-L-L-D) sequence can also take β-turn only when Pro is at the 2nd position, especially the tetrapeptide sequence with L-Pro at the 2nd and D-amino acids at the 1st and 4th has the highest potential to take β-turn conformation [98, 104]. Based on these results, they designed a series of cyclic peptides with "sidedness" structure formed by the β-turn preferring D-Leu-L-Pro-Xaa-D-Leu sequence or its antipodal sequence, L-Leu-D-Pro-Xaa-L-Val, as schematically illustrated in Figure 7 [98, 104].

They also supposed that not only the conformation but the hydrophobicity of the whole molecule was important for the activity of GS. Thus, the designed analog

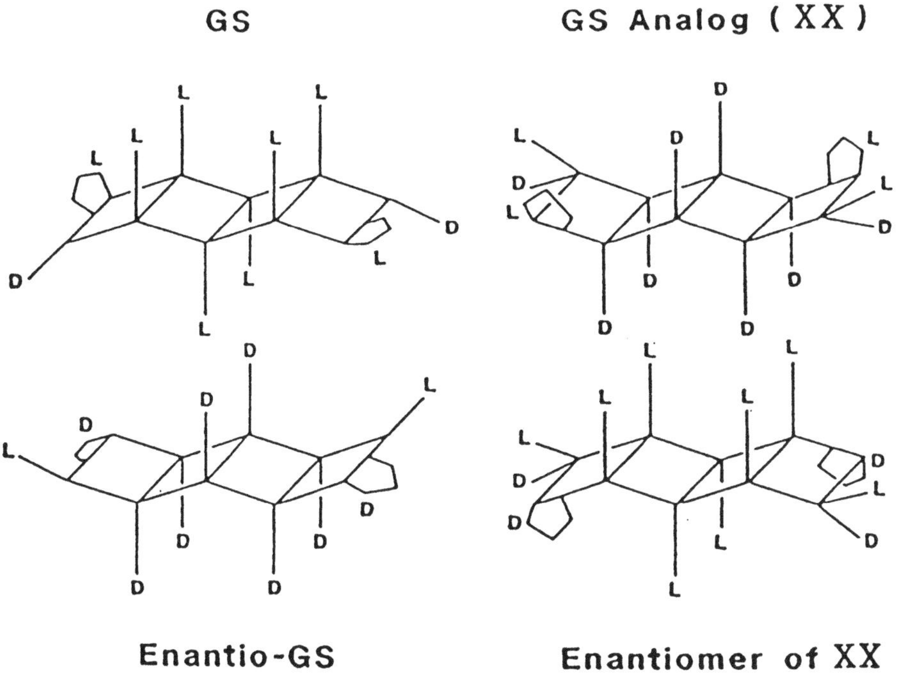

Figure 7 Conformational models of gramicidin S and designed analogs.

Table 5 Antimicrobial activities and retention time on HPLC of designed analogs of gramicidin S.

No.	Cyclic decapeptide (XX)[1]	Minimum inhibitory concentration (μg/ml)[2]		Retention time (min) on HPLC[3]
		B. subtilis	*S. aureus*	
	1,1′ 2,2′ 3,3′ 4,4′ 5,5′			
1	cyclo(-D-Leu-D-Lys-D-Leu-Pro-Trp-)$_2$	10–20	10–40	>26
2	cyclo(-Cha-)$_2$	2.5–10	2.5–10	8
3	cyclo(-Phe-)$_2$	5–10	5–10	5
4	cyclo(-Leu-)$_2$	40–80	160–310	4.6
5	cyclo(-Ala-)$_2$			3.8
6	cyclo(-Asn-)$_2$			4
7	cyclo(-Gly-)$_2$	40–80		4.9
8	cyclo(-D-Ala-)$_2$	40–80		5.6
9	cyclo(-D-Leu-)$_2$	10–20	40–80	7.6
10	cyclo(-D-Phe-)$_2$	2.5–10	5–20	8.5
11	cyclo(-D-Cha-)$_2$	1.25–5	5–20	13.6
12	cyclo(-D-Trp-)$_2$	5–20	5–20	>26
13	Gramicidin S (GS)	5	5	18.6

[1] Only the varied residues are shown except in the case of No. 1 and GS, and all optically active amino acids are of the L-configuration unless otherwise indicated. [2] The figures are derived from the data (Reference 98) by calculation on the basis of those for GS as a standard. [3] The figures are derived from the data (Reference 98). Column: Develosil ODS-5, 4.6 × 5 mm; eluent: MeOH: 5% $NaClO_4$ (5 : 1); flow rate: 1 ml/min.

(XX), cyclo(-D-Leu-D-Lys-D-Leu-Pro-Xaa-)$_2$, with the D-Leu-L-Pro-Xaa-D-Leu sequence contains a variety of amino acids at Xaa residues with different hydrophobicity, and their antimicrobial activities are summarized in Table 5. In the Table, the retention time on HPLC as an index of the hydrophobicity of the whole molecule is also shown [188].

As expected, the activity increased with increasing hydrophobicity of Xaa residues in XX and the highest activity was observed when Xaa was D-Cha (No. 11 in Table 5). This analog shows a strong and even surpassed activity compared with that of the parent GS. The bioactivity of the analogs with Cha, Phe, D-Phe and D-Trp (Nos. 2, 3, 10 and 12 in Table 5) is comparable to that of GS. The mirror image conformation of GS-like *β*-sheet structure for the analog XX with D-Xaa was suggested by CD studies (see Figure 7). Another designed analog (XXI), cyclo (-Val-Orn-Leu-D-Pro-Xaa-)$_2$, that is, the enantiomer of the analog XX, exhibited similar activity and HPLC profile to those of XX (Nos. 51–54 in Table 1). The CD profile of the analog XXI with L-Xaa was similar to that of GS, as expected from the conformational model shown in Figure 7.

Stabilization of bioactive conformation of peptides by restriction of conformational freedom would provide a powerful strategy for designing an active analog. Fixation of conformationally important elements such as *β*-turn structure in GS is one of the way. Based on this idea, a bicyclic dipeptide, (3*S*,6*S*,9*R*)-2-oxo-3-amino-7-thia-

1-azabicyclo[4.3.0]nonane-9-carboxylic acid, was designed and synthesized simulating the backbone conformation of the two central amino acid residues, D-Ala-Pro, in β-turn (type II′) part of [D-Ala$^{4,4'}$]GS [104, 189]. Structural models of derivative (XXII) of the bicyclic dipeptide, named bicyclic β-turn dipeptide (BTD), and the corresponding part (XXIII) of the analog, [D-Ala$^{4,4'}$]GS, are shown in Figure 8.

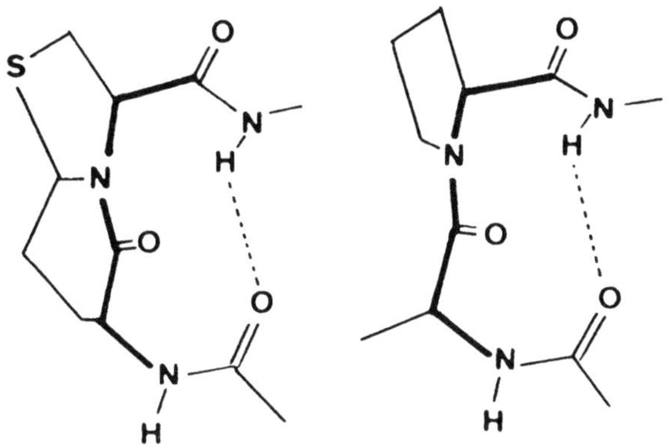

XXII XXIII

Figure 8 Structural models of XXII: the acylated BTD amide and XXIII: the corresponding part of the gramicidin S analog. Dotted lines, H-bonding; bold lines, the central part of backbone in β-turn, which is fixed in XXII.

The usefulness of fixation of β-turn by the BTD unit in bioactive peptides has been demonstrated by the incorporation of BTD into GS at the two β-turns in place of the D-Phe-Pro sequences [56, 104]. The resulting analog, [BTD$^{4-5,4'-5'}$]GS (No. 69 in Table 1), showed the same antimicrobial activity and closely similar CD profile to that of GS, indicating that there are two β-turns in the bioactive conformation of GS.

Meanwhile, Shimohigashi et al. [60] and Kodama et al. [101] paid attention to α,β-dehydroamino acid (Δamino acid) that elicits some conformational changes. Presence of Δamino acid residue would increase a rigidity due to limited bond angles of the residue and a restriction of side chain orientations when such Δamino acid is incorporated into peptides. In fact, the incorporation of ΔPhe into GS in place of the D-Phe$^{4,4'}$ resulted in the reinforcement of the β-sheet backbone conformation by the strengthening of the hydrogen bond between Val-NH and Leu-CO. As we discussed repeatedly, this conformation is the most important structural element for GS activity, and the reinforcement led to the exhibition of strong antimicrobial activity (see No. 66 in Table 1) [60, 101].

Similarly, the antimicrobial activity of a GS analog containing ΔAla at 4,4′ positions, [ΔAla$^{4,4'}$]GS (No. 65 in Table 1), is comparable to that of the [ΔPhe$^{4,4'}$]GS described above and 4-fold higher than that of the saturated [D-Ala$^{4,4'}$]GS (No. 44 in Table 1). This result also indicates possible stabilization effect of ΔAla residues on β-turn structure [52, 55].

7. Active Analogs of Gramicidin S Against Gram-Negative Bacteria

GS, with a high antimicrobial activity against Gram-positive bacteria but not against Gram-negative ones, has hydrophobic part on one side and hydrophilic part on the opposite side to form amphiphilic structure, but the hydrophobic D-Phe side chains at 4,4′ positions are on the hydrophilic side (see Figure 6). In order to investigate the effect of increase of cationic charges on the hydrophilic side on activity, Ando et al. [52, 55] synthesized the analog VIII, [D-Dpr$^{4,4'}$]GS, containing two additional basic D-Dpr residues instead of the D-Phe$^{4,4'}$ and evaluated its antimicrobial activities (see No. 56 in Table 1 and No. 2 in Table 6).

Table 6 Antimicrobial activities of gramicidin S analogs and gramicidin S-like analogs.

No.	Cyclic peptide[1]	Minimum inhibitory concentration (μg/ml)			Ref.
		B. subtilis	*E. coli*	*S. typhosa*	
1	Gramicidin S (GS)	3.13	>100	>100	52, 55
2	[D-Dpr$^{4,4'}$]GS (VIII)	12.5	25	25	52, 55
3	[Lys$^{1,1'}$]GS		>100		80
4	cyclo(-Orn-Leu-Orn-D-Phe-Pro-)$_2$ (XXV)	50	>100	>100	62
5	cyclo(-Orn-Leu-Orn-Phe-Pro-)$_2$ (XXVI)	12.5	50	25	62
6	cyclo(-Orn-Leu-Orn-Phe-D-Pro-)$_2$	12.5	>100	100	62
7	cyclo(-Orn-Leu-Orn-Leu-D-Phe-Pro-)$_2$	50	>100	>100	190
8	cyclo(-Leu-Orn-Leu-Orn-D-Phe-Pro-)$_2$	12.5	100	>100	190
9	cyclo(-Leu-Orn-Leu-Orn-Leu-D-Phe-Pro-)$_2$ (XXIV)	6.25	12.5	25	61, 191
10	cyclo(-Leu-Orn-Leu-Orn-Leu-D-Phe-Sar-)$_2$	25	100	100	61
11	cyclo(-Leu-Orn-Leu-Orn-Leu-D-Phe-Gly-)$_2$	100	>100	>100	61

[1] Optically active amino acids shown in this Table are of the L-configuration unless otherwise indicated.

Surprisingly, the analog VIII was found to show different antimicrobial spectrum pattern from that of parent GS. Namely, VIII had an appreciable activity against several Gram-negative bacteria including *Escherichia coli* and *Salmonella typhosa* but negligible activity against Gram-positive ones such as *Staphylococcus aureus* [52, 55]. Meanwhile, introduction of cationic charges on the hydrophobic side by Lys residues at 1,1′ positions resulted in an inactive analog against both types of bacteria (Nos. 11 and 13 in Table 1 and No. 3 in Table 6) [80].

Stimulated by these facts, Ando et al. designed such GS analogs or GS-like macroring analogs (Nos. 4–11 in Table 6) that contain four basic amino acid residues and possibly hold β-sheet conformation (see structure of a cyclic tetradecapeptide (XXIV) in Figure 9) [61, 62, 190, 191].

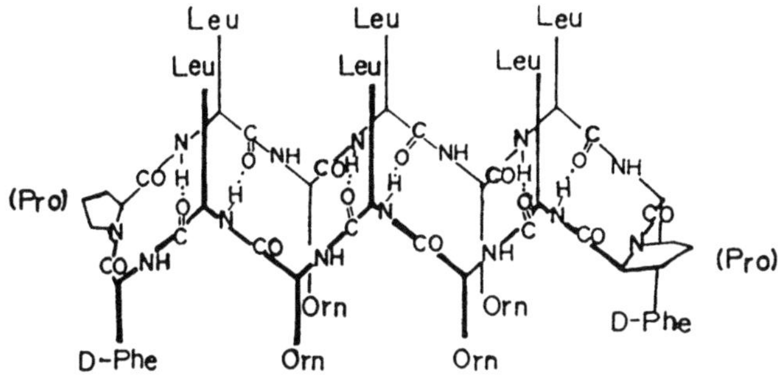

Figure 9 Possible structure of gramicidin S-like analog (XXIV).

Interestingly, some of them (Nos. 5 and 9 in Table 6) have a marked inhibitory potency against Gram-negative bacteria comparable to that of [D-Dpr$^{4,4'}$]GS. It seems that the increase of basicity on the hydrophilic side without change of the amphiphilic property induced by GS-like β-sheet structure is one of the important factors for revealing the activity against Gram-negative ones. However, as judged by CD evidence, the GS analog (XXV) (No. 4 in Table 6) satisfying such requirements does not show any activity against Gram-negative ones, whereas the analog (XXVI) (No. 5 in Table 6) holding a certain conformation other than that of GS exhibits the activity against Gram-negative ones.

So far several poly cationic antibacterial peptides have been reported. Macro-ring analogs of GS, sesquiGS (cyclic pentadecapeptide) and diGS (cyclic eicosapeptide), with increased cationic charges showed weak activity against Gram-positive bacteria, but no data on Gram-negative ones [192]. Amphiphilic basic model peptides, e. g., Ac-(Leu-Ala-Lys-Leu)$_3$-NHCH$_3$ (XXVII), that mimic the prepiece moieties of mitochondrial protein precursors were found to be as active as GS against Gram-positive bacteria, but inactive against Gram-negative ones [193, 194]. Meanwhile, a copolymer (XXVIII) [195] composed of equimolar amounts of Orn and Leu, and poly-Lys [196] are strongly active against *Escherichia coli.*

At present time, we can not give definite explanation why some peptides such as the analog XXV or the model peptide XXVII with basic and amphiphilic properties show no activity against Gram-negative bacteria while some basic peptides such as the analog XXVI or the copolymer XXVIII without GS-like conformation show the activity. In this connection, it is noteworthy that polymyxins, known for their strong inhibitory activity against Gram-negative bacteria, contain five or six basic amino acid residues out of 10 amino acid components together with fatty acid attached to the peptides [197]. It is also of interest to note that natural amphiphilic basic peptides recently isolated and sequenced can inhibit the growth of *E. coli*; lepidopterans from silkworm [198, 199], cecropins from cecropia moth [200–203], sarcotoxins from flesh fly [204–206], and magainins from the skin of the African clawed frog [207].

8. Mechanism of Antimicrobial Action of Gramicidin S

Interaction of GS with membranes of sensitive bacteria is the first and important event for triggering its antibacterial action [11]. Recent progress in the studies on the interaction of GS with model membranes and bacterial membranes will be described in this Section to shed light on the mechanism of antimicrobial action of this antibiotic.

8.1 Interactions with model membranes

Datema et al. [155] investigated the interaction of deuterated GS with dipalmitoylphosphatidylcholine (DPPC) liposomes by ^{2}H-NMR as well as ^{31}P-NMR and differential scanning calorimetry (DSC). Their results supported a previous proposal that electrostatic interaction of GS with phospholipid head group may be the first step for its binding to the bacterial membranes. It meant that the protonated amino groups of the Orn side chains in GS molecule interact with the negatively charged phosphate moiety of the phospholipid [7, 84, 208–211].

In contrast, Higashijima et al. [154] have demonstrated the importance of the hydrophobic interaction in the binding of GS with phospholipid membranes. As mentioned in Section 5.1, they utilized active and inactive GS analogs to analyze the interaction of GS with phospholipid membrane by CD and ^{1}H-NMR spectroscopy. Then, they came to propose that, in the binding of GS with phospholipid membrane, the main ring (and hydrophobic side) of GS molecule is trapped in the hydrophobic interior of phospholipid membrane while the side chain NH_3^+ groups of the Orn residues are exposed in the aqueous phase.

Importance of such hydrophobic interaction in the binding of GS with model membranes has also been pointed out by many experiments. These include lateral diffusion experiment of *N*-4-nitrobenz-2-oxa-1,3-diazole GS in PC, DPPC and dimyristoylphosphatidylcholine (DMPC) multibilayers using fluorescence recovery after photobleaching and DSC [212], laser-Raman and DSC investigations of GS-DMPC liposome interaction [213] and DSC study on the interaction between GS and DPPC or dipalmitoylphosphatidylethanolamine [214].

8.2 Interactions with bacterial membranes

To elucidate the actual mode of antimicrobial action by GS, Yonezawa et al. studied GS distribution on bacterial cells and on their membranes using synthetic ^{14}C-labeled GS ([^{14}C]GS) with the same activity as that of parent GS (No. 79 in Table 1) [45, 46, 215]. [^{14}C]GS was incubated with cells of sensitive Gram-positive *Bacillus subtilis* or insensitive Gram-negative *Escherichia coli*, and the amount of [^{14}C]GS

Table 7 Amounts of adsorbed GS and components on *B. subtilis* cell[1].

Amounts of adsorbed GS and components on cells	Natural cell	Protoplast membrane	Sonicated cell (cell wall)
Adsorbed GS (dpm)	1083	1105	193
Glucosamine (μg)	13.5	4.0	8.4
Phosphorous (μg)	1.82	1.7	0.3

[1] The values are corrected to 1.6×10^9 as the cell number.

Table 8 Amounts of adsorbed GS and components on *E. coli* cells[1].

Amounts of adsorbed GS and components on cells	Natural cell	Spheroplast membrane
Adsorbed GS (dpm)	100	322
Glucosamine (μg)	2.9	1.7
Phosphorous (μg)	1.1	1.1

[1] See the footnote (1) in Table 7.

adsorbed on the cells was measured (Tables 7 and 8) [215]. Tables 7 and 8 show the easy adsorption of [^{14}C]GS on sensitive *B. subtilis*, but not so on insensitive *E. coli*. Remarkable drop of the amount of the adsorbed [^{14}C]GS on the sonicated cell (cell wall) of *B. subtilis* compared with that of [^{14}C]GS adsorbed on the natural cells and the protoplast membrane clearly indicates that GS is adsorbed on the cell membrane of sensitive Gram-positive bacteria. Interestingly GS molecules covered most of the cell surface of *B. subtilis* at the minimum inhibitory concentration of 1.5 μg/ml; the number of the molecule adsorbed per cell was 1.3×10^6 [45].

Inhibitory profile of [^{14}C]GS-adsorption on cells of *B. subtilis* by several GS analogs indicated that degree of the inhibition paralleled the antibacterial activity. While, inactive [$Hnv^{2,2'}$]GS (No. 32 in Table 1) and [$Gly^{1,1'}$]GS (No. 2 in Table 1) had no longer such inhibitory activity [46].

Summarizing up the data obtained, Yonezawa et al. concluded that GS may be adsorbed on the phospholipid bilayer of the cell membrane of sensitive bacteria by electronic interaction between positive charges of Orn residues of GS and negative charges of phospholipid. Then, the hydrophobic side chains of Val and Leu residues enter the internal hydrophobic portion of the phospholipid bilayer. These interactions result in damages of the phospholipid bilayer of the cell membrane and in prevention of functioning of the cell membrane [215]. This is compatible with the previous proposal by Kaprel'yants et al. [216], Eremin et al. [217] and Dergunov et al. [218] that the electrostatic interaction and subsequent hydrophobic interaction are important processes for the binding of GS to the bacterial membrane.

Very recently, Katsu et al. presented an interesting another model for GS-phospholipid interaction following the observation of structure-activity relationship of GS analogs with regard to membrane permeability. They employed various cellular systems such as *Staphylococcus aureus* as Gram-positive bacteria, *Escherichia coli* as Gram-negative ones, human erythrocytes, rat liver mitochondria and artificial liposomal membranes [219, 220]. Two types of GS analogs were used [220]: (1) cyclo(-D-Leu-D-Lys-D-Leu-Pro-Xaa-)$_2$ XX (see Figure 7), where Xaa = Gly, D-Leu and D-Cha (Nos. 7, 9 and 11 in Table 5) and (2) [Orn(Ac)$^{2,2'}$]GS (XXIX) (No. 5 in Table 2) which lacks cationic moiety of GS. All the analogs have a β-sheet conformation as GS.

It was found that the antimicrobial activity (see Nos. 7, 9, 11 and 13 in Table 5) of the peptides against *S. aureus* coincided well with their ability to increase the permeability of the cell as judged by K^+ efflux (Table 9) [220]. Disruption of the barrier function of the outermembrane by the cationic compound 48/80 [221] forced the *E. coli* cells more sensitive to GS and XXa (XXa = D-Cha) (Table 9). These results indicated that the lower antimicrobial activity of GS against *E. coli* is due to the low ability of the peptide to disrup the outermembrane structure existing in Gram-negative bacteria. The cytoplasmic membrane of *S. aureus* cells seems to be rich in negatively charged lipid at neutral pH used in the K^+ efflux experiments judging from the phospholipid compositions of the membrane [222]. Therefore, *S. aureus* cells might be more sensitive to cationic peptides, and even the analogs, XXb (XXa = D-Leu) and XXc (Xaa = Gly), stimulated the efflux of K^+ to some extent.

Table 9 Percentage of peptide-induced K^+ efflux from cells[1].

	GS	XXa (Xaa = D-Cha)	XXb (Xaa = D-Leu)	XXc (Xaa = Gly)	[Orn(Ac)$^{2,2'}$]GS (XXIX)
S. aureus	100	100	30	10	20
E. coli	30	20	0	0	0
E. coli[2]	100	100	0	0	0
Mitochondria	100	100	0	0	100
Erythrocytes	100	100	0	0	0

[1] Percentage effused within 5 min is calculated. Cells were suspended in pH 7.2 buffer. [2] Pre-treated with compound 48/80 to disrup the outer membrane structure.

In order to examine the membrane fluidity of liposomes induced by GS and its analogs, changes of the phase transition of DPPC liposomes in the presence of GS and its analogs were also measured by the fluorescence polarization technique or DSC [219, 220]. GS and XXa (Xaa = D-Cha) decreased the phase transition temperature of neutral DPPC liposomes, while XXb (Xaa = D-Leu) and XXc (Xaa = Gly) did not so to a significant extent. These results show a good correlation between the ability of the peptides to decrease the phase transition temperature of DPPC liposomes and to increase the K^+ permeability of the cytoplasmic membranes.

Based on the results obtained, Katsu et al. proposed a hypothetical model on the GS-phospholipid interaction (Figure 10) [220] considering characteristic "sidedness" structure of GS which is essential for revealing the antimicrobial activity [130, 133]. The model is based on the previous one [154] that the hydrophobic interaction is

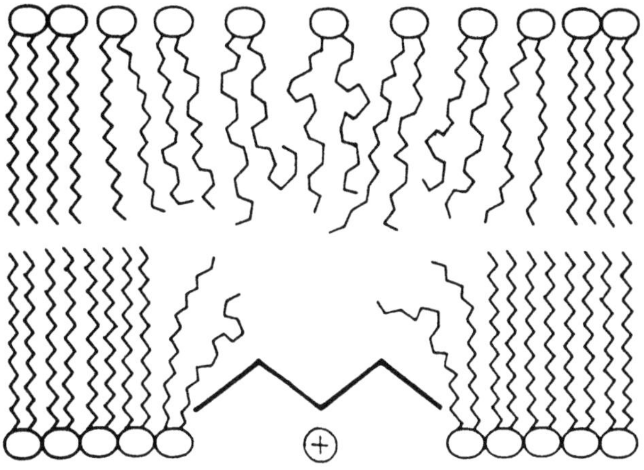

Figure 10 Schematic illustration of binding of gramicidin S with phospholipid membrane. Circles, charged head groups of phospholipid; zigzag lines, fatty acid chains; straight lines with cationic charge, gramicidin S molecule.

primarily important in the binding of GS with phospholipid membranes [154, 188]. This model is slightly revised to illustrate the induced disturbance.

Following the model depicted in Figure 10, they explain the mode of interaction of GS with phospholipid membrane as follows. In phospholipid bilayer, two cationic side chains of Orn residues in GS are exposed in aqueous phase, while the hydrophobic side in GS is penetrated into the hydrophobic parts of phospholipid molecules. The partial penetration of GS molecules makes a certain space in the interior of membrane, resulting in an increase in the movement of the acyl chains of phospholipid molecules to decrease the phase transition temperature. The appearance of these disordered regions in an ordered membrane might trigger the efflux of K^+. Thus, the appropriate hydrophobicity arising from the characteristic "sidedness" structure in GS causes the expansion of the membrane structure of neutral lipids which constitute main parts of the biological membrane. The expansion results in stimulation of the permeability of the membrane, which brings about various biological activities of GS including the antimicrobial activity.

All of the observations described in this Section support the view that GS permeates the outer membrane of sensitive bacteria and interacts with the inner membrane, altering the permeability of the membranes and releasing intracellular components. In contrast, GS cannot permeate the cell membranes of resistant bacteria. However, as mentioned above, there is difference of opinion as to the details of the mode of interaction of GS with phospholipid membrane. Thus, we have to say that the precise mechanism of antimicrobial action of GS is still unclear at the present time.

9. Concluding Remarks

GS has a long history since its discovery in 1944, but still has attracted continuing interest to chemists, biochemists and biologists because of its unique structure and bioactivity. Numerous structure-activity studies on this antibiotic have been carried out for clarifying the functional elements necessary for its biological activity, leading to the recognition that the specific "sidedness" structure of GS is the most important structural element for the exhibition of strong antimicrobial activity. Recent synthetic studies clearly indicate that the conformational considerations based on a whole molecule architecture are increasingly of importance for the design of strongly active GS analogs against Gram-positive bacteria or Gram-negative ones and for the understanding of GS-phospholipid interactions. Particularly, it is important to know bioactive conformation of GS and its analogs in the environment where they actually act.

Even now, the mechanisms of antibacterial action and other biological action of GS [223, 224] and its biofunction [12, 225] in the producer microorganisms have not been completely elucidated. However, good use of specifically designed GS analogs of various biological interests based on such conformational considerations will be a great help for the better and profound understanding of them. Further development of membrane sciences, especially with respect to the structure and function of the cell membrane of the target sensitive bacteria or insensitive bacteria, would be also needed.

Note

Very recently Shimohigashi et al. [226] have demonstrated that the two Orn side chains of GS are not always necessary for exhibiting the antimicrobial activity. Thus, they designed and synthesized two analogs of GS, cyclo(-Val-Ala-Leu-ΔPhe-Pro-)$_2$ (des-philic GS) lacking the hydrophilic Orn side chains and cyclo(-Ala-Orn-Ala-ΔPhe-Pro-)$_2$ lacking the hydrophobic Val and Leu side chains, to elucidate the role of the amphiphilic property of GS in biomembrane-interaction. The $\Delta Phe^{4,4'}$ residues were incorporated for stabilizing the GS bioactive conformation [60, 101]. Interestingly, des-philic GS without amino groups showed as high antimicrobial activity as GS for *B. subtilis* PCI 219 and *S. aureus* 1840, while its activity for different strains of *S. aureus* was weaker than that of GS. For *S. flexneri*, des-philic GS exhibited no activity in spite of a fairly high activity of GS. Meanwhile, des-phobic GS showed no activity for any bacteria. To get an insight into the mechanism of peptide-membrane interactions, analysis of interactions of GS and these unique analogs with liposomes by carboxyfluorescein-leakage has been initiated [227]. Further progress in a series of structure-function studies of GS using such uniquely designed GS analogs with specific bioactivity has been greatly expected.

References

1. Gause, G. F., and Brazhnikova, M. G., Gramicidin "S": its origin and mode of action, *Am. Rev. Soviet Med. 2*, 134, 1944; *Chem. Abstr. 39*, 1893, 1945.
2. Gause, G. F., and Brazhnikova, M. G., Gramicidin S and its use in the treatment of infected wounds, *Nature 154*, 703, 1944.

3. Brazhnikova, M. G., Obituary of Prof. G. F. Gause, *J. Antibiot. 40*, 1079, 1987.
4. Kato, T., and Izumiya, N., Chemistry and biochemistry of gramicidin S and related compounds, in *Chemistry and Biochemistry of Amino Acids, Peptides and Proteins*, Vol. 2, Weinstein, B., Ed., Marcel Dekker, New York, 1974, p. 1.
5. Izumiya, N., Chemistry of gramicidin S, in *Biologically Active Peptides Produced by Microorganisms* (in Japanese), Umezawa, H., Shiba, T., and Takita, T., Eds., Kyoritsu Shuppan, Tokyo, 1976, p. 166.
6. Kleinkauf, H., and von Döhren, H., Nucleic acid independent synthesis of peptides, *Curr. Top. Microbiol. Immunol. 91*, 129, 1981.
7. Ovchinnikov, Yu. A., and Ivanov, V. T., The cyclic peptides: structure, conformation, and function, in *The Proteins*, Vol. 5, 3rd ed., Neurath, H., and Hill, R. L., Eds., Academic Press, Inc., New York, 1982, p. 307.
8. Izumiya, N., Kato, T., and Waki, M., Synthesis of biologically active cyclic peptides, *Biopolymers 20*, 1785, 1981.
9. Aoyagi, H., Lee, S., and Izumiya, N., Design and synthesis of bioactive cyclic peptides, AM-toxin I and gramicidin S, *J. Mol. Graph. 5*, 35, 1987.
10. Aoyagi, H., Ando, S., Lee, S., and Izumiya, N., Synthesis of antibacterial peptides, gramicidin S analogs and designed amphiphilic oligopeptides, *Tetrahedron 44*, 877, 1988.
11. Izumiya, N., Kato, T., Aoyagi, H., Waki, M., and Kondo, M., *Synthetic Aspects of Biologically Active Cyclic Peptides – Gramicidin S and Tyrocidines*, Kodansha Ltd., Tokyo, and Wiley, New York, 1979.
12. *Peptide Antibiotics – Biosynthesis and Functions: Enzymatic Formation of Bioactive Peptides and Related Compounds*, Kleinkauf, H., and von Döhren, H., Eds., de Gruyter, Berlin, 1982.
13. Schröter-Kermani, C., von Döhren, H., and Kleinkauf, H., Active thiol-directed binding and adsorption of gramicidin S-synthetase 1 to disulfide-containing matrices, *Biochim. Biophys. Acta 883*, 345, 1986.
14. Vater, J., Schlumbohm, W., Palacz, Z., Salnikow, J., Gadow, A., and Kleinkauf, H., Formation of D-Phe-Pro-Val-cyclo-Orn by gramicidin S synthetase in the absence of L-leucine, *Eur. J. Biochem. 163*, 297, 1987.
15. Danders, W., Marahiel, M. A., Krause, M., Kosui, N., Kato, T., Izumiya, N., and Kleinkauf, H., Antibacterial action of gramicidin S and tyrocidines in relation to active transport, in vitro transcription, and spore outgrowth, *Antimicrob. Agents Chemother. 22*, 785, 1982.
16. Vostroknutova, G. N., Simakova, I. M., Kharat'yan, E. F., Bulgakova, V. G., Udalova, T. P., and Ostrovskii, D. N., Study of the structuro-functional peculiarities of the bacterium *Bacillus brevis* in connection with the accumulation of gramicidin S in cells, *Biokhimiya 48*, 818, 1983.
17. Krauss, E. M., and Chan, S. I., Complexation and phase transfer of nucleic acids by gramicidin S, *Biochemistry 23*, 73, 1984.
18. Frangou-Lazaridis, M., and Seddon, B., Effect of gramicidin S on the transcription system of the producer *Bacillus brevis* Nagano, *J. Gen. Microbiol. 131*, 437, 1985.
19. Ristow, H., and Paulus, H., Induction of sporulation in *Bacillus brevis* 1. Biochemical events and modulation of RNA synthesis during induction by tyrocidine, *Eur. J. Biochem. 129*, 395, 1982.
20. Pschorn, W., Paulus, H., Hansen, J., and Ristow, H., Induction of sporulation in *Bacillus brevis* 2. Dependence on the presence of the peptide antibiotics tyrocidine and linear gramicidin, *Eur. J. Biochem. 129*, 403, 1982.
21. Bohg, A., and Ristow, H., DNA-supercoiling is affected *in vitro* by the peptide antibiotics tyrocidine and gramicidin, *Eur. J. Biochem. 160*, 587, 1986.
22. Abbreviations used are according to IUPAC-IUB Commissions: Nomenclature and symbolism for amino acids and peptides, Recommendations 1983, *Eur. J. Biochem. 138*, 9, 1984. All optically active amino acids are of the L-configuration unless otherwise indicated. Other abbreviations are shown in the text where they firstly appear.
23. Schwyzer, R., and Sieber, P., Die Synthese von Gramicidin S, *Helv. Chim. Acta 40*, 624, 1957.
24. Nozaki, S., and Muramatsu, I., Natural homologs of gramicidin S, *J. Antibiot. 37*, 689, 1984.
25. Nozaki, S., and Muramatsu, I., Natural homologs of gramicidin S. II. Synthesis of gramicidin S-2 and S-3, *Bull. Chem. Soc. Jpn. 58*, 331, 1985.
26. Katz, E., and Demain, A. L., The peptide antibiotics of *Bacillus*: chemistry, biogenesis, and possible functions, *Bacteriol. Rev. 41*, 449, 1977.

27. Altmann, M., Koischwitz, H., Salnikow, J., and Kleinkauf, H., (3,3′-Leu)-gramicidin S formation by gramicidin S synthetase, *FEBS Lett. 93*, 247, 1978.
28. Aarstad, K., Zimmer, T. L., and Laland, S. G., The fidelity of gramicidin S synthetase, *Eur. J. Biochem. 112*, 335, 1980.
29. Aarstad, K., Zimmer, T. L., and Laland, S. G., Replacement of phenylalanine in gramicidin S by other amino acids, *FEBS Lett. 103*, 118, 1979.
30. Hotchkiss, R. D., and Dubos, R. J., Bactericidal fractions from an aerobic sporulating *Bacillus, J. Biol. Chem. 136*, 803, 1940.
31. Battersby, A. R., and Craig, L. C., The chemistry of tyrocidine. I. Isolation and characterization of a single peptide, *J. Am. Chem. Soc. 74*, 4019, 1952.
32. Ruttenberg, M. A., and Mach, B., Studies on amino acid substitution in the biosynthesis of the antibiotic polypeptide tyrocidine, *Biochemistry 5*, 2864, 1966.
33. Fujikawa, K., Sakamoto, Y., Suzuki, T., and Kurahashi, K., Biosynthesis of tyrocidine by a cell-free enzyme system of *Bacillus brevis* ATCC 8185 II. Amino acid substitution in tyrocidine, *Biochim. Biophys. Acta 169*, 520, 1968.
34. Ohno, M., Kato, T., Makisumi, S., and Izumiya, N., Studies of peptide antibiotics. IV. The synthesis of tyrocidine A, *Bull. Chem. Soc. Jpn. 39*, 1738, 1966.
35. Okamoto, K., Nonaka, K., and Izumiya, N., Studies of peptide antibiotics. XXXIV. Syntheses of tyrocidine A and its analogs containing glycine, *Bull. Chem. Soc. Jpn. 50*, 231, 1977.
36. Kuromizu, K., and Izumiya, N., Studies of peptide antibiotics. XX. Synthesis of tyrocidine B, *Bull. Chem. Soc. Jpn. 43*, 2199, 1970.
37. Kuromizu, K., and Izumiya, N., Studies of peptide antibiotics. XXI. Synthesis of tyrocidine C, *Bull. Chem. Soc. Jpn. 43*, 2944, 1970.
38. Mitsuyasu, N., Matsuura, S., Waki, M., Ohno, M., Makisumi, S., and Izumiya, N., Studies of peptide antibiotics. XVIII. The synthesis of tyrocidine E, *Bull. Chem. Soc. Jpn. 43*, 1829, 1970.
39. Zharikova, G. G., Myaskovskaya, S. P., and Silaev, A. B., Antibiotic gratisin derived from the cells of the mutant *Bacillus brevis* strain Y-33, *Vestn. Mosk. Univ., Biol., Pochvoved. 27*, 110, 1972.
40. Myaskovskaya, S. P., Zharikova, G. G., and Silaev, A. B., Amino acid composition and some of the physicochemical properties of gratisin, *Vestn. Mosk. Univ., Biol., Pochvoved. 28*, 123, 1973.
41. Tamaki, M., Takimoto, M., Sofuku, S., and Muramatsu I., Synthetic studies on gratisin. II, *J. Antibiot. 36*, 751, 1983.
42. Tamaki, M., Studies on peptide antibiotic "gratisin", *Bull. Chem. Soc. Jpn. 57*, 3210, 1984.
43. Kopple, K. D., Synthesis of cyclic peptides, *J. Pharm. Sci. 61*, 1345, 1972.
44. Hardy, P. M., and Ridge, B., Recent advances in the chemistry of cyclic peptides, *Prog. Org. Chem. 8*, 129, 1973.
45. Yonezawa, H., Kaneda, M., Tominaga, N., Higashi, S., and Izumiya, N., Adsorption of ^{14}C-labeled gramicidin S on cell of bacteria, *J. Biochem. (Tokyo) 90*, 1087, 1981.
46. Yonezawa, H., Okamoto, K., Kaneda, M., Tominaga, N., and Izumiya, N., Studies of peptide antibiotics. XLIV. Syntheses and biological activities of gramicidin S analogs containing δ-hydroxy-L-norvaline or glycine, *Int. J. Peptide Protein Res. 22*, 573, 1983.
47. Brady, S. F., Varga, S. L., Freidinger, R. M., Schwenk, D. A., Mendlowski, M., Holly, F. W., and Veber, D. F., Practical synthesis of cyclic peptides, with an example of dependence of cyclization yield upon linear sequence, *J. Org. Chem. 44*, 3101, 1979.
48. Minematsu, Y., Waki, M., Suwa, K., Kato, T., and Izumiya, N., Facile synthesis of gramicidin S via cyclization of a linear pentapeptide, *Tetrahedron Lett. 21*, 2179, 1980.
49. Pastuszak, J., Gardner, J. H., Singh, J., and Rich, D. H., Improved synthesis of [4-alanine]chlamydocin: cyclization studies of tetrapeptides containing five α-substituents, *J. Org. Chem. 47*, 2982, 1982.
50. Matsuura, S., Takiguchi, H., Waki, M., and Izumiya, N., Studies of peptide antibiotics. XXXII. Cyclization reactions of linear peptides with sequence of gramicidin S, *Mem. Fac. Sci., Kyushu Univ., Ser. C 9*, 277, 1975.
51. Sato, K., and Nagai, U., Studies on the β-turn of peptides. VII. Syntheses and antibiotic activities of gramicidin S analogs with L-Pro-L-Asn or L-Pro-D-Ala sequence at the β-turn part, *Bull. Chem. Soc. Jpn. 56*, 3329, 1983.

52. Ando, S., Aoyagi, H., Shinagawa, S., Nishino, N., Waki, M., Kato, T., and Izumiya, N., [4,4′-D-Diaminopropionic acid] gramicidin S: a synthetic gramicidin S analog with antimicrobial activity against Gram-negative bacteria, *FEBS Lett. 161*, 89, 1983.
53. Ando, S., Aoyagi, H., Waki, M., Kato, T., and Izumiya, N., Studies of peptide antibiotics. XLIII. Syntheses of gramicidin S analogs containing D-serine or dehydroalanine in place of D-phenylalanine and asymmetric hydrogenation of the dehydroalanine residue, *Int. J. Peptide Protein Res. 21*, 313, 1983.
54. Tanimura, K., Kato, T., Waki, M., Lee, S., Kodera, Y., and Izumiya, N., Facile synthesis of (2*R*,3*R*)-phenylalanine-2,3-d_2 and NMR study on deuterated gramicidin S, *Bull. Chem. Soc. Jpn. 57*, 2193, 1984.
55. Ando, S., Kato, T., and Izumiya, N., Studies of peptide antibiotics. XLVI. Syntheses of gramicidin S analogs containing D-α,β-diaminopropionic acid or α,β-dehydroalanine, *Int. J. Peptide Protein Res. 25*, 15, 1985.
56. Sato, K., and Nagai, U., Synthesis and antibiotic activity of a gramicidin S analogue containing bicyclic β-turn dipeptides, *J. Chem. Soc., Perkin Trans. 1*, 1231, 1986.
57. Sato, K., Kato, R., and Nagai, U., Studies on β-turn of peptides. XII. Synthetic confirmation of weak activity of [D-Pro$^{5,5'}$]-gramicidin S predicted from β-turn preference of its partial sequence, *Bull. Chem. Soc. Jpn. 59*, 535, 1986.
58. Kondo, M., Kimura, M., Sato, K., and Horimoto, H., Syntheses and properties of [1,1′-bis(α-aminoisobutyric acid)] gramicidin S and [1-α-aminoisobutyric acid] semigramicidin S, *Bull. Chem. Soc. Jpn. 60*, 1391, 1987.
59. Ono, S., Lee, S., Kodera, Y., Aoyagi, H., Waki, M., Kato, T., and Izumiya, N., Environment-dependent conformation and antimicrobial activity of a gramicidin S analog containing leucine and lysine residues, *FEBS Lett. 220*, 332, 1987.
60. Shimohigashi, Y., Kodama, H., Imazu, S., Horimoto, H., Sakaguchi, K., Waki, M., Uchida, H., Kondo, M., Kato, T., and Izumiya, N., [4,4′-(*Z*)-Dehydrophenylalanine]gramicidin S with stabilized bioactive conformation and strong antimicrobial activity, *FEBS Lett. 222*, 251, 1987.
61. Ando, S., Takiguchi, H., and Izumiya, N., Studies of peptide antibiotics. XLV. Syntheses of gramicidin S-like analogs with macro-ring structures, *Bull. Chem. Soc. Jpn. 56*, 3781, 1983.
62. Ando, S., Nishikawa, H., Takiguchi, H., and Izumiya, N., Syntheses of cyclic decapeptides with four ornithyl residues related to gramicidin S, *Bull. Chem. Soc. Jpn. 59*, 1201, 1986.
63. Tamaki, M., Takimoto, M., and Muramatsu, I., Studies of a peptide antibiotic "gratisin". II, *Bull. Chem. Soc. Jpn. 60*, 2101, 1987.
64. Sofuku, S., Sugiyama, Y., and Muramatsu, I., Synthesis of gramicidin S (GS) analogs, [3-4-δAva]-GS and [3-4,3′-4′-bis(δAva)]-GS (δAva = 5-aminovaleric acid), *Bull. Chem. Soc. Jpn. 59*, 185, 1986.
65. Ji, A.-X., and Bodanszky, M., Cyclization studies with a model pentapeptide, *Int. J. Peptide Protein Res. 22*, 590, 1983.
66. Shioiri, T., Ninomiya, K., and Yamada, S., Diphenylphosphoryl azide. A new convenient reagent for a modified Curtius reaction and for the peptide synthesis, *J. Am. Chem. Soc. 94*, 6203, 1972.
67. Brady, S. F., Freidinger, R. M., Paleveda, W. J., Colton, C. D., Homnick, C. F., Whitter, W. L., Curley, P., Nutt, R. F., and Veber, D. F., Large-scale synthesis of a cyclic hexapeptide analogue of somatostatin, *J. Org. Chem. 52*, 764, 1987.
68. Mukaiyama, T., Kamekawa, K., and Watanabe, Y., One-pot cyclization of a peptide by the use of (5-nitropyridyl)diphenyl phosphinate: the synthesis of cyclic decapeptide gramicidin S, *Chem. Lett.*, 1367, 1981.
69. Schwyzer, R., and Sieber, P., Verdoppelungsreaktionen beim Ringschluß von Peptiden. I. Synthese von Gramicidin S und von bis-homo-Gramicidin S aus den Pentapeptid-Einheiten, *Helv. Chim. Acta 41*, 2186, 1958.
70. Waki, M., and Izumiya, N., Studies of peptide antibiotics. X. Syntheses of cyclosemigramicidin S and gramicidin S, *Bull. Chem. Soc. Jpn. 40*, 1687, 1967.
71. Schwyzer, R., and Sieber, P., Ein synthetisches Analoges des Gramicidin S: c-(Val-Lys-Leu-Phe-Pro)$_2$, (L-L-L-D-L)$_2$, *Helv. Chim. Acta 41*, 1582, 1958.
72. Stepanov, V. M., Silaev, A. B., and Polin, A. N., Preparation and antibiotic properties of N^{δ}-peptides of gramicidin S, *Antibiotiki 3*, 49, 1958.

73. Kondo, M., and Izumiya, N., Studies of peptide antibiotics. XI. Analogs of gramicidin S and cyclosemigramicidin S containing glycine or alanine in place of valine, *Bull. Chem. Soc. Jpn. 40*, 1975, 1967.
74. Kondo, M., Aoyagi, H., Kato, T., and Izumiya, N., Studies of peptide antibiotics. VI. Syntheses of cyclic penta and decapeptides with a glycyl-L-ornithyl-L-leucyl-D-phenylalanylglycyl sequence, *Bull. Chem. Soc. Jpn. 39*, 2234, 1966.
75. Takiguchi, H., Nishikawa, H., Ando, S., and Izumiya, N., Studies of peptide antibiotics. XXXVII. Synthesis of [1,1′,3,3′-L-alanine]-gramicidin S, *Bull. Chem. Soc. Jpn. 51*, 297, 1978.
76. Kondo, M., and Izumiya, N., Studies of peptide antibiotics. XIX. Syntheses of 1,1′-leucine-gramicidin S and 1-leucine-cyclosemigramicidin S, *Bull. Chem. Soc. Jpn. 43*, 1850, 1970.
77. Iwai, M., Nakajima, K., Uno, A., Hase, S., Takeuchi, I., and Okawa, K., Studies on hydroxy amino acids. III. The synthesis of cyclic peptide containing L-serine, *Bull. Chem. Soc. Jpn. 43*, 3246, 1970.
78. Ando, S., Nishikawa, H., and Takiguchi, H., Syntheses of [1,1′-L-Phe]-gramicidin S, [3,3′-L-Phe]-gramicidin S, and [1,3′-L-Phe]-gramicidin S, *Fukuoka Univ. Sci. Rep. 11*, 109, 1981.
79. Takiguchi, H., Nishikawa, H., and Izumiya, N., Synthesis of [1,1′,3,3′-L-phenylalanine]-gramicidin S, *Fukuoka Univ. Sci. Rep. 4*, 107, 1974.
80. Hashiguchi, A., Ando, S., Aoyagi, H., Waki, M., Kato, T., and Izumiya, N., Studies of peptide antibiotics. XLVII. Syntheses and biological activities of [L-Lys$^{1,1'}$] gramicidin S and related peptides, *Mem. Fac. Sci., Kyushu Univ., Ser. C 15*, 91, 1985.
81. Tamaki, M., Okitsu, T., Araki, M., Sakamoto, H., Takimoto, M., and Muramatsu, I., Synthesis and properties of gramicidin S analogs containing D-Phe-L-Pro-D-Val or L-Phe-L-Pro-D-Val sequences in place of D-Phe-L-Pro-L-Val sequence in the β-turn part of the antibiotic, *Bull. Chem. Soc. Jpn. 58*, 531, 1985.
82. Abe, H., Sato, K., Kato, T., and Izumiya, N., Studies of peptide antibiotics. XXXIII. Syntheses of gramicidin S analogs containing *N*-methyl-L-valine in place of L-valine, *Bull. Chem. Soc. Jpn. 49*, 3113, 1976.
83. Zhuze, A. L., Kogan, G. A., Krit, N. A., Andronova, T. M., Filatova, M. P., Senyavina, L. B., Meshcheryakova, E. A., Ryabova, I. D., Radvel, G. A., and Shchukina, L. A., Depsipeptide modification as an approach to elucidation of intramolecular hydrogen bond contributions to the conformational stability of peptides (gramicidin S), *Mol. Biol. (Moscow) 8*, 84, 1974.
84. Ovchinnikov, Yu. A., and Ivanov, V. T., Conformational states and biological activity of cyclic peptides, *Tetrahedron 31*, 2177, 1975.
85. Krit, N. A., Zhuze, A. L., Andronova, T. M., Filatova, M. P., Ravdel, G. A., and Shchukina, L. A., Synthesis of the gramicidin S depsipeptide analogs, *Bioorg. Khim. 1*, 247, 1975.
86. Sato, K., Ueda, K., Kondo, M., Aoyagi, H., and Izumiya, N., Studies of peptide antibiotics. XXXVIII. Synthesis of *S,S′*-bi([1-L-hemicystine]-gramicidin S), a dimerized analog of gramicidin S, *Bull. Chem. Soc. Jpn. 51*, 1830, 1978.
87. Tamaki, M., Takimoto, M., Sōfuku, S., and Muramatsu, I., A new synthetic antibiotic analog of gramicidin S, [2,2′-L-serine]-gramicidin S, *J. Antibiot. 33*, 105, 1980.
88. Waki, M., Abe, O., Okawa, R., Kato, T., Makisumi, S., and Izumiya, N., Studies of peptide antibiotics. XII. Syntheses of [2,2′-*a*, γ-diaminobutyric acid]- and [2,2′-lysine]-gramicidin S, *Bull. Chem. Soc. Jpn. 40*, 2904, 1967.
89. Rothe, M., and Eisenbeiss, F., Synthesis of all-L-gramicidin S, *Angew. Chem. Int. Ed. Engl. 7*, 883, 1968.
90. Nonaka, K., Sato, K., Terada, S., Kato, T., and Izumiya, N., Studies of peptide antibiotics. XXXIX. Syntheses of gramicidin S analogs containing L-histidine in place of L-ornithine, *Bull. Chem. Soc. Jpn. 51*, 2127, 1978.
91. Abe, O., and Izumiya, N., Studies of peptide antibiotics. XVII. Analogs of gramicidin S containing glycine or alanine in place of leucine, *Bull. Chem. Soc. Jpn. 43*, 1202, 1970.
92. Nishikawa, H., Takiguchi, H., and Izumiya, N., Synthesis and properties of [3,3′-L-valine]-gramicidin S, *Fukuoka Univ. Sci. Rep. 7*, 103, 1977.
93. Sugano, H., Abe, H., Miyoshi, M., Kato, T., and Izumiya, N., Studies of peptide antibiotics. XXX. Syntheses of gramicidin S analogs containing *N*-methylleucine in place of leucine, *Bull. Chem. Soc. Jpn. 47*, 698, 1974.
94. Kawai, M., and Nagai, U., Comparison of conformation and antimicrobial activity of synthetic analogs of gramicidin S: stereochemical consideration of the role of D-phenylalanine in the antibiotic, *Biopolymers 17*, 1549, 1978.

95. Lee, S., Ohkawa, R., and Izumiya, N., Studies of peptide antibiotics. XXIV. Synthesis of 4,4′-D-alanine-gramicidin S, *Bull. Chem. Soc. Jpn. 44*, 158, 1971.
96. Aoyagi, H., Kato, T., Waki, M., Abe, O., Okawa, R., Makisumi, S., and Izumiya, N., Studies of peptide antibiotics. XIV. Analogs of gramicidin S containing D-valine or D-leucine in place of D-phenylalanine, *Bull. Chem. Soc. Jpn. 42*, 782, 1969.
97. Ando, S., Aoyagi, H., Waki, M., Kato, T., Izumiya, N., Okamoto, K., and Kondo, M., Syntheses of gramicidin S analogs containing D-serine or a dehydroalanine residue, and asymmetric hydrogenation of the dehydroalanine, *Tetrahedron Lett. 23*, 2195, 1982.
98. Sato, K., and Nagai, U., Design and synthesis of gramicidin S analogs with high antibiotic activity, in *Peptide Chemistry 1985*, Kiso, Y., Ed., Protein Res. Found., Osaka, 1986, p. 217.
99. Snezhkova, L. G., Shepel, E. N., Ryabova, I. D., Miroshnikov, A. I., Ivanov, V. T., and Ovchinnikov, Yu. A., The synthesis and antimicrobial activity of new gramicidin S analogs, *Bioorg. Khim. 1*, 347, 1975.
100. Tamaki, M., Takimoto, M., and Muramatsu, I., Synthesis and properties of gramicidin S analogs containing Pro-D-Phe sequence in place of D-Phe-Pro sequence in the β-turn part of the antibiotic, *Bull. Chem. Soc. Jpn. 58*, 1469, 1985.
101. Kodama, H., Shimohigashi, Y., Imazu, S., Sakaguchi, K., Waki, M., Kato, T., and Izumiya, N., Conformational stabilization of gramicidin S by α,β-dehydrogenation of D-Phe$^{4,4'}$ residues, in *Peptide Chemistry 1986*, Miyazawa, T., Ed., Protein Res. Found., Osaka, 1987, p. 115.
102. Sōfuku, S., Synthesis of a gramicidin S analog containing δ-aminovaleric acid, [4-5-δ-aminovaleric acid]-gramicidin S, *Bull. Chem. Soc. Jpn. 46*, 968, 1973.
103. Sōfuku, S., Muramatsu, I., Okada, K., and Hagitani, A., Synthesis of [4-5,4′-5′-bis(δ-aminovaleric acid)]-gramicidin S, *Bull. Chem. Soc. Jpn. 48*, 2888, 1975.
104. Nagai, U., Bioactivity and conformation of peptides – molecular design based on stereochemistry, *Kagaku To Seibutsu* (in Japanese) *24*, 822, 1986.
105. Aoyagi, H., Kato, T., Ohno, M., Kondo, M., Waki, M., Makisumi, S., and Izumiya, N., Studies of peptide antibiotics. III. Cyclo-(L-valyl-L-ornithyl-L-leucyl-D-phenylalanylglycyl)$_2$, *Bull. Chem. Soc. Jpn. 38*, 2139, 1965.
106. Ovchinnikov, Yu. A., Kiryushkin, A. A., and Kozhevnikova, I. V., Synthesis of peptides on a polymer carrier in solution. II. Synthesis of L-valyl-L-tyrosyl-L-valyl-L-histidyl-L-prolyl-L-phenylalanine and Gly^5-Gly^{10}-gramicidin S, *Zh. Obshch. Khim. 38*, 2636, 1968.
107. Abe, O., Utsumi, Y., and Izumiya, N., Studies of peptide antibiotics. XXXVI. Synthesis and biological activity of [5,5′-leucine]-gramicidin S, *Bull. Chem. Soc. Jpn. 50*, 2341, 1977.
108. Nishikawa, H., Takiguchi, H., Kimoto, E., and Izumiya, N., Studies of peptide antibiotics. XL. Synthesis of gramicidin S analog containing L-isoleucine in place of L-proline, *Mem. Fac. Sci., Kyushu Univ., Ser. C 12*, 253, 1980.
109. Takiguchi, H., Nishikawa, H., and Izumiya, N., Synthesis of [5,5′-L-phenylalanine]-gramicidin S, *Fukuoka Univ. Sci. Rep. 5*, 139, 1975.
110. El-Maghraby, M. A., Synthesis of a cyclic decapeptide analogue of gramicidin S containing 4-fluoroproline in 5–10-positions, *Indian J. Chem. Sect. B 14*, 988, 1976.
111. Takiguchi, H., Nishikawa, H., Matsudai, K., and Izumiya, N., Synthesis of [5,5′-α-aminoisobutyric acid]-gramicidin S, *Fukuoka Univ. Sci. Rep. 3*, 31, 1974.
112. Aoyagi, H., and Izumiya, N., Studies of peptide antibiotics. V. Syntheses of cyclic penta- and decapeptides with the L-valyl-L-ornithyl-L-leucyl-D-phenylalanylsarcosyl sequence, *Bull. Chem. Soc. Jpn. 39*, 1747, 1966.
113. Matsuura, S., Waki, M., Makisumi, S., and Izumiya, N., Studies of peptide antibiotics. XVI. Analogs of gramicidin S containing β-alanine in place of L-proline, *Bull. Chem. Soc. Jpn. 43*, 1197, 1970.
114. Matsuura, S., Waki, M., and Izumiya, N., Studies of peptide antibiotics. XXVII. Synthesis of an analog of gramicidin S containing a glycine residue in place of a proline, *Bull. Chem. Soc. Jpn. 45*, 863, 1972.
115. Sōfuku, S., Yoshida, A., Baba, H., and Muramatsu, I., Syntheses of gramicidin S analogs containing δ-aminovaleric acid. [5-1′-δ-Aminovaleric acid]-gramicidin S and [5-1′,5′-1-bis-(δ-aminovaleric acid)]-gramicidin S, *Bull. Chem. Soc. Jpn. 50*, 2143, 1977.
116. Takiguchi, H., and Izumiya, N., Studies of peptide antibiotics. XXXI. Syntheses of *iso*-gramicidin S and *iso*-[2,2′-lysine]-gramicidin S, *Bull. Chem. Soc. Jpn. 47*, 221, 1974.
117. Waki, M., and Izumiya, N., Studies of peptide antibiotics. XIII. Synthesis of retrogramicidin S, *Bull. Chem. Soc. Jpn. 41*, 1909, 1968.

118. Shemyakin, M. M., Ovchinnikov, Yu. A., Ivanov, V. T., and Ryabova, I. D., Topochemical approach to the structure-activity relation. Retroenantio-Gly^{5,10}-gramicidin S, *Experientia 23*, 326, 1967.
119. Srinivasa, B. R., Chemical and biological properties of formyl gramicidin S, *J. Antiobiot. 35*, 571, 1982.
120. Nagamurthi, G., and Rambhav, S., Gramicidin-S: structure-activity relationship, *J. Biosci. 7*, 323, 1985.
121. Silaev, A. B., Relation between structure, functional groups, and activity of polypeptide antibiotics, in *Humboldt-Symp. Grundfragen Biol.*, Berlin, 1960, p. 156; *Chem. Abstr. 59*, 6635e, 1963.
122. Hiramatsu, S., Studies on the derivatives of gramicidin J_1, *J. Osaka City Med. Cent.* (in Japanese) 8, 2017, 1959.
123. Hashimoto, Y., Studies on the mode of the action of gramicidin J_1. I. On the site of action of fluorescent derivatives of gramicidin J_1, *J. Osaka City Med. Cent.* (in Japanese) 9, 4019, 1960.
124. Ohnishi, M., Ando, S., Waki, M., and Izumiya, N., Studies of peptide antibiotics. XLII. Preparations and antibacterial properties of derivatives of natural gramicidin S, *Mem. Fac. Sci., Kyushu Univ., Ser. C 13*, 143, 1981.
125. Stepanov, V. M., and Silaev, A. B., Preparation of gramicidin-C derivatives containing carboxy groups, *Zh. Obshch. Khim. 31*, 3811, 1961.
126. Kawai, M., Fukuta, N., Ito, N., Ohya, M., Butsugan, Y., Maruyama, M., and Kudo, Y., Preparation and properties of trimethylammonium group-containing analogues of [D-Ala^2, Leu^5]enkephalin and gramicidin S, *J. Chem. Soc., Chem. Commun.* 955, 1986.
127. Kawai, M., Ohya, M., Butsugan, Y., Saito, K., and Higashijima, T., Selective *N*-methylation of peptide bond. Preparation and properties of [MeOrn$^{2,2'}$,D-MePhe$^{4,4'}$] gramicidin S, *Chem. Lett.*, 1835, 1984.
128. Yukawa, T., Studies on some cyclic heptapeptides derived from gramicidin J, *J. Osaka City Med. Cent.* (in Japanese) 6, 285, 1957.
129. Hodgkin, D. C., and Oughton, B. M., Possible molecular models for gramicidin S and their relationship to present ideas of protein structure, *Biochem. J. 65*, 752, 1957.
130. Schwyzer, R., Synthesis of cyclic polypeptides, in *Amino Acids and Peptides with Antimetabolic Activity*, (CIBA Foundation Symposium), Churchill, New York, 1958, p. 171.
131. Stern, A., Gibbons, W. A., and Craig, L. C., A conformational analysis of gramicidin S-A by nuclear magnetic resonance, *Proc. Natl. Acad. Sci. U.S.A. 61*, 734, 1968.
132. Ohnishi, M., and Urry, D. W., Temperature dependence of amide proton chemical shifts: the secondary structures of gramicidin S and valinomycin, *Biochem. Biophys. Res. Commun. 36*, 194, 1969.
133. Kato, T., and Izumiya, N., Conformations of di-*N*-methyl-leucine gramicidin S and *N*-methylleucine gramicidin S compatible with the sidedness hypothesis, *Biochim. Biophys. Acta 493*, 235, 1977.
134. Okamoto, K., Yonezawa, H., and Izumiya, N., Preparation of pure peptides from a mixture of gramicidins and tyrocidines by droplet countercurrent chromatography, *J. Chromatogr. 92*, 147, 1974.
135. Kosui, N., Kato, T., and Izumiya, N., Studies of peptide antibiotics. XLI. Syntheses of 6-L-valine-tyrocidine A and 7-L-ornithine-tyrocidine A, *Int. J. Peptide Protein Res. 18*, 127, 1981.
136. Ando, S., Takiguchi, H., Nishikawa, H., and Izumiya, N., Syntheses of [8,9-L-phenylalanine]-tyrocidine E and [8,9-L-leucine]-tyrocidine E, *Fukuoka Univ. Sci. Rep. 10*, 7, 1980.
137. Kuo, M.-C., and Gibbons, W. A., Determination of individual side-chain conformations, tertiary conformations, and molecular topography of tyrocidine A from scalar coupling constants and chemical shifts, *Biochemistry 18*, 5855, 1979.
138. Tamaki, M., Takimoto, M., Sofuku, S., and Muramatsu, I., Synthetic studies on gratisin, *J. Antibiot. 34*, 1227, 1981.
139. Tamaki, M., Takimoto, M., Nozaki, S., and Muramatsu, I., HPLC behavior of peptides related to gratisin and gramicidin S, in *Peptide Chemistry 1985*, Kiso, Y., Ed., Protein Res. Found., Osaka, 1986, p. 327.
140. Rae, I. D., Stimson, E. R., and Scheraga, H. A., Nuclear Overhauser effect and the conformation of gramicidin S, *Biochem. Biophys. Res. Commun. 77*, 225, 1977.

141. Hull, S. E., Karlsson, R., Main, P., Woolfson, M. M., and Dodson, E. J., The crystal structure of a hydrated gramicidin S-urea complex, *Nature 275*, 206, 1978.
142. Schwyzer, R., and Ludescher, U., Conformational study of gramicidin S using the phthalimide group as nuclear magnetic resonance marker, *Biochemistry 7*, 2519, 1968.
143. Ovchinnikov, Yu. A., Ivanov, V. T., Bystrov, V. F., Miroshnikov, A. I., Shepel, E. N., Abdullaev, N. D., Efremov, E. S., and Senyavina, L. B., The conformation of gramicidin S and its N,N′-diacetyl derivative in solutions, *Biochem. Biophys. Res. Commun. 39*, 217, 1970.
144. Dygert, M., Gō, N., and Scheraga, H. A., Use of a symmetry condition to compute the conformation of gramicidin S, *Macromolecules 8*, 750, 1975.
145. Kuo, M.-C., and Gibbons, W. A., Total assignments, including four aromatic residues, and sequence confirmation of the decapeptide tyrocidine A using difference double resonance, Qualitative nuclear Overhauser effect criteria for β turn and antiparallel β-pleated sheet conformations, *J. Biol. Chem. 254*, 6278, 1979.
146. Kuo, M.-C., Drakenberg, T., and Gibbons, W. A., A study of proton relaxation mechanisms, stereochemistry, and dynamics of the decapeptide tyrocidine A, *J. Am. Chem. Soc. 102*, 520, 1980.
147. Kuo, M.-C., and Gibbons, W. A., Nuclear Overhauser effect and cross-relaxation rate determinations of dihedral and transannular interproton distances in the decapeptide tyrocidine A, *Biophys. J. 32*, 807, 1980.
148. Zhou, N., Mascagni, P., Gibbons, W. A., Niccolai, N., Rossi, C., and Wyssbrod, H., Confirmation of the solution structure of tyrocidine A using perturbation of proton relaxation rates by nitroxide spin labels, *J. Chem. Soc., Perkin Trans. 2*, 581, 1985.
149. Jones, C. R., Kuo, M.-C., and Gibbons, W. A., Multiple solution conformations and internal rotations of the decapeptide gramicidin S, *J. Biol. Chem. 254*, 10307, 1979.
150. Kuo, M.-C., Jones, C. R., Mahn, T. H., Miller, P. R., Nicholls, L. J. F., and Gibbons, W. A., Simplification and spin-spin analysis of the side chain proton magnetic resonance spectrum of the decapeptide gramicidin S using difference scalar decoupling and biosynthesis of specifically deuterated analogs, *J. Biol. Chem. 254*, 10301, 1979.
151. Tanimura, K., Kato, T., Waki, M., and Izumiya, N., Facile synthesis of (2*R*,3*R*)-phenylalanine-2,3-d_2 and its application to conformational analysis of gramicidin S, *Tetrahedron Lett. 24*, 3737, 1983.
152. Niccolai, N., Valensin, G., Rossi, C., and Gibbons, W. A., The stereochemistry and dynamics of natural products and biopolymers from proton relaxation spectroscopy: spin-label delineation of inner and outer protons of gramicidin S including hydrogen bonds, *J. Am. Chem. Soc. 104*, 1534, 1982.
153. Fischman, A. J., Live, D. H., Wittbold, W. M. Jr., and Wyssbrod, H. R., The assignment of the resonances of the backbone amide protons of arginine vasopressin and gramicidin S in D_2O by decoupling during exchange, *J. Magn. Reson. 40*, 527, 1980.
154. Higashijima, T., Miyazawa, T., Kawai, M., and Nagai, U., Gramicidin S analogs with a D-Ala, Gly, or L-Ala residue in place of the D-Phe residue: molecular conformations and interactions with phospholipid membrane, *Biopolymers 25*, 2295, 1986.
155. Datema, K. P., Pauls, K. P., and Bloom, M., Deuterium nuclear magnetic resonance investigation of the exchangeable sites on gramicidin A and gramicidin S in multilamellar vesicles of dipalmitoylphosphatidylcholine, *Biochemistry 25*, 3796, 1986.
156. Howarth, O. W., and Lian, L. Y., Resolved librational motions of gramicidin S and glutathione dimer as studied by nuclear magnetic resonance spectroscopy, *J. Chem. Soc., Perkin Trans. 2*, 263, 1982.
157. Saitō, H., and Tabeta, R., Conformational characterization of biologically active peptides, Pro-Leu-Gly-NH_2 and gramicidin S, in the solid and solution as revealed by conformation-dependent ^{13}C chemical shifts, *Chem. Lett.* 83, 1985.
158. Niccolai, N., Rossi, C., Mascagni, P., Neri, P., and Gibbons, W. A., ^{1}H-^{13}C Selective NOE studies of the decapeptide gramicidin S, *Biochem. Biophys. Res. Commun. 124*, 739, 1984.
159. Live, D. H., Davis, D. G., Agosta, W. C., and Cowburn, D., Long range hydrogen bond mediated effects in peptides: ^{15}N NMR study of gramicidin S in water and organic solvents, *J. Am. Chem. Soc. 106*, 1939, 1984.
160. Förster, H. G., Müller, D., and Kricheldorf, H. R., Secondary structure of peptides: 5. ^{15}N n.m.r. CP/MAS spectroscopy of solid polypeptides and gramicidin S, *Int. J. Biol. Macromol. 5*, 101, 1983.

161. Krauss, E. M., and Chan, S. I., Intramolecular hydrogen bonding in gramicidin S. 2. Ornithine, *J. Am. Chem. Soc. 104*, 6953, 1982.
162. Kricheldorf, H. R., ^{15}N NMR spectroscopy 30-structure/shift relationships of oligopeptides and copolypeptides, including gramicidin S, *Org. Magn. Reson. 15*, 162, 1981.
163. Hawkes, G. E., Randal, E. W., Hull, W. E., and Convert, O., Use of natural-abundance ^{15}N-NMR spectroscopy to investigate the secondary structure of peptides: gramicidin S, *Biopolymers 19*, 1815, 1980.
164. Gondol, D., and van Binst, G., Distance determination by a two-dimensional NOE NMR study on the medium-sized peptide gramicidin S, *Biopolymers 25*, 977, 1986.
165. Roques, B. P., Rao, R., and Marion, D., Use of nuclear Overhauser effect in the study of peptides and proteins, *Biochimie 62*, 753, 1980.
166. Huang, D.-H., Walter, R., Glickson, J. D., and Krishna, N. R., Solution conformation of gramicidin S: an intramolecular nuclear Overhauser effect study, *Proc. Natl. Acad. Sci. U.S.A. 78*, 672, 1981.
167. Jones, C. R., Sikakana, C. T., Hehir, S., Kuo, M.-C., and Gibbons, W. A., The quantitation of nuclear Overhauser effect methods for total conformational analysis of peptides in solution, Application to gramicidin S, *Biophys. J. 24*, 815, 1978.
168. Bothner-By, A. A., and Johner, P. E., Specificity of interproton nuclear Overhauser effects in gramicidin-S dissolved in deuterated ethylene glycol, *Biophys. J. 24*, 779, 1978.
169. Jones, C. R., Sikakana, C. T., Kuo, M.-C., and Gibbons, W. A., Interproton distances for the β-turn residues of the peptide gramicidin S determined from nuclear Overhauser effect ratios, *J. Am. Chem. Soc. 100*, 5960, 1978.
170. Krauss, E. M., and Cowburn, D., Anomalous exchange kinetics of peptide amide protons in aqueous solutions, *Int. J. Peptide Protein Res. 17*, 42, 1981.
171. Liquori, A., and De Santis, P., X-Ray structure *versus* predicted conformation of gramicidin S, *Int. J. Biol. Macromol. 2*, 112, 1980.
172. Rackovsky, S., and Scheraga, H. A., Intermolecular antiparallel β sheet: comparison of predicted and observed conformations of gramicidin S, *Proc. Natl. Acad. Sci. U.S.A. 77*, 6965, 1980.
173. Némethy, G., and Scheraga, H. A., Hydrogen bonding involving the ornithine side chain of gramicidin S, *Biochem. Biophys. Res. Commun. 118*, 643, 1984.
174. Naik, V. M., Krimm, S., Denton, J. B., Némethy, G., and Scheraga, H. A., Vibrational analysis of peptides, polypeptides and proteins XXVII. Structure of gramicidin S from normal mode analyses of low-energy conformations, *Int. J. Peptide Protein Res. 24*, 613, 1984.
175. Hong, S. B., Kim, S. K., Kim, M. S., and Suh, S. W., Preliminary X-ray crystallographic and Raman studies on gramicidin S · 2HCl, *Arch. Biochem. Biophys. 243*, 563, 1985.
176. Magdoff-Fairchild, B., and White, J. G., Tyrocidine A: crystal growth and preliminary X-ray diffraction data, *J. Mol. Biol. 135*, 757, 1979.
177. Fox, J. A., Tu, A. T., Hruby, V. J., and Mosberg, H. I., Raman study of crystalline peptides containing β turns, *Arch. Biochem. Biophys. 211*, 628, 1981.
178. Krauss, E. M., and Chan, S. I., Spectroscopic studies of intramolecular hydrogen bonding in gramicidin S, *J. Am. Chem. Soc. 104*, 1824, 1982.
179. Sankaram, M. B., and Easwaran, K. R. K., CD and NMR Studies on the interaction of lithium ion with valinomycin and gramicidin S, *Biopolymers 21*, 1557, 1982.
180. Gierasch, L. M., Lacy, J. E., Thompson, K. F., Rockwell, A. L., and Watnick, P. I., Conformations of model peptides in membrane-mimetic environments, *Biophys. J. 37*, 275, 1982.
181. Kondo, M., Okamoto, K., Nishi, I., Yamamoto, M., Kato, T., and Izumiya, N., A new method for conformational analysis of peptides by hydrogen-chlorine replacement reaction, *Chem. Lett.*, 703, 1980.
182. Kondo, M., Yada, Y., Okamoto, K., Yonezawa, H., Kato, T., and Izumiya, N., Structure and activity of chlorinated gramicidin S, in *Peptide Chemistry 1982*, Sakakibara, S., Ed., Protein Res. Found., Osaka, 1983, p. 257.
183. Wilkins, C. L., Weil, D. A., Yang, C. L. C., and Ijames, C. F., High mass analysis by laser desorption Fourier transform mass spectrometry, *Anal. Chem. 57*, 520, 1985.
184. Cody, R. B. Jr., Amster, I. J., and McLafferty, F. W., Peptide mixture sequencing by tandem Fourier-transform mass spectrometry, *Proc. Natl. Acad. Sci. U.S.A. 82*, 6367, 1985.

185. Sato, K., Kawai, M., and Nagai, U., β-Turn preference of tetrapeptide sequences as analyzed by CD spectra of their Dnp-pNA derivatives, *Biopolymers 20*, 1921, 1981.
186. Sato, K., Kawai, M., and Nagai, U., Studies on the β-turn of peptides. II. Syntheses and conformational properties of *N*-(2,4-dinitrophenyl)tetrapeptide *p*-nitroanilides related to the β-turn part of gramicidin S, *Bull. Chem. Soc. Jpn. 56*, 1527, 1983.
187. Sato, K., Nagai, U., and Higashijima, T., Studies on the β-turn of peptides. III. Syntheses and conformational properties of *N*-(2,4-dinitrophenyl)tetrapeptide *p*-nitroanilides related to β-turn preferring sequences in proteins, *Bull. Chem. Soc. Jpn. 56*, 1657, 1983.
188. Tamaki, M., Takimoto, M., Nozaki, S., and Muramatsu, I., Adsorption of cyclic peptides analogous to gramicidin S and gratisin onto octadecylsilica stationary phase and bacterial cells, *J. Chromatogr. 413*, 287, 1987.
189. Nagai, U., and Sato, K., Synthesis of a bicyclic dipeptide with the shape of β-turn central part, *Tetrahedron Lett. 26*, 647, 1985.
190. Nomiyama, K., Ando, S., Nishikawa, H., Takiguchi, H., and Izumiya, N., Syntheses and properties of cyclic dodecapeptides with four ornithyl residues, in *55th Autumn Natl. Mtg. Chem. Soc. Jpn.*, Fukuoka, 1987, Abstr. (in Japanese), II, p. 775.
191. Ando, S., Nishikawa, H., Takiguchi, H., and Izumiya, N., Synthesis of cyclo-(Leu-Orn-Leu-Orn-Leu-D-Phe-Pro-)$_2$, *Fukuoka Univ. Sci. Rep. 13*, 49, 1983.
192. Matsuura, S., Waki, M., Kato, T., and Izumiya, N., Studies of peptide antibiotics. XXVIII. Syntheses of sesquigramicidin S and digramicidin S, *Bull. Chem. Soc. Jpn. 46*, 977, 1973.
193. Lee, S., Mihara, H., Aoyagi, H., Kato, T., Izumiya, N., and Yamasaki, N., Relationship between antimicrobial activity and amphiphilic property of basic model peptides, *Biochim. Biophys. Acta 862*, 211, 1986.
194. Mihara, H., Kanmera, T., Yoshida, M., Lee, S., Aoyagi, H., Kato, T., and Izumiya, N., Design and synthesis of amphiphilic basic peptides with antibacterial activity and their interaction with model membrane, *Bull. Chem. Soc. Jpn. 60*, 697, 1987.
195. Bichowsky-Slomnicki, L., Berger, A., Kurtz, J., and Katchalski, E., The antibacterial action of some basic amino acid copolymers, *Arch. Biochem. Biophys. 65*, 400, 1956.
196. Shima, S., Matsuoka, H., Iwamoto, T., and Sakai, H., Antimicrobial action of ε-poly-L-lysine, *J. Antibiot. 37*, 1449, 1984.
197. Storm, D. R., Rosenthal, K. S., and Swanson, P. E., Polymyxin and related peptide antibiotics, *Ann. Rev. Biochem. 46*, 723, 1977.
198. Teshima, T., Ueki, Y., Nakai, T., and Shiba, T., Structure determination of lepidopteran, self-defense substance produced by silkworm, *Tetrahedron 42*, 829, 1986.
199. Teshima, T., Nakai, T., Kitazawa, M., and Shiba, T., Total synthesis of lepidopteran A, self-defence substance produced by silkworm, *Tetrahedron Lett. 28*, 4705, 1987.
200. Steiner, H., Hultmark, D., Engström, Å., Bennich, H., and Boman, H. G., Sequence and specificity of two antibacterial proteins involved in insect immunity, *Nature 292*, 246, 1981.
201. Hultmark, D., Engström, Å., Bennich, H., Kapur, R., and Boman, H. G., Insect immunity: isolation and structure of cecropin D and four minor antibacterial components from Cecropia pupae, *Eur. J. Biochem. 127*, 207, 1982.
202. Qu, X.-m., Steiner, H., Engström, Å., Bennich, H., and Boman, H. G., Insect immunity: isolation and structure of cecropins B and D from pupae of the Chinese oak silk moth, *Antheraea pernyi, Eur. J. Biochem. 127*, 219, 1982.
203. van Hofsten, P., Faye, I., Kockum, K., Lee, J.-Y., Xanthopoulos, K. G., Boman, I. A., Boman, H. G., Engström, Å., Andreu, D., and Merrified, R. B., Molecular cloning, cDNA sequencing, and chemical synthesis of cecropin B from *Hyalophora cecropia, Proc. Natl. Acad. Sci. U.S.A. 82*, 2240, 1985.
204. Okada, M., and Natori, S., Purification and characterization of an antibacterial protein from haemolymph of *Sarcophaga peregrina* (flesh-fly) larvae, *Biochem. J. 211*, 727, 1983.
205. Okada, M., and Natori, S., Primary structure of sarcotoxin I, an antibacterial protein induced in the hemolymph of *Sarcophaga peregrina* (flesh fly) larvae, *J. Biol. Chem. 260*, 7174, 1985.
206. Matsumoto, N., Okada, M., Takahashi, H., Ming, Q. X., Nakajima, Y., Nakanishi, Y., Komano, H., and Natori, S., Molecular cloning of a cDNA and assignment of the *C*-terminal of sarcotoxin IA, a potent antibacterial protein of *Sarcophaga peregrina, Biochem. J. 239*, 717, 1986.

207. Zasloff, M., Magainins, a class of antimicrobial peptides from *Xenopus* skin: isolation, characterization of two active forms, and partial cDNA sequence of a precursor, *Proc. Natl. Acad. Sci. U.S.A. 84*, 5449, 1987.
208. Finer, E. G., Hauser, H., and Chapman, D., Nuclear magnetic resonance studies of interactions of phospholipids with cyclic antibiotics, *Chem. Phys. Lipids 3*, 386, 1969.
209. Pache, W., Chapman, D., and Hillaby, R., Interaction of antibiotics with membranes: polymyxin B and gramicidin S, *Biochim. Biophys. Acta 255*, 358, 1972.
210. Semenov, S. N., Melnik, E. I., Snezkova, L. G., Miroshnikov, A. I., and Ivanov, V. T., The mode of gramicidin S action on model lipid systems, *Bioorg. Khim. 3*, 1055, 1977.
211. Pink, D. A., Lookman, T., MacDonald, A. L., Zuckermann, M. J., and Jan, N., Lateral diffusion of gramicidin S, M-13 coat protein and glycophorin in bilayers of saturated phospholipids, mean field and Monte Carlo studies, *Biochim. Biophys. Acta 687*, 42, 1982.
212. Wu, E.-S., Jacobson, K., Szoka, F., and Portis, A. Jr., Lateral diffusion of a hydrophobic peptide, *N*-4-nitrobenz-2-oxa-1,3-diazole gramicidin S, in phospholipid multibilayers, *Biochemistry 17*, 5543, 1978.
213. Susi, H., Sampugna, J., Hampson, J. W., and Ard, J. S., Laser-Raman investigation of phospholipid-polypeptide interactions in model membranes, *Biochemistry 18*, 297, 1979.
214. Nakagaki, M., Handa, T., and Sehara, C., Thermal studies on interactions between gramicidin S and phospholipid membranes, *Yakugaku Zasshi 101*, 774, 1981.
215. Yonezawa, H., Okamoto, K., Tomokiyo, K., and Izumiya, N., Mode of antibacterial action by gramicidin S, *J. Biochem. (Tokyo) 100*, 1253, 1986.
216. Kaprel'yants, A. S., Nikiforov, V. V., Miroshnikov, A. I., Snezhkova, L. G., Eremin, V. A., and Ostrovskii, D. N., Bacterial membranes and mechanism of action of gramicidin S, *Biokhimiya 42*, 329, 1977.
217. Eremin, V. A., Bulgakova, V. G., Kaprel'yants, A. S., Polin, A. N., and Ostrovskii, D. N., Accommodation of gramicidin S in membranes of bacteria sensitive to it and protein-lipid interactions, *Biokhimiya 44*, 548, 1979.
218. Dergunov, A. D., Kaprel'yants, A. S., and Ostrovskii, D. N., Changes in structural state of boundary lipids in bacterial membrane induced by membranotropic antibiotic gramicidin S, *Biokhimiya 46*, 1499, 1981.
219. Katsu, T., Kobayashi, H., and Fujita, Y., Mode of action of gramicidin S on *Escherichia coli* membrane, *Biochim. Biophys. Acta 860*, 608, 1986.
220. Katsu, T., Kobayashi, H., Hirota, T., Fujita, Y., Sato, K., and Nagai, U., Structure-activity relationship of gramicidin S analogues on membrane permeability, *Biochim. Biophys. Acta 899*, 159, 1987.
221. Katsu, T., Yoshimura, S., and Fujita, Y., Increase in permeability of *Escherichia coli* outer membrane induced by polycations, *FEBS Lett. 166*, 175, 1984.
222. White, D. C., and Frerman, F. E., Extraction, characterization, and cellular localization of the lipids of *Staphylococcus aureus, J. Bacteriol. 94*, 1854, 1967.
223. Okamoto, K., Tomita, Y., Yonezawa, H., Hirohata, T., Ogura, R., and Izumiya, N., Inhibitory effect of gramicidin S on the growth of murine tumor cells in vitro and in vivo, *Oncology 41*, 43, 1984.
224. Okamoto, K., Sakanashi, T., Nakagawa, T., Tomita, Y., Yonezawa, H., Uemura, Y., Ogura, R., and Izumiya, N., Interaction of gramicidin S and tumor cell, in *Peptide Chemistry 1986*, Miyazawa, T., Ed., Protein Res. Found., Osaka, 1987, p. 195.
225. Krauss, E. M., and Chan, S. I., Complexation and phase transfer of nucleotides by gramicidin S, *Biochemistry 22*, 4280, 1983.
226. Shimohigashi, Y., Yoshitomi, H., Ono, S., Sakaguchi, K., and Waki, M., Structure function studies of gramicidin S. I. Ornithine side chains are not necessary to the antimicrobial activity, in *Peptide Chemistry 1988*, Ueki, M., Ed., Protein Res. Found., Osaka, 1989, p. 313.
227. Yoshitomi, H., Shimohigashi, Y., Ono, S., Waki, M., and Inoue, T., Structure function studies of gramicidin S. II. Interaction between liposomes and highly bioactive gramicidin S analog lacking the two ornithine side chains, in *Peptide Chemistry 1988*, Ueki, M., Ed., Protein Res. Found., Osaka, 1989, p. 317.

Chapter 10
Cyclosporine: Synthetic Studies, Structure-Activity Relationships, Biosynthesis and Mode of Action

Hans G. Fliri and Roland M. Wenger

1. Introduction
2. Cyclosporine
   2.1 Isolation
   2.2 Structure determination
   2.3 Other naturally occurring cyclosporines. Nomenclature
3. Total Synthesis
   3.1 Synthesis of MeBmt
      3.1.1 The Wenger synthesis
      3.1.2 The Evans synthesis
      3.1.3 The Seebach synthesis
      3.1.4 The syntheses of Rich
      3.1.5 The Schmidt synthesis
   3.2 Synthesis of cyclosporine
4. Chemical Modification of Cyclosporine
5. Structure-Activity Relationships
6. Conformation of Cyclosporine in the Crystal, in Aprotic Solvents and in Biological Fluids
   6.1 Crystal and solution conformation
   6.2 Cyclosporine conformation in biological fluids
   6.3 Immunochemical study of the cyclosporine conformation in aqueous medium
   6.4 Immunological studies of cyclosporine and their interpretation in terms of the conformation of the compound in aqueous media
   6.5 Calculations using molecular dynamics programs indicate that the crystal structure conformation is in water more stable than any others
7. Biosynthesis
8. Mechanism of Action of Cyclosporine
   8.1 Effects on T cells
   8.2 Effects on B cells
   8.3 Possible intracellular targets for cyclosporine
9. Conclusions and Outlook
References

1. Introduction

Microorganisms are the source of a variety of cyclic peptides and depsipeptides. Many of these natural products incorporate either unusual (non proteinogenic) amino acids or chemically modified versions of proteinogenic precursors such as *N*-methyl amino acids, dehydro amino acids or heterocyclic structures derived by cyclization of peptides containing serine, threonine or cysteine. Most of these compounds have been discovered on the basis of remarkable biological properties. Some examples of structures and biological properties are given below.

Table 1 Examples of bioactive peptides.

Compound name	No.	Producing organism	Activity
Cyclosporine	1	Tolypocladium inflatum GAMS	immunosuppressive
Chlamydocine	2	Diheterospora chlamydosporia	cytostatic
Amanitine	3	Amanita phalloides	toxic (protein synthesis inhibitor)
Dolastatine	4	Dolabella auricularia	cell growth inhibitory
Patellamide B	5	Lissoclinum patella	cell growth inhibitory
Bleomycin	6	Streptomyces verticillus	antineoplastic
Verticillin	7	Verticillium spp.	toxic
Vancomycin	8	Streptomyces orientalis	antibacterial (against grampositive bacteria, esp. against penicillin resist.)

The biosynthesis of these molecules does not occur on ribosomes but employs a mechanism which resembles rather the biosynthesis of fatty acids in which acetate units activated as thiol esters are assembled stepwise. In this chapter emphasis will be placed on the cyclosporines, a group of cyclic peptides currently of high interest to the medical and biochemical community due to their immunosuppressive properties [1]. The first member of this group, originally named "Cyclosporin A" [2] and throughout this paper called cyclosporine, has been marketed since 1983 for the prevention of graft rejection in organ transplantations under the registered trade mark SANDIMMUN® (E) [3] (for nomenclature see also Section 2.3). The discovery and structure determination of cyclosporine and its natural congeners, its total synthesis and some chemical modifications will be discussed together with various approaches to determine its conformation in solution, as well as its biosynthesis and biological properties.

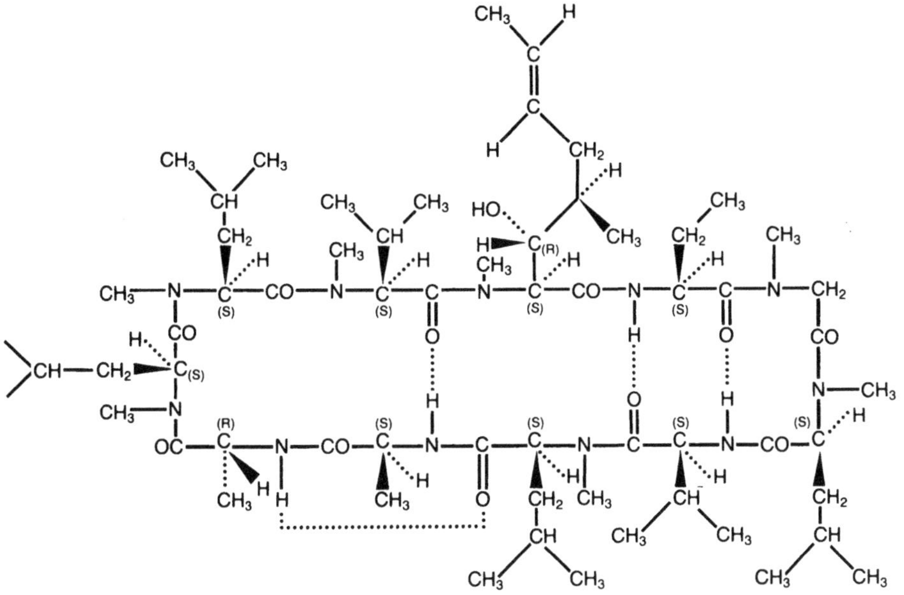

1

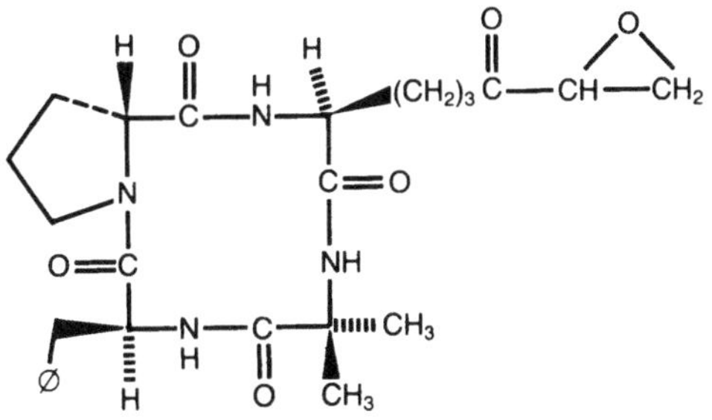

2

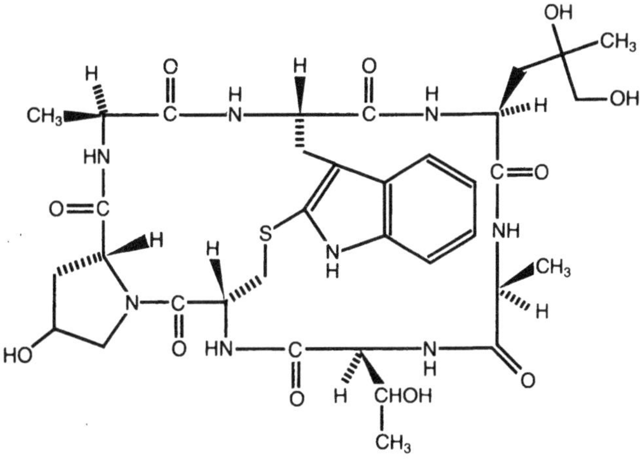

Figure 1a

3

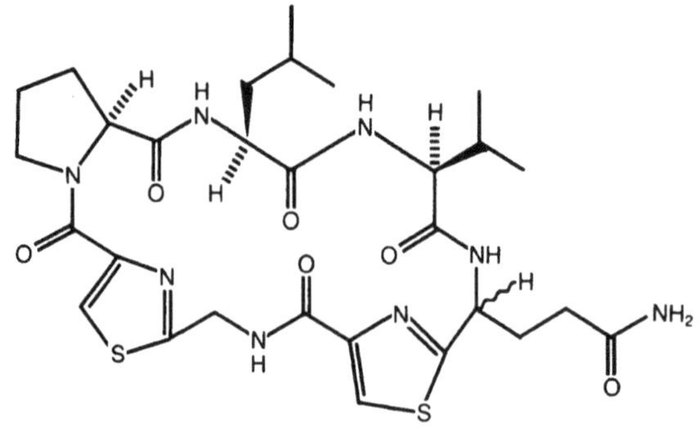

4

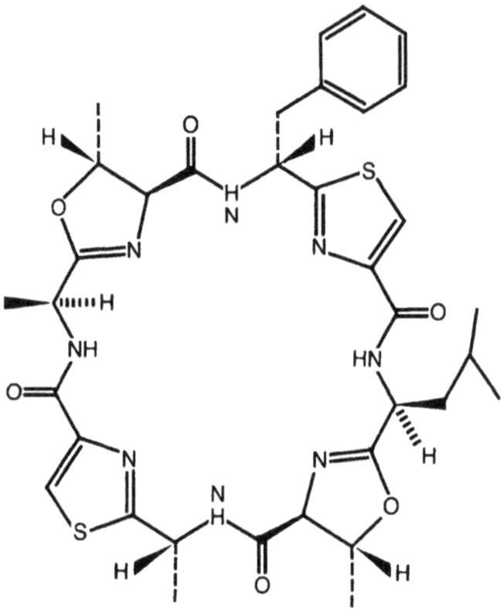

5

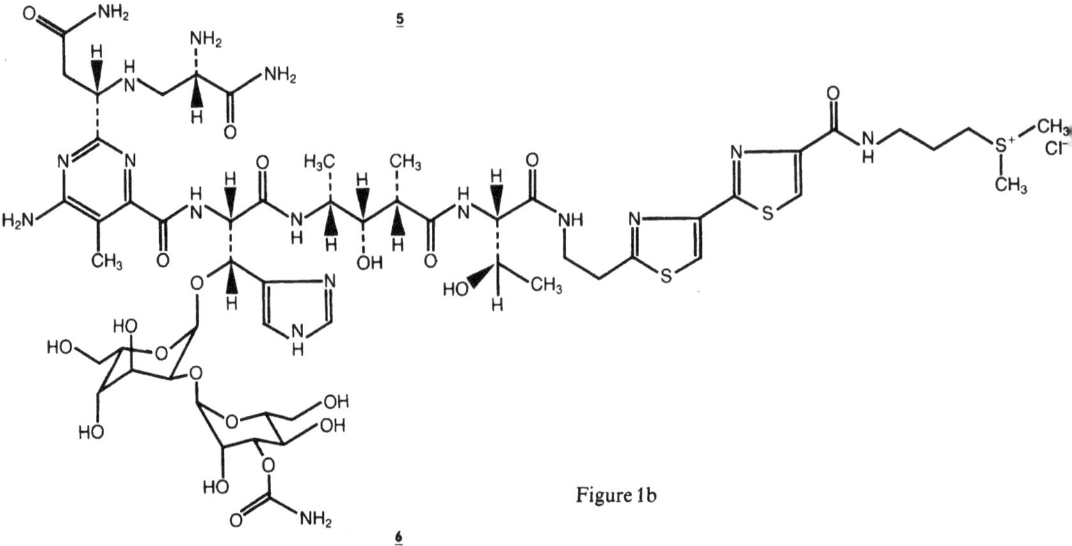

Figure 1b

6

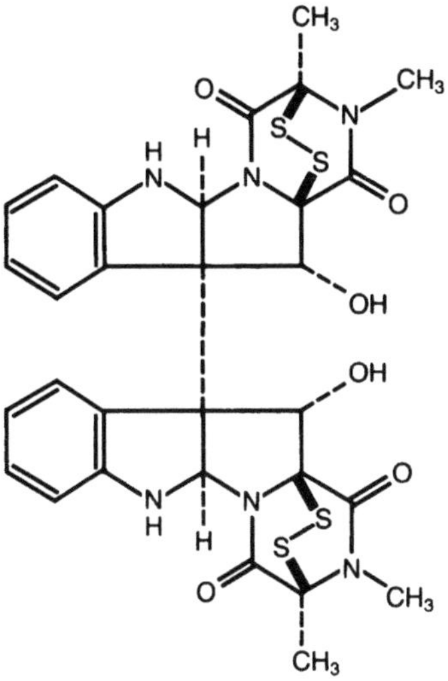
CH3
O
CH3
N
S
H
H
S
N
N
O
OH
OH
O
N
N
S
H
H
S
N
O
CH3
CH3

7

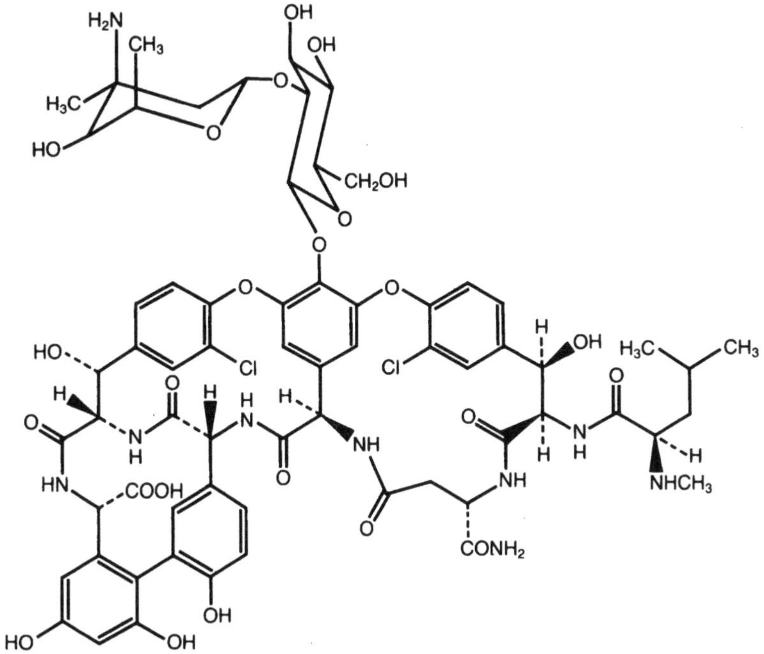
H2N
CH3
OH
OH
H3C
O
HO
O
CH2OH
O
O
O
O
H
HO
OH
H3C
CH3
Cl
Cl
O
O
H
H
H
H
O
O
N
N
NH
N
H
H
H
H
H
O
HN
COOH
NH
NH
NHCH3
O
CONH2
OH
HO
OH

8

Figure 1c

2. Cyclosporine

2.1 Isolation

Microorganisms which were isolated from soil samples collected at Hardanger Vidda in Norway and Wisconsin (USA) were found to produce a mixture of metabolites exhibiting antifungal properties. These microorganisms were later identified as *Tolypocladium inflatum* GAMS and *Cylindrocarpon lucidum* BOOTH (fungi imperfecti). The isolation, culture conditions and taxonomy of these microorganisms have been described by Dreyfuss et al. [4]. From the mixture of metabolites the main component was isolated and found to be cyclosporine **1**. The narrow range of antifungal activity was not sufficient to warrant its development in this direction; however, the compound was found to be of very low toxicity and was thus subjected to a variety of pharmacological assays. In the course of this program it was observed that **1** inhibited humoral and cellular immune responses and had antiinflammatory properties while being devoid of general cytotoxicity [5].

2.2 Structure determination

The structure of cyclosporine (Figure 1) was elucidated by a combination of chemical degradation and X-ray crystal structure analysis [2, 6]. Microtitration and electrophoretic mobility indicated a neutral compound while IR- and NMR-spectroscopy suggested the presence of amide groups, a disubstituted double bond of trans geometry and an alcoholic hydroxyl group of which an acetyl derivative could be formed only under forcing conditions. Total hydrolysis of cyclosporine and identification of the fission products yielded 4 moles of *N*-methyl-L-leucine, sarcosine, *N*-methyl-L-valine, L-valine, L-2-aminobutyric acid, racemic alanine (2 moles) and artefacts of a previously unknown amino acid consisting of 9 carbon atoms. At that time this amino acid could not be isolated from cyclosporine hydrolysis. Only after catalytic hydrogenation followed by hydrolysis could a bis-trifluoroacetyl-dihydroderivative [2] be obtained and the structure assigned as 3-hydroxy-4-methyl-2-methylamino-6-octenoic acid. The configuration of this amino acid became known from the X-ray analysis of an iodo-derivative of cyclosporine [6] and its isolation as (4R)-4((E)-2-butenyl)-4,N-dimethyl-L-threonine (abbreviated MeBmt) **9** (Figure 4) only became possible after total synthesis [22] (Section 3.1). The sequence of amino acids in cyclosporine **1** could not be determined directly by Edman degradation since no free amino group was available in the molecule. However, treating cyclosporine with methane sulfonic acid in methanol effected an *N*,*O*-acyl migration furnishing iso-cyclosporine **10** containing a free methylamino group (Figure 2). Treatment of **10** with methyl isothiocyanate in pyridine followed by trifluoroacetid acid in 1-chlorobutane afforded the anhydro-methylthiohydantoine **12** as the only product derived from **9** and established it as the first amino acid in the peptide sequence. Repetitive Edman degradation under special conditions (Edman degradation of *N*-methylated peptides was previously unknown; it had, however, been shown that phenyl isothiocyanate reacted with sarcosine to give

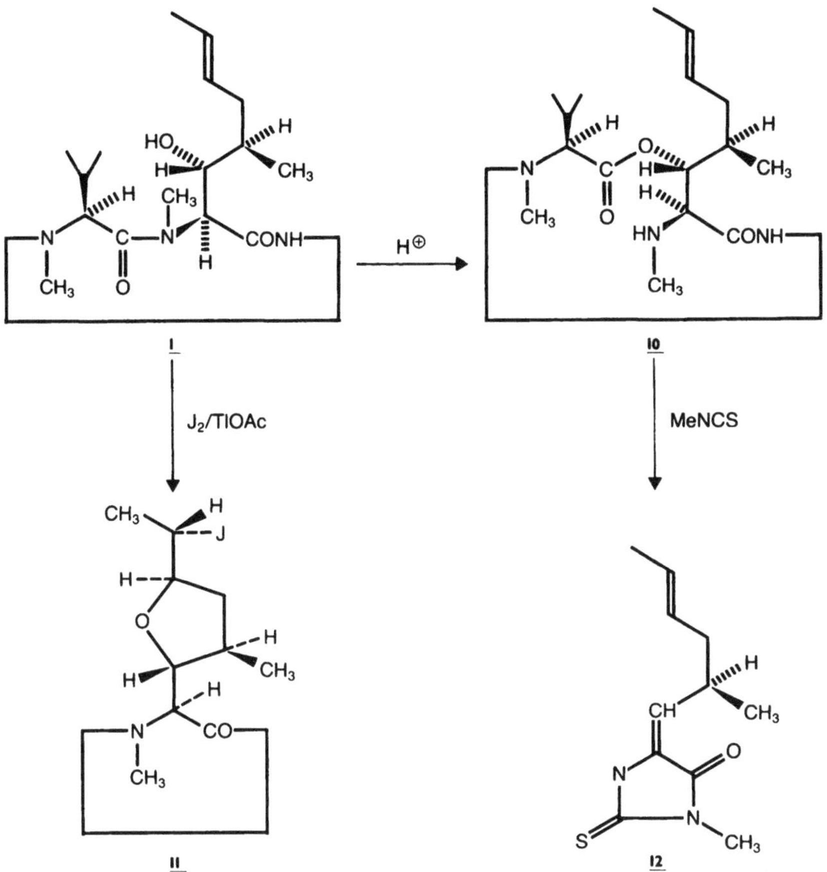

Figure 2

mainly undesired products [7]) using methyl isothiocyanate established the complete sequence except for the alanines (the formation of thiohydantoines is accompanied by racemisation and does not provide information regarding the configuration of the parent amino acid). The absolute configuration of MeBmt, the amino acid in position 1 of the sequence and the configuration of the Ala-Ala sequence in pos. 7 and 8 were determined by X-ray crystal structure analysis of an iodo ether derivative 11 (Figure 2), which had been obtained by treatment of 1 with iodine and thallium acetate [6]. The structural assignment has been later confirmed by X-ray crystal structure determination of cyclosporine itself and by two dimensional nuclear magnetic resonance studies [8, 9].

The crystal structure of 1 is schematically shown in Figure 3. The cyclic peptide can be divided into two structurally different moieties: a) A β-fragment, consisting of amino acids 11–7, which form an antiparallel β-pleated sheet with a right handed type-II′ β-turn [10] at residues 2–5. Three H-bonds bridge the two short β strands and contribute to the stability of the fragment. b) Residues 7–11 form

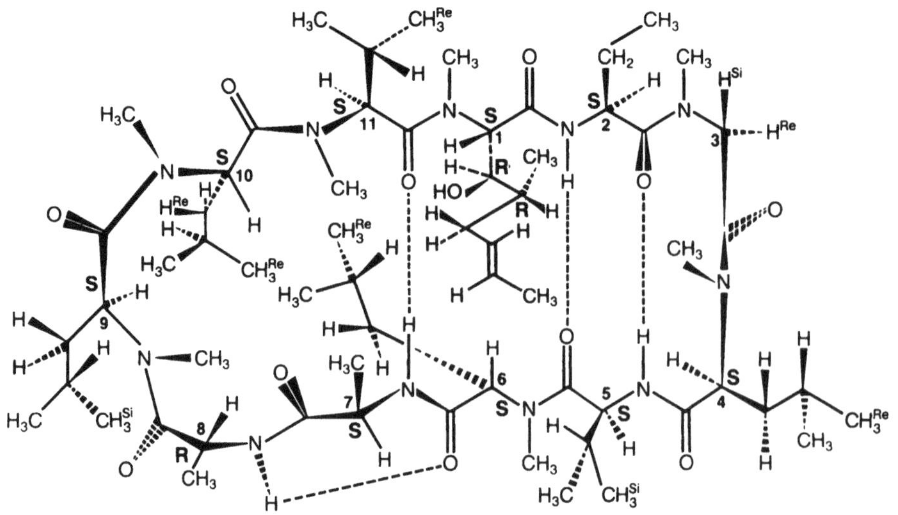

Figure 3

the second fragment, called the loop. Several interesting features are found in this fragment: The fourth intramolecular H-bond bridging N_8 and O_6, D-Ala in position 8 and a cis-amide bond between residues 9 and 10. Another remarkable feature is the position of the *N*-methyl group of MeVal-11, which projects into the center of the loop and makes a number of close contacts to backbone atoms. The lipophilic side chains of MeLeu, Val and Abu residues are in the staggered conformation with MeLeu side chains being extended. The side chain of MeBmt is folded back into the ply of the *β*-pleated sheet, allowing the molecule to adopt a compact globular shape. The conformation of 1 in various solvents will be discussed later in this chapter (Section 6) together with experimental approaches to obtain this information.

2.3 Other naturally occurring cyclosporines. Nomenclature [3, 14]

The name *Cyclosporin A* is the scientific name used in the literature for the drug commercially available as SANDIMMUN® (E) [3a]. The international nonproprietary name (INN) adopted by the World Health Organization (WHO) is *Ciclosporin* and the United States Adopted Name (USAN) is *Cyclosporine*. The British Approved Name (BAN) is *Cyclosporin*, the French name *Ciclosporine* and in Spain it is *Ciclosporina*. For practical reasons, it was proposed to define *cyclosporin(e)* as the cyclic undecapeptide with the structure of *Cyclosporin A* [3b]. The other fungal metabolites *cyclosporins B, C ...* have then, according to peptide nomenclature [11] the following names: Ala^2-cyclosporine (for cyclosporin B), Thr^2-cyclosporine (for cyclosporin C) and so on. A total of 25 other natural products have

Table 2 Naturally occurring cyclosporines.

Name	Amino acid in cyclosporine *1*	pos.	Amino acid in metabolite/analogue	No.	Name
CS A	Abu	2		*1*	cyclosporine (CS)
CS B	Abu	2	Ala	*13*	Ala^2-CS
CS C	Abu	2	Thr	*14*	Thr^2-CS
CS D	Abu	2	Val	*15*	Val^2-CS
CS E	MeVal	11	Val	*16*	Val^{11}-CS
CS F	MeBmt	1	(3′-desoxy)MeBmt	*17*	(3′-desoxy)MeBmt^1-CS
CS G	Abu	2	Nva	*18*	Nva^2-CS
CS H	MeVal	11	D-MeVal	*19*	D-MeVal^{11}-CS
CS I	Abu^{2}MeLeu^{10}	2,10	Val^{2}Leu^{10}	*20*	Val^{2}Leu^{10}-CS
CS K	MeBmt^{1}Abu^2	1,2	(3′-desoxy)MeBmt^{1}Val^2	*21*	(3′-desoxy)MeBmt^{1}Val^2-CS
CS L	MeBmt	1	Bmt	*22*	Bmt^1-CS
CS M	Abu^{2}Val^5	2,5	Nva^{2}Nva^5	*23*	Nva^{2,5}-CS
CS N	Abu^{2}MeLeu^{10}	2,10	Nva^{2}Leu^{10}	*24*	Nva^{2}Leu^{10}-CS
CS O	MeBmt^{1}Abu^2	1,2	MeLeu^{1}Nva^2	*25*	MeLeu^{1}Nva^2-CS
CS P	MeBmt^{1}Abu^2	1,2	Bmt^{1}Thr^2	*26*	Bmt^{1}Thr^2-CS
CS Q	MeLeu	4	Val	*27*	Val^4-CS
CS R	MeLeu^{6}MeLeu^{10}	6,10	Leu^{6}Leu^{10}	*28*	Leu^{6,10}-CS
CS S	Abu^{2}MeLeu^4	2,4	Thr^{2}Val^4	*29*	Thr^{2}Val^4-CS
CS T	MeLeu	10	Leu	*30*	Leu^{10}-CS
CS U	MeLeu	6	Leu	*31*	Leu^6-CS
CS V	Ala	7	Abu	*32*	Abu^7-CS
CS W	Abu^{2}MeVal^{11}	2,11	Thr^{2}Val^{11}	*33*	Thr^{2}Val^{11}-CS
CS X	Abu^{2}MeLeu^9	2,9	Nva^{2}Leu^9	*34*	Nva^{2}Leu^9-CS
CS Y	Abu^{2}MeLeu^6	2,6	Nva^{2}Leu^6	*35*	Nva^{2}Leu^6-CS
CS Z	MeBmt^1	1	L-2-methylamino-octanoic acid (= MeAoc)	*36*	MeAoc^1-CS

been published so far and been named "cyclosporines A to Z" [12]. Table 2 gives their structures and abbreviated names (**13** to **36**). The most common variation concerns pos. 2, Abu being replaced by alanine, threonine, valine or norvaline. Some modifications are due to a lower degree of *N*-methylation and replacement of MeBmt by (3′-deoxy)MeBmt, MeLeu or L-2-methylamino-octanoic acid (MeAoc) has also been observed. No variation at positions 3 and 8 has yet been found. Some of the observed variations resulted in considerable conformational changes (as evidenced by NMR) which in most cases had profound consequences for the biological activities of these derivatives.

3. Total Synthesis

When the therapeutic potential of cyclosporine **1** was fully recognized it became also clear that access to derivatives which could not be obtained from natural sources or by chemical modification of the natural products would be required in order to better understand structure activity relationships and also for immunochemical investigations [13]. In addition, since the biochemical mode of action of cyclosporine occurred by a novel (and still incompletely understood) mechanism it was necessary to provide tools for biochemical investigations such as affinity labels or derivatives suitable for crosslinking with solid supports for affinity chromatography. The first total synthesis of cyclosporine was reported by Wenger [14]. Subsequently several other groups have documented their interest. Stereospecific syntheses for MeBmt have been published by Evans [15], Seebach [16], Rich [17, 18], and by Schmidt [19]. Rich also announced recently the synthesis of the MeBmt-analogues **81** and **82** [20] (see Figure 9) and in addition the total synthesis and biological activities of some novel cyclosporine derivatives [21].

The synthesis of cyclosporine can be dissected into three distinct tasks: a) The classical problem of forming a macrocycle, b) Stereospecific construction of MeBmt and c) Coupling of *N*-methylated amino acids without racemisation.

3.1 Synthesis of MeBmt

3.1.1 The Wenger synthesis

This amino acid was first synthesized by Wenger [22]. His approach is illustrated in Figure 4.

Starting from (R, R)-(+)-tartaric acid in a first series of operations the optically active triol **38** was obtained. The key step in the formation of **38** is the introduction of the methyl group by reacting the (S,S)-epoxide **37** with methyl lithium cuprate which proceeds in excellent yield (89%). Due to the C_2 symmetry of **37** only one product is obtained; the methyl group is introduced with inversion of configuration and triol **38** is obtained in a yield of 56% over 7 steps.

Introduction of the unsaturated C_3 unit with a trans double bond was achieved by selectively protecting two of the three hydroxyl groups in **38** (acetonide formation), replacing the third by an aldehyde function and then subjecting **39** to a Wittig reaction. Deprotection yields the olefinic diol **40** (42% from triol **38**). Conversion of **40** to the hydroxy aldehyde **41** proved difficult. Direct oxidation of **40** (Pfitzner-Moffat method) gave **41** in poor yield. Only after selective protection of the secondary hydroxyl group via **42** and **43** were satisfactory yields of **41** obtained (70% over 5 steps).

The final Strecker synthesis was performed on the unprotected aldehyde **41** (the methylamino nitrile **44** was acid labile and tended to reconvert back into **41** under deprotection conditions). A diastereomeric mixture of methylamino nitriles was obtained which, after conversion into the oxazolidinones **46** (1,1′-carbonyl-diimidazole) furnished a single [23] carboximidate **48** upon treatment with potassium carbonate/ethanol. This reaction proceeds through a ketene imine intermediate **47** (IR 2230 cm^{-1}) which adds ethanol to give the more stable trans oxazolidinone **48** corresponding to the desired threo-configuration of hydroxy and methylamino groups. Final hydrolysis using acidic followed by basic conditions afforded enantiomerically pure MeBmt **9** in 90% yield.

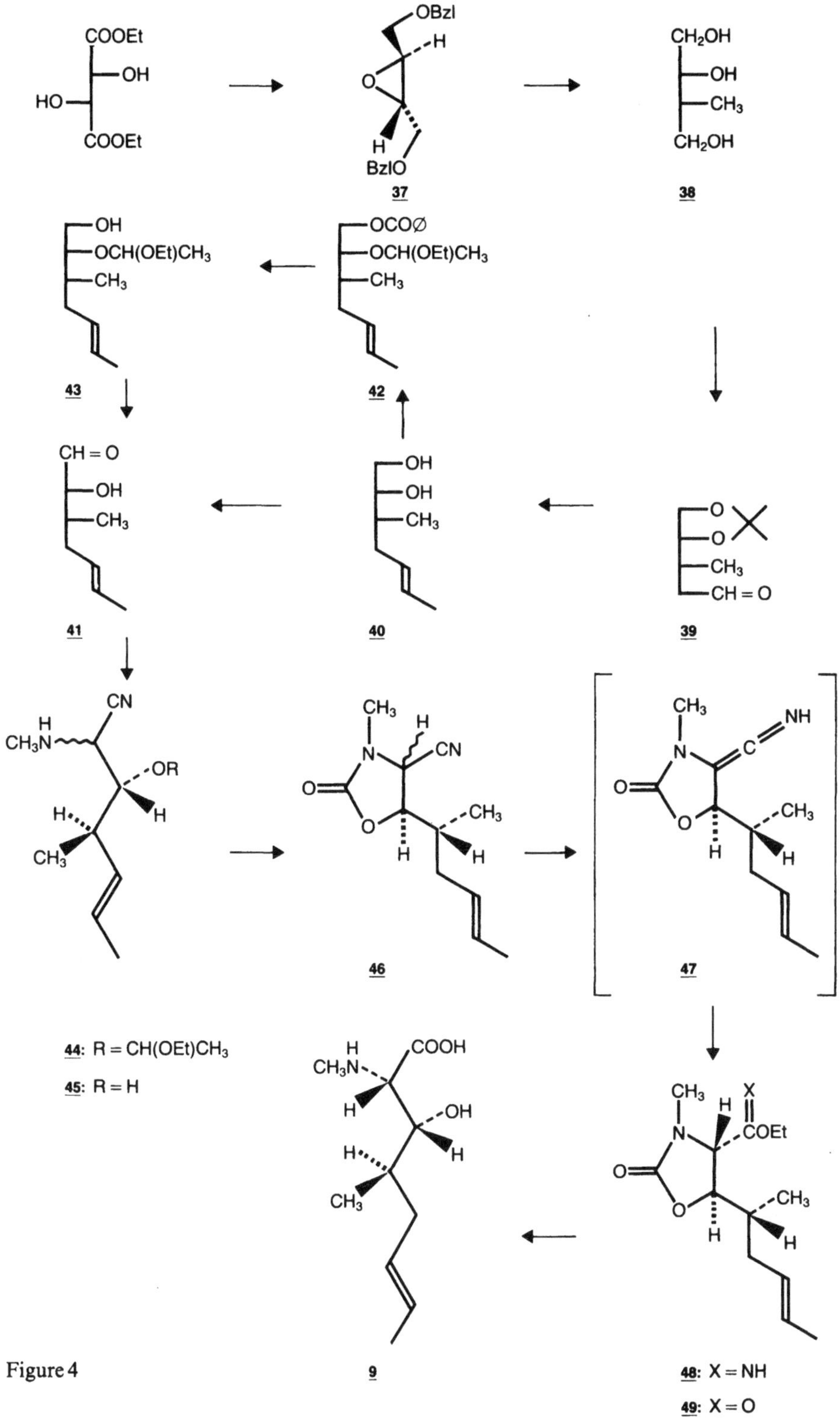

COOEt
OH
HO
COOEt
OBzl
H
O
H
BzlO
37
CH2OH
OH
CH3
CH2OH
38
OH
OCH(OEt)CH3
CH3
43
OCOØ
OCH(OEt)CH3
CH3
42
CH = O
OH
CH3
41
OH
OH
CH3
40
O
O
CH3
CH = O
39
CN
H
CH3N
OR
H
H
CH3
CH3
N
H
CN
O
O
CH3
H
H
46
CH3
NH
N
C
O
O
CH3
H
H
47
44: R = CH(OEt)CH3
45: R = H
H
COOH
CH3N
H
OH
H
H
CH3
9
CH3
X
H
N
COEt
O
O
CH3
H
H
48: X = NH
49: X = O

Figure 4

3.1.2 The Evans synthesis [15]

This synthesis is illustrated in Figure 5

Using the chiral enolate 51 derived from the isothiocyanate 50 (stannous triflate, N-ethyl piperidine) in an aldol reaction with the optically active aldehyde 52 the optically active 2-thiono-oxazolidine 53 was obtained in 73% yield. Transesterifi-

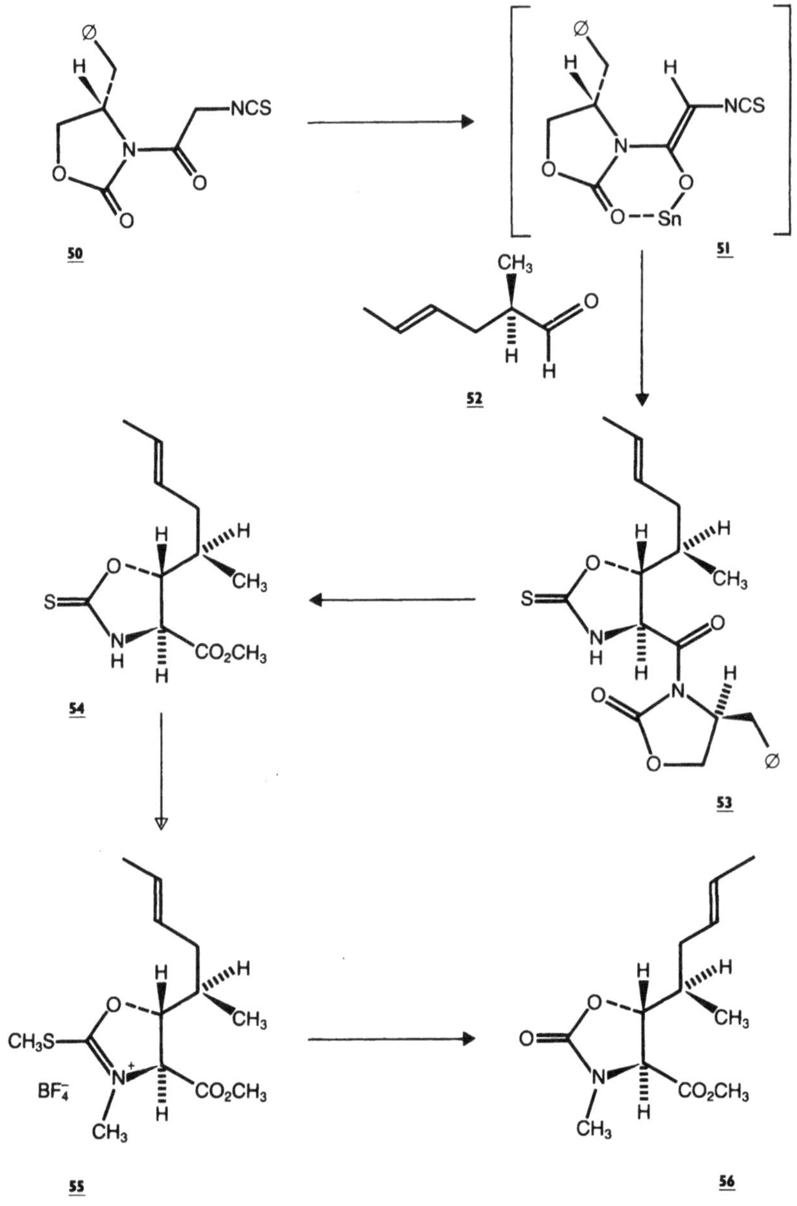

Figure 5

cation of **53** to the corresponding methyl ester **54** was accomplished with a solution of magnesium methoxide in methanol. Trimethyloxonium tetrafluoroborate converted **54** into the oxazolidinium salt **55**, which was hydrolyzed to give the oxazolidinone **56**. Hydrolysis of **56** as reported by Wenger for **49** gave enantiomerically pure MeBmt. Evans also reported a synthesis of allo-MeBmt, the (2S,3S,4R)-epimer **65** of MeBmt [24] (Figure 6). Whereas the methodology illustrated above provides

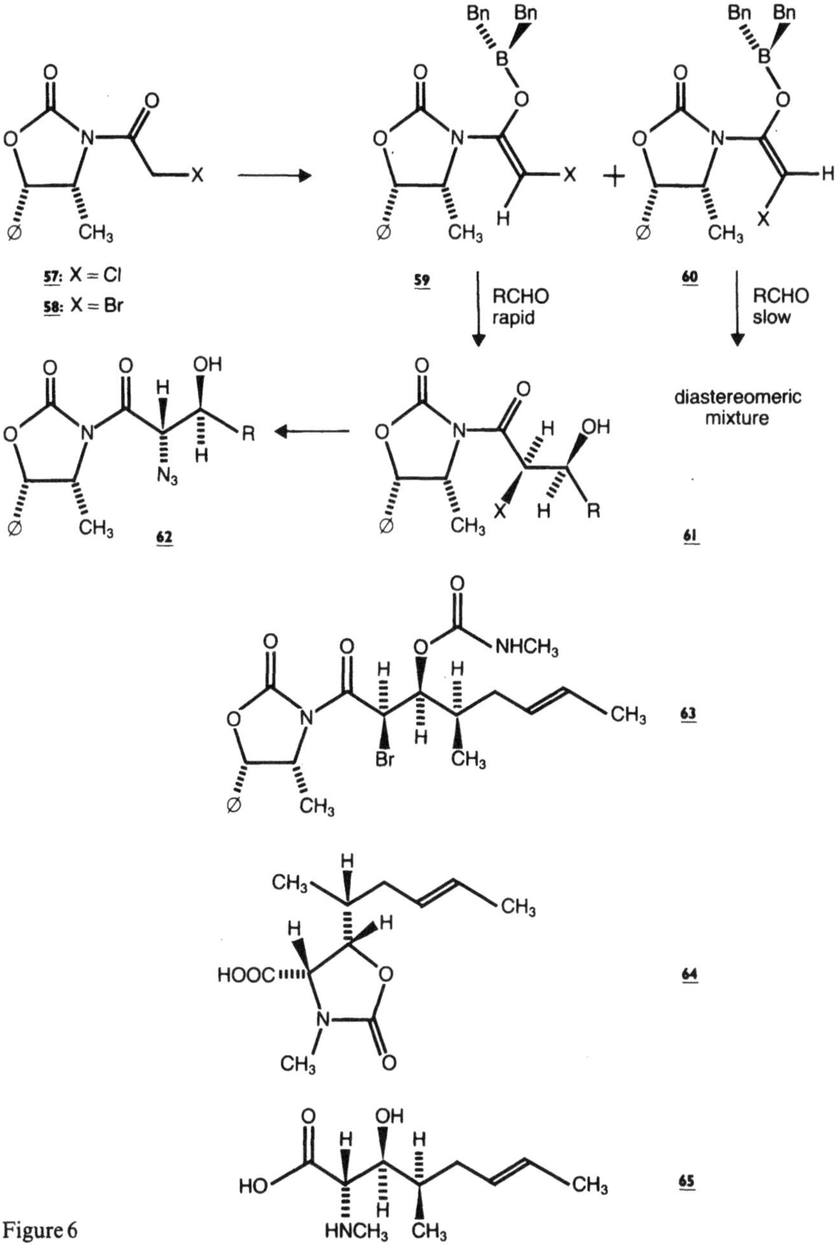

Figure 6

rapid access to enantiomerically pure syn-β-hydroxy-α-amino acids (threonine derivatives) the following complementary protocol was developed to produce anti-β-hydroxy-α-amino acids (allo-threonine derivatives).

Haloacetyl oxazolidinones **57** or **58** were converted into a mixture of boron Z- and E-enolates **59** and **60** (NMR-assignment). When aldehydes were added to this mixture the Z-enolate **59** was consumed much more rapidly and aldol adducts derived from it were obtained with high levels of diastereoselectivity. The E-isomer **60** required prolonged reaction times at room temperature to be completely consumed and showed only poor diastereoselectivity. This different reactivity of the isomeric enolates allowed the use of excess enolate as a mixture of Z- and E-isomers **59** and **60** in aldol reactions. Diastereomerically pure aldol products were obtained in high yields by preferential reaction of the Z-enolate **59**. The products are α-halo-β-hydroxy-acid derivatives **61** (R = CH_3, C_6H_{11}) with threo configuration. Subsequent displacement of the halogen atoms by azide proceeds with inversion of configuration (when using the chlorides formation of some retention products was also observed) to give α-azido-β-hydroxy acid derivatives **62** of erythro configuration. Using this protocol L-allo-threonine as well as the above mentioned (3S)-epimer **65** of MeBmt were successfully prepared via carbamates **63** and oxazolidinone **64**.

3.1.3 The Seebach synthesis [16a, b]

Similarly to the strategy developed by Evans the Seebach synthesis is also based upon an aldol reaction of a chiral enolate derived from glycine with the optically active aldehyde **52**. Rather than linking glycine to a chiral auxiliary, however, Seebach prepared the racemic imidazolidinone **66** and separated the enantiomers by crystallization of the mandelic acid salts. The desired optically active glycine-synthon **67** (S-configuration) was then obtained by benzoylation. The lithium enolate of this synthon adds to aldehydes with high diastereoselectivity to give, after acidic hydrolysis, threo-α-amino-β-hydroxy acid derivatives of high enantiomeric purity. Figure 7 illustrates this approach.

The initial aldol adducts undergo *N,O* acyl migration and the products obtained are benzoic acid esters of hydroxy amino acid *N*-methyl amides. The MeBmt intermediate **68** after *N*-methylation was converted to the *N*-methyl amide **69**. The final hydrolysis to MeBmt **9** was achieved using dilute aqueous hydrochloric acid [16b].

3.1.4 The syntheses of Rich [17]

The first synthesis of MeBmt by Rich (Figure 8) involved a protected hydroxy aldehyde similar to that used by Wenger (Figure 4). Sharpless oxidation of the dienol **70** and reaction of the optically active protected epoxide **71** with lithium dimethyl cuprate in the presence of boron trifluoride etherate furnished the mono-protected diol **72** (80%) along with the isomer **73** (14%). Methyl isocyanate followed

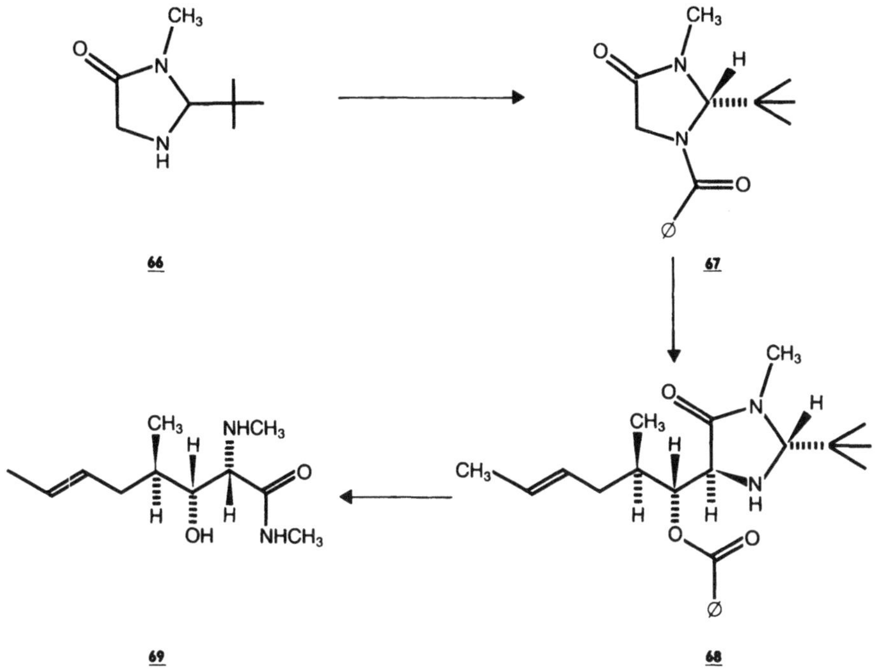

Figure 7

by deprotection yielded the primary alcohol which was oxidized and acetylated to give the latent aldehyde **74**. Cyanotrimethylsilane replaced the acetoxy group in **74** to give the nitrile **46**, which was converted into MeBmt using Wenger's method.

Subsequently an additional nonstereoselective route was reported by Rich [18] (Figure 9). Addition of the optically active aldehyde **52** to the Li-enolate of the sarcosine derivative **76** yielded an approximately 1 : 1 mixture of the diastereomeric oxazolidinones **77** and **78**. Deprotection and separation by fractional crystallization of the ephedrine salts gave the enantiomerically and diastereomerically pure oxazolidinone acids **79** and **80** (R = H, Figure 9) which were further transformed according to the Wenger procedure. This route using **79** furnishes MeBmt **9** in 4 steps from aldehyde **52** in an overall yield of ca. 20%. Using this approach the MeBmt derivatives **81**–**83** were also prepared. **83** is the starting material for 4′-desmethyl-MeBmt[1]-cyclosporine which has been previously reported [25].

3.1.5 The Schmidt synthesis [19]

This approach is based on the erythro-selective addition of the lithium enolate of *N,N,O*-tris-trimethylsilyl-glycine **84** to the optically active aldehyde **52** (Figure 10).

The (2S,3S,4R)- and (2R,3R,4R)-diastereomers **85** and **86** obtained were converted into the benzamide methyl esters **87** and **88** and separated by chromatography.

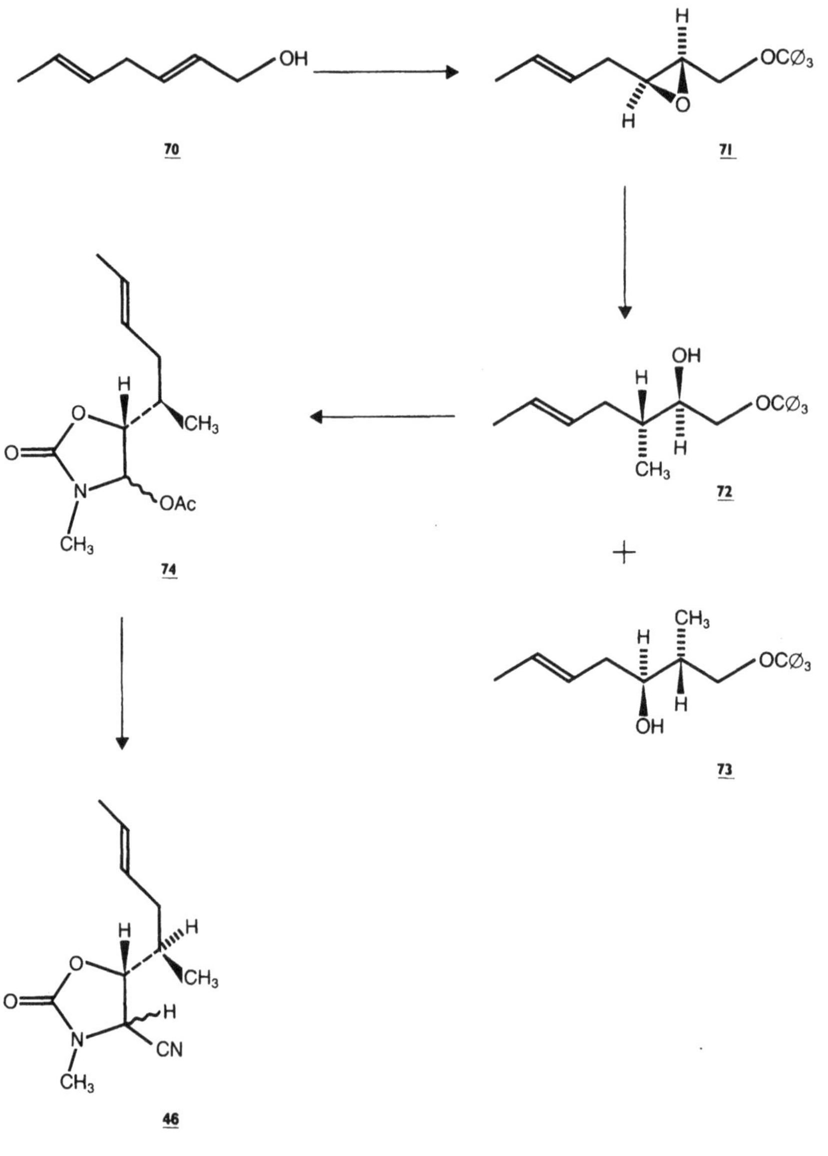

Figure 8

Thionyl chloride cyclized the former isomer **87** with inversion of the configuration at C_3 to the oxazoline **89** in which all the asymmetric centers have the configuration present in MeBmt. *N*-methylation and hydrolysis converted **89** into MeBmt. Schmidt also reported conversion of the other diastereomer **88** by an analogous sequence into the (2S,3S,4R)-epimer of MeBmt, a compound which has also been prepared by Evans [24]. The initial adduct, however, is most likely of (2R,3R,4R)-

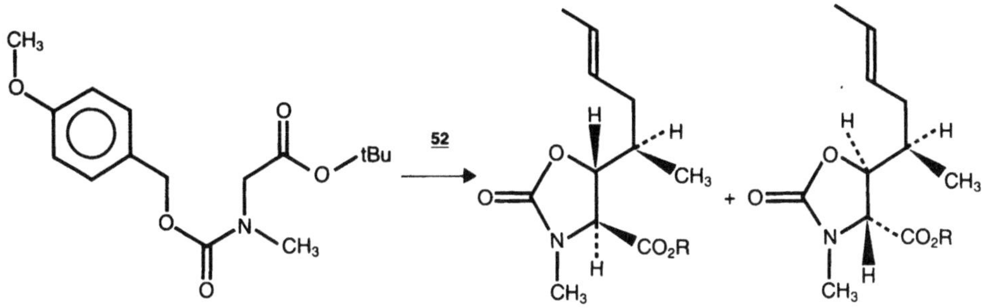

76 77 R = t−Bu 78

80 R = H 79

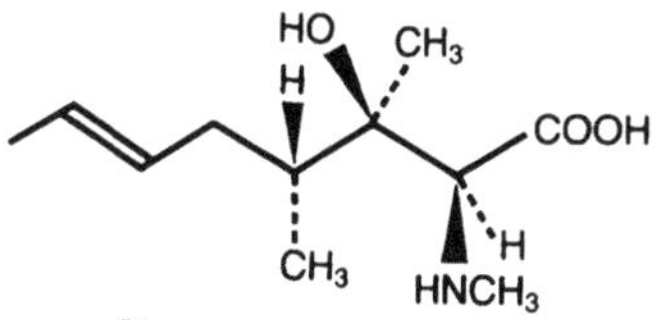

81

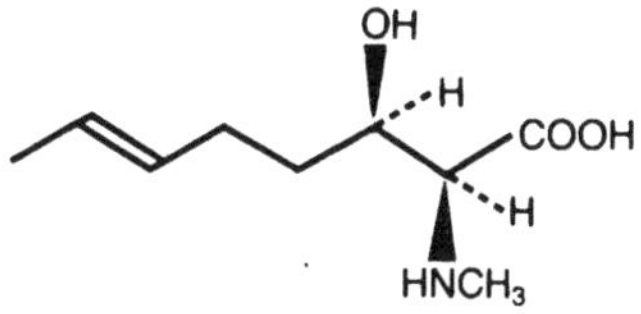

83

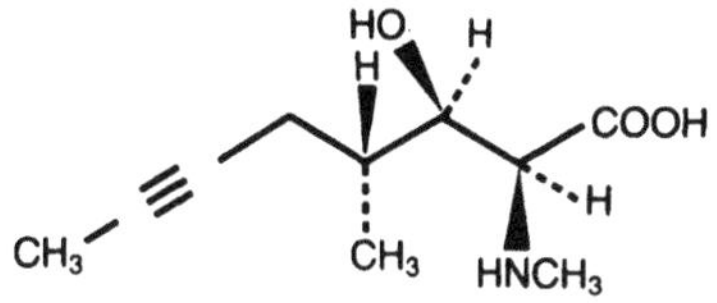

82

Figure 9

configuration and consequently the oxazoline **90** obtained with thionyl chloride should have (2R,3S,4R) configuration with trans arrangement of the ester group and alkyl side chain. This stereochemistry makes subsequent epimerisation at C_2 very unlikely and the final product obtained from **90** is probably the (2R,3S,4R)-diastereomer of MeBmt (a D-threonine derivative).

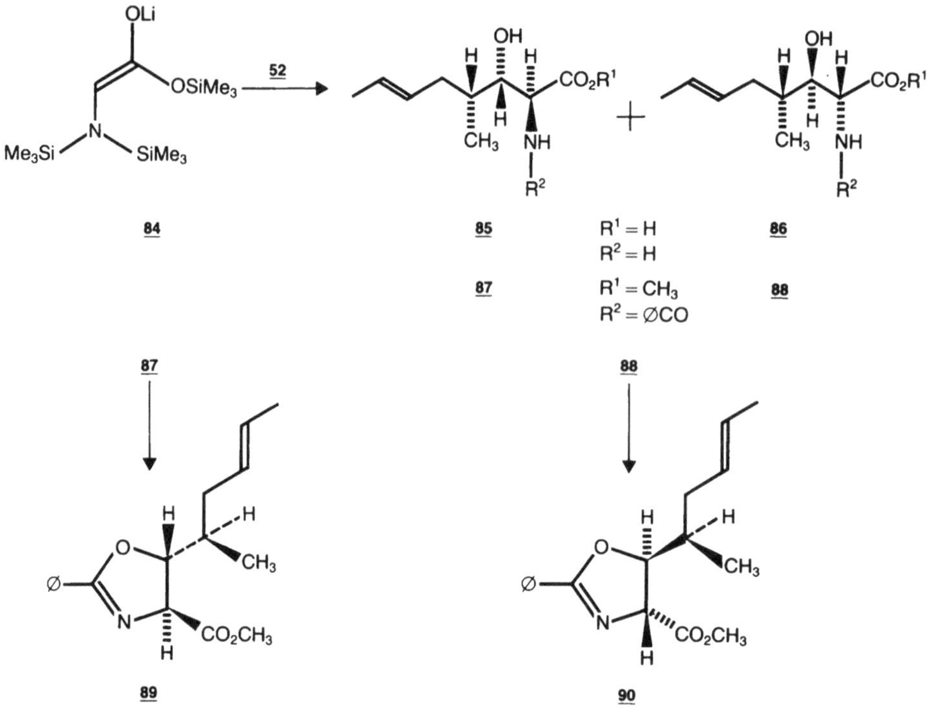

Figure 10

3.2 Synthesis of cyclosporine

The first total synthesis of cyclosporine and some derivatives which do not occur naturally has been reported by Wenger [14, 25]. Recently Rich [21], closely following Wenger's work reported the synthesis of some derivatives modified in position 1. The strategy as developed by Wenger is discussed below. The peptide bond between amino acids 7 (L-Ala) and 8 (D-Ala) was chosen for the cyclization step for several reasons:

1. peptide bond formation between *N*-methyl amino acids presents more difficulties than is the case with non-*N*-methylated derivatives.
2. intrachain H-bonds might stabilize the linear undecapeptide in a folded conformation similar to the macrocycle and thus facilitate ring closure.
3. ample precedence in the literature suggests a preference for ring closure over side reactions if the linear peptide has a D-amino acid at the *N*- or *C*-terminus [26].

For the synthesis of the linear undecapeptide, a fragment condensation technique introducing MeBmt at the end was chosen in order to minimize the number of steps after introduction of this amino acid. The peptide fragments were built up in

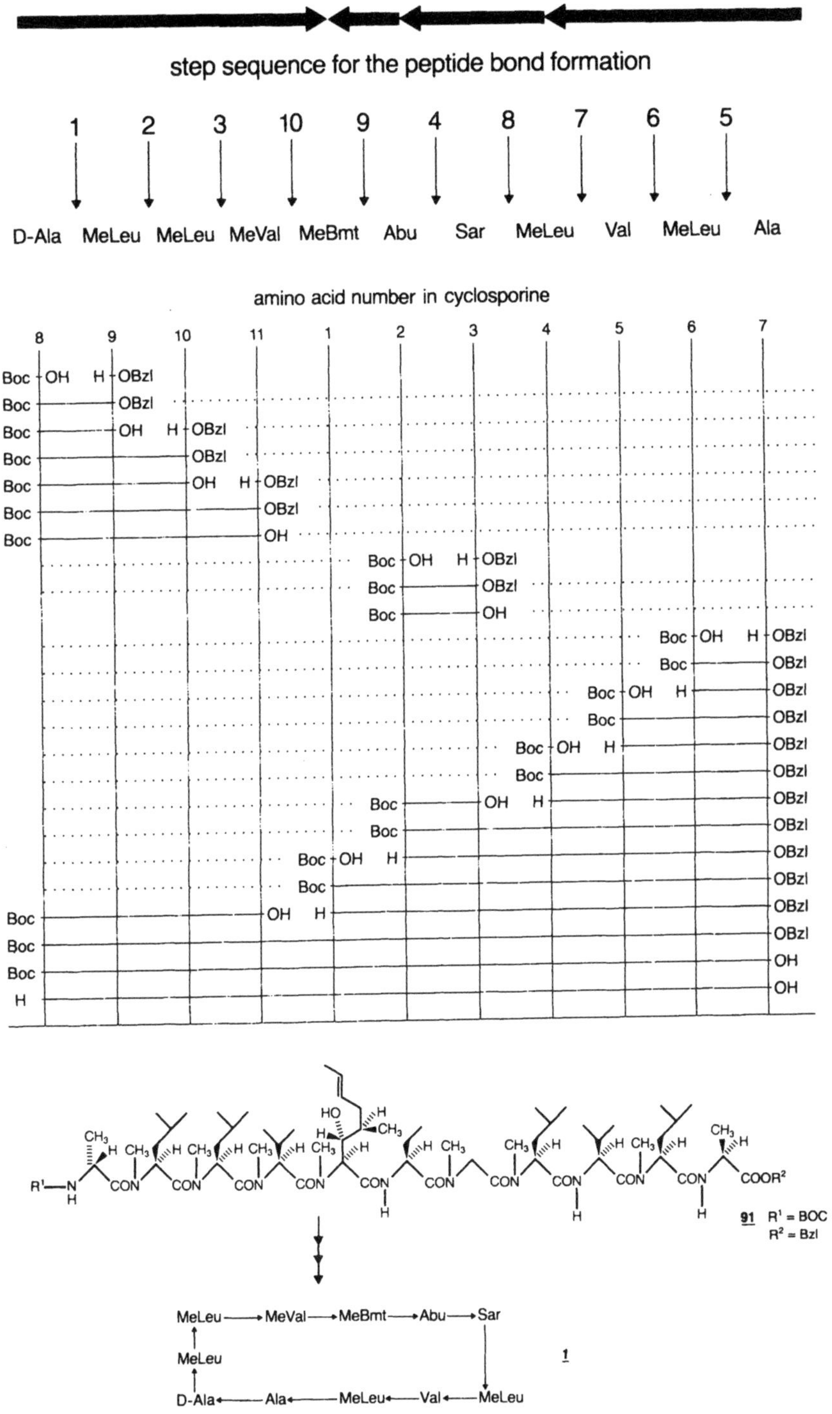

step sequence for the peptide bond formation
1 2 3 10 9 4 8 7 6 5
D-Ala MeLeu MeLeu MeVal MeBmt Abu Sar MeLeu Val MeLeu Ala
amino acid number in cyclosporine
8 9 10 11 1 2 3 4 5 6 7
Boc OH H OBzl
Boc OH
91 R¹ = BOC
R² = Bzl
MeLeu → MeVal → MeBmt → Abu → Sar
MeLeu
D-Ala ← Ala ← MeLeu ← Val ← MeLeu
1

Figure 11

the direction shown by the arrows in Figure 11 using the step sequence which is indicated numerically. The protective groups chosen were generally tert-butoxy-carbonyl (Boc) for the amino groups and benzyl (Bzl) for the carboxy groups.

Activation of carboxyl groups for the formation of peptide bonds between *N*-methyl-amino acids presents considerable difficulties. Increased steric hindrance and slow reactions contribute to extensive epimerisation in coupling reactions [27]. An additional factor facilitating racemisation of *N*-methyl-amino acids upon carboxyl activation is the increase in acidity of the *a* hydrogen atom. A detailed theoretical analysis by Veber [28] has identified three carbanion stabilizing factors:

1. *N*-alkylation increases the proportion of cis peptide present [29]. While an unalkylated cis-peptide is less stable by 1.6 kcal than the trans peptide [30] the methylated analogues are of approximately equal stability.
2. A deprotonated peptide is isoelectronic with the butadienyl dianion. CNDO/2 calculations suggest the transoid conformation of this dianion is approximately 6–7 kcal more stable than the cisoid. A cis-peptide upon deprotonation gives a carbanion corresponding to a transoid butadienyl system whereas deprotonation of the trans-peptide creates the corresponding butadienyl system in the cisoid conformation (fig. 12). Therefore a cis-peptide is more acidic at the *a* position than the corresponding trans-peptide and epimerization occurs more easily.
3. There is another factor by which *N*-methylation can contribute to an increase in acidity at the *a* position: The 'versatile methyl group' [31] can not only support anionic and cationic hyperconjugation, i. e. stabilization of positive and negative charges in adjacent positions, but it may also stabilize a negative charge formally located two atoms away.

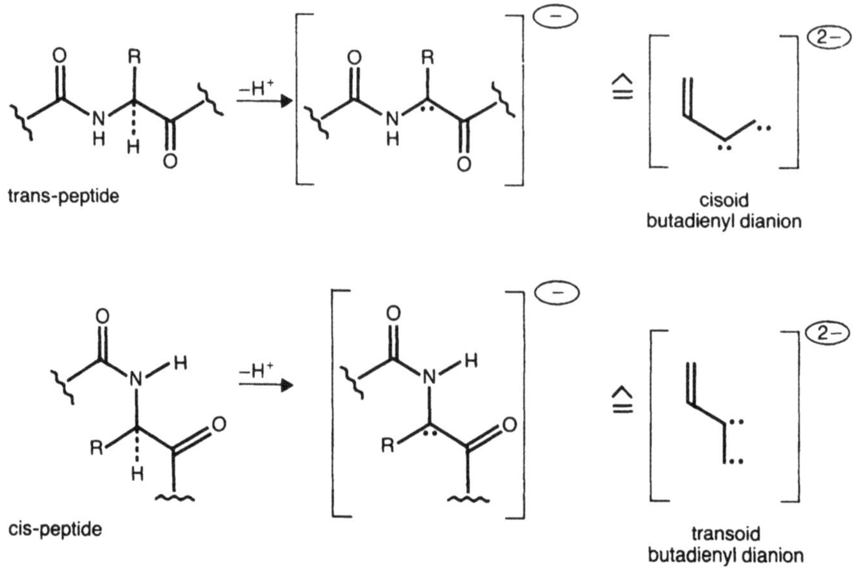

Figure 12

Thus carefully controlled experimental procedures had to be developed to minimize racemisation/epimerisation during coupling reactions. Wenger used a variation of the mixed pivalic anhydride method reported by Zaoral [32]. To adopt this method for *N*-methyl-amino acid derivatives he allowed for slow anhydride formation with pivaloyl chloride at −20 °C in chloroform in the presence of 2 equivalents of a tertiary base such as *N*-methyl morpholine before adding the *O*-protected amino acid or peptide to be coupled as the free base. The coupling reaction was then also performed at −20 °C in order to slow down epimerization via enolization of the activated carboxyl group. A different protocol was recently reported by Rich [33] who synthesized the cyclosporine 8−11 tetrapeptide fragment using bis(2-oxo-3-oxazolidinyl)phosphinic chloride (BOP-Cl) as a coupling reagent and the 9-fluorenyl-methoxycarbonyl group for *N*-protection. Wenger built up the 8−11 tetrapeptide from the left to the right as indicated in Figure 12 making the peptide bond 1 first. This tetrapeptide could not be made in the reverse direction starting with MeVal-OBzl by making bond 3 first because of instantaneous diketo-piperazine formation of the dipeptide ester H·MeLeu·MeVal·OBzl. In contrast, this sequence is fully stable as the *tert*-butyl ester H·MeLeu·MeVal·OtBu [33], an observation first made by Wenger [27]. The fragment 4−7 Boc · MeLeu · Val · MeLeu · Ala · OBzl was synthesized starting from the right forming bonds 5,6,7 in that order. Bond 4 was made by coupling Boc·Abu·OH to Sar-OBzl to give the 2−3 fragment. In the preparation of these fragments there was no coupling of consecutive *N*-methyl amino acids which reflected in the excellent yields of each coupling reaction. Joining the free acid of the 2−3 fragment to the *N*-deprotected 4−7 tetrapeptide using the pivalic anhydride method afforded the hexapeptide 2−7 in 88% yield. After *N*-deprotection (trifluoroacetic acid, −20 °C) MeBmt was added to this fragment using dicyclohexyl carbodiimide (DCC) in the presence of *N*-hydroxy-benztriazol (BtOH). During the coupling reaction the OH- and NH-groups of MeBmt were protected as an isopropylidene derivative; the presence of this five-membered ring ensured to maintain the thermodynamically more stable trans configuration of the carboxy and alkenyl substituents during the coupling reaction. Later experiments by Wenger have shown that N-Boc-MeBmt can equally be used to introduce this amino acid into the molecule [3b]. Removal of the isopropylidene (1 equiv. 1N HCl in methanol) or Boc (trifluoro acetic acid at −20 °C) protective group gave the heptapeptide H · MeBmt · Abu · Sar · MeLeu · Val · MeLeu · Ala · OBzl. The Boc-protected 8−11 tetrapeptide was then joined to the 1−7 heptapeptide benzyl ester using (1H-benzo[d][1,2,3]triazol-1-yloxy)tris(dimethylamino)-phosphonium hexafluorophosphat ($BtOP^+(NMe_2)_3PF_6$) [34] to give the bis-protected undecapeptide **91**. The ester group of **91** was removed by alkaline hydrolysis and the Boc group cleaved with trifluoroacetic acid. Cyclisation was effected in dilute methylene chloride solution (0.0002 M) using again the reagent developed by Castro [34] in the presence of *N*-methylmorpholine at room temperature. Cyclosporine **1** was obtained in 62% yield. Using the mixed phosphonic anhydride method described by Wissmann and Kleiner [35] or the pentafluorophenol-DCC complex of Kovacs [36] gave similar results.

Having established the methodology to synthesize cyclosporine potentially any amino acid of the sequence can be varied. Total synthesis thus becomes a complementary source of cyclosporine derivatives providing compounds not found in nature and the effects of a wide variety of specific variations on the biological activity can be studied. Some of these variations will be described below.

4. Chemical Modification of Cyclosporine

(Figure 13). The natural metabolites have been subjected to a wide variety of chemical modifications, many of them focussing on the olefinic and hydroxyl function of MeBmt. Some of this work has recently been published by Wenger [37].

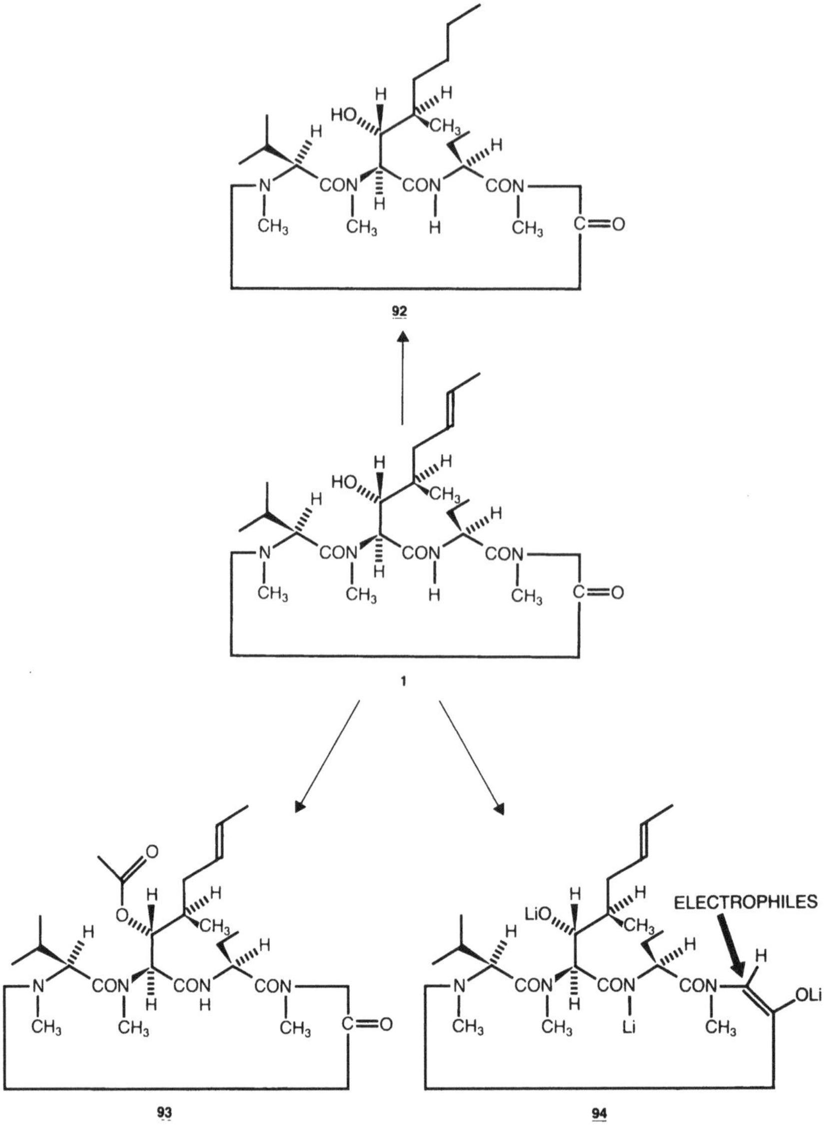

Figure 13

Catalytic hydrogenation gives dihydro compounds such as dihydro-cyclosporine **92** which generally show diminished immunosuppresive activity. Acetylation to **93** or removal of the hydroxyl group of MeBmt producing **17** (Cyclosporin F) dramatically reduce the immunosuppressive activity of these derivatives. Cyclosporine can be converted into a polyanion **94** using an excess of lithium diisopropylamide **38**. The resulting enolate originates by removal of a proton from sarcosine in pos. 3 and can be trapped with a wide variety of electrophiles such as carbonyl compounds, alkyl halides, or disulfides. The reaction with these electrophiles occurs from the bottom side giving products with D-configuration at this center; specifically, methyl iodide gives (D-MeAla)^3-cyclosporine, methyl chloroformate affords the 3-α-carbomethoxy derivative, formaldehyde (D-MeSer)^3-cyclosporine and reaction of **94** with allyl bromide gives (D-allylsarcosine)^3-cyclosporine.

5. Structure-Activity Relationships

A detailed and extensive discussion of structure-activity relationships is given in refs. 25 and 37. Earlier work [6] had indicated that it might be of particular interest to investigate changes in the immediate environment of the unusual amino acid, MeBmt. The relevance of the alkenyl side chain of MeBmt could be assessed by substituting *N*-methyl-L-threonine for MeBmt; this amino acid was expected to retain the polar features of MeBmt but not the lipophilic side chain. One of the natural metabolites contains threonine in place of α-amino-butyric acid in pos. 2 (cyclosporine C, **14**) and is a relatively good immunosuppressant. To examine whether or not this additional hydroxy group was involved in some special receptor interactions the corresponding serine derivative **96** was prepared. In the preparation of this compound the novel protected serine derivative **95** was used [39].

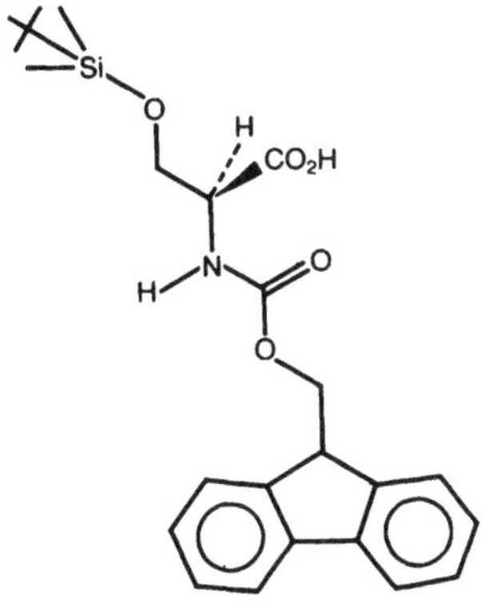

95

As mentioned previously sarcosine in pos. 3 participates in a type II′ β-turn (see Section 2.2 and 3.5). Replacement of sarcosine by D-proline would stabilize this structural element and make it more rigid. A natural isomer of cyclosporine has D-MeVal in place of MeVal in pos. 11 (cyclosporine H, **19**. This natural metabolite was synthesized along with other variations of amino acid 11: MeLeu would have similar lipophilic properties as MeVal and at the same time somewhat different steric bulk. Other variations in this position, more closely resembling the structure of cyclosporine itself, consisted in the introduction of MeIle (*N*-methyl-isoleucine) **105** and aMeIle (*N*-methyl-allo-isoleucine) **106**. The necessity to have an isopropyl group in this position could be evaluated by synthesizing a derivative containing MeAla in this position (D-MeAla^{11}-cyclosporine, **107**). The importance of the configuration of amino acid 11 is illustrated by the low level of activity of cyclosporine H (D-MeVal^{11}-cyclosporine, **19**). Except for L-Pro^{3}-cyclosporine **98** and D-MeVal^{11}-cyclosporine **19** all of the above modifications could be expected to retain the peptide ring conformation of cyclosporine and changes in biological activity should indicate whether or not these amino acids are involved in interactions with the receptor. The immunosuppressive activities of some natural metabolites together with some derivatives obtained either by chemical modification of natural product(s) or by total synthesis are listed in Table 3.

Table 3 Immunosuppressive activities of cyclosporine analogs and derivatives.

Amino acid in **1**	Pos.	Amino acid of derivative	Source	Comp. no.	Activity
MeBmt	1	dihydro-MeBmt	ss	**92**	+
MeBmt	1	(3-O-Ac)-MeBmt	ss	**94**	–
MeBmt	1	MeThr	s	**111**	–
Abu	2	Ser	s	**96**	–
Sar	3	D-Pro	s	**97**	–
Sar	3	L-Pro	s	**98**	–
Sar	3	D-Ala	s	**99**	+
Sar	3	L-Ala	s	**100**	–
Sar	3	D-Phe	ss	**101**	–
Ala	7	D-Ala	s	**102**	–
D-Ala	8	des-8-D-Ala	s	**103**	–
MeVal	11	MeLeu	s	**104**	–
MeVal	11	MeIle	s	**105**	+
MeVal	11	aMeIle	s	**106**	–
MeVal	11	MeAla	s	**107**	–
des-(8,9,10)-γ-Abu^{11}			s	**108**	–
des-(8,9,10)-γ-MeAbu^{11}			s	**109**	–
MeBmt	1	(8′-hydroxy)MeBmt		**110** [45]	+

ss ... semisynthetic compound
s ... synthetic compound
\+ ... active immunosuppressant
– ... no significant immunosuppressive activity

From these results it is clear that the lipophilic carbon chain of MeBmt is important for biological activity. Of equal importance seems to be the hydroxyl group of this amino acid since acetylation or complete removal result in substantial decrease in immunosuppressive activity. On the other hand the framework of the macrocycle seems also important as evident from the lack of activity of either MeBmt alone or of cyclic peptides with a shortened sequence. Some variations are tolerated in pos. 2 where Abu can be replaced by either Nval **18** or Thr **14** without significant loss of activity. Shortening the carbon chain, however as in Ser^2-Cs **96** or Ala^2-Cs **13** results in decreased activity. Interesting results are obtained with MeLeu^{11}-cyclosporine **104** and D-Pro^3-cyclosporine **97**. Although NMR indicates for both of these derivatives solution conformations very similar to cyclosporine and the X-ray crystal structure of **104** is very similar to that of cyclosporine, **97** as well as **104** are nearly devoid of activity, suggesting that in these cases the increased steric bulk in positions 3 and 11 probably prevents efficient binding of these derivatives to the cyclosporine receptor. Less surprising is the lack of activity of the L-Pro^3 derivative **98** since its conformation in the region of the β-turn is very different from cyclosporine (NMR studies and X-ray crystal structure determination). Another major conformational change is effected by epimerization of MeVal^{11} (cyclosporine H, **19**) as indicated by NMR and X-ray analysis of an iodo derivative. Again, the conformational change resulted in loss of activity. Since replacement of this amino acid in position 11 by MeAla **107** also leads to greatly diminished activity it seems evident that the isopropyl group of MeVal in this position plays an important role. This hypothesis is substantiated by the loss of activity of the derivatives having MeIleu **105** or allo-MeIleu **106** in position 11, the loss of activity of **105** being less pronounced as that of **106**. In this latter derivative the *re*-methyl group of *N*-methyl valine which, according to the X-ray crystal structure of cyclosporine would appear to be closer to the *N*-methyl group of MeBmt, carries the additional methyl group. Thus it appears that this methyl group is more sensitive to variation than the *si*-methyl group of MeVal^{11}, which is more distant from the *N*-methyl group of MeBmt. Major conformational changes can also be effected by introducing additional H-bridges into the molecule, eg. by omitting some of the *N*-methyl groups eg. Val^{11}-Cs **16** at amino acid 11, Leu^6-Cs **31** at amino acid 6 or Leu^{10}-Cs **30** at amino acid 10. The importance of the loop is demonstrated by the lack of activity of derivatives in which the 8–11 segment has been replaced by 4-amino-butanoyl (**108**) or 4-methylamino-butanoyl units (**109**).

In summary evidence seems to accumulate that the biological activity of cyclosporine is associated with a large part of the molecule and variations are permitted only in a few positions without significant loss in immunosuppressive activity. It is clear that the unusual amino acid MeBmt is intimately involved in the biological actions of cyclosporine but by itself this amino acid is not sufficient for immunosuppressive activity. The biological activity of cyclosporine can, from the results presented above, be associated with a large portion of the cyclosporine structure, that includes amino acids 1, 2, 3, and 11. These amino acids are clustered together on the surface of the cyclosporine molecule. Further derivatives will have to be made and more research has to be done to answer many of the remaining questions concerning the structure-activity relationships of cyclosporine.

6. Conformation of Cyclosporine in the Crystal, in Aprotic Solvents and in Biological Fluids

6.1 Crystal and solution conformation

As described under Section 2.2 the cyclosporine conformation observed in the crystal [9] is a rather rigid conformation containing a β-fragment (residues 11 – 7) and a loop (residues 8 – 11). As a consequence of this rather rigid conformation of the cyclosporine skeleton, six amino acids have their carbon chains oriented quasi perpendicular to the 'plane' of the peptide ring. They are Abu[2], Val[5], Ala[7], and MeVal[11], projecting up and MeBmt[1] and MeLeu[6] projecting down as shown in Figure 3. The remaining amino acid residues have their carbon chain more or less directed into the 'plane' of the peptide ring. In solution (Figure 14) in apolar solvents ($CDCl_3$, C_6D_6, CD_2Cl_2) the backbone conformation of cyclosporine in the most stable arrangement is very similar to the X-ray structure [9]. A β-pleated sheet involving a β-II′-bend about Abu-2 to Val-5 is built up via 3 H-bridges. As in the crystal, there is a single peptide bond in the cis conformation between MeLeu-9 and MeLeu-10. The main difference between crystal and solution conformations

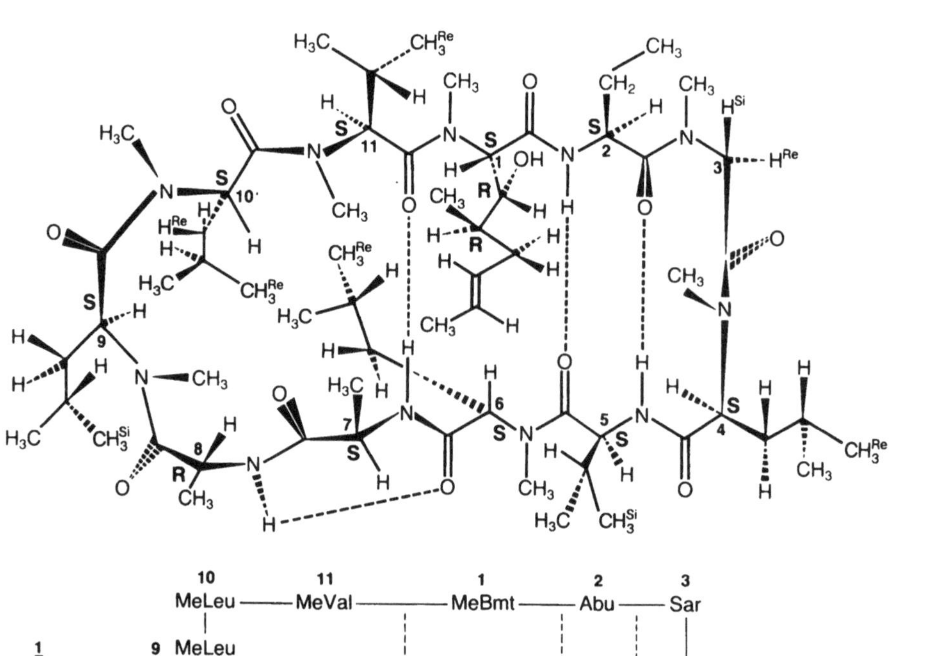

Figure 14 Solution structure of cyclosporine in nonpolar solvents (schematic representation). As compared to the crystal conformation the carbon chain of MeBmt has slightly rotated around the $C_\alpha - C_\beta$-bond and the OH-group points up towards the carbonyl group of MeBmt.

lies in the orientation of the carbon chain of MeBmt-1. Its carbon chain has rotated around the C_α-C_β-sigma bond by about 120° as compared to the conformation in the crystal. This rotation allows the formation of an intramolecular H-bond between the β-hydroxy group and the carbonyl oxygen atom of MeBmt itself replacing the intermolecular H-bond found in the crystal. The carbon chain of MeLeu-10 has equally rotated by 120°, this change not effecting much the position of this residue in comparison to the crystal conformation (Figure 3).

6.2 Cyclosporine conformation in biological fluids

In the presence of water it is unclear to what extent the conformation of cyclosporine observed in the crystalline state (Figure 3) or the conformation obtained by NMR measurements in solution in aprotic solvents (Figure 14) are relevant for biological activity. Results obtained with monoclonal antibodies and by extensive molecular dynamics calculations, as described below suggest that in aqueous environment and possibly on the biological receptor the conformation of cyclosporine may resemble more the one found in the crystalline state.

6.3 Immunochemical study of the cyclosporine conformation in aqueous medium

The rigid structure of the cyclosporine molecule provides a conformationally highly restricted antigen. Cyclosporine **1** and its derivatives with their rigid ring structures make it possible to study subtle structural parameters that govern the antigen-antibody interactions. Cyclosporine coupled to a protein carrier was used to raise polyclonal [40] and monoclonal [13, 41, 42] antibodies against cyclosporine. The fine specificity of the monoclonal antibodies for cyclosporine was determined by studying the crossreactivities of the antibodies for different cyclosporine analogues showing preserved skeletal conformation. When the capacity of a cyclosporine derivative to inhibit the reaction between cyclosporine and antibody was the same as that of unmodified cyclosporine it was concluded that the modified residue was not involved in antibody binding. Conversely, when the inhibiting capacity of the derivative was decreased, it was assumed that the unmodified amino acid was a contact residue involved in the epitope recognized by the monoclonal antibody.

6.4 Immunological studies of cyclosporine and their interpretation in terms of the conformation of the compound in aqueous media

Using some especially designed derivatives of **1** it was possible to attach the molecule at two different positions to protein carriers for immunisation and to prepare a large number of monoclonal antibodies recognizing different epitopes on the surface of 1 [42]. Immunisation with D-Lys^{8}-cyclosporine coupled via a succinate spacer to ovalbumin led mainly to antibodies recognizing a contiguous surface including amino acid residues 1 to 6 and 10 and 11 (Figure 15a) whereas antibodies resulting from immunisation with succinate coupled to Thr^{2}-cyclosporine recognized mainly residues 6, 8, and 9 (Figure 15b). The ability of these antibodies to discriminate

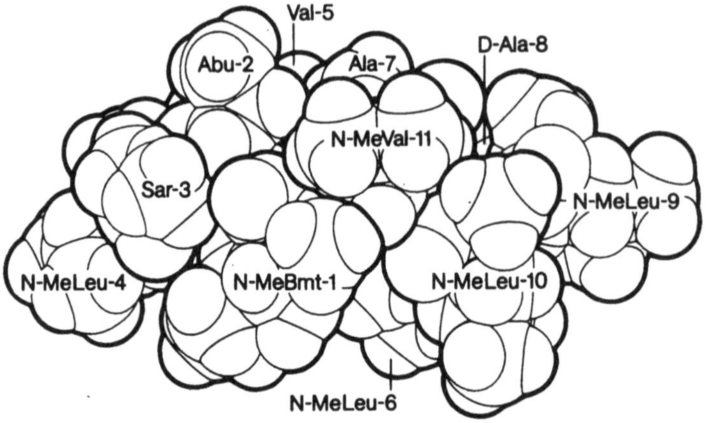

Figure 15a "Active face" of cyclosporine mainly showing amino acids 3,2,1,11 and 10 projecting to the observer. This region is part of the surface of the molecule in the crystal conformation.

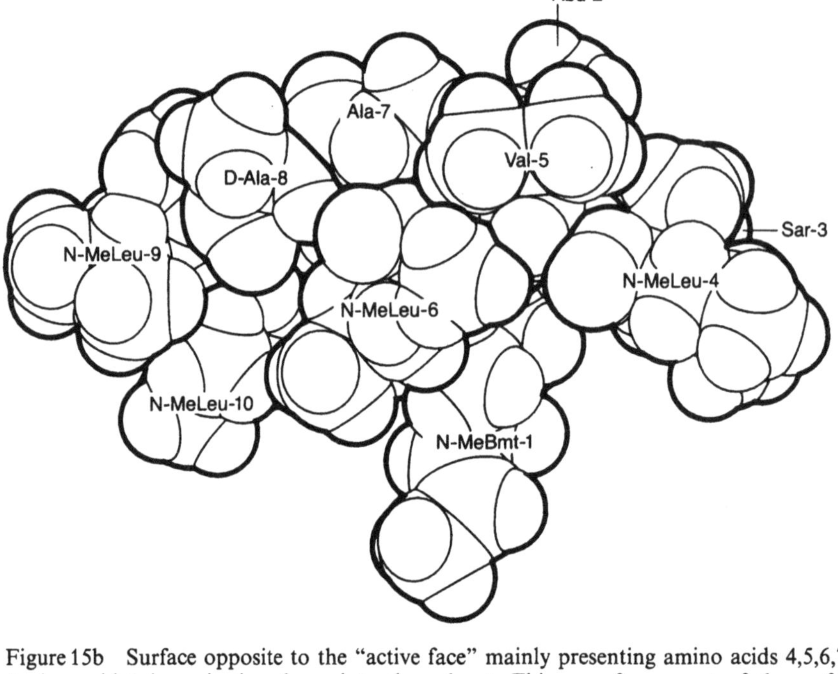

Figure 15b Surface opposite to the "active face" mainly presenting amino acids 4,5,6,7,8 and 9. Amino acid 1 is projecting down into the solvent. This area forms part of the surface of the molecule in the conformation in nonpolar solvents.

between structural changes along the length of the MeBmt carbon chain in a series of derivatives was then examined [43, 44]. It was found that monoclonal antibodies recognizing amino acid residues 6, 8 and 9 also recognized the terminal carbon atoms of the MeBmt chain but not the carbon atoms 3 and 4. By contrast, those monoclonal antibodies recognizing residues 1, 2, 3 and 10 and 11 did distinguish between structural modifications on the carbon atoms 2 and 3 but not on the terminal two carbon atoms (as seen e. g. with metabolite 17 110 [45]). These findings suggest that when 1 is in aqueous solution the terminal carbon atoms of MeBmt form part of a contiguous surface including amino acids 6, 8, 9. The conformation of cyclosporine as determined by X-ray crystallography fulfills these conditions but not that determined by NMR in aprotic solvents (Figure 15b). It would therefore appear that in the presence of water, the lipohilic carbon chain of MeBmt[1], unlike in aprotic solvents, folds back into the ply of the β-pleated sheet so that the molecule forms a globular, compact form similar to that found in the crystal.

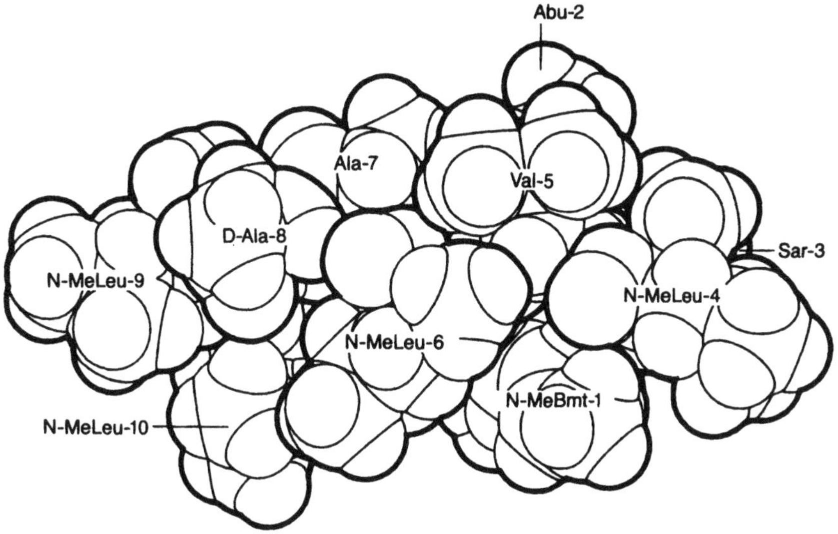

Figure 15c Expected solution conformation of cyclosporine in aqueous environment. The part of the molecule seen corresponds to Figure 15b. This surface is part of the molecule in the crystal conformation.

6.5 Calculations using molecular dynamics programs indicate that the crystal structure conformation is in water more stable than any others

Extensive molecular dynamics calculations (MD) have been done using the MD programs GROMOS (University of Groningen) and DISCOVER (Biosym Technologies, Inc., San Diego) [46]. MD simulations of the cyclosporine crystal structure and dynamic modelling of the solution structure based on NMR have shown that in water the crystal structure conformation is the most stable one. If the solution conformation based on NMR is put into a hypothetical water box the cyclosporine molecule instantly adopts the conformation observed in the crystal. In other words the crystal structure conformation of cyclosporine is stable and the most probable one in aqueous environment and should be a good model for interpreting relationships between biological activity and structure.

7. Biosynthesis

Kobel et al. [47] could demonstrate by feeding ^{3}H and ^{13}C labelled precursors that the 7 *N*-methyl groups and the methyl group at C_4 of MeBmt originate from methionine and that the linear carbon chain of MeBmt is formed by head-to-tail coupling of 4 acetate units. The formation of the macrocycle is effected by a large multienzyme complex as shown recently by Zocher et al. [48] and represents another example of the nonribosomal mechanism widely used by microorganisms to synthesize cyclic peptides and depsipeptides. This mechanism bears close resemblance to the biosynthesis of fatty acids and has been elucidated by the group of Lipmann [49]. Amino acids are activated as thiol esters in a two-step reaction (Figure 16). First the amino acid is transformed into an aminoacyl adenylate by reaction with ATP; this step also occurs in the ribosomal mechanism. Whereas in the latter the second step consists in a transfer of the aminoacyl group to a reactive hydroxyl at the *O*-terminal adenosine of a specific tRNA, in the non-ribosomal mechanism

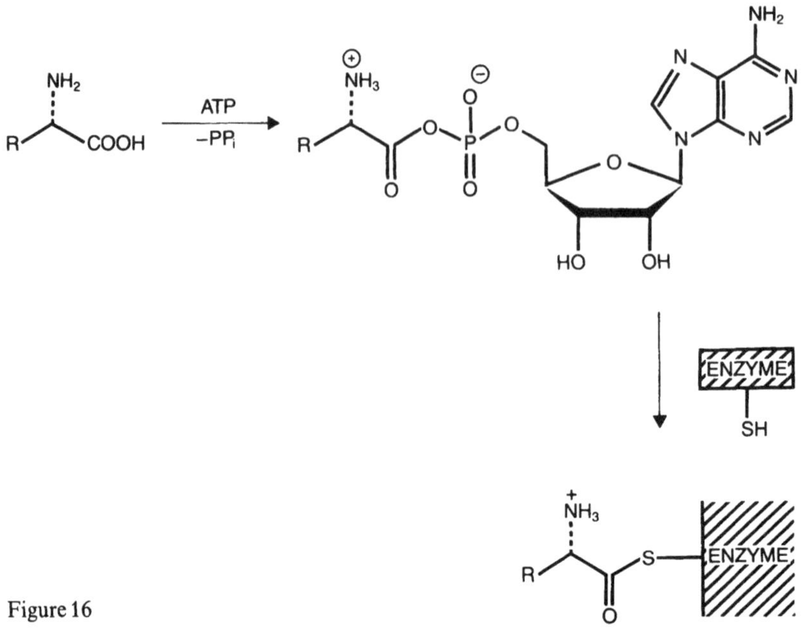

Figure 16

aminoacyl transfer to an enzymebound thiol occurs. Thiol esters are much more powerful acylating agents than normal esters and non-ribosomal peptide synthesis represents a beautiful example of how nature makes use of this difference in reactivity. This has also been recognized for a long time by chemists in the early days of peptide synthesis [50]. In the nonribosomal mechanism the sequence of the peptide is guided by a protein template laid out on a polyenzyme system (Figure 17). This situation contrasts sharply to the ribosomal mechanism where the genetic information is transferred by codon-anticodon interaction on the mRNA template (Figure 18). The ribosomal mechanism ensures maximal accuracy of transmission of information to produce specific sequences of long polypeptide chains with high reproducibility. Substitution of amino acids at specific sites in the sequence requires appropriate changes at the level of the gene. This is not the case with protein templates where the limit may not exceed sequences of perhaps twenty amino acids [49] and where amino acids with similar physicochemical and steric characteristics

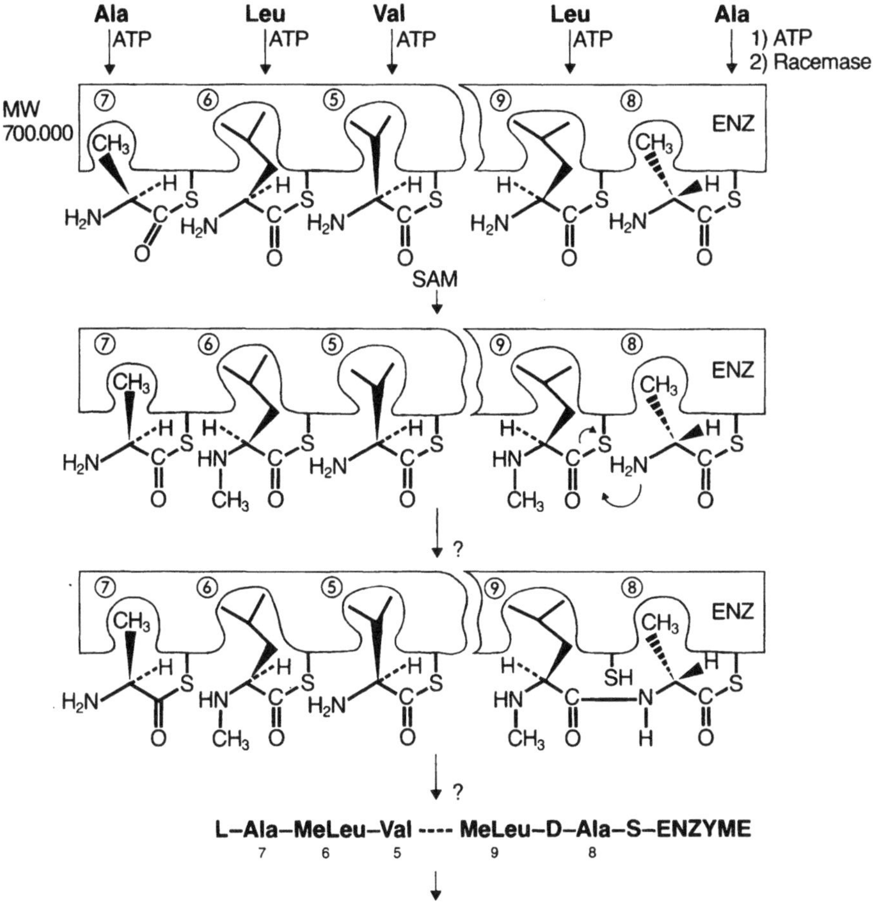

Figure 17 **Cyclosporine**

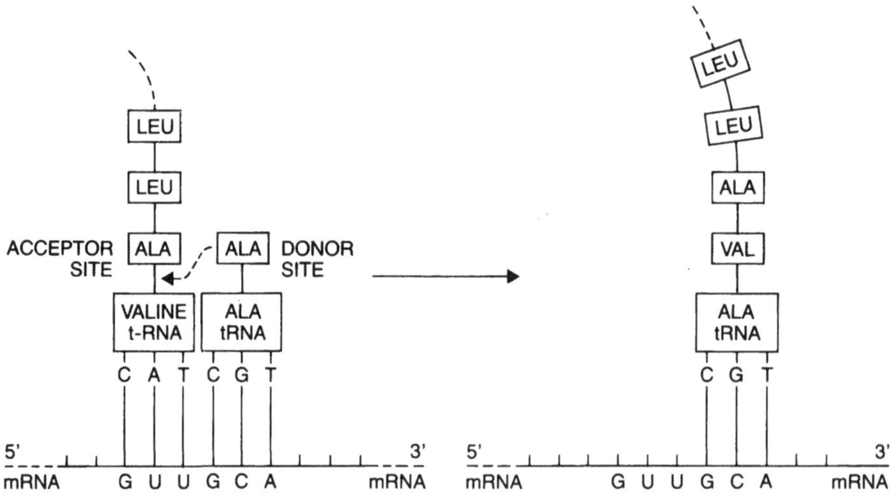

Figure 18

can replace each other. Phenylalanine can substitute for tyrosine in a cell-free system of tyrocidin synthase. In some cases it has been possible to separate the activating activities for individual amino acids from the multienzyme complex [49]. *N*-methylation in the biosynthesis of cyclosporine occurs at the level of thioester-linked amino acids [48]. Recently, Billich and Zocher [51] reported the isolation of an enzyme from *Tolypocladium inflatum* capable to synthesize cyclosporine from the constituent amino acids with concomitant consumption of ATP and *S*-adenosylmethionine (SAM). This contrasts with previous findings of the same group [48] where an enzyme preparation (molecular weight approx. 700,000) from a different strain had been shown to activate the individual amino acids of cyclosporine and to perform the *N*-methylation but de novo synthesis of **1** with the cell-free preparation had not been observed; the only isolable product had been the diketopiperazine cyclo-(D-Ala · MeLeu). The reason for this discrepancy is still not clear. However, using this enzyme preparation Billich and Zocher were able to synthesize various cyclosporines in vitro [51].

8. Mechanism of Action of Cyclosporine

The first observation that cyclosporine suppresses certain functions in the immune system (cell mediated immune responses and antibody formation) was reported in 1976 by Borel [52]. The compound acts reversibly and selectively on lymphocytes, mainly T cells and it affects the induction phase rather than cell proliferation. A discussion of current concepts of activation of lymphocytes in context with a hypothetical model of action of cyclosporine at the cellular level together with general mechanisms of cell activation and signal transduction is given in ref. 1.

8.1 Effects on T cells

Mature T cells can be distinguished into two phenotypes: those that express the surface glycoprotein CD4 and those that express the CD8 molecule. The CD4 molecule is expressed primarily on cells with helper/inducer function whereas CD8 ist found on cytotoxic and suppressor cells [53–55]. A stringent association exists between these T cell subsets and the major histocompatibility complex (MHC) gene products recognized by the T cell receptors (TCR). $CD8^+$ cells recognize antigen associated with class I MHC gene products whereas $CD4^+$ cells interact with class II MHC bearing targets. Interaction of the TCR with the antigen:MHC complex displayed on the surface of accessory cells (antigen-presenting cells (APC)) such as macrophages, dendritic cells or B cells forms a ternary complex which in turn represents one necessary requirement for cell activation [57–59]. Normal T cells, however, require at least two signals for cell activation, IL-1 being secreted from the accessory cell usually regarded as the second signal. The mechanism by which these two signals are transmitted into the interior of the cells is not fully understood. In some instances, IL-1 can be replaced by phorbol esters suggesting a role for protein kinase C in this event [60].

The cell interaction described above triggers a series of events on both cell types. Both macrophages and T cells newly express and amplify the expression of existing surface structures and release soluble mediators, so called lymphokines. Of prime importance in the context of this chapter is the expression of the receptor for the T cell growth factor (interleukin 2, IL-2), the IL-2 receptor on the surface of T cells and the production of IL-2. The expression of the receptor occurs early in the activation (after 6–10 hours) and is associated with transition from G_0 into the G_1 phase of the cell cycle. This transition and the concomitant IL-2 receptor expression can be triggered by phorbol esters alone in normal T cell populations [63]. In contrast, phorbol esters alone are generally not sufficient (with the exception of certain tumor cell lines) to induce production of IL-2, indicating a more stringent signal requirement for this event [60]. The expression of the IL-2 receptor has been called stage 1 of T cell activation whereas IL-2 production is associated with stage 2. For stage 2 the concomitant action of IL-1 seems to be necessary. The interaction of IL-2 with its receptor then initiates stage 3, T cell growth. The signal transduction of the interaction of IL-2 with its receptor seems to be mediated by G-proteins [64, 65]. A schematic representation of the events leading to cell activation and proliferation as well as of the G-protein mediated signal transduction mechanism is given in Figures 19 and 20.

As a consequence of activation T cells release a variety of factors affecting other cells such as e. g. macrophages or B cells. These factors include interferons, macrophage activating factors, colony stimulating factors, B cell differentiation factor (BSF-2, identical to β-2-interferon [66]) or also factors affecting bone homeostasis such as osteoclast activating factor [67]. IL-2 will act on any cell which expresses the IL-2 receptor. This means that any interference with T cell activation or IL-2 production will have serious consequences for IL-2 dependent cell-cell interactions. Reports on the effects of cyclosporine on the individual stages of T cell activation

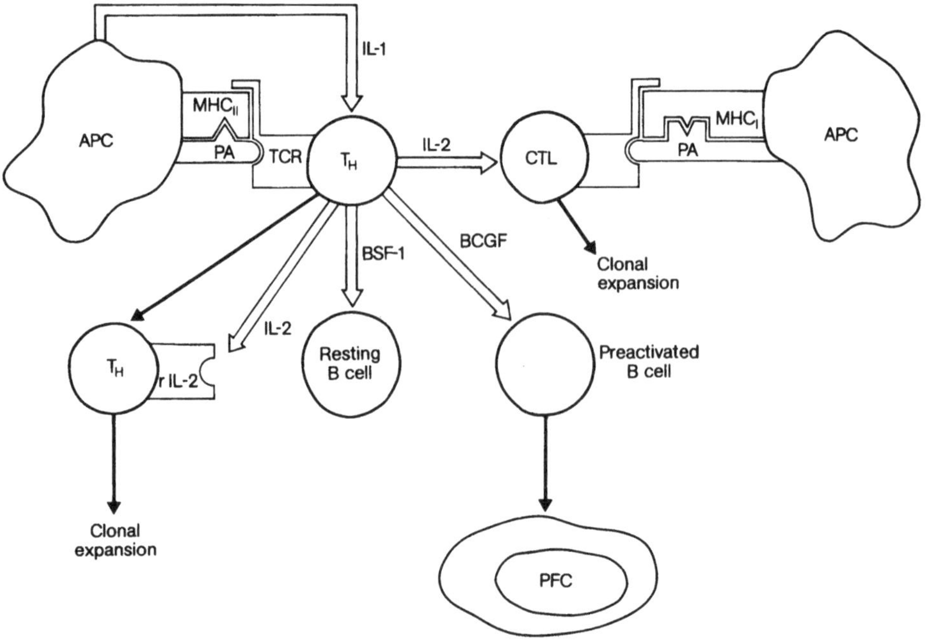

Figure 19 An antigen presenting cell (APC) displays on its surface the complex formed by binding of processed antigen (PA) to an MHC class II (MHC_{II}) molecule. Binding of this complex to the T cell receptor (TCR) represents one necessary signal to activate the T helper cell (T_H). The second signal required is provided by IL-1 release from the APC. The T_h cell responds by expressing the receptor for IL-2 (rIL-2) on its surface and by releasing soluble mediators such as IL-2 and factors affecting B cells. Interaction of IL-2 with the rIL-2 the leads to clonal expansion of the T_H as well as cytotoxic T lymphocytes (CTL) which have received the activation signal provided by binding of the PA/MHC class I complex displayed on the surface of an APC.BSF-1 activates resting B cells and BCGF acts on preactivated B cells which transform into plaque forming cells (PFC).

have been controversial. Borel had shown [68] that cyclosporine could inhibit mitogen-induced T cell proliferation when added after 2 hours of cell culture but not after 48 hours. Once a cell has progressed beyond the early G_1 phase addition of cyclosporine is without effect [69]. Substantial evidence points to inhibition by cyclosporine of IL-2 production [70–74], probably at the level of transcription. The effects of cyclosporine on the expression of the IL-2 receptor are in part conflicting. While some authors [71–73, 75] reported no effects of cyclosporine on the expression of the receptor others [70, 76] have obtained strong evidence that this event is inhibited. Specifically, it seems to be the early activation of the transcription of the genes encoding the IL-2 receptor and some other lymphokines which is sensitive to inhibition by cyclosporine [76]. Since IL-2 plays a key role in the proliferation of T cells, inhibitory concentrations of cyclosporine will prevent clonal expansion of T-helper independent cytotoxic T-lymphocytes (which produce their own IL-2) as well as cytotoxic T cell precursors which depend on IL-2 provided by helper T cells [77–79]. The cytolytic potential of these cells, however, remains unimpaired.

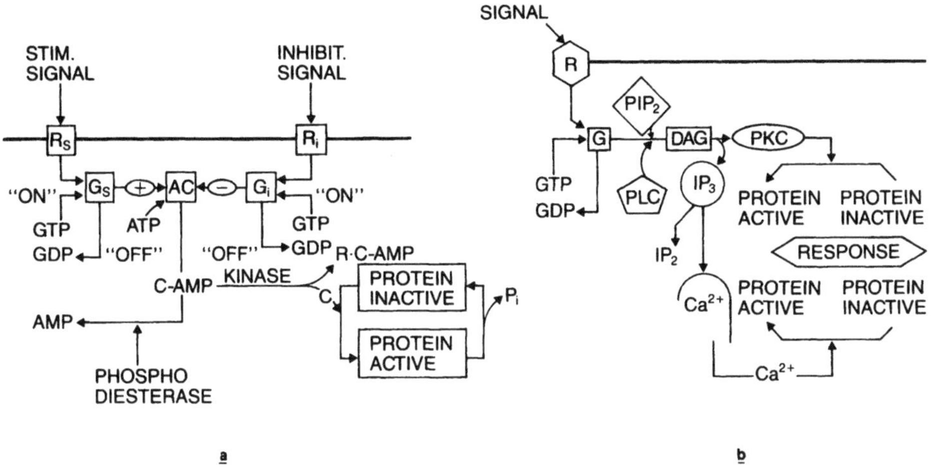

Figure 20a Upon stimulation of a stimulatory receptor (R_s) or an inhibitory receptor (R_i) a conformational change is induced in the receptor which enables it to bind G_s (G_i). This in turn leads to binding of GTP to G_s (G_i) inducing again a conformational change of the G-protein into a structure which enables it to activate/inactivate adenylate cyclase (AC). This constitutes the "on"-reaction. c-AMP then binds to the regulatory subunit of a protein kinase (R.c-AMP) thereby liberating the catalytic subunit C. This in turn leads to activation of an inactive protein and to the cellular response. The GTP which is bound to G_s/G_i gets hydrolyzed to GDP; this event represents the "off"-reaction, because the activation signal for AC is turned off.

Figure 20b Stimulation of the receptor (R) causes the G-protein (G) to release GDP and to bind GTP. This in turn leads to activation of phospholipase C; the mechanism is similar to the c-AMP pathway depicted in Figure 20a. Upon activation of phospholipase C (PLC) the signalling pathway bifurcates: hydrolysis of phosphatidylinositol-4,5-diphosphate (PIP_2) generates two second messengers, diacyl glycerol (DAG) and inositol triphosphate (IP_3). DAG activates protein kinase C (PKC) which in turn converts an inactive protein into an active by phosphorylating it. IP_3 causes release of Ca^{2+} from intracellular membranous caverns causing an increase of intracellular Ca^{2+}. This leads to activation of a Ca^{2+}-dependent protein kinase which in turn converts an inactive protein into the active phosphorylated form.

8.2 Effects on B cells

Cyclosporine can affect B lymphocyte function by two ways, either by directly acting on the B cell or indirectly by inhibiting lymphokine production by T cells. As had been shown initially by Borel [68] cyclosporine strongly reduced the formation of antibodies in mice. Similarly, in guinea pigs a single dose of the drug can suppress the production of antibodies to hapten and carrier in primary and secondary responses [80]. Polyclonal activation using the B cell mitogen LPS (lipopolysaccharide) or mitogenic stimulation by phorbol esters was, however not inhibited. Interestingly, these activators have in common that they do not mobilize intracellular calcium [81–83]. There is substantial evidence that B cell activation, proliferation and differentiation is also controlled by T cell-derived lymphokines [84–88]. Various factors have been proposed to act on B cells and their interrelationship is still uncertain. Two different types of T cell derived factors have been described: B cell stimulatory factor 1 (BSF-1) acts on resting B cells and drives them to a transitional activated state characterized by the expression of high levels of Ia antigens [88, 89]. In combination with antibodies to surface Ig receptors BSF-1 stimulates the cell to synthesize DNA [87, 90–92]. The second type of factor (BCGF-2) has been detected on the basis of its activity to induce proliferation of the murine T cell lymphoma BCL1 in vitro [93, 94]. This factor also induces differentiation in eosinophils and has later been named interleukin-4 [95, 96]. This lymphokine does not affect resting B cells but rather induces DNA synthesis and Ig secretion in naturally occurring large B cells and also stimulates proliferation of B cells which have been preactivated by anti-Ig-antibodies [97]. It has recently been shown by the group of Klaus [98–100] that activation of resting B cells by BSF-1 is resistant to cyclosporine whereas activation by anti-Ig-antibodies alone or in combination with BSF-1 is drug susceptible. The stimulation of large (preactivated) B cells by IL-4 to synthesize DNA and to secrete immunoglobulins is inhibited by low concentrations of cyclosporine.

8.3 Possible intracellular targets for cyclosporine

Various hypotheses have been put forward regarding the precise mechanism by which cyclosporine inhibits cell activation. Colombani and Hess [101] suggested calmodulin as a candidate for inhibition by the drug. This hypothesis has been questioned by LeGrue et al. [102] as well as by Handschumacher [103] who had previously isolated a cytosolic protein from human spleen and bovine thymus cytosol and named it cyclophilin [104, 105]. Cyclophilin binds cyclosporine in a 1 : 1 ratio with a binding constant of approx. 10^{-7} M. It has a molecular weight of 17,737 (163 amino acid residues) and no apparent homology with known protein sequences. Major (pI 9.1) and minor (pI 7.4) isoforms have been identified without, however, providing the structural basis for this difference. The affinity of a series of cyclosporine derivatives to cyclophilin shows some correlation with their immunosuppressive activity [104, 121, 122]. Cyclophilin occurs abundantly in many cells other than T cells or B cells and it is presently not known whether it has any function in cell activation. It was recently suggested by Harding [106] that cyclophilin has serine/threonine protein kinase activity and that this activity can be inhibited by cyclosporine but not by a non-immunosuppressive analog, $MeLeu^{11}$-cyclosporine. Whether cyclophilin has any relevance for the mechanism of action of cyclosporine is presently uncertain but will very likely be clarified in the near future.

9. Conclusions and Outlook

The present view of the mechanism of immunosuppressive activity of cyclosporine is consistant with the abrogation of production of interleukine 2 and other lymphokines. The drug-sensitive event seems to be the transcription of the respective gene. Two DNA-sequence elements are primarily involved in eukaryotic gene transcription: promoters and enhancers. These DNA-sequences are required for efficient and accurate transcription and consist usually of an AT-rich region. Promoter sequences are located at a close distance upstream of the sequence to be transcribed whereas enhancers can act over long distances. Promoters and enhancers both contain DNA-elements which specifically interact with proteins, so called trans-acting factors (a recent discussion of this concept in the context of regulation of class II MHC gene expression is found in ref. 108. For a general review the reader is referred to an article by Yamamoto [107]. At the present time few of the proteins regulating gene transcription are understood in detail. On the basis of their activating or inhibiting activities on gene transcription one can distinguish between positive and negative trans-acting factors. The best understood positive trans-acting factors include the glucocorticoid [109, 110], estrogen [111, 112] and progesterone [113, 114] receptors. These proteins have recently been cloned and provide an opportunity to investigate at the molecular level how transcription is regulated in higher eukaryotes. Of negative regulation even less is understood. Glucocorticoids have been reported to inhibit (among others, such as e. g. the gene of prolactin [115] or proopiomelanocortin [116]) the transcription of the TNF-*a*/cachectin gene [117] on one hand but also to activate transcription of the gene coding for macrocortine/lipomoduline [118] and represent an example for a dual role of a single hormone. Another example of negative regulation is represented by cyclosporine. What is known with certainty is that the compound enters the cell and represses the transcription of the gene coding for IL-2, gamma-interferon and other lymphokines. Whether this inhibition is effected by the cyclosporine-cyclophilin complex is presently not known. However, several aspects contradict this assumption; compared to positive or negative transcription factors cyclophilin has a wide tissue distribution and is a relatively abundant protein. Furthermore, its structure lacks some of the typical characteristics of DNA-binding proteins (e. g. the 'zinc-finger', [119]).

The highly specific structure activity relationships of cyclosporine derivatives suggest a unique binding site for the drug. Cyclosporine might inhibit the binding of transactivator protein(s) to DNA by either forming a complex with the activator which is unable to bind to the promoter/enhancer sequence on the DNA or by inactivating this protein. Other mechanisms are conceivable ultimately resulting in inhibition of gene transcription. Undoubtedly future will bring to light novel insights concerning the precise mechanisms underlying the regulation of gene transcription and how cyclosporine exerts its activities. The compound continues to fascinate scientists of various disciplines due to the various intriguing biological activities. On the other hand the availability of this drug has opened new prospects in the field of organ transplantation. Only recently the potential of this compound in the treatment of various autoimmune diseases has been widely accepted and been the subject of a recent symposium [120]. It will be very interesting to follow the future developments triggered by the discovery of the cyclosporines in the various disciplines of biochemistry as well in the therapy of human diseases.

References

1. Wenger, R. M., Payne, T. G., Schreier, M., Cyclosporine: Chemistry, structure-activity relationships and mode of action, *Progress in Clinical Biochemistry*, Vol. 3, Springer-Verlag, Berlin, Heidelberg 1986, p. 157.
2. Ruegger, A., Kuhn, M., Lichti, H., Loosli, H. R., Huguenin, R., Quiquerez, C., and von Wartburg, A., Cyclosporin A, ein immunsuppressiver Pilzmetabolit aus Trichoderma polysporum (Link ex Pers.) Rifai, *Helv. Chim. Acta 59*, 1075, 1976.
3. For the nomenclature of cyclosporins see: a) Borel, J. F., Editorial: Ciclosporin and its future, in *Progr. Allergy 38*, 9, 1986. b) Wenger, R. M., Synthesis of ciclosporin and analogues: Structural and conformational requirements for immunosuppressive activity, in *Progr. Allergy 38*, 46, 1986.
4. Dreyfuss, M., Haerri, E., Hofmann, H., Kobel, H., and Tscherter, H., Cyclosporine A and C: New metabolites from Trichoderma polysporum (Link ex Pers.) Rifai, *Eur. J. Appl. Microbiol. 3*, 1072, 1976.
5. Borel, J. F., Editorial: Ciclosporin and its future, *Progr. Allergy 38*, 9, 1986. See also Staehelin, H., Historical background, *Progr. Allergy 38*, 19, 1986.
6. Petcher, T. J., Weber, H. P., Ruegger, A., Crystal and molecular structure of an iodo-derivative of the cyclic undecapeptide cyclosporin A, *Helv. Chim. Acta 59*, 1480, 1976.
7. Beyermann, H. C., Maat, L., Sinnemar, A., von Veen, Conversions of sarcosine and pipecolic acid with phenyl and methyl isothiocyanate, *Rec. Trav. chim. Pays-Bas 87*, 11, 1968.
8. Kessler, H., Loosli, H. R., Oschkinat, H., Assignment of the ^{1}H-, ^{13}C- and ^{15}N-NMR spectra of cyclosporin A in $CDCl_3$ and C_6D_6 by a combination of homo- and heteronuclear two-dimensional techniques, *Helv. Chim. Acta 68*, 661, 1985.
9. Loosli, H. R., Kessler, H., Oschkinat, H., Weber, H. P., Petcher, T. J., Widmer, A., The conformation of cyclosporin A in the crystal and in solution, *Helv. Chim. Acta 68*, 682, 1987.
10. a) Chou, P. Y., Fasman, G. D., β-Turns in proteins, *J. Molec. Biol. 115*, 135, 1977. b) Smith, J. A., Pease, L. G., Reverse turns in peptides and proteins, *CRC Crit. Revs. Biochem. 8*, 315, 1980.
11. IUPAC-IUB Joint Commission on Biochemical Nomenclature. Nomenclature and symbolism for amino acids and peptides, recommendations 1983. *Eur. J. Biochem. 138*, 9, 1984.
12. Von Wartburg, A., Traber, R., Chemistry of the natural cyclosporin metabolites, *Progr. Allergy 38*, 28, 1986.
13. a) Quesniaux, V. F. J., Tees, R., Schreier, M. H., Wenger, R. M., Donatsch, P., Van Regenmortel, M. H. V., Monoclonal antibodies to cyclosporin, *Progr. Allergy 38*, 108, 1986. b) Quesniaux, V. F. J., The use of monoclonal antibodies to probe the surface of cyclosporines, *Transplantation Proceedings 18*, Suppl. 5, 111, 1986.
14. Wenger, R. M., Synthesis of cyclosporine, *Helv. Chim. Acta 67*, 502, 1984.
15. Evans, D. E., and Weber, A. E., Asymmetric glycine enolate aldol reactions: Synthesis of cyclosporine's unusual amino acid, MeBmt. *J. Amer. Chem. Soc. 108*, 6757, 1986.
16. a) Seebach, D., Juaristi, E., Miller, D. D., Schickli, C., and Weber, T., Addition of chiral glycine, methionine, and vinylglycine enolate derivatives to aldehydes and ketones in the preparation of enantiomerically pure α-amino-β-hydroxy acids, *Helv. Chim. Acta 70*, 237, 1987. b) Seebach, D., personal communication to R. W., 1987.
17. Tung, R. D., and Rich, D. H., Total synthesis of the unusual cyclosporine amino acid MeBmt, *Tetrahedron Lett. 28*, 1139, 1987.
18. Aebi, J. D., Dhaon, M. K., and Rich, D. H., A Short synthesis of enantiomerically pure (2S,3R,4R,6E)-3-hydroxy-4-methyl-2-(methylamino)-6-octenoic acid, the unusual C-9 amino acid found in the immunosuppressive peptide cyclosporine, *J. Org. Chem. 52*, 2881, 1987.
19. Schmidt, U., and Siegel, W., Synthesis of (4R)-4-(E)-2-butenyl-4, N-dimethyl-L-threonine (MeBmt), the characteristic amino acid of cyclosporine, *Tetrahedron Lett. 28*, 2849, 1987.
20. Deyo, D., Tung, R. D., Rich, D. H., Enantioselective syntheses of β-hydroxy, α-amino acids for cyclosporine analogs, *Amer. Chem. Soc. Meeting,* New Orleans, Louisiana, Aug. 30, 1987, Abstract No. 104.
21. Rich, D. H., Dhaon, M. K., Dunlap, B., and Miller, S. P. F., Synthesis and antimitogenic activities of four analogues of cyclosporin A modified in the 1-position, *J. Med. Chem. 29*, 978, 1986.

22. Wenger, R. M., Synthesis of enantiomerically pure (2S,3R,4R,6E)-3-hydroxy-4-methyl-2-methylamino-6-octenoic acid starting from tartaric acid, *Helv. Chim. Acta 66*, 2308, 1983.
23. **48** is obtained as a single diastereomer with trans orientation of the substituents on the oxazolidinone ring. The compound is, however a mixture of Z- and E-isomers with respect to the imino group. Rich[17] obtained this compound as a 8:1 trans/cis mixture and Schmidt[19] has also questioned Wenger's result. It has, however been repeatedly confirmed by Wenger, that careful use of the published procedure gives only the transoxazolidinone.
24. Evans, D. A., Sjogren, E. B., Weber, A. E., Conn, R. E., Asymmetric synthesis of anti-β-hydroxy-α-amino acids, *Tetrahedron Lett. 28*, 39, 1987.
25. a) Wenger, R. M., Synthesis of cyclosporine and analogues: Structural requirements for immunosuppressive activity, *Angew. Chem. Int. Ed. Engl. 24*, 77, 1985. b) Wenger, R. M., Cyclosporine and analogues: Structural requirements for immunosuppressive activity, *Transplantation Proc. 28*, Suppl. 5, 213, 1986.
26. Brady, S. F., Varga, S. L., Freidinger, R. M., Schwenk, D. A., Mendlowski, M., Holly, F. W., Veber, D. F., Practical synthesis of cyclic peptides, with an example of dependence of cyclization yield upon linear sequence. *J. Org. Chem. 44*, 3101, 1979, and references cited therein.
27. McDermott, J. R., Benoiton, N. L., N-Methylamino acids in peptide synthesis. IV. Racemization and yields in peptide-bond formation. *Can. J. Chem. 51*, 2562, 1973. For a detailed discussion of the problems concerning the synthesis of N-methyl peptides see Wenger, R. M., Synthesis of cyclosporine, Part II: Synthesis of Boc-D-Ala-MeLeu-MeLeu-MeVal-OH, a part of the peptide sequence of cyclosporine, by different strategic ways and synthesis of its isomers Boc-D-Ala-MeLeu-D-MeLeu-MeVal-OH, Boc-D-Ala-MeLeu-D-MeLeu-D-MeVal-OH, and Boc-D-Ala-MeLeu-MeLeu-D-MeVal-OH as reference compounds, *Helv. Chim. Acta 66*, 2672, 1983.
28. Gund, P., and Veber, D. F., On the ease of base-catalyzed epimerization of N-methylated peptides and diketopiperazines, *J. Amer. Chem. Soc. 101*, 1885, 1979.
29. a) Goodman, M., and Fried, M., Conformational aspects of polypeptide Structure. XX. Helical poly-N-methyl-L-alanine. Experimental results. *J. Amer. Chem. Soc. 89*, 1264, 1967. b) Mark, E., and Goodman, M., Conformational aspects of polypeptide structure. XXI. Helical poly-N-methyl-L-alanine. *J. Amer. Chem. Soc. 89*, 1267, 1967.
30. a) Tonelli, A. E., On the stability of cis and trans amide bond conformations in polypeptides. *J. Amer. Chem. Soc. 93*, 7153, 1971. b) LaPlanche, L. A., and Rogers, M. T., Cis and trans configurations of the peptide bond in N-monosubstituted amides by nuclear magnetic resonance. *J. Amer. Chem. Soc. 86*, 337, 1964.
31. DeFrees, D. J., Bartmess, J. E., Kim, J. K., McIver, Jr., R. T., and Hehre, W. J., In support of anionic hyperconjugation. The versatile methyl group. *J. Amer. Chem. Soc. 99*, 6451, 1977.
32. Zaoral, M., Amino acids and peptides. Pivaloyl chloride as a reagent in the mixed anhydride synthesis of peptides, *Collect. Czech. Chem. Commun. 27*, 1273, 1962.
33. Tung, R. D., Dhaon, M. K., and Rich, D. H., BOP-Cl Mediated synthesis of the cyclosporine A 8–11 tetrapeptide fragment, *J. Org. Chem. 51*, 3350, 1986.
34. Castro, B., Dormoy, J. R., Elvin, J. G., Selve, C., Reactifs de couplage peptidique IV. (1) – L'hexafluorophosphate de benzotriazolyl N-oxytrisdimethylamino phosphonium (B. O. P.). *Tetrahedron Lett. 14*, 1219, 1975.
35. Wissmann, H., Kleiner, H. J., Neue Peptidsynthese, *Angew. Chem. 92*, 129, 1980.
36. Kovacs, J., *The Peptides*, Vol. 2, Gross, E., Meienhofer, J., eds., Academic Press, New York – London, 1980, p. 485.
37. Wenger, R. M., Cyclosporine and analogues – isolation and synthesis – mechanism of action and structural requirements for pharmacological activity. *Fortsch. Chem. Org. Naturst. 50*, 123, 1986.
38. Seebach, D., Murtiashaw, W., Naef, R., Shoda, S. I., Krieger, M., Bollinger, P., Leutwiler, A., Wenger, R. M., Aktive Cyclosporinderivate durch C-Alkylierung unter Ersatz von H^{re} der Sarkosin-Einheit. Erzeugung polylithiierter Peptide. *Swiss Chemical Society Meeting*, Berne, Oct. 18, 1985.
39. Wenger, R. M., Synthesis of cyclosporine and analogues: Structure, activity, relationships of new cyclosporine derivatives, *Transplantation Proceedings 15*, Suppl. 1, 2230, 1983.

40. Donatsch, P., Abisch, E., Homberger, M., Traber, R., Trap, R., and Voges, R., A radioimmunoassay to measure cyclosporin A in plasma and serum samples, *J. Immunoassay 2*, 19, 1981.
41. Quesniaux, V. F. J., Himmelspach, K., and Van Regenmortel, M. H. V., An enzyme immunoassay for the screening of monoclonal antibodies to cyclosporin, *Immunol. Lett. 9*, 99, 1985.
42. Quesniaux, V. F. J., Tees, R., Schreier, M. H., Wenger, R. M., and Van Regenmortel, M. H. V., Fine specificity and cross-reactivity of monoclonal antibodies to cyclosporine. *Mol. Immunol. 24*, 1159, 1987.
43. Quesniaux, V. F. J., Wenger, R. M., Schreier, M. H., and Van Regenmortel, M. H. V., Immunochemical study of cyclosporine conformation in aqueous medium. *Protides of the Biological Fluids*, Peeters, ed., Pergamon Press, Oxford, Vol. 35, p. 507, 1987.
44. Quesniaux, V. F. J., Wenger, R. M., Schmitter, D., Van Regenmortel, M. H. V., A study of the conformation of cyclosporine in aqueous medium by means of monoclonal antibodies, *Int. J. Pept. Prot. Res. 29*, 1987, in press.
45. Maurer, G., Loosli, H. R., Schreier, E., Keller, B., Disposition of cyclosporine in several animal species and man. I. Structural elucidation of its metabolites. *Drug Metabolism and Disposition 12*, 120, 1984.
46. Weber, H. P., Lautz, J., and van Gunsteren, W., communication at the *XXXVth Annual Colloquium: Protides of the Biological Fluids*, Brussels, Belgium, April 27–29, 1987.
47. Kobel, H., Loosli, H. R., and Voges, R., Contribution to knowledge of the biosynthesis of cyclosporin A, *Experientia 39*, 873, 1983.
48. Zocher, R., Nihira, T., Paul, E., Madry, N., Peeters, H., Kleinkauf, H., and Keller, U., Biosynthesis of cyclosporin A: Partial purification and properties of a multifunctional enzyme from Tolypocladium inflatum, *Biochemistry 25*, 550, 1986.
49. Lipmann, F., Attempts to map a process evolution of peptide biosynthesis, *Science 173*, 875, 1971.
50. Wieland, Th., Schafer, W., Bokelmann, E., Über Peptidsynthesen V: Über eine bequeme Darstellungsweise von Acylthiophenolen und ihre Verwendung zu Amid- und Peptid-Synthesen. *J. Liebigs Ann. d. Chem. 573*, 99, 1951.
51. Billich, A., and Zocher, R., Enzymatic synthesis of the immune suppressor cyclosporin A, *J. Biol. Chem.*, 1987, in press.
52. Borel, J. F., Feurer, C., Gubler, H. U., Staehelin, H., Biological effects of cyclosporin A: A new antilymphocytic agent, *Agents Actions 6*, 468, 1976.
53. Reinherz, E., Kung, P., Goldstein, G., and Schlossmann, S., Separation of functional subsets of human T cells by a monoclonal antibody. *Proc. Nat. Acad. Sci. U.S.A. 76*, 4061, 1979.
54. Goldstein, P., Goridis, C., Schmitt-Verhulst, A.-M., Hayot, B., Pierres, A., Van Agthoven, A., Kaufmann, Y., Eshhar, Z., Pierres, M., Lymphoid cell surface interaction structures detected using cytolysis-inhibiting monoclonal antibodies. *Immunol. Rev. 68*, 5, 1982.
55. Engleman, E., Benike, C., Glickman, E., and Evans, R., Antibodies to membrane structures that distinguish suppressor/cytotoxic and helper T lymphocyte subpopulations block the MLR in man, *J. Exp. Med. 153*, 193, 1981.
56. Gay, D., Maddon, P., Sekaly, R., Talle, M. A., Godfrey, M., Long, E., Goldstein, G., Chess, L., Axel, R., Kappler, J., and Marrack, P., Functional interaction between human T cell protein CD4 and the major histocompatibility complex HLA-DR antigen, *Nature 328*, 1987.
57. A recent update of the current concept is given by Schwartz, R. H., Fugue in T-lymphocyte recognition, *Nature 326*, 738, 1987. See also refs. 54, 55.
58. Guillet, J. G., Lai, M. Z., Briner, T. J., Buus, S., Sette, A., Grey, H. M., Smith, J. A., Gefter, M. L., Immunological self, nonself discrimination, *Science 235*, 865, 1987.
59. Buus, S., Sette, A., Colon, S. M., Miles, C., Grey, H. M., The relation between major histocompatibility complex (MHC) restriction and the capacity of Ia to bind immunogenic peptides, *Science 235*, 1353, 1987.
60. Droge, W., Protein kinase C in T cell regulation, *Immunology Today 7*, 340, 1986.
61. Hara, T., and Fu, S. M., Human T cell activation. I. Monocyte-independent activation and proliferation induced by anti-T3 monoclonal antibodies in the presence of tumor promoter 12-O-tetradecanoyl phorbol-13-acetate. *J. Exp. Med. 161*, 641, 1985.

62. Depper, J. M., Leonard, W. J., Kroenke, M., Noguchi, P. D., Cunningham, R. E., Waldmann, T. A., Greene, W. C., Regulation of interleukin 2 receptor expression: effects of phorbol diester, phospholipase C, and reexposure to lectin or antigen. *J. Immunol. 133*, 3054, 1984.
63. Ando, I., Hariri, G., Wallace, D., Tumor promoter phorbol esters induce unresponsiveness to antigen and expression of interleukin 2 receptor on T cells. *Eur. J. Immunol. 15*, 196, 1985.
64. Evans, S. W., Beckner, S., and Farrar, W. L., Stimulation of specific GTP binding and hydrolysis activities in lymphocyte membrane by interleukin-2, *Nature 325*, 166, 1987.
65. For a review on G-proteins see Dohlmann, H. G., Caron, M. G., Lefkowitz, R. J., A family of receptors coupled to guanine nucleotide regulatory proteins, *Biochemistry 26*, 2657, 1987.
66. Sehgal, P. B., May, L. T., Tamm, I., Vilcek, J., Human β-2 interferon and B cell differentiation factor BSF-2 are identical, *Science 235*, 731, 1987.
67. Horowitz, M., Baron, R., Mart, J., Andreoli, M., Vigrery, A., Osteoclast activating factor production by murine spleen cells is sensitive to cyclosporin A and is distinct from IL-2 and Interferon. *Calcif. Tiss. Int. 36*, 462, 1984.
68. Borel, J. F., Feurer, C., Magnee, C., Staehelin, H., Effects of the new antilymphocytic peptide cyclosporin A in animals, *Immunology 32*, 1017, 1977.
69. Koponen, M., Grieder, A., Loor, F., The effects of cyclosporins on the cell cycle of T-lymphoid cell lines. *Exp. Cell Res. 140*, 237, 1982.
70. Palacios, R., Concanavalin A triggers T lymphocytes by directly interacting with their receptors for activation. *J. Immunol. 128*, 337, 1982.
71. Bunjes, D., Hardt, C., Roellinghoff, M., Wagner, H., Cyclosporin A mediates immunosuppression of primary cytotoxic T cell responses by impairing the release of interleukin 1 and interleukin 2. *Eur. J. Immunol. 11*, 657, 1981.
72. Dos Reis, G. A., Shevach, E. M., Effect of Cyclosporin A on T cell function in vitro: The mechanism of suppression of T cell proliferation depends on the nature of the T cell stimulus as well as the differentiation state of the responding T cell. *J. Immunol. 129*, 2360, 1982.
73. Kroenke, M., Leonard, W. J., Depper, J. M., Arya, S. K., Worg-Staal, F., Gallo, R. O., Waldmann, T. A., Greene, W. O., Cyclosporin A inhibits T-cell growth factor gene expression at the level of mRNA transcription. *Proc. Nat. Acad. Sci. U.S.A. 81*, 5214, 1984.
74. Nordmann, R., Andersen, E., Trussardi, R., Mazer, N. A., Kinetics of interleukin 2 mRNA and protein produced in the human T-cell line Jurkat and effect of cyclosporin A, Biochemistry, 28, 1971, 1989.
75. Cammisuli, S., The effect of cyclosporin A on cell interactions within the immune system, in *Cyclosporin A*, White, D. J. G., ed., Elsevier Biomedical Press, Amsterdam, 1982, p. 243.
76. Gauchat, J. F., Khandijan, E. W., and Weil, R., Cyclosporin A prevents induction of the interleukin 2 receptor gene in cultured murine thymocytes, *Proc. Nat. Acad. Sci. U.S.A. 83*, 6430, 1986.
77. Bunjes, D., Hardt, C., Solbach, W., Deusch, K., Rollinghoff, M., Wagner, H., Studies on the mechanism of action of cyclosporin A in the murine and human T cell response in vitro, in *Cyclosporin A*, White, D. J. G., ed., Elsevier Biomedical Press, Amsterdam, 1982, p. 261.
78. Andrus, L., Lafferty, K. J., Inhibition of T-cell activity by cyclosporin A, *Scand. J. Immunol. 15*, 449, 1982.
79. Morris, P. J., Mason, D. W., Hutchinson, I. V., The effect of cyclosporine A on lymphocytes in animal models of tissue transplantation, *Transplant. Proc. 15*, 2287, 1983.
80. Parker, D., Droessler, K., Turk, J. L., Kinetics of the effect of a single dose of cyclosporin-A on antibody and cell mediated immune responses in the guinea pig. *Int. J. Immunopharmacol. 6*, 67, 1984.
81. Bijsterbosch, M. K., and Klaus, G. G. B., Cyclosporine does not inhibit mitogen-induced inositol phospholipid degradation in mouse lymphocytes. *Immunology 56*, 435, 1985.
82. Klaus, G. G. B., Bijsterbosch, M. K., and Holman, M. J., Activation and proliferation signals in mouse B cells. VII. Calcium ionophores are non-mitogenic polyclonal B cell activators. *Immunology 56*, 321, 1985.
83. Kay, J. E., Benzie, R. T. C. R., and Borghetti, A. F., Effect of cyclosporin A on lymphocyte activation by the calcium ionophore A23187. *Immunology 50*, 441, 1983.
84. Muraguchi, L., Butler, J. L., Kehrl, J. H., Falkoff, R. J. M., and Fauci, A. S., Selective suppression of an early step in human B cell activation by cyclosporine A, *J. Exp. Med. 158*, 690, 1983.

85. Howard, M., and Paul, W. E., Regulation of B cell growth and differentiation by soluble factors, *Annu. Rev. Immunol. 1*, 307, 1983.
86. Kishimoto, T., Factors affecting B cell growth and differentiation, *Annu. Rev. Immunol. 3*, 133, 1985.
87. Howard, M., Farrar, J., Hilfiker, M., Johnson, B., Takatsu, K., Hamaoka, T., and Paul, W. E., Identification of a T cell-derived B cell growth factor distinct from interleukin 2. *J. Exp. Med. 155*, 914, 1982.
88. Roehm, N. W., Leibson, H. J., Zlotnik, A., Kappler, J., Marrack, P., and Cambier, J. C., Interleukin-induced increase in Ia expression by normal mouse B cells. *J. Exp. Med. 160*, 679, 1984.
89. Noelle, R. P., Krammer, P. H., Ohara, J., Uhr, J., and Vitetta, J. S., Increased expression of Ia antigens on resting B cells: an additional role for B cell growth factor. *Proc. Nat. Acad. Sci. U.S.A. 81*, 6149, 1984.
90. Yoshizaki, K., Nakagawa, T., Fukunaga, K., Kaieda, T., Maruyama, S., Kishimoto, S., Yamamura, Y., and Kishimoto, T., Characterization of human B cell growth factor [BCGF] from cloned T cells or mitogen stimulated T cells. *J. Immunol. 130*, 1241, 1983.
91. Muraguchi, A., Kasahara, T., Oppenheim, J. J., Fauci, A. S., B cell growth factor and T cell growth factor produced by mitogen-stimulated normal peripheral blood lymphocytes are distinct molecules. *J. Immunol. 129*, 2486, 1982.
92. Maizel, A. L., Morgan, J. W., Mehta, S. R., Kouttab, N. M., Bator, J. M., and Sahasrabuddhe, C. G., Long-term growth of human B cells and their use in a microassay for B cell growth factor. *Proc. Nat. Acad. Sci. U.S.A. 80*, 5047, 1983.
93. Swain, S. L., and Dutton, R. W., Production of a B cell growth promoting activity, (DL)BCGF, from a cloned T cell line and its assay on the BCL1 B cell tumor, *J. Exp. Med. 156*, 1821, 1982.
94. Swain, S. L., Howard, M., Kappler, J., Marrack, P., Watson, J., Booth, R., Wetzel, G. D., and Dutton, R. W., Evidence for two distinct classes of murine B cell growth factors with activities in different functional assays. *J. Exp. Med. 158*, 822.
95. Warren, D. J., and Sanderson, C. J., Production of a T cell hybrid producing a lymphokine stimulating eosinophil differentiation. *Immunology 54*, 615, 1985.
96. Sanderson, C. J., O'Garra, A., Warren, D. J., and Klaus, G. G. B., Eosinophil differentiation factor has also B cell growth factor activity. Proposed name interleukin-4. *Proc. Nat. Acad. Sci. U.S.A. 83*, 437, 1986.
97. O'Garra, A., Warren, D. J., Holman, M., Popham, A. M., Sanderson, J. C., and Klaus, G. G. B., Interleukin-4 (B cell growth factor-II/eosinophil differentiation factor) is a mitogen and differentiation factor for preactivated murine B lymphocytes. *Proc. Nat. Acad. Sci. U.S.A. 83*, 5228, 1986.
98. O'Garra, A., Warren, D. J., Holman, M., Popham, A. M., Sanderson, C. J., and Klaus, G. G. B., Effects of cyclosporine on responses of murine B cells to T cell derived lymphokines. *J. Immunol. 137*, 2220, 1986.
99. Dongworth, D. W., Klaus, G. G. B., Effects of cyclosporine A on the immune system of the mouse. I. Evidence for a direct selective effect of cyclosporine A on B cells responding to anti-immuno-globulin antibodies, *Eur. J. Immunol. 12*, 1018, 1982.
100. Klaus, G. G. B., Hawrylowicz, C. M., Activation and proliferation signals in mouse B cells. II. Evidence for activation (G0 and G1) signals differing in sensitivity to cyclosporine. *Eur. J. Immunol. 14*, 250, 1984.
101. Colombani, P. M., Robb, A., and Hess, A. D., Cyclosporin A binding to calmodulin: A possible site of action on T lymphocytes. *Science 228*, 337, 1985.
102. LeGrue, S. J., Turner, R., Weisbrodt, N., and Dedman, J. R., Does the binding of cyclosporine to calmodulin result in immunosuppression? *Science 234*, 68. 1986.
103. Hait, W. N., Harding, M. W., Handschumacher, R. E., Calmodulin, cyclophilin, and cyclosporin A. *Science*, 233, 987, 1986.
104. Handschumacher, R. E., Harding, M. W., Rice, J., Drugge, R. J., Cyclophilin: A specific cytosolic binding protein for cyclosporine A. *Science 226*, 544, 1984.
105. Harding, M. W., Handschumacher, R. E., and Speicher, D. W., Isolation and amino acid sequence of cyclophilin. *J. Biol. Chem. 261*, 8547, 1986.

106. Harding, M. W., Gorelick, F. S., and Handschumacher, R. E., Cyclophilin represents a novel class of protein kinases. *Fed. Proc. 46*, 2161, abstract no. 1377, 1987.
107. Yamamoto, K., Steroid receptor regulated transcription of specific genes and gene networks, *Ann. Rev. Genet. 19*, 209, 1985.
108. Dorn, A., Bollekens, J., Staub, A., Benoist, C., and Mathis, D., A multiplicity of CCAAT box-binding proteins, *Cell 50*, 863, 1987.
109. Miesfeld, R., Okret, S., Wikstrom, A. C., Wrange, O., Gustafsson, J. A., Yamamoto, K., Characterization of a steroid hormone receptor gene and mRNA in wild-type and mutant cells. *Nature 312*, 779, 1984.
110. Hollenberg, S. M., Weinberger, C., Ong, E. S., Cerelli, G., Oro, A., Lebo, R., Thompson, E. B., Rosenfeld, M. G., Evans, R. M., Primary structure and expression of a functional human glucocorticoid receptor cDNA. *Nature 318*, 635, 1985.
111. Green, S., Walter, P., Kumar, V., Krust, A., Bornert, J. M., Argos, P., Chambon, P., Human oestrogen receptor cDNA: Sequence, expression and homology to v-*erb*-A. *Nature 320*, 134, 1986.
112. Krust, A., Green, S., Argos, P., Kumar, V., Walter, P., Bornert, J. M., Chambon, P., The chicken oestrogen receptor sequence: Homology with v-*erb*A and the human oestrogen and glucocorticoid receptors. *EMBO J. 5*, 891, 1986.
113. Jeltsch, J. M., Krozowski, Z., Quirin-Stricker, C., Gronemeyer, H., Simpson, R. J., Garnier, J. M., Krust, A., Jacob, F., and Chambon, P., Cloning of the chicken progesteron receptor. *Proc. Nat. Acad. Sci. U.S.A. 83*, 5424, 1986.
114. Conneely, O. M., Sullivan, W. P., Toft, D. O., Birnbaumer, M., Cook, R. G., Maxwell, L. B., Zarucki-Schulz, T., Greene, G. L., Schrader, W. T., O'Malley, B. W., Molecular cloning of the chicken progesteron receptor. *Science 233*, 767, 1986.
115. Potter, E., Nicolaisen, A., Ong, E., Evans, R., Rosenfeld, M., Thyrotropin releasing hormone exerts rapid nuclear effects to increase production of the primary prolactin mRNA transcript. *Proc. Nat. Acad. Sci. U.S.A. 78*, 6662, 1981.
116. Eberwine, J. H., Roberts, J. C., Glucocorticoid regulation of proopiomelanocortin gene transcription in the rat pituitary, *J. Biol. Chem. 259*, 2166, 1984.
117. Caput, D., Beutler, B., Hartog, K., Thayer, R., Brown-Shimer, S., Cerami, A., Identification of a common nucleotid sequence in the 3′-untranslated region of mRNA molecules specifying inflammatory mediators. *Proc. Nat. Acad. Sci. U.S.A. 83*, 1670, 1986.
118. Blackwell, G. J., Carnuccio, R., Di Rosa, M., Flower, A. J., Parente, L., Persico, P., Macrocortin: a polypeptide causing the anti-phospholipase effect of glucocorticoids. *Nature 287*, 147, 1980.
119. Alexis, M. N., Glucocorticoids: new insights into their mechanism. *Trends in Pharm. Sci. 8*, 10, 1987.
120. Schindler, R., ed, *Ciclosporin in Autoimmune Diseases, First International Symposium*, Basle, March 18–20, 1985.
121. Quesniaux, V. F. J., Schreier, M. H., Wenger, R. M., Hiestand, P. C., Harding, M. W., Van Regenmortel, M. H. V., Cyclophilin binds to the region of cyclosporine involved in its immunosuppressive activity. *Eur. J. Immunol. 17*, 1359, 1987.
122. Quesniaux, V. F. J., Schreier, M. H., Wenger, R. M., Hiestand, P. C., Van Regenmortel, M. H. V., Molecular characteristics of cyclophilin-cyclosporine interaction, *Transplantation*. Vol. 46, 235–275, Suppl. August 1988.

Chapter 11
Biosynthesis and Chemical Synthesis of Bleomycin

Tomohisa Takita and Yasuhiko Muraoka

1. Introduction
2. Biosynthesis and Semisynthesis of Unnatural Congeners
   2.1 Biosynthesis
   2.2 Semisynthesis
3. Biosynthesis and Chemical Synthesis of Building Blocks of Bleomycin Molecule
   3.1 Biosynthesis of unusual amino acids
   3.2 Chemical synthesis of unusual amino acids
      3.2.1 Synthesis of pyrimidine-containing amino (PBA)
      3.2.2 Synthesis of *erythro-β*-hydroxy-L-histidine
      3.2.3 Synthesis of (2S,3S,4R)-4-amino-3-hydroxy-2-methylpentanoic acid (AHM)
      3.2.4 Synthesis of 2′-(2-aminoethyl)-2,4′-bithiazole-4-carboxylic acid (ABC)
4. Biosynthetic and Chemical Construction of Bleomycin Molecule
   4.1 Biosynthesis of peptide part of bleomycin: deglycobleomycin
   4.2 Chemical synthesis of deglycobleomycin
   4.3 Chemical synthesis of the disaccharide moiety for total synthesis of bleomycin
   4.4 Total synthesis of bleomycin
5. Conclusion
References

1. Introduction

In 1956, Umezawa and his associates discovered phleomycin during the course of a study of water-soluble basic antibiotics [1]. This antibiotic was found to inhibit Ehrlich carcinoma with a high therapeutic index, but it showed strong renal toxicity toward dogs precluding its clinical study.

Bleomycin was discovered by Umezawa and his associates after an intensive search for phleomycin-like antibiotics [2]. Bleomycin did not cause renal damage in dogs, and the clinical study started in 1965. Bleomycin is currently used in the treatment of squamous cell carcinomas, Hodgkin's lymphomas, testis tumor etc.

Bleomycin is produced by *Streptomyces verticillus* ATCC 15003. The antibiotic is a mixture of congeners and is obtained as the blue-copper complex. The structure

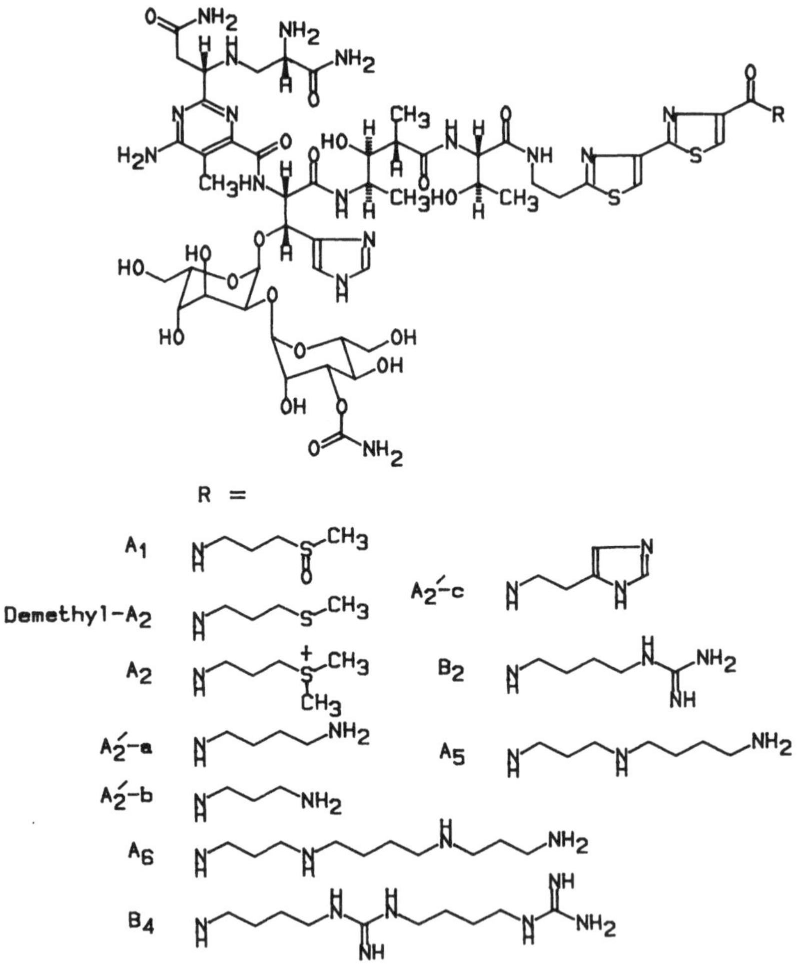

Figure 1 Structure of bleomycins.

was determined as a glycopeptide amide shown in Figure 1 [3]. The congeners are different from each other in the carboxy terminal amine moiety. The glycopeptide part common to these congeners was named bleomycinic acid (Figure 1: R = OH)

Bleomycinic acid consists of a linear pentapeptide and an O-carbamoyl disaccharide. The peptide part is composed from pyrimidoblamic acid (PBA) [1], *erythro-β*-hydroxy-L-histidine (OH-His) [2], (2S,3S,4R)-4-amino-3-hydroxy-2-methylpentanoic acid (AHM) [3], L-threonine (Thr), 2′-(2-aminoethyl)-2, 4′-bithiazole-4-carboxylic acid (ABC) [4]. And the sugar part is composed from 3-O-carbamoyl-D-mannose and L-gulose. The saccharide connects to the peptide part by *O*-glycosidic linkage to the hydroxy group of *β*-hydroxy-histidine. The structure of copper complex of bleomycin was also proposed by Takita et al. (Figure 2) [4]. On the basis of this study, it became possible that the mechanism of action of bleomycin can be discussed on molecular level (Figure 3) [4].

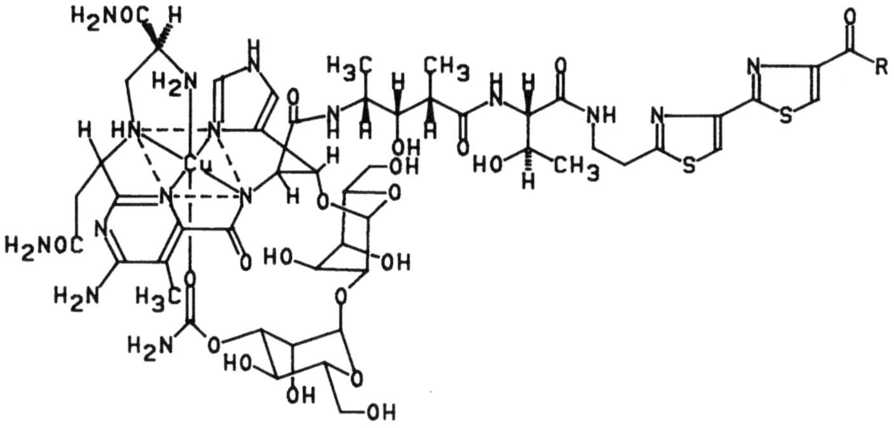

Figure 2 Structure of copper complex of bleomycin.

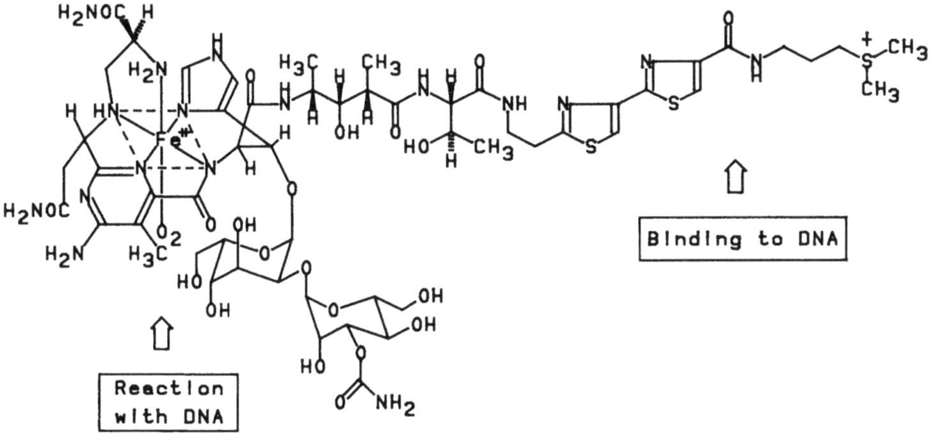

Figure 3 Interaction of bleomycin with DNA.

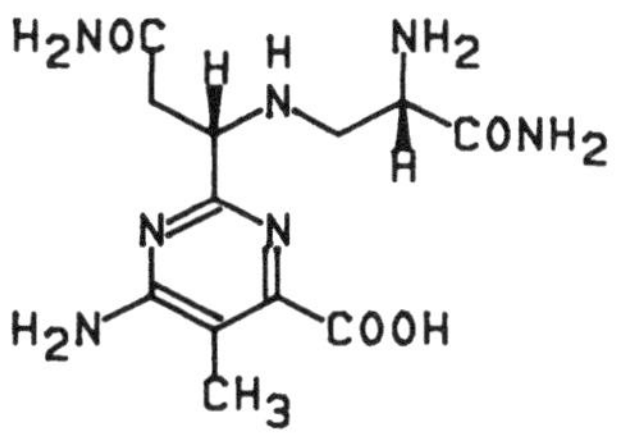

(PBA) [1]

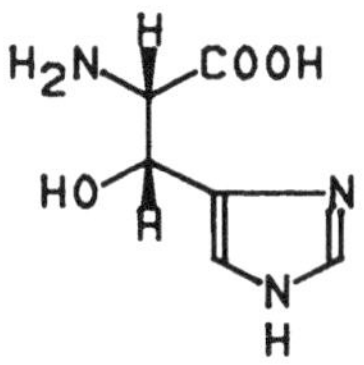

(HO-His) [2]

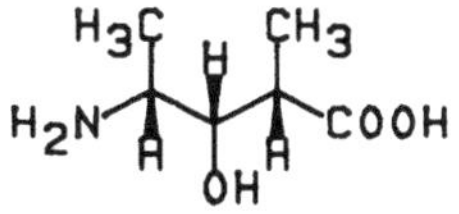

(AHM) [3]

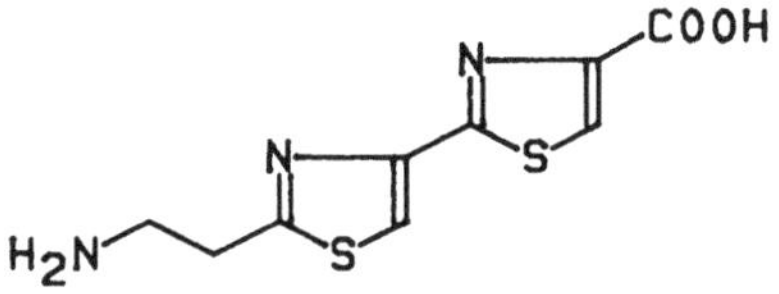

(ABC) [4]

2. Biosynthesis and Semisynthesis of Unnatural Congeners

2.1 Biosynthesis

The bleomycin-producing strain (*Streptomyces verticillus* ATCC 15003) was isolated from a soil sample collected at a coal mine in Fukuoka Prefecture, Japan. The productivity of bleomycin has been improved by selection of the highly productive strain among the mutants, which were induced by conventional physical and chemical methods. A CM-Sephadex C-25 chromatographic separation of bleomycin mixture produced by an improved strain under an ordinary culture is shown in Figure 4a.

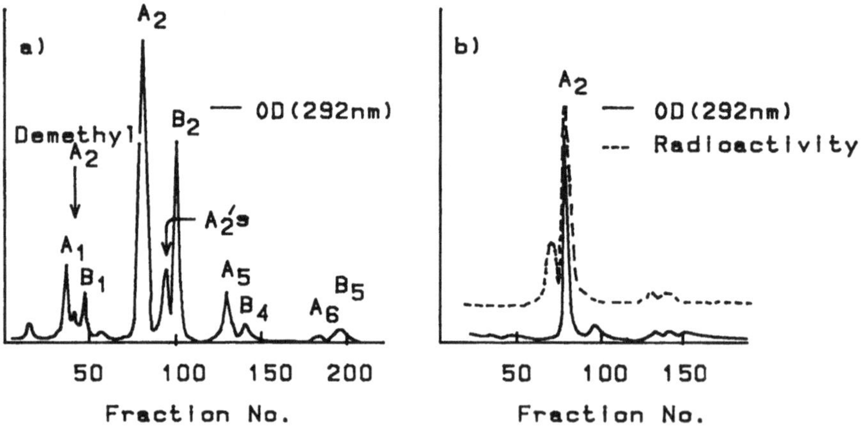

Figure 4 Analysis of bleomycin mixture produced by fermentations by CM-Sephadex C-25 chromatography: (a) produced without addition of amine (b) produced by addition of 3-aminopropyldimethylsulfonium chloride (^{14}C CH_3).

The incorporation of a specific amine into the carboxy terminal amine moiety of bleomycin was first proved by a culture experiment, in which radioactive 3-aminopropyl-dimethylsulfonium chloride ([^{14}C]methyl) was added to the culture medium (Fig. 4b) [5]. In this experiment, it became evident that labeled bleomycin A_2 was produced and further that the addition of the amine of A_2 suppressed the production of other congeners. This precursor effect was also found in the other terminal amines of natural bleomycins, except for those of bleomycins A_1, demethyl-A_2, and A_6. The result are summarized in Table 1. The terminal amines of bleomycins A_1 and demethyl-A_2 were not incorporated. These bleomycins are derived from bleomycin A_2 by spontaneous demethylation and oxidation. Addition of A_6 amine (spermine) caused the selective production of bleomycin A_5, whose terminal amine is spermidine, a degradation product of spermine.

Table 1 Bleomycins produced by addition of terminal amines of natural bleomycins.

Terminal amine added	Produced BLMs (%)		Control (%)[a]
A_1	Almost same as control		9.2
Demethyl-A_2	Almost same as control		2.4
A_2	A_2	82	54.5
A_2'-a	A_2'-a	25	
A_2'-b	A_2'-b	85	3.5
A_2'-c	A_2'-c	93	
B_2	B_2 45, new BLM 34[b]		26.7
B_4	B_4 38, new BLM 40[b]		2.4
A_5	A_5 100		1.3
A_6	A_5 100, A_6 trace		Trace

[a] Composition of BLMs produced without addition of amine.
[b] Deaminated products of the terminal guanido group of BLMs B_2 and B_4, respectively.

Table 2 Unnatural amines well incorporated into the terminal amine moiety of bleomycin.

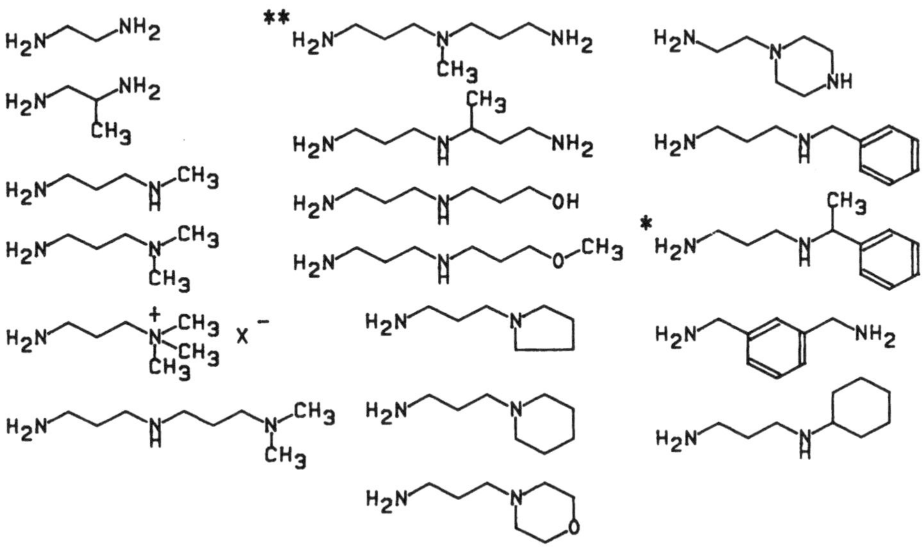

* Terminal amine of peplomycin.
** Terminal amine of the biosynthetic bleomycin, which is used as the starting material of liblomycin.

The above results suggested that alkylamines which have at least one more additional basic function such as amine, guanidine, imidazole, sulfonium etc. could be incorporated into the terminal amine moiety of bleomycin. In fact, unnatural amines shown in Table 2 were well incorporated into the terminal amine moiety of bleomycin, and a large number of new biosynthetic bleomycins have been prepared. Among them, peplomycin, whose terminal amine is *N*-[(S)-1-phenylethyl]-1,3-diaminopropane, was selected as a second-generation bleomycin and has been used clinically in Japan since 1981. Liblomycin is a third-generation bleomycin under clinical trial in Japan. It is derived from a biosynthetic bleomycin, of which the terminal amine is bis(3-aminopropyl)-methylamine.

2.2 Semisynthesis

Chemical methods to replace the terminal amine moiety of natural bleomycin were developed. Bleomycinic acid, the key intermediate of artificial bleomycin can be obtained by two methods.

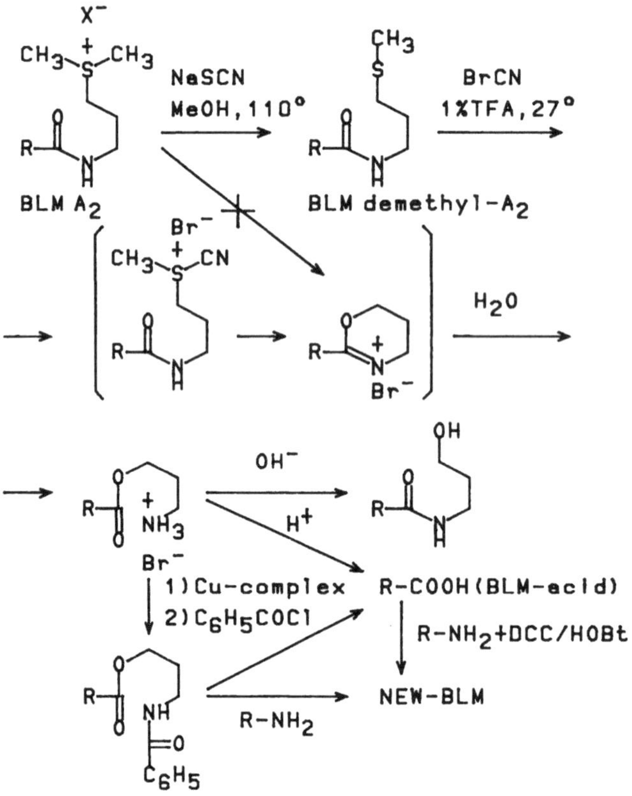

Figure 5 Scheme of semisynthetic bleomycin.

The first one is hydrolysis of bleomycin B_2 by an enzyme produced by a strain of Fungi, *Fusarium anguioides* [6]. This enzyme, acyl-agmatine hydrolase, is a newly found enzyme and recognizes acyl-agmatines as substrates hydrolyzing the amide bond between the acyl group of bleomycinic acid and the amino group of agmatine in bleomycin B_2. The second one is BrCN mediated selective cleavage of 3-mercaptopropylamide of bleomycin demethyl A_2, which is derived from bleomycin A_2 by pyrolysis or NaNCS treatment [7]. By the use of BrCN in dilute trifluoroacetic acid, bleomycin demethyl A_2 is converted to bleomycinic acid 3-aminopropyl ester, which can be hydrolyzed to bleomycinic acid or transformed directly to bleomycin derivative, via bleomycinic acid 3-benzamidepropyl ester. Thus, various bleomycins can be prepared from bleomycinic acid by the use of peptide-forming-reagent such as dicyclohexylcarbodiimide plus 1-hydroxybenztriazole or aminolysis of bleomycinic acid esters (Figure 5).

To get bleomycin analogues, semisynthetic methods are more complicated than fermentation methods but are valuable in their ability to permit the incorporation of a broader range of terminal amines; amines which have no additional basic function (e. g. A_1-amine, demethyl A_2 amine, other monoamine and aromatic amines), amines which are easily metabolized by the producing microorganism (e. g. A_6 amine), amines which are toxic to the microorganism, or insoluble amines in the culture medium etc. This semisynthetic procedure is also suitable for small scale but large variety of preparation of new bleomycins, in the early stage of screening of the useful analogues.

3. Biosynthesis and Chemical Synthesis of Building Blocks of Bleomycin Molecule

3.1 Biosynthesis of unusual amino acids

Bleomycin contains three novel amino acids, PBA [1], AHM [3], and ABC [4], which have unusual skeletons. The biosynthesis of these amino acids have been studied by incorporation of ^{14}C-labeled compounds [5]. The results are shown in Table 3. The PBA moiety of bleomycin is decomposed into a pyrimidine-containing amino acid [5] and β-aminoalanine [6] by competitive β-elimination under total acid hydrolysis condition. Thus it was found that serine was incorporated into [6] and asparagine (but not asparatic acid) and methionine-CH_3 were incorporated into [5].

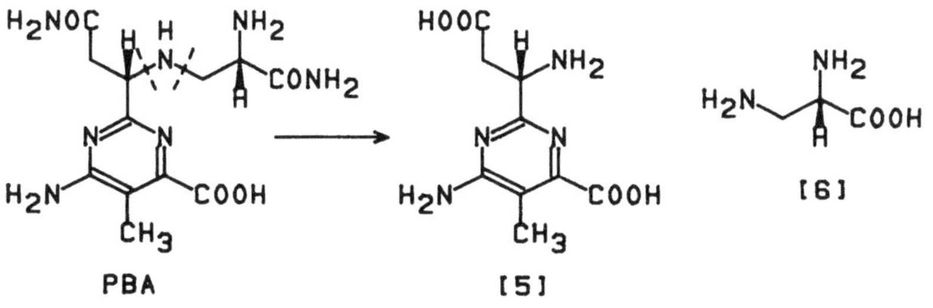

Table 3 Incorporation of ^{14}C-labeled compounds into novel amino acid constituents of bleomycin.

Novel amino acid	Significantly incorporated labeled compound
PBA	L-Asn(U)[a], L-Met(CH_3), L-Ser(U)
AHM	L-Ala(U), L-Met(CH_3), acetic acid(2)
ABC	β-Ala(1), DL-Cys(3)

[a] Position of labeling; U, uniformly labeled.

Therefore PBA appears to be biosynthesized from 2 moles of asparagine and one mole each of serine and methionine-CH_3 (Figure 6). As described later (Section 6), demethyl PBA(DM-PBA) [7] is first synthesized from serine and asparagine, and the methyl group is introduced after the pyrimidine ring formation. It is interesting to note that DM-PBA is contained in a bleomycin family antibiotic SF1771 [9] and SF1961 [10] in place of PBA.

AHM was found to be biosynthesized from alanine, acetic acid, and methionine-CH_3 by incorporation of ^{14}C-labeled compounds [5] (Figure 6). The incorporation of methionine-CH_3 into the α-methyl of AHM was confirmed further by the incorporation study using L-methionine-methyl-^{13}C analyzed by ^{13}C-nuclear magnetic resonance spectrometry. In this study, the incorporation of methionine-CH_3 into the methyl on pyrimidine nucleus of PBA was also confirmed. The incorporation of alanine into the C_3-C_5 moiety of AHM was confirmed by the incorporation of DL-alanine-3-^{13}C [11].

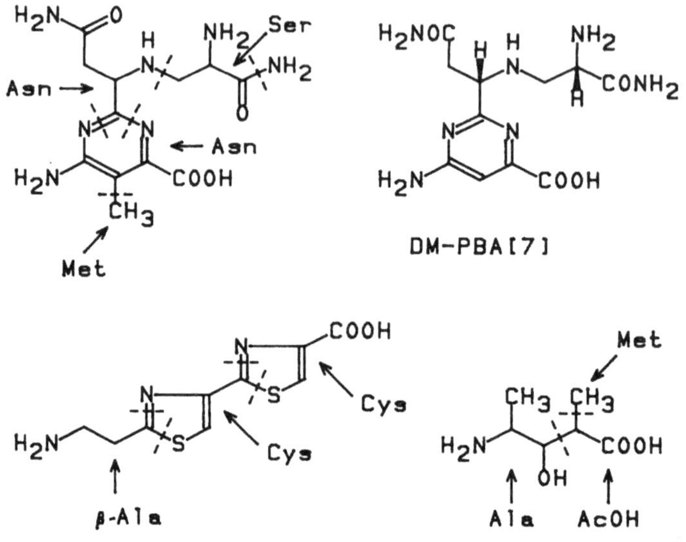

Figure 6 Biosynthesis of PBA[1], AHM[2] and ABC[3].

ABC was found to be biosynthesized from 1 mole of β-alanine and 2 moles of cystein. The ^{3}H-labeled ABC was not incorporated into bleomycin. This suggested that ABC is formed after incorporation of β-alanine and cystein stepwisely into the peptide chain. In phleomycin, a bleomycin family antibiotic, 2-[2-(2-aminoethyl)-Δ^2-thialolin-4-yl]-thiazole-4-carboxylic acid [7] (dihydro-ABC) is contained in place of ABC. Dehydrogenation of dihydro-ABC in phleomycin to ABC was performed not only by active MnO_2 [12] but also by a cell free enzyme preparation from the bleomycin producing organism [13].

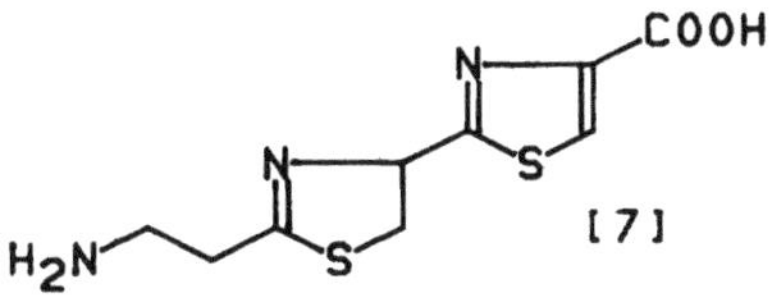

[7]

erythro-β-Hydroxy-L-histidine, another novel amino acid was found to be derived from histidine by oxygenation after the incorporation into the peptide chain from the biosynthetic studies that will be described later (Section 4.1).

3.2 Chemical synthesis of unusual amino acids

3.2.1 Synthesis of pyrimidine-containing amino acid (PBA)

PBA [5] was first synthesized from 4-carboxy-2,5-dimethyl-6-hydroxypyrimidine via pyrimidine acrylic acid [8], [14]. Racemic [5] was easily resolved to the optical pure form by chromatographic separation [15]. [5] was transformed to epimeric mixture of tricarboxylic acid [9], (formerly named as pseudo-dipeptide and later as pyrimidobleonic acid) by treatment with methyl acetamideacrylate followed by deblocking [16].

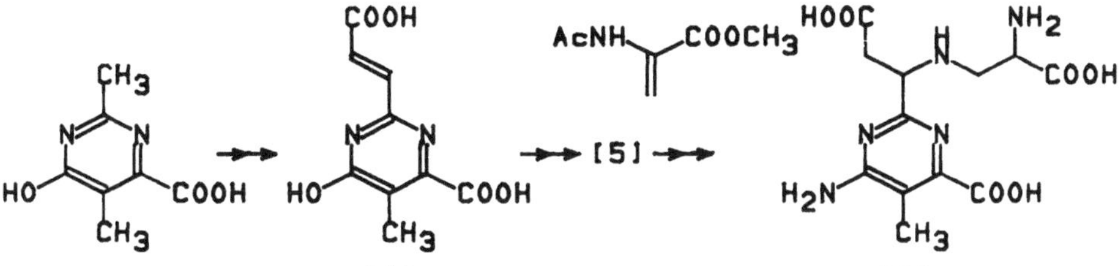

Tricarboxylic acid [9] was sophisticatedly converted to Boc PBA. Trimethyl ester of [9] was derived to the copper complex. By the complex formation, selective cleavage of the ester on the pyrimidine ring was achieved. Protection of the primary amine with Boc-S reagent followed by ammonolysis gave Boc PBA [10], [17].

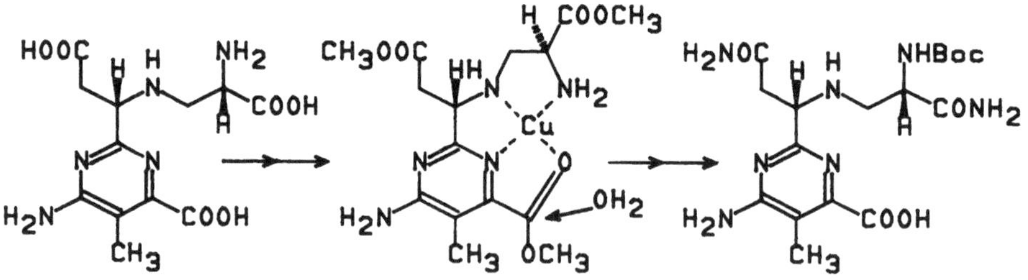

[9] [10]

Direct synthesis toward Boc-PBA were carried out by two groups [17, 18]. Both strategies are similar: first preparing substituted pyrimidin-2-aldehyde [11] [17] or pyrimidine acrylic acid [8] [18] and then forming the complete amino acid amide. In both routes, the asymmetric carbon at the α-aminocarboxamide part was derived from the chiral synthon prapared from the L-α-amino acid. However, the chiral center of the α-position of the pyrimidine side chain could not be controlled.

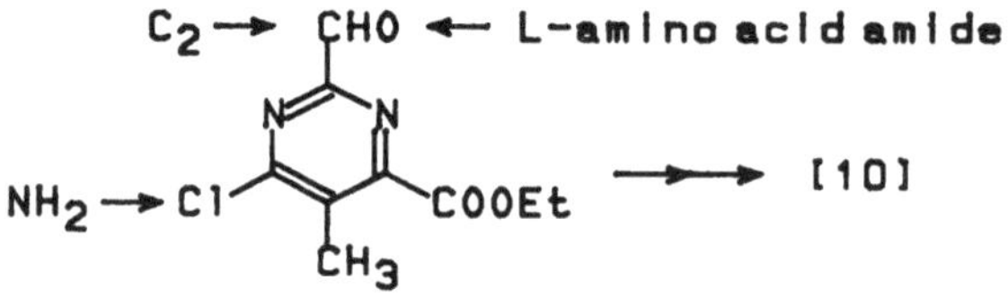

[11]

3.2.2 Synthesis of *erythro*-β-hydroxy-L-histidine

erythro-β-Hydroxy-DL-histidine was first synthesized from 4-formylimidazole by conventional Akabori method in stereoisomeric mixture. *threo*-β-Hydroxyhistidine, diastereoisomer of the natural product, was separated by the use of cellulose column chromatography [19]. Later, stereoselective synthesis of the *erythro*-isomer was achieved by the use of *N*-pyruvylidenglycinatocopper (Section 2). This reaction is highly stereospecific, and the stereospecificity can be reasonably explained by coordination of the imidazole to the copper ion. This seems to be the best procedure to get *erythro*-β-hydroxy-histidine to date, because synthetic racemic *erythro*-β-hydroxy-histidine is readily resolved by co-crystalization with D(-)tartaric acid [20, 21].

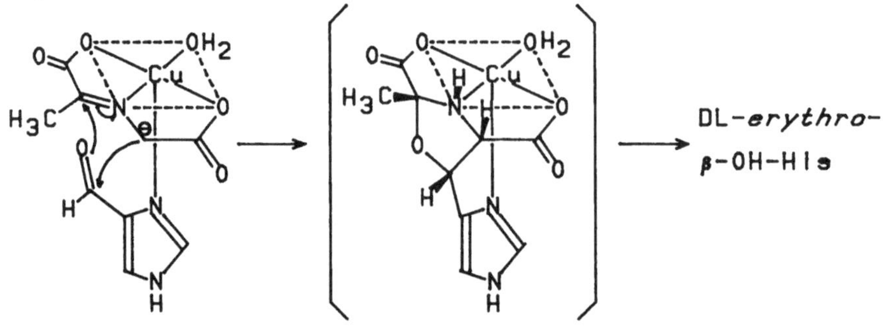

A multi-step stereospecific synthesis of *erythro-β*-hydroxy-L-histidine from D-glucosamine was published [22]. The synthesis was attained by retaining stereochemistry of 2-amino and 3-hydroxy groups of the starting material.

3.2.3 Synthesis of (2S,3S,4R)-4-amino-3-hydroxy-2-methylpentanoic acid (AHM)

AHM was first synthesized in 1974 from D-alanine derivative [12] and half ester of methylmalonic acid. Although the stereochemistry at C4 in the synthesized AHM was derived from D-alanine, the stereochemistry at 2 and 3 were not controlled. The natural stereoisomer was only 12.2% among the products, and was separated from others by ion exchange chromatography [23]. Later, the conversion of the (2R,3S,4R) stereoisomer to the desired (2S,3S,4R)-isomer was accomplished by cyclization and base treatment [24].

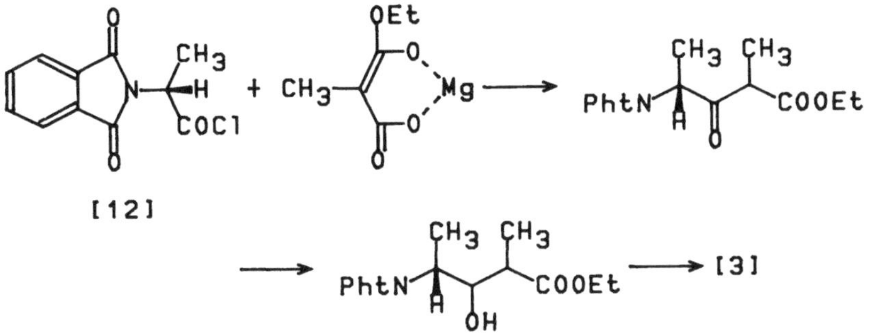

Stereocontrolled syntheses of AHM from D-alanine were achieved by two methods. One is utilization of stereocontrolled aldol type condensation of E-vinyloxyboran [13] with D-alaninal derivatives [14], in which two asymmetric centers are selectively introduced in one step reaction [25]. In this method, *N*-masked AHM active ester [15] was prepared, and it was used directly in the peptide synthesis [32].

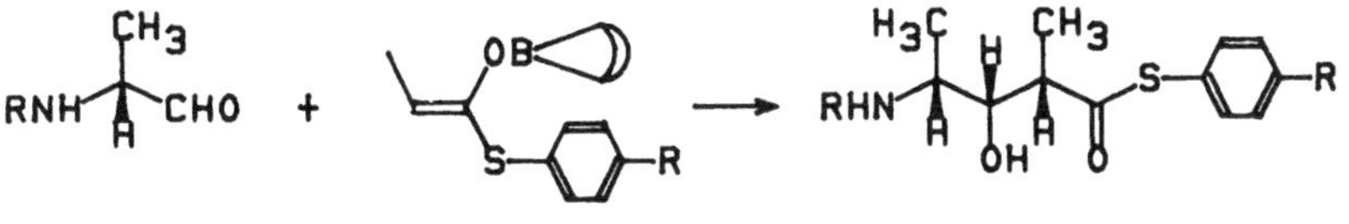

[14] [13] [15]

The other involves acylation of *N*-propionyl oxazolidinone [16] with Boc-D-alanine anhydride, followed by reduction of the resulting ketone. These two reactions are well stereo-controlled giving the desired stereoisomer [26].

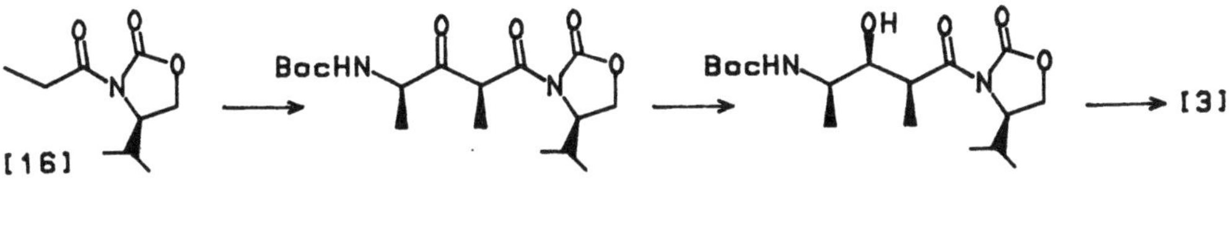

In these synthetic works, D-alanine was used as the chiral synthon for three carbon unit from C_3 to C_5 of AHM like as biogenetic route. A multi-step synthesis of AHM starting from L-rhamnose was also published [27].

3.2.4 Synthesis of 2′-(2-aminoethyl)-2,4′-bithiazole-4-carboxylic acid (ABC)

By the use of conventional thiazole synthesis, ABC was synthesized by two different routes. The first one is: *N*-benzoyl-*β*-alaninethioamide was coupled with ethyl bromopyruvate, and the resulting thiazole ester was converted to thioamide, and then coupled again with bromopyruvate [28]. The second one is: ethyl bromopyruvate was coupled with *O*-benzoylthiolactamide, and the resulting thiazole ester was converted to 2-bromoacetyl-thiazole-4-carboxylate, which was coupled with *N*-benzoyl-*β*-alaninethioamide [29]. Biomimetic synthesis of ABC was published in 1977, in which cyclization, dehydration, and dehydrogenation of *β*-alanyl-cysteinyl-cystein are involved [30]. The dehydrogenation process of thiazoline to thiazole is first introduced to conversion of phleomycin to bleomycin in 1972 [12].

4. Biosynthetic and Chemical Construction of Bleomycin Molecule

4.1 Biosynthesis of peptide part of bleomycin: deglycobleomycin

The nine biosynthetic intermediates (P-3 ~ deglycobleomycin) have been isolated from the fermentation broth of bleomycin [13]. The structures of these intermediates were determined by degradation study and ^{13}C-NMR spectrometry. These peptides were arranged in the order by their structures shown in Figure 7. The Figure clearly demonstrates the biosynthetic steps of the peptide part of bleomycin. The smallest one is P-3, which is composed from demethyl-PBA and histidine. The second smallest one is P-3A, which has additional alanine at the C-terminal of P-3. In figure 7, the tilted arrow means peptide chain elongation and the asterisk means modification at this stage. The third one, P-4, has AHM at the C-terminal. The biosynthetic processes from P-3 to P-4 is described more in detail later. P-5 has threonine at the C-terminal. At this stage, methylation occurs at the pyrimidine ring. The next one, P-5mB, has *β*-alanine at the C-terminal. Between P-5mB and P-6m, two moles of cystein are incorporated and modification occur to yield ABC at the C-terminal of P-6m. At this stage, hydroxylation occurs at the *β*-position of histidine. Finally, the terminal amine is incorporated to give deglycobleomycin.

The detailed biosynthetic processes from P-3 to P-4 are shown in Figure 8. These processes can be justified by isolation of a key intermediate named P-3K. DM-PBA-His ~ S-Enz on the top line means P-3 is connected to the synthetase with thioester. The nucleophilic attack of the amino group of alanine on the synthetase

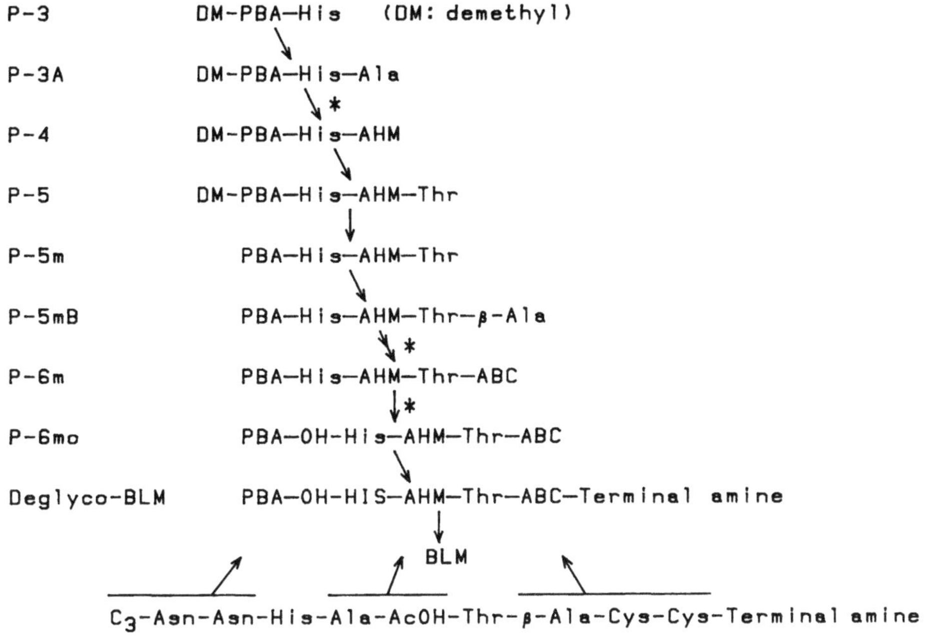

Figure 7 Biosynthetic intermediates of bleomycin. The tilted arrow indicates elongation of peptide chain. Modification occurs at the step marked by an asterisk.

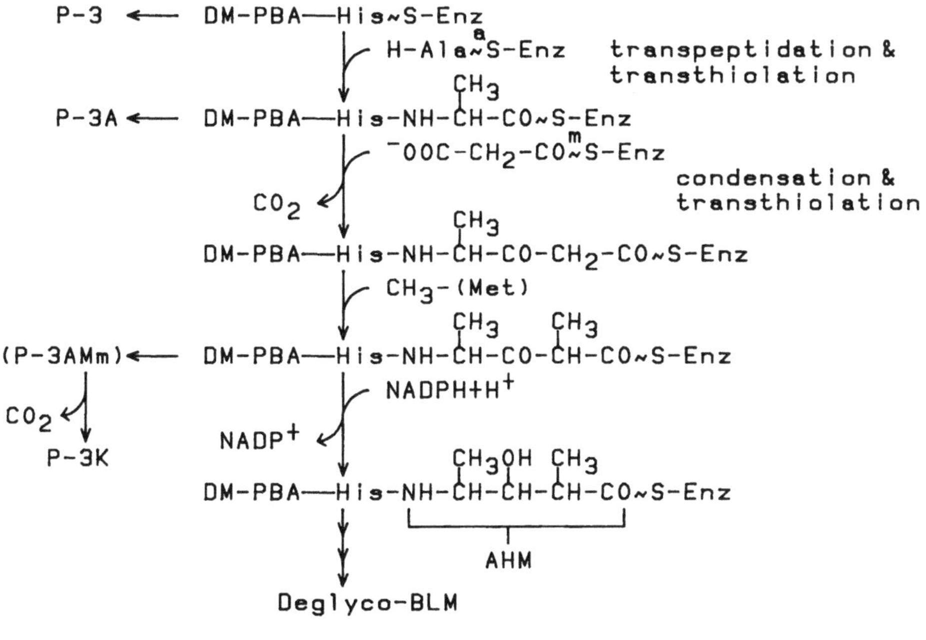

Figure 8 Biosynthesis of AHM.

to the thioester of P-3 forms a new peptide bond. This process is called transpeptidation. The formed P-3A is transferred to the peptide site of the synthetase. This process is called transthiolation. These processes are proposed by analogy from the biosynthetic study of gramicidin S [31]. In the next step, malonate, but not an amino acid, reacts with P-3A to form β-keto-thioester. At this stage, methylation occurs at the α-methylene. The spontaneous hydrolysis of the thioester liberates the β-keto-acid, which is immediately decarboxylated to yield the ethyl ketone. This is just P-3K. The isolation of P-3K suggests that these biosynthetic processes actually occur during the peptide chain elongation of bleomycin. The biological reduction of the keto group at the thioester yields AHM of the C-terminal of P-4.

Thus, biosynthesis of peptide antibiotic including bleomycin is much more similar to fatty acid synthesis rather than ordinary protein synthesis. The formation of AHM, of which the components are alanine and malonate, during the peptide chain elongation is symbolic of the biosynthesis of peptide antibiotic. Three biosynthetic processes are compared in Figure 9.

After the biosynthesis of deglycobleomycin, the disaccharide moiety seems to be introduced to the peptide aglycon, but the details of this process are not yet known.

```
Protein synthesis
      amino acid + ATP ⟶ aminoacyl-adenylate +PPi
      aminoacyl-adenylate + tRNA ⟶ aminoacyl-tRNA + AMP
       (5'-P-acid anhydride)          (3'-O-ester)
                                           ↓
                                        Protein
Fatty acid synthesis
      acetate + ATP ⟶ acetyl-adenylate + PPi
      acetyl-adenylate + CoASH ⟶ acetyl-CoA + AMP
       (5'-P-acid anhydride)        (thio-ester)
                                         ↓
                                     Fatty acid
Peptide antibiotic synthesis
      amino acid + ATP ⟶ aminoacyl-adenylate + PPi
      aminoacyl-adenylate + Enz-SH ⟶ aminoacyl-S-Enz + AMP
       (5'-P-acid anhydride)            (thio-ester)
                                             ↓
                                      Peptide antibiotic
```

Figure 9 Comparison of biosynthesis of protein, fatty acid and peptide antibiotic.

4.2 Chemical synthesis of deglycobleomycin

In contrast to the above described biosynthesis, of which the peptide chain is elongated from the N-terminal to the C-terminal, chemical synthesis of peptide is generally carried out by stepwise elongation from C-terminal to N-terminal to avoid racemization.

Deglycobleomycin was synthesized in three ways. In the first work, protected HO-His-AHM dipeptide fragment was condensed with tripeptide S (Thr-ABC-amine of A2), that was synthesized by stepwise elongation from the C-terminal. Boc-PBA was finally condensed to the resulting pentapeptide [32]. In the second work, tetrapeptide S of A_2 was synthesized by stepwise elongation from the C-terminal, and then the two hydroxy groups were protected as acetate. The diacetyl tetrapeptide S was condensed with Boc-HO-His(Dnp) to give protected pentapeptide S which has only one free hydroxy group for the glycosidation. This pentapeptide was used in the first total synthesis of bleomycin described later. Finally, pentapeptide S was coupled with Boc-PBA followed by stepwise removal of the *O*-acetate and Boc group at the N-terminal to give deglycobleomycin [33]. In the third work, N^{α},N^{im}-di-Boc-HO-His-AHM dipeptide was condensed with tripeptide S of demethyl A_2 like as the first work, and then coupled with Boc-PBA followed by deblocking [34].

4.3 Chemical synthesis of the disaccharide moiety for total synthesis of bleomycin

Although there are no experimental results that L-gulose in bleomycin is derived from D-glucose, this speculation is quite natural if we assume that L-gulose has the same stereochemistry at the C2-C4 as that of D-glucose C4-C2, that is, the stereochemistry of L-gulose is upside-down of that of D-glucose (Figure 10).

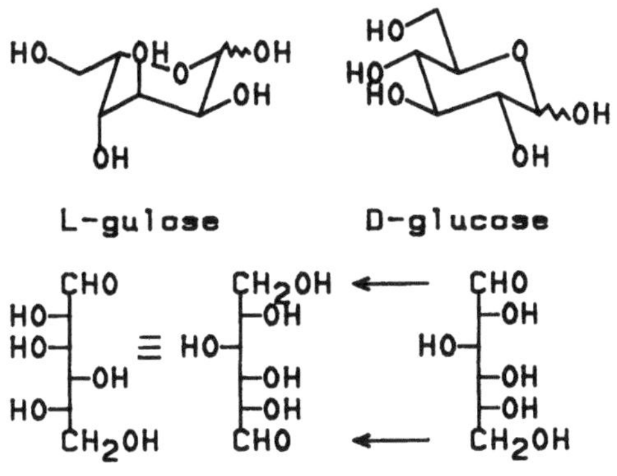

Figure 10 Relation between L-glucose and D-glucose.

In two synthetic studies of the sugar part of bleomycin, D-glucose was used as the starting material for the L-gulose moiety. Tsuchiya et al. used a D-glucose derivative [**17**] which has azide group at the 6 position and free hydroxy group at the 5-position, to which a D-mannose derivative [**18**] was glycosidized. The aldehyde group at the 1 position of glucose of the resulting disaccharide [**19**] was reduced to alcohol [**20**] and the azide methylene group was converted to aldehyde by photolysis to convert D-glucose to L-gulose in the disaccharide [**21**] [35]. This disaccharide was converted to cyclic carbonate [22] by carbonyldiimidazole after protection of the primary alcohols by silylation. After removal of the silyl groups followed by acetylation, the cyclic carbonate ester was converted to carbamoyl group by ammonolysis. Peracetylated carbamoyl disaccharide [23, X = OAc] thus obtained was used for total synthesis of bleomycin after conversion to the glycosyl halide [23, X = Br] [36, 37].

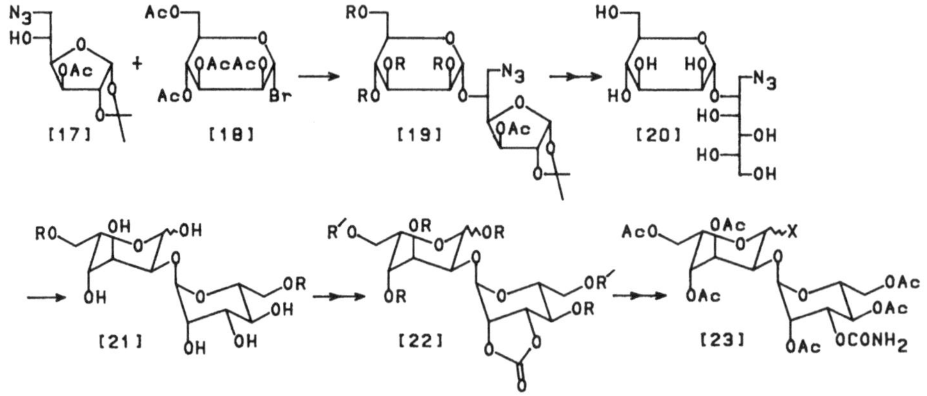

Later, American chemists synthesized carbamoyldisaccharide halide [24] from D-glucose via 1,6-anhydrogulose [25] [38], glycosydation by peracetyl 3-*O*-carbamoyl mannopyranosyl bromide [39], and ring opening of anhydrogulose [40]. This disaccharide halide differed from the previous one [23] in 3,4-protective group of L-gulose, and acetyl group on carbamoyl nitrogen.

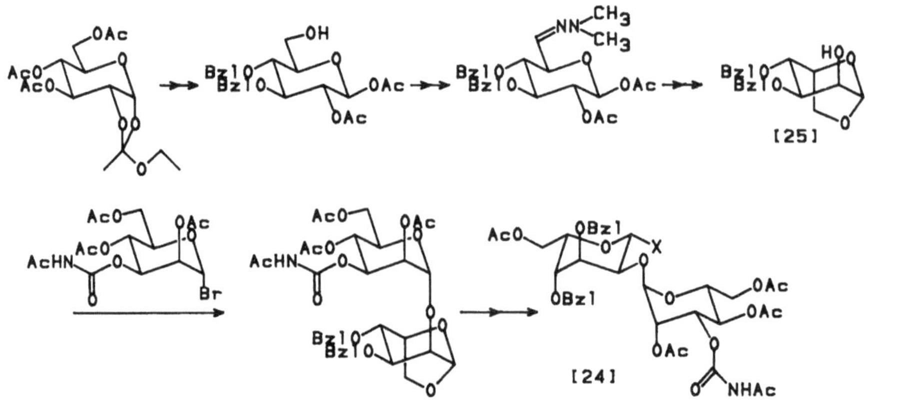

4.4 Total synthesis of bleomycin

Bleomycin seems to be biosynthesized by glycosydation of deglycobleomycin. We tried direct glycosydation of deglycobleomycin and some of its protected derivatives, but did not succeed. This seems to be due to the hydroxy groups of OH-His in deglycobleomycin being highly sterically hindered by neighboring substituents.

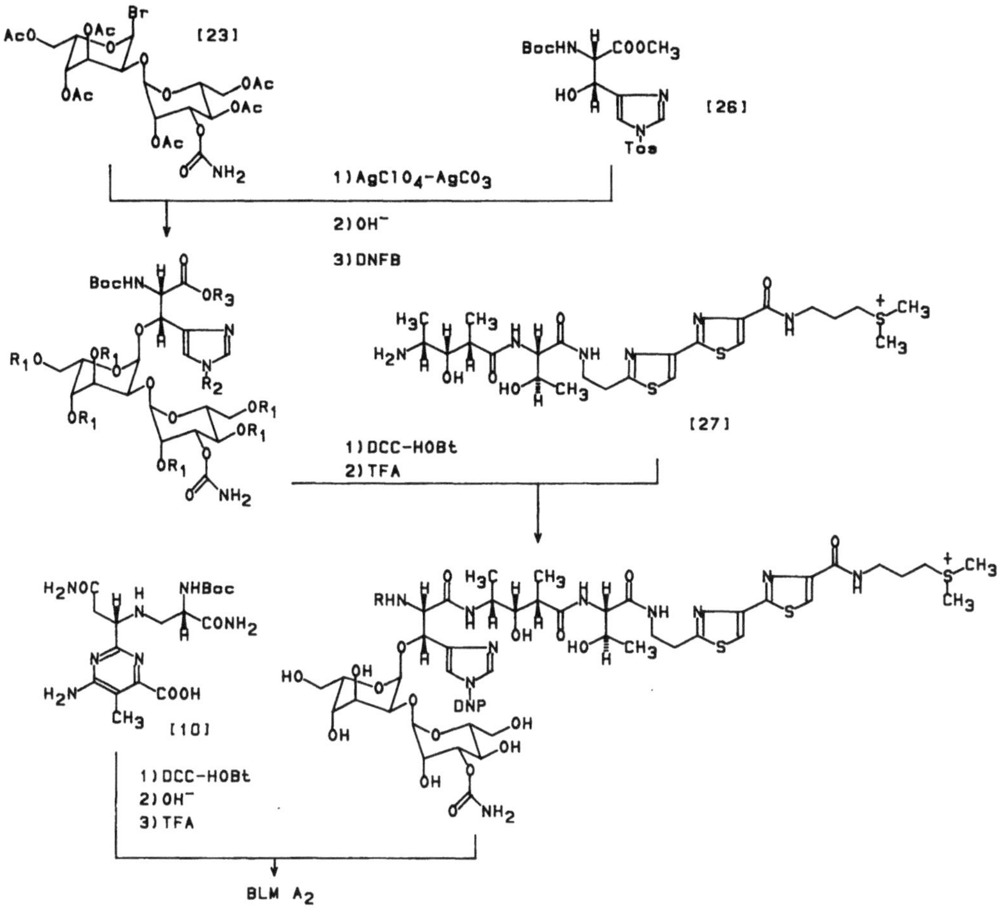

Figure 11 Total synthesis of bleomycin A_2.

In the first successful total synthesis, the disaccharide was coupled with a protected pentapeptide S (Boc-HO-His(Dnp)-AHM(Ac)-Thr(Ac)-ABC-Amine) and then Boc-PBA was coupled. In this case, a large amount of N^{im}-glycosydated products was formed. But the first synthesis of bleomycin was achieved after removal of the protective groups [36, 37].

In our improved total synthesis, the disaccharide [23] was first condensed with Boc-HO-His(Tos)-OMe [26], then tetrapeptide S [27] and Boc-PBA [16] were coupled with HO-His successively [21].

This improved synthetic method was applied to the synthesis of several bleomycin analogues in which AHM is replaced by other γ-amino acids. The studies on the structure-activity relationships showed that AHM is not only the connecting chain between the reaction site and the binding site to DNA in bleomycin molecule (see Figure 3), but also has some role on the manifestation of bleomycin activity (Table 4)

Table 4 Bioactivities of synthetic analogues of bleomycin.

Analog	Antibacterial activity (*B. subtilis*, %)	IC_{50} (μg/ml) against HeLa cell	DNA cleavage activity (pBR322) *in vitro*
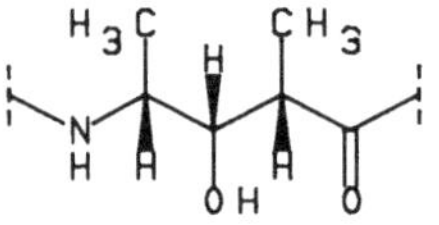	100	1.18	100
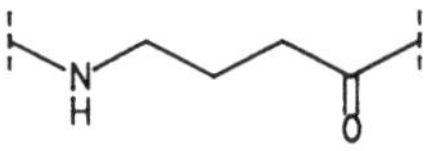	0	≫10	~10
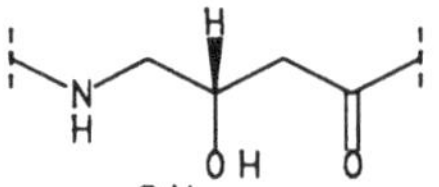	11	≫10	~10
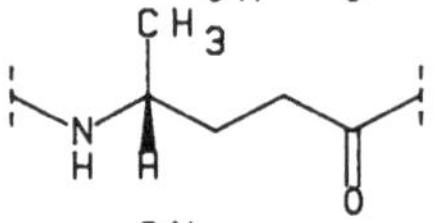	16	9.0	~200
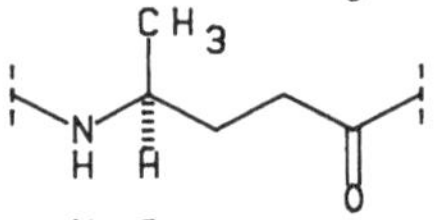	0	≫10	~10
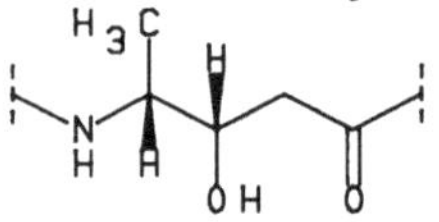	100	4.07	~100

[41]. The glycosydation between HO-His and disaccharide was further studied, and it was found that the protection of OH-His with N^{im}-Tos, N^{α}-Z and methyl ester is the best for the glycosydation [42].

An American group used $N^{\alpha,Nim}$-diBoc HO-His benzyl ester for glycosidation with carbamoyl disaccharide. They coupled it with AHM benzyl ester, tripeptide S and Boc-PBA successively. But details of the processes such as reaction yield and side reaction were not described [40].

5. Conclusion

Bleomycin is one of the most complicated natural products produced by microorganism. However, the biosynthetic process is not special, but rather similar to that of other peptide antibiotics such as gramicidin S and bacitracin. One of the characteristics of biosynthesis of bleomycin is that after the incorporation of amino acids into the peptide chain, a part of the peptide is modified in various ways such as cyclization and dehydrogenation to give pyrimidine and thiazole ring etc. This indicates the potentiality of microorganisms to produce a vast number of modified peptides.

We have been successful in the total synthesis of bleomycin. On the basis of this study, we established the procedure to rebuild bleomycin molecule from its fragments, which are efficiently obtained by selective cleavage of the molecule. We now get a variety of bleomycins and their analogues by biosynthesis and chemical modification and reconstruction. By the use of these compounds, we can expect to understand the mechanism of action of bleomycin in more detail and also to get a more effective antitumor agent.

References

1. Maeda, K., Kosaka, H., Yagishita, K., Umezawa, H., A new antibiotic, phleomycin. *J. Antibiotics, Ser. A 9*, 82, 1956.

2a. Umezawa, H., Maeda, K., Takeuchi, T., Okami, Y., New antibiotics, bleomycin A and B, *J. Antibiotics 19*, 200, 1966.

2b. Umezawa, H., Suhara, Y., Takita, T., Maeda, K., Purification of bleomycins, *J. Antibiotics 19*, 210, 1966.

3. Takita, T., Muraoka, Y., Nakatani, T., Fujii, A., Umezawa, Y., Naganawa, H., Umezawa, H., Chemistry of bleomycin. XIX. Revised structures of bleomycin and phleomycin, *J. Antibiotics 31*, 801, 1978.
4. T. Takita, Y. Muraoka, T. Nakatani, A. Fujii, Y. Iitaka, H. Umezawa, Chemistry of bleomycin. XXI. Metal-complex of bleomycin and its implication for the mechanism of bleomycin action, *J. Antibiotics 31*, 1073, 1978.
5. Fujii, A., Takita, T., Shimada, N., Umezawa, H., Biosyntheses of new bleomycins, *J. Antibiotics 27*, 73, 1974.
6. Umezawa, H., Takahashi, Y., Fujii, A., Saino, T., Shirai, T., Takita, T., Preparation of bleomycinic acid: Hydrolysis of bleomycin B_2 by a *Fusarium* acylagmatine amidohydrolase., *J. Antibiotics 26*, 117, 1973.
7. Takita, T., Fujii, A., Fukuoka, F., Umezawa, H., Chemical cleavage of bleomycin to bleomycinic acid and synthesis of new bleomycins, *J. Antibiotics 26*, 252, 1973.
8. Nakatani, T., Studies on the biosynthesis of antitumor antibiotic, bleomycin, *Ph. D. thesis, University of Tokyo*, 1981.
9. Ohba, K., Shomura, T., Tsuruoka, T., Omoto, S., Kojima, M., Hisamatsu, T., Inoue, S., Niida T., SF-1771, A new antibiotic related to bleomycin, *J. Antibiotics 33*, 1236, 1980.
10. Shomura, T., Omoto, S., Ohba, K., Ogino, H., Kojima, M., Inoue, S., SF-1961, a new antibiotic related to bleomycin, *J. Antibiotics 33*, 1243, 1980.
11. Nakatani, T., Fujii, A., Naganawa, H., Takita, T., Umezawa, H., Chemistry of bleomycin. XXVI. Biosynthetic study using ^{13}C-enriched precursors, *J. Antibiotics 33*, 717, 1980.
12. Takita, T., Muraoka, Y., Fujii, A., Itoh, H., Maeda, K., Umezawa, H., The structure of the sulfur-containing chromophore of phleomycin, and chemical transformation of phleomycin to bleomycin, *J. Antibiotics 25*, 197, 1972.
13. Fujii, A., Biogenetic aspects of bleomycin-phleomycin group antibiotics, in *Bleomycin: Chemical, Biochemical and Biological Aspects*, Hecht, S. M., Ed., Springer-Verlag, New York, 1979, p. 75.
14. Yoshioka, Y., Muraoka, Y., Takita, T., Maeda, K., Umezawa, H., Chemistry of bleomycin. VII. Synthesis of β-amino-β-(4-amino-6-carboxy-5-methylpyrimidin-2-yl)-propionic acid, an amine component of bleomycin, *J. Antibiotics 25*, 625, 1972.
15. Muraoka, Y., Takita, T., Maeda, K., Umezawa, H., Chemistry of bleomycin. IV. The structure of amine component II of bleomycin A_2, *J. Antibiotics 23*, 252, 1970.

16. Takita, T., Muraoka, Y., Maeda, K., Umezawa, H., Selective cleavage of bleomycin, in *Proc. 8th Symp. on Peptide Chem.* Osaka, Shiba, T., Ed., Osaka, Japan, 1970, p. 179.
17. Umezawa, H., Morishima, H., Saito, S., Takita, T., Umezawa, H., Kobayashi, S., Otsuka, M., Narita, M., Ohno, M., Synthesis of the pyrimidine moiety of bleomycin, *J. Am. Chem. Soc. 102*, 6630, 1980.
18. Arai, H., Hagmann, W. K., Suguna, H., Hecht, S. M., Synthesis of the pyrimidine moiety of bleomycin and epibleomycin, *J. Am. Chem. Soc. 102*, 6631, 1980.
19. Takita, T., Yoshioka, T., Muraoka, Y., Maeda, K., Umezawa, H., Chemistry of bleomycin. V. Revised structure of an amine component of bleomycin A_2., *J. Antibiotics 24*, 795, 1971.
20. Yoshioka, T., Chemical studies on bleomycin: Syntheses of constituent amino acids and related compounds (Japanese), *Ph. D. thesis, Faculty of pharmaceutical Science, Univ. of Tokyo*, p. 151, 1977.
21. Saito, S., Umezawa, Y., Yoshioka, T., Takita, T., Umezawa, H., Muraoka, Y., An improved total synthesis of bleomycin, *J. Antibiotics 36*, 92, 1983.
22. Hecht, S. M., Rupprecht, K. M., Jabobs, P. M., Synthesis of L-*erythro*-β-hydroxyhistidine from D-glucosamine, *J. Am. Chem. Soc. 101*, 3982, 1979.
23. Yoshioka, T., Hara, T., Takita, T., Umezawa, H., Chemistry of bleomycin. XIV. Biogenetic-like synthesis of (2S,3S,4R)-4-amino-3-hydroxy-2-methyl-*n*-valeric acid, an amine component of bleomycin, *J. Antibiotics 27*, 356, 1974.
24. Levin, M. D., Subrahamanian, K., Katz, H., Smith, M. B., Burlett, D. J., Hecht, S. M., Bleomycin: synthesis and structural verification of the tripeptide S and Tetrapeptide S moieties, *J. Am. Chem. Soc. 102*, 1452, 1980.
25. Narita, M., Otsuka, M., Kobayashi, S., Ohno, M., Umezawa, Y., Morishima, H., Saito, S., Takita, T., Umezawa, H., Stereoselective synthesis of (2S,3S,4R)-4-amino-3-hydroxy-2-methylpentanoic acid, an amino acid constituent of bleomycin, by aldol condensation, *Tetrahedron Lett. 23*, 525, 1982.
26. DiPardo, R. M., Bock, M. G., (2S,3S,4R)-4-amino-3-hydroxy-2-methylpentanoic acid. Enantioselective synthesis of an amino acid constituent of bleomycin, *Tetrahedron Lett. 24*, 4805, 1983.
27. Ohgi, T., Hecht, S. M., (2S,3S,4R)-4-amino-3-hydroxy-2-methylvalerate. Synthesis of anamino acid constituent of bleomycin from L-rhamnose, *J. Org. Chem. 46*, 1232, 1981.
28. Zee-Cheng, K. Y., Cheng, C. C., Synthesis of 2′-(2-aminoethyl)-2,4′-bithiazole-4-carboxylic acid, A component of the antitumor antibiotic bleomycin(1), *J. Heterocycl. Chem. 7*, 1439, 1970.
29. Sakai, T. T., Riordan, J. M., Booth, T. E., Glickson, J. D., Synthesis and DNA binding of [3-[2′-(2-acetamidoethyl)-2,4′-bithiazole-4-carboxamido]propyl]dimethylsulfonium chloride, a fragment of bleomycin-A_2, *J. Med. Chem. 24*, 279, 1981.
30. McGowan, D. A., Jordis, U., Minster, D. K., Hecht, S. M., A biomimetic synthesis of the bithiazole moiety of bleomycin, *J. Am. Chem. Soc. 99*, 8078, 1977.
31. Lipmann, F., Attempts to map a process evolution of peptide biosynthesis. Synthesis of peptide antibiotics from thiol-linked amino acids parallels fatty acid synthesis, *Science 173*, 875, 1971.
32. Takita, T., Umezawa, Y., Saito, S., Morishima, H., Umezawa, H., Muraoka, Y., Suzuki, M., Otsuka, M., Kobayashi, S., Ohno, M., Total synthesis of deglyco-bleomycin A_2, *Tetrahedron Lett. 22*, 671, 1981.
33. Saito, S., Umezawa, Y., Morishima, H., Takita, T., Umezawa, H., Narita, M., Otsuka, M., Kobayashi, S., Ohno, M., A new synthesis of deglyco-bleomycin A_2 aiming at the total synthesis of bleomycin, *Tetrahedron Lett. 23*, 529, 1982.
34. Aoyagi, Y., Suguna, H., Murugesan, N., Ehrenfeld, G. M., Chang, L.-H., Ohgi, T., Shekhani, M. S., Kirkup, M. P., Hecht, S. M., Deglycobleomycin: Total synthesis and oxygen transfer properties of an active bleomycin analogue, *J. Am. Chem. Soc. 104*, 5237, 1982.
35. Tsuchiya, T., Miyake, T., Kageyama, S., Umezawa, S., Umezawa, H., Takita, T., Total synthesis of the disaccharide of bleomycin, 2-O-(α-D-mannopyranosyl)-L-gulopyranose, *Tetrahedron Lett. 22*, 1413, 1981.
36. Takita, T., Umezawa, Y., Saito, S., Morishima, H., Umezawa, H., Muraoka, Y., Suzuki, M., Otsuka, M., Kobayashi, S., Ohno, M., Tsuchiya, T., Miyake, T., Umezawa, S., Total synthesis of bleomycin, in *Peptides: Synthesis – Structure – Function* (Proceedings of the 7th American Peptide Symposium), Rich, D. H., and Gros, E., Eds., Pierce Chem. Co., 1981, p. 29.

37. Takita, T., Umezawa, Y., Saito, S., Morishima, H., Naganawa, H., Umezawa, H., Tsuchiya, T., Miyake, T., Kageyama, S., Umezawa, S., Muraoka, Y., Suzuki, M., Otsuka, M., Narita, M., Kobayashi, S., Ohno, M., Total synthesis of bleomycin A2, *Tetrahedron Lett.* *23*, 521, 1982.
38. Minster, D.K., Hecht, S.M., Synthesis of L-gulose from D-glucose via aldose interchange, *J. Org. Chem.* *43*, 3987, 1978.
39. Pozsgay, V., Ohgi, T., Hecht, S.M., Synthesis of the carbohydrate moiety of bleomycin, *J. Org. Chem.* *46*, 3761, 1981.
40. Aoyagi, Y., Katano, K., Suguna, H., Primeau, J., Chang, L.-H., Hecht, S.M., Total synthesis of bleomycin, *J. Am. Chem. Soc.* *104*, 5537, 1982.
41. Umezawa, H., Takita, T., Saito, S., Muraoka, Y., Takahashi, K., Ekimoto, H., Minamide, S., Nishikawa, K., Fukuoka, T., Nakatani, T., Fujii, A., Matsuda, A., New analogs and derivatives of bleomycin, in *Bleomycin Chemotherapy*, Sikic, B.I., Rozencweig, M., and Carter, S.K., Eds., Academic Press, 1985, Chapter 23, p. 289.
42. Miyake, T., Tsuchiya, T., Umezawa, S., Saito, S., Umezawa, H., Studies on glycosylation of *erythro-β*-hydroxy-L-histidine. A key step of bleomycin total synthesis, *Bull. Chem. Soc. Jpn.* *59*, 1387, 1986.

Chapter 12
Small Molecular Protease Inhibitors and Their Biological Effects

Takaaki Aoyagi

1. Introduction
2. Protease Inhibitors
3. Inhibitors Against Endopeptidases
4. Enzymatic Activities of Cellular Membrane
5. Inhibitors Against Cell Surface Enzymes
   5.1 Inhibitors against exopeptidases
   5.2 Inhibitors against enzymes on cellular membrane
6. Enzymatic Changes in Various Pathologic Conditions
   6.1 Roles of catabolic enzymes in muscular dystrophy
      6.1.1 Murine muscular dystrophy
      6.1.2 Duchenne muscular dystrophy
   6.2 Spontaneously hypertensive rats and aminopeptidases
   6.3 Immunologic diseases and hydrolytic enzymes
7. Modification of Metabolic Homeostasis Caused by Low-Molecular-Weight Enzyme Inhibitors
   7.1 Enzymatic oscillations caused by bestatin
   7.2 Influence of angiotensin-converting enzyme inhibitor, foroxymithine
   7.3 Effects of bestatin on hydrolytic enzymes in progressive muscular dystrophy
8. Pharmacologic Applications of Enzyme Inhibitors
9. Summary
Acknowledgements
References

1. Introduction

Proteases are known to play important roles in various processes and functions of the living organisms such as inflammation, immune response, complement activation, cell proliferation, cell fusion, carcinogenesis, metastasis, virus infection, autoimmune diseases, muscular dystrophy, demyelination diseases and so forth. They are closely linked to biological phenomena in the individual from ontogeny to death. Living organisms are maintained with an extensive network of closely linked enzymatic reactions. Enzyme inhibitors can thus be used as powerful tools for analyzing various aspects of the homeostasis in living organisms. Moreover, they are of great value in elucidating disease processes and seem to have a potential usefulness in treatment of various diseases. However, insufficient data are yet available for a full systematization of our knowledge on protease inhibitors.

For many years, we have been screening for inhibitors against various proteases including cellular surface enzyme and have found various kinds of inhibitors. The protease inhibitors which have been discovered in microbial metabolites are quite distinguished for their peculiar characteristics: they are all small molecular substances with low toxicity and with unique chemical structure. These inhibitors are useful in the analysis of reaction mechanisms and steric conformation of the corresponding enzymes, they facilitate purification of enzymes as the most suitable ligands in affinity chromatography and thus contribute to the discovery of new enzymes and proenzymes.

We found recently that certain inhibitors against the enzymes located on plasma membrane can modify cellular functions including immune reaction. The finding provides a new approach to the studies of regulatory mechanisms of the cells.

The inhibitors are widely used in a variety of fields in medicine, pharmacology, biochemistry and biology. They are used for the recognition or identification of functioning enzymes in organs, cells, body fluids and blood; they are used as the reagents to elucidate the roles of enzymes in various aspects of the physiological functions; they are used as the reagents to prevent the enzymatic degradation of various kinds of biologically active substances during the course of their isolation and purification.

Furthermore, we found recently that administration *in vivo* of one of these small molecular inhibitors affect not only the level of the corresponding enzyme *in vivo* but also the level of a wide range of enzymes which are closely linked with each other to form networks of enzymes *in vivo*. We have demonstrated that they help to provide new approaches to the elucidation of the mechanism of homeostasis and also to the treatment of various types of disease processes including intractable ones. Thus it is expected that the studies on enzyme inhibitors will contribute greatly to the progress of biological sciences.

2. Protease Inhibitors

Proteolytic enzymes, which are widely distributed in animals, plants, and microbes, are classified as endopeptidases and exopeptidases based on their mode of actions on synthetic substrates. Moreover, endopeptidases are divided into four categories according to their catalytic residues: serine proteinases, cysteine proteinases, aspartic proteinases, and metal proteinases. Serine proteinases include trypsin, plasmin, chymotrypsin, elastase, thrombin, kallikrein, cathepsin G, and so forth. Cysteine proteinases include papain, ficin, bromelin, cathepsin B1, B2, H, and L, and so forth. Aspartic proteinases include pepsin, gastricsin, chymosin, renin, cathepsin

D, and E, and so forth. Metal proteinases include thermolysin, collagenase, aminopeptidases, carboxypeptidases, and so forth.

On the other hand, exopeptidases are divided into four groups: aminopeptidases, dipeptidylaminopeptidases, carboxypeptidases, and dipeptidylcarboxypeptidases. Aminopeptidases include aminopeptidase A and B, leucine aminopeptidase, proline iminopeptidase, methionine aminopeptidase, cathepsin H, and so forth. Dipeptidylaminopeptidases include dipeptidylaminopeptidase I, II, III, and IV, and so forth. Carboxypeptidases include carboxypeptidase A, B, and Y, cathepsin A, and so forth. Dipeptidylcarboxypeptidases include angiotensin-converting enzyme, cathepsin B, and so forth.

For many years, we have been searching for inhibitors against various proteases including cellular surface enzymes and have found more than 60 kinds of enzyme inhibitors of small molecular size from the culture filtrate of microbes [1 – 6]. These inhibitors, in contrast to those discovered in animals and plants, are characterized in common by their low-molecular-weight and their novel structures. The structures of these new inhibitors were completely unforeseen, and hence they stimulated works on the mechanisms of inhibition and the structure-activity relationships. Also in many other fields of biology and medicine, enzyme inhibitors have attracted the attention of many researchers. Especially, inflammations, immune responses, complement reactions, tumorigenesis, metastasis of tumors, and several intractable diseases including muscular dystrophy, autoimmune diseases, and so forth are research topics to which enzyme inhibitors are now often applied as powerful tools.

3. Inhibitors Against Endopeptidases

Inhibitors of serine and cysteine proteinases such as plasmin, trypsin, kallikrein, chymotrypsin, elastase, papain and cathepsins, are of considerable interest because these enzymes are involved in blood coagulation, fibrinolysis, kinin formation, pulmonary emphysema, immunity, tumorigenesis, metastasis, and so forth. Such enzyme inhibitors also have obvious applications to studies on the related diseases.

Screening on culture filtrates of microbes for inhibitory activities against trypsin, plasmin, papain, chymotrypsin, and elastase led to the discovery of the following inhibitors; leupeptin, antipain, chymostatin, and elastatinal (Figure 1).

Leupeptin, which contains L-argininal, inhibits trypsin, plasmin, papain, cathepsin B, and so forth. Antipain, which also contains L-argininal, inhibits papain, trypsin, cathepsin B, and so forth. Chymostatin, which contains L-phenylalaninal, inhibits chymotrypsins. Elastatinal, which contains L-alaninal, inhibits pancreas elastase. These inhibitors have all been found from culture filtrates of *Streptomyces* strain (Table 1) [7 – 10].

As mentioned above, leupeptin, antipain, chymostatin, and elastatinal contain C-terminal aldehyde groups, and if these aldehyde groups are oxidized or reduced to

Leupeptin

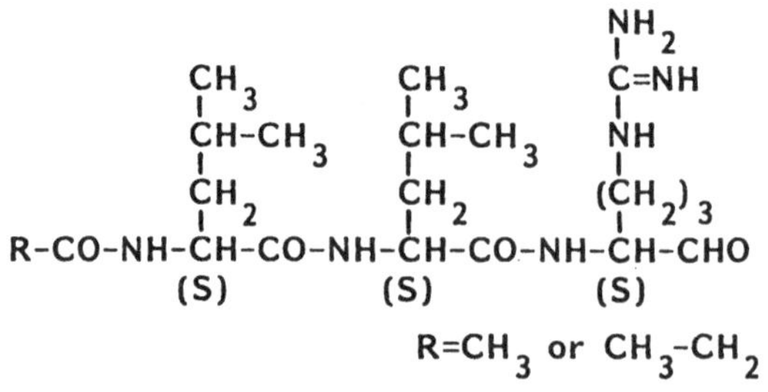

Antipain

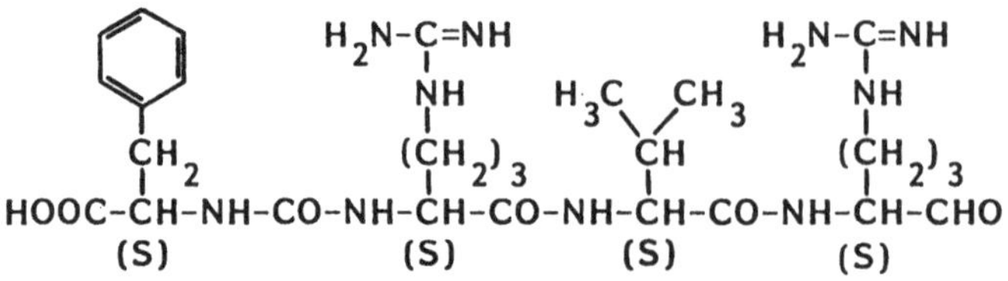

Chymostatin

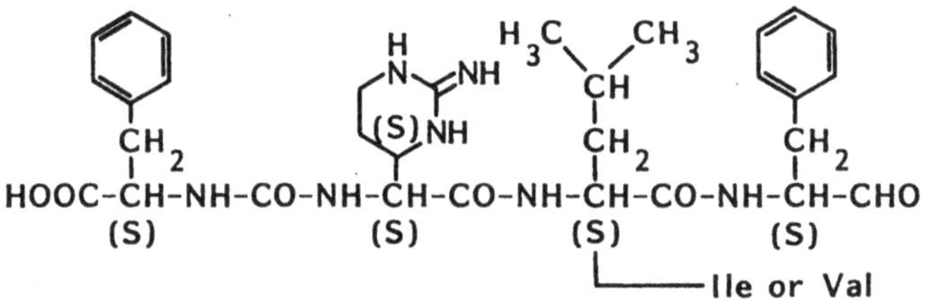

Elastatinal

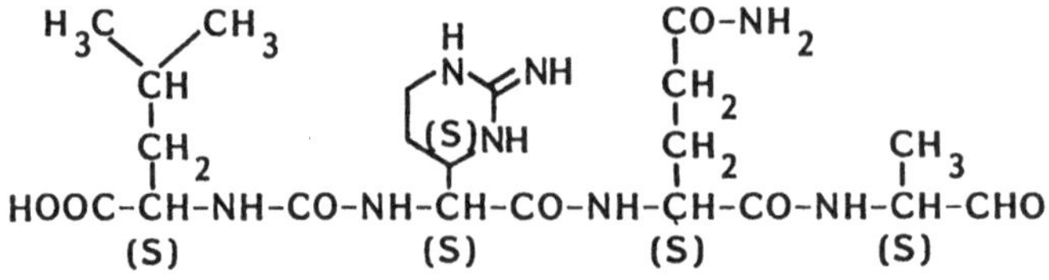

Figure 1 Inhibitors of serine and cysteine proteinases.

give an acid or alcohol, the compounds lose their activity. Leupeptin and antipain have the argininal residue, chymostatin the phenylalaninal residue, and elastatinal the alaninal residue at their terminal corbon. This indicates that α-acylamino aldehydes related to the acylamines portion of susceptible substrates have strongly inhibitory structures. This finding is very useful for understanding the relationship between the catalytic site of the enzyme and the essential structure of its inhibitor. The kinetic constants of the above mentioned inhibitors are shown in Table 2.

Table 1 Strains of *Streptomyces* producing endopeptidase inhibitors.

Inhibitors	Inhibited enzymes	Strains
Leupeptin	Trypsin, Plasmin, Papain, Cathepsin B	*Streptomyces roseus, S. roseochromogenes, S. lavendulae, S. albireticuli, S. thioluteus, S. chartleusis, S. noboritoensis* and more than 11 other species
Antipain	Trypsin, Papain, Cathepsin B	*S. michiganensis, S. violascens, S. mauvecolor, S. yokosukanensis*
Chymostatin	Chymotrypsins, Papain, Cathepsin B	*S. hygroscopicus, S. lavendulae*
Elastatinal	Elastase	*S. griseoruber* and more than 2 other species
Pepstatin	Pepsin, Gastricsin, Chymosin, Cathepsin D, Renin	*S. testaceus, S. argenteolus* var. toyonakensis, *S. naniwaensis, S. parvisporogenes* and more than 5 other species
Phosphoramidon	Thermolysins Collagenase	*S. tanashiensis* and more than 5 other species

Table 2 Kinetic constants of inhibitors against endopeptidases.

Inhibitor	Enzyme	Substrate	K_m ($\times 10^{-4}$M)	K_i ($\times 10^{-7}$M)	Type of inhibition
Leupeptin	Trypsin	TAME[2]	7.1	3.4	Competitive
		BAEE[3]	4.8	1.3	Competitive
Elastatinal	Elastase	Ac-$(Ala)_3$ · NA[4]	3.2	2.4	Competitive
		Ac-$(Ala)_3$ · ME	4.3	2.1	Competitive
Pepstatin	Pepsin	Phe-Gly-His-Phe(NO_2)-Phe-Ala-Phe · ME	0.4	0.001	Competitive
Phosphoramidon	Thermolysin	Z-Gly-Leu · NH_2[5]	20	0.28	Competitive
	B. subtilis	Z-Gly-Leu · NH_2	33	65.0	Competitive
	B. griseus	Z-Gly-Leu · NH_2	40	0.2	Competitive
P-Leu-Trp[1]	Thermolysin	Z-Gly-Leu · NH_2	20	0.02	Competitive
	B. subtilis	Z-Gly-Leu · NH_2	33	3.0	Competitive
	B. griseus	Z-Gly-Leu · NH_2	40	0.01	Competitive

[1] L-Leucyl-L-tryptophan *N*-Phosphate.
[2] *a*-*N*-(*p*-toluene sulfonyl)-L-arginine methyl ester HCl.
[3] *a*-*N*-benzoyl-L-arginine ethyl ester HCl.
[4] Acetyl-L-alanyl-L-alanyl-L-alanine *p*-nitroanilide.
[5] Carbobenzoxyglycyl-L-leucineamide.

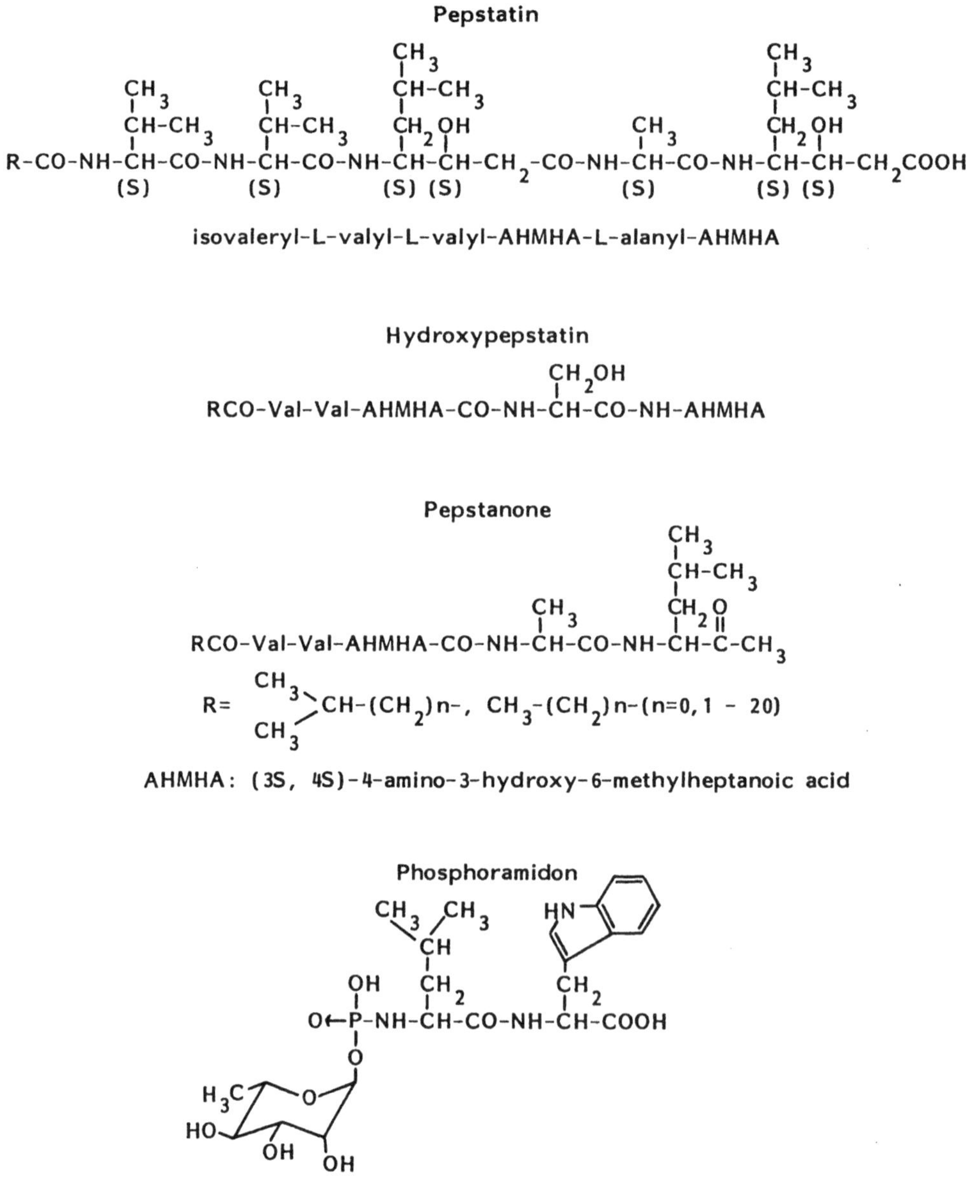

Figure 2 Inhibitors of carboxyl and metal proteinases.

Specific inhibitors of pepsin long remained unknown, although such inhibitors would have obvious application in the treatment of gastric ulcers. Pepstatin, pepstanone, and hydroxypepstatin have been found by our screening for inhibitors of pepsin (Figure 2) [11–13]. They inhibit pepsin, cathepsin D, chymosin, renin, and so forth. Pepstatin, which has been used widely in biochemical studies, contains the isovaleryl group. The structure of pepstatin was determined as *iso*-valeryl-L-valyl-L-valyl-AHMHA-L-alanyl-AHMHA (AHMHA or statin: 4-amino-3-hydroxy-6-methylheptanoic acid). Pepstatin-producing strains of *Streptomyces testaceus* afforded several kinds of pepstatin which differed from one another in the fatty acid moiety (C_2-C_{20}) or amino acid moiety [14]. Pepstatins, pepstanones, and hydroxypepstatins were found to show an almost equal activity against pepsin and cathepsin D [15, 16]. However, pepstatins are more effective against renin than either pepstanone or hydroxypepstatin, and the activity of pepstatin increases with increasing number of carbon atoms in the fatty acid moiety. The hydroxyl group in the inner AHMHA moiety is necessary for activity. The activity of dehydroacetylpepstatin against pepsin is 100 times weaker than that of pepstatin [15]. These results suggest that not only the hydrophobic binding to the enzyme, but also hydrogen bonding through one of the hydroxyl groups, may be necessary for pepstatin to inhibit the activity of pepsin. Moreover, as minor components, pepstanone containing (3 S)-3-amino-5-methylhexane-2-one instead of the C-terminal AHMHA and hydroxypepstatin containing L-serine instead of L-alanine were isolated.

The tight binding of pepstatin to pepsin is confirmed by kinetic studies. The Kcat value for the hydrolysis of the synthetic substrate, Phe-Gly-His-Phe(NO_2)-Phe-Ala-PheOMe, by pepsin is very large, and kinetic data can be obtained at very low pepsin concentrations with a short reaction time. Pepstatin was shown to be competitive, with a K_i of 9.7×10^{-11} M [17]. The kinetic constants of the inhibitor is shown in Table 2.

Pepstatin is very effective against Shay rat ulcer, which is caused by ligation of the pylorus, indicating that secreted pepsin may be the main cause of this ulcer. Pepstatin exhibits a very low toxicity.

As an inhibitor of metal-proteinases, we obtained the thermolysin inhibitor, phosphoramidon, from *Streptomyces tanashiensis* in our screening studies (Figure 2) [18]. The same compound was also isolated by the screening of Ehrlich positive products [19]. The structure of this inhibitor was determined as *N*-(*a*-L-rhamnopyranosyloxyhydroxyphosphinyl)-L-leucyl-L-tryptophan (Rha-P-Leu-Trp). Several analogs have been prepared and their activities have been compared with that of phosphoramidon. Phosphoryl-L-leucyl-L-tryptophan (P-Leu-Trp) was found to be more active than phosphoramidon. The kinetic constants of the above mentioned inhibitors are shown in Table 2.

4. Enzymatic Activities of Cellular Membrane

Inhibition of growth or locomotion of mammalian cells, their morphological changes, their metastasis to specific tissues and other phenomena are mediated through their mutual recognition, aggregation (association), adhesion, and so forth. These phenomena are the most fundamental ones in mammalian cells and have been the matters of absorbing great interest in biology. Especially, much attention has been focused on the relationship between the phenomena in the cellular membrane and the corresponding cellular functions. Many reports have been published in this field. Studies on the mutual actions of cells are called "cell sociology" and studies on the relationship between a certain specific function of the cellular membrane and the corresponding structure are called "ektobiology" [20]. The relationship between the phenomena in the cellular membrane and the corresponding cellular functions seems to be an important field of research which molecular biology hereafter should challenge in order to broaden our knowledge on life.

When cells are treated with proteolytic enzymes, DNA synthesis in the cells is promoted and also the cellular proliferation is shown to be stimulated. Furthermore, enhancement of the cellular agglutinability with lectins [21] and release of several glycoproteins from the cellular membrane are also known to occur [22]. On the other hand, it is known that plasminogen activator is located on the surface of transformed cells and also that proteolytic activities can be detected on the surface of human erythrocyte [23]. Development of esterolytic activity on the surface of mouse B lymphocytes which are stimulated with anti-immunoglobulin antibody was also recently reported [24]. The relationship between the phenomena in the cellular membrane and the corresponding cellular functions has attracted attention of many researchers and many related reports have been published.

During the early stage of infection with influenza virus, we noticed that the rearrangement of sialic acid-containing-glycoconjugates, a constituent of plasma membrane of the host cells, take place. At the same time, the fluidity of plasma membrane is decreased when virion is absorbed on cell membrane, and this causes a drastic change in the activities of the enzymes which are located on the cell membrane [25–27].

Furthermore, we determined the enzymatic activities on mixovirus virion grown in chorioallantoic membrane (CAM) cells and in monkey kidney (MK) cells. In case of such virion that is grown in CAM cells, the activity of aminopeptidase B (AP-B) is higher than AP-A as is the case with the surface of its host cells shown in Table 3. In the virion that is grown in MK cells, however, the activity of AP-A is higher than AP-B as is the case with the surface of its host cells [28, 29]. This suggests that the aminopeptidase in the envelope of virion came from that located on the plasma membrane of the host cells.

Table 3 Enzymatic activities on myxovirus virion grown in CAM and in MK Cells.

Enzyme	Specific activity ($nmol \cdot min^{-1} \cdot mg^{-1}$ protein)							
	CAM				MK			
	Cell	PR-8	NDV	HVJ	Cell	PR-8	FM-1	HVJ
AP-A	0.6	2.3	2.1	2.2	0.7	4.0	2.9	1.9
AP-B	4.7	6.3	9.9	21.3	0.2	1.5	0.8	0.8

AP: aminopeptidase, CAM: chorioallantoic membrane cells, MK: monkey kidney cells.

In order to get further proof that the enzyme activities so far described were actually located on the cell surface, the enzyme activities of intact cells were determined with use of L-Leu-Gly-Gly or L-Leu-Gly-Gly · glass beads as the substrate. As shown in Table 4, at least Leu-AP is distributed on the cell surface. When the cells were incubated with glass beads only, under similar conditions, the enzyme activities were not released into the supernatant [30, 31]. These results are taken to indicate that the surface of plasma membrane has various enzymes located on it.

Table 4 Release of leucine from substrate and glass beads-bound substrate.

Substrate	Leucine released (μmol/10^6 cells)				
	SV40-C3H	Ehrlich	L-1210	MK	MDCK
L-Leu-Gly-Gly	1.33	1.38	1.52	1.30	1.21
L-Leu-Gly-Gly-glass beads	0.25	0.30	0.38	0.25	0.22
Glass beads	0	0	0	0	0

We have found the activities of aminopeptidases, alkaline phosphatase, and esterase on the surface of lymphocytes, macrophages, normal or virus-transformed cells, tumor cells, and so forth (Table 5). Considering the biological role of these enzymes located on the cell membrane, we undertook a search for inhibitors against the cell surface enzymes. The inhibitors against these enzymes form an enzyme-inhibitor

Table 5 Comparison of enzyme activities on surface of various cells.

Cell	Specific activity (nmol · min^{-1} · mg^{-1} protein)			
	AP-A	AP-B	Alk-Phosphatase	Esterase
Monkey kidney	0.7	0.2	2.2	113.5
Macrophage	0.8	4.8	2.7	46.0
Polymorphonuclear leukocyte	0.2	3.6	25.8	14.0
Lymphocyte (spleen) mouse	0.1	0.7	4.2	26.5
Lymphocyte (spleen) rat	0.0	0.8	3.0	22.8
Thymocyte	0.6	2.1	2.6	14.0
SN 1051	<0.1	1.0	3.6	29.1
SN 1054	<0.1	1.8	5.5	30.3
Bri-7	0.1	2.6	2.1	15.8
Bri-8	0.1	2.5	3.0	43.9
C3H2K normal	0.9	0.6	1.9	33.2
C3H2K transformed	0.1	0.3	2.0	31.1
3Y1 normal	2.4	0.3	0.6	24.0
3Y1 transformed	0.2	0.5	0.6	23.5
L	0.3	2.1	1.0	45.1
FM3A	0.3	2.7	3.7	15.8
L1210	1.1	0.7	2.9	17.0
AH66	3.5	8.4	1.4	57.0
AH66F	0.5	6.5	1.2	17.8
Ehrlich	0.6	14.7	4.0	54.8
Gardner lymphosarcoma	0.4	2.3	2.2	29.6

complex on the cellular membrane. Studies on the effect of this complex on the cellular function will make a new approach to elucidate the relationship between the membrane phenomena and the corresponding cellular functions.

5. Inhibitors Against Cell Surface Enzymes

5.1 Inhibitors against exopeptidases

We have found the activities of aminopeptidases, carboxypeptidases, alkaline phosphatase, esterase, and so forth, on the surface of lymphocytes, macrophages, normal or virus-transformed cells, tumor cells, and so on. Searching for specific inhibitors against these cell surface enzymes, we have discovered inhibitors of exopeptidases, alkaline phosphatase, esterase, and so forth in culture filtrates of streptomycetes. As will be described later, inhibitors of cellular surface enzyme enhanced immune responses, and this effect stimulated interest in studies on inhibitors of these enzymes.

As inhibitors against enzymes including aminopeptidases, dipeptidylaminopeptidases, carboxypeptidases, and dipeptidylcarboxypeptidases, we have amastatin, bestatin, arphamenines A and B, α-aminoacylarginine, actinonin, probestin, prostatin, formestins A and B, ebelactones A and B, leupeptin, Ac-Leu-Argal, diprotins A and B, benzylmalic acid, histargin, chymostatin, foroxymithine, and so forth. They specifically inhibit the corresponding enzymes. The microbes producing these inhibitors as well as chemical structure of the latter, are shown in Tables 6 and 7 and Figures 3, 4 and 5. Inhibitors against exopeptidases were discovered as those against the plasma-membrane-located enzymes as follows.

Table 6 Strains of microbes producing aminopeptidase inhibitors.

Inhibitors	Inhibited enzymes	Strains
Amastatin	AP-A, Leu-AP, AP-M, tri or tetra AP	*Streptomyces* sp. ME98-M3 and more than 3 other species
Bestatin	AP-B, Leu-AP, AP-M, tri or tetra AP	*S. olivoreticuli* and more than 5 other species
Arphamenines A and B	AP-B	*Chromobacterium violaceum*
α-Aminoacylarginine	AP-B	*Actinomycetes* (MF931-A2)
Actinonin	Leu-AP, AP-M	*Streptomyces sp.* MG848-hF6
Probestin	Leu-AP, AP-M	*S. azureus*
Formestins A and B	fMet-AP	*Actinomadura reseoviolacea*
Ebelactones A and B	fMet-AP	*S. aburaviensis* (MG7-G1)
Leupeptin	DAP-III	*S. roseus* and more than 11 other strains
Diprotins A and B	DAP-IV	*Bacillus cereus* BMF673-RF1

Amastatin strongly inhibits aminopeptidase A (AP-A), Leu-AP, Tyr-AP, and tripeptidyl and tetrapeptidyl aminopeptidases [32]. Amastatin has five asymmetric

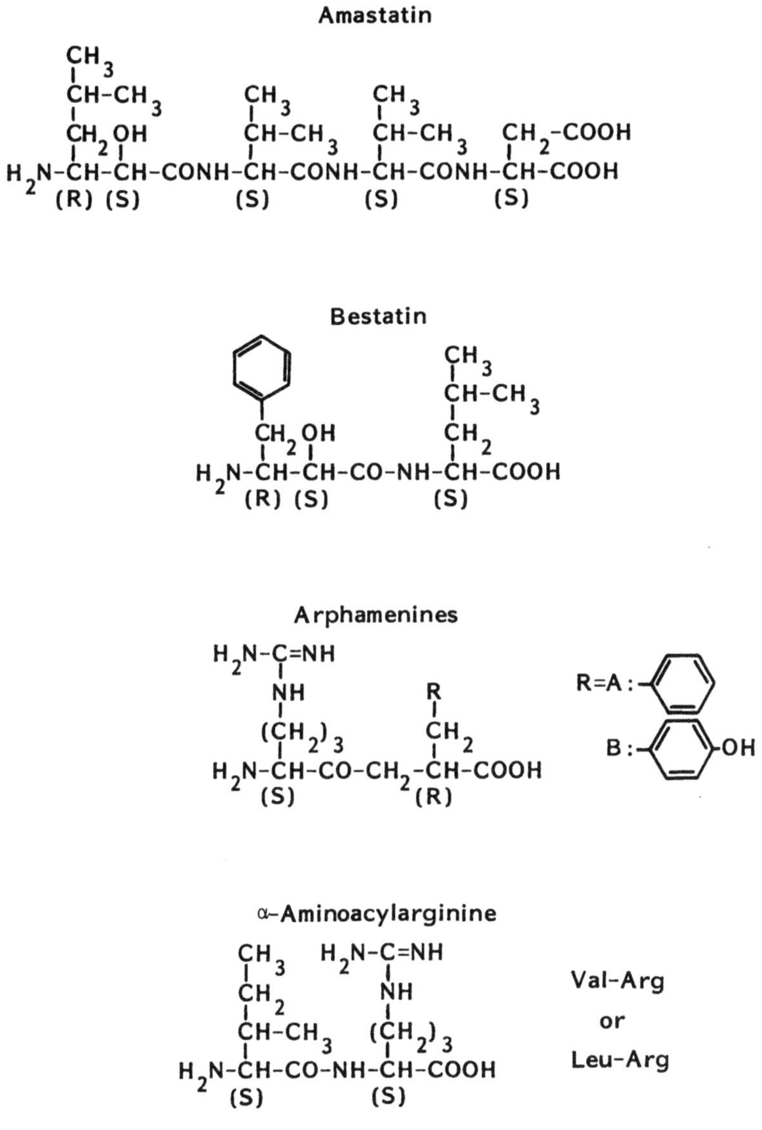

Figure 3 Inhibitors of aminopeptidases-1.

Table 7 Strains of microbes producing carboxypeptidase inhibitors.

Inhibitors	Inhibited enzymes	Strains
(S)-*a*-Benzylmalic acid	CP-A, CP-B	*S. hygroscopicus* (MG368-CF16)
Histargin	CP-B, ACE	*S. roseoviridis* (MF118-A5)
Chymostatin	CP-Y	*S. hygroscopicus, S. lavendulae*
Foroxymithine	ACE	*S. nitrosporeus* (MG329-CF56)

CP-A: carboxypeptidase A, ACE: angiotensin-converting enzyme.

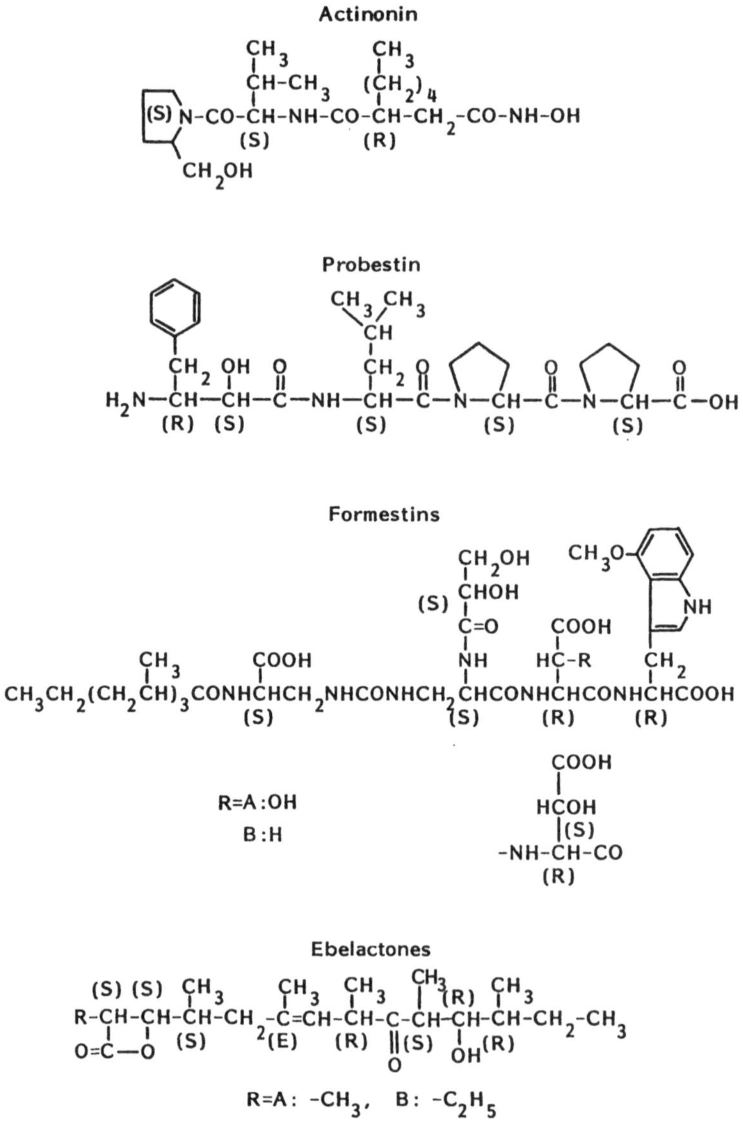

Figure 3 Inhibitors of aminopeptidases-2.

carbon atoms in its molecules. Studies on the inhibitory activity of its stereoisomers obtained by chemical synthesis indicated the following points to be important for its developing activity: (2S,3R)-3-amino-2-hydroxy-5-methylhexanoic acid is located at the N-terminal; the configuration at C2 of this new amino acid is (S); a hydroxyl group is located C2; the amino acid at the C-terminal is aspartic acid or glutamic acid; the amino acid located in the middle of the molecule is hydrophobic (Figure 3) [33].

Ac-Leu-Argal

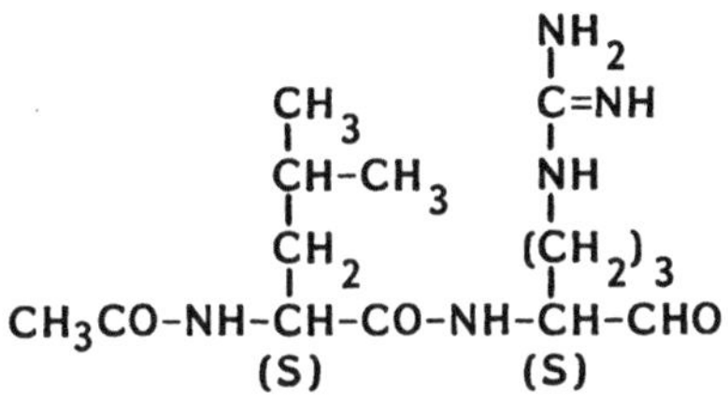

Diprotins

A

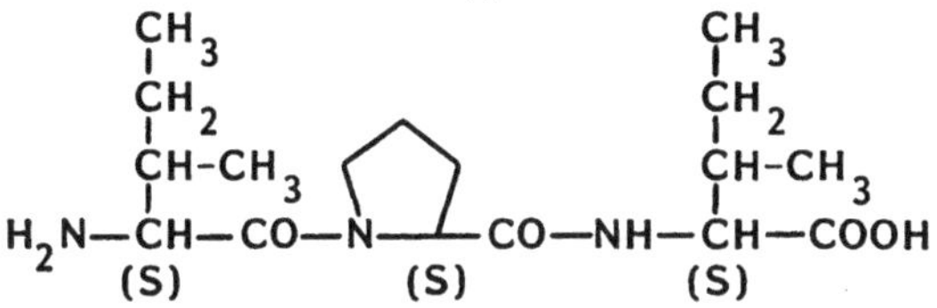

B

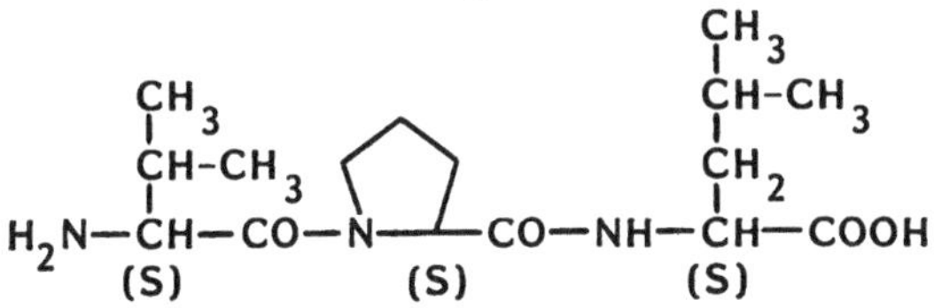

Figure 4 Inhibitors of dipeptidylaminopeptidases.

Bestatin strongly inhibits AP-B, Leu-AP, Ala-AP, and tripeptidyl and tetrapeptidyl aminopeptidases [34]. Bestatin has 3 asymmetric centers and its 8 stereoisomers have been synthesized. The S-configuration of the C-2 of the (2S,3R)-3-amino-2-hydroxy-4-phenylbutanoyl moiety (AHPA) is the key stereochemistry for activity. All the 4 stereoisomers of bestatin with 2S-configuration display inhibitory activites against AP-B and Leu-AP, whereas the 4 other stereoisomers with 2R-configuration have no or only slight activities (Figure 3) [35].

It is important to determine how bestatin and its stereoisomers can modify the cell functions by inhibiting the aminopeptidase activity located on the cell membrane. Ishizuka *et al.* have studied the effect of these compounds on immunoresponsive cells and found that bestatin analogs with the 2 S-configuration can enhance the delayed type hypersensitivity (DTH) [36]. As shown in Table 8, a stimulatory effect of bestatin on DTH establishment is observed in the dose range of 0.1 to 100 μg, but with lower doses showing a marked action of enhancing the DTH, it does not increase the number of antibody-forming cells. On the other hand, isomers having

(S)-α-Benzylmalic acid

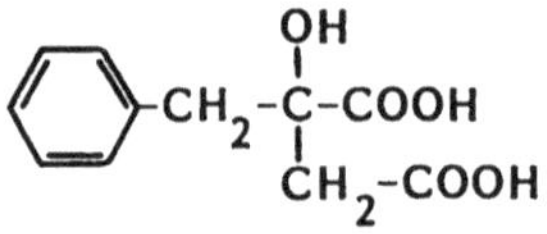

Histargin

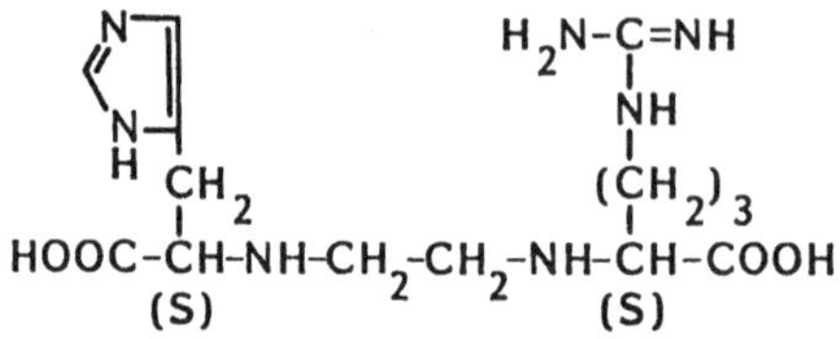

Foroxymithine

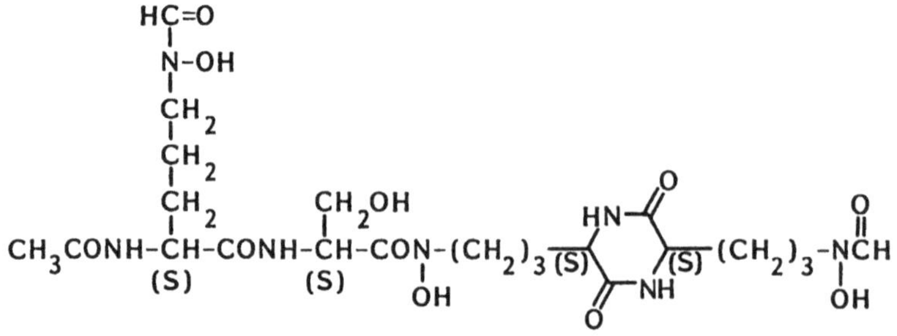

Figure 5 Inhibitors of carboxypeptidases and dipeptidylcarboxypeptidases.

the 2R-configuration have no or only slight activities to inhibit these enzymes and to enhance the DTH. This suggests that the action of bestatin to enhance the immune responses is due to its interaction with aminopeptidases.

Among the inhibitors which enhance immune responses, bestatin has been studied in the most detail. The results of the studies testing the effects of bestatin on the mouse immune system are summarized as follows: (1) Low doses of bestatin enhance the delayed-type hypersensitivity and increase thymidine incorporation into spleen cells. (2) High doses of bestatin increase the number of mouse spleen cells producing antibodies. (3) Bestatin has antitumor effects, enhances the effect of antitumor antibiotics, and promotes the production of interleukins 1 and 2. (4) Bestatin restores the decreased resistance against bacterial infections. In clinical studies, the

Table 8 Effect of bestatin and its stereoisomers on the DTH response and on the aminopeptidases in mice.

Stereoisomer	Dose (μg/mouse)	DTH Enhancement (%)	IC_{50} (μg/ml) AP-B	IC_{50} (μg/ml) Leu-AP
(2S,3R)-AHPA-S-Leu	10	55		
	1	50	0.05	0.01
	0.1	38		
(2S,3R)-AHPA-R-Leu	10	56		
	1	51	0.56	3.4
	0.1	41		
(2R,3R)-AHPA-S-Leu	10	8		
	1	13	>100	>100
	0.1	0		
(2R,3R)-AHPA-R-Leu	10	24		
	1	5	>100	>100
	0.1	9		
(2R,3S)-AHPA-R-Leu	10	0		
	1	2	>100	>100
	0.1	0		

(2S,3R)-AHPA-S-Leu (bestatin): [(2S,3R)-3-amino-2-hydroxy-4-phenylbutanoyl]-(S)-leucine.

oral administration of 30 or 60 mg of bestatin daily has been confirmed to increase the reduced T-cell percentage. It increased the reduced NK-cell activity in cancer patients and improved the bone marrow picture of cancer patients. The suppressor activity in cancer patients' serum inhibiting IgG formation is nullified by bestatin both *in vitro* and *in vivo* [37]. The effects of bestatin on the immune system do not seem to be solely due to the inhibition of aminopeptidases but also to its binding to cells (probably macrophages) involved in immune responses. Bestatin is now considered to be a promising biological response modifiers (BRM) and it is clinically used for the therapy of leukemia.

We have further continued the screening for inhibitors of AP-B, and discovered another group of the inhibitors which we named arphamenines A and B [38]. Arphamenines are completely specific inhibitors of AP-B. The structures were determined to be 5-amino-8-guanidino-4-oxo-2-phenylmethyl-octanoic acid for arphamenine A and 5-amino-8-guanidino-2-(4-hydroxyphenylmethyl)-4-oxo-octanoic acid for arphamenine B [39]. These inhibitors have a methylene ketone ($-CO-CH_2-$) in place of the scissile peptide bond ($-CO-NH-$) of L-arginyl-L-phenylalanine, the substrate of AP-B. There have been many papers reporting the biosynthesis of the peptide bonds of various peptide compounds, but none reporting the biosynthesis of a methylene ketone group linking two amino acids. The biosynthetic pathway of arphamenine A is shown in Figure 6 [40]. Formation of the methylene ketone moiety represents a new biosynthetic pathway. Furthermore, we also found L-isoleucyl-L-arginine, L-leucyl-L-arginine, and L-valyl-L-arginine by screening for AP-B [41].

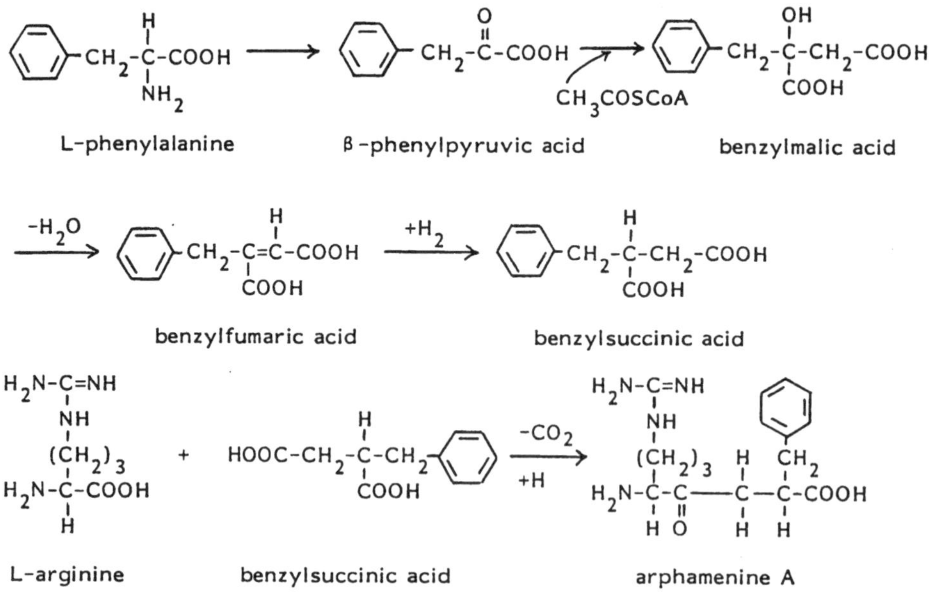

Figure 6 Biosynthesis of arphamenine A.

Table 9 Kinetic constants of aminopeptidase inhibitors.

Inhibitor	Enzyme	Substrate	K_m ($\times 10^{-4}$M)	K_i ($\times 10^{-7}$M)	Type of inhibition
Amastatin	AP-A[1]	Glu · NA[4]	8.0	2.5	Competitive
Amastatin	Leu-AP	Leu · NA	37.0	16.0	Competitive
Bestatin	AP-B	Arg · NA	1.0	0.6	Competitive
Bestatin	Leu-AP	Leu · NA	5.8	0.2	Competitive
Arphamenine A	AP-B	Arg · NA	1.0	0.025	Competitive
Arphamenine B	AP-B	Arg · NA	1.0	0.008	Competitive
Ile-Arg	AP-B	Arg · NA	1.0	5.0	Competitive
Val-Arg	AP-B	Arg · NA	1.0	21.0	Competitive
Actinonin	AP-M	Leu · NA	0.8	1.7	Competitive
Formestin A	fMet-AP[2]	fMet · NA	2.0	0.39	Competitive
Formestin B	fMet-AP	fMet · NA	2.0	1.23	Competitive
Ebelactone A	fMet-AP	fMet · NA	2.0	1.73	Noncompetitive
Ebelactone B	fMet-AP	fMet · NA	2.0	0.63	Noncompetitive
Diprotin A	DAP-IV[3]	Gly-Pro · NA	4.0	22.0	Competitive
Diprotin B	DAP-IV	Gly-Pro · NA	4.0	76.0	Competitive

[1] aminopeptidase A, [2] formylmethionine aminopeptidase, [3] dipeptidyl aminopeptidase IV, [4] glutamic acid β-naphthylamide.

It is known that aminopeptidase M (AP-M) is located in the microsome fraction of brush boader of renal tubules in mammals. We discovered actinonin as a specific inhibitor against AP-M [42], but actinonin in itself had been known as an antibiotic and its structure had also been reported [43]. Furthermore, we found probestin and prostatin, which strongly inhibit AP-M and Leu-AP (Figure 3) [44].

As a part of our continuous effort toward the discovery of novel enzyme inhibitors from microbial cultures, we have searched for inhibitors against N-formylmethionine aminopeptidase (fMet-AP). This enzyme is also located on the cell surface and has been reported to play an important role in chemotaxis. Hence, we were interested in this enzyme and its inhibitors. We were successful in obtaining an inhibitor, which we named formestins A and B [45]. On the other hand, ebelactones A and B, which were found by screening for inhibitors of esterase, inhibited not only esterase but also fMet-AP (Figure 3) [46].

Furthermore, we screened for inhibitors against dipeptidylaminopeptidase (DAP). Ac-Leu-Argal and leupeptin, specific inhibitors against DAP-III, and diprotins A and B, specific inhibitors against DAP-IV, were discovered (Figure 4) [47, 48].

An inhibitor of carboxypeptidase A (CP-A) was identified as (S)-α-benzylmalic acid [49]. It is a specific inhibitor of CP-A, and it also weakly inhibits CP-B. A specific inhibitor of CP-B was found and was named histargin [50]. It inhibits dipeptidyl carboxypeptidase including angiotensin-converting enzyme (ACE) (Figure 5). In the process of screening for ACE, we found foroxymithine [51]. It inhibits dipeptidyl carboxypeptidase including ACE (Figure 5).

Table 10 Kinetic constants of carboxypeptidase inhibitors.

Inhibitor	Enzyme	Substrate	K_m ($\times 10^{-4}$M)	K_i ($\times 10^{-7}$M)	Type of inhibition
(S)-α-Benzylmalic acid	CP-A[1]	Hip-Phe[3]	10.0	6.7	Competitive
Histargin	CP-B	Hip-Lys	62.0	320.0	Competitive
Foroxymithine	ACE[2]	Hip-His-Leu	0.5	–	Uncompetitive

[1] carboxypeptidase A, [2] angiotensin-converting enzyme, [3] hippuryl-L-phenylalanine.

Chart 1 Proteases and their low-molecular-weight inhibitors.

	Peptidases	*Proteinase*	*Inhibitors*
Protease	Endo-	Serine & Cysteine-	Leupeptin, Antipain, Chymostatin, Elastatinal, Ac-Leu-Argal, Lystatin Poststatin
		Aspartic-	Pepstatin, Pepstanone, Hydroxypepstatin
		Metal-	Phosphoramidon, Steffimycins B & D
		Peptidase	
	Exo-	Amino-	Amastatin, Actinonin, Arphamenines A & B, Bestatin, Ebelactones A & B, Formestin, Probestin, Prostatin, Leuhistin
		Dipeptidylamino	Ac-Leu-Argal, Antipain, Leupeptin, Diprotins A & B, Octastatins A & B
		Carboxy-	(S)-α-Benzylmalic acid, Histargin
		Dipeptidylcarboxy-	EDDS, Foroxymithine, Histargin

The type of inhibition of these inhibitors against exopeptidases is shown in Tables 9 and 10. All the inhibitors we found are listed in Chart 1 in relation to the classification of the proteases.

5.2 Inhibitors against enzymes on cellular membrane

Our previous studies demonstrated the activities of exopeptidases, carboxypeptidases, alkaline phosphatase, esterase, and so forth, on the surface of lymphocytes, macrophages, normal or virus-transformed cells, and tumor cells. Searching for

Forphenicine

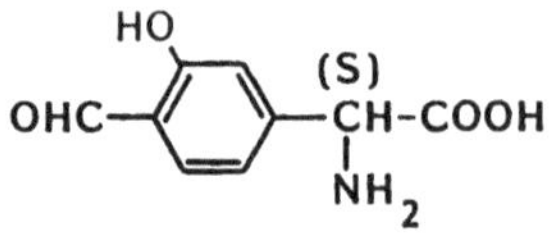

Esterastin

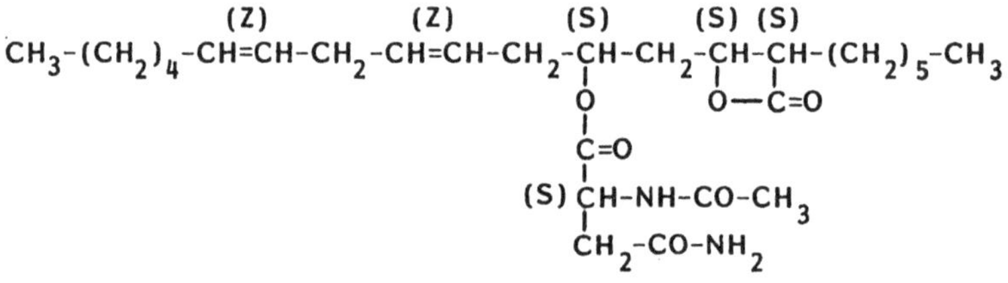

Plipastatins

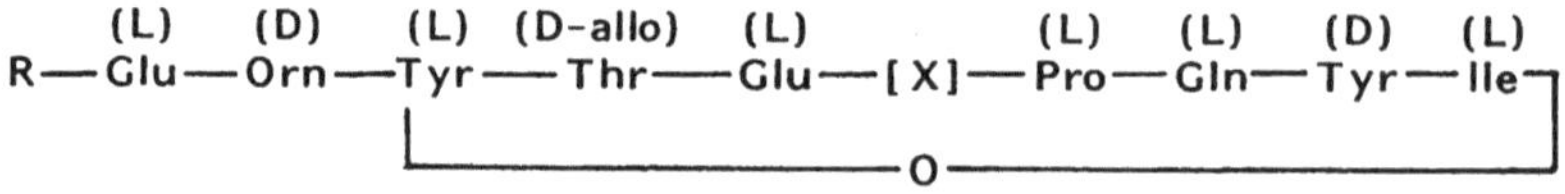

Plipastatins	R		[X]
A1	$CH_3(CH_2)_{12}CH(OH)CH_2CO-$	3-(R)-hydroxy hexadecanoic acid	D-Ala
A2	$CH_3CH_2CH(CH_3)(CH_2)_{10}CH(OH)CH_2CO-$	14-(S)-methyl-3-(R)-hydroxy hexadecanoic acid	D-Ala
B1	$CH_3(CH_2)_{12}CH(OH)CH_2CO-$	3-(R)-hydroxy hexadecanoic acid	D-Val
B2	$CH_3CH_2CH(CH_3)(CH_2)_{10}CH(OH)CH_2CO-$	14-(S)-methyl-3-(R)-hydroxy hexadecanoic acid	D-Val

Figure 7 Inhibitors of plasma-membrane-located-enzymes.

specific inhibitors against exopeptidases, we have discovered various kinds of inhibitors as described before (Figures 3–5). We also have discovered inhibitors of alkaline phosphatase, esterase and phospholipase such as forphenicine, esterastin, ebelactones A and B, and plipastatins A1, A2, B1, and B2. All these inhibitors enhanced immune responses, except for an esterase inhibitor, esterastin, which suppresses immune responses. Chemical structures of these inhibitors are shown in Figure 7.

Forphenicine strongly inhibits alkaline phosphatase prepared from chicken intestine [52]. Formyl group is essential for the molecule to develop its inhibitory activity. Esterastin, which was discovered as an inhibitor against esterase, shows a strong inhibitory activity against esterase, lipase, cholesterol esterase and it also inhibits *N*-formylmethionine aminopeptidase (fMet-AP) weakly [53]. Ebelactones A and B, which were discovered as an inhibitor against esterase, strongly inhibit not only esterase, lipase, cholesterol esterase but also fMet-AP [46]. The compounds are very interesting also in the field of studies on chemotaxis mechanism. In both esterastin and ebelactones A and B, *β*-lactone part of the molecules plays an important role in developing their inhibitory activity. The compound which is obtained by decyclization of the *β*-lactone ring of esterastin shows a very similar structure to mycolic acid, which is usually found in mycobacteria. It does not inhibit esterase any more. In contrast to mycobacteria, actinomycetes has a biosynthetic pathway by which *β*-lactone is formed. The type of inhibition of the above compounds is shown in Table 11.

Table 11 Kinetic constants of inhibitors against plasma-membrane-located-enzymes.

Inhibitor	Enzyme	Substrate	K_m ($\times 10^{-4}$M)	K_i ($\times 10^{-7}$M)	Type of inhibition
Forphenicine	Alkaline phosphatase	NPP[1]	4.6	–	Uncompetitive
Esterastin	Esterase (pancreas)	NPA[2]	4.0	0.002	Competitive
Ebelactone A	Esterase (liver)	NPA	6.7	0.92	Competitive
Ebelactone B	Esterase (liver)	NPA	6.7	0.005	Competitive

[1] *p*-nitrophenyl phosphate, [2] *p*-nitrophenyl acetate.

6. Enzymatic Changes in Various Pathologic Conditions

6.1 Roles of catabolic enzymes in muscular dystrophy

6.1.1 Murine muscular dystrophy

There have been several studies suggesting that the increased activities of intramuscular endopeptidases are related to muscular dystrophy in mice [54–59]. These observations suggest a possible application of protease inhibitors for the treatment of muscular dystrophy [60–62]. However, it is too early to assume that the destruction of muscle in dystrophy is only due to the increased activities of intramuscular endopeptidases.

Table 12 Various enzymatic activities in the muscle of control and dystrophic mice.

Enzyme	Activities* ± SD (n = 4)	
	Control	Dystrophy
AP-A	0.73 ± 0.08	1.68 ± 0.32[c]
AP-B	5.02 ± 1.11	5.38 ± 1.14
Gly-AP	0.65 ± 0.17	2.10 ± 0.27[d]
Ala-AP	5.77 ± 0.96	14.13 ± 0.50[d]
Ser-AP	0.51 ± 0.18	0.85 ± 0.17[b]
Pro-IP	1.10 ± 0.50	1.08 ± 0.43
Leu-AP	3.59 ± 0.31	7.70 ± 0.20[d]
Met-AP	4.64 ± 0.51	11.80 ± 0.60[d]
fMet-AP	5.32 ± 1.31	4.58 ± 0.34
Phe-AP	4.97 ± 0.70	12.40 ± 0.80[d]
Trp-AP	0.96 ± 0.15	3.18 ± 0.05[d]
Gly-Pro-AP	0.52 ± 0.03	0.85 ± 0.13[c]
Gly-Pro-Leu-AP	0.50 ± 0.14	0.80 ± 0.14[b]
Gly-His-Lys-AP	0.09 ± 0.04	0.40 ± 0.00[d]
Trypsin-like	17.00 ± 10.70	8.40 ± 3.74
Chymotrypsin-like	12.20 ± 4.80	18.60 ± 2.40[a]
Elastase-like	8.12 ± 8.95	15.60 ± 6.90
Cathepsin C	0.41 ± 0.09	1.53 ± 0.36[c]
Cathepsin D**	0.00 ± 0.00	0.01 ± 0.00[b]
α-D-Glucosidase	0.20 ± 0.10	0.43 ± 0.05[c]
β-D-Galactosidase	0.04 ± 0.05	0.20 ± 0.08[b]
α-D-Mannosidase	0.08 ± 0.09	0.20 ± 0.00[b]
N-Ac-β-D-Glucosaminidase	0.17 ± 0.05	0.98 ± 0.10[d]
Phosphatase	1.25 ± 0.30	1.78 ± 0.53
Esterase	14.10 ± 16.00	80.10 ± 45.40[b]
RNase (free)***	1.08 ± 0.17	5.53 ± 0.70[d]

* Enzyme activities are expressed as nmol/min/mg protein. ** Cathepsin D activity is expressed as absorbance (at 280 nm)/min/mg protein. *** RNase activities are expressed as unit/mg protein. 1 unit is equivalent to the activity of 1 ng standard RNase from bovine pancreas.
[a] $P < 0.1$; [b] $P < 0.05$; [c] $P < 0.01$; [d] $P < 0.001$.

In order to elucidate the metabolic abnormalities in dystrophic mice and chickens, we investigated a spectrum of enzymes in muscles and other organs [63, 64]. We demonstrated that the activities of aminopeptidases, endopeptidases, glycosidases, phosphatase, esterase, and ribonuclease are abnormally elevated in various organs of dystrophic mice and chickens when compared to the controls.

In Table 12 the enzyme activities are expressed in terms of specific activity per minute per mg protein in the muscle. In the muscle of dystrophic mice, the most significant differences were seen in glycyl aminopeptidase (Gly-AP), Ala-AP, Leu-AP, Met-AP, Phe-AP, Trp-AP, Gly-His-Lys-AP, N-Ac-β-D-glucosaminidase (GlcNAc-as)eand free RNase ($P < 0.001$). A highly significant difference was seen with aminopeptidase A (AP-A), cathepsin C, α-D-galactosidase and total RNase ($P < 0.01$). A significant difference was seen with Ser-AP, Gly-Pro-Leu-AP, cathepsin D, β-D-galactosidase, α-D-mannosidase, and esterase ($P < 0.05$).

Table 13 shows the inhibitory effects (IC_{50}) of amastatin, bestatin, esterastin, and pepstatin on tissue homogenates. As can be seen, the IC_{50} values of amastatin on AP-A activities in muscle homogenates of mice were 0.4 µg/ml for the controls and 1.4 µg/ml for the dystrophic preparations. This larger value for the dystrophic preparations was seen in bone as well. The IC_{50} values of this agent on Leu-AP activities were 11 µg/ml for controls and 4.5 µg/ml for dystrophic preparations, while they were 4.6 µg/ml and 3.5 µg/ml, respectively, in bone.

Table 13 Effects of various inhibitors on their corresponding enzyme activities in organ homogenates of mice.

Inhibitor	Enzyme	Muscle		Bone	
		C	D	C	D
Amastatin	AP-A	0.4	1.4	0.6	2
	Leu-AP	11	4.5	4.6	3.5
Bestatin	AP-B	0.4	0.8	0.2	0.3
	Leu-AP	28	50	22	50
Esterastin	Esterase	0.8	0.8	0.6	1.0
Pepstatin	Cathepsin D	0.005	0.003	0.002	0.002

Each figure represents the 50% inhibition concentration (µg/ml).
C: control mice, D: dystrophic mice, AP: aminopeptidase.

The IC_{50} values of bestatin on AP-B activities in muscle were 0.4 µg/ml for control and 0.8 µg/ml for dystrophic animals, while 0.2 µg/ml and 0.3 µg/ml in bone. The IC_{50} values of bestatin in muscle on Leu-AP were 28 µg/ml for control and 50 µg/ml for dystrophy. Similar tendencies were seen in bone. The IC_{50} values of esterastin on esterase activities were similar for control and for dystrophy. The inhibitory action of pepstatin on cathepsin D activities was extremely high, the IC_{50} being 0.002–0.005 µg/ml for controls and 0.002–0.003 µg/ml for dystrophic preparations.

Table 14 Changes in enzymatic activity of hindlimb muscle of dystrophic mice induced by low-molecular-weight inhibitors.

Enzyme	Specific activity ± SD (n = 5)				
	None (n = 10)	Leupeptin	Bestatin	Forphenicinol	Forphenicine
AP-A	1.60 ± 0.23	1.44 ± 0.22	1.54 ± 0.09	1.28 ± 0.38[a]	1.36 ± 0.48
AP-B	10.07 ± 1.35	9.60 ± 1.55	10.78 ± 0.56	9.76 ± 0.66	8.35 ± 1.06[b]
Pro-IP	3.63 ± 0.78	3.14 ± 0.84	3.48 ± 0.22	4.22 ± 0.56	3.99 ± 0.44
Leu-AP	4.56 ± 0.37	6.46 ± 0.20[d]	7.12 ± 0.37[d]	6.15 ± 1.49[c]	5.93 ± 1.44[b]
fMet-AP	3.60 ± 0.31	4.87 ± 0.88[c]	4.70 ± 0.27[d]	5.24 ± 1.40[d]	3.82 ± 0.83
Phe-AP	13.41 ± 2.12	11.36 ± 0.78[a]	12.11 ± 0.70	11.86 ± 1.24	9.38 ± 2.69[c]
Gly-Pro-Leu-AP	1.28 ± 0.17	1.38 ± 0.11	1.32 ± 0.09	1.15 ± 0.15	1.08 ± 0.25[a]
Trypsin-like	36.40 ± 10.94	52.35 ± 8.58[b]	38.05 ± 5.01	33.81 ± 12.51	22.51 ± 4.33[b]
Chy-try-like	40.17 ± 8.53	50.92 ± 14.82[a]	40.90 ± 6.09	51.95 ± 12.75[a]	25.21 ± 4.03[c]
Elastase-like	29.68 ± 4.94	36.06 ± 14.86	34.20 ± 3.64[a]	31.55 ± 5.29	26.00 ± 6.00
Cathepsin D	0.0042 ± 0.0005	0.0056 ± 0.0005[d]	0.0047 ± 0.0006	0.0043 ± 0.0009	0.0034 ± 0.0011[a]
Glucosidase	0.49 ± 0.08	0.52 ± 0.09	0.52 ± 0.03	0.45 ± 0.09	0.37 ± 0.15[a]
Mannosidase	0.21 ± 0.05	0.21 ± 0.02	0.23 ± 0.02	0.18 ± 0.03	0.19 ± 0.04
GlcNAc-ase	0.86 ± 0.15	1.20 ± 0.19[c]	0.98 ± 0.19	0.96 ± 0.09	0.71 ± 0.26
Creatine kinase	35.68 ± 4.56	32.88 ± 4.69	34.83 ± 1.78	35.33 ± 10.41	25.71 ± 2.43[d]
Phosphatase	1.90 ± 0.79	1.43 ± 0.50	1.71 ± 0.26	1.44 ± 0.63	2.71 ± 0.51[a]
Esterase	46.42 ± 7.50	49.95 ± 5.95	49.11 ± 6.36	50.71 ± 9.04	34.65 ± 9.29[b]

[a] $P < 0.1$; [b] $P < 0.05$; [c] $P < 0.01$; [d] $P < 0.001$.

Table 15 Changes in enzymatic activity of spleen of dystrophic mice induced by low-molecular-weight inhibitors.

Enzyme	Specific activity ± SD (n = 5)				
	None (n = 10)	Leupeptin	Bestatin	Forphenicinol	Forphenicine
AP-A	0.52 ± 0.18	0.43 ± 0.05	0.35 ± 0.03[a]	0.50 ± 0.09	0.64 ± 0.16
Ap-B	16.74 ± 1.15	15.78 ± 2.60	18.40 ± 1.35[b]	15.69 ± 2.07	12.53 ± 3.58[c]
Pro-IP	1.47 ± 0.04	1.60 ± 0.15[b]	1.65 ± 0.17[c]	1.79 ± 0.27[c]	1.48 ± 0.40
Leu-AP	4.38 ± 0.98	5.11 ± 1.48	5.94 ± 0.63[c]	7.13 ± 1.03[d]	5.02 ± 0.82
fMet-AP	11.94 ± 2.13	12.96 ± 2.97	10.85 ± 0.75	10.86 ± 1.29	11.51 ± 0.92
Phe-AP	4.29 ± 0.12	5.73 ± 1.10[d]	4.70 ± 0.46[b]	5.77 ± 0.12[d]	5.47 ± 1.00[c]
Gly-Pro-Leu-AP	0.55 ± 0.01	1.03 ± 0.35[d]	0.68 ± 0.05[d]	0.87 ± 0.09[d]	0.94 ± 0.41[c]
Trypsin-like	231.33 ± 52.17	160.73 ± 73.59[b]	170.03 ± 30.56[b]	175.03 ± 10.93[b]	179.99 ± 39.37[a]
Chy-try-like	351.13 ± 113.35	299.83 ± 119.69	198.37 ± 27.66[b]	209.97 ± 120.64[b]	248.64 ± 68.27[a]
Elastase-like	147.43 ± 97.42	126.57 ± 55.77	103.60 ± 31.54	134.92 ± 13.84	195.18 ± 64.61
Cathepsin D	0.0303 ± 0.0129	0.0384 ± 0.0134[b]	0.0476 ± 0.0174	0.0366 ± 0.0129	0.0293 ± 0.0202
Glucosidase	1.05 ± 0.32	0.76 ± 0.34	0.85 ± 0.19	0.15 ± 0.16[d]	0.69 ± 0.05[b]
Mannosidase	0.74 ± 0.30	0.49 ± 0.34	0.48 ± 0.12[a]	0.0 ± 0.0[d]	0.10 ± 0.14[d]
GlcNAc-ase	9.11 ± 9.81	3.43 ± 1.16	3.92 ± 0.96	2.27 ± 0.13	2.04 ± 0.22
Phosphatase	1.94 ± 0.32	2.43 ± 0.48[b]	2.02 ± 0.33	2.54 ± 0.95[b]	2.67 ± 0.45[c]
Esterase	15.27 ± 1.10	16.52 ± 1.27[a]	16.50 ± 1.02[a]	34.66 ± 2.05[d]	24.33 ± 3.77[d]

[a] $P < 0.1$; [b] $P < 0.05$; [c] $P < 0.01$; [d] $P < 0.001$.

Of the substances tested, amastatin, bestatin, and pepstatin showed strong inhibitory actions. These inhibitors may be worthy of testing their effects on muscular dystrophy. We compared the therapeutic effects of these inhibitors on dystrophic mice. Leupeptin, bestatin, forphenicinol, and forphenicine significantly affected the enzymatic activities in the dystrophic muscles [65].

Table 14 shows the effects of the inhibitors on enzymes in hindlimb muscle. The enzymatic changes which occurred were much different from the ones in forelimb muscle. The activity of Leu-AP was increased by the effects of any of the inhibitors tested. The activity of fMet-AP was increased in most cases. The activity of trypsin-like was elevated by the effect of leupeptin but was suppressed by that of forphenicine. The activities of cathepsin D and GlcNAc-ase were elevated by leupeptin. The activities of AP-B, Phe-AP, trypsin-like, Chy-try-like, creatine kinase and esterase were suppressed by the effect of forphenicine.

Table 15 lists the enzymatic changes in spleen. Leupeptin significantly suppressed trypsin-like but elevated Pro-IP, Phe-AP, Gly-Pro-Leu-AP, cathepsin D, and phosphatase. Bestatin suppressed trypsin-like and Chy-try-like but elevated AP-B, Pro-IP, Leu-AP, Phe-AP, and Gly-Pro-Leu-AP. The effects of forphenicinol were similar to those of bestatin in the fact that it suppressed the activities of trypsin-like, Chy-try-like, glucosidase, and mannosidase but elevated Pro-IP, Leu-AP, Phe-AP, Gly-Pro-Leu-AP, phosphatase, and esterase. Forphenicine suppressed AP-B, and glucosidase but elevated many others including Phe-AP, Gly-Pro-Leu-AP, mannosidase, phosphatase, and esterase.

Table 16 compares the patterns of enzymatic changes caused by each of the agents tested in the present study. It is clearly seen that the effects of these agents resembled each other in the muscles of forelimb and hindlimb. However, forphenicinol differed from the others in its effect on the enzyme networks in spleen. This may be related to the peculiar effects of this inhibitor on immunologically responsive cells.

Table 16 Correlations among enzymatic changes induced by low-molecular-weight inhibitors.

Organs	x	Forphenicine		
	y	Leupeptin	Bestatin	Forphenicinol
Forelimb		0.69[a]	0.94[b]	0.92[b]
Hindlimb		0.75[b]	0.84[b]	0.86[b]
Spleen		0.92[b]	0.89[b]	0.08

[a] $P < 0.05$; [b] $P < 0.01$.

The present results demonstrated the similarity of various inhibitors on enzyme networks in dystrophic mice. Thus it was suggested that low-molecular-weight enzyme inhibitors would be useful in the treatment of muscular dystrophy. Furthermore, it is noteworthy that forphenicinol presented a peculiarity in its effect on the enzyme networks in spleen. This may be related to its unique immunological actions [66] and may warrant its potential usefulness for the therapy of some intractable diseases.

6.1.2 Duchenne muscular dystrophy

Previously, we demonstrated that the activities of aminopeptidases, serine proteinases, carboxyl proteinases, glycosidases, cathepsin C, phosphatase, esterase and RNase are abnormally elevated in various organs of dystrophic mice and chickens when compared with the controls [63, 64, 67].

Although serum enzyme activities of patients with Duchenne muscular dystrophy (DMD) are well investigated, in routine clinical practice only a few of the wide range of possible tests are used regularly, namely those for creatine kinase (CK), aspartate aminotransferase (AST), alanine aminotransferase (ALT), lactate dehydrogenase (LDH), and fructose-bisphosphate aldolase (aldolase) [68–70]. Hence it seems important to investigate the behavior in serum of the above mentioned catabolic enzyme activities and to evaluate their relation to the routine clinical tests.

Enzyme activities were tested in serum from normal subjects and from patients with Duchenne muscular dystrophy (DMD) and asthma [71]. When compared with normal controls, the patients of Grade I showed a significant increase in the activities of most aminopeptidases. On the contrary, the activities of trypsin, cathepsin C, esterase, and RNase were significantly decreased (Table 17).

However, the increased enzyme activities in these patients tended to decrease toward the control values when the severity of the disease increases to Grade II and to Grade III. A typical example is seen in the case of Gly-Pro-AP, which showed an increase from 15.10 in normal subjects to 27.86 in Grade I but again decreased to 17.69 in Grade II and finally to 7.19 $\text{nmol} \cdot \text{min}^{-1} \cdot \text{ml}^{-1}$ in Grade III. Aminopeptidases represented by the activities of AP-A, Gly-AP, Ala-AP, Ser-AP, Leu-AP, Met-AP, Phe-AP, Trp-AP as well as Conv-Enz, sulfatase and phosphatase showed similar tendencies.

On the other hand, such enzyme activities that were significantly decreased in the Grade I patients, tended to increase toward the control levels with the severity of the disease. The activities of trypsin, for example, once reduced from 892 in normal subjects to 780 but again increased to 953 and finally to 967 $\text{nmol} \cdot \text{min}^{-1} \cdot \text{ml}^{-1}$. The activities of cathepsin C, esterase and RNase showed similar tendencies.

In Table 18 the coefficients of correlation (r) and the statistical degree of significance are shown. Sixteen out of the 24 enzyme activities tested showed significant logarithmic regression on the increase of patients' age. Most of these 16 significant regressions were associated with negative correlations ($r = -0.418$ to -0.884). It is to be noted that all of the clinically applied enzymes (i. e. CK, AST, ALT, LDH, and aldolase) were the ones exhibiting negative correlations. In contrast to these negative correlations of the clinically used parameters and of most aminopeptidases, trypsin, cathepsin C, and RNase revealed positive correlations. It should be noted that these enzyme activities were in nearly every instance those which tended to be decreased in Grade I patients as compared with the normal controls (Table 17). It is noteworthy that these enzymes are generally the ones which showed special tendencies when analyzed in relation to the severity of the disease.

Table 17 Comparison of serum enzyme activities between normal subjects and dystrophic patients.

Enzyme	Specific activity ± SD ($nmol \cdot min^{-1} \cdot ml^{-1}$)			
		muscular dystrophy		
	normal ($n = 9$)	Grade I ($n = 9$)	Grade II ($n = 10$)	Grade III ($n = 5$)
AP-A	4.09 ± 1.78	12.50 ± 2.15[d]	7.46 ± 3.70[b]	2.80 ± 0.73
AP-B	15.69 ± 7.44	15.14 ± 2.65	13.89 ± 2.13	12.41 ± 1.95
Gly-AP	4.56 ± 2.21	13.08 ± 2.66[d]	7.76 ± 3.24[b]	5.86 ± 0.78
Ala-AP	61.27 ± 20.50	79.19 ± 16.74[a]	65.77 ± 16.92	56.96 ± 10.91
Ser-AP	1.81 ± 0.85	4.56 ± 0.86[d]	3.06 ± 0.78[c]	2.59 ± 0.20[a]
Leu-AP	23.72 ± 9.80	34.77 ± 6.09[b]	27.84 ± 8.54	22.99 ± 4.69
Met-AP	42.90 ± 14.79	46.15 ± 8.99	41.64 ± 8.19	41.23 ± 9.65
Phe-AP	14.69 ± 3.66	48.28 ± 7.05[d]	40.10 ± 8.46[d]	36.82 ± 4.44[d]
Trp-AP	6.55 ± 3.04	18.23 ± 3.01[d]	15.17 ± 3.70[d]	11.75 ± 1.05[c]
Gly-Pro-AP	15.10 ± 4.90	27.86 ± 6.63[d]	17.69 ± 8.43	7.19 ± 2.32[c]
Gly-Pro-Leu-AP	8.73 ± 5.14	5.10 ± 1.81[a]	5.22 ± 1.89[a]	4.64 ± 2.15
Trypsin	892 ± 86	780 ± 132[b]	953 ± 113	967 ± 233
Cathepsin C	8.65 ± 4.64	1.65 ± 0.72[d]	4.00 ± 1.98[c]	6.82 ± 4.13
Conv-Enz	8.67 ± 2.23	10.37 ± 3.95	7.14 ± 1.85	4.22 ± 1.15[c]
Sulfatase	0.52 ± 0.45	0.61 ± 0.66	0.40 ± 0.34	0.44 ± 0.27
Phosphatase	2.05 ± 1.90	2.85 ± 1.30	2.67 ± 1.06	1.54 ± 0.51
Ac-ch-esterase	13.79 ± 2.16	10.31 ± 10.30	7.93 ± 6.85[b]	12.11 ± 5.09
Esterase	3232 ± 631	1379 ± 588[d]	1674 ± 455[d]	1936 ± 634[c]
RNase	58.62 ± 13.14	40.43 ± 21.28[b]	47.91 ± 12.88[a]	55.13 ± 5.09

[a] $P < 0.1$; [b] $P < 0.05$; [c] $P < 0.01$; [d] $P < 0.001$. P always for comparison with 'normal'.
Grade I patients: with the average age of 8, ranging from 4 to 11.
Grade II patients: with the average age of 14, ranging from 7 to 28.
Grade III patients: with the average age of 19, ranging from 15 to 23.
asthma: with the average age of 10, ranging from 2 to 20.
normal: with the average age of 30, ranging from 22 to 40.

Table 18 Relationship between serum enzyme activities and age of patients with duchenne muscular dystrophy.

Enzyme	Correlation coefficients	Enzyme	Correlation coefficients
AP-A	-0.732^{d}	Cathepsin C	0.773^{d}
AP-B	-0.143	Conv-Enz	-0.493^{b}
Gly-AP	-0.519^{c}	Sulfatase	-0.437^{b}
Ala-AP	-0.217	Phosphatase	-0.578^{c}
Ser-AP	-0.473^{b}	Ac-ch-esterase	-0.025
Leu-AP	-0.355^{a}	Esterase	0.281
Met-AP	-0.004	RNase	0.469^{b}
Phe-AP	-0.299	CK	-0.884^{d}
Trp-AP	-0.418^{b}	AST	-0.841^{d}
Gly-Pro-AP	-0.603^{c}	ALT	-0.755^{d}
Gly-Pro-Leu-AP	0.183	LDH	-0.792^{d}
Trypsin	0.643^{d}	Aldolase	-0.722^{d}

^{a} $P < 0.1$; ^{b} $P < 0.05$; ^{c} $P < 0.01$; ^{d} $P < 0.001$.
Sample numbers are 24 for all the enzyme activities except LDH and aldolase for which sample numbers are 23.

The activities of many hydrolytic enzymes are known to be increased in the muscular tissues of dystrophic mice and chicken [63, 64]. Our serial studies suggested the possibility that not only the abnormalities in endopeptidases but also the abnormalities in exopeptidases play some roles in pathogenesis of muscular dystrophy. Thus it was suggested that low-molecular-weight enzyme inhibitors would be useful in the treatment of this disease [72].

6.2 Spontaneously hypertensive rats and aminopeptidases

Although peptidergic mechanisms seem to be involved in blood pressure regulation, there is still controversy on the roles of physiologically active peptides including the renin-angiotensin system in genetic hypertension.

In order to understand the role of peptide metabolism in spontaneously hypertensive rats (SHR), we performed a longitudinal study for 20 weeks on SHR and Wistar-Kyoto rats (WKR) to determine the relationship between peptide metabolism and the age-dependent increase in blood pressure [73].

Figure 8 compares the time courses of the plasma levels of AP-A and AP-B in SHR with those in WKR. The AP-A level gradually decreased with age, making very similar hyperbola-like curves for the two strains. In regard to the AP-B level, however, such a hyperbola-like curve was found only in WKR. In SHR it remained fluctuating in the range of 5 to 7 $nmol \cdot min^{-1} \cdot ml^{-1}$ throughout the period of observation. When these data were replotted on semilogarithmic paper, statistically significant regression lines were obtained for AP-A in both strains and AP-B in WKR.

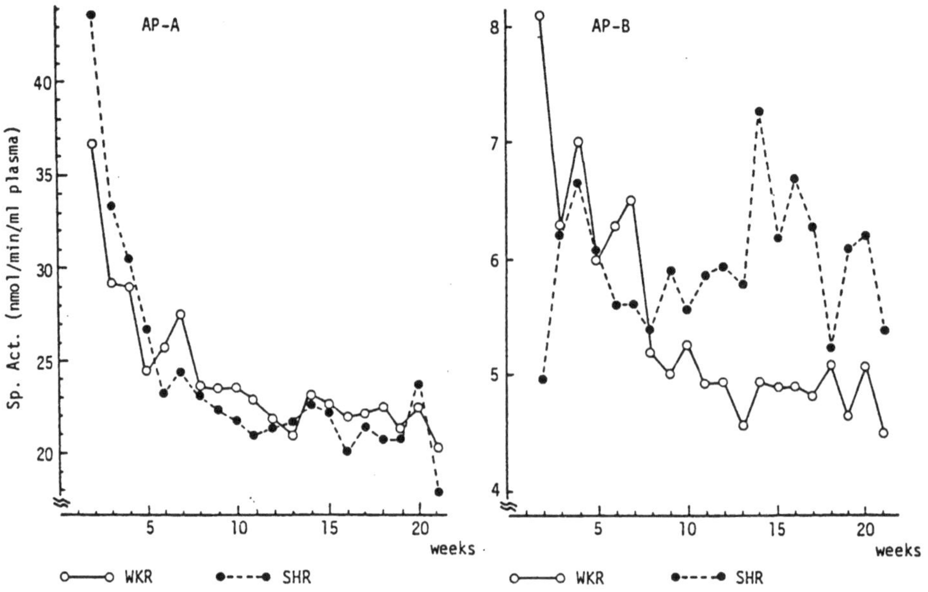

Figure 8 Age-dependent changes of plasma aminopeptidase levels in SHR and WKR.

Table 19 Comparison of plasma protease levels and renin-angiotensin systems in WKR and SHR.

Enzyme	Specific Activity ± SD (n = 20)	
	WKR	SHR
AP-A[1]	24.26 ± 3.87	24.13 ± 5.92
AP-B[1]	5.45 ± 0.94	5.94 ± 0.57
Kallikrein[1]	1232.31 ± 228.61	1009.51 ± 195.19[a]
ACE[1]	1.26 ± 0.15	1.11 ± 0.45
PRA[2]	14.25 ± 3.02	14.15 ± 3.73
IRAI[2]	3.32 ± 0.76	3.72 ± 1.01
Blood pressure[3]	130.83 ± 11.53[4]	164.84 ± 24.64[4,b]

[1] Specific activity is expressed in nmol/min/ml plasma. [2] Specific activity is expressed in ng/hr/ml plasma. [3] Blood pressure is expressed in mm Hg. [4] $n = 18$. ACE: angiotensin-converting enzyme, PRA: plasma renin activity, IRAI: immunoreactive angiotensin I.
[a] $P < 0.05$; [b] $P < 0.001$.

Table 19 lists all of the enzymatic activities examined in plasma, as well as the levels of plasma renin activity (PRA) and immunoreactive angiotensin I (IRAI) in WKR and SHR. Only the level of kallikrein was significantly decreased in SHR when compared with WKR.

In order to test the modes of age-related changes in all of the parameters checked in this study, we calculated their regression lines (Table 20). When plotted on semilogarithmic scales, statistically significant regression lines were obtained for

Table 20 Age-dependency of plasma levels of proteases, renin, and angiotensin I in WKR and SHR.

Enzyme	WKR ($n = 20$)		SHR ($n = 20$)	
	Regression lines	Correlation (r)	Regression lines	Correlation (r)
AP-A	$y = -5.24 \log x + 36.2$	-0.90^{c}	$y = -7.59 \log x + 41.5$	-0.87^{c}
AP-B	$y = -1.28 \log x + 8.4$	-0.91^{c}	$y = 0.15 \log x + 5.6$	0.18
Kallikrein	$y = 305 \log x + 540$	0.89^{c}	$y = 257 \log x + 427$	0.88^{c}
ACE	$y = -0.19 \log x + 1.69$	-0.83^{c}	$y = -0.53 \log x + 2.32$	-0.79^{c}
PRA	$y = -0.76 \log x + 16.0$	-0.17	$y = 3.31 \log x + 6.6$	0.59^{b}
IRAI	$y = 0.13 \log x + 3.03$	0.12	$y = 0.73 \log x + 2.05$	0.48^{a}
Blood pressure[1]	$y = 21.0 \log x + 79.9$	0.91^{c}	$y = 45.1 \log x + 55.7$	0.91^{c}

[1] $n = 18$. [a] $P < 0.05$; [b] $P < 0.01$; [c] $P < 0.001$.

the levels of AP-A ($r = -0.90$), AP-B ($r = -0.91$), kallikrein ($r = 0.89$), and angiotensin-converting enzyme (ACE, $r = -0.83$) in WKR, whereas in SHR, AP-A ($r = -0.87$), kallikrein ($r = 0.88$), ACE ($r = -0.79$), PRA ($r = 0.59$), and IRAI ($r = 0.48$) were significant. It is noteworthy that PRA and IRAI, having no significant correlations with age in WKR, showed significant correlations in SHR. The opposite was true of the level of AP-B, whose correlation coefficient decreased from -0.91 in WKR to 0.18 in SHR. The blood pressure showed highly significant correlations ($r = 0.91$) with age in both strains.

In order to compare the metabolic networks represented by interrelations of the various parameters in SHR and WKR, we performed a multivariate analysis [74]. Figure 9 shows a bird's-eye view of the relations among the parameters tested in

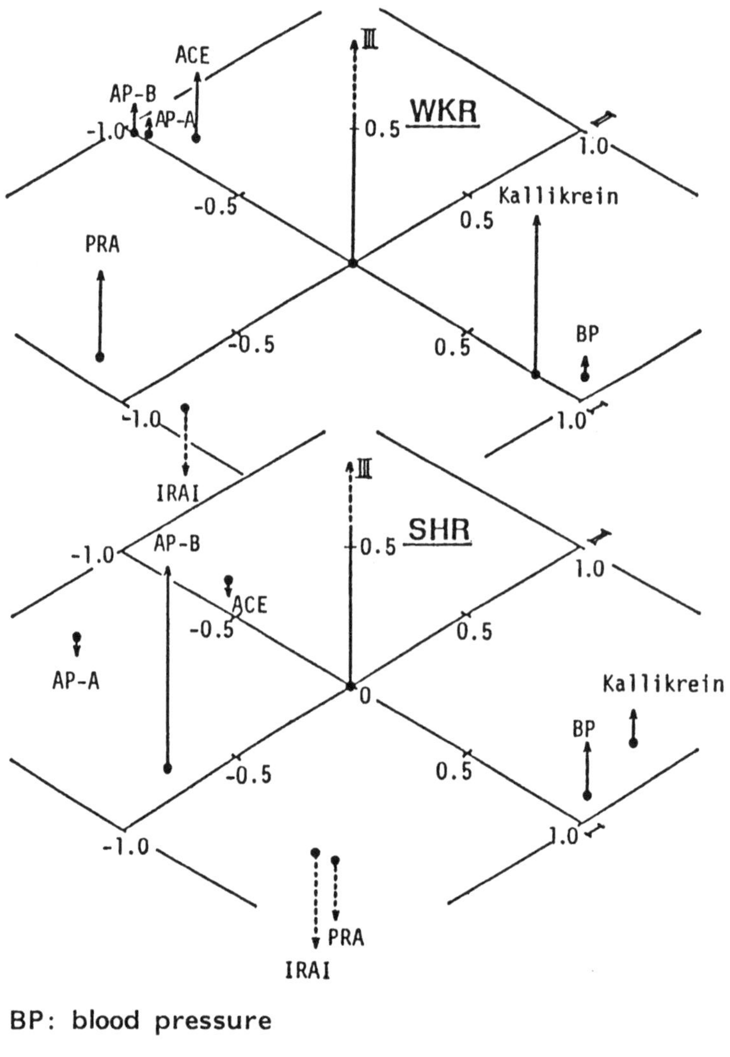

Figure 9 Bird's-eye view of the enzyme networks in SHR and WKR.

the two strains. The first principal component is the one explaining the largest portion of sample variance, namely 56.4% in WKR and 48.7% in SHR. The second and the third components explained 22.8% and 8.0%, respectively, of the sample variance in WKR, 25.7% and 10.9%, respectively, in SHR. By paying attention to the factor loadings, namely, the correlations of each enzyme to each principal component, one can know the character of that particular component. The scales of the three axes, I, II, and III, represent the sizes of such correlation coefficients. One should note that this is a three-dimensional display, expressing the enzymes, PRA, and IRAI as arrows standing on a two dimensional plane which is composed of the axes for the first and the second components. The height of the arrows represents the scale for the third principal component.

In both WKR and SHR, it can be seen that the first principal component is closely related to the blood pressure (BP). Note that in WKR the arrows representing the enzymes, AP-B, AP-A, and ACE, are standing in close connection. However, this interrelationship was broken in SHR. The arrows representing PRA and IRAI are standing in close proximity with each other in SHR when compared with WKR. This result suggested abnormalities in the networks of proteolytic enzymes and in the peptide metabolism surrounding the renin-angiotensin system in SHR. These abnormalities may play some important role in pathophysiological mechanisms of hypertension in SHR.

In order to confirm the pathophysiological significance of the dissociation between the plasma levels of AP-A and AP-B in SHR, we tested the effect of chronic administration of several enzyme inhibitors including inhibitors of AP-B on the course of hypertension in these rats [75].

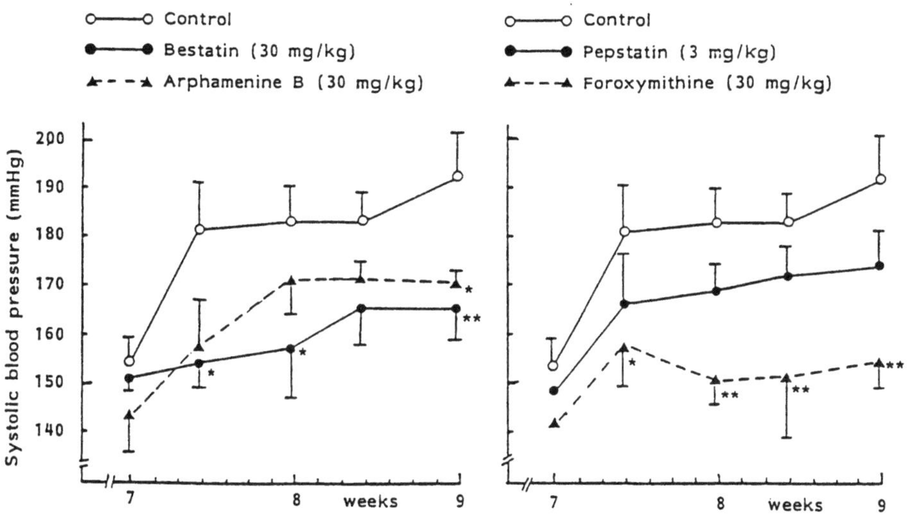

The statistical significance is as to the difference between the controls and the treated. *: $p < 0.05$, **: $p < 0.01$

Figure 10 Suppressive effects of several enzyme inhibitors on the blood pressure in SHR.

Figure 10 shows the effects of daily injection of these inhibitors on age-related changes in blood pressure in SHR. As can be seen, the blood pressure in the control animals gradually increased with age, showing an increase of 40 mmHg at 9 weeks of age. It is a matter of course that foroxymithine, an ACE inhibitor, showed the most striking hypotensive effect. However, it is noteworthy that bestatin also significantly suppressed the development of hypertension. Arphamenine B also tended to suppress the blood pressure, but a statistically significant effect was seen only at the age of 9 weeks. The effect of pepstatin, which is an inhibitor of renin, was not significant. This was possibly due to the fact that the absorption of this agent was not complete because of its insufficient solubility.

The observation that bestatin and arphamenine B showed significant hypotensive effects in SHR favors the possibility that the increase in the plasma level of AP-B, which was found in SHR in our previous study, really plays a pathogenetic role [73]. The stronger effect of bestatin, as compared with that of arphamenine B may be taken to indicate that not only AP-B but also Leu-AP and triaminopeptidase are somehow associated with the development of hypertension in SHR. Regardless of the precise mechanisms involved, the present results may warrant further investigation of the peptide metabolism in this model of hypertension.

6.3 Immunologic diseases and hydrolytic enzymes

Previous studies indicated the importance of hydrolytic enzymes in pathogenesis of autoimmune diseases. Increased activities of lysosomal enzymes have been detected in areas of ongoing inflammation and tissue injury [76–81]. Proteases with acidic pH optima are found in the azurophil granules of polymorphonuclear cells, and they are frequently postulated as accelerators of inflammatory processes [58]. From the standpoint of the application of low-molecular-weight enzyme inhibitors as therapeutic agents, it was of interest to investigate the abnormalities of enzyme networks in a representative experimental model of an autoimmune disease [82]. For this purpose a strain of New Zealand mice was studied as a model for systemic lupus erythematosus (SLE) [78].

Table 21 compares the splenic enzyme activities between the NZB/W mice and their controls. At 10 weeks of age, aminopeptidases such as Leu-AP and Phe-AP showed significantly decreased activities in NZB/W mice. Of the endopeptidases, chymotrypsin-like and elastase-like ones showed significantly increased activities, whereas post-proline-cleaving enzyme (Post-Pro-Enz) and cathepsin D activities were decreased. Glycosidases, such as glucosidase and mannosidase, showed significantly increased activities in NZB/W mice. At 20 weeks of age, the activities of AP-A, AP-B, Pro-IP, fMet-AP, glucosidase, and mannosidase were significantly increased, while those of Gly-Pro-Leu-AP, chymotrypsin-like, elastase-like, and esterase were decreased in NZB/W mice. At 30 weeks of age, the activities of AP-B, Pro-IP, fMet-AP, Post-Pro-Enz, glucosidase, mannosidase, GlcNAc-ase, and phosphatase were all significantly increased, while those of trypsin-like and elastase-like were lower [82].

Table 21 Comparison of spleen enzyme activities between NZB/W and control mice.

Enzyme	Specific activity ± SD (nmol · min^{-1} · mg^{-1} protein)					
	10 weeks		20 weeks		30 weeks	
	Control (n = 5)	NZB/W (n = 7)	Control (n = 5)	NZB/W (n = 5)	Control (n = 5)	NZB/W (n = 4)
AP-A	0.90 ± 0.04	0.95 ± 0.19	1.01 ± 0.07	1.28 ± 0.14[b]	0.77 ± 0.16	0.59 ± 0.13
AP-B	8.13 ± 1.30	9.17 ± 1.03	7.81 ± 0.93	10.63 ± 1.44[b]	7.33 ± 1.09	9.69 ± 1.04[a]
Pro-IP	1.35 ± 0.29	1.38 ± 0.16	1.14 ± 0.13	1.57 ± 0.23[b]	0.97 ± 0.09	1.25 ± 0.16[a]
Leu-AP	8.75 ± 0.77	7.58 ± 0.69[a]	8.64 ± 0.63	9.31 ± 0.84	7.25 ± 0.62	8.80 ± 1.51
fMet-AP	11.51 ± 0.87	11.67 ± 2.01	8.03 ± 1.57	12.24 ± 0.81[c]	10.14 ± 1.06	12.57 ± 1.04[a]
Phe-AP	10.46 ± 1.47	8.07 ± 0.98[b]	9.76 ± 0.80	9.33 ± 1.18	8.68 ± 0.79	8.09 ± 1.11
Gly-Pro-Leu-AP	2.12 ± 0.24	1.85 ± 0.28	2.39 ± 0.32	1.78 ± 0.12[b]	1.78 ± 0.15	1.58 ± 0.19
Ang-Conv-Enz	0.56 ± 0.11	0.52 ± 0.12	0.68 ± 0.14	0.53 ± 0.15	0.45 ± 0.10	0.35 ± 0.07
Trypsin-like	86.04 ± 7.35	92.99 ± 26.52	101.83 ± 16.43	95.42 ± 11.35	82.13 ± 23.90	48.85 ± 10.51[a]
Chymotrypsin-like	70.32 ± 8.11	97.79 ± 16.62[b]	121.72 ± 10.61	78.13 ± 28.12[a]	87.18 ± 19.47	59.73 ± 17.29
Elastase-like	55.94 ± 10.98	76.79 ± 15.11[a]	97.51 ± 13.74	71.64 ± 5.84[b]	69.32 ± 13.52	46.58 ± 14.33[a]
Post-Pro-Enz	3.39 ± 0.44	1.94 ± 0.37[c]	2.89 ± 0.34	2.95 ± 0.44	2.40 ± 0.39	3.36 ± 0.41[b]
Cathepsin D	0.030 ± 0.004	0.021 ± 0.007[a]	0.030 ± 0.004	0.030 ± 0.006	0.030 ± 0.004	0.026 ± 0.007
Glucosidase	1.28 ± 0.11	1.62 ± 0.19[b]	1.20 ± 0.13	1.93 ± 0.07[c]	0.99 ± 0.09	1.46 ± 0.19[b]
Mannosidase	0.37 ± 0.03	0.54 ± 0.07[c]	0.22 ± 0.03	0.34 ± 0.05[b]	0.23 ± 0.03	0.54 ± 0.12[c]
GlcNAc-ase	3.34 ± 0.43	3.06 ± 0.51	3.22 ± 0.57	3.61 ± 0.43	2.41 ± 0.33	3.17 ± 0.26[b]
Phosphatase	5.41 ± 1.05	5.15 ± 0.71	5.18 ± 0.89	5.68 ± 0.36	3.46 ± 0.88	6.35 ± 1.25[b]
Esterase	32.71 ± 4.23	54.27 ± 45.50	39.71 ± 3.74	33.10 ± 2.40[a]	37.32 ± 10.40	46.48 ± 2.89

[a] $P < 0.05$; [b] $P < 0.01$; [c] $P < 0.001$.

Because of these diversities of enzymatic changes with increasing age of the animals, we performed a multivariate analysis to assess the role of each enzyme in the pathophysiology of this SLE model. Since a particular role for Post-Pro-Enz was indicated by this study, we focused our attention on this enzyme.

The activity of Post-Pro-Enz showed a highly positive correlation with that of Pro-IP in all the organs tested from the control animals. We tested the relationship between the activities of the above two enzymes in spleen. In the control animals, there was a significant correlation ($r = 0.85$) between the two enzymatic activities, but the correlation was not significant ($r = 0.03$) in NZB/W mice. Furthermore, it is to be noted that the relationship of these two enzymes to the aging process completely changes in this pathologic model in comparison with the control. Namely, it is clear that both of the enzymatic activities decreased with increasing age in the control animals. In NZB/W mice, however, the activity of Post-Pro-Enz increased, rather than decreased, with age; and no significant trend was seen for Pro-IP (Table 21).

In order to know the general relationship between the two enzymatic activities, we examined their activities in six organs: forelimb muscle, hindlimb muscle, heart muscle, spleen, liver, and kidney. As can be seen in Table 22, their correlations were as high as 0.80 to 0.97 in all of the organs, except liver, of the control animals, while in half of the organs of the lupus mice they ranged from −0.25 to 0.77 [82].

Table 22 Correlation between the activities of Post-Pro-Enz and Pro-IP in six organs.

Animals	Muscle			Spleen	Liver	Kidney
	Forelimb	Hindlimb	Heart			
Control	0.92^{c}	0.80^{c}	0.97^{c}	0.85^{c}	0.42	0.82^{c}
NZB/W	−0.25	0.35	0.77^{b}	0.03	−0.07	0.23

Note. Numbers refer to coefficients of correlation obtained.
^{b} $P < 0.01$; ^{c} $P < 0.001$.

This close relationship between the two enzymes was more or less distorted in various organs of NZB/W mice. In spleen, especially, the correlation coefficients were dramatically lowered from 0.85 to 0.03. This suggests that the metabolism somehow related to proline is extensively altered in the spleens of these animals. Moreover, it is to be noted that the change in the time course of Post-Pro-Enz has been completely reversed only in the spleen of NZB/W mice. Thus it is likely that this enzyme plays some critical role in the pathogenesis of immunological abnormalities in this experimental model. If this hypothesis is valid, a specific inhibitor of this enzyme should have some effect on this disease in NZB/W mice. Trial experiments along these lines are now being done in this laboratory.

From the standpoint of applying enzyme inhibitors as immunomodifiers, it seems important to know the enzymatic profiles in immunologically competent cells. Hence we studied the relation of various enzymes to subpopulations of lymphocytes in man [83]. As shown in Table 23, the fraction of total mononuclear cells had

Table 23 Enzymatic profiles of mononuclear cell fraction and T-cell rich fraction.

Enzyme	Specific activity ± SD (nmol · min^{-1} · mg^{-1} protein)	
	Mononuclear fract. ($n = 4$)	T-cell rich fract. ($n = 10$)
AP-A	1.20 ± 0.29	0.78 ± 0.24[a]
AP-B	34.68 ± 3.88	24.92 ± 4.30[b]
Pro-IP	10.42 ± 1.60	9.81 ± 2.99
Leu-AP	19.20 ± 6.89	10.45 ± 1.29[b]
fMet-AP	2.01 ± 0.14	2.35 ± 0.33
DAP-II	14.33 ± 5.14	33.13 ± 18.18
DAP-IV	4.09 ± 1.79	6.46 ± 1.30[a]
Post-Pro-Enz	9.38 ± 2.59	14.86 ± 2.85[b]
Cathepsin B	20.62 ± 8.67	9.32 ± 3.57[b]
GlcNAc-ase	2.72 ± 0.66	4.88 ± 0.88[c]
Sialidase	0.013 ± 0.011	0.0 ± 0.0[b]
Poly(ADP-ribose)Syn.	0.035 ± 0.019	0.006 ± 0.008[c]
Esterase	46.93 ± 22.79	35.15 ± 8.06

[a] $P < 0.05$; [b] $P < 0.01$; [c] $P < 0.001$.

Table 24 Discriminant function scores for differentiating mononuclear cell fraction and T-cell rich fraction.

Positive correlation		Negative correlation	
Variables	Γ	Variables	Γ
T-cells	0.84	B-cells	−0.94
GlcNAc-ase	0.79	Macrophage	−0.91
Post-Pro-Enz	0.69	Poly(ADP-ribose)Syn	−0.78
DAP-IV	0.63	Leu-AP	−0.76
		AP-B	−0.75
		Sialidase	−0.75
		Cathepsin B	−0.72
		AP-A	−0.63

significantly higher activities of AP-A, AP-B, cathepsin B, sialidase, and poly(ADP-ribose)synthetase, whereas the T-cell rich fraction had higher activities of dipeptidylaminopeptidase IV, prolyl endopeptidase (Post-Pro-Enz), and N-Ac-β-D-glucosaminidase (GlcNAc-ase).

In order to know the relations of these enzymes to lymphocytes subpopulations more precisely, we tried to mathematically differentiate the two fractions both from cellular and enzymatic aspects using the discriminant function analysis. As shown in Table 24, the multivariate axis for maximally separating the two groups had positive high correlations to the percentage of T-cells, and the activities of GlcNAc-ase, Post-Pro-Enz, DAP-IV and so on. But the same axis had high negative correlations to the percentage of B-cells as well as macrophages, and the activities of poly(ADP-ribose)synthetase, Leu-AP, AP-B, sialidase, cathepsin B, and AP-A. The enzymes with lower correlations than 0.60 were omitted in the table.

Judging from the correlations of each variable to such an axis it seemed reasonable for us to conclude that the enzymes characterizing T-cells were GlcNAc-ase, Post-Pro-Enz, and DAP-IV, whereas those characterizing B cells and/or macrophages were poly(ADP-ribose)synthetase, Leu-AP, AP-B, cathepsin B, and AP-A.

Although from these results alone it is hard to differentiate between the features of B-cells and macrophages, they may give an important clue for the future study in this field.

7. Modification of Metabolic Homeostasis Caused by Low-Molecular-Weight Enzyme Inhibitors

7.1 Enzymatic oscillations caused by bestatin

Our previous studies demonstrated the complex relationships between the actions *in vivo* and *in vitro* of enzyme inhibitors [84–86]. In order to understand the effects of immunomodulators *in vivo*, we performed time series analysis on mice given an aminopeptidase inhibitor, bestatin. The study was undertaken to compare the enzymatic oscillation induced in three major organs, spleen, kidney, and liver of mice by the active [(2S,3R)AHPA-(S)-Leu, bestatin] and inactive [(2R,3R)AHPA-(S)-Leu] forms of bestatin. Table 8 compares the *in vitro* actions of bestatin and its stereoisomers. [(2S,3R)AHPA-(S)-Leu, bestatin] inhibits the activities of AP-B, Leu-

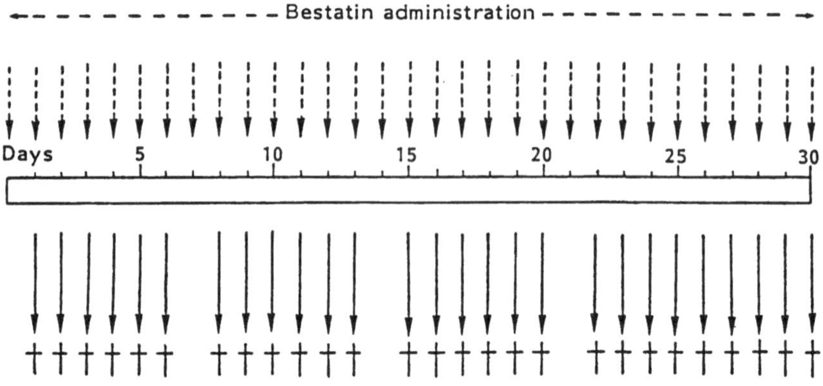

Experimental animals: ICR mice, i.p. injection

Inhibitors: (2S,3R)-bestatin, (2R,3R)-bestatin

Organs: forelimb muscle, hindlimb muscle, heart muscle, spleen, liver, kidney

Enzymatic assays: aminopeptidase (5), endopeptidase (7), GlcNAc-ase, phosphatase, esterase

Figure 11 Effects of bestatin stereoisomers on enzyme networks in murine organs.

AP, and Gly-Pro-Leu-AP, whereas the (2S,3S)-compound had a slight inhibitory action. Two other stereoisomers, (2R,3S)- and (2R,3R)-compounds, had no inhibitory action. Thus, the configuration of the C-2 position seems to be the most important for the inhibitory action of bestatin.

As shown in Figure 11, two hundred micrograms of bestatin [(2S,3R)AHPA-(S)-Leu] or stereoisomer [(2R,3R)AHPA-(S)-Leu] dissolved in 0.2 ml saline were given daily for 30 days. Three mice at a time were killed every 24 to 48 hours, and we tested various enzymatic activities in extracts of spleen, kidney, and liver [87, 88]. These included 7 aminopeptidases, 5 endopeptidases, 3 glycosidases, phosphatase, and esterase. Figure 12 exhibits the time courses of the enzymatic activities in spleen during treatment with either one of the two forms of AHPA-Leu. Leu-AP is one of the enzymes directly inhibited by bestatin *in vitro*. As fas as the splenic enzymes are concerned, the activity of Leu-AP showed two peaks of activity at 1 and 3 weeks, respectively, after the initiation of the treatment with (2S,3R)AHPA-(S)-Leu. However, the treatment with (2R,3R)AHPA-(S)-Leu gave rise to more random fluctuations of this enzyme [88]. Although the time course of Gly-Pro-Leu-AP

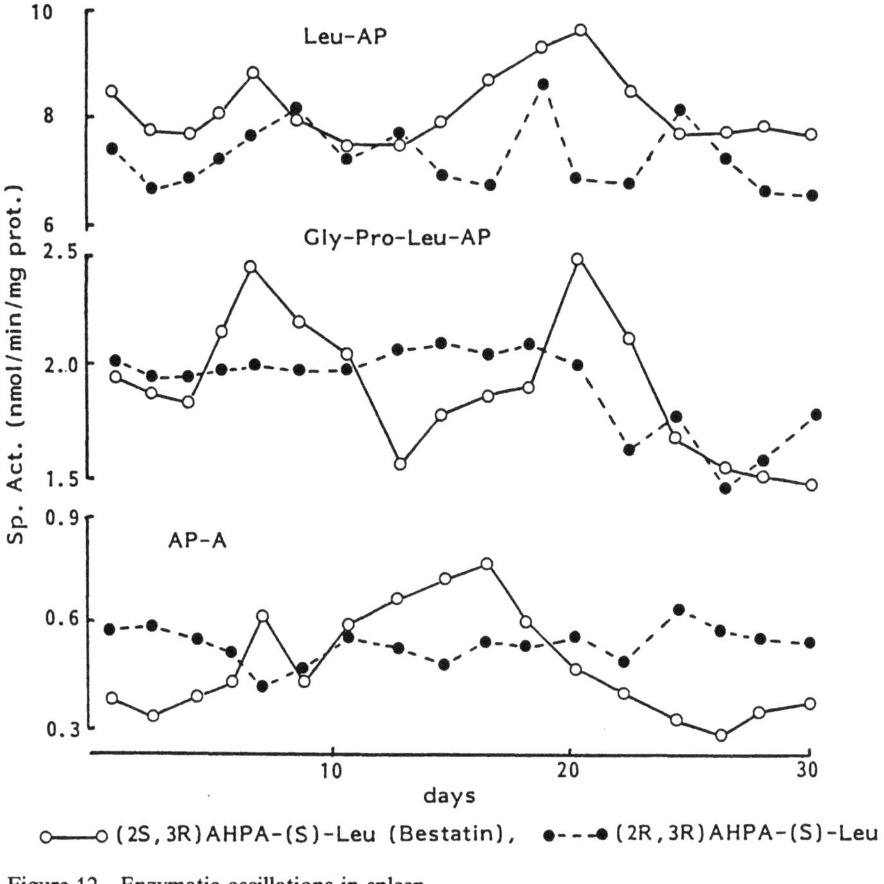

Figure 12 Enzymatic oscillations in spleen.

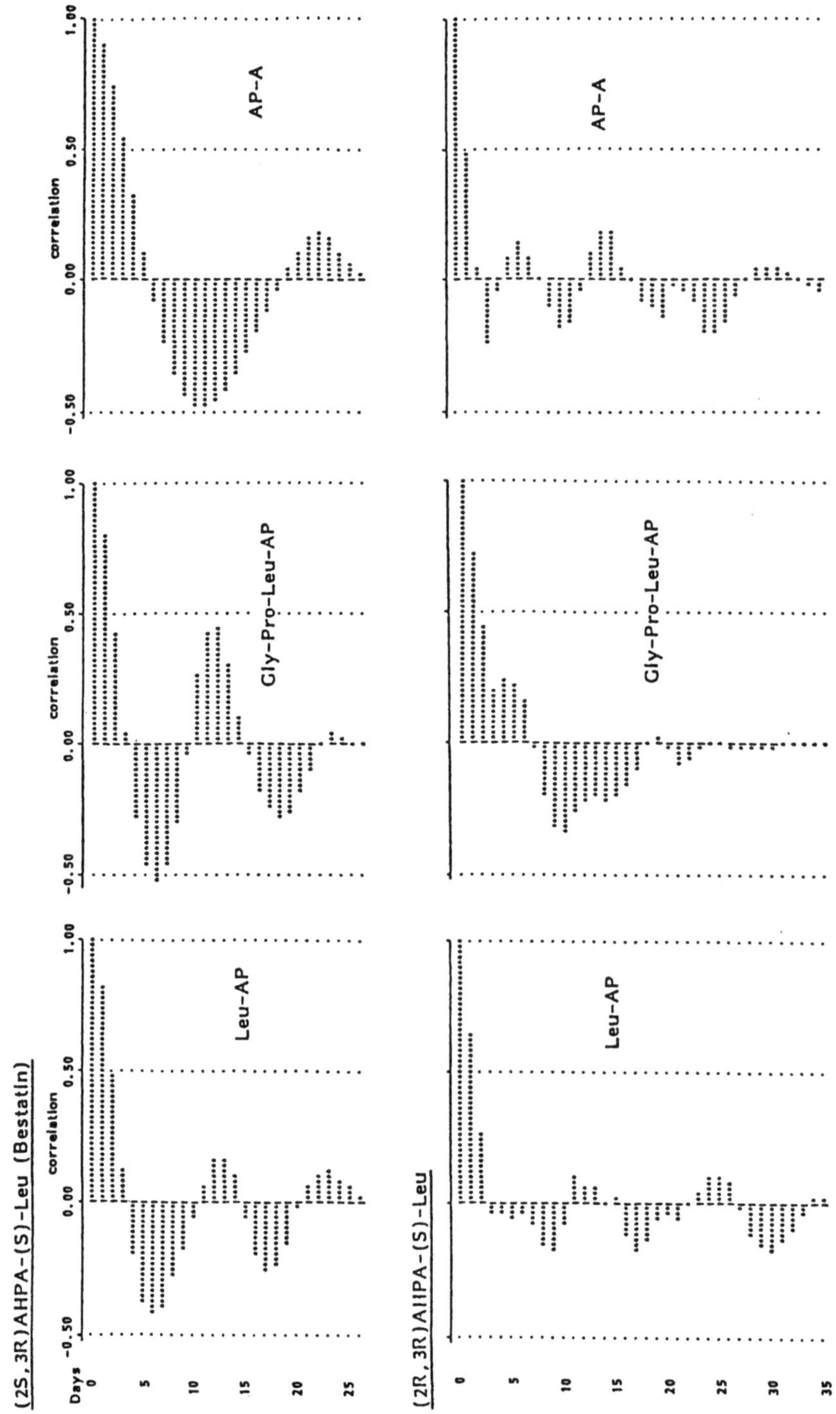

Figure 13 Autocorrelation curves for enzymatic activities in spleen.

much resembled that of Leu-AP in either group of animals, that of AP-A showed somewhat different tendencies. Namely, the activity of this enzyme showed only one peak during treatment with the active form of bestatin. When the inactive form was used, however, AP-A activity showed only minor fluctuations. In order to study the significance of these enzymatic fluctuations, we made autocorrelation curves from the raw data of enzymatic activities (Figure 12) [89].

Figure 13 shows the curves for spleen. Judging from the type of the oscillations resembling sine curves, it is likely that the metabolic processes involving those enzymes are more synchronized in the bestatin-treated animals than in the controls [88]. Bestatin is a chemically well-defined immunomodifier. Of the bestatin-related compounds, all 2S-isomer compounds but not all 2R-isomerones, showed the hypersensitivity-enhancing activity. This correlation between the antienzyme activity and the immunity-enhancing action was postulated to provide the evidence for the binding of bestatin with the enzymes and/or similar proteins of immunoresponsive cell. The above results also suggested the association between the enzymatic oscillations and the biological activity of bestatin. Irrespective of their precise mechanisms, it seems certain that the oscillative phenomena discovered by us are somehow related to the regulatory movement of the dynamic system composed by multiple enzymes in cells. In fact, our recent data obtained by use of a new statistical approach for analyzing feedback systems in the body supported such a view.

7.2 Influence of angiotensin-converting enzyme inhibitor, foroxymithine

The observation that bestatin, an inhibitor of arginine aminopeptidase (AP-B), triggers rhythmical oscillations of various enzymatic activities in spleen and kidney [87, 88] prompted us to test the effects of other inhibitors. Since such observations may give an important clue for the *in vivo* application of enzyme inhibitors, we performed a long-term study using an inhibitor of angiotensin-converting enzyme (ACE), foroxymithine [51]. The results showed that the inhibitor blocks the normal regulatory mechanism of the renin-angiotensin system *in vivo* [90].

Figure 14 shows the time course of the plasma levels of ACE in the animals with or without the administration of foroxymithine. In the control animal, the ACE level gradually decreased throughout the whole period of observation. In the experimental animal, it tended to decrease for the first 70 to 80 days but then increased thereafter. This finding in the experimental animal was accompanied by an abrupt increase in the plasma renin activity (PRA) around the 70th day in the same animal, as shown in Figure 14. In contrast, PRA tended to decrease throughout the whole period in the control animal. Although the initial value, obtained just before the start of treatment with the drug, was much lower in this animal than in the control, it was probably due to the natural diversity among the individual animals. However, it is to be stressed that the most important point of the present study is the network, as a whole, of metabolism rather than the absolute level of each variable. Taking the above observations together, we conclude that the metabolism in the experimental animal changed drastically around the 70th day. The tendency toward an increasing ACE level after this point could be explained by the feedback mechanism through the action of renin.

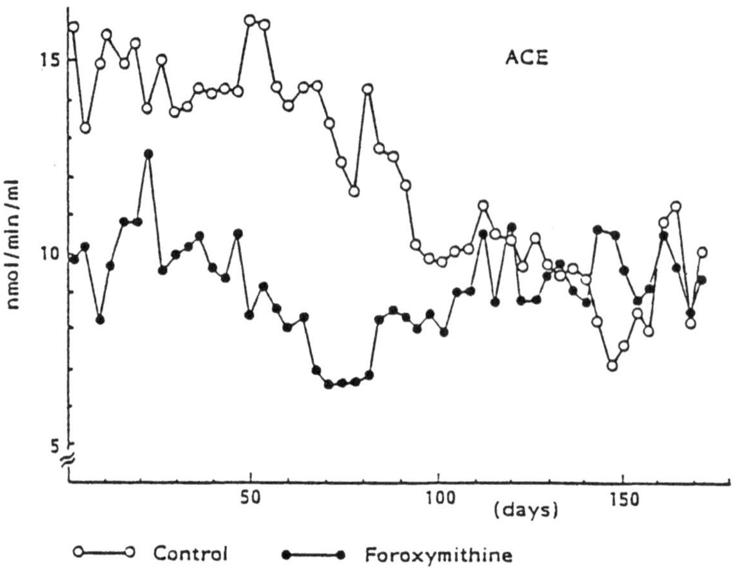

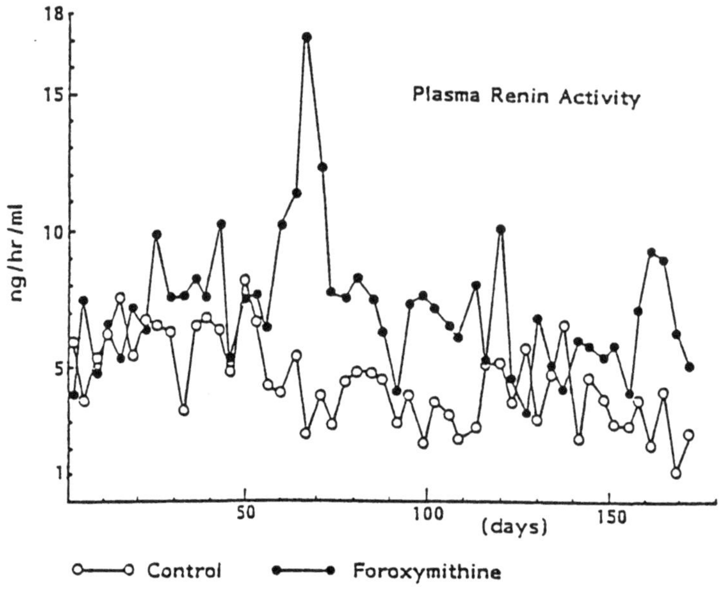

Figure 14 Time course of plasma levels of ACE and renin activity (PRA) in the rabbits with or without foroxymithine administration.

Table 25 Correlations among renin, angiotensin I, and exopeptidases in serum.

	IRAI	PRA	ACE	AP-A	AP-B
Control:					
IRAI		0.37[b]	0.03	0.06	−0.01
PRA	0.37[b]		0.61[c]	0.47[c]	0.47[c]
ACE	0.03	0.61[c]		0.81[c]	0.82[c]
AP-A	0.06	0.47[c]	0.81[c]		0.88[c]
AP-B	−0.01	0.47[c]	0.82[c]	0.88[c]	
Experimental:					
IRAI		0.32[a]	−0.31[a]	−0.01	−0.23
PRA	0.32[a]		−0.24	−0.16	0.08
ACE	−0.31[a]	−0.24		0.59[c]	0.12
AP-A	−0.01	−0.16	0.59[c]		0.03
AP-B	−0.23	0.08	0.12	0.03	

[a] $P < 0.05$; [b] $P < 0.01$; [c] $P < 0.001$.

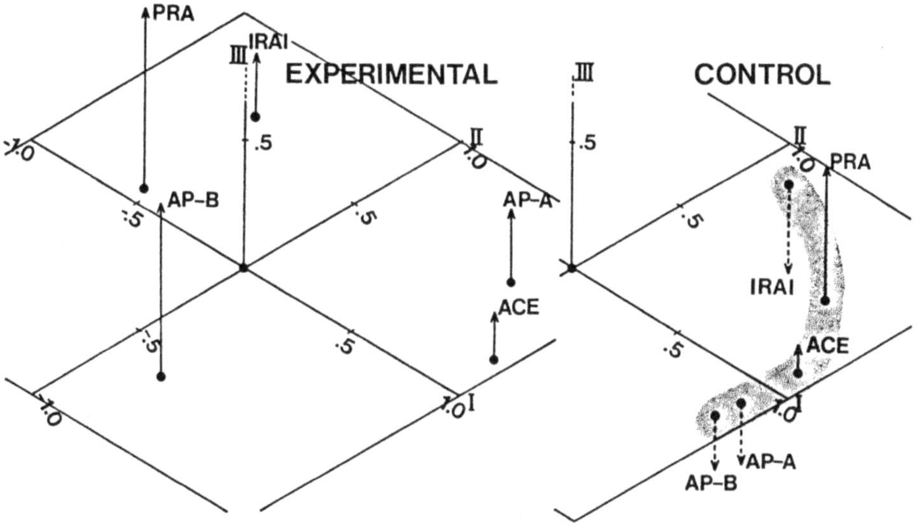

Figure 15 Bird's-eye view showing the relationships among the variables examined.

Table 25 shows correlations among all the variables selected for this study. The control animal showed significant and relatively high correlations among the levels of PRA, immunoreactive angiotensin I (IRAI), ACE, AP-A, and AP-B when compared with the experimental animal.

To compare these relationships in the two animals more systematically, we adopted principal-component analysis, one of the methods of multivariate analysis [74]. Three principal components were extracted from the correlation matrix consisting of the correlation coefficients between the various pairs of enzymes or other variables tested. The first, i. e., the most important principal component, was that explaining the largest portion of sample variance (61.6% in the control and 37.4%

in the experimental). The second component explained 23.5 and 23.2% and the third, 9.3 and 21.1% in the control and experimental, respectively. In Figure 15, the degree of correlation of each enzyme or other variable to these components is equivalent to the coordinates on the three axes drawn. As a result, each variable was expressed as an arrow standing vertically on the diagonal plane, framed by the axes of the first and the second component. The height of each arrow expresses the degree of correlation of that particular variable to the third component. Thus it can be seen that the first component in the control was closely correlated to ACE ($r = 0.93$), AP-A ($r = 0.92$), and AP-B ($r = 0.92$), but that the degree of correlation decreased in the experimental as follows: $r = 0.85$ for ACE, $r = 0.68$ for AP-A, and $r = 0.23$ for AP-B.

From this figure, one can see that the close relationships among the metabolic parameters seen in the control animal almost disappeared in the experimental animal. Also, the observation is compatible with our experiences with various enzyme inhibitors, which showed that exogenous inhibitors have rather more extensive secondary effects than would be expected from their actions *in vitro* [84, 85, 88]. From now on, the dynamic aspects of systems in the body should be taken into consideration in interpretation of the hemodynamic and endocrine alterations induced by enzyme inhibitors.

7.3 Effects of bestatin on hydrolytic enzymes in progressive muscular dystrophy

It is known that the low-molecular-weight enzyme inhibitors produce various pharmacological effects simply not explainable by their direct inhibitory actions [84–86]. Since a preliminary report suggested that bestatin has a suppressive effect on the development of murine muscular dystrophy [72], we performed a long-term study using this drug on a patient with progressive muscular dystrophy [91].

Figures 16 and 17 show the time course of several enzymes during the control, as well as experimental, period. The fluctuations in the enzymatic levels were unexpectedly wide in both of these periods. One can see, however, that the remarkable fluctuations in the AP-A level seen at round the 30th and 90th day were also evident in the levels of Pro-IP and carboxypeptidase B (CP-B). As for other enzymes examined, such fluctuations were not apparent. Although the above fluctuations were not rhythmical, they seemed to somehow reflect some variation in metabolic networks [91].

In spite of these fluctuations, it seems apparent that the level of AP-B abruptly decreased a few weeks after the initiation of the bestatin treatment, although such movements were not seen with regard to AP-A and Pro-IP. Of the three enzymatic levels presented in Figure 17, the levels of esterase apparently tended to decrease during the experimental period. After the initiation of bestatin treatment there was a sharp drop in CP-B activity, but the level returned to the control range within 10 days.

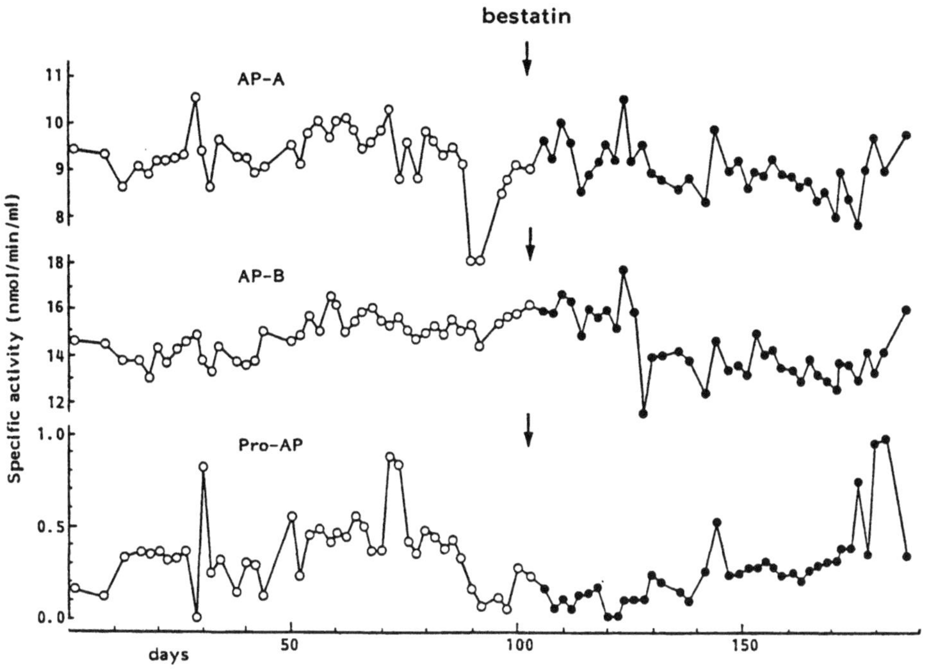

Figure 16 Time course of serum aminopeptidase activities in muscular dystrophy patient before and during bestatin administration.

Table 26 lists the means and standard deviations of all the serum enzyme levels measured as well as the statistical significance of their difference between the two periods. It was found that the levels of AP-A, AP-B, CP-B, alkaline phosphatase, and esterase were all significantly decreased during the administration of bestatin. While CK, LDH, GOT, and GPT activities showed mild to moderate elevation, there were no significant differences between the two periods. Although other enzymes did not show any significant difference, it seems possible that some changes occurred in the interrelationships among the various enzymes between the two periods. In order to test this possibility, a multivariate test [74] was done.

Three principal components were extracted from the correlation matrix which includes the correlation coefficients between the various pairs of enzymes examined. The first i. e., the most important principal component was the one explaining the largest portion of sample variance (24.7% in the control period and 29.2% in the experimental period). The second component explained 18.9% and 21.0%, while the third one, 12.7% and 19.8%, in the control and experimental periods, respectively.

The degree of correlation of these components to each enzymes is described in Figure 18. During the control period, AP-A, for example, had correlation coefficients of 0.61, −0.54, and 0.15 to the above three principal components, respectively. During the experimental period, in contrast, the correlation co-efficients of AP-A

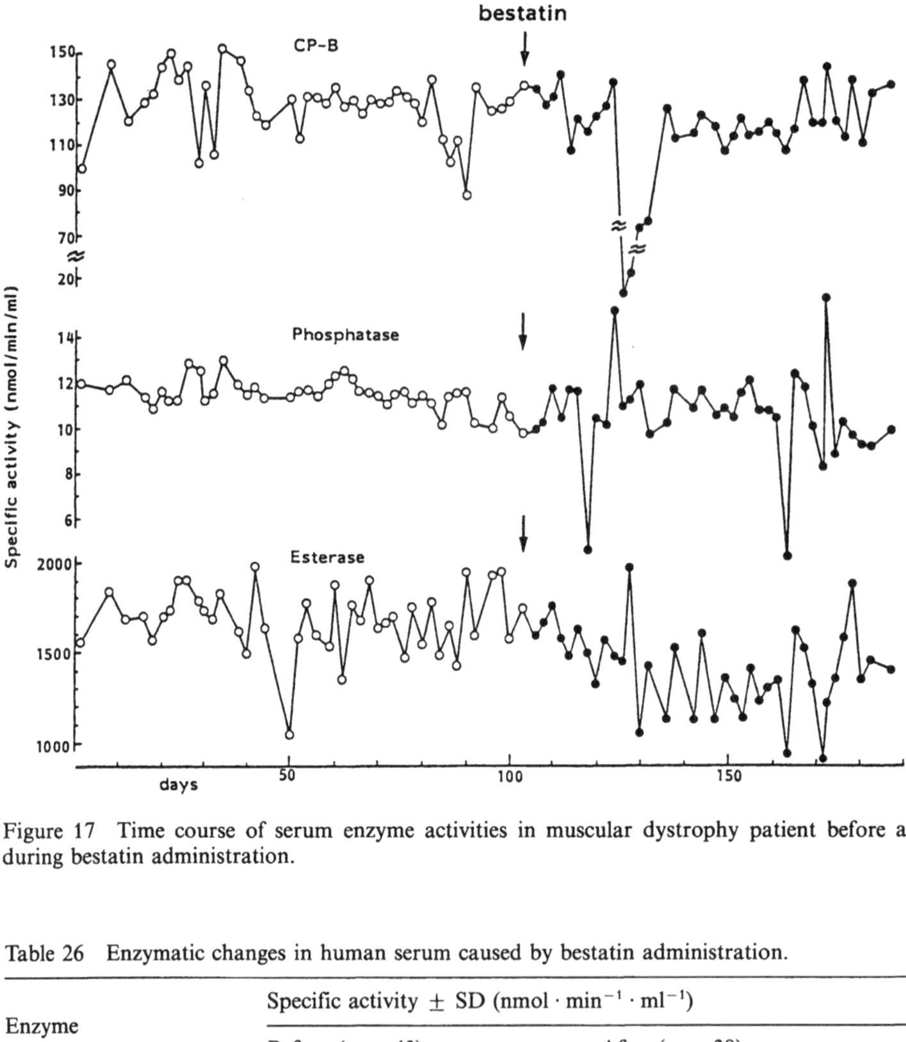

Figure 17 Time course of serum enzyme activities in muscular dystrophy patient before and during bestatin administration.

Table 26 Enzymatic changes in human serum caused by bestatin administration.

Enzyme	Specific activity ± SD (nmol · min^{-1} · ml^{-1})	
	Before (n = 43)	After (n = 38)
AP-A	9.26 ± 0.71	8.97 ± 0.56[a]
AP-B	14.84 ± 0.84	14.22 ± 1.37[a]
Pro-IP	0.35 ± 0.19	0.27 ± 0.21
Leu-AP	27.57 ± 1.27	26.83 ± 2.15
CP-A	0.33 ± 0.56	0.64 ± 1.12
CP-B	127.68 ± 13.77	115.09 ± 28.78[a]
ACE	1.69 ± 1.76	1.53 ± 0.64
Post-Pro-Enz	0.04 ± 0.07	0.01 ± 0.09
Kallikrein	812.03 ± 129.07	760.96 ± 135.04
Cathepsin D	0.0018 ± 0.0014	0.0018 ± 0.0016
Phosphatase	11.45 ± 0.66	10.59 ± 2.08[a]
Esterase	1,682.92 ± 181.78	1,408.33 ± 231.92[c]
CK	642.27 ± 152.60 (n = 22)	632.86 ± 138.67 (n = 28)

[a] $P < 0.05$; [c] $P < 0.001$.

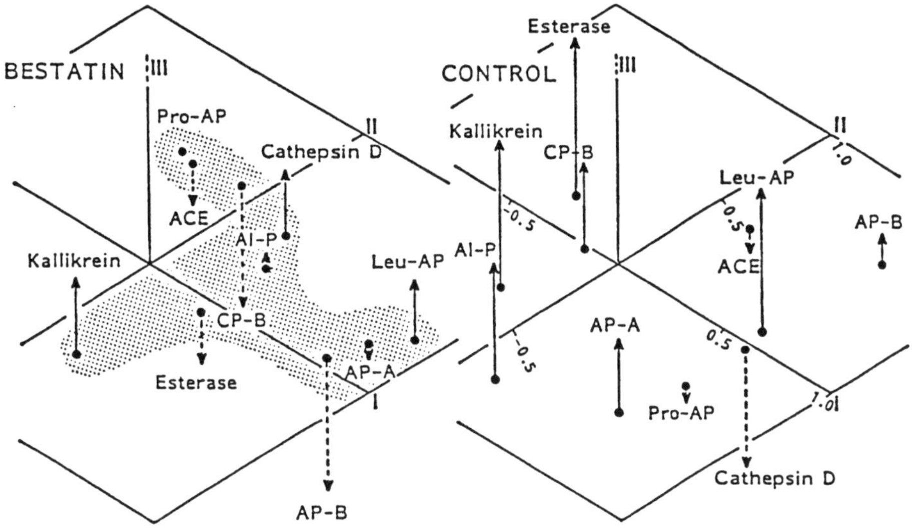

Figure 18 Bird's-eye view showing the relationships among various serum enzymes

were 0.85, 0.25, and −0.06, respectively. The correlation coefficients obtained for all the enzymes were plotted in this figure to enable us to see the enzymatic relations stereographically. As a result of this, each enzyme was drawn as an arrow standing vertically on the diagonal plane, framed by the axes of the first and the second component. The height of each arrow expresses the degree of correlation of that particular enzyme to the third component. During the control period, the correlation coefficients between the pairs of three representative enzymes were as follows: 0.40 for AP-A *vs.* AP-B, 0.47 for AP-B *vs.* Leu-AP, and 0.36 for Leu-AP *vs.* AP-A. During bestatin treatment, these correlation coefficients rose to 0.66, 0.79, and 0.63, respectively. Reflecting the closer relationships among the three enzymes, the arrows representing these enzymes are closer together in the left figure than in the right one. Generally, the correlation coefficients between the various pairs of enzymes were higher in the period of bestatin treatment than in the control period. Thus the arrows tended to be more clustered during the period of bestatin treatment. These findings are compatible with the notion that the enzyme networks are more synchronized during bestatin treatment than during the control period [91].

In order to rule out the possibility that the changes in enzyme levels represent those in enzymatic sensitivities between the two periods, we compared the degree of inhibition of several enzymes by means of exogenous enzyme inhibitors. As seen in the Table 27, no significant changes occurred in the IC_{50} values for any of the enzymes tested [91].

These observations lead us to conclude that bestatin exerts a synchronizing effect on the movements of enzyme networks. The results of multivariate analysis in the present study are also compatible with such a notion. Recently, considerable attention has been aroused concerning the treatment of muscular dystrophy with

Table 27 Inhibitory activities of various inhibitors against enzymes in human serum.

Inhibitor	Enzyme	IC_{50} (μg/ml) Bestatin administration without	with
Amastatin	AP-A	0.056	0.056
Bestatin	AP-B	4.2	4.2
	Leu-AP	3.5	3.7
	Gly-Pro-Leu-AP	13.0	11.0
Leupeptin	Kallikrein	31.0	45.0
Esterastin	Esterase	≧100.0	≧100.0
Ebelactone B	Esterase	≧100.0	≧100.0

protease inhibitors [62]. This is based on the findings that the activities of various endopeptidases and exopeptidases are increased in dystrophic muscles [63]. Specifically, a preliminary report suggested that this drug suppresses the development of murine muscular dystrophy [72]. It is regrettable that the serum levels of CK, LDH, GOT, and GPT in this patient did not show any significant difference between the control and experimental periods. However, one might predict that inhibitor therapy would not be effective in such a middle-aged patient as this with an established disease. Since the serum levels of proteolytic enzymes were significantly suppressed by treatment with bestatin, it may be worth while to try to use this agent on younger patients with initial symptoms of the disease.

8. Pharmacologic Applications of Enzyme Inhibitors

Of the low-molecular-weight enzyme inhibitors discovered in our institute, those acting against the enzymes located on cellular surfaces, such as aminopeptidases, carboxypeptidases, esterase, alkaline phosphatase, and so forth, proved to have immunomodifying effects [92]. One of the alkaline phosphatase inhibitors, levamisole, is known to have favorable effects in patients with autoimmune diseases in spite of its immunopotentiating action [93]. Thus it seemed worth while to test the effects of immunomodifiers on the development of experimental allergic encephalomyelitis (EAE), which is known to be a useful model for studies of cell-mediated autoallergic immune diseases. We tested the effects of low-molecular-weight enzyme inhibitors with immunomodifying agents for suppression of EAE [94].

As can be seen in Table 28, all of the 10 untreated guinea pigs and those treated only with phosphate buffered saline (PBS) died within 3 weeks. The most effective inhibitor used was phosphoramidon, Z-diprotin A and forphenicine. Also, arphamenine A and B, formestin A, and esterastin saved three out of the five animals. Pathological findings also confirmed the efficacy of the inhibitors. Muscular structures were severely damaged by the disease processes and the development of the

Table 28 Effects of various inhibitors on survival of animals injected with EAE antigen.

Inhibitors	Dose (μg/GP)	No. of animals	Average day of onset	Mean day of death	Survivors (35 day after)	Statistical significance
None	–	10	11.5	13.3	0/10	
Phosphoramidon	10	5	12.8	>30.0	5/5	$p < 0.01$
	500	5	12.6	>27.8	4/5	$p < 0.01$
Z-Diprotin A	100	5	12.6	>27.0	4/5	$p < 0.01$
	400	5	12.2	>24.4	3/5	$p < 0.05$
Arphamenine A	10	5	12.2	>24.4	3/5	$p < 0.05$
	500	5	12.0	>23.4	3/5	$p < 0.05$
Arphamenine B	10	5	14.5	>27.0	3/5	$p < 0.05$
Formestin A .	10	5	10.2	>22.4	3/5	$p < 0.05$
Forphenicine	500	5	16.0	>31.4	4/5	$p < 0.01$
Esterastin	100	5	14.8	>28.0	3/5	$p < 0.05$

NS: Actinonin, Diprotin A, Amastatin, Forphenicinol, Ebelactone A, etc.

pathological changes was suppressed by the effects of phosphoramidon, Z-diprotin A and forphenicine. Also, the findings in cerebral tissues were compatible with these results. Demyelination of the central nerves occurred in the diseases animals but not in the above agents-treated ones, although meningeal infiltration of mononuclear cells persisted in some of the latter animals. The findings in the animals saved by the other inhibitors were similar to these.

Previously, Brosnan *et al.* reported that inhibitors of plasminogen activators and other neutral proteinases gave significant protection against the clinical expression of EAE in Lewis rats [95]. The present study, using other inhibitors, clearly demonstrated their suppressive effects on pathological changes in the central nervous tissues and muscles. Our previous studies suggested that this disease model presents unexpectedly extensive changes involving multiple organs [80, 81]. Accordingly, the beneficial effects of these inhibitors may have a broader significance, encompassing even systemic autoimmune diseases. The effects of the inhibitors against alkaline phosphatase, aminopeptidase B and esterase were remarkable.

Finally, let me summarize the biological and therapeutic effects of enzyme inhibitors found in our institute (Chart 2). The actions of immunopotentiation were found to be possessed by amastatin, actinonin, bestatin, arphamenines A and B, diprotins A and B, benzylmalic acid, forphenicine, forphenicinol, ebelactones A and B, and histargin. Fertilization is known to be suppressed by leupeptin, antipain, and chymostatin. The analgesic action is exerted by bestatin, arphamenines A and B, amastatin, and actinonin. Carrageenin edema is suppressed by leupeptin, antipain, chymostatin, elastatinal, pepstatin, phosphoramidon, esterastin, whereas ascites and pleural fluid are suppressed by pepstatin. The blister formation after burn is suppressed by leupeptin. The released serine proteinases by pancreatitis can be inhibited by leupeptin. The high plasma renin activity in renal vascular hypertension is inhibited by pepstatin. Therapeutic effects on muscular dystrophy were suggested by the use of leupeptin, antipain, bestatin, or forphenicinol. Chemical carcinogenesis can be suppressed by leupeptin and antipain. Malignant diseases in man can be

Chart 2 Biological and therapeutic effects of enzyme inhibitors.

Immunopotentiation	Ac, Am, Ar(A,B), Bm, Bs, Dp(A,B), Es, Eb(A,B), Fl, Fn, Fo, Hs
Fertilization	Ap, Cs, Lp
Analgesic action	Ac, Am, Ar(A,B), Bs
Inflammation	Ap, Cs, El, Es, Lp, Ph, Ps
Ascites and pleural fluid	Ps
Chemical carcinogenesis	Ap, Lp
Burn	Lp
Pancreatitis	Lp
Hypertension	Bs, Fo, Hs, Ps
Muscular dystrophy	Ap, Bs, Fl, Lp
Malignant diseases	Bs, Fl, Fn
Autoimmune diseases	Ar(A), Es, Fn, Lp

Am: amastatin, Ac: actinonin, Ap: antipain, Ar: arphamenine, Bm: (S)-α-benzylmalic acid, Bs: bestatin, Cs: chymostatin, Dp: diprotin, Eb: ebelactone, El: elastatinal, Es: esterastin, Fl: forphenicinol, Fn: forphenicine, Fo: foroxymithine, Lp: leupeptin, Hs: histargin, Ph: phosphoramidon, Pr: probestin, Ps: pepstatin.

suppressed by bestatin, forphenicine, and forphenicinol. Demyelinative diseases can be suppressed by phosphoramidon, Z-diprotin A, arphamenines A and B, formestin A, forphenicine, and esterastin. I would like to close this section hoping that the application of these enzyme inhibitors will, some day, lead to elucidation of the cause of diseases and even to their therapy.

9. Summary

In organisms a large number of enzymes are working in complicated networks to express various biological functions. In order to analyze such functions from various aspects, specific enzyme inhibitors are likely to become useful tools. They are also useful for studies of the reaction mechanisms and analysis of three-dimensional structures of enzymes. Moreover, they are of great values in elucidating disease processes and seem to have usefulness in treatment of various diseases.

In order to elucidate the cause of a disease process, we consider it useful to analyze the enzyme-network as the function of time and follow it for a rather long period. It is important to study the kinetics of activities of a variety of enzymes and examine the data with multivariate analysis. In this way we can identify a certain enzyme that is the most pertinent to the disease process. It will be useful to search for a specific inhibitor against such an enzyme and to study the effect of the inhibitor in animal models of the disease process.

Searching for inhibitors in culture filtrate of microbes, we discovered many substances which specifically inhibit various enzymes such as serine-, cysteine-, aspartic-, and metal-proteinases, aminopeptidases, dipeptidyl aminopeptidases, carboxypeptidases, dipeptidyl carboxypeptidases, and so forth. These inhibitors have low-molecular-weights and unique structures. We found significant activities of exopeptidases, such as aminopeptidases, carboxypeptidases, alkaline phosphatase, esterase, and so forth, on surface membranes of various mammalian cells. Searching for specific inhibitors against these cell surface enzymes, we have discovered inhibitors such as amastatin, bestatin, arphamenines A and B, actinonin, probestin, prostatin, formestins A and B, diprotins A and B, histargin, foroxymithine, forphenicine, ebelactones A and B, esterastin, and so forth. These inhibitors proved to bind to the cellular surface and to modify the functions of cells involved in immune responses.

Thus the studies on these enzyme inhibitors may well afford important keys to understand various aspect of biological phenomena and diseases: inflammation, immune response, complement reaction, carcinogenesis, metastasis, viral infection, autoimmune diseases, muscular dystrophy, demyelination diseases, and so forth. Because of their interesting pharmacological activities, some of the inhibitors are now under clinical evaluation for their uses as medical drugs. Enzyme inhibitors seem to propose a new promising field of science.

Acknowledgements

I wish to express my heartfelt thanks to the late Professor Hamao Umezawa and Dr. Tomio Takeuchi, the Director General of the Institute of Microbial Chemistry. Also, it is needless to say that these performances were made with cooperation of my many collaborators. I sincerely hope that this tree of works, which was planted by the late Professor Umezawa, will bear fruit in many fields of medicine, pharmacology, and biochemistry.

References

1. Umezawa, H., *Enzyme Inhibitors of Microbial Origin*, University of Tokyo Press, Tokyo, 1972.
2. Aoyagi, T., and Umezawa, H., Structures and activities of protease inhibitors of microbial origin, in *Proteases and Biological Control*, Reich, E., Riffkin, D. B., and Show, E., Eds., Cold Spring Harbor Laboratory, New York, 1975, p. 429.
3. Umezawa, H., and Aoyagi, T., Activities of proteinase inhibitors of microbial origin, in *Proteinases in Mammalian Cells and Tissues*, Barrett, A. J., Ed., North-Holland, Amsterdam, 1977, chap. 15.
4. Aoyagi, T., Structure and activities of proteinase inhibitors of microbial origin, in *Bioactive Peptides Produced by Microorganisms*, Umezawa, H., Takita, T., and Shiba, T., Eds., Kodansha, Tokyo & Wiley, New York, 1978, cvap. 7.
5. Umezawa, H., Structures and activities of protease inhibitors of microbial origin, in *Methods in Enzymology*, Vol. 45, Lorand, L., Ed., Academic Press, New York, 1976, chap. 55.
6. Aoyagi, T., Biological significance of proteases and their inhibitors, *Proc. Jpn. Soc. Invest. Dermatol. 10*, 77, 1986.
7. Aoyagi, T., Takeuchi, T., Matuzaki, A., Kawamura, K., Kondo, S., Hamada, M., Maeda, K., and Umezawa, H., Leupeptins, new proteinase inhibitors from actinomycetes, *J. Antibiot. 22*, 283, 1969.

8. Suda, H., Aoyagi, T., Hamada, M., Takeuchi, T., and Umezawa, H., Antipain, a new protease inhibitor isolated from actinomycetes, *J. Antibiot. 25*, 263, 1972.
9. Umezawa, H., Aoyagi, T., Morishima, H., Kunimoto, S., Matsuzaki, M., Hamada, M., and Takeuchi, T., Chymostatin, a new chymotrypsin inhibitor produced by actinomycetes, *J. Antibiot. 23*, 425, 1970.
10. Umezawa, H., Aoyagi, T., Okura, A., Morishima, H., Takeuchi, T., and Okami, Y., Elastatinal, a new elastase inhibitor produced by actinomycetes, *J. Antibiot. 26*, 787, 1973.
11. Umezawa, H., Aoyagi, T., Morishima, H., Matsuzaki, M., Hamada, M., and Takeuchi, T., Pepstatin, a new pepsin inhibitor produced by actinomycetes, *J. Antibiot. 23*, 259, 1970.
12. Miyano, T., Tomiyasu, M., Iizuka, H., Tomisaka, S., Takita, T., Aoyagi, T., and Umezawa, H., New pepstatins, pepstatins B and C, and pepstanone A, produced by streptomyces, *J. Antibiot. 25*, 489, 1972.
13. Umezawa, H., Miyano, T., Murakami, T., Takita, T., Aoyagi, T., Takeuchi, T., Naganawa, H., and Morishima, H., Hydroxypepstatin, a new pepstatin produced by streptomyces, *J. Antibiot. 26*, 615, 1973.
14. Aoyagi, T., Yagisawa, Y., Kumagai, M., Hamada, M., Morishima, H., Takeuchi, T., and Umezawa, H., New pepstatins, pepstatins BU, PR and AC produced by streptomyces, *J. Antibiot. 26*, 539, 1973.
15. Aoyagi, T., Kunimoto, S., Morishima, H., Takeuchi, T., and Umezawa, H., Effect of pepstatin on acid protease, *J. Antibiot. 24*, 687, 1971.
16. Aoyagi, T., Morishima, H., Nishizawa, R., Kunimoto, S., Takeuchi, T., and Umezawa, H., Biological activity of pepstatins, pepstanone A and partial peptides on pepsin, cathepsin D and renin, *J. Antibiot. 25*, 689, 1972.
17. Kunimoto, S., Aoyagi, T., Nishizawa, R., Komai, T., Takeuchi, T., and Umezawa, H., Mechanism of inhibition of pepsin by pepstatin, *J. Antibiot. 27*, 413, 1974.
18. Suda, H., Aoyagi, T., Takeuchi, T., and Umezawa, H., A thermolysin inhibitor produced by actinomycetes: phosphoramidon, *J. Antibiot. 26*, 621, 1973.
19. Umezawa, S., Tatsuta, K., Izawa, O., Tsuchiya, T., and Umezawa, H., A new microbial metabolite phosphoramidon, *Tetrahedron Letters 1*, 97, 1972.
20. Kalckar, H. M., Galactose metabolism and cell "sociology", *Science 150*, 305, 1965.
21. Burger, M. M., Proteolytic enzymes initiating cell division and escape from contact inhibition of growth, *Nature 227*, 170, 1970.
22. Cook, G. M. W., Heard, D. H., and Seaman, G. V. F., A sialomucopeptide liberated by trypsin from the human erythrocyte, *Nature 188*, 1011, 1960.
23. Goldberg, A. R., Wolf, B. A., and Lefebvre, P. A., Plasminogen activators of transformed and normal cells, in *Proteases and Biological Control*, Reich, E., Rifkin, D. B., and Shaw, E., Eds., Cold Spring Harbor Laboratory, New York, 1975, p. 857.
24. Becker, E. L., and Unanue, E. R., The requirement for esterase activation in the anti-immunoglobulin-triggered movement of B lymphocytes, *J. Immunol. 117*, 27, 1976.
25. Aoyagi, T., Suzuki, J., Nerome, K., Nishizawa, R., Takeuchi, T., and Umezawa, H., Sialic acid residues exposed on mammalian cell surface: the effect of adsorption of denatured virus particles, *Biochem. Biophys. Res. Commun. 57*, 271, 1974.
26. Aoyagi, T., Nerome, K., Suzuki, J., Takeuchi, T., and Umezawa, H., Change of enzyme activities during the early stage of influenza virus infection, *Biochem. Biophys. Res. Commun. 60*, 1178, 1974.
27. Aoyagi, T., Komiyama, T., Nerome, K., Takeuchi, T., and Umezawa, H., Characterization of myxovirus sialidase, *Experientia 31*, 896, 1975.
28. Aoyagi, T., and Umezawa, H., Hydrolytic enzymes on the cellular surface and their inhibitors found in microorganisms, in *Industrial and Clinical Enzymology*, Vitale, L., and Simon, V., Eds., Pergamon Press, Oxford and New York, 1980, p. 89.
29. Aoyagi, T., and Umezawa, H., The relationships between enzyme inhibitors and function of mammalian cells, *Acta Biol. Med. Germ. 40*, 1523, 1981.
30. Aoyagi, T., Suda, H., Nagai, M., Ogawa, K., Suzuki, J., Takeuchi, T., and Umezawa, H., Aminopeptidase activities on the surface of mammalian cells, *Biochim. Biophys. Acta 452*, 131, 1976.
31. Aoyagi, T., Nagai, M., Iwabuchi, M., Liaw, W. S., Andoh, T., and Umezawa, H., Aminopeptidase activities on the surface of mammalian cells and their alterations associated with transformation, *Cancer Res. 38*, 3505, 1978.

32. Aoyagi, T., Tobe, H., Kojima, F., Hamada, M., Takeuchi, T., and Umezawa, H., Amastatin, an inhibitor of aminopeptidase A, produced by actinomycetes, *J. Antibiot. 31*, 636, 1978.
33. Tobe, H., Morishima, H., Aoyagi, T., Umezawa, H., Ishiki, K., Nakamura, K., Yoshioka, T., Shimauchi, Y., and Inui, T., Synthesis and structure-activity relationships of amastatin analogues, inhibitors of aminopeptidase A, *Agric. Biol. Chem. 46*, 1865, 1982.
34. Umezawa, H., Aoyagi, T., Suda, H., Hamada, M., and Takeuchi, T., Bestatin, an inhibitor of aminopeptidase B, produced by actinomycetes, *J. Antibiot. 29*, 97, 1976.
35. Nishizawa, R., Saino, T., Takita, T., Suda, H., Aoyagi, T., and Umezawa, H., Synthesis and structure-activity relationships of bestatin analogues, inhibitors of aminopeptidase B, *J. Med. Chem. 20*, 510, 1977.
36. Umezawa, H., Ishizuka, M., Aoyagi, T., and Takeuchi, T., Enhancement of delayed-type hypersensitivity by bestatin, an inhibitor of aminopeptidase B and leucine aminopeptidase, *J. Antibiot. 29*, 857, 1976.
37. Umezawa, H., Small molecular weight immunomodifiers produced by microorganisms: their screening and discoveries, and the genetics of microbial secondary metabolites, in *Small Molecular Immunomodifiers of Microbial Origin-Fundamental and Clinical Studies of Bestatin*, Umezawa, H., Ed., Japan Sci. Soc. Press, Tokyo/Pergamon Press, Oxford, 1980, p. 1.
38. Umezawa, H., Aoyagi, T., Ohuchi, S., Okuyama, A., Suda, H., Takita, T., Hamada, M., and Takeuchi, T., Arphamenine A and B, new inhibitors of aminopeptidase B, produced by bacteria, *J. Antibiot. 36*, 1572, 1983.
39. Ohuchi, S., Suda, H., Naganawa, H., Takita, T., Aoyagi, T., Umezawa, H., Nakamura, H., and Iitaka, Y., The structure of arphamenine A and B, *J. Antibiot. 36*, 1576, 1983.
40. Okuyama, A., Ohuchi, S., Tanaka, T., Aoyagi, T., and Umezawa, H., Cell-free biosynthesis of arphamenine A, *Biochem. Int. 12*, 485, 1986.
41. Yamamoto, K., Suda, H., Ishizuka, M., Takeuchi, T., Aoyagi, T., and Umezawa, H., Isolation of *a*-aminoacyl arginines in screening of aminopeptidase B inhibitors, *J. Antibiot. 33*, 1597, 1980.
42. Umezawa, H., Aoyagi, T., Tanaka, T., Suda, H., Okuyama, A., Naganawa, H., Hamada, M., and Takeuchi, T., Production of actinonin, an inhibitor of aminopeptidase M, by actinomycetes, *J. Antibiot. 38*, 1629, 1985.
43. Ollis, W. D., East, A. J., Gordon, J. J., and Sutherland, I. O., The constitution of actinonin-structural and synthetic studies, in *I. A. M. Symposia on Microbiology*, No. 6, Chemistry of Microbial Product, The Microbial Research Foundation, Tokyo, 1964, p. 204.
44. Aoyagi, T., Yoshida, S., Nakamura, Y., Naganawa, H., Hamada, M., and Takeuchi, T., in preparation.
45. Aoyagi, T., Ogawa, K., Iinuma, H., Naganawa, H., Hamada, M., and Takeuchi, T., in preparation.
46. Umezawa, H., Aoyagi, T., Uotani, K., Hamada, M., Takeuchi, T., and Takahashi, S., Ebelactone, an inhibitor of esterase, produced by actinomycetes, *J. Antibiot. 33*, 1594, 1980.
47. Nishikiori, T., Kawahara, F., Naganawa, H., Muraoka, Y., Aoyagi, T., and Umezawa, H., Production of acetyl-L-leucyl-L-argininal, inhibitor of dipeptidyl aminopeptidase III by bacteria, *J. Antibiot. 37*, 680, 1984.
48. Umezawa, H., Aoyagi, T., Ogawa, K., Naganawa, H., Hamada, M., and Takeuchi, T., Diprotins A and B, inhibitors of dipeptidyl aminopeptidase IV, produced by bacteria, *J. Antibiot. 37*, 422, 1984.
49. Tanaka, T., Suda, H., Naganawa, H., Takeuchi, T., Aoyagi, T., and Umezawa, H., Production of (S)-*a*-benzylmalic acid, inhibitor of carboxypeptidase A by actinomycetes, *J. Antibiot. 37*, 682, 1984.
50. Umezawa, H., Aoyagi, T., Ogawa, K., Iinuma, H., Naganawa, H., Hamada, M., and Takeuchi, T., Histargin, a new inhibitor of carboxypeptidase B, produced by actinomycetes, *J. Antibiot. 37*, 1088, 1984.
51. Umezawa, H., Aoyagi, T., Ogawa, K., Obata, T., Iinuma, H., Naganawa, H., Hamada, M., and Takeuchi, T., Foroxymithine, a new inhibitor of angiotensin-converting enzyme, produced by actinomycetes, *J. Antibiot. 38*, 1813, 1985.
52. Aoyagi, T., Yamamoto, T., Kojiri, K., Kojima, F., Hamada, M., Takeuchi, T., and Umezawa, H., Forphenicine, an inhibitor of alkaline phosphatase, produced by actinomycetes, *J. Antibiot. 31*, 244, 1978.

53. Umezawa, H., Aoyagi, T., Hazato, T., Uotani, K., Kojima, F., Hamada, M., and Takeuchi, T., Esterastin, an inhibitor of esterase, produced by actinomycetes, *J. Antibiot. 31*, 639, 1978.
54. Iodice, A. A., The inhibition by pepstatin of cathepsin D and autolysis of dystrophic muscle, *Life Sciences 19*, 1351, 1976.
55. Toyooka, T., Shimizu, T., and Masaki, T., Inhibition of proteolytic activity of calcium activated neutral protease by leupeptin and antipain, *Biochem. Biophys. Res. Commun. 82*, 484, 1978.
56. Ishiura, S., Murofuchi, H., Suzuki, K., and Imahori, K., Studies of a calcium-activated neutral protease from chicken skeletal muscle, *J. Biochem. 84*, 225, 1978.
57. Goll, D. E., Okitani, A., Dayton, W. R., and Reville, W. J., A Ca-activated muscle protease in myofibrillar protein turnover, in *Protein Turnover and Lysosome Function*, Segel, H. L., and Doyle, D. J., Eds., Academic Press, New York, 1978, p. 587.
58. Bird, J. W. C., Spanier, A. M., and Schwartz, W. N., Cathepsins B and D: proteolytic activity and ultrastructural localization in skeletal muscle, in *Protein Turnover and Lysosome Function*, Segal, H. L., and Doyle, D. J., Academic Press, New York, 1978, p. 589.
59. Sanada, Y., Yasogawa, N., and Katsunuma, N., Serine protease in mice with hereditary muscular dystrophy, *J. Biochem. 83*, 27, 1978.
60. McGowan, E. B., Shafiq, S. A., and Stracher, A., Delayed degeneration of dystrophic and normal muscle cell cultures treated with pepstatin, leupeptin and antipain, *Exp. Neurol. 50*, 649, 1976.
61. Libby, P., and Goldberg, A. L., Leupeptin, a protease inhibitor, decreases protein degradation in normal and diseased muscles, *Science 199*, 534, 1978.
62. Stracher, A., McGowan, E. B., and Shafiq, S. A., Muscular dystrophy: inhibition of degeneration *in vivo* with protease inhibitors, *Science 200*, 50, 1978.
63. Aoyagi, T., Wada, T., Kojima, F., Nagai, M., and Umezawa, H., Various enzyme activities in muscle and other organs of dystrophic mice, *J. Clin. Invest. 67*, 51, 1981.
64. Wada, T., Kojima, F., Nagai, M., Aoyagi, T., and Umezawa, H., Intramuscular enzyme abnormalities of dystrophic chickens as compared to dystrophic mice, *J. Pharm. Dyn. 4*, 184, 1981.
65. Aoyagi, T., Wada, T., Ohuchi, S., Kojima, F., Nagai, M., Harada, S., and Umezawa, H., Similar effects of various low-molecular-weight enzyme inhibitors on enzyme networks in dystrophic mice, *J. Pharm. Dyn. 7*, 681, 1984.
66. Ishizuka, M., Ishizeki, S., Masuda, T., Momose, A., Aoyagi, T., Takeuchi, T., and Umezawa, H., Studies on effects of forphenicinol on immune responses, *J. Antibiot. 35*, 1042, 1982.
67. Aoyagi, T., Wada, T., and Umezawa, H., Catabolic enzymes in dystrophic muscle, in *Muscular Dystrophy*, Ebashi, S., Ed., University of Tokyo Press, Tokyo, 1982, p. 239.
68. Ricker, H., and Shimrigk, K., Enzyme diagnostics of muscular disorders, in *Advances in Clinical Enzymology*, Schmidt, E., Schmidt, F. W., Trautschold, I., and Hannover, R. F., Eds., S. Karger, Basel, 1979, p. 124.
69. Heyck, H., and Laudahn, G., Fermentchemische Serum-Befunde bei Myopathien, in *Myopathien. Genetik, Biochemie, Pathologie, Klinik und Therapie unter besonderer Berücksichtigung des Kindesalters*, Bockmann, R., Ed., Thieme, Stuttgart, 1965, p. 176.
70. Thomson, W. H. S., Sweetin, J. C., and Elton, R. A., The neurogenic and myogenic hypotheses in human (Duchenne) muscular dystrophy, *Nature 249*, 151, 1974.
71. Aoyagi, T., Wada, T., Kojima, F., Nagai, M., Miyoshino, S., and Umezawa, H., Two different modes of enzymatic changes in serum with progression of Duchenne muscular dystrophy, *Clin. Chim. Acta 129*, 165, 1983.
72. Matsushita, H., Study on the clinical applications of secondary metabolites from microbes, *Annual Report of Muscular Dystrophy Committee*, Supported by Ministry of Health and Welfare, 1980, p. 55 (Japanese).
73. Aoyagi, T., Wada, T., Kojima, F., Harada, S., Hachisu, M., Murata, S., and Umezawa, H., The dissociation between the plasma levels of aminopeptidases A and B in spontaneously hypertensive rats, *Biotech. and Appl. Biochem. 8*, 32, 1986.
74. Kendall, M., *Multivariate Analysis*, Charles Griffin and Co., London, 1975.
75. Murata, S., and Umezawa, H., Inhibitors of aminopeptidase B suppress the development of hypertension in spontaneously hypertensive rats, *Chem. Pharm. Bull. 34*, 4852, 1986.
76. Kerby, G. P., and Taylor, S. M., Enzymatic activity in human synovial fluid from rheumatoid and nonrheumatoid patients, *Proc. Soc. Exp. Biol., (NY) 126*, 865, 1967.

77. Karan, A., Saatci, Y., and Bakkaloglu, A., The role of cathepsin D in pathogenesis of acute poststreptococcal glomerulonephritis, *Acta Pediat. Scand.* *56*, 355, 1976.
78. Bennett, R. M., and Skosey, J. L., Lactoferrin and lysozyme levels in synovial fluid. Differential indices of articular inflammation and degradation, *Arthr. Rheum.* *20*, 84, 1977.
79. Aoyagi, T., Wada, T., Yamamoto, K., Hayashi, Y., Nagai, Y., and Umezawa, H., Various enzyme activities in synovial fluid of patients with osteoarthritis and rheumatoid arthritis, *Biochem. Int.* *2*, 637, 1981.
80. Aoyagi, T., Wada, T., Ishikawa, Y., Kojima, F., Nagai, M., Osanai, T., Nagai, Y., and Umezawa, H., Systemic enzymatic changes in guinea pigs suffering from experimental allergic encephalomyelitis, *J. Pharm. Dyn.* *6*, 963, 1983.
81. Aoyagi, T., Wada, T., Ishikawa, Y., Kojima, F., Nagai, M., Osanai, T., Nagai, Y., and Umezawa, H., Role of intramuscular enzymatic changes in the development of muscular weakness in rats with experimental allergic neuritis, *Exp. Neurol.* *84*, 326, 1984.
82. Aoyagi, T., Wada, T., Kojima, F., Nagai, M., Okubo, M., Masaki, Y., and Umezawa, H., Abnormality of the post-proline-cleaving enzyme activity in mice with systemic lupus erythematosus-like syndrome, *J. Appl. Biochem.* *7*, 273, 1985.
83. Aoyagi, T., Wada, T., Kojima, F., Sasakawa, S., Tamura, M., and Umezawa, H., Multiple enzymes related to differentiation of lymphocyte subpopulations in man, *Biochem. Int.* *18*, 383, 1989.
84. Aoyagi, T., Wada, T., Iwabuchi, M., Kojima, F., Nagai, M., and Umezawa, H., Changes in enzymatic activity induced by low molecular enzyme inhibitors in various organs of dystrophic mice, *Biochem. Int.* *5*, 97, 1982.
85. Aoyagi, T., Wada, T., Umezawa, K., Kojima, F., Nagai, M., and Umezawa, H., Relation between *in vivo* effects and *in vitro* effects of serine and thiol proteinase inhibitors, *J. Pharm. Dyn.* *6*, 643, 1983.
86. Aoyagi, T., Wada, T., Ohuchi, S., Kojima, F., Nagai, M., Harada, S., and Umezawa, H., Similar effects of various low-molecular-weight enzyme inhibitors on enzyme networks in dystrophic mice, *J. Pharm. Dyn.* *7*, 681, 1984.
87. Aoyagi, T., Wada, T., Ohuchi, S., Kojima, F., Nagai, M., Kawahara, F., and Umezawa, H., Oscillation of enzyme networks in spleen triggered by an immunopotentiator, bestatin, *Biochem. Int.* *9*, 405, 1984.
88. Aoyagi, T., Wada, T., Yamamoto, K., Kojima, F., Nagai, M., Harada, S., and Umezawa, H., Different enzymatic oscillations *in vivo* caused by the stereoisomers of an aminopeptidase inhibitor, bestatin, *J. Appl. Biochem.* *6*, 212, 1984.
89. Bendat, J. S., and Plersol, A. G., *Random Data: Analysis and Measurement Procedures* (Japanese translation), Wiley, New York, 1971, p. 11.
90. Aoyagi, T., Wada, T., Iinuma, H., Ogawa, K., Kojima, F., Nagai, M., Kuroda, H., Obayashi, A., and Umezawa, H., Influence of angiotensin-converting enzyme inhibitor, foroxymithine, on dynamic equilibrium around the renin-angiotensin system *in vivo*, *J. Appl. Biochem.* *7*, 388, 1985.
91. Aoyagi, T., Wada, T., Kojima, F., Nagai, M., Harada, S., Kinoshita, M., Yamada, N., and Umezawa, H., Bestatin suppresses the serum level of several hydrolytic enzymes in a patient with progressive muscular dystrophy, *J. Clin. Biochem. Nutr.* *1*, 65, 1986.
92. Umezawa, H., Fundamental studies on bestatin: a small molecular microbial product enhancing immune responses, in *Recent Results in Cancer Research*, Vol. 76, Carter, S. K., Sakurai, Y., and Umezawa, H., Eds., Springer Verlag, Berlin, 1981, p. 209.
93. Holborow, E. J., and Swannel, A. J., Rheumatoid and sero-negative arthritides, in *Immunology in Medicine*, Holborow, E. J., and Reeves, W. G., Eds., Academic Press, London, 1977, p. 541.
94. Aoyagi, T., Wada, T., Nagai, M., Sakaguchi, H., Osanai, T., Nagai, Y., and Umezawa, H., Low-molecular-weight enzyme inhibitors suppress the development of experimental allergic encephalomyelitis, *Experientia* *40*, 1405, 1984.
95. Brosnan, C. F., Cammer, W., Norton, W. T., and Bloom, B. R., Proteinase inhibitors suppress the development of experimental allergic encephalomyelitis, *Nature* *285*, 235, 1980.

Chapter 13
Directed Biosynthesis of Neoviridogriseins

Yasushi Okumura

1. Introduction
2. Selective Production of Neoviridogriseins by Directed Biosynthesis
   2.1 Effect of exogenous amino acid on the production of neoviridogriseins
   2.2 Specific accumulation of neoviridogrisein II by inhibition of proline hydroxylation reaction
   2.3 Specific accumulation of neoviridogrisein II by the mutant blocked in *trans*-4-hydroxy-L-proline formation from L-proline
3. Summary
References

1. Introduction

Neoviridogriseins [1, 2], depsipeptide antibiotics belonging to the virginiamycin B group [3–5], are viridogrisein [6] analogs produced by *Streptomyces griseoviridus* P8648 [7]. The organism produces neoviridogriseins I, II and III and viridogrisein under undirected fermentation conditions. However, the main product of the organism is viridogrisein and the other three compounds are very minor products. On the other hand, biological evaluation has revealed that neoviridogrisein II is the most active compound among viridogrisein analogs and seems to be a suitable compound for practical use [8]. Consequently, production improvement for neoviridogrisein II has become necessary. For this purpose, the directed biosynthesis of neoviridogrisein II has been attempted by supplementation of various compounds [9] and then by a certain mutant [10]. Furthermore, the production of more active analogs of neoviridogrisein has been attempted by the same methods [11, 12].

Structures of neoviridogriseins belonging to virginiamycin B-II group are shown in Figure 1 and for comparison, structures of virginiamycin B-I group are presented in Figure 2 [3, 5, 7].

Viridogrisein (etamycin [13], compound 6613 [14]) was a sole member in virginiamycin B-II group until neoviridogriseins have been discovered, while there have been many analogs reported in the virginiamycin B-I group.

Generally speaking, the biosyntheses of small peptide antibiotics such as gramicidin S, tyrocidine and actinomycin are directed by the thiotemplate system and not by the ribosome system of protein synthesis [15–17]. Because of the broad substrate

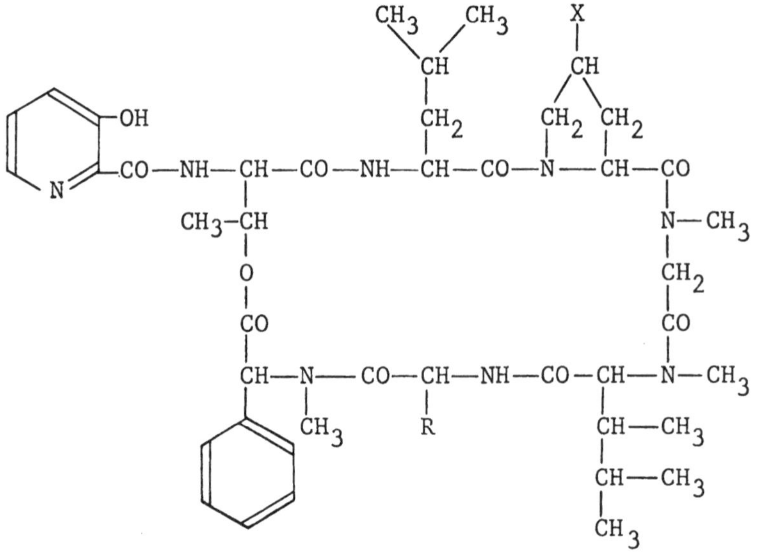

Antibiotic	R	X
neoviridogrisein I	C_2H_5	H
neoviridogrisein II	CH_3	H
neoviridogrisein III	C_2H_5	OH
viridogrisein	CH_3	OH
neoviridogrisein-MP	CH_3	CH_3
neoviridogrisein-Cl-c	CH_3	Cl(cis)
neoviridogrisein-Cl-t	CH_3	Cl(trans)

Figure 1 Structures of virginiamycin B-II group of antibiotics. Reproduced with permission of The Japan Antibiotic Research Association (Ref. 7).

specificity of the thiotemplate system, there remain many possibilities to biochemically replace the constituent amino acid with structurally related compounds and to produce novel and desirable analogs. On this line of approach, many analogs of gramicidin S, tyrocidine, actinomycin and quinoxaline have been produced [15–18]. From these facts, the directed biosyntheses of neoviridogriseins, especially the selective production of neoviridogrisein II, seemed to be possible.

Neoviridogrisein II differs from viridogrisein only at the imino acid moiety. That is, neoviridogrisein II contains D-proline instead of *cis*-4-hydroxy-D-proline in viridogrisein [1]. It has been reported that the precursor of *cis*-4-hydroxy-D-proline in viridogrisein is *trans*-4-hydroxy-L-proline which originates from L-proline [19, 20], and that D-proline in neoviridogrisein II is also derived from L-proline [9]. Based on these reports, the inhibition of this hydroxylation reaction seems to be effective for selective production of neoviridogrisein II. In this review, the latest information about the directed biosyntheses of neoviridogriseins, especially neoviridogrisein II, is described.

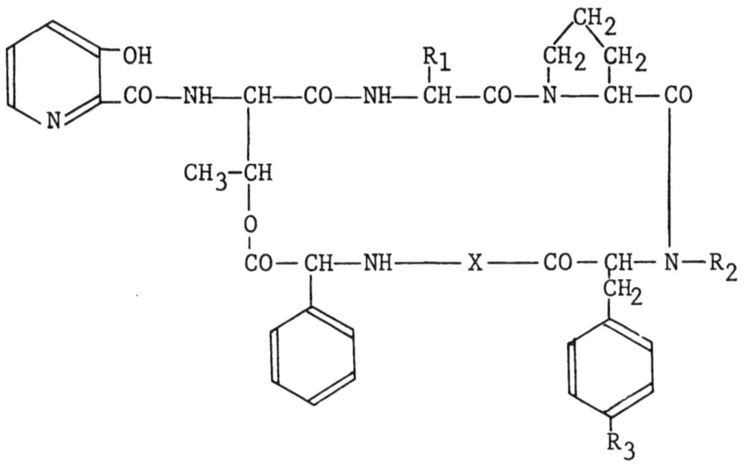

Antibiotic	R_1	R_2	R_3	X
virginiamycin S	C_2H_5	CH_3	H	4-oxopipecolic acid
virginiamycin S_2	C_2H_5	H	H	4-hydroxypipecolic acid
virginiamycin S_3	C_2H_5	CH_3	H	3-hydroxy-4-oxopipecolic acid
virginiamycin S_4	CH_3	CH_3	H	4-oxopipecolic acid
vernamycin Bα	C_2H_5	CH_3	$N(CH_3)_2$	4-oxopipecolic acid
vernamycin Bβ	C_2H_5	CH_3	$NHCH_3$	4-oxopipecolic acid
vernamycin Bγ	CH_3	CH_3	$N(CH_3)_2$	4-oxopipecolic acid
vernamycin Bδ	CH_3	CH_3	$NHCH_3$	4-oxopipecolic acid
vernamycin C	C_2H_5	CH_3	$N(CH_3)_2$	aspartic acid
ostreogrycin B_3	C_2H_5	CH_3	$N(CH_3)_2$	3-hydroxy-4-oxopipecolic acid
patricin A	C_2H_5	CH_3	H	proline
patricin B	C_2H_5	CH_3	H	pipecolic acid

Figure 2 Structures of virginiamycin B-I group of antibiotics. Reproduced with permission of The Japan Antibiotic Research Association (Ref. 7).

2. Selective Production of Neoviridogriseins by Directed Biosynthesis

Streptomyces griseoviridus P8648 isolated from a soil sample produces a large amount of viridogrisein together with its analogs, neoviridogriseins I, II and III. Since the novel antibiotics are very minor products under undirected fermentation conditions, the yield improvement of them has been attempted by directed biosynthesis.

2.1 Effect of exogenous amino acid on the production of neoviridogriseins

Various amino acids structurally related to each amino acid residue have been supplemented to the culture of *S. griseoviridus* P8648, and the effects are summarized in Table 1.

Table 1 Effects of exogenous amino acids on the production of neoviridogriseins.

Constituent amino acid	Supplemented amino acid	Effects	
		Productivity	New product
L-Thr	L-Thr	+	−
	D-Thr	−	−
	L-Ser	±	−
	D-Ser	−	−
D-Leu	L-Leu	+	−
	D-Leu	−	−
	L-Val	±	−
	D-Val	−	−
	L-Ileu	±	−
	D-Ileu	±	−
D-Hyp(cis)	L-Pro	+	+(I,II)
	D-Pro	−	−
	L-Hyp(trans)	+	−
	D-Hyp(cis)	−	−
	L-3-MePro	−	−
	L-4-MePro	±	+(MP)
	L-5-MePro	−	−
	L-4-ClPro	±	+(Cls)
	L-ThioPro	−	−
	L-Az	−	−
Sar	Sar	+	−
	Gly	+	−
L-β,N-DML	L-Leu	+	−
	L-N-MeLeu	±	−
L-Ala	L-Ala	+	−
	D-Ala	+	−
	L-Abu	+	+(I,III)
	D-Abu	−	−
	L-Val	±	−
	D-Val	−	−
L-PheSar	L-PheGly	+	−
	D-PheGly	−	−
	L-Phe	±	−
	D-Phe	−	−
(Met)	L-Met	+	−
	D-Met	−	−
	L-Eth	−	−

Abbreviations: L-3-MePro: L-3-methylproline, L-5-MePro: L-5-methylproline, L-4-Cl-Pro: L-4-chloroproline, L-ThioPro: L-thioproline, L-Az: L-azetidine-2-carboxylic acid, L-β,*N*-DML: L-β,*N*-dimethylleucine, L-N-MeLeu: L-*N*-methylleucine, L-PheSar: L-phenylsarcosine, L-PheGly: L-phenylglycine, L-Eth: L-ethionine. +: enhanced, ±: not changed, −: suppressed.

Since all effective amino acids except alanine are in the L-configuration, the precursors of the individual amino acid moieties are in the L-form. In addition, the formation and the incorporation of the D-amino acids may coincidentally occur or the D-amino acids formed may be immediately incorporated into the antibiotics, because there have never been the free D-forms of constituent amino acids detected in the cells and medium. This phenomenon is similar to those of gramicidin S, tyrocidine and actinomycin. The supplementation of L-amino acid corresponding to each moiety increases the production of neoviridogriseins. Furthermore, the addition of proline analogs has caused the production of new neoviridogriseins named neoviridogriseins-MP, -Cl-c and -Cl-t [11, 12].

In neoviridogriseins I and III, the alanine moiety of viridogrisein is replaced with L-α-amino-n-butyric acid. This type of substitution seems common in the virginiamycin B-I group. For example, under undirected fermentation conditions (in the absence of exogenous L-α-amino-n-butyric acid), virginiamycins S, S_2 and S_3 which contain D-α-amino-n-butyric acid are abundantly produced together with a small amount of virginiamycin S_4 which contains D-alanine instead of D-α-amino-n-butyric acid [21]. Another example is the case of vernamycins [22]. The normal composition of the vernamycin B complex is far more favorable for vernamycins Bα and Bβ which contain D-α-amino-n-butyric acid than for vernamycins Bγ and Bδ which contain D-alanine. The typical production ratio of the vernamycin B complex (Bα : Bβ : Bγ : Bδ) is 53 : 24 : 7 : 4. It seems probable that the virginiamycin B-I group producers such as *S. virginiae* [23], *S. loidensis* [24] and *S. ostreogriseus* [25] have a fairly large intracellular pool of α-amino-n-butyric acid, because the addition of α-amino-n-butyric acid is unnecessary for the biosyntheses of these antibiotics containing α-amino-n-butyric acid. On the other hand, since *S. griseoviridus* P8648 produces almost an undetectable amount of neoviridogriseins I and III under undirected fermentation conditions, the organism seems to have a very small intracellular pool of the amino acid. Another plausible reason for the negligible production of neoviridogriseins I and III under no supplementation of α-amino-n-butyric acid may stem from the higher affinity of the responsible enzyme system for alanine than that for α-amino-n-butyric acid. Additionally, since the addition of D-α-amino-n-butyric acid to the culture of *S. griseoviridus* P8648 inhibits the production of neoviridogriseins, this organism has no racemase (or epimerase) activity for α-amino-n-butyric acid.

The mechanism of hydroxyproline biosynthesis by *S. griseoviridus* PD-04955 [19, 20], one of viridogrisein producers, differs from that characterized in animal systems [26]. In animal systems, hydroxyproline, found in a certain protein such as collagen, is synthesized only in a proline-rich polypeptide by prolyl hydroxylase. It is in an animal system the free form of hydroxyproline can be never incorporated into the hydroxyproline position of the polypeptide. On the other hand, free hydroxyproline is readily and specifically incorporated into the antibiotic in the microbial system [9, 10, 19, 20]. In addition, hydroxyproline in viridogrisein is in the D-cis form, while the hydroxyproline moiety is in the L-trans form in the animal system. It has been clarified that a direct precursor of *cis*-4-hydroxy-D-proline in viridogrisein is the L-trans form which is converted from L-proline before incorporation [9, 10, 19, 20]. The biosynthesis of *trans*-4-hydroxy-L-proline from L-proline may be one of the rate-limiting steps in the antibiotic biosynthesis and, as a result, L-proline added in excess to the culture may be directly accepted by the antibiotic synthesizing system. Since the affinity of the system for L-proline may be much lower than that for *trans*-4-hydroxy-L-proline, a large amount of L-proline may be necessary for the direct incorporation.

In the meanwhile, the replacement of the imino acid moiety in viridogrisein has been attempted by the supplementation with proline analogs and new analogs named neoviridogriseins-MP, -Cl-c and -Cl-t have been obtained [11, 12]. Since the incorporated amount of *trans*-4-chloro-L-proline into the new neoviridogrisein was greater than that of the L-cis form and there was no difference of the permeability into the cells, it has been suggested that the affinity of neoviridogrisein synthesizing system for proline analogs may be affected by the configuration at the C-4 position. In the case of methylprolines, *S. griseoviridus* P8648 can utilize only 4-methylproline. 3-Methyl and 5-methylproline can not be incorporated into the antibiotics. Furthermore, 3-methyl and 5-methylproline inhibit the antibiotic biosynthesis (Table 2).

Table 2 Concentrations of methylproline analogs required for 50% inhibition of neoviridogrisein synthesis.

Compound	Concentration (mM)
3-Methyl-DL-proline (cis + trans)	0.9
5-Methylproline (D-cis + L-trans)	8.8
4-Methyl-DL-proline (cis)	no inhibition
4-Methylproline (D-cis + L-trans)	no inhibition

Each methylproline was added at 24 h after inoculation to the culture of *S. griseoviridus* P8648. Production of neoviridogrisein was measured at 120 h after the supplementation. Reproduced with permission of The Agricultural Chemical Society of Japan (Ref. 11).

In actinomycins, proline analogs such as methylproline [27, 28], pipecolic acid [29], sarcosine [30] and azetidine-2-carboxylic acid [31] affect the synthesizing system and some of them are directly incorporated into the actinomycin molecule instead of proline. On the other hand, when various proline analogs are supplemented to *S. griseoviridus* P8648, only 4-methyl- and 4-chloroprolines are incorporated into the antibiotics while the other compounds inhibit the formation of neoviridogriseins. This suggests that these proline analogs may bind to the antibiotic synthesizing system at the same site for proline or hydroxyproline, and that these proline analogs bound to the antibiotic synthesizing system can not be utilized for further peptide bond elongation steps.

There are invariable moieties. That is, 3-hydroxy-picolinic acid, L-threonine, D-leucine, sarcosine, *β,N*-dimethyl-L-leucine and L-phenylsarcosine can not be replaced by structurally related compounds. Of course, if different strains or different cultivation conditions will be used, there may be some possibilities to replace these positions.

Antimicrobial activities of various neoviridogriseins are summarized in Table 3 [32].

Table 3 Minimum inhibitory concentrations of neoviridogriseins.

Microorganism	MIC (μg/ml)						
	I	II	III	VG	MP	Cl-c	Cl-t
Staphylococcus aureus 209P	0.2	0.1	0.2	0.2	0.2	0.2	0.4
S. aurues Smith	0.2	0.2	0.4	0.4	0.4	0.4	0.8
S. aureus (TC, CP, PC)^r	0.2	0.1	0.4	0.4	ND	ND	ND
Streptococcus pneumoniae	0.4	0.4	0.4	0.4	0.4	0.8	0.8
Bacillus subtilis ATCC6633	0.4	0.2	0.8	0.8	0.8	0.8	0.8
Sarcina lutea	0.2	0.2	0.2	0.2	0.2	0.8	0.8
Escherichia coli K-12	>25	>25	>25	>25	>25	>25	>25
Pseudomonas aeruginosa	>25	>25	>25	>25	>25	>25	>25
Candida albicans	>25	>25	>25	>25	>25	>25	>25

Abbreviations: I: neoviridogrisein. I, II: neoviridogrisein II, III: neoviridogrisein III, VG: viridogrisein, MP: neoviridogrisein-MP, Cl-c: neoviridogrisein Cl-c, Cl-t: neoviridogrisein-Cl-t, TC: tetracycline, CP: chloramphenicol, PC: penicillin G, ( )^r: resistant, ND: not determined.

2.2 Specific accumulation of neoviridogrisein II by inhibition of proline hydroxylation reaction

As described in the previous section, neoviridogrisein II is the most active compound among neoviridogriseins including viridogrisein and seems to be more the desirable compound for practical use. Neoviridogrisein II can be produced by supplementation of L-proline to the culture, but this method is unsuitable for industrial production. Consequently, another method is required for the large scale production.

It has been known that the precursor of *cis*-4-hydroxy-D-proline in viridogrisein is *trans*-4-hydroxy-L-proline and also L-proline. The former is a direct precursor rather than the later [19, 20]. Based on the facts, it has been expected that if the proline hydroxylation reaction can be inhibited, the production of viridogrisein will decrease and the biosynthesis of neoviridogrisein II will increase. From this viewpoint, selective inhibition of the proline hydroxylation reaction has been attempted.

In animal system, prolyl hydroxylase requires *a*-ketoglutarate, ascorbate, ferrous ion and atmospheric oxygen as co-factors [33]. Among these co-factors, the most controlable compound, ferrous ion, has been tested.

The supplementation of ferrous or ferric ion has suppressed the biosynthesis of neoviridogrisein II and enhanced the production of viridogrisein even under conditions favouring neoviridogrisein II production. Ferrous ion has been more effective than ferric ion (Figure 3).

Furthermore, *a,a'*-dipyridyl, a well-known inhibitor for prolyl hydroxylase from animal systems, reduces the effect of ferrous ion (Figure 4).

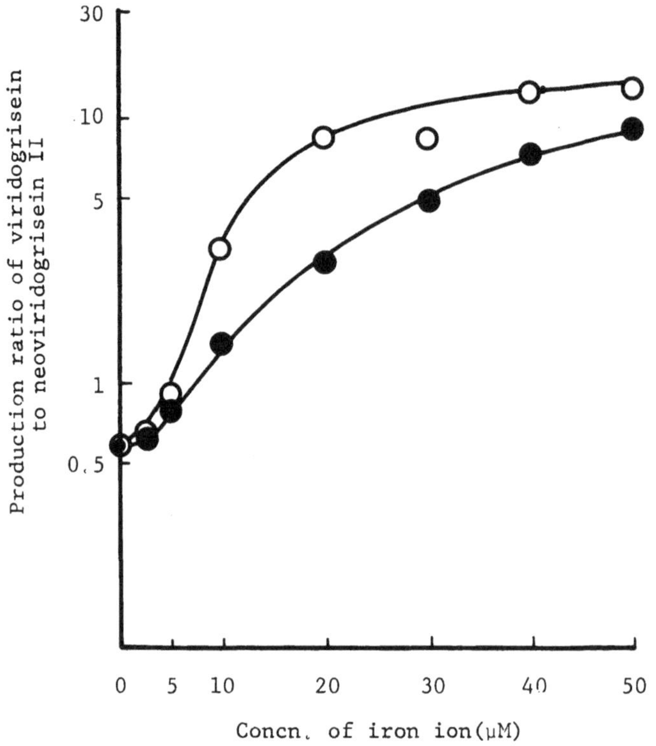

Figure 3 Effect of ferrous or ferric ion on the production of viridogrisein and neoviridogrisein II. *S. griseoviridus* P8648 was cultivated in a chemically defined medium containing 0.5% L-proline. Ferrous (○) or ferric (●) ion was added to the culture at the initial time. Reproduced with permission of The Agricultural Chemical Society of Japan (Ref. 9).

From these results, it has been clarified that the biosynthesis of *trans*-4-hydroxy-L-proline from L-proline in *S. griseoviridus* P8648 requires ferrous ion and the hydroxylation is inhibited by *a,a'*-dipyridyl which suppresses the reaction by removing ferrous ion from the system, and that the supplementation of the inhibitor may be one of the selective production methods for neoviridogrisein II. However, since *a,a'*-dipyridyl is highly toxic in the producing organism, a low or nontoxic effector is desired for the selective production of neoviridogrisein II.

It has been reported that prolyl hydroxylase from animal systems is inhibited by zinc ion [34]. In the microbial system, effect of zinc ion has been tested. Zinc ion shows the selective inhibition of proline hydroxylation reaction even in the microbial system (Figure 5 and Table 4).

Omission of ferrous ion and trans-4-hydroxy-L-proline from and supplementation of zinc ion to the culture of *S. griseoviridus* P8648 cause increased production of neoviridogrisein II and decreased production of viridogrisein. This is one of the selective production methods for neoviridogrisein II on a large scale.

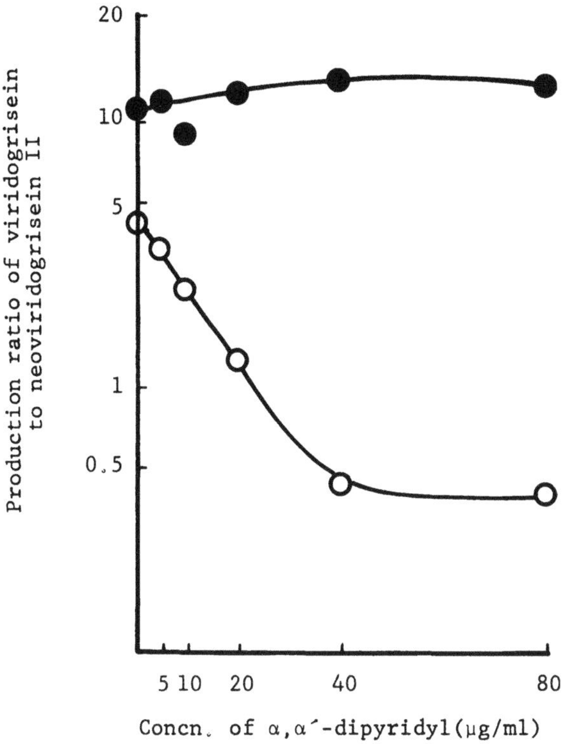

Figure 4 Effect of *a,a'*-dipyridyl on the production of viridogrisein and neoviridogrisein II. In the presence of 0.5% L-proline (○) or 0.05% (●) trans-4-hydroxy-L-proline, *S. griseoviridus* P8648 was cultivated in a chemically defined medium containing 10 μM of ferrous ion. Reproduced with permission of The Agricultural Chemical Society of Japan (Ref. 9).

Table 4 Effect of zinc ion on the incorporation of L-[^{14}C-(U)]-proline or *trans*-[^{3}H-(G)]-4-hydroxy-L-proline into viridogrisein and/or neoviridogrisein II.

Labeled compound	% Incorporation			
	Zn^{2+} 0 μM		Zn^{2+} 30 μM	
	VG*	NVG II**	VG*	NVG II**
L-[^{14}C-(U)]-proline	7.6	2.1	3.5	6.2
trans-[^{3}H-(G)]-4-hydroxy-L-proline	8.0	0	7.9	0

S. griseoviridus P8648 was cultured in a chemically defined medium. The indicated concentration of zinc ion was added at the initial time. Each radioactive compound was added to the culture at 72 h after inoculation to a concentration of 0.02 μCi/ml (^{14}C compound) and 0.2 μCi/ml (^{3}H compound). The incorporation was carried out further 24 h cultivation. VG*: viridogrisein and NVG II**: neoviridogrisein II. Reproduced with permission of The Agricultural Chemical Society of Japan (Ref. 9).

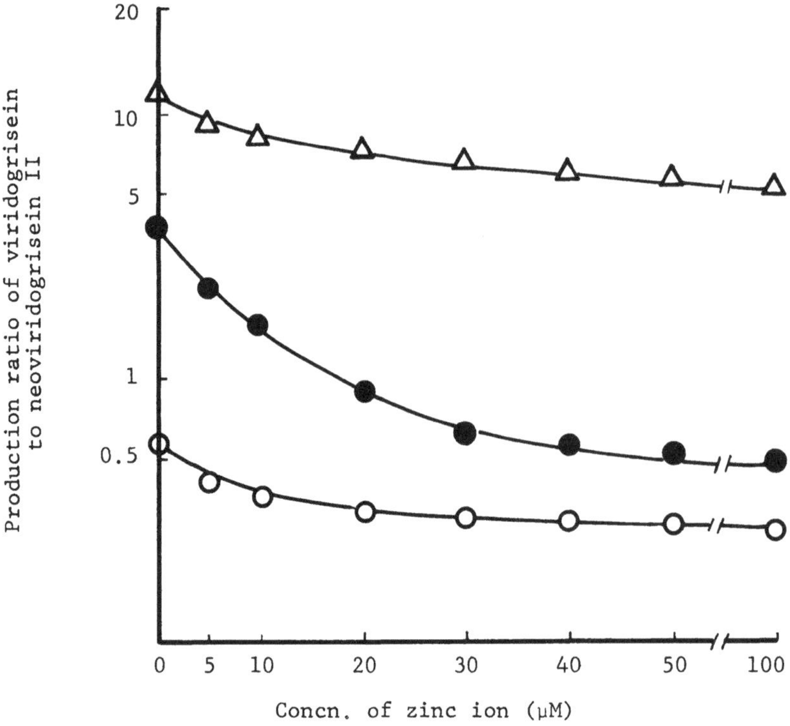

Figure 5 Effect of zinc ion on the production of viridogrisein and neoviridogrisein II. In the presence of ferrous ion (○; 0 μM, ●; 10 μM, △; 50 μM), *S. griseoviridus* P8648 was cultured in a chemically defined medium containing 0.5% L-proline. Reproduced with permission of The Agricultural Chemical Society of Japan (Ref. 9).

2.3 Specific accumulation of neoviridogrisein II by the mutant blocked in *trans*-4-hydroxy-L-proline formation from L-proline

As described in previous sections, directed biosynthesis of neoviridogrisein II has been established. Supplementation of L-proline and zinc ion and omission of *trans*-4-hydroxy-L-proline and ferrous ion can direct to favorite production of neoviridogrisein II. Especially, addition of zinc ion is more suitable for the large scale production. However, ferrous ion easily formed from ferric ion is commonly present in medium and water and the effect of zinc ion is drastically reduced by ferrous ion. Consequently, this directed biosynthetic method is not satisfactory for the industrial production. The most desirable method may be the use of a certain mutant which is blocked in the proline hydroxylation reaction.

Mutation work has been carried out for obtaining desirable mutants and finally, a mutant named strain G-89 has been isolated [10]. This strain does not produce viridogrisein in spite of the presence of ferrous ion and the absence of L-proline

Table 5 Property of strain G-89 on the incorporation of L-proline or *trans*-4-hydroxy-L-proline.

Radioactive imino acid	Zn^{2+} (μM)	% Incorporation			
		Parent strain		G-89	
		VG*	NVG II**	VG*	NVG II**
L-[^{14}C-(U)]-proline	0	7.6	2.1	0	8.5
	30	3.5	6.2	0	8.0
trans-[^{3}H-(G)]-4-hydroxy-L-proline	0	7.9	0	8.0	0
	30	8.0	0	8.2	0

The cultivation conditions were same as those of Table 4. VG*: viridogrisein and NVG II**: neoviridogrisein II. Reproduced with permission of The Agricultural Chemical Society of Japan (Ref. 10).

and zinc ion, while in the presence of trans-4-hydroxy-L-proline this strain can produce viridogrisein. These results suggest that strain G-89 has lost the L-proline hydroxylation ability and retained neoviridogriseins synthesizing ability. This has been confirmed by the use of radiolabeled L-proline and trans-4-hydroxy-L-proline (Table 5).

The possible formation processes of the imino acid moieties in viridogrisein and neoviridogrisein II are summarized in Figure 6.

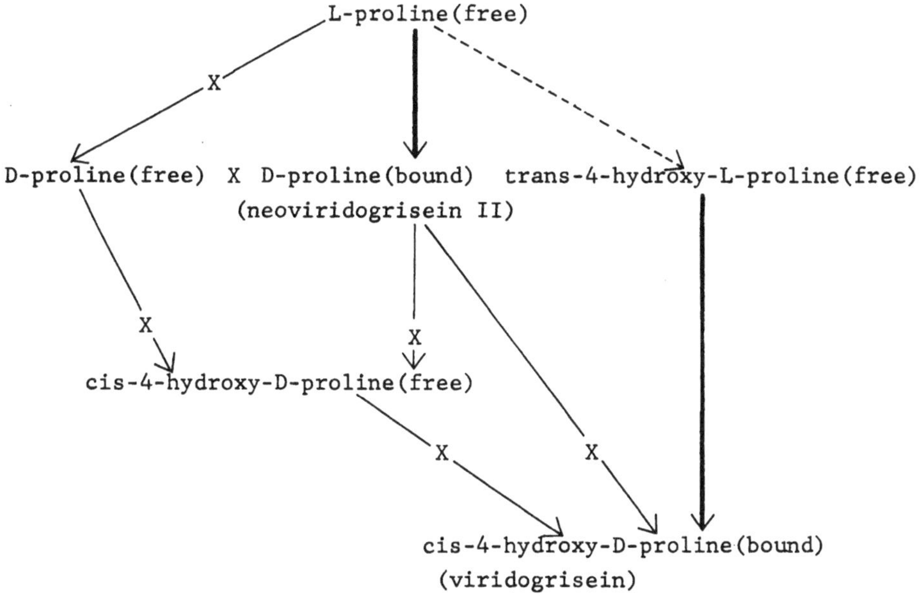

Figure 6 Possible biosynthetic pathway for the imino acid moiety in viridogrisein or neoviridogrisein II.
—×—→: These reactions don't occur in *S. griseoviridus* P8648 and its mutant G-89.
------→ This reaction is blocked in strain G-89.
Reproduced with permission of The Agricultural Chemical Society of Japan (Ref. 10).

The imino acid moieties in both antibiotics are derived from L-proline, and the direct precursors are free L-forms of the corresponding imino acid. Furthermore, since the free D-forms can not be detected in the mycelia and the cultured broth of the organism, racemization (or epimerization) of free L-imino acids may proceed concurrently with incorporation of the formed D-epimers to the antibiotic synthesizing system.

For the sake of clarification of the hydroxyproline formation from L-proline in this streptomycete, isolation of the proline hydroxylase has been attempted [35]. The cell free extract from *S. griseoviridus* P8648 exhibites proline hydroxylase activity. The optimum pH and temperature are 7.5 and 25°C, respectively. Effects of co-factors and inhibitors are presented in Table 6.

Table 6 Effects of co-factors and inhibitors on the proline hydroxylation.

Factor deleted or Inhibitor added	Concentration (mM)	Relative activity (%)
Complete	–	100
Fe^{2+}	–	0
α-Ketoglutarate	–	0
Ascorbate	–	78
Catalase	–	100
Zn^{2+}	0.03	39
	0.25	31
	0.50	0
α,α'-Dipyridyl	0.01	82
	0.05	57
	1.00	29
	5.00	22
	10.0	0
EDTA	0.25	0

The effects of co-factors were measured by deletion method. After each factor was omitted from the complete reaction mixture, the reaction was carried out. The inhibitors were added to the indicated concentration at the initial time and the reaction was performed. Reproduced with permission of Academic Press (Ref. 35).

This reaction essentially requires ferrous ion and α-ketoglutarate. Ferric ion can not replace ferrous ion [36]. Ascorbate also contributes to the reaction but not essentially. The reaction is inhibited by zinc ion, α,α'-dipyridyl and EDTA which are known as inhibitors of prolyl hydroxylase of animal systems. Since these co-factors and inhibitors are same as those of the animal systems, the reaction mechanisms may be very similar. However, there is one quite different point between the two systems. That is, in animal system only peptidyl proline can be hydroxylated, while in the microbial system free proline can be hydroxylated to the free form of hydroxyproline. From these characteristics of this enzyme, it has been clarified that target point of directed biosynthesis of neoviridogrisein II is this enzyme.

3. Summary

Nowadays, only small numbers of peptide antibiotics are accepted for practical use. The reasons may be due to their biological properties such as toxicity, solubility in water, narrow antimicrobial activity and low productivity. If these disadvantages are overcome, the utilization of peptide antibiotics will be expanded, because there are many kinds of peptide antibiotics having multifarious biological properties.

It may be the most important point about their practical use that peptide antibiotics are usually produced as complex forms and new analogs can be produced by replacement of the constituent moieties and, as a result, their biological properties can be improved. The biological evaluation or structure-activity relationships may able to lead the most desirable compound. After that, selective production of the compound can be attempted by various methods. Directed biosynthesis is one of the most suitable techniques for this purpose as described in this chapter.

References

1. Okumura, Y., Takei, T., Sakamoto, M., Ishikura, T., and Fukagawa, Y., Controlled biosynthesis of neoviridogriseins, new homologues of viridogrisein. II. Production, biological properties and structure of neoviridogrisein II, *J. Antibiot. 32*, 584, 1979.
2. Okumura, Y., Takei, T., Sakamoto, M., Ishikura, T., and Fukagawa, Y., Controlled biosynthesis of neoviridogriseins, new homologues of viridogrisein. III. Production, structures and biological properties of neoviridogriseins I and III, *J. Antibiot. 32*, 1002, 1979.
3. Cocito, C., Antibiotics of the virginiamycin family, inhibitors which contain synergistic components, *Microbiol. Rev. 43*, 145, 1979.
4. Okumura, Y., Peptidolactones, in *Biochemistry and Genetic Regulation of Commercially Important Antibiotics*, Vining, L. C., Ed., Addison-Wesley, Massachusetts, 1983, p. 147.
5. Vazquetz, D., The streptogramin family of antibiotics, in *Antibiotics* Vol. III, Corcoran, J. W., and Hahn, F. E., Eds., Springer-Verlag, Berlin, 1975, p. 521.
6. Haskell, T. H., Maretzki, A., and Bartz, Q. R., Viridogrisein: Chemical studies, *Antibiot. Annual* 1954/1955, p. 784, 1955.
7. Okumura, Y., Takei, T., Kouno, K., Lein, J., Ishikura, T., and Fukagawa, Y., Controlled biosynthesis of neoviridogriseins, new homologues of viridogrisein. I. Taxonomy and fermentation, *J. Antibiot. 32*, 575, 1979.
8. Okumura, Y., Sakamoto, M., Takei, T., Ishikura, T., and Fukagawa, Y., Controlled biosynthesis of neoviridogriseins, new homologues of viridogrisein. IV. In vitro synergism between neoviridogrisein II and the antibiotics of the mikamycin A group, *J. Antibiot. 32*, 1130, 1979.
9. Okumura, Y., Onishi, M., Okamoto, R., and Ishikura, T., Some effectors on the biosynthesis of cis-4-hydroxy-D-proline in viridogrisein, *Agric. Biol. Chem. 46*, 3063, 1982.
10. Okumura, Y., Onishi, M., Nagato, N., Okamoto, R., and Ishikura, T., A certain mutant blocked in trans-4-hydroxy-L-proline formation from L-proline during neoviridogrisein biosynthesis, *Agric. Biol. Chem. 47*, 1087, 1983.
11. Okumura, Y., Fukagawa, Y., Okamoto, R., and Ishikura, T., Influence of methylproline isomers upon neoviridogrisein synthesis, *Agric. Biol. Chem. 46*, 731, 1982.
12. Okumura, Y., Okamoto, R., and Ishikura, T., Neoviridogrisein-Cls, new analogs of viridogrisein, produced under the supplementation of 4-chloro-L-proline, *Agric. Biol. Chem. 48*, 543, 1984.
13. Sheehan, J. C., Zachan, H. G., and Lawson, W. B., The structure of etamycin, *J. Am. Chem. Soc. 80*, 3349, 1957.

14. Kruglyak, E. B., Konstantinova, N. V., and Sulavko, L. A., A new method of isolation of 6613 antibiotic and comparison of the drug with etamycin, *Antibiotiki 6*, 298, 1961.
15. Kleinkauf, H., Die Biosynthese von Antibiotica-Peptiden, *Chemie in unserer Zeit 14*, 105, 1980.
16. Lipmann, F., Gevers, W., Kleinkauf, H., and Roskoski, R., Polypeptide synthesis on protein templates: The enzymatic synthesis of gramicidin S and tyrocidine, *Adv. Enzymol. 35*, 1, 1971.
17. Katz, E., Actinomycins, in *Antibiotics*, Vol. II, Gottlieb, D., and Shaw, P. D., Eds., Springer-Verlag, Berlin, 1967, p. 276.
18. Katagiri, K., Yoshida, T., and Sato, K., Quinoxaline Antibiotics, in *Antibiotics* Vol. III, Corcoran, J. W., and Hahn, F. E., Eds., Springer-Verlag, Berlin, 1975, p. 521.
19. Hook, D. J., and Vining, L. C., Biosynthetic precursors of etamycin, a peptidolactone antibiotic from *Streptomyces griseoviridus, Can. J. Biochem. 51*, 1630, 1973.
20. Katz, E., Kamal, F., and Mason, K., Biosynthesis of trans-4-hydroxy-L-proline by *Streptomyces griseoviridus, J. Biol. Chem. 254*, 6684, 1979.
21. Vanderhaeghe, H., Janssen, G., and Compernolle, F., The structure of minor components of virginiamycin S, *Tetrahedron Lett.*, 2687, 1971.
22. Bodanszky, M., and Ondetti, M. A., Structures of the vernamycin B group of antibiotics, *Antimicrob. Agents Chemother.*, 360, 1964.
23. Vanderhaeghe, H., and Parmentier, G., The structures of factor S of staphylomycin, *J. Am. Chem. Soc. 82*, 4414, 1960.
24. Donovick, R., Dutcher, J. D., Heuser, L. J., and Pagano, J. F., *U.S. Patent 2,990,325* 1961.
25. Ball, S., Boothroyd, B., Less, K. A., Raper, H., and Smith, E. L., Preparation and properties of an antibiotic complex E129, *Biochem. J. 68*, 24, 1958.
26. Cardinale, G. J., and Udenfirend, S., Prolylhydroxylase, *Adv. Enzymol. 41*, 245, 1974.
27. Katz, E., Williams, W. K., Mason, K. T., and Mauger, A. B., Novel actinomycins formed by biosynthetic incorporation of cis- and trans-4-methylproline, *Antimicrob. Agents Chemoter. 11*, 1056, 1977.
28. Yoshida, T., Mauger, A. B., Wikop, B., and Katz, E., Influence of methylproline isomers upon actinomycin biosynthesis, *Biochem. Biophys. Res. Commun. 25*, 66, 1966.
29. Formica, J. V., and Katz, E., Isolation, purification and characterization of pipecolic acid-containing actinomycins, Pip2, Pip1α and Pip1β, *J. Biol. Chem. 248*, 2066, 1973.
30. Katz, E., and Goss, W. A., Controlled biosynthesis of actinomycin with sarcosine, *Biochem. J. 73*, 458, 1959.
31. Formica, J. V., and Apple, M. A., production, isolation and properties of azetomycins, *Antimicrob. Agents & Chemother. 9*, 214, 1976.
32. Okumura, Y., Personal communication, 1983.
33. Myllylä, R., Tuderman, L., and Kivirikko, K. I., Mechanism of the prolyl hydroxylase reaction. 2. Kinetic analysis of the reaction sequence, *Eur. J. Biochem. 80*, 349, 1977.
34. Tuderman, L., Myllylä, R., and Kivirikko, K. I., Mechanism of the prolyl hydroxylase reaction. 1. Role of co-substrates, *Eur. J. Biochem. 80*, 341, 1977.
35. Onishi, M., Okumura, Y., Okamoto, R., and Ishikura, T., Proline hydroxylation by cell free extract of a streptomycete, *Biochem. Biophs. Res. Commun. 120*, 45, 1984.
36. Okumura, Y., Personal communication, 1984.

Chapter 14
Biochemical, Genetical and Biotechnical Aspects of Antibiotic Production via Immobilised Biocatalysis

Erick J. Vandamme

1. Novel Antibiotic Compounds
2. Novel Production Ways for Antibiotics
3. Total Enzymatic Synthesis of Antibiotics
4. Partial Enzymatic Synthesis of Antibiotics
5. Antibiotic Fermentation via Immobilised Viable Cells
6. β-Lactam-Antibiotic Bioconversions With Immobilised Biocatalysts
   6.1 Penicillin bioconversions
   6.2 Cephalosporin bioconversions
7. Chiral Side Chain Production for β-Lactam Antibiotics Synthesis
8. Bioconversion of Novel β-Lactam Antibiotics
9. Bioconversion of Other Important Antibiotics
10. Perspectives
References

1. Novel Antibiotic Compounds

About 150 microbial antibiotic compounds are produced today on large scale by fermentation. This is a minority as compared to the already more than 7000 known different natural microbial antibiotics discovered by ingenious screening methods. These economic antibiotics are mainly used in medicine and agriculture as antibacterial, antifungal or antiviral agents. Fewer compounds are applied as antitumour agents, immunomodulators, as anthelmintical, herbicidal, antiprotozoal, piscicidal or anticoccidial agents, feed additives or food preservatives (Demain, 1983; Vandamme, 1984) (Table 1).

Apart from continued intensive screening for new antibiotic producing microorganisms, various approaches have been sought to force well-known antibiotic producer strains to synthesize novel antibiotics (Table 2).

Feeding unnatural precursors to penicillin fermentations is in this respect a historical example (Behrens et al., 1948) with lasting industrial impact; precursor-toxicity and permeability barriers pose heavy limits to a generalisation of this approach.

Table 1 Use of Antibiotic Compounds.

Antibacterials	Antitumors
Antifungals	Antivirals
Coccidiostats	Feed supplements
Antiprotozoals	Food preservatives
Plant-disease controllers	Piscicidals
Anthelmintics	Herbicidals
Immuno-modulators	

Table 2 Current approaches to third novel antibiotic compounds.

- screening
- precursored fermentation (wild type + natural precursor):
  penicillin G, V, ...
- mutasynthesis (idiotrophs + unnatural precursors):
  aminoglycosides (streptomutins)
- hybrid biosynthesis (wild type + enzyme inhibitor + unnatural precursor):
  macrolides (= chimeramycins)
- interspecies mating:
  aminoglycosides, anthracyclines, ...
- protoplast fusion:
  macrolides, β-lactams, ...
- cloning and transfer of antibiotic biosynthesis genes:
  mederrhodins, actinomycins, ...
- bioconversion:
  semi-synthetic β-lactams, ...

Antibiotic production by mutasynthesis (Shier et al., 1969) or hybrid biosynthesis (Omura et al., 1983) are also interesting techniques to yield new compounds. However, such fermentations proceed via at least two stages:

1. cultivation of an idiotrophic mutant or of the antibiotic producer in the presence of an enzyme inhibitor,
2. the addition of an analogue of a biosynthetic intermediate.

A further approach to obtain novel antibiotic compounds is interspecific mating (Schlegel and Fleck, 1980) and protoplastfusion (Ikeda et al., 1984); however, here the origin of the novel compounds is not absolutely clear, since an unexpressed capacity for antibiotic formation in one strain could be expressed upon transfer into the other strain. The direct transfer of cloned antibiotic biosynthesis genes isolated from one producer into a recipient strain offers the possibility of generating hybrid strains, producing true novel antibiotic compounds by simple one step fermentation (Hopwood et al., 1985).

The production of the new hybrid antibiotics mederrhodin A and B by a genetically engineered *Streptomyces* strain via introducing the genes for actinorhodin biosynthesis from *S. coelicolor* is an example of this technique (Omura et al., 1986).

All previous techniques deal with novel compounds to be produced by conventional fermentation, involving complex pathways and multi-step enzymatic reactions.

Several new compounds are in turn derived from such fermentation products by chemical or enzymatic conversion (Vandamme, 1983, 1984; Savidge, 1984). In these processes, usually only one enzymatic step is involved to obtain a new product, the so called semi-synthetic antibiotic derivative with superior therapeutic properties.

2. Novel Production Ways for Antibiotics

Apart from conventional batch fermentation or bioconversion processes, novel ways for production of known or novel antibiotic compounds include

a) multi-step total or partial enzymatic antibiotic synthesis via immobilised biocatalysts,
b) antibiotic fermentation via immobilised viable cells,
c) antibiotic bioconversions with immobilised biocatalysts.

Indeed, today several antibiotic bioconversion processes – involving single step bioreactions – are run industrially with immobilised enzymes or cells as biocatalysts, where they replace laborious or uneconomical chemical processes. The replacement of conventional fermentation – involving invariably a complex multistep reaction sequence – with a total or partial sequence of (immobilised) enzyme-reactions or with immobilised viable cell technology also offers great potential to arrive at novel compounds or improved processes.

The recent cloning of genes, coding for antibiotic biosynthesis enzymes or for bioconversion enzymes, opens an exciting new area leading towards improved production of indirect gene products such as antibiotics via r-DNA technology (Table 3).

Table 3 Novel approaches in the production of antibiotics.

- Conventional and NON-conventional fermentation or bioconversion
- multi-step enzymic total or partial synthesis
- cloning of antibiotic biosynthesis-genes (gramicidin S-synthetase, tyrocidine synthetase, bacitracin synthetase, factor A, isopenicillin N synthase, expandase, hydroxylase, tylosin genes, erythromycin genes, macrolide genes, ...)
- immobilised viable cell fermentation
- bioconversion with immobilised biocatalysts
- cloning of bioconversion enzyme genes (penicillin G acylase, penicillin V acylase, glutaryl-7-ACA acylase, HABA-acylase

3. Total Enzymatic Synthesis of Antibiotics

Despite the fact that chemical routes are known for the total synthesis of many important antibiotics only few compounds – chlooramphenicol, thienamycin, cellocidin, relatively simple antibiotic molecules –, are produced industrially by chemical synthesis. Most antibiotics are very complex biological molecules virtually excluding large scale chemical production from an economic point of view.

However, although universally applied, conventional fermentation displays several drawbacks, asking for improved production ways.

As secondary metabolites, their synthesis is usually delayed until growth declines; the enzymes involved in antibiotic synthesis – named the antibiotic "synthetases" – are unstable and rapidly inactivated; substrate conversion into antibiotic is inefficient (10% in the case of penicillin) and strain degeneration presents a universal problem. In this respect, several attempts have been made to replace fermentation or cellular synthesis by acellular processes, i. e. total enzymic antibiotic synthesis in vitro. In such a process, it is the ultimate aim to use isolated (and immobilised) enzymes, which, in sequential reactions, perform the total synthesis of an antibiotic upon addition of its precursors, ATP and cofactors. This concept has been studied intensively with the oligopeptide antibiotics, including gramicidin S, tyrocidine,

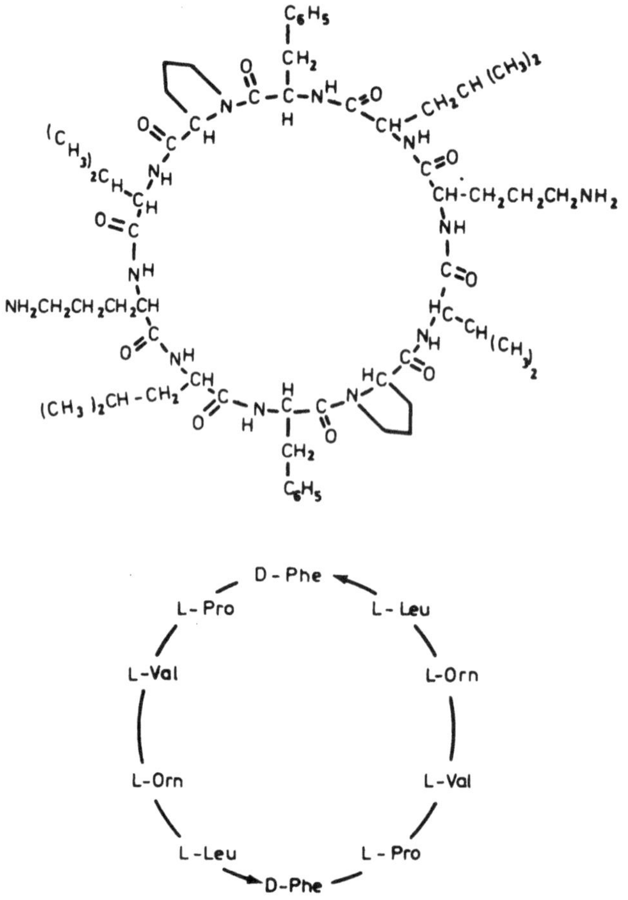

Figure 1 Structure and schematical representation of gramicidin S.

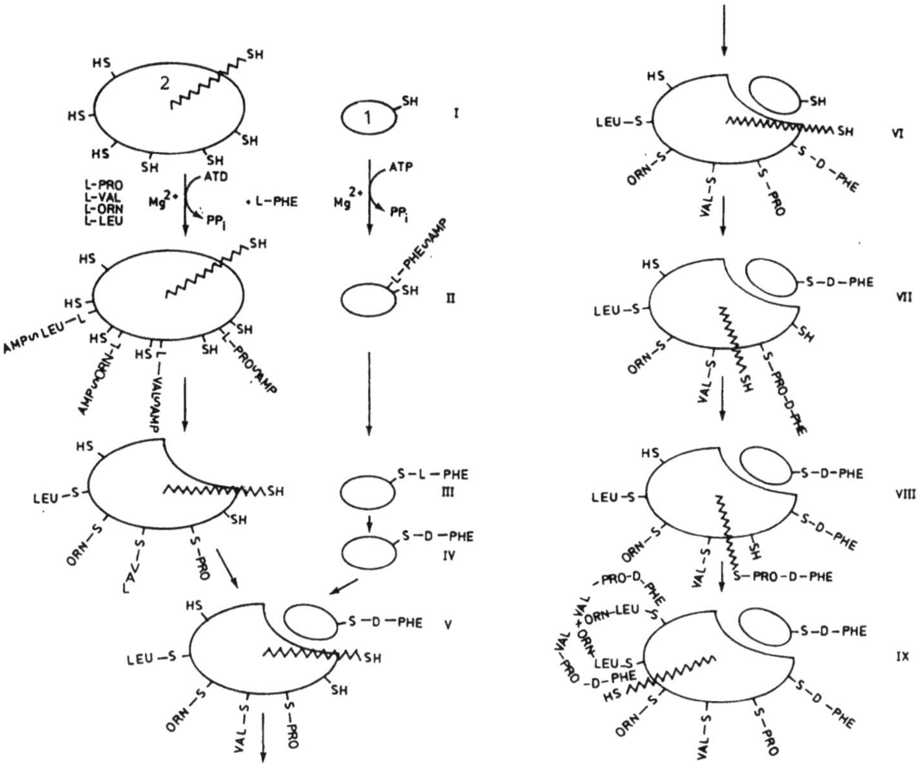

Figure 2 Model for the functioning of the multi-enzyme thiotemplate mechanism in gramicidin S biosynthesis (1 = GS-synthetase 1, 2 = GS-synthetase 2)
I, Schematic representation of GS synthetases 1 and 2. The zig-zag line represents 4′-phosphopantetheine. II, Activation of amino acids and binding on GS synthetases 1 and 2. III, Thioester binding of amino acids and conformational change of GS synthetase 2. IV, Racemization of L-Phe to its D-isomer. V, GS synthetases 1 and 2 complex, loaded with thioester-bound amino acids. VI, Transfer of D-phenylalanine on GS synthetase 1 to a thiol site of GS synthetase 2. VII, Formation of the first intermediate dipeptide. VIII, Transfer of the dipeptide to the 4-phosphopantetheine arm. IX, Two pentapeptides cyclize 'head to tail' in an intramolecular reaction of GS synthetase 2 [Reproduced from Vandamme, E. J. in *Topics in Enzyme and Fermentation Biotechnology* (Wiseman, A., ed.) Ellis Horwood, Chichester, 1981, vol. 5, p. 185, by permission of Ellis Horwood Limited©] (Ref. 79).

bacitracin, enniatins, ... One process that is understood in detail is the biosynthesis of gramicidin S (GS), a cyclic decapeptide antibiotic produced by certain *Bacillus brevis* strains (Figure 1). Only two enzymes take part in the biosynthetic pathway, i. e. GS synthetase 1 and 2. The two enzymes, 5-amino acid substrates and cofactor phosphopantetheine, involved, have been characterized and the biosynthetic mechanism has been resolved (Vandamme, 1981; Vater, 1989) (Figure 2).

The mechanism involved – which seems to occur frequently in peptide – antibiotic producing bacteria and fungi –, has been named "multi-enzyme thiotemplate" biosynthesis, and it requires, in addition to the two enzymes, only the building block amino acids or their analogues as precursors, ATP as an energy source, and Mg^{2+} ions. The overall biosynthesis scheme for GS can be simplified as follows:

$$\left.\begin{array}{l}2\ \text{Phe}\\ 2\ \text{Pro}\\ 2\ \text{Val}\\ 3\ \text{Orn}\\ 2\ \text{Leu}\end{array}\right. + 10\ \text{ATP} \xrightarrow[\text{Mg}^{2+}]{\text{GS synthetase 1 and 2}} 1\ \text{GS} + 10\ \text{AMP} + 10\ \text{PP}_i$$

Using the two GS synthetases, isolated from high-yield gramicidin S-producing *B. brevis* ATCC 9999 cells, gram quantities of gramicidin S have been produced in vitro by this total enzymic method (Demain et al., 1976).

The soluble GS synthetases are only transitorily formed during the *B. brevis* growth cycle which makes it difficult to harvest large quantities of this labile enzymes (Vandamme et al., 1982) (Figure 3).

Particulate GS synthetases are also present, which might be easier to recover and to stabilize outside the cells (Demain et al., 1981; Poirier and Demain, 1981). In this respect, the cloning of the structural genes for the GS synthetases seems important to be able to produce the enzymes abundantly; the recent cloning of an

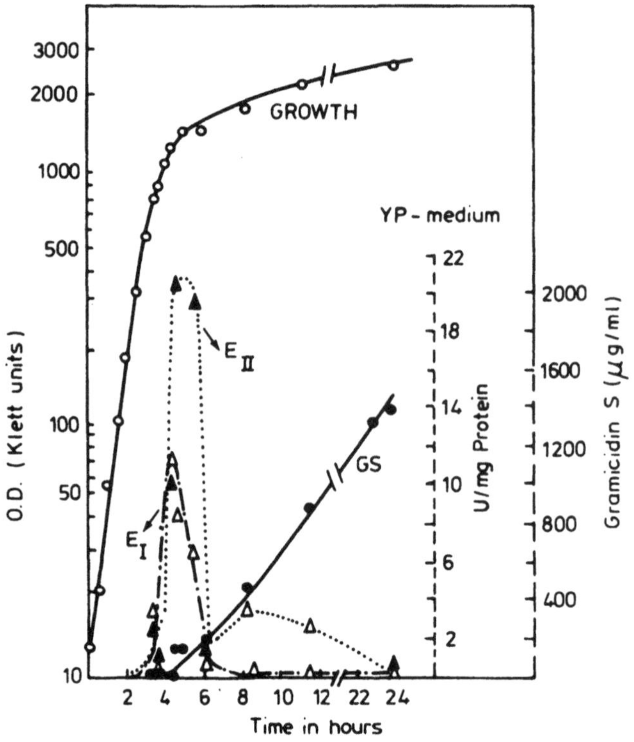

Figure 3 Fermentation dynamics of gramicidin S and transient formation of soluble GS-synthetase 1 (=E_1) and 2 (=E_2).

ornithine activating fragment of GS-synthetase 2 from *B. brevis* and its expression in *E. coli* is a valuable step in this direction (Krause et al., 1985). The thus produced GS synthetases could be coimmobilised to construct a bioreactor, coupled to an enzymatic ATP-regeneration system which may allow for a continuous process for GS production. Also instead of using expensive pure amino acids as substrates, protein hydrolysates could serve as a cheap amino acid source. The successful screening of *B. brevis* strains, which overproduce extracellular proteins (up to 20 g/l) (Udaka, 1987) is also important, in view of using them as hosts for homologous cloning and production of *Bacillus* oligopeptide antibiotic-synthetases (Tsukagoshi et al. 1985).

The depsipeptide antibiotic enniatin (Figure 4), produced by a fungus – *Fusarium oxysporum* – is also produced via the thiotemplate mechanism, but here only one multifunctional enzyme, enniatin synthetase (MW 250,000) is involved. It catalyses the synthesis of enniatins in the presence of Mg^{2+}, ATP, hydroxyisovaleric acid, S-adenosyl-L-methionine and a branched chain L-amino acid such as isoleucine (= enniatin A), valine (= enniatin B) or leucine (= enniatin C). Addition of isoleucine to an enniatin-producing culture resulted in a complex mixture of homologues and hybrid enniatins. The purified multi-enzyme also synthesizes in vitro different enniatins depending on the added amino acid, but with less hybrids formation. The partially purified enzyme has been immobilised by adsorption to propylagarose, with 45% retention of original activity and an operational half-life of 15 h. In vitro

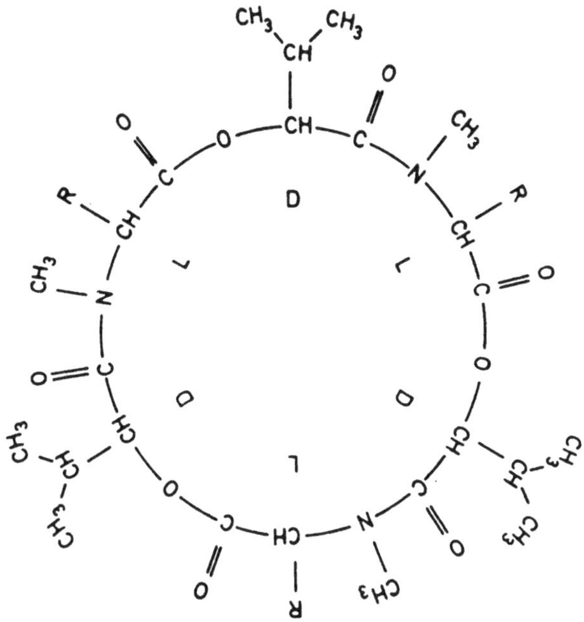

Enniatin A, R = $-CH(CH_3)CH_2CH_3$; enniatin B, R = $-CH(CH_3)_2$; enniatin C, R = $-CH_3CH(CH_3)_2$.

Figure 4 Structure of enniatin homologues.

selective synthesis – however with low yield – of several enniatin homologues was achieved, including non-natural depsipeptides, depending on the type of non-proteinogenic amino acid supplied (norvaline, norleucine or α-aminobutyric acid) (Madry et al., 1984). The immobilised enzyme-system proved thus more advantageous than the classical fermentation in view of obtaining specific enniatins. An important observation here is also that the specificity of the purified synthetase is broader than that of the cells as enzyme source.

Membrane- or organelle-localized enzyme complexes have also been found to be able to catalyse total antibiotic synthesis. Kurzatkowsky et al. (1982) isolated vesicles (40 – 200 nm in diameter) from *P. chrysogenum* PQ-96 protoplasts which displayed penicillin G total biosynthesis capacity from the added building block amino acids and an activated side chain.

Immobilisation of these vesicles in a calcium alginate gel left 44% of the activity of native vesicles. This activity remained stable after 240 h storage at 4°C, whereas the native vesicles had lost all biosynthetic activity within 60 h. Such a stabilized organelle-biocatalyst in a continuous bioreactor might result in improved penicillin production by providing a natural compartimentalized enzyme complex, with minimal substrate-transport limitations.

Potential applications of total enzymic synthesis are:

- improved production of currently used and known antibiotics,
- the production of biosynthetic intermediates as starting materials for chemical or bioconversion into new antibiotics by omitting or adding certain enzymes in such systems,
- total enzymic synthesis of novel antibiotics since the in vitro specificities of antibiotic-forming enzymes are not very strict; substrates can be used which could otherwise not be incorporated by whole cells because of permeability constraints, degradation or toxicity,
- the synthesis of other complex molecules of medical, nutritional, agricultural or industrial importance.

4. Partial Enzymatic Synthesis of Antibiotics

Total enzymatic in vitro synthesis of antibiotics is only possible if all enzymes, substrate-compounds and cofactors involved in the pathway are known and available; this is only the case for a few compounds, mainly oligopeptide antibiotics as described above.

On the contrary, particular enzymic steps and intermediates involved in economic antibiotic synthesis are rather well studied. Applying the concept of enzymic synthesis here implicates the addition of natural intermediates of the desired antibiotic to the bioreactor. Microbial enzymes useful in this respect are those involved in the late steps of penicillin and cephalosporin biosynthesis (Figure 5).

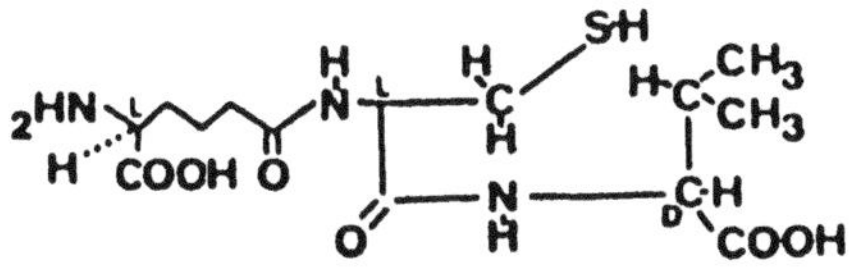

δ-(L-α-aminoadipyl)-cysteinyl-D-valine (ACV)

cyclase; isopenicillin N sythase (IPNS)

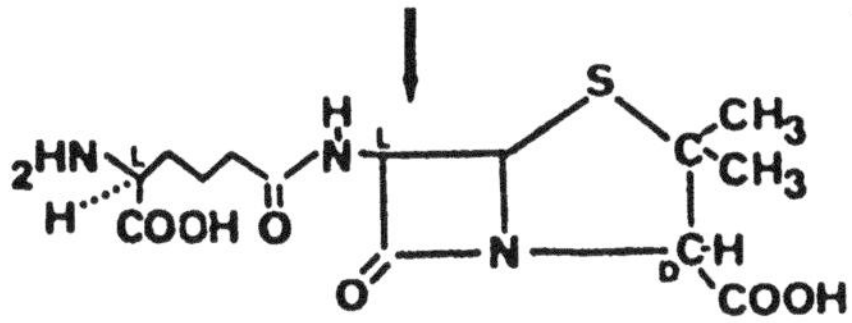

isopenicillin N

epimerase

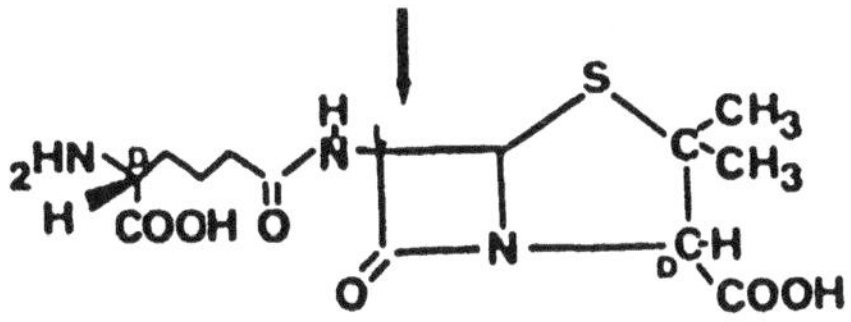

penicillin N

expandase; deacetoxycephalosporin C synthase (DAOCS)

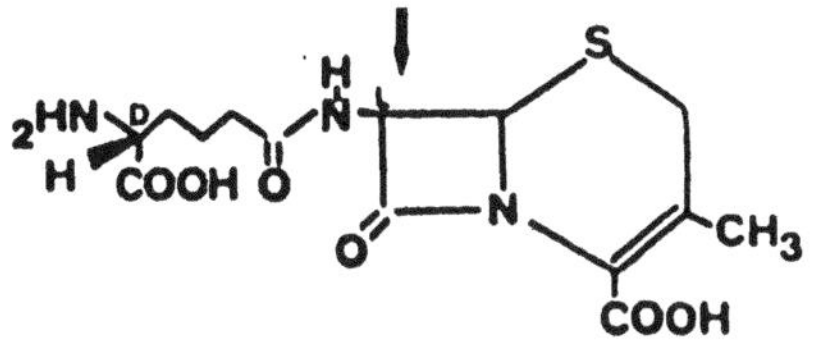

deacetoxycephalosporin C

hydroxylase; deacetylcephalosporin C synthase (DACS)

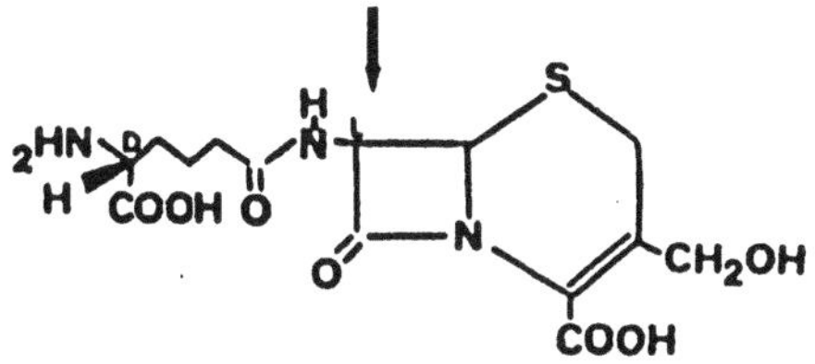

deacetylcephalosporin C

Figure 5 Biosynthesis pathway to penicillins and cephalosporins in *Streptomyces clavuligerus*.

The common natural intermediate here is the tripeptide δ-(L-α-amino-adipyl)-L-cysteinyl-D-valine (ACV), which is further converted by isopenicillin synthetase (cyclase) into isopenicillin N, which is then transformed into natural penicillins by a transferase activity in penicillin producing fungi (*Penicillium chrysogenum*).

The early enzymes carrying out the condensation of the precursor amino acids (α-aminoadipic acid, cysteine, valine) into ACV are only now being characterized, preventing a total enzymatic in vitro synthesis.

In cephalosporin producing fungi (*Cephalosporium acremonium*) and actinomycetes (*several Streptomyces sp.*), isopenicillin N is further epimerised (epimerase) into penicillin N, which in turn undergoes ring expansion, effected by the deacetoxy-cephalosporin C synthetase (expandase) into deacetoxycephalosporin C (Dotzlaf and Yeh, 1987). This intermediate is further enzymatically oxidized (hydroxylase) into deacetylcephalosporin C, from which cephalosporin C or cephamycins are formed (Figure 5). Jensen et al. (1982, 1984) reported on the immobilisation in a DEAE-tris-acrylgel of the four soluble enzymes (cyclase, epimerase, expandase, hydroxylase), present in a cell free extract of the prokaryotic *Streptomyces clavuligerus*; they demonstrated continuous multistep conversion of the natural intermediate δ-(L-α-aminoadipyl)-L-cysteinyl-D-valine (ACV)-tripeptide into a mixture of penicillin and cephalosporin β-lactam antibiotics.

They speculated on the fact that any one of these biosynthesis enzymes might be an appropriate biocatalyst (in free or immobilised form) for the production of unnatural β-lactams providing unnatural intermediates: indeed, by altering chemically the amino acids of the natural peptide-intermediate ACV, unnatural intermediates (ACV with L-α-aminoadipic acid replaced by L-carboxymethylcysteine) were obtained which could be converted into novel antibiotics by the *S. clavuligerus* enzymes (Bahadur et al., 1981; Wolfe et al., 1982). Of these four enzymes, especially, the isopenicillin N synthetase (cyclase) has been subjected to detailed study as well in *Penicillium chrysogenum* (Ramos et al., 1985) as in *Cephalosporium acremonium* (Hollander et al., 1984) and in *S. clavuligerus* (Jensen et al., 1982). The cyclase displays similar properties, whether derived from fungi or actinomycetes: it needs ACV and O_2 as substrates; it is stimulated by dithiothreitol, Fe^{2+} and ascorbate, but is inhibited by Co^{2+}, Mn^{2+} and glutathione; optimal temperature is 25°C and optimal pH 7,8; a molecular weight of 39,000 – 42,000 is typical. The cyclase is also transitorily formed, peaking at the onset of antibiotic detection in the fermentation cycle. High-penicillin producer strains displayed high and stable levels of cyclase during the fermentation cycle as compared to low penicillin producers (Ramos et al., 1985). Shen et al. (1986) also detected higher levels of cyclase and expandase in high-producer cephalosporin C strains of *C. acremonium* and found the expandase of the superior strain to be deregulated with respect to carbon source repression.

An exciting observation – in fact related to the substrate specificity of the cyclase – whether originating from fungi or Streptomycetes – but with important industrial relevance – was made by Martin and his collaborators (Luengo et al., 1986 a; Castro et al., 1986): by supplying the partially purified cyclases of high producing *P. chrysogenum, C. acremonium* and *S. lactamdurans* strains with the synthetic linear

ACV-analogue phenylacetyl-L-cysteinyl-D-valine (PCV), direct enzymatic synthesis of penicillin G was obtained although with much lower efficiency (Figure 6). The use of immobilised cyclases to produce penicillin G from PCV is now the next logical step. Already, the gene encoding for cyclase protein has been isolated, sequenced and expressed in *E. coli* (Samson et al., 1985). In the recombinant *E. coli* strain, the active cyclase-protein was a major (≈20%) component of the cell protein (Carr et al., 1986; Jensen et al., 1987).

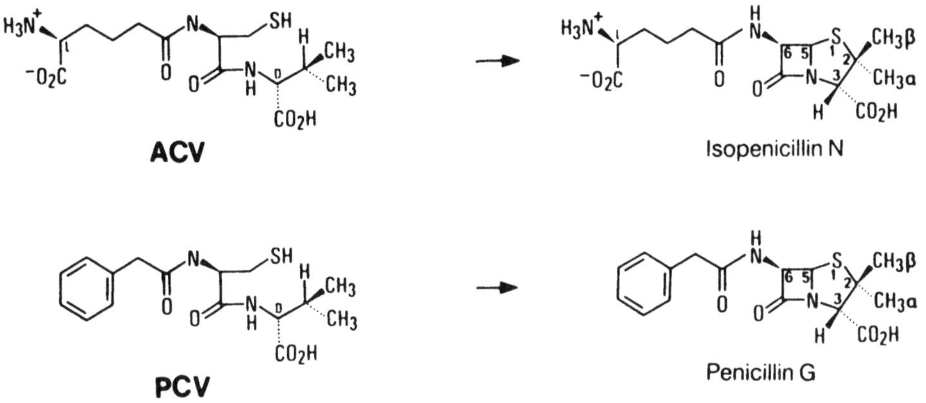

Figure 6 Structure of ACV and PCV and their cyclization products isopenicillin N and penicillin G.

The last enzyme believed to be involved in natural penicillin fermentations (no precursors added) is an acyl CoA:6-amino-penicillanic acid acyltransferase, which has been purified from *P. chrysogenum* AS-P-78 (Luengo et al., 1986 b). This enzyme could be used for in vitro synthesis of different natural penicillins (hexanoyl-, heptanoyl- and octanoylpenicillin) by direct acylation of the penicillin nucleus, 6-aminopenicillanic acid (6-APA) with the corresponding fatty acid-CoA-derivatives; only aliphatic side chain precursors with a 6 to 8 atoms-carbon length are accepted as a substrate; this acylation is not observed with the conventional bacterial penicillin acylases; supplying phenylacetyl CoA resulted in formation of penicillin G.

Screening for strains or mutants, able to accept higher concentrations of common or strange fatty acids should yield acylCo-6-APA-acyltransferases, which in soluble or immobilised form, could be used to synthesize new aliphatic penicillins.

It is obvious that large scale fermentation of the above mentioned enzymes and their hyperproduction via deregulation or genetic techniques is essential to arrive at economical enzymatic partial synthesis of natural and unnatural β-lactam antibiotics (Fayerman, 1986; Garcia-Dominguez et al., 1987).

5. Antibiotic Fermentation via Immobilised Viable Cells

The difficulties still encountered with complex total or partial enzymic synthesis processes for antibiotic production using cell-free (immobilised) enzyme preparations might be circumvented by using immobilised whole viable cells (Vandamme, 1976; Chibata and Tosa, 1981; Karube et al., 1984; Brodelius and Vandamme, 1987). The rationale for viable cell immobilisation can be summarized as follows: it obviates enzyme extraction/purification, lowering effective enzyme cost; generally higher operational stability is obtained; high yield of enzyme activity on immobilisation; cofactor regeneration in situ is possible; retention of structural and conformational integrity, great potential for multi-step processes, greater resistance to environmental perturbations. This concept would be particularly valuable for those antibiotics which are, or can be, excreted into the fermentation broth. Immobilised viable cell-fermentations of penicillin G, cephamycin, thienamycin, candicidin, nisin, bacitracin, colistin, tylosin, nikkomycin, patulin, cyclosporin and several other compounds have already been tested in this respect (Veelken and Pape, 1982; Vandamme, 1983; Arcuri et al., 1986).

Immobilised viable cell systems also offer several important advantages over conventional fermentation:

1. placement of fermentation on heterogeneous catalysis basis.
2. ability to run continuous or plug flow, fluidized bed, ... operations as opposed to traditional batch fermentation.
3. operation at high dilution rates without culture wash-out.
4. ability to recharge the system by controlled reproduction of resting cells.
5. decrease or elimination of lag and growth phases for product accumulation associated with the non-growth phase of the fermentation.
6. possibility of accelerated reaction rates and higher yields due to increased cell density.
7. improved rheological control.
8. simplified antibiotic recovery from a cell-free liquid.

So far, none of the mentioned compounds is produced commercially via immobilised cell technology because of serious problems related with: long-term stability of the (immobilised) enzymatic enzyme complexes involved, the low specific activity and low volumetric productivity, cell leakage.

An important observation was made by Freeman and Aharonowitz (1981) (Figure 7): they compared the cephalosporin production capacity of *S. clavuligerus* cells in resting condition (a situation also occuring in immobilised cell systems) as a function of culture age with that of the free growing cells; it was found that resting cells, prepared from the early log-phase – a growth stage where antibiotic formation could not be detected in the fermentation process – were highly active as antibiotic producers, also upon immobilisation.

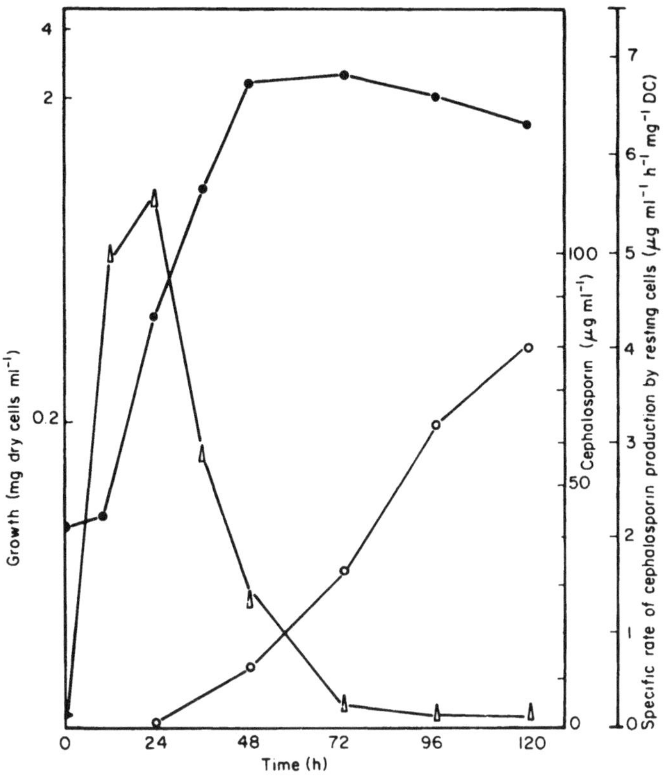

Figure 7 Capacity of *Streptomyces clavuligerus* resting cells (△) to produce cephalosporins as a function of culture age as compared to the antibiotic formation pattern (○) during growth (●).

Conversely, resting cells, prepared from stationary growth – where high antibiotic levels were present in the fermentation process – were poor antibiotic producers.

This implicates that the physiological state of the cells or mycelium to be immobilised is of utmost importance to obtain high antibiotic yields in immobilised viable cell systems: or the cells are to be harvested at that particular growth stage and fixed there or have to be forced to reach that physiological state by controlled in situ reproduction in the immobilised biocatalyst system.

Viable cells capable of (secondary) metabolism are usually also capable of undesirable growth within and out of immobilised biocatalyst-preparations. In this respect, nutritional and environmental conditions need to be selected which allow secondary product synthesis with no growth. Otherwise, strains should be used where growth can be stopped "at will" e. g. by temperature shift or by limitation of an auxotrophic requirement, without interfering with antibiotic synthesis.

Several enzymes in the β-lactam antibiotic pathways require oxygen (isopenicillin N synthetase, expandase, hydroxylase, ...): the immobilisation potential for β-lactam antibiotic producing viable micro-organisms is hence influenced by the availability of oxygen to such cells; this limits the use of cell-entrapment techniques, which are best studied otherwise. Alternative ways to supply oxygen to immobilised cells include co-immobilisation with oxygen-generating algae, increased oxygen supply using perfluorochemicals or haemoglobin or *in situ* oxygen generation from hydrogen peroxide with catalase activity.

6. *β*-Lactam Antibiotic Bioconversions with Immobilised Biocatalysts

6.1 Penicillin bioconversions

Precursored penicillins G and V produced by directed fermentation can be hydrolysed into their nucleus 6-aminopenicillanic acid (6-APA) and their side chain acid, phenylacetic and phenoxyacetic acid respectively, by the action of microbial penicillin acylases (E.C. 3.5.1.11) (Figure 8) or by chemical methods.

Coupling new side chains to 6-APA yields the semi-synthetic penicillins, which find extensive use in therapy today.

Industrially, chemical as well as enzymatical routes are used to produce 6-APA. By introduction of the immobilised biocatalyst-principle, the enzymatic process currently displays better economics and is increasingly used.

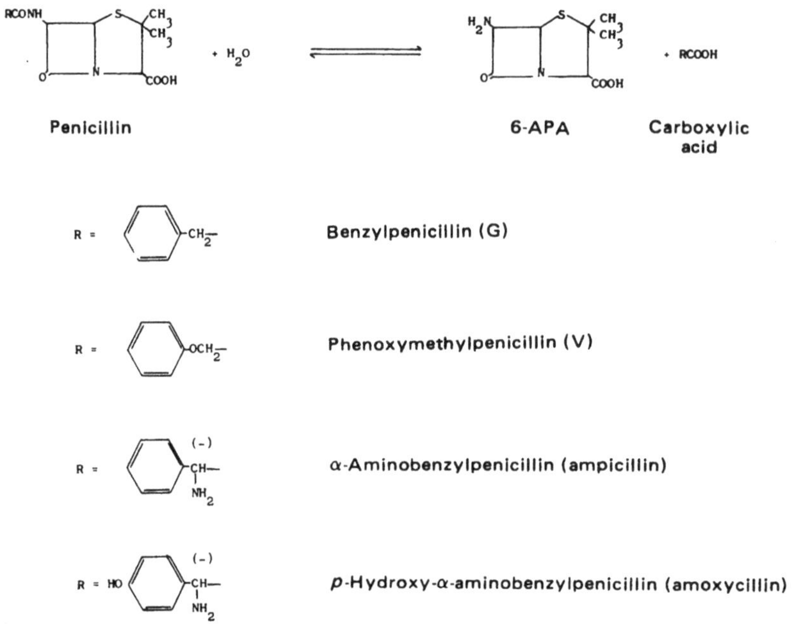

Figure 8 Formation and reacylation of 6-aminopenicillanic acid (6-APA).

The penicillin acylases can be derived from bacteria, actinomycetes, yeasts or fungi (Vandamme, 1980; Mahajan, 1984; Lowe, 1985). Their properties have been thoroughly studied and are reviewed elsewhere (Vandamme, 1980; Mahajan, 1984; Daumy et al., 1985). Highly productive and genetically engineered strains are available for industrial bioconversion of penicillin G or V into 6-APA (Vandamme, 1983; Schomer et al., 1984; Olsson et al., 1985; Daumy et al., 1986); two types of enzyme – penicillin G and V acylase – are produced by several companies and used in-house or sold on the market (Table 4). Almost every possible immobilisation procedure has been applied to the penicillin acylase enzyme or its producing cells. A comprehensive survey of these aspects has been compiled by the author (Vandamme, 1980, 1983).

Table 4 Industrial enzyme production for β-lactam antibiotic-bioconversions.

1. Penicillin G acylase (E.C. 3.5.1.11) Bayer, West Germany Beecham, U. K. Gist-Brocades, NL Snam Progetti, IT Pfizer, USA Toyo Jozo, Japan	4000 tons of 6-APA per year
2. Penicillin V acylase Biochemie, Austria Novo Industri, DK Bristol-Myers, USA	500 tons of 6-APA per year
3. Cephalosporin C acylase: ?	
4. Glutaryl-7-ACA acylase + D-amino acid oxidase (E.C. 1.4.3.3) Asahi Chemical Ind. Co., Japan	7-ACA
5. Hydantoinase (E.C. 3.5.22): Kanegafuchi, Japan, Singapore Recordati, (Snam Progetti), IT	50 tons of D-phenyl-glycine per year

Initially in industry, cell suspensions of penicillin acylase active strains could only be used once and the productivity was low, about 0,5 to 1 kg of 6-APA per kg. *E. coli* suspension. Introducing methods to immobilize *E. coli* cells increased productivity up to 50 kg of 6-APA per kg of immobilised catalyst. Bayer applied a poly(methacryl)glutaraldehyde method, while Pfizer used binding to glucidyl methacrylate polymers. Beecham used a cyanogen bromide-activated dextran/sephadex methode, Astra a cyanogen bromide activated Sephadex method and Toyo Jozo glutaraldehyde-activated porous polyacrylonitrile fibres. Current productivities are within the range 100–1000 kg 6-APA/kg; however, few data have been published by the companies involved (Table 5 and 6). It can only be assumed that ±70% of all 6-APA produced today is made by the immobilised catalyst route. This means that about 5000 ton of 6-APA is produced by 15–30 ton of immobilised penicillin acylase catalyst (Poulsen, 1984). Industrial 6-APA producers are listed in Table 7.

Table 5 Typical process data for immobilised penicillin G acylase (Poulson, 1984).

Enzyme product	Rigid granules or Dextran/Sephadex; *E. coli* acylase	Acylase (*B. megaterium*) covalently bound to glutaraldehyde-activated porous poly-acrylonitrile fibres
Reactor	Column	Parallel column system
Substrate	Penicillin G or certain semisynthetic cephalosporins (4–15% w/v)	Penicillin G (10% w/v)
Temperature	35–40°C	30–36°C
Inlet pH	7–8	8.4 ± 0.1
Operating life	2000–4000 hours	Approx. 1200 hours
Productivity	1000–2000 kg/kg enzyme	500–700 kg/kg enzyme
	According to Beecham, U.K.	According to Toyo Jozo, Japan

Table 6 Process data for immobilised penicillin V acylase (Poulson, 1984).

Enzyme product	Penicillin V acylase cell homogenates covalently bound with glutaraldehyde
Reactor	Stirred batch or column
Substrate	Penicillin V (4–12% w/v)
Temperature	35°C
Inlet pH	7.5–8.0
Operating life	1000–2000 hours
Productivity	200–600 kg/kg enzyme
	Data from Novo Industri, Denmark

Table 7 Industrial production of 6-APA.

Antibioticos, Spain	Kyowa Hakko, Co., Japan
Asahi Chem., Japan	Novo Industri, DK
Astra, Sweden	Ranbaxy, India
Bayer, West Germany	Rohm, West Germany
Beecham, U. K.	Snam-Progetti, IT
Biochemie, Austria	Spofa, Czechoslovakia
Bristol-Myers, USA	Toyo Jozo, Japan
Cipan, Portugal	Wyeth, USA
Gist-Brocades, The Netherlands	Yung-Jin, S-Korea
Glaxo, U. K.	Companies in China, USSR
Hoechst, West Germany	

Despite its wide industrial application, the immobilised penicillin acylase reaction is still confronted with severe problems (Savidge, 1984). Upon penicillin hydrolysis into 6-APA at pH 7,0–8,0 at 37°C, the sidechain acid is liberated, which causes a drop in pH, resulting in a slower reaction rate. A higher starting pH is not desirable because of β-lactam ring hydrolysis and inactivation of penicillin. Strict pH control is thus necessary during this bioconversion process, which is therefore difficult to run in a continuous packed-bed reactor (Figure 9).

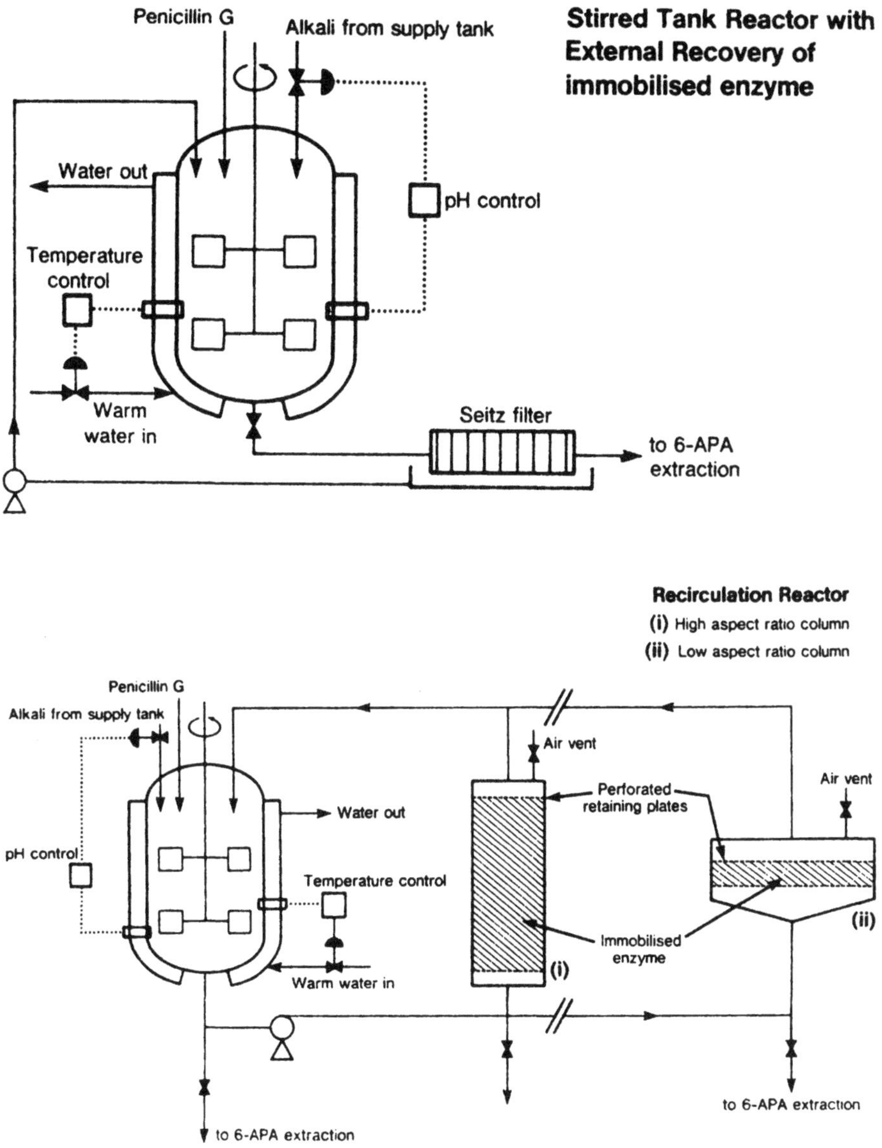

Figure 9 Industrial immobilised biocatalyst reactors for 6-APA production (reproduced with kind permission of T. A. Savidge).

Internal in situ pH-regulation by enzymatic action has been achieved by co-immobilisation of urease and adding urea in the inlet stream (Rousseau et al., 1984). Ammonia produced by urease action neutralizes the acid formed by side chain splitting of penicillin. Penicillin acylase has a more alkaline pH-optimum (pH 7,5 – 8,0) than urease. Thus if the pH rises, the activity of the acylase (acid forming enzyme) increases relatively to the acid consuming (urease) enzyme; this results in a pH-drop (and vice versa). This is a self-regulating system! In industry, a batch reactor is generally used. Such a process is then faced with another aspect of the catalytic properties of the enzyme, namely end-product inhibition. So, in this particular case, it is in fact the nature of the biochemical reaction that determines the type of reactor and reactor controls to be used.

Conventional stirred-tank reactors are used with adaptations for immobilised biocatalyst retention or recuperation; a less shear-intensive agitation system is usually also needed to avoid mechanical damage to the enzyme or cell carrier as well as an efficient system for the addition and distribution of alkali to avoid local alkaline inactivation of enzyme and penicillin, and to obtain strict pH control (Figure 9). If this reactor type still fails because of biocatalyst-carrier limitations, a recycle reactor-column combination could be used. In such a system, pH adjustment occurs separately from the immobilised enzyme (Savidge, 1984). These bottlenecks claim for a search for penicillin acylases with an acid pH optimum for hydrolysis, insensitivity to substrate of end product inhibition and tolerance to water miscible solvents. Ingenious screening and protein engineering techniques should be able to solve the problem.

In this respect, interesting penicillin V acylases have already been detected in *Erwinia aroideae* and in *Bacillus sphaericus* with optimal hydrolytic action at pH 6,0 and 5,8, respectively (Vandamme, 1980; Carlsen and Emborg, 1981).

Several bacterial penicillin acylases can also direct the synthesis of penicillins from 6-APA and an appropriate side chain at acid pH values.

Relatively little information is available on the acylation of 6-APA to produce (semisynthetic) penicillins using (immobilised) biocatalysts (Vandamme, 1983; Nam and Ryu, 1984).

The penicillin G acylases from *E. coli* ATCC 9637, *Bacillus circulans, B. megaterium* and the ampicillin acylase from *Pseudomonas melanogenum* are able to synthesize ampicillin or amoxycillin from 6-APA and D-phenylglycine methyl ester, or p-hydroxyphenylglycine methyl ester, respectively.

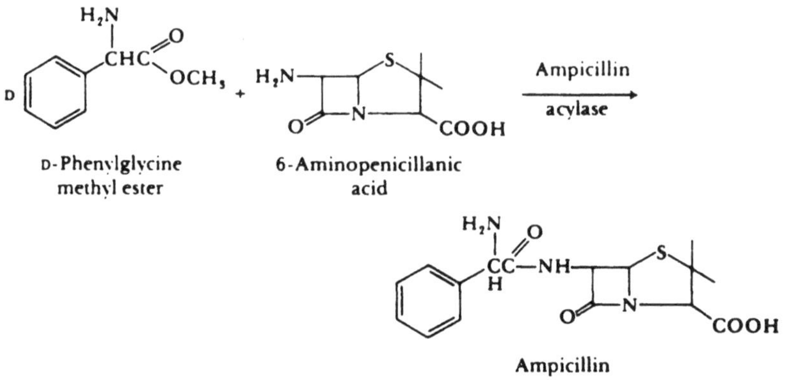

Figure 10 Synthesis of Ampicillin from 6-APA and D-phenylglycine methylester by Ampicillin acylase.

Using ampicillin acylase for ampicillin synthesis, there is no need to separate the side chain, liberated during the 6-APA formation step with penicillin G or V acylase (Figure 10).

The development of a commercially feasible penicillin acylase process for enzymatic synthesis of semisynthetic penicillins from 6-APA is slow due to the low conversion ratios and the availability of relatively simple chemical acylation procedures.

6.2 Cephalosporin bioconversions

Cephalosporins are penicillin analogues that are synthesized by fermentation with *Cephalosporium acremonium* (*Acremonium chrysogenum*), or by chemical ring expansion of penicillin. The major fermentation product, cephalosporin C, contains the 7-aminocephalosporanic acid (7-ACA) nucleus and α-aminoadipic acid as the side chain (Figure 11). 7-ACA derivatives with other side chains have not been obtained by precursored fermentation contrary to the 6-APA derivatives obtained with *Penicillium chrysogenum* fermentations. The responsible enzyme(s) for side chain exchange – an acyltransferase system in *Penicillium chrysogenum* – is not found in *Cephalosporium acremonium*.

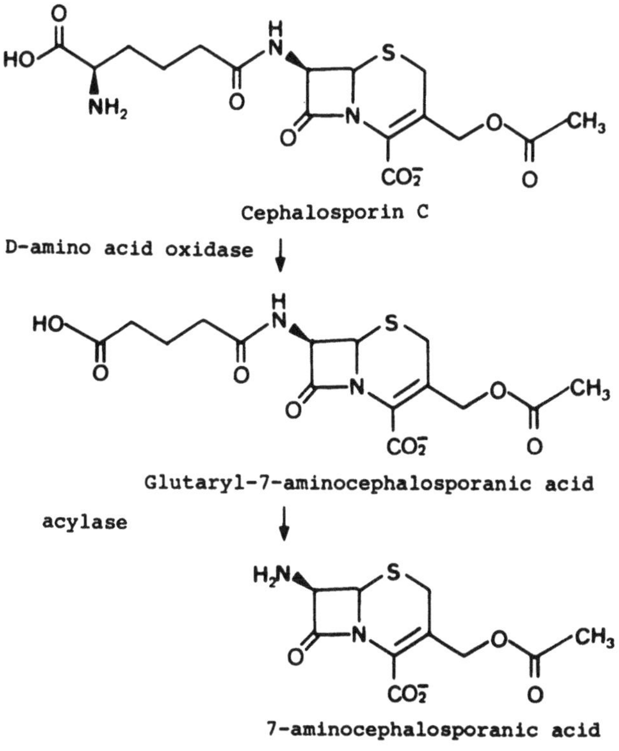

Figure 11 Bioconversion of cephalosporin C via 7-β-(4-carboxybutanamido)-cephalosporanic acid (glutaryl-7-ACA) into 7-ACA by D-amino acid oxidase and glutaryl-7-ACA-acylase.

Cloning of the acyltransferase gene(s) from the penicillin producer into cephalosporin C producer strains might result in fermentations yielding new precursored cephalosporins.

So far, obtaining the 7-ACA nucleus for the synthesis of semisynthetic cephalosporins, depended upon chemical removal of the natural α-aminoadipic acid side chain.

Although Japanese patents claim the direct enzymatic hydrolysis of cephalosporin C into 7-ACA and α-aminoadipic acid with *Pseudomonas putida* cephalosporin C acylase, a two-step enzymatic process for 7-ACA production seems to attract wider interest. In such a process cephalosporin C is first oxidized and deaminated by yeast (*Trigonopsis variabilis* CBS 4095 or *Rhodotorula gracilis* ATCC 26217) D-amino acid oxidase (E.C. 1.4.3.3) activity into keto-adipic-7-aminocephalosporanic acid, which spontaneously yields glutaryl-7-ACA (7-β-(4-carboxybutane-amido)cephalosporanic acid), which is then hydrolysed by glutaryl-7-ACA acylase of *Comamonas* or *Pseudomonas* sp. into the desired 7-ACA (Figure 11). This process was commercialized recently by ASAHI Chemical Industry, Co., Japan (Table 8).

Table 8 Two-step bio-process for 7-ACA production.

Cephalosporin C

↓ D-amino acid oxidase (*Trigonopsis variabilis*, *Rhodotorula gracilis*, *Citrobacter*)

Keto-Adipic-7-ACA

↓ Spontaneously

Glutaryl-7-ACA

↓ Glutaryl-7-ACA-acylase (*Pseudomonas*, *Arthrobacter*)

7-ACA

The intracellular yeast D-amino acid oxidase catalyses the oxidative deamination of several D-amino acids (D-ALA, METH, TYR, LEU, PRO, SER, LYS) with formation of the corresponding α-keto acid and ammonia, but is also active on cephalosporin C (Simonetta et al., 1982). The purified enzyme is a FAD-flavoprotein with MW of 86,000 daltons (Szwajcer and Mosbach, 1985). Toluene-ethanol permeabilized cells displayed high activity (Kubicek-Pranz and Rohr, 1985). Genetic transfer of the D-amino acid oxidase activity into *Cephalosporium* could lead to direct fermentation of glutaryl-7-ACA.

Shibuya et al. (1981) and Ichikawa et al. (1981) reported on the screening and mutation of *Pseudomonas* (*Pseudomonas putida* ATCC 950, *Pseudomonas* SY-77-1) strains, able to deacylate glutaryl-7-ACA into 7-ACA (Figure 11). These strains specifically hydrolysed cephalosporin-compounds having alipathic dicarboxylic acid in the acyl side chain; cephalosporin C was not hydrolysed.

β-Lactamase negative and constitutive mutants were obtained, yielding high enzyme levels. The cloning of the *Pseudomonas* glutaryl-7-ACA acylase gene in *E. coli* has recently been reported (Matsuda en Komatsu, 1985).

Direct fermentation of glutaryl-7-ACA rather than cephalosporin C formation and use of glutaryl-7-ACA-acylase could thus provide an alternative economic route for 7-ACA-production.

Immobilised enzyme or cell technology is now being applied to this new type of amino acid oxidase and acylase and might compete with conventional chemical methods for 7-ACA production. An immobilised cell system has been described where cells of *Pseudomonas* sp. or *Comamonas* sp. entrapped in cellulose acetate capsules hydrolyse glutaryl-7-ACA into 7-ACA. This compound is converted chemically into the clinically important cephaloglycin and cephalothin (Figure 12). These

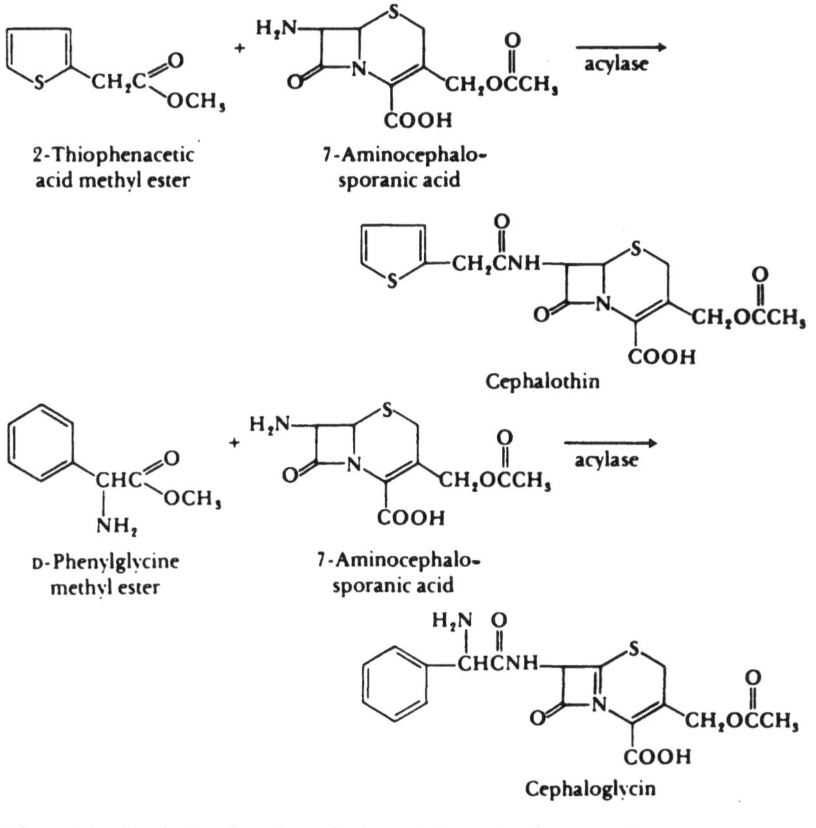

Figure 12 Synthesis of semi-synthetic cephalosporins from 7-ACA.

coupling reactions have also been performed with the Celite-adsorbed penicillin G-acylase from *B. megaterium* and with *Xanthomonas citri* cells (Nam et al., 1985; Kim et al., 1983; Ryu and Nam, 1983).

Another useful bioconversion is the deacylation of 7-ACA and cephalosporins (Figure 13). Acetate removal by enzymatic hydrolysis (acetylesterase) is the only efficient method to prepare deacetyl-7-ACA. In this respect, *Bacillus subtilis* esterase was absorbed onto bentonite or contained within an ultrafiltration device and used for multiple batch reactions to deacetylate 7-ACA-solutions (Abbott et al., 1976).

Cephalosporins can also be produced from the precursored penicillin G or penicillin V by a series of chemical reactions that expand the five-membered thiazolidine ring of the penicillins into the six-membered dihydrothiazine ring of the cephalosporins.

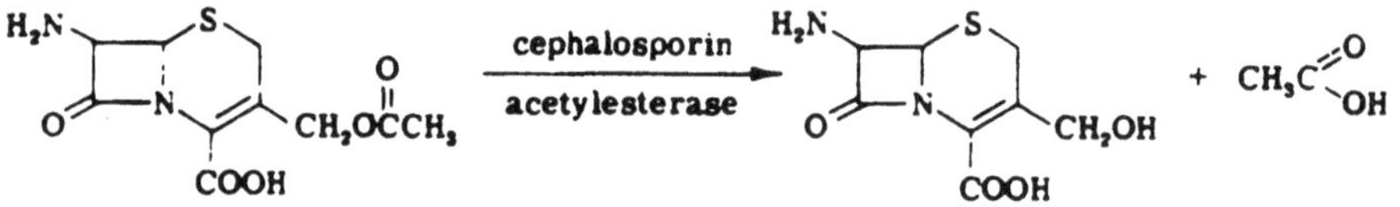

Figure 13 Conversions of 7-ACA into deacetyl-7-ACA by cephalosporin acetylesterase.

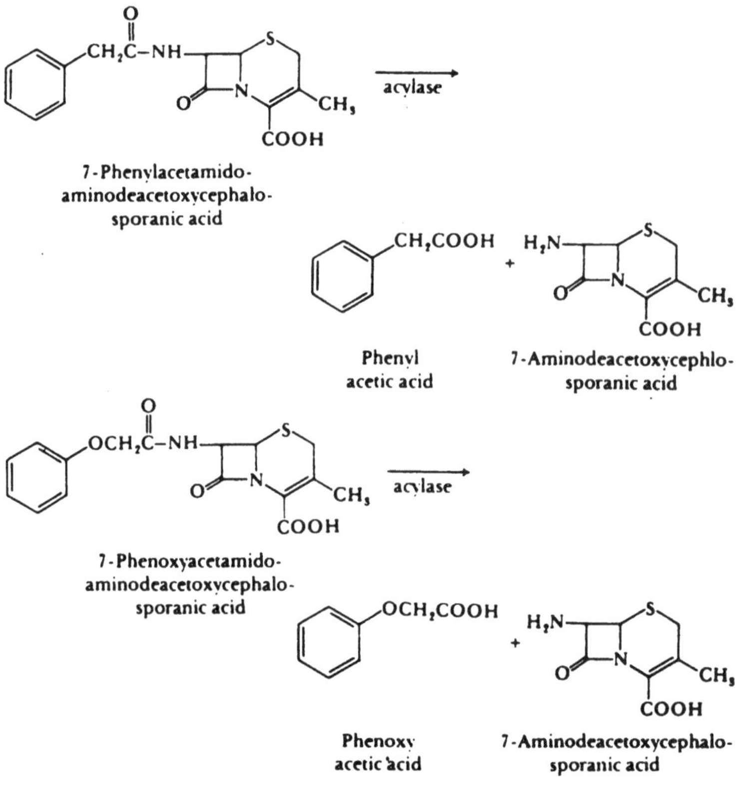

Figure 14 Hydrolysis of cephalosporins into 7-aminodeacetoxy-cephalosporanic acid (7-ADCA).

The cephalosporins obtained contain the penicillin G or V side chain, which can then be removed enzymically by conventional penicillin G or V acylase action. The cephalosporin nucleus thus obtained is 7-aminodesacetoxycephalosporanic acid (7-ADCA) (Figure 14).

Direct conversion of 6-APA into 7-ADCA has not been reported; enzymes able to hydrolyse 7-ACA into its desacetoxy-derivative 7-ADCA or to form 7-ACA through coupling of an acetoxy-group to 7-ADCA have not yet been described.

In this respect, immobilised biocatalysts have been used for the deacylation (or acylation) of such penicillin-derived cephalosporin compounds. Extracellular penicillin G acylases from *E. coli, B. megaterium* and *P. rettgeri*, absorbed onto Celite, activated carbon, carboxymethylcellulose or an Amberlite ion-exchange resin hydrolysed 7-phenylacetamido-ADCA.

The partially purified penicillin V-acylase as well as intact cells of *Erwinia aroideae* were entrapped in cellulose triacetate fibres at Glaxo Laboratories, UK; a fibre-packed column hydrolysed 7-phenoxyacetamido-ADCA to 7-ADCA in 58% yield after 3 h of substrate circulation at 37°C (Fleming et al., 1974).

Acylation of 7-ADCA with D-phenylglycine derivatives produces the useful antibiotic cephalexin (Figure 15). Several immobilised enzyme or cell systems for the synthesis of this clinically useful antibiotic have been developped (Kato et al., 1980; Kim et al., 1983). The enzyme form *Acetobacter turbidans* ATCC 9325 has been immobilised to curdlan and used for continuous synthesis of cephalexin. Whole cells of *Xanthomonas citri*, immobilised in carrageenan, produced continuously cephalexin at pH 6,2 at 25°C with 83% conversion rate (Kim et al., 1983).

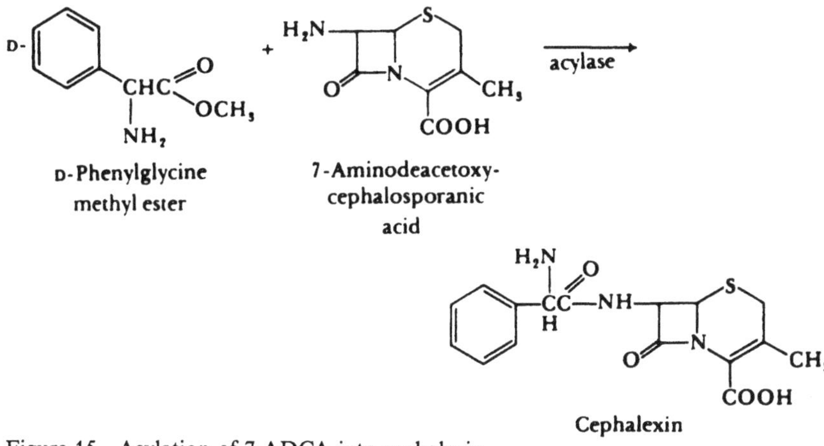

Figure 15 Acylation of 7-ADCA into cephalexin.

Chemical acylation of the cephalosporin nucleus needs blocked derivatives of the reactants, while enzymatic acylation does not. This fact, combined with the advantages of immobilised enzyme or cell technology favours the use of enzymatic procedures for semisynthetic cephalosporin synthesis.

7. Chiral Side-Chain Production for β-Lactam Antibiotics Synthesis

Microbial (immobilised) enzymes and cells are also used for the manufacture of side chain molecules to be coupled to 6-APA or to the cephalosporin nucleus to obtain semi-synthetic β-lactams. Especially, the necessity for D-chiral side chains has favoured the introduction of enzymatic processes in industry (Table 4). D-p-Hydroxyphenylglycine is such a non-proteineous chiral compound needed for coupling to 6-APA to produce amoxycillin. Several enzymatic routes have been proposed to produce this valuble compound. One industrial route utilizes immobilised cells, catalysing the stereoselective cleavage of the DL-hydroxyhydantoin precursor into D-carbamoyl-p-hydroxyphenylglycine, which can be converted chemically (or enzymatically) into the desired product D-p-hydroxyphenylglycine with retention of chiral configuration.

A microbial hydantoinase (*Pseudomonas, Bacillus, Agrobacterium*) has been screened for, which is sufficiently stable and active at slight alkaline pH values, since only in this pH range is the speed of the spontaneous chemical racemisation of the untouched L-hydantoin precursor into the DL-form fast enough to allow for a complete conversion into the chiral end product. The stereoselective selective cleavage by the microbial hydantoinase acts as a sink to pull the racemisation, thereby replenishing substrate for the enzyme (Figure 16).

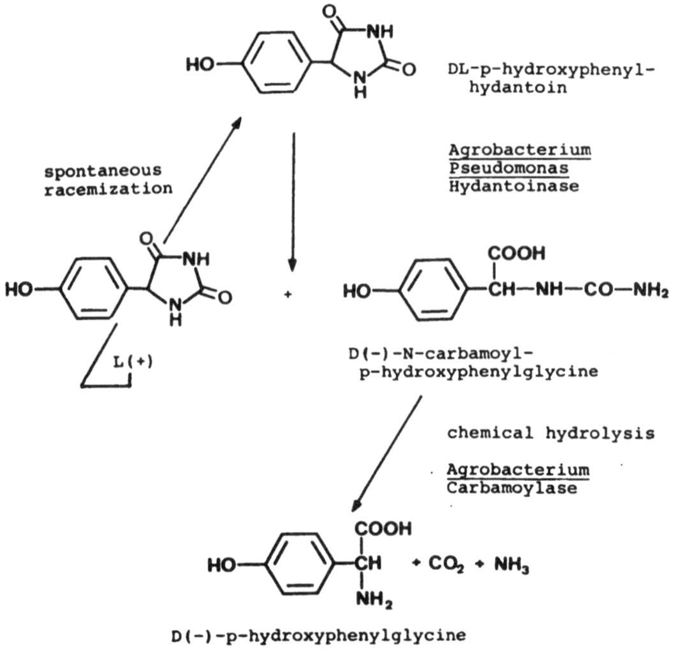

Figure 16 Enzymatic resolution of DL-p-hydroxyphenylhydantoin into D(−)-p-hydroxy-phenylglycine with hydantoinase and carbamoylase from *Agrobacterium* sp.

D-p-Phenylglycine is a similar non-proteineous amino acid used for the synthesis of ampicillin from 6-APA, cephaloglycin from 7-ACA and cephalexin from 7-ADCA.

Further developments led to the isolation of strains (i. e. *Pseudomonas* sp. AJ-11 220, *Agrobacterium radiobacter* NRRL-B-11 291) with both D-hydantoinase as well as D-carbamoylase activity, thereby avoiding the need for the chemical hydrolysis of the D-carbamoyl derivative (Olivieri et al., 1981). Such a process allows the complete conversion of the racemic substrate to the desired chiral D-amino acid in a single reactor (Figure 16) (Yokozeki and Kubota, 1987). Use of immobilised cells could further improve the process.

8. Bioconversion of Novel β-Lactam Antibiotics

Several new natural β-lactam antibiotics have been recently described, including bicyclic compounds, i. e. cephamycins, clavulanic acid, thienamycins, olivanic acids, asparenomycin, cephabacins, chitinovorins, but also monocyclic β-lactams, i. e. nocardicins, formadicins, sulfazecins, SQ 28 332 (Imada et al., 1981; Singh et al., 1983).

Many of these natural compounds have been reported to undergo bioconversion by microbial cell or enzyme preparations. *Pseudomonas schuylkillensis* acylase hydrolysed nocardicin C at pH 8 and 37°C into its nucleus 3-aminonocardicinic acid (3-ANA). The olivanic acid related PS-5 antibiotic was deacylated by *Pseudomonas* sp. 1158 cells, immobilised in polyacrylamide gel at pH 7,4 and 30°C. An *E. coli* acylase has been reported to acylate thienamycin reversibly (Figure 17).

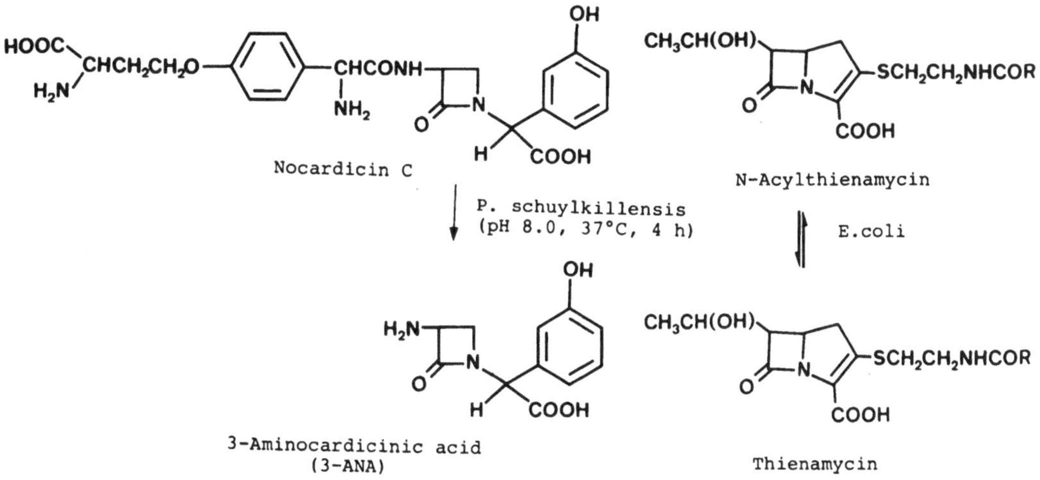

Figure 17 Bioconversion of Nocardicin C into its nucleus 3-aminonocardicinic acid (3-ANA) and of thienamycin.

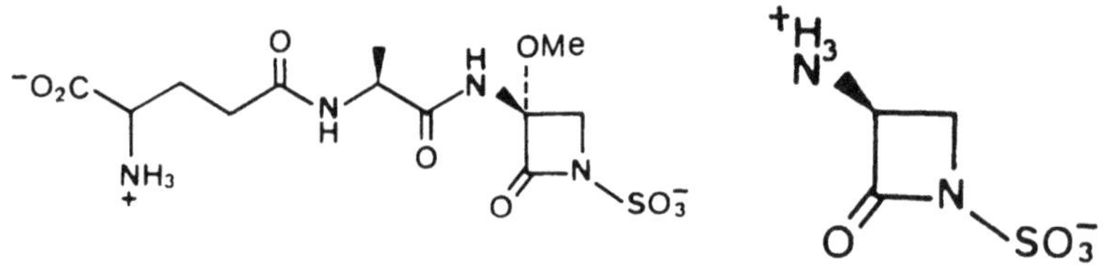

Figure 18 Structures of the monobactam Sulfazecin and of its nucleus, 3-amino-monobactamic acid (3-AMA).

The monocyclic N-sulfonated *β*-lactams isolated from bacteria (*Pseudomonas, Gluconobacter, Chromobacter, Agrobacterium, Flexibacter* (Imada et al., 1981; Singh et al., 1983) are N-acyl derivatives of the nucleus 3-aminomonobactamic acid (3-AMA) (Figure 18). These natural monobactams did not serve as substrates for *E. coli* penicillin acylase; the chemically prepared nucleus 3-AMA as well as 3-amino-4*α*-methylmonobactamic acid could be acylated at pH 4,5 at 45°C using a polyacrylamide immobilised *E. coli* penicillin G acylase. Unnatural acyl residues which could be coupled were readily deacylated at pH 7,5 (O'Sullivan and Aklonis, 1984). Intentional screening for specific monobactam acylases could lead the way to immobilised biocatalyst technology for semi-synthetic monobactam-production.

9. Bioconversion of Other Important Antibiotics

The macrolide feed-antibiotic tylosin, commercially produced by *Streptomyces fradiae*, can be converted into 4-acyltylosin by *Streptomyces thermotolerans*; cloning of this enzyme into *S. fradiae* would yield 4-acyltylosin directly; coimmobilisation of the two *Streptomyces* species could also yield 4-acyltylosin in one run (Figure 19).

The formation of the broad spectrum aminoglycoside antibiotic amikacin from kanamycin A, produced by *Streptomyces kanamyceticus*, and L(−)*α*-hydroxy-*γ*-aminobutyric acid (L-HABA) with a non producer mutant of the butirosin pro-

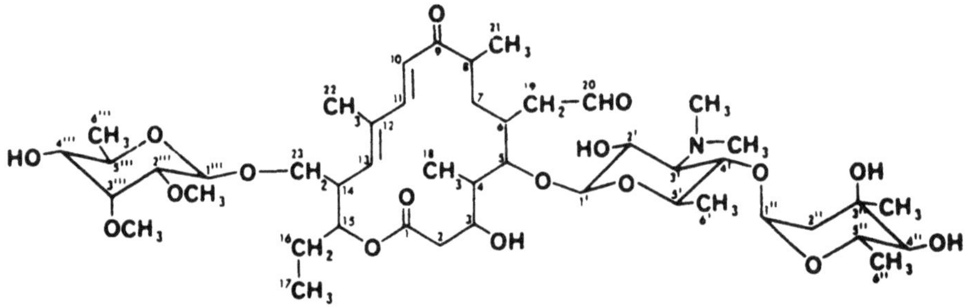

Figure 19 Structure of tylosin.

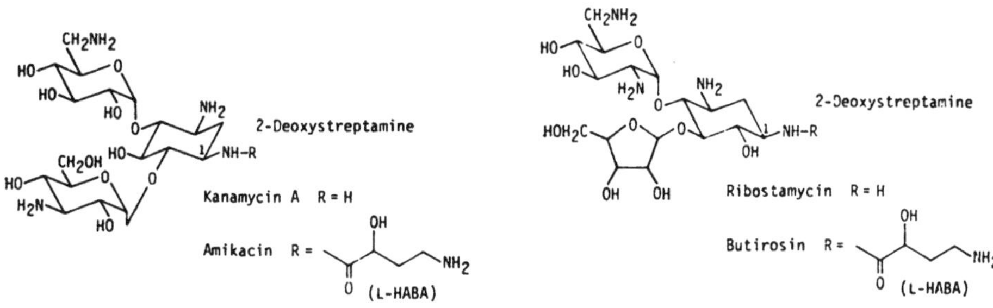

Figure 20 Structures of butirosin and ribostamycin produced by *Bacillus circulans* and of kanamycin A, produced by *Streptomyces kanamyceticus*, and its semi-synthetic derivative, amikacin.

ducing *Bacillus circulans* could replace its expensive chemical synthesis; furthermore a crude fermentation broth from *S. kanamyceticus* could be used as kanamycin A substrate for bioconversion (Cappelletti and Spagnoli, 1983) (Figure 20) with the hydroxyaminobutyric acid (HABA)-acylase present in *Bacillus circulans.*

Alternatively, cloning of this HABA-acylase gene from the butirosin-producing *Bacillus circulans* into *Streptomyces kanamyceticus* could lead to the direct and efficient synthesis of the important semi-synthetic aminoglycoside, amikacin.

Rifamycin B, one of the major metabolic products of *Nocardia mediterranei*, can be converted into rifamycin O, rifamycin S and rifamycin SV by chemical methods. Especially, rifamycin S is of great therapeutical importance as the key intermediate for the synthesis of rifamycin-derivatives, including rifampicin, used as an antituberculosic and antilepra-drug (Figure 21).

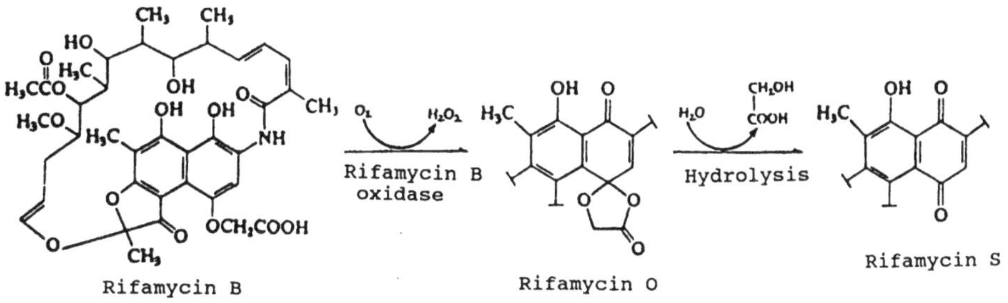

Figure 21 Proposed mechanism of rifamycin oxidase on rifamycin B to form rifamycin S.

Chemical conversion of rifamycin B to rifamycin S requires an organic solvent system, artificial oxidant and strong acidic conditions, which results in unsatisfactory yields.

A bioconversion process of rifamycin B to rifamycin S was recently described by Han et al. (1983) and Seong et al. (1983, 1985). Polyacrylamide gel or cellulose acetate entrapped acetone-treated cells of *Humicola* ATCC 20 620 and *Monocillium*

ATCC 20621 in a fluidised bed reactor or rotating disk reactor at pH 7,8 and 45 – 50°C were capable of catalysing the conversion of rifamycin B to rifamycin O by rifamycin B-oxidase (E.C. 1.10.3.36) action, a new type of diphenol oxidase; the thus formed rifamycin O was spontaneously hydrolysed to rifamycin S in neutral aqueous conditions. It was found that aeration – since oxygen is a cosubstrate –, cell-leakage and precipitation of rifamycin S were important factors in the continuous reactions (Chung et al., 1985, 1986; Lee et al., 1985).

10. Perspectives

In contrast to immobilised mono-step bioconversions for antibiotic production, which are already operational on a large scale (6-APA and 7-ACA production, chiral side chain resolution), industrial biocatalysis by multi-step enzyme systems or by immobilised cell fragments or viable cells remains a goal to achieve. The complexity of the pathways involved and the many basic aspects yet to be well understood hamper a quick translation of this concept into economic reality. Indeed, microbiological, physiological and technical problems inherent to the immobilised biocatalyst principle, need further study to arrive at an optimal performance of immobilised catalyst bioreactors.

For most complex multi-enzyme processes, there is as yet no economic incentive to replace well-established fermentation processes by immobilised enzyme or cell reactor technology; for bioconversion reactions, emphasis is clearly focused upon switching from free or immobilised purified single-enzyme reactions to immobilised (viable) cell processes.

However, cloning of genes, coding for antibiotic-biosynthesis enzymes such as acyltransferase (Luengo et al., 1986 b); isopenicillin N synthetase (cyclase) (Samson et al., 1985; Carr et al., 1986; Jensen et al., 1987; Skatrud et al., 1987), expandase (Rollins et al., 1987), HABA-acylase, ... or bioconversion enzymes such as penicillin G and V acylases (Schomer et al., 1984; Lowe et al., 1987; Olsson et al., 1985; Okash et al., 1989), ampicillin acylase, glutaryl-7-ACA acylase (Matsuda and Komatsu 1985), hydantoinase, D-amino acid oxidase, tylosin-acylase, rifamycin B-oxidase, ... should receive more attention so as to produce these industrial enzymes more efficiently and abundantly (Marahiel et al., 1987): otherwise such bioconversion-enzyme-genes could be introduced into antibiotic producer strains so as to extend or deviate their antibiotic pathways with an additional useful last step reaction (Vournakis and Elander, 1983; Kurth and Demain, 1984) and to obtain at the end of the fermentation process the "semi-synthetic" derivative directly!

It is obvious from the examples taken here from industry and research laboratories that an integration of microbiological, genetical and process-technological knowledge is needed to arrive at a successful and economic performance of novel antibiotic fermentations and bioconversions.

References

Abbott, B. J., Cerimele, B., and Fukuda, D. S. (1976). Immobilisation of a cephalosporin acetylesterase by containment within an ultrafiltration device. *Biotechnol. Bioeng. 19*, 1033–1042.

Arcuri, E. J., Slaff, G., and Greasham, R. (1986). Continuous Production of Thienamycin in Immobilized Cell Systems. *Biotechnol. Bioeng. 28*, 842–849.

Bahadur, G., Baldwin, J. E., Wan, T., Jung, M., Abraham, E. P., Huddleston, J. A., and White, R. L. (1981). On the proposed intermediary of β-hydroxyvaline and thiazepinone-containing peptides in penicillin biosynthesis. *J. Chem. Soc. Chem. Commun. 93, 1146–1147.*

Behrens, O. K., Corse, J., Edwards, J. P., Garrison, L., Jones, R. G., Soper, Q. F., van Abeele, F. R., and Whitehead, C. W. (1948). Biosynthesis of penicillin. IV. New crystalline biosynthetic penicillin. J. Biol. Chem. 175, 793–809.

Brodelius, P., and Vandamme, E. J. (1987). Immobilised Cell Systems. In: *Biotechnology*. Reed, G., and Rehm, H. J. (eds.). Vol. 7a (405–464). VCH-Verlag, Weinheim.

Cappelletti, L. M., and Spagnoli, R. (1983). Biological Transformation of Kanamycin A to Amikacin (BBK-8). *J. Antibiotics 36*, 328–330.

Carlsen, F., and Emborg, C. (1981). *Bacillus sphaericus* V-penicillin acylase. *Biotechnol. Lett. 3*, 375–378.

Carr, L. G., Skatrud, P. L., Scheets, M. E., Queener, S. W., and Ingolia, T. D. (1986). Cloning and expression of the isopenicillin N synthetase gene from *Penicillium chrysogenum. Gene 48*, 257–266.

Castro, J. M., Liras, P., Cortes, J., and Martin, J. F. (1986). Conversion of phenylacetyl-cysteinyl-valine in vitro in penicillin G by isopenicillin N synthase of *Streptomyces lactamdurans. FEMS Microbiol. Lett. 34*, 349–353.

Chibata, I., and Tosa, T. (1981). Use of Immobilised Cells. *Annu. Rev. Biophys. Bioeng. 10*, 197–216.

Chung, B. H., Chang, H. N., and Han, M. H. (1985). Immobilisation and characterisation of Rifamycin B oxidase in Cellulose Acetate beads. *Korean J. Appl. Microb. Bioeng. 13*, 115–118.

Chung, B. M., Chang, H. N., and Han, M. H. (1986). Enzymatic Conversion of Rifamycin B in a Rotating Packed Disk Reactor. *J. Ferment. Technol. 64*, 343–345.

Daumy, G. O., Danley, D., and McColl, A. S. (1985). Role of protein subunits in *Proteus rettgeri* Penicillin G acylase. *J. Bacteriol. 163*, 1279–1281.

Daumy, G. O., Williams, J. A., McColl, A. S., Zuzel, T. J., and Danley, D. (1986). Expression and Regulation of the Penicillin G Acylase Gene from *Proteins rettgeri* cloned in *Escherichia coli. J. Bacteriol. 168*, 431–433.

Demain, A. L. (1983). New Applications of Microbial Products. *Science 219*, 709–714.

Demain, A. L., Piret, J. M., Friebel, T. O. E., Vandamme, E. J., and Matteo, C. C. (1976). Studies on *Bacillus brevis* directed towards the cell free synthesis of Gramicidin S. In: *Microbiology-1976*. Schlessinger, D. (ed.). American Society for Microbiology, Washington D.C., p. 437–443.

Demain, A. L., Poirier, A., Agathos, S., and Nimi, O. (1981). Appearance and disappearance of Gramicidin S Synthetase. *Developm. Industr. Microbiol. 22*, 233–244.

Dotzlaf, J. E., and Yeh, W. K. (1987). Copurification and Characterisation of deacetoxycephalosporin C synthetase/hydroxylase from *C. acremonium. J. Bacteriology 169*, 1611–1618.

Fayerman, J. F. (1986). New Developments in Gene Cloning in Antibiotic Producing Microorganisms. *Bio/Technology 4*, 786–789.

Fleming, I. D., Turner, M. K., and Napier, E. J. (1974). *German Patent 2422374.*

Freeman, A., and Aharonowitz, Y. (1981). Immobilisation of Microbial Cells in Crosslinked Prepolymerised, linear Polyacrylamide Gels: Antibiotic Production by Immobilized *Streptomyces clavuligerus* Cells. *Biotechnol. Bioengin. 23*, 2747–2759.

Garcia-Dominguez, M., Martin, J. F., Mahro, B., Demain, A. L., and Liras, P. (1987). Efficient Plasmid Transformation of the β-lactam producer *Streptomyces clavuligerus. Appl. Environm. Microb. 53*, 1376–1381.

Han, M. H., Seong, B. L., Son, H. J., and Mheen, T. I. (1983). Rifamycin B oxidase from *Monocillium spp.*, a new type of diphenol oxidase. *FEBS-Lett. 151*, 36–40.

Hollander, I. J., Shen, Y. Q., Heim, J., Demain, A. L., and Wolfe, S. (1984). A pure enzyme catalyzing penicillin biosynthesis. *Science 224*, 610–612.

Hopwood, D. A., Malpartida, F., Kieser, H. M., Ikeda, H., Duncan, J., Fujii, I., Rudd, B. A., Floss, H. G., and Omura, S. (1985). Production of 'hybrid' antibiotics by genetic engineering. *Nature (London) 314*, 642–644.

Ichikawa, S., Shibuya, Y., Matsumoto, K., Fujii, T., Komatsu, K., and Kodaira, R. (1981). Purification and Properties of 7-β-(4-carboxybutanamido) cephalosporic acid Acylase produced by Mutants derived from *Pseudomonas*. *Agric. Biol. Chem. 45*, 2231–2236.

Ikeda, H., Inoue, M., Tanaka, H., and Omura, S. (1984). Interspecific protoplast fusion among macrolide-producing streptomycetes. *J. Antibiot. 37*, 1224–1230.

Imada, A., Kitano, K., Kintaka, K., Muroi, M., and Asai, M. (1981). Sulfazecin and Isosulfazecin, novel β-lactam antibiotics of bacterial origin. *Nature 289*, 590–591.

Jensen, S. E., Leskiw, B. K., and Westlake, D. W. S. (1987). Cloning and Sequence Analysis of the Isopenicillin N Synthetase Gene from *Streptomyces clavuligerus*. *SIM-News, 37* (4), p. 77. Abstract p. 48, Ann. Mtg. SIM, Baltimore, 9–14 aug.

Jensen, S. E., Westlake, D. W. S., Bowers, R. J., and Wolfe, S. (1982). Cephalosporin formation by cell free extracts from *Streptomyces clavuligerus*. *J. Antibiotics 35*, 1351–1360.

Jensen, S. E., Westlake, D. W. S., and Wolfe, S. (1984). Production of penicillins and cephalosporins in an immobilized enzyme reactor. *Appl. Microbiol. Biotechnol. 20*, 155–160.

Karube, I., Suzuki, S., and Vandamme, E. J. (1984). Antibiotic Production with Immobilized Cells. In: *Biotechnology of Industrial Antibiotics*. E. J. Vandamme (ed.). Marcel Dekker Inc., New York & Basel, USA, p. 761–780.

Kato, K., Kawahaka, K., Takahashi, T., and Kakinuma, A. (1980). Substrate specificity of α-amino acid ester hydrolase from *Xanthomonas citri*. *Agric. Biol. Chem. 44*, 1075–1081.

Kim, J. K., Nam, D. H., and Ryu, D. D. Y. (1983). Cephalexin Synthesis using Immobilised *Xanthomonas citri* Cells. *Appl. Biochem. Biotechnol. 8*, 195–202.

Krause, M., Marahiel, M. A., von Döhren, H., Kleinkauf, H. (1985). Molecular cloning of an Ornithine-Activating Fragment of the Gramicidin S Synthetase 2 Gene from *Bacillus brevis* and its expression in *Escherichia coli*. *J. Bacteriol. 162*, 1120–1125.

Kubicek-Pranz, E. M., and Rohr, M. (1985). D-Amino acid oxidase from *Trigonopsis variabilis*: comparison of enzyme specificity *in vivo* and *in vitro*. *Biotechnol. Lett. 7*, 9–14.

Kurth, R., and Demain, A. L. (1984). The Impact of New Genetics on Antibiotic Production. In: *Biotechnology of Industrial Antibiotics*. E. J. Vandamme (ed.). Marcel Dekker Inc., New York & Basel, USA, p. 781–789.

Kurzatkowski, W., Kurylowics, W., and Paszkiewics, A. (1982). Penicillin G production by immobilised fungal vesicles. *Eur. J. Appl. Microb. Biotechnol. 5*, 211–213.

Lee, G. M., Choi, C. Y., Park, J. M., and Han, M. H. (1985). Biotransformation of Rifamycin B to Rifamycin S using Immobilised Whole Cells of *Humicola sp.* in a Fluidised Bed Reactor. *J. Chem. Technol. Biotechnol. 35B*, 3–10.

Lowe, D. A. (1985). Industrial importance of biotransformations of β-lactam antibiotics. *Dev. Ind. Microbiol. 26*, 143–155.

Lowe, D. A., Romancik, G., and Elander, R. P. (1987). Enzymic Hydrolysis of p-Hydroxypenicillin V by a *Thielaviopsis species*. Abstracts of the 1987 Annual Meeting, SIM, Baltimore, 9–14 aug., P-67. *SIM-News 37*, (4), p. 80.

Luengo, J. M., Alemany, M. T., Salto, F., Ramos, F., Lopez-Nieto, M. J., and Martin, J. F. (1986 a). Direct Enzymatic Synthesis of Penicillin G using Cyclases of *Penicillium chrysogenum* and *Acremonium chrysogenum*. *Bio/Technology 4*, 44–47.

Luengo, J. M., Iriso, J. L., and Lopez-Nieto, M. J. (1986 b). Direct enzymatic synthesis of natural penicillins using phenylacetyl-CoA: 6-APA Phenylacetyl transferase of *Penicillium chrysogenum*: minimal and maximal side chain length requirements. *J. Antibiotics 39*, 1754–1759.

Madry, N., Zocher, R., Grodzki, K., and Kleinkauf, H. (1984). Selective Synthesis of depsipeptides by the immobilized multi-enzyme enniatin synthetase. *Appl. Microbiol. Biotechnol. 20*, 83–86.

Mahajan, P. B. (1984). Penicillin Acylases, an Update. *Appl. Biochem. Biotechnol. 9*, 537–554.

Marahiel, M. A., Zuber, P., Czekay, G., and Losick, R. (1987). Identification of the Promoter for a Peptide Antibiotic Biosynthesis Gene from *Bacillus brevis* and its Regulation in Bacillus subtilis. *J. Bacteriol. 169*, 2215–2222.

Matsuda, A., and Komatsu, K. (1985). Molecular Cloning and Structure of the Gene for 7-β-(4-carboxybutanamido)-cephalosporanic acid Acylase from a *Pseudomonas* strain. *J. Bacteriol. 163*, 1222–1228.

Nam, D. H., and Ryu, D. D. Y. (1984). Enzymatic Synthesis of Phenoxymethylpenicillin using *Erwinia aroideae* enzyme. *J. Antibiotics 37*, 1217–1223.

Nam, D. H., Sohn, H. S., and Ryu, D. D. Y. (1985). Enzymatic Synthesis of Cephaloglycin. *Bull. Korean Chem. Soc. 4*, 72–76.

Ohash, !, Katsuta, Y., Nagashima, M., Kamei, T., and Yano, M. (1989). Expression of the *Arthrobacter viscosius Penicillin G Acylase Gene in E. coli* and *B. subtilis* Appl. Anvironm. Microbiol. *55*, p. 1351–1356.

Olivieri, R., Fascetti, E., Angelini, L., and Degen, L. (1981). Microbial Transformation of Racemic Hydantoins to D-Amino Acids. *Biotechnol. Bioeng. 23*, 2173–2183.

Olsson, A., Hagstrom, T., Nilsson, B., Umlen, M., and Gatenbeck, S. (1985). Molecular Cloning of *Bacillus sphaericus* Penicillin V acylase Gene and its Expression in *Escherichia coli* and *Bacillus subtilis. Appl. Environm. Microbiol. 49*, 1084–1089.

Omura, S., Ikeda, H., Malpartida, F., Kieser, H. M., and Hopwood, D. A. (1986). Production of New Hybrid Antibiotics, Mederrhodins A and B, by a Genetically Engineered Strain. *Antimicrob. Agents Chemother. 29*, 13–19.

Omura, S., Sadakane, N., Tanaka, Y., and Matsubara, H. (1983). Chimeramycins: new macrolide antibiotics produced by hybrid biosynthesis. *J. Antibiotics 36*, 927–930.

O'Sullivan, J., and Aklonis, C. A. (1984). Enzymatic Acylation of Monobactams. *J. Antibiotics 37*, 804.

Poirier, A., and Demain, A. L. (1981). Arginine Regulation of Gramicidin S Biosynthesis. *Antimicrob. Ag. Chemother. 20*, 506–514.

Poulsen, P. B. (1984). Current Applications of Immobilised enzymes for manufacturing purposes. *Biotechn. Gen. Eng. Revs. 1*, 121–140.

Ramos, F. R,, Lopez-Nieto, M. J., and Martin, J. F. (1985). Isopenicillin N Synthetase of *Penicillium chrysogenum*, an enzyme that converts δ-(L-α-Aminoadipyl)-L-Cysteinyl-D-Valine to Isopenicillin N. *Antimicrob. Ag. Chemother. 27*, 380–387.

Rollins, M. J., Carmichael, R. D., Jensen, S. E., and Westlake, D. W. S. (1987). Effect of Aeration and Nutriënt Supply on Desacetoxycephalosporin C Synthetase Activity in *Streptomyces clavuligerus*. Abstracts of the 1987 Annual Meeting, SIM, Baltimore, 9–14 aug., P-69. *SIM-News 37 (4)*, p. 80.

Rousseau, I., Roth-Van Eijk, C., and Liou, J. K. (1984). Internal pH-control of Immobilized Enzyme Particles: Penicillin Acylase-Urease System. *Progr. Ind. Microbiol. 20*, 73–80.

Ryu, D. D. Y., and Nam, D. H. (1983). Enzymatic Synthesis of Semisynthetic β-lactam antibiotics. *Jap. Enz. Eng. News 9*, 30–62.

Samson, S. M., Belagaje, R., Blankenship, D. T., Chapman, J. L., Perry, D., Skatrud, P. L., Vanfrank, R. M., Abraham, E. P., Baldwin, J. L., Queener, S. W., and Ingolia, T. D. (1985). Isolation, sequence determination and expression in *Escherichia coli* of the isopenicillin N synthetase gene from *Cephalosporium* acremonium. *Nature 318*, 191–194.

Savidge, T. A. (1984). Enzymatic conversions used in the production of penicillins and cephalosporins. In: *Biotechnology of Industrial Antibiotics*. E. J. Vandamme (ed.), Marcel Dekker Inc., New York & Basel, p. 171–224.

Schlegel, B., and Fleck, W. F. (1980). New anthracycline antibiotics produced by interspecific recombinants of streptomycete. I. Selection of *Streptomyces violaceus subsp. iremyceticus*, an iremycin-producing subspecies. *Z. Allg. Mikrobiol. 20*, 527–530.

Schomer, U., Segner, A., and Wagner, F. (1984). Penicillin Acylase from the Hybrid Strain *Escherichia coli* 5 K (pHM 12): Enzyme Formation and Hydrolysis of β-lactam Antibiotics with Whole Cells. *Appl. Environm. Microbiol. 47*, 307–312.

Seong, B. L., Son, H. J., Mheen, T. I., and Han, M. H. (1983). Microbial transformation of Rifamycin B: a new synthetic approach to rifamycin derivatives. *J. Antibiotics 36*, 1402–1404.

Seong, B. L., Son, H. J., Mheen, T. I., Park, Y. H., and Han, M. H. (1985). Enzymatic Oxidation of Rifamycins by a microorganism of the Genus *Humicola. J. Ferment. Technol. 63*, 515–522.

Singh, P. D., Johnson, J. H., Ward, P. C., Sott-Wells, J., Treso, W. H., and Sykes, R. B. (1983). SQ 28, 332, a new monobactam produced by a *Flexibacter sp. J. Antibiotics 36*, 1245–1251.

Shen, Y. Q., Wolpe, S., and Demain, A. L. (1986). Levels of Isopenicillin N, Synthetase and Deacetoxycephalosporin C-Synthetase in *Cephalosporium acremonium* producing high and low levels of Cephalosporin C. *Bio/Technology 4*, 61–64.

Shibuya, Y., Matsumoto, K., and Fujii, T. (1981). Isolation and Properties of 7-β-(4-Carboxybutanamido)cephalospranic acid Acylase producing bacteria. *Agric. Biol. Chem. 45*, 1561–1567.
Shier, W. T., Rinehart, K. L., and Gottlieb, D. (1969). Preparation of four new antibiotics from mutant of *Streptomyces fradiae*. *Proc. Natl. Acad. Sci. USA 63*, 198–204.
Simonetta, M. P., Vanoni, M. A., and Curti, B. (1982). D-Amino acid oxidase activity in the yeast *Rhodotorula gracilis*. *FEMS Microbiol. Lett. 15*, 27–31.
Skatrud, P. L., Queener, S. W., Fisher, D. L., and Chapman, J. L. (1987). Strain Improvement Studies in *Penicillium chrysogenum* using the cloned *P. chrysogenum* Isopenicillin N Synthetase Gene and the asmd S Gene of *Aspergillus nidulans*. Abstracts of the 1987 Annual Meeting, SIM, Baltimore, 9–14 aug., P-69. *SIM-News 37 (4)*, p. 77.
Szwajcer, E., and Mosbach, K. (1985). Isolation and Partial Characterisation of a D-amino acid oxidase active against cephalosporin C from the yeast *Trigonopsis variabilis*. *Biotechnol. Letters 7*, 1–7.
Tsukagoshi, N., Iritani, S., Sasaki, T., Takemura, T., Ihara, H., Idota, Y., Yamagata, H., and Udaka, S. (1985). Efficient Synthesis and Secretion of a Thermophilic α-Amylase by Protein-Producing *Bacillus brevis* 47 Carrying the *Bacillus stearothermophilus* Amylase Gene. *J. Bacteriol. 164*, 1182–1187.
Udaka, S. (1987). From Amino Acid Fermentation to Protein Fermentation: Extracellular Production of Proteins by *Bacillus brevis*. *Proceedings 4th ECB-1987*, Vol. 3, p. 32 (Elsevier Sc. Publ.), Amsterdam.
Vandamme, E. J. (1976). Immobilised microbial cells as Industrial Biocatalysts. *Chem. Ind. 24*, 1070–1072.
Vandamme, E. J. (1980). Penicillin Acylases and β-lactamases. In: *Economic Microbiology*. Rose, A. H. (ed.). Academic Press, New York – London, vol. 5, p. 467–522.
Vandamme, E. J. (1981). Properties, Biogenesis and Fermentation of the Cyclic Decapeptide Antibiotic Gramicidin S. In: *Topics in Enzyme and Fermentation Biotechnology*. Wiseman, A. (ed.). Ellis Horwood, Chichester, vol. 5, p. 185–261.
Vandamme, E. J. (1983). Peptide Antibiotic Production through Immobilised Biocatalyst Technology. *Enz. Microb. Technol. 5*, 403–416.
Vandamme, E. J. (ed.) (1984). *Biotechnology of Industrial Antibiotics*. Marcel Dekker Inc., New York & Basel, pp. 832.
Vandamme, E. J., Leyman, D., Debuyser, D., de Visscher, P., Spriet, J., Nimi, O., Poirier, A., and Demain, A. L. (1982). Environmental Influences on the Dynamics of the Gramicidin S Fermentation. In: *Peptide Antibiotics Biosynthesis and Functions*. Kleinkauf, H., and von Döhren, H. (eds.). de Gruyter, Berlin, pp. 117–135.
Vater, J. (1989) this volume p. 33–56
Veelken, M., and Pape, H. (1982). Production of Tylosin and Nikkomycin by Immobilised *Streptomyces* cells. *Eur. J. Appl. Microb. Biotechn. 15*, 206–210.
Vournakis, J. N., and Elander, R. P. (1983). Genetic Manipulation of Antibiotic Producing Microorganisms. *Science 219*, 703–709.
Wolfe, S., Demain, A. L., Jensen, S. E., and Westlake, D. W. S. (1984). Enzymatic Approach to Syntheses of Unnatural β-Lactams. *Science 226*, 1386–1392.
Yokozeki, K., and Kubota, K. (1987). Mechanism of Asymmetric Production of D-Amino Acids from the corresponding Hydantoins by *Pseudomonas* sp. *Agr. Biol. Chem. 51*, 721–728.

Chapter 15
Compilation of Peptide Structures – A Biogenetic Approach

Hans von Döhren

Introduction
- General biosynthetic mechanism
- Compound data

Abbreviations and Symbolism
- Configuration
- Bonds
- Abbreviated nomenclature used
- Cyclic structures
- Side chain modifications
- Chain length
- Combinations of chains
- Piperidine-type cyclizations
- Unusual constituents

Compilation

Index of table

List of abbreviations
- a) compounds
- b) sources

Introduction

The aim of this compilation of these mainly nonribosomally formed peptide metabolites is to evaluate for comparative purpose a structural classification scheme, that is based only on the number of peptide bonds involved, size of cyclic structures, and mode of cyclization or modification. It includes peptides of nonribosomal origin and heavy modified structures originating from mRNA-templates.

<u>General biosynthetic mechanism</u>. As has been evaluated before [1], the general mechanism of peptide biosynthesis is head growth [2]. Each cyclic structure has a linear precursor. A general enzymic mechanism of peptide formation may lead to structures of up to 20 (or more) amino acids in defined sequence. Enzyme templates catalyzing sequential steps have been found to be organized in multifunctional structures with fused catalytic functions. These systems are of high interest since the substrate specificity does not match that of the ribosomal system involving proof-reading steps. Thus peptide analogs may be synthesized by feeding procedures *in vivo,* or enzymatically *in vitro*. In addition, the enzyme systems might be modified with respect to their template function by genetic manipulations. Enzyme templates for peptide synthesis and modification are expected to be structurally related. Thus certain functional assays or structural probes are useful to identify unknown systems. In addition, structural comparisons of peptides try to identify homologies pointing to functional properties. Such homologies could suggest similar structural properties with regard to molecular interactions on their respective targets.

Compound data. The reader is referred for additional information to general data files like CABS and the NIH antibiotic file of Bérdy [3], or the Peptide Institute of the Protein Research Foundation, Osaka [4], where the compounds can be retrieved by various modes.
For classification of compounds a scheme is used as essentially outlined in a 1980-review [5], that has already been used in two compilations[6]. This table now is an update including sequence data. Compounds have been mostly termed according to recommendation of the IUPAC-IUB nomenclature commision [7]. Since from biosynthetic evidence no major distinction is found between amino and hydroxy acids (both are activated as acyladenylates, and cyclic structures are frequently closed by amino- or hydroxy-groups), peptidolactones and depsipeptides have been included.

Abbreviations and Symbolism

Configurations of residues
Generally the prefix L is omitted, and only the prefix D is used. For reasons of simplicity, D is placed directly before the abbreviation of the residue without a hyphen. Data given in the RS-system have been changed to DL in this table(except for cysteine and cystine which correspond to LD).

Types of bonds involved
Peptide bonds are written as →, indicating CO-N(H) bond. Ester bonds are indicated by -, with no direction given. In all cases the respective carboxyl and hydroxy groups can be unambiguously assigned. Modes of cyclization of branched peptides and lactones are indicated by a bar (|) at the acceptor amino or hydroxy function. If other groups than α-carboxyl/α-amino- are involved, this has been indicated like -β→, when useful.

Abbreviated Nomenclature of Compounds involved
Carboxylic acids are written as C_n with *n* giving the number of carbons in a respective linear (C_n) or terminally branched chain (iso: iC_n, anteiso: aC_n). Substitutions like other methyl functions are indicated as *nm,* with n referring to the position of the methyl group m. Hydroxyl functions are given as *nh*, with n referring to the position of the hydroxyl h. Double bonds are indicated by n,n+1*d,* with the numbers indicating the position of the bond *d.*
Amino acids are written in the recommended three-letter abbreviations. Frequently new abbreviations are introduced, also with four or five letters. Substitutions are indicated as follows:
Methylations of amino groups, as occurring most frequently in methylated peptide bonds, are shown as *Me*. The common prefix N-Me is omitted. Other methylations are indicated with *m*, as in the case of carboxylic acids.Hydroxylations are indicated by *h*. The recommended nomenclature for hydroxy-Lys (Hyl) and hydroxy-Pro (Hyp) is not followed, since the general use of HyX for the amino acid Xxx is difficult to maintain, consider e.g. hydroxy-leucine. Instead, the prefix *h* is used like hXxx, or in the case of a defined hydroxyl *n*hXxx, for two substitutions n,mhXxx. Dehydroamino acids are given as *d*Xxx if the position of the double bond is 2,3. If the position is different, then it is indicated as in the case of carboxylic acids.

Cyclic structures

Cyclizations are indicated by "c". In cyclodipeptides no direction of peptide bonds can be assigned. Instead of "A B" the terminology "cA→B" has been used, with alphabetically ordered sequences. In cases where actually information on the direction of biosynthesis was available, this order may be reversed. Generally only few starting poiunts of cyclic peptide biosynthesis are known. In such unknown cases, searches for peptide sequence homologies should consider a given sequence in repetition, like

$$cA_1A_2A_3 \longrightarrow A_1A_2A_3A_1A_2A_3$$

Side chain modifications

Side chain cyclizations have been indicated with the following symbolisms:

a) side chain to carboxyl-function
thiazolidine/thiazol-rings   oxazolidine/oxazole-rings

```
      ┌─┐                         ┌─┐
→Xxx→Cys→                    →Xxx→Ser→
```

b) disulfides

```
 ┌─S───S─┐
→Cys→  →Cys→
```

c)thioether

```
 ┌──────S──────┐
→Cys→        →Xxx→
```

d)ether bonds, direct links between side chains (C-C, C-N)

```
 ┌─────────────O┐
→Xxx→        →Yyy→   (hydroxyl-function on Yyy)

α┌─────────────┐β
→Xxx→        →Yyy→   (with α,β indicating position of bond)
```

Chain length

To define the number of residues is not always simple. Besides amino acids containing α,α-peptide bonds, compounds containing carboxyl and amino(imino)-functions have been treated as single residues. These have been differentiated here from simple carboxyls, termed *acyl-residues*. However, frequently hydroxy acids takle their place in peptide structures, in depsipeptides, or lactones. Their mode of incorporation does not differ from the mode of amino acid incorporation. So generally hydroxy acids have been considered equivalent to amino acids in the description of chain length. Chains containing both types of residues have been termed depsipeptides.

Combination of chains

Chain combinations are indicated by different directions of peptide bonds within a given compound. Prominent examples are
(1) a urea-type of bond containing proteinase inhibitors, and
(2) cephabacins, where a side chain is added step-by-step to a modified cephalosporin.

The cases have not yet been solved generally, but the respective peptides have been grouped in the following way:

(1) the length has been defined by the total number of amino acids, the presence of the urea-type bond indicated by -(U)-.

(2) since there are considerable variations in the side chains indicating a step-wise mechanism of addition, the compounds have been treated as modified tripeptides, with no attempts to group the peptide side chains.

Piperidine-type cyclizations

In the *thiostreptone/nosiheptide*-group of peptides an unusual type of cyclization has been found leading to different types of piperidine-rings. This strucutural type has been indicated by the addition -(PIP)- in the general structural description. The recent studies by Floss et al.[8] support the biosynthetic scheme proposed by [9] with a cyclization mechanism involving three amino acid residues.

Unusual constituents

Since there is a mechanistically close link of polyketide and polypeptide formation with respect to elongation principle involving the pantetheine-cofactor, it is not surprising to detect residues apparently formed by condensation of amino acid and acetate within metabolites. According to the definition of the number of residues, such a compound would be treated as a single entity, and not as two compounds, corresponding to two sequential enzymic steps. In the structural description the possible precursors have been indicated, like *[Leu,OAc]*. This is a proposal that the corresponding residue originates from condensation of L-leucine and acetate.

References

[1] H.Kleinkauf and H. von Döhren (1987) Annu. Rev. Microbiol. **41**, 259 (1987)

[2] F. Lipmann (1968) Essays Biochem. **4**,1

[3] Database of bioactive natural products, Miklós Kertész, SZENZOR Management Consulting Co., H-1134 Budapest, Lehel u. 11

[4] Peptide Institute, Protein Research Foundation, 4-1-2 Ina, Minoh-Shi, Osaka 562

[5] H.Kleinkauf and H.v.Döhren (1981) Curr. Top. Microbiol. Immunol. **91**,129

[6] H.Kleinkauf and H.v.Döhren (1985) *Peptide Antibiotics*. In Comprehensive Biotechnology (M.Moo-Young, Ed.),Vol.2, p.95-135, Pergamon, Oxford;(1986) *Peptide Antibiotics*. In Biotechnology (H.-J. Rehm and H. Reed,eds.), Vol.4 (H.Pape and H.-J.Rehm,eds.),pp.283-307, Verlag Chemie, Weinheim

[7] Eur. J. Biochem. **138**, 9-37 (1984)

[8] H.G.Floss (1988) *Biosynthesis of the antibiotics nosiheptide and thiostrepton*. In Biology of Actinomycetes '88 (Y. Okami, T. Beppu, and H. Ogawara, eds.) Japan Scientific Society Press Tokyo 1988, pp.401-405

[9] B.W.Bycroft and M.S.Gowland (1978) J. Chem. Soc. Chem. Comm. 256-257

Type cyclodipeptides
subtype piperazinedione C-2
modified piperazinediones C-2-M
branched piperazinediones n-P-C-2

C-2, C-2-M

Gliotoxin	*Gliocladium fimbriatum, Aspergillus fumigatum, Trichoderma viride, Penicillium sp.*		J.Med.Chem. 21,796,799 (78)
Tryptophane-dehydrobutyrine-piperazinedione **(CWT)**	*Streptomyces spectabilis*	RLV reverse transcriptase (weak)	JA 27,733 (74)
Spori**d**esmolides	*Pithomyces dartamm*	G+,AV	JCS Perkin I 1978,1476
Ver**t**icillin	*Verticillium sp.*	G+,MY,AT,AV	JA 23,420 (70),JCS Perkin I 1973,1819
Cairomycin	*Streptomyces sp.AS-C-19*	G+,G-	AAC 11,373 (77), 19, 941(81)B,
Bicyclomycin	*Streptomyces sapporoensis S.irabensis*	G-	JA 25,569, 576,582(72) 27,976(74) 29,155(76)
Cyclo-leucyl-N-methyl-phenylalanine **(CLF)**	*Alternaria alternata*		Phytochem. 27,357(88)
Cyclo-**l**eucyl-**p**roline (Gancidin)	*Streptomyces gancidicus*	antibacterial (weak), AT	Heterocycl. 7,341(77)
Cyclo-D-phenylalanyl-proline **(CFP)**	*Bacillus brevis*		this vol., p.
Hyalodendrin	*Hyalodendron sp.*	G+,AF	JCS Perkin I 1973,2600 US Pat. 3,715,352
Albonoursin	*Streptomyces noursei* *Actinomyces albus*	AF	Ant.8,201 (63),TL 1963,1881
(3Z,6E)-1-N-methyl-albonoursin **(ALBm)**	*Streptomyces albus* ATCC 3004	G+,weak,dermatophytes in vitro	Phyt.23,200 (84), J.Chem.Res. 1984,283
Fumitremorgin B	*Aspergillus fumigatus A.caespitosus Penicillium lanosum*	tremorganic toxin	JCS CC 1974 408
Verruculogen	*Penicillium verruculosum*	tumor producing	JACS 96, 6785(74)

3-Benzyl-6-ethyl-3,6-bis(methylthio)piperazinedione	*Gliocladium deliquescens*		JCS CC 1983 810
Cyclo-prolyl-tyrosine **(CPY)**	*Streptomyces gelaticus*	cytotoxic	ABC 45,2613 (81)
Cyclo-prolyl-valine	*Streptomyces tanashiensis*	plant growth retardant, insecticidal	
Piperazinedione 593A	*Streptomyces griseoluteus*	AV	JACS 98, 6742(76), US Pat. 3,718,651 Ger.Pat. 2,029,708
Chaetocin	*Chaetomium minutum*	G+,G-,AV cytostatic	HCA 53,1061 (70)
11,11-Dihydroxychaetocin Melinacidin **(CHAh)**	*Verticillium tererum* *Acrostalagmus cinnabarinus*	G+,AV	JA 30,468 (77),US Pat 3,857,936 Br.Pat. 1,229,297
Chaetomin	*Chaetomium cochloides*	G+,G-,AV	JACS 98, 6741(76)

```
GLI     c[Abu→Phe]
CWT     cAbu→MeTrp
                 ┌─┐
SPD     cMeAla→[Trp]
             └─Sn─┘
                 ┌─┐
VET     (MeAla→Trp-)2
             └S-S┘
CAI     cAsp→Lys
BIC     c2,5h3mdIle→3,4,5hLeu
         └──────────────────┘2
CLF     cLeu→MePhe
CLP     cLeu→Pro
CF2     cPhe→Phe
CFP     cDPhe→Pro
CPF     cPhe→Pro
HYA     cMeDPhe→MeDSer
              └──Sn──┘-
ALB     cdPhe→dVal
ALBm    cdPhe→MedVal
FUM
VER     cMeAla→[2hTrp]  c[2hTrp]→MeAla
            └──Sn──┘        └──Sn──┘
BEP     c[Phe→Ser]
CPY     cPro→Tyr
CPV     cPro→Val
PIZ     c(5ClPig)2
                  ┌────────┐
CHA     cMeSer→[Trp] c[Trp]→MeSer
```

```
CHAm     cMeSer→[2hTrp] c[2hTrp]→MeSer
               └─S2─┘           └─S2─┘
                    N
CHE      cMeSer→Trp  c[Trp]→MeSer
               └S2┘       └─S2─┘
```

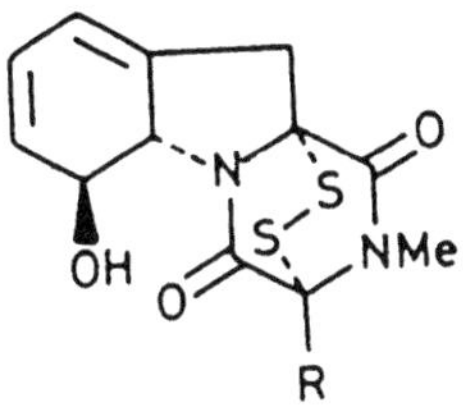

GLI

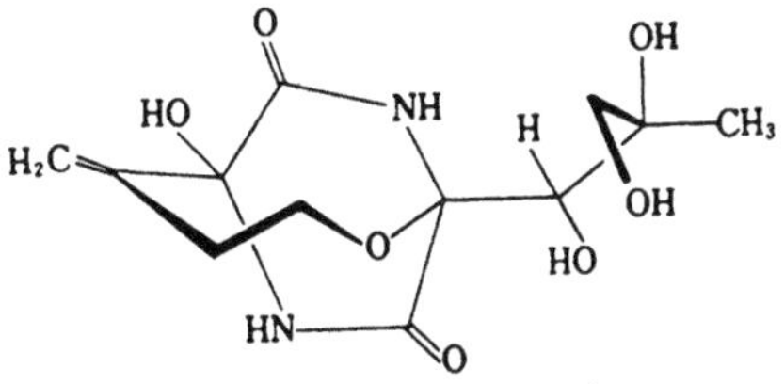

BIC

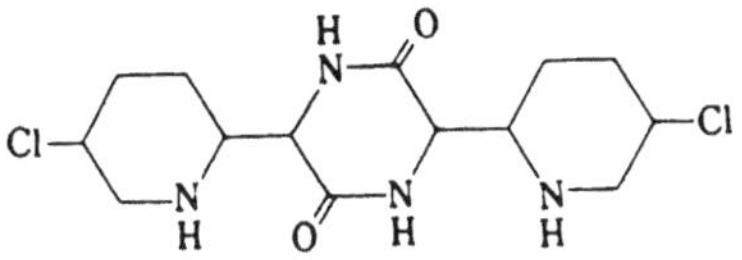

PIZ

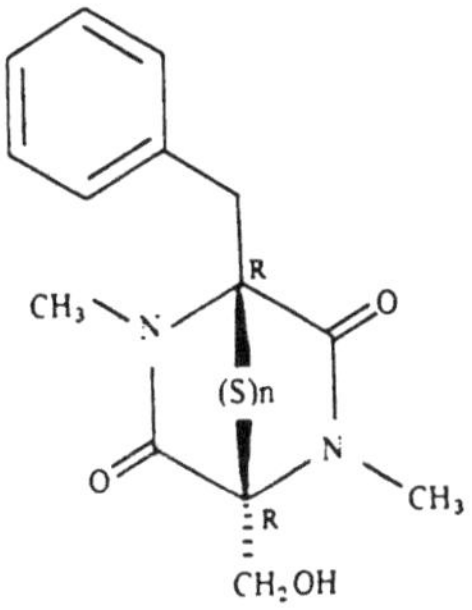

HYA (n=2,3,4)

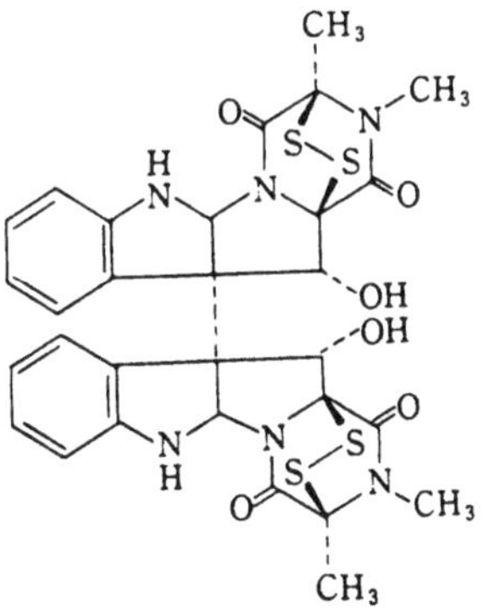

CHAm

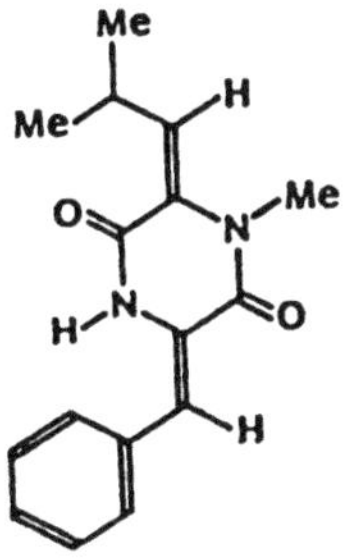

ALMm

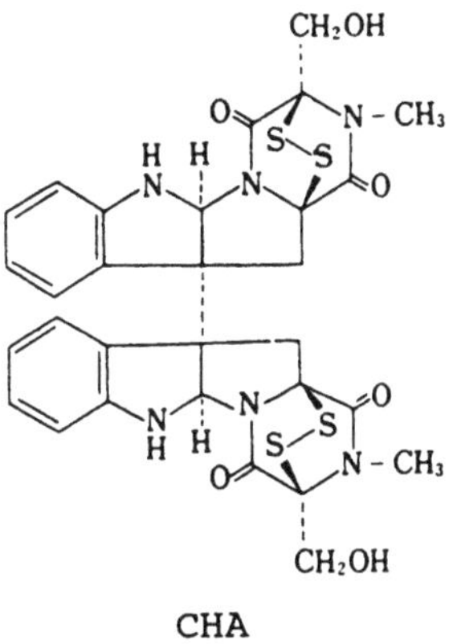

CHA

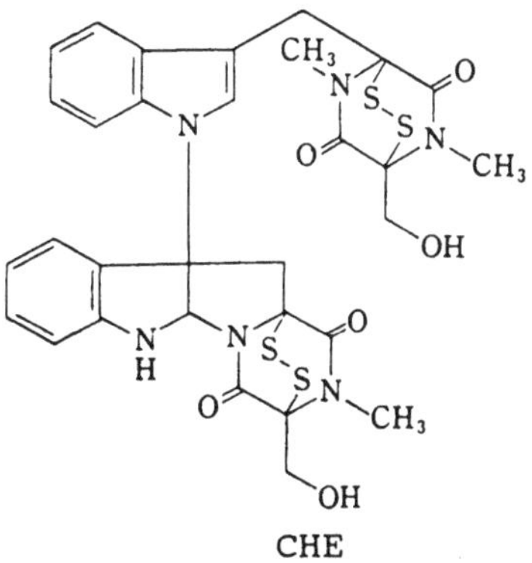

CHE

R-4-P-C-2

Foroxymithine	*Streptomyces nitrosporeus*	ACE inh.	JA 38,1813 (85)

FOR OAc→NfhOrn→Ser→c(NhOrn→NfhOrn)

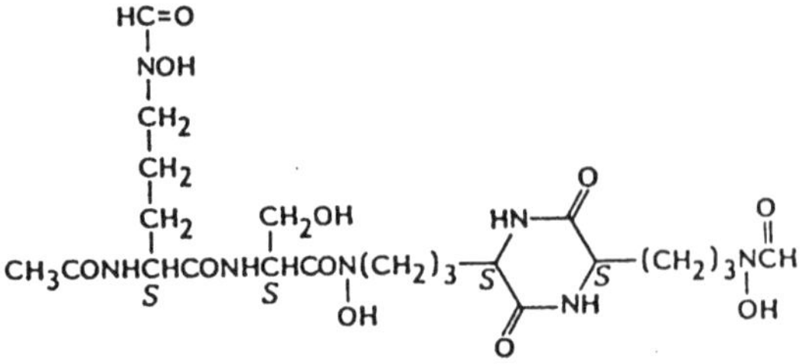

FOR

R-11-P-C-2

Amphomycin	*Streptomyces canus* *S.violaceus* *S.albogriseolus*	inh. peptido-glycan synth. feed additive	JACS 95, 2352(73) Br.Pat. 736,325 Ger.Pat. 1,021,538 B.20,1561 (81)

AMH aC_{13}/iC_{12}→Asp→*threo*3mAsp→Asp→Gly→Asp→Gly→D*erythro*D*bu*→[8]Val→
→Pro→c(D*threo*Dbu→DPip)

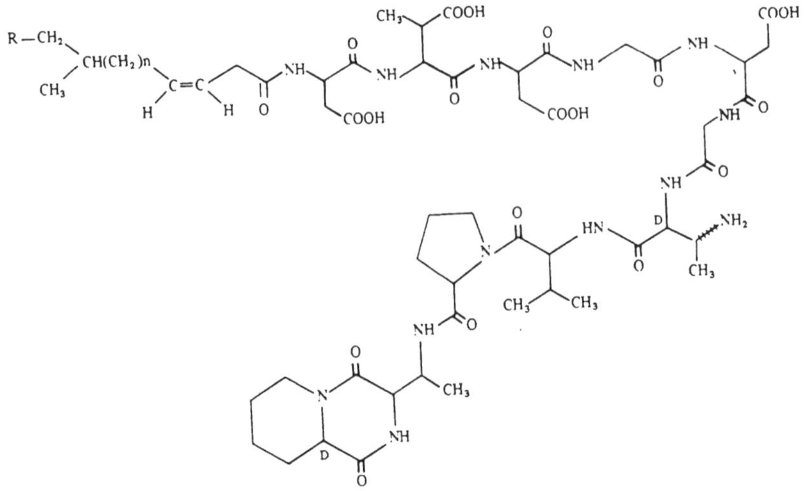

AMH (R=H,CH_3, n=4,5)

Type 3-cycles
subtype cyclopeptide R-C-3
cycloester P-C-E-3
acyl-cyclodepsipeptide

R-C-3

Aspochracin	*Aspergillus ochraceus*	insecticide	ABC 33,1491, 1501(69)

P-C-E-3

Enterobactin	*Escherichia coli* *Aerobacter aerogenes* *Salmonella typhimurium*	iron transport	V.Braun, this volume, p.103

R-C-D-3

Pyridomycin	*Streptomyces pyridomyceticus* *S.albidofuscus*	G-,My,AT	Bull.Chem.Soc 48,2081(75)

ASP 2,3d4,5d6,7dC$_8$→Lys-δ→MeVal→MeAla→

ENT c{(Dhb)→Ser}$_3$

PYR HyPic→Thr→Ahmpp-4mdHiv-

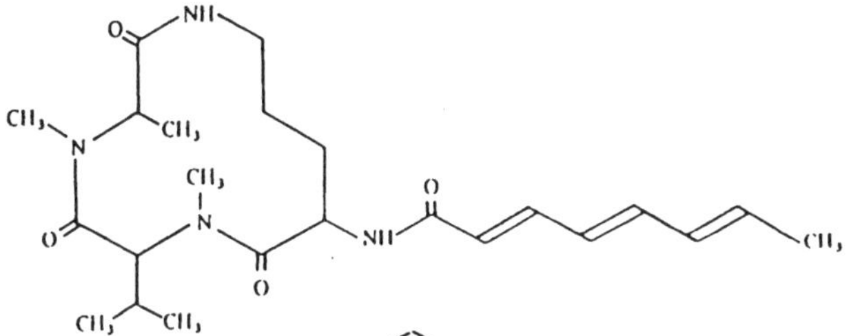

ASP

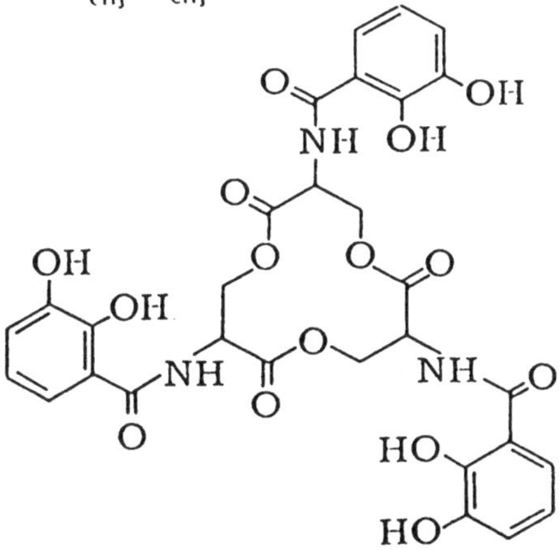

ENT

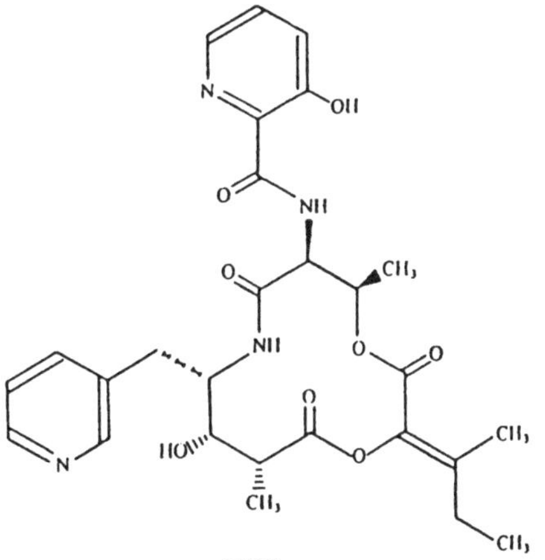

PYR

Type 4-cycles
subtype cyclopeptide C-4
branched cyclopeptide n-P-C-4
lactone L-4
branched lactone n-P-L-4
depsipeptide D-4
others

C-4

Chlamydocin	*Diheterospora chlamydospora*	cytostatic	HCA 57,533(74) TL 24,5305(83) Belg.Pat. 747,441 Brit.Pat. 1,300,137
HC-toxin I HC-toxin II	*Helminthosporium carbonum*	plant pathogen	T 38,45(82) TL 24,5309(83) 28,3419(87) J.Med.Chem. 30,71(87)
FR 900261 WF 3161	*Petriella guttulata*	plant pathogen	JA 36,478(83)
Cyl-1 Cyl-2	*Cylindrocladium scoparium*	plant growth inhibitory	ABC 37,643, 1185(73)
Tentoxine	*Alternaria tenuis, A. mali*	chlorosis inducing toxin	TL 1982,5263
Dihydro-tentoxine	*A. alternata*		Phyt.27,357 (88)

CHL cAoe→Aib→Phe→DPro
HCI cAoe→DPro→Ala→Gly
HCII cAoe→DPro→Ala→DAla
FR9 cAoe→Phe→Leu→Pip
CY1 cAoe→DOMeTyr→Ile→Pip
CY2 cAoe→DOMeTyr→Ile→Pro
TEN cMeAla→Ile-MedPhe→Gly
DIH cMeAla→Leu→MePhe→Gly

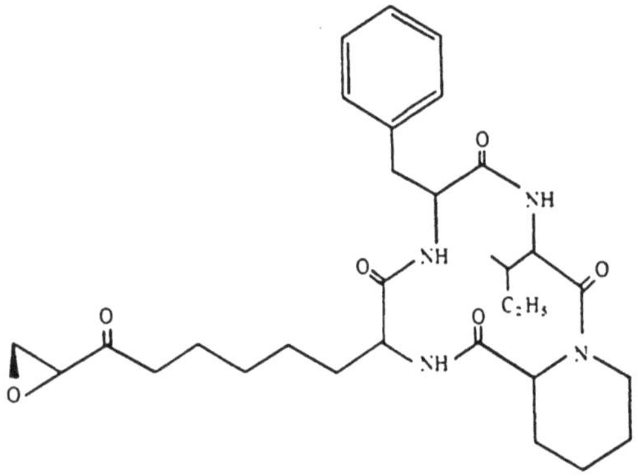

CHL

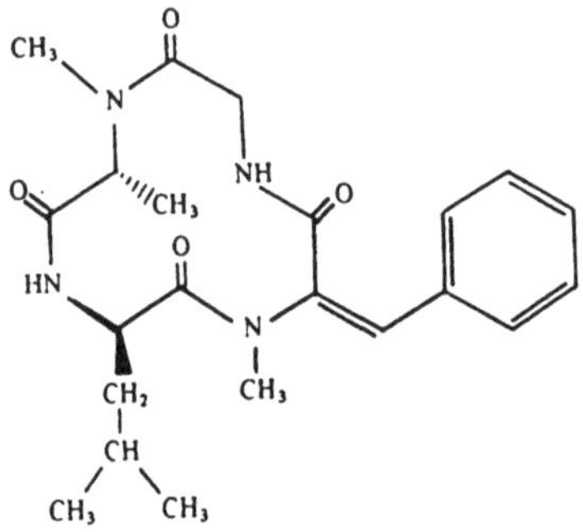

TEN

n-P-C-4
7-P-C-4

Bottromycin A2	*Streptomyces bottropensins* *Kitasatoa kanaensis*	antimycoplasma	JA 36,1076(83) Brit.Pat. 762,746

BOT cGly→3mPro→Val→tLeu→tLeu→3mPhe→(Cys)OMe (Cys) = Cy2 linkage see structure

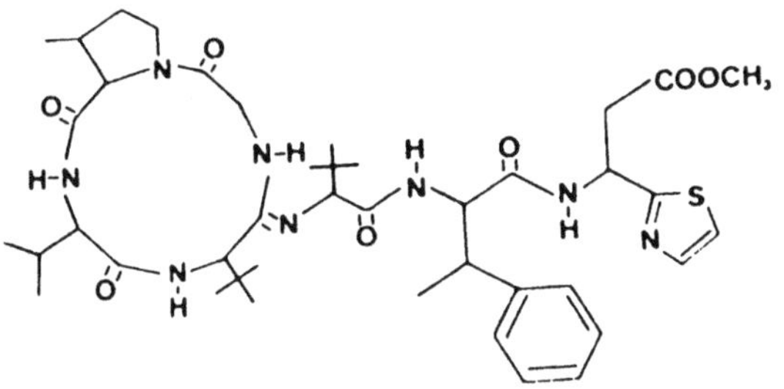

BOT

L-4

Alternarolide AM-toxin	*Alternaria mali*	G+,AF	Chem.Lett.1977 1411, Jpn.Pat. 75-123,881
Beauverolide	*Beauveria bassiana*		JCS Perkin I, 2878(80)
Jaspamide Jasplakinolide	*Jaspis sp.*	AF,insecticidal, cytotoxic, antinematodial	JACS 108,3123 (86) TL 27,2797(86)

ALT cHiv→App→dAla→Ala-
BEA c3h4mC_8→Trp→Phe→DaIle-
JAS 2,4,6m4,5dC_8→Ala→MeBrTrp→βTyr-

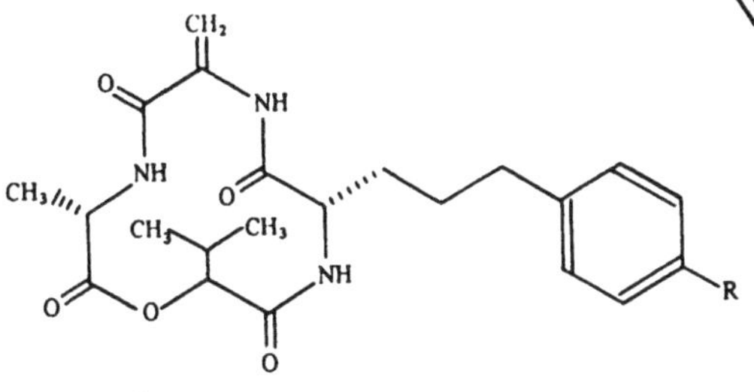

ALT (R : I-OMe, II-H, III-OH)

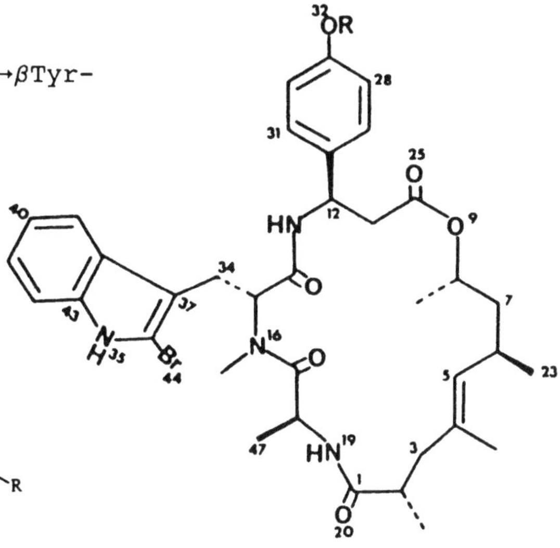

JAS (R=H;Ac)

n-P-L-4
R-7-P-L-4

A-3302-A TL-119	*Bacillus subtilis*	G+,My	JA 28.,126(75) Jpn.Pat.75-106,492

A33 OAc/OPr→DPhe→DLeu→Phe→Thr→Val→Ala→*d*Abu-

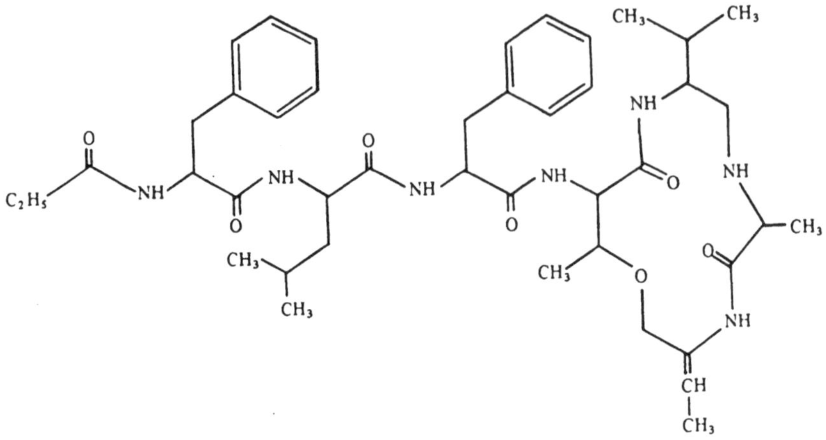

A-3

D-4

Angolide	*Hitomyces sacchari*	protection of spore surface	TL 1964,3313 BJ 98,8(66)
Serratamolide	*Serratia marcescens* *Bacterium prodigiosum*	G+,AF	TL 1969,47

ANG cHiv→Ile-Hiv→***erythro***DIle-
SER c3hC$_{10}$→Ser-3hC$_{10}$→Ser-

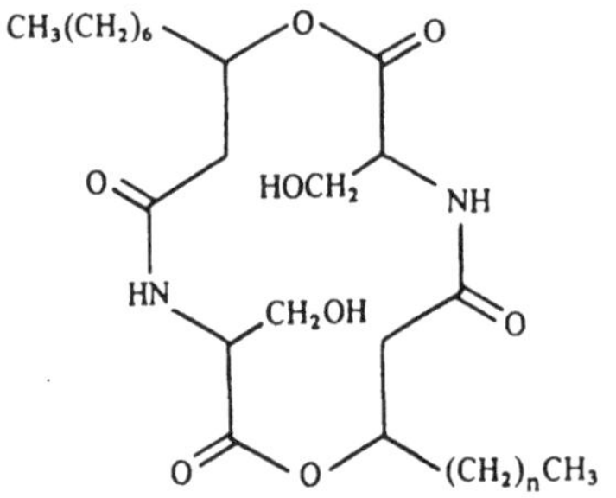

SER (A : n=6, B : n=8)

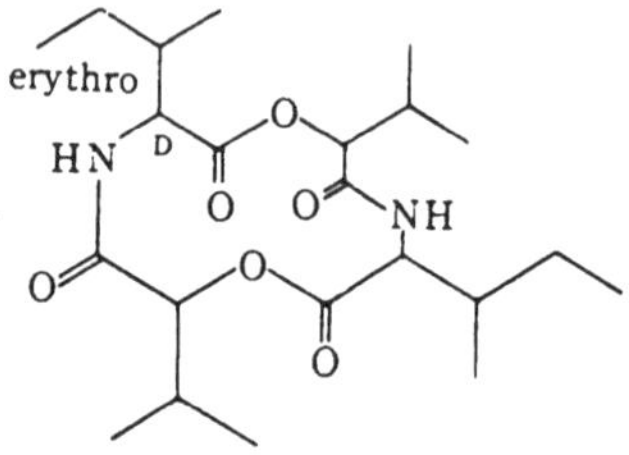

ANG

type 5-cycles
subtype cyclopeptide C-5
cyclopeptide, β-linkage C-5(β)
cyclopeptide, modified C-5-M
branched cyclopeptide n-P-C-5
acyl-lactone R-L-5
branched lactone, type 1 n-P-L-5
, type 2 n-P*-L*-5
diacyl-dilactone R(C-5)$_2$

C-5(β)

Cyclochlorotine	*Penicillium islandicum*	toxin	Chem.Lett.1973,1319

CCT cβPhe→Ser→3,4ClPro→Abu→Ser

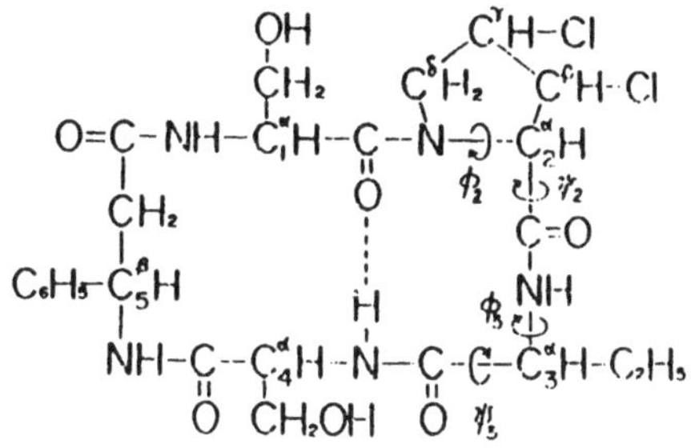

CCT

C-5-M

Malformin	*Aspergillus niger*	G+,G-,	Int.J.Pept.Prot.Res 20,16(82),Biopolym. 23,2513(84)

MA1 cVal→DCys→DCys→Val→DLeu
└S—S┘

MA2 cIle→DCys→DCys→Val→DLeu
└S—S┘

MA3 cLeu→DCys→DCys→Val→DLeu
└S—S┘

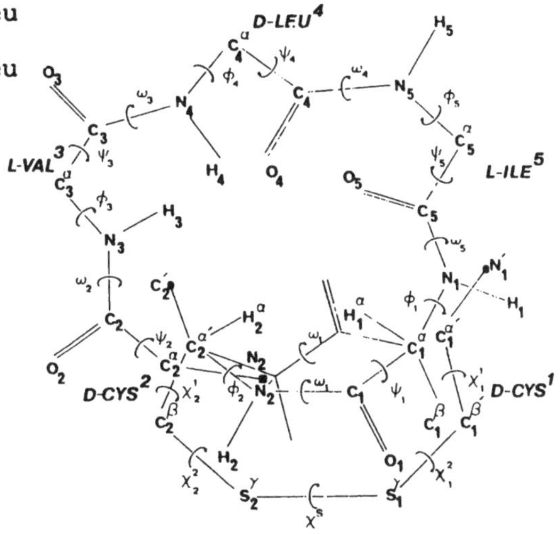

MA

6-P-C-5(β)-M

Capreomycin	*Streptomyces capreolys* *Dactylosporangium variesporum*	My	T 39,921(78) US Pat 3,143,468 Brit.Pat. 920,563
LL-BM-547	*Nocardia sp.*	G+,My	JOC 421,1282(77)
Tuberactinomycin	*Streptomyces griseoverticillatus-tuberacticus, S.sp.*	My	JA 30,1008,1073 (77),US Pat. 2,828,245, Jpn. Pat.73-30,396
Viomycin (Tuberactinomycin B)	*Streptomyces vinaceus, S. puniceus, S.floridae, S.abikoensis, S.olivoreticuli, S.californicus, S. griseus-purpureus*	G+,G-,My	JA 26,528(73) US Pat.2,828,245 Brit.Pat.651,269 Jpn.Pat.55-3096 Hung.Pat.153,683

CAP*IA* cDpr→Ala→Dpr→*d*Dpr→Cap→ (βLys ↓ at Dpr; ↑) IB cDpr→Ser→Dpr→*d*Dpr→Cap→ (βLys ↓ at Dpr; ↑)

LLB MeβArg→Dpr→Ser→Ser→*d*Dpr($CONH_2$)→Cap→ ↑

TUB*O* βLys→Dpr→Ser→Ser→*d*Dpr($CONH_2$)→Cap→ ↑

TUB*N* 4hβLys→Dpr→Ser→Ser→*d*Dpr($CONH_2$)→Cap→ ↑

TUB*A* 4hβLys→Dpr→Ser→Ser→*d*Dpr($CONH_2$)→5hCap→ ↑

VIO (TUB*B*) βLys→Dpr→Ser→Ser→*d*Dpr($CONH_2$)→5hCap→ ↑

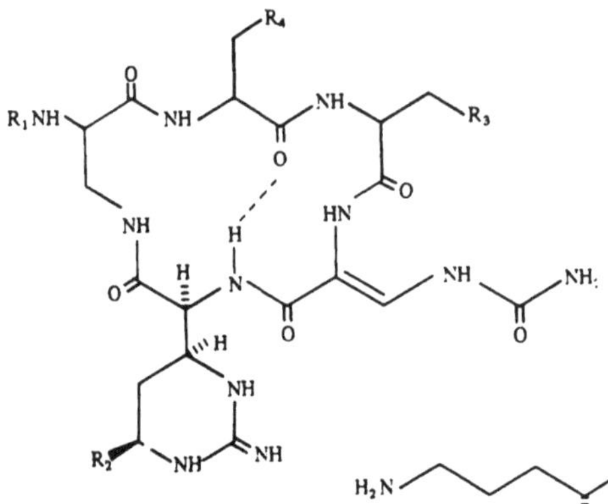

	R1	R2	R3	R4
VIO	X	OH	OH	OH
TUB*A*	Y	OH	OH	OH
TUB*N*	Y	H	OH	OH
TUB*O*	X	H	OH	OH
LLB	Z	OH	OH	OH
CAP*IA*	H	H	NHX	OH
CAP*IB*	H	H	NHX	H
CAP*IIA*	H	H	NH_2	OH?
CAP*IIB*	H	H	NH_2	H?

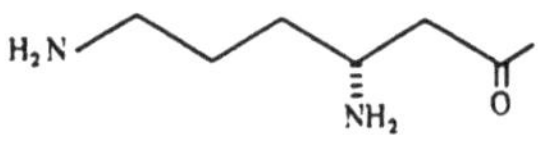

X

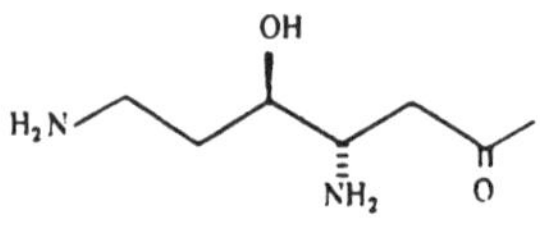

Y

R-L-5

Bre**v**igellin	*Penicillium brevicompactum*	gelatinous substance	T 37,1795 (82)

BRV Bz→Thr→Ala→Pro→Cpp→βAbu-

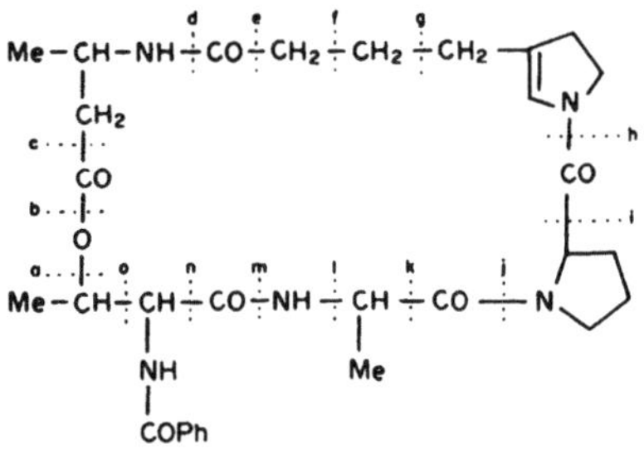

BRV

n-P-L-5
R-10-P-L-5

Imacidine	*Streptomyces olivaceus*	inh cell wall formation (Actinomyces)	Ann.1982,28

IMA iC_{13}2h→Cya→Leu→γAbu→His→Ala→aThr→Thr→His→DAla→DaThr-

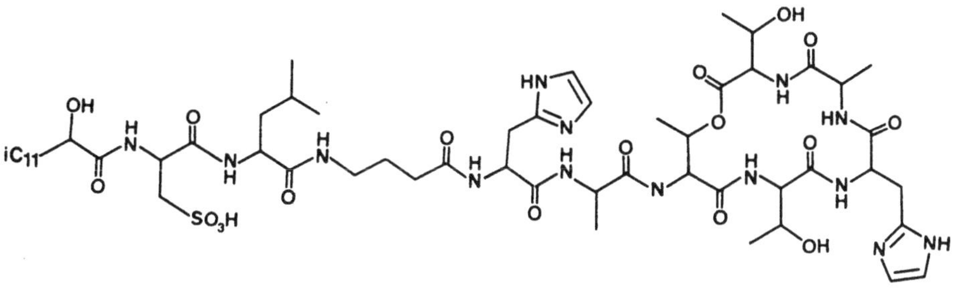

IMA

R-5-L
R-(5-L)

Actinomycins	*Streptomyces antibioticus* *S.chrysomallus* *S.parvulus* and various *S.*	G+, AT	Adv.Appl. Microbiol.17, 203(74), JBC 259,8226(84), ABC 51,803(87)
Actinomycin, monofunctional	*S.chrysomallus*		JBC 259,8226 (84)
Actinomycin, acidic	*S.antibioticus*		ABC 51,803(87)

acyl-residues: monofunctional 4m3hA, 4m3hB,
bifunctional Act (monosubstitution possible)

R→[Thr-1]→[Daa-2]→[imino 3]→[Sar-4]→[Meaa-5]-

1- Thr,MeThr,4hThr
2- DVal,DaIle,DLeu,DIle
3- Pro,trans4hPro,4oPro,allo(cis)4hPro,cis5mPro,4o5mPro, 3h4o5mPro,3h5mPro,trans4mPro,allo(cis)4mPro,Pip,4oPip, trans4hPip,Aze,Sar
4- Sar,Gly,Avl,5mPro
5- MeVal,MeaIle,MeAla

ACT R→Thr→DVal→Pro→Sar→MeVal-

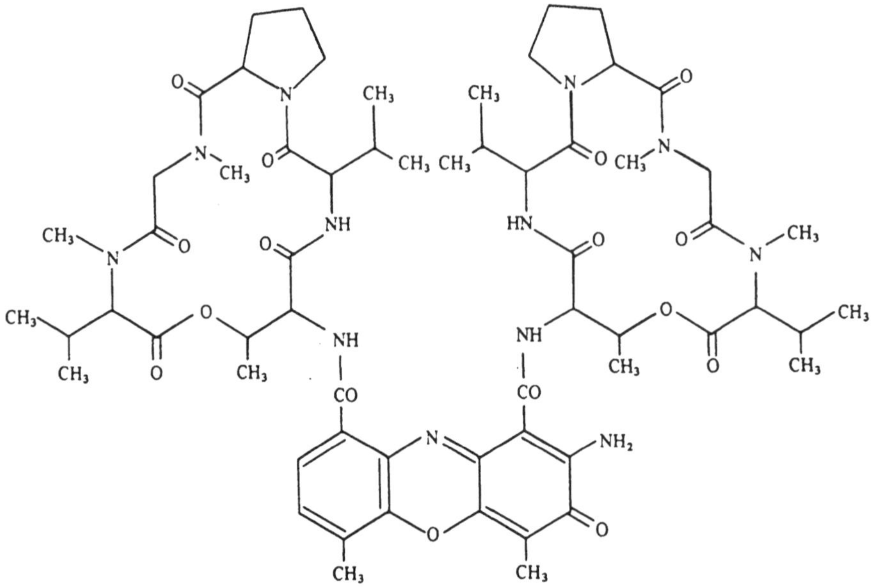

ACT

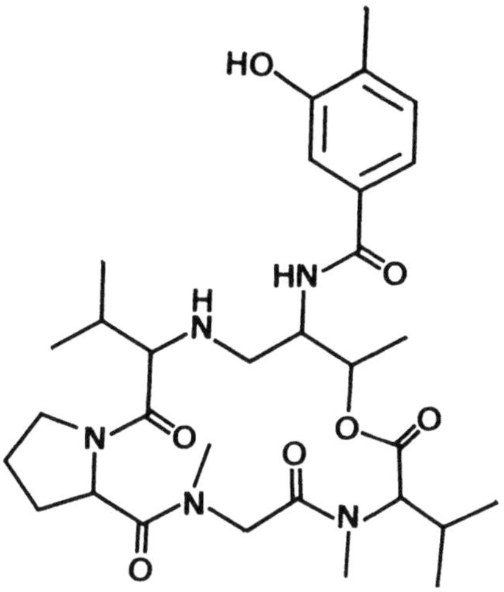

ACT*m*

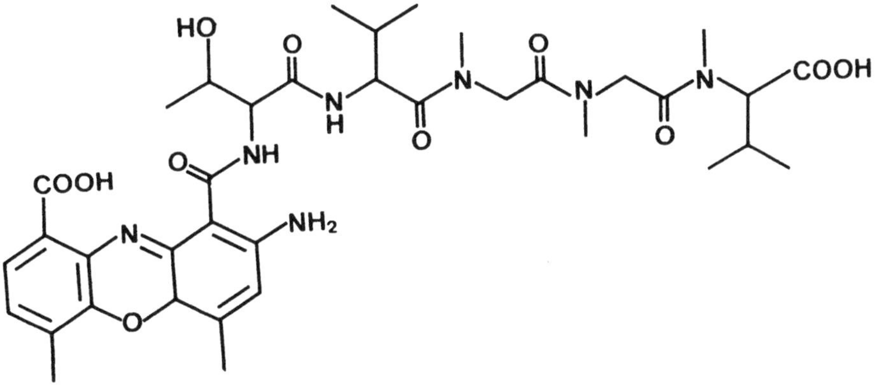

ACT*a*

Type 6-cycles
subtype cyclohexapeptide C-6
-modified C-6-M
lactone L-6
branched lactone n-P-L-6
depsipeptide D-6

C-6

Asperchrome	*Aspergillus ochraceus*	G+, iron transport	J.Bac.158,683 (84)
Ferrichrome	*Penicillium* *Aspergillus* *Neurospora* *Ustilago*		Biotechnology Vol.4,p.216
Ferrichrysine	*Aspergillus*		
Ferricrocin	*Aspergillus*		
Ferrirhodin	as ferrichrome		
Ferrirubin	*Fusarium roseum*		
Malonichrome	*Fusarium roseum*		
Phomopsin	*Phomopsis leptostromiformis*	mycotoxin (lupinosis)	JCS CC 1983, 1259
RA	*Rubiae radix*	AT	Chem.Pharm.Bull 31,1424(83), 32,284(84)
Bouvardin	*Bouvardia terniflora*	AT, no immune response	

ASP cSer→Ser→Gly→(N^{6}hOAc/hipOrn)$_3$→
FERc cGly→Gly→Gly→(N^{6}hOAcOrn)$_3$→
FERy cSer→Ser→Gly→(N^{6}hOAcOrn)$_3$→

FERcC cSer→Ala→Ser→(N^{6}OAcOrn)$_3$→
FERo cGly→Ser→Gly→(N^{6}OAcOrn)$_3$→

FERh cSer→Ser→Gly→(N^{6}hcishipOrn)$_3$→

FERu cSer→Ser→Gly→(N^{6}htranshipOrn)$_3$→
MAL cGly→Ala→Gly→(N^{6}hmalOrn)$_3$→
PHO c3,4*d*Pro→*d*Ile→*d*Asp→Me3Cl5hTyr→3,4*d*Val→3hIle→
RAIII cMeOmTyr→Ala→Me2hTyr→MeOmTyr→Ala→Ala→
└──O──┘

IV cMeOmTyr→Ala→MeTyr→Me2hOmTyr→Ala→Ala→
└──O──┘

V cMeOmTyr→Ala→MeTyr→MeTyr→Ala→Ala→
└──O──┘

V-23 cMeOmTyr→Ala→MeTyr→MeOAcTyr→Ala→Ala→
└──O──┘

VII cMeOmTyr→Ala→MeTyr→MeOmTyr→Ala→Ala→
└──O──┘

BOU cDAla→Ala→MeOmTyr→Ala→Me3hTyr→MeTyr→
└──O──┘

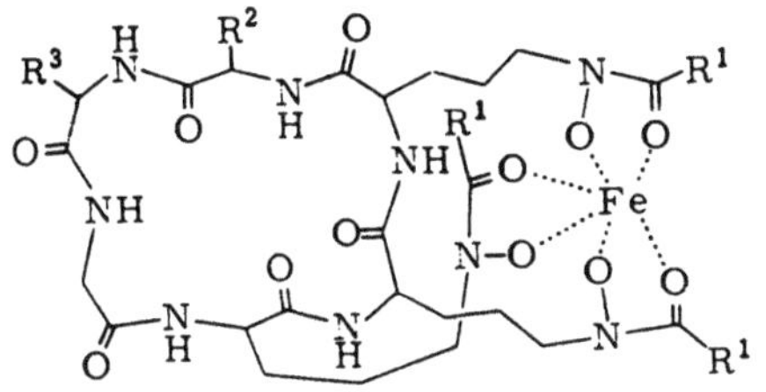

	R1	R2	R3
FER	Me	H	H
FER*o*	Me	H	CH_2OH
FER*c*	Me	H	Me
FER*y*	Me	CH_2OH	CH_2OH
ASP	Me/$CH{=}C(CH_3)CH_2CH_2OH$	CH_2OH	CH_2OH
MAL	CH_2CO_2H	H	Me
FER*A*	*trans*$CH{=}C(CH_3)CH_2CO_2H$	CH_2OH	CH_2OH
FER*h*	*trans*$CH{=}C(CH_3)CH_2CH_2OH$	CH_2OH	CH_2OH
FER*u*	*cis*$CH{=}C(CH_3)CH_2CH_2OH$	CH_2OH	CH_2OH

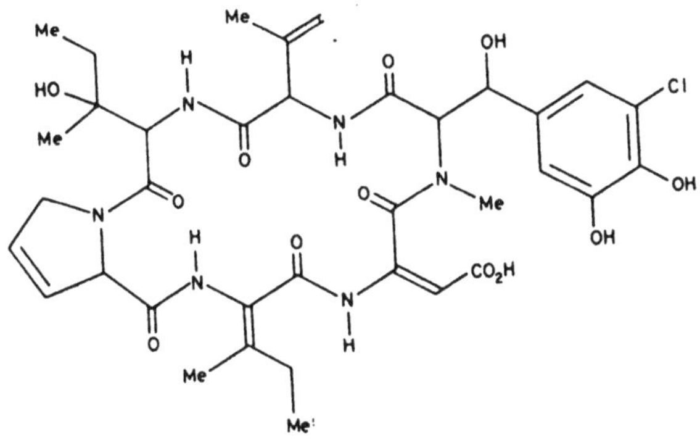

PHO

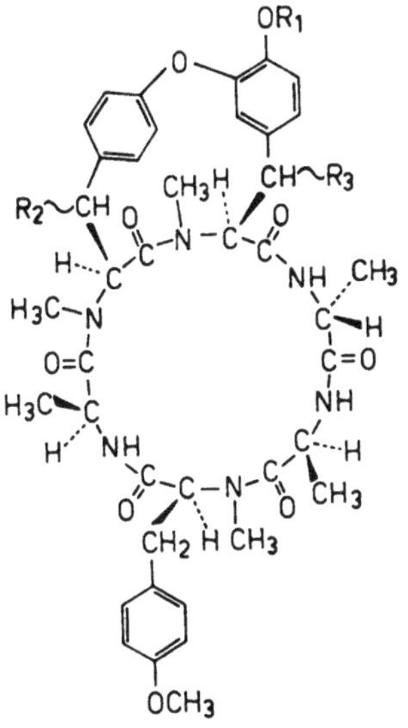

RA	R1	R2	R3
III	Me	OH	H
IV	Me	H	OH
V	H	H	H
V-23	Ac	H	H
VII	Me	H	H

R-C-6

Mulundocandin	*Aspergillus sydowi*		JA 41,281(87)
Echinocandin	*A.rugulosis*	AF	HCA 62,1252(79)
	A.nidulans-echinolatus		Swiss Pat. 568,386
S-41062-F	*Cryptosporiopsis ATCC 20594*	AF	Eur.Pat.57,724 (82)
Aculeacin	*Aspergillus aculeatus*	AF	JA 30,297,303 (77),Can.Pat 1,041,446(78)

MUL mC_{14}→4h5hOrn→Thr→4hPro→3,4hHty→Ser→3h4mPro→
↑

ECH 6,7*d*9,10*d*C_{18}→4h5hOrn→Thr→4hPro→3,4hHty→Thr→3h4mPro→
↑

S-4 C_{16}→4h5hOrn→Ser→3hPro→3hHty→hAsn→3h4mPro→
↑

ACU C_{17}→4h5hOrn→Thr→3hPro→3,4hHty→Thr→3h4mPro→
↑

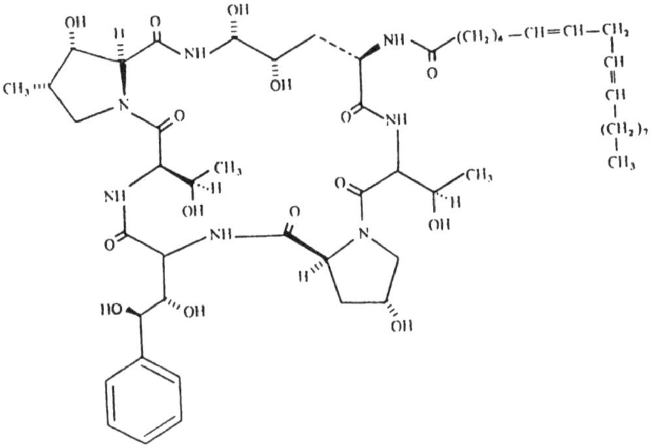

ECH

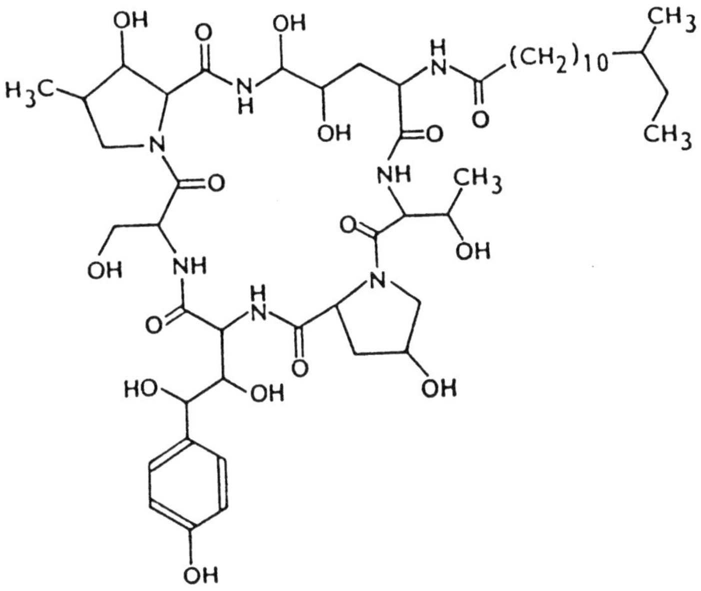
OH
O
OH
H
N
(CH2)10
CH3
H3C
N
H
OH
O
O
CH3
N
O
NH
CH3
O
OH
OH
NH
N
H
N
O
O
HO
OH
OH
OH

MUL

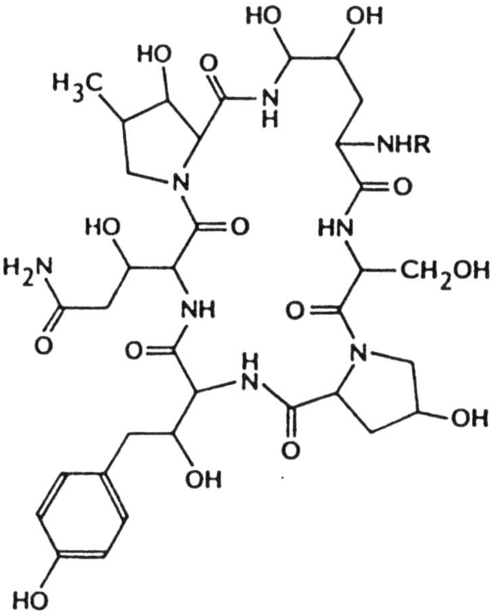
HO
OH
HO
O
H3C
N
H
NHR
N
O
HO
O
HN
H2N
CH2OH
NH
O
O
N
O
H
N
OH
O
OH
HO

S41 ($R=C_{16}$)

L-6

Isariin	*Isaria cretacea*	weak G+,AF	JCS PerkinI 1974,802,Chem. Pharm.Bull.22, 2136(74)
SF-1902-A2	*Streptomyces hygroscopicus*	G-	JA 34,1416(81)
Globomycin	*Streptomyces halstedii* *Streptoverticillium cinnamoneum*	G-,inh. cell wall synthesis, prolipoprotein-processing	JA 31,421, 426(78)
Destruxin	*Metarhizium anisopliae* *Aspergillus ochraceus* *Oospora destructer*	insecticidal	Phytochem.20, 715(81)
Roseotoxin	*Trichothecium roseum*	toxin	JACS 106,2388 (84)
Monamycin	*Streptomyces jamaicensis*	G+,surface active,ionophore M^+	JCS PerkinI 1977,2369

SF1 $2m3hC_7$→MeLeu→Ile→Ser→Thr→Gly-

GLO $3hC_{8/10}$→MeLeu→aIle→Ser→aThr→Gly-

DES*A* 4,5*d*2hC$_5$/4m2hC$_5$→Pro→Ile→MeVal→MeAla→βAla-

ROS 4,5*d*2hC$_5$→Pro→Ile→MeVal→MeAla→βAla-

ROS*2* 4,5*d*2hC$_5$→*trans*3mPro→Ile→MeVal→MeAla→βAla-

ISA $3hC_{12}$→Gly→Val→DLeu→Ala→Val-

MON*B2* Hiv→DPia→4hPia→MeLeu→Pro→DIle-

MON*B3* 4mHiv→DPia→4hPia→MeLeu→Pro→DVal-

MON*H2* Hiv→D4ClPia→4hPia→MeLeu→4mPro→DIle-

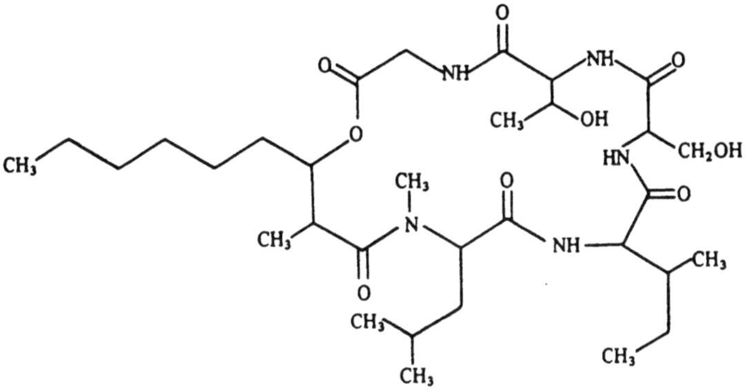

SF-1

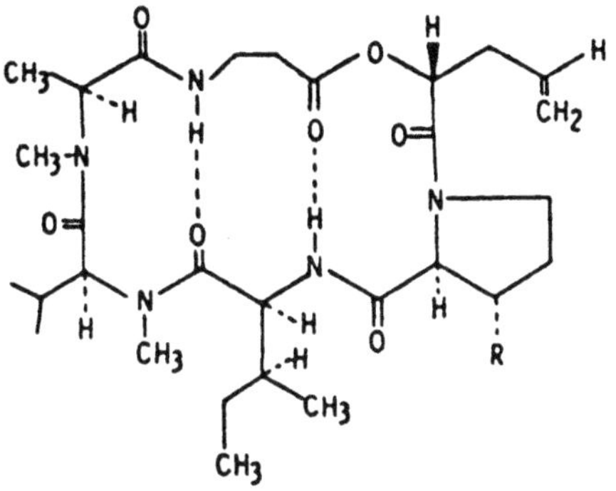

DESA R=H
ROS2 R=Me

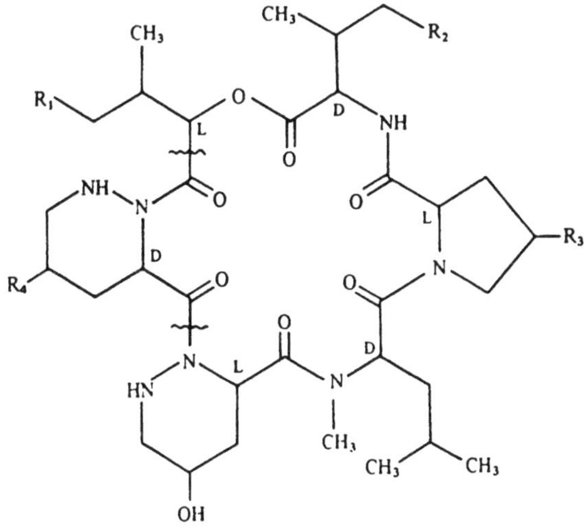

MON	R1	R2	R3	R4
A[a]	H	H	Me	H
B1	H	H	Me	H
B2	H	Me	H	H
B3	Me	H	H	H
C[a]	Me	H	Me	H
D1	Me	H	Me	H
D2	H	Me	Me	H
E[a]	Me	Me	Me	H
F	Me	Me	Me	H
G1	H	H	Me	Cl
G2	Me	Me	H	Cl
G3	Me	H	H	Cl

a

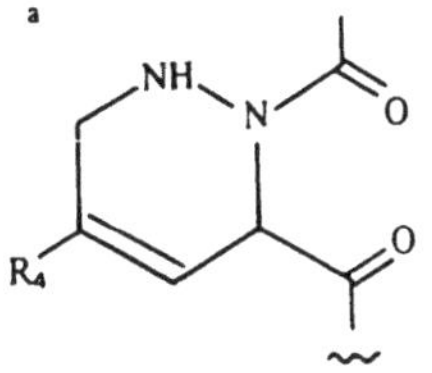

R-6-L-M

Doricin	*Streptomyces loidensis*	G+	TL 1966,4231 US Pat.3,299,047
Ostreogrycin B Pristinamycin Vernamycin Mikamycin	*S.ostreogriseus* *S.mitakaensis* *S.pristinae,* *S.piralis,* *S.olivaceus* *S.graminofaciens*	G+ ribosome binding(70s)	JA 30,665(77) US Pat. 2,787,580 2,990,325 3,137,640 3,154,475 3,299,047 3,311,538 Brit.Pat. 819,872 Fr.Pat. 1,301,837
Staphylomycin Virginiamycin	*S.virginiae*		TL 1971,2687 JA 30,141(77)
Yakusimycin	*Streptomyces antibioticus*	G+	Jpn.Pat. 73-30,398 73-10,294
Azinothricin	*Streptomyces sp.*		JA 39,17(86)
A83586C	*Streptomyces karantakensis*	G+(in vitro)	JA 41,726(88)

DOR 3hPic→Thr→DAbu→Pro→MePhe(pNMe$_2$)→Asp→Phg-

OST*B* 3hPic→Thr→DAbu→Pro→MePhe(pNMe$_2$)→4oPip→Phg-

OST*B1* 3hPic→Thr→DAla→Pro→MePhe(pNMe$_2$)→4oPip→Phg-

OST*B2* 3hPic→Thr→DAbu→Pro→MePhe(pNHMe)→4oPip→Phg-

OST*B3* 3hPic→Thr→DAbu→Pro→MePhe(pNHMe)→4o5hPip→Phg-

STA*S* 3hPic→Thr→DAbu→Pro→MePhe→4oPip

STA*S*1 3hPic→Thr→DAbu→Pro→Phe→4hPip

STA*S*2 3hPic→Thr→DAbu→Pro→MePhe→4o5hPip

STA*S*3 3hPic→Thr→DAla→Pro→MePhe→4oPip

A83 2,3h3,7e11o6,8,10,12m2et8,9d12,13dC_{14}→3hLeu→Pia→^{3}NhAla→
→^{4}MeAla→Pia→DThr→

AZI 2,3h3,7e11o6,8,10,12m2et8,9d12,13dC_{14}→3hLeu→Pia→NhOmSer→
→^{4}MeAla→Pia→DThr→

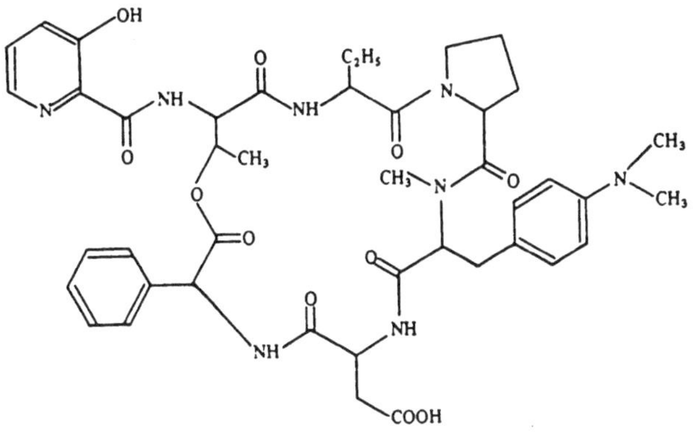

DOR

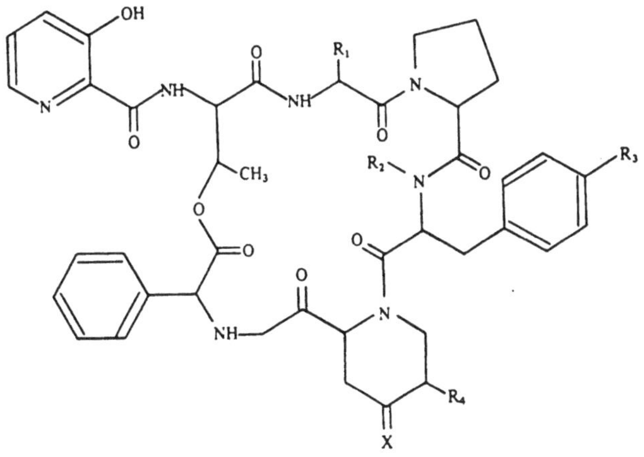

OST/STA

	R1	R2	R3	R4	X
OST*B*	Et	Me	NMe_2	H	O
OST*B1*	Me	Me	NMe_2	H	O
OST*B2*	Et	Me	NHMe	H	O
OST*B3*	Et	Me	NHMe	OH	O
STA*S*	Et	Me	NHMe	H	O
STA*S1*	Et	H	NHMe	H	H,OH
STA*S2*	Et	Me	NHMe	OH	O
STA*S3*	Me	Me	NHMe	H	O

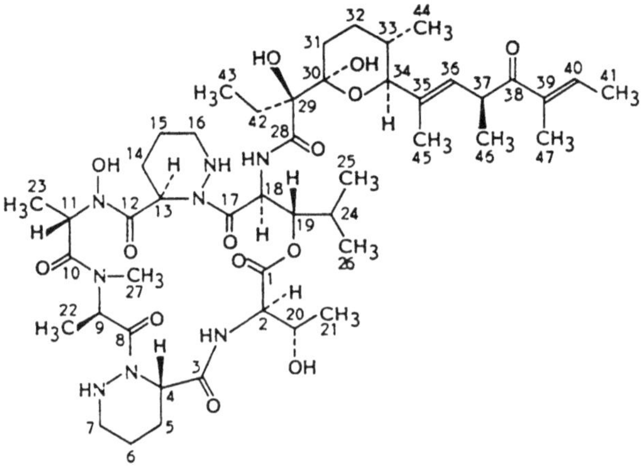

A83

7-P-L-6

Cyclo**he**ptamycine	*Streptomyces sp.*	G+,My	T 26,4931(70)

8-P-L-6

Esperin	*Bacillus mesentericus*	My	T 25,1985(79)

CYH fVal→Thr→DOmTyr→DAla→MeIle→3hAvl→5OmTrp-

ESP $2hC_{13/14/15}$→Glu→Leu→DLeu→Val→Asp→DLeu→Leu

ESP2 $2hC_{13/14/15}$→Glu→Leu→DLeu→Val→Asp→DLeu→Val

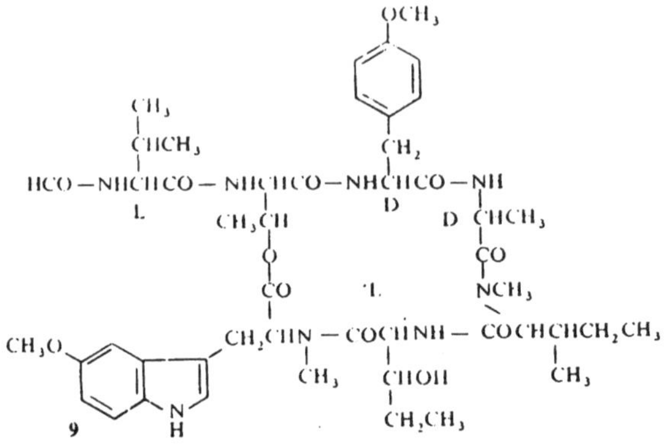

CYH

D-6

Enniatin	*Fusarium oxysporum, F. orthoceras, F. sambucinum, F. latentium*	G+,My,AF,AV M^+-binding (Li^+,K^+)	Chem.Pharm.Bull. 21,1171,1175, 1184(73), this vol.,p.
Beauvericin	*Beauveria bassiana Paecilomyces fumoroseus, Polyporus sulphureus*	G+,G-, insecticidal	Aust.J.Chem. 31,1397(78)
Sporidesmolide	*Pithomyces dartamm*	G+,AT Protection of spore surface	JCS Perkin I 1476(78)
Neurosporin	*Neurospora crassa*	Siderophore	JACS 106,1285 (84)

ENN c(MeVal-DHiv→)$_3$
BEA c(MePhe-DHiv→)$_3$
SPO cHiv→DVal→DLeu-2hiC$_{6/7}$→Val→MeLeu-
SPOII cHiv→DIle→DLeu-2hiC$_{6/7}$→Val→MeLeu-
NEU cN^{1}OAcN^{6}hDOrn→Hbu-δ→N^{6}hN^{1}OAcDOrn-Hbu-δ→DOrn-Hbu-δ→

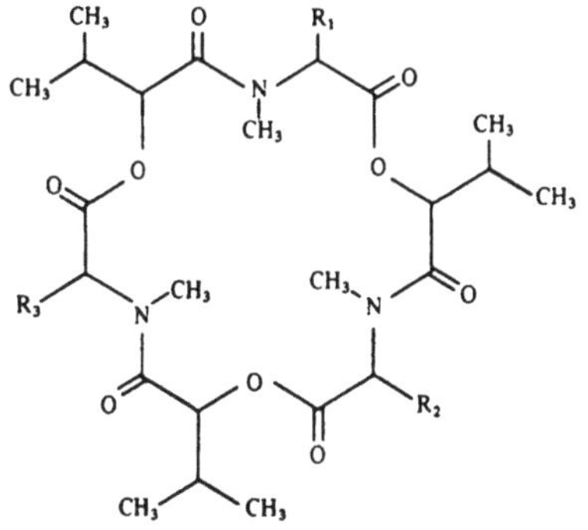

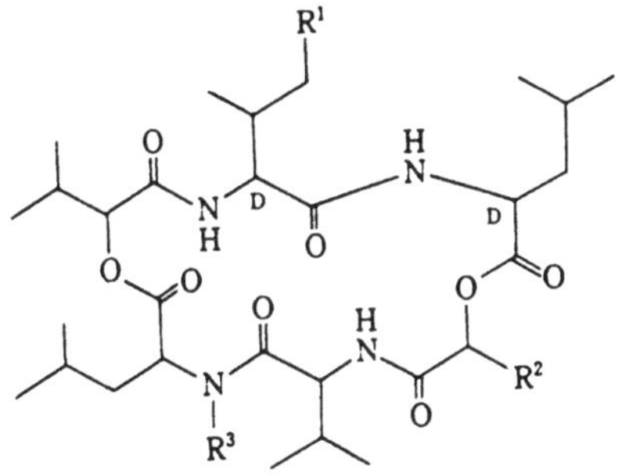

	R corresponding to		
	1	2	3
ENNA	Ile	Ile	Ile
ENNB	Val	Val	Val
ENNC	Leu	Leu	Leu
BEV	Phe	Phe	Phe

	R corresponding to		
	1	2	3
SPOI	DVal	Hib	Me
SPOII	DIle	Hib	Me
SPOIII	DVal	Hib	Me
SPOIV	DVal	Hip	Me

n-D-L-6

Didemnin	*Didemnum solidum*	AT	JACS 109,6846 (87),PNAS 85, 4119(88)

DIDA MeDLeu→Thr→Ist-Hip→Leu→Pro→MeOmTyr- D-7-L-6

DIDB L-Lac→Pro→MeDLeu→Thr→Ist-Hip→Leu→Pro→MeOmTyr- D-8-L-6

DIDC L-Lac→MeDLeu→Thr→Ist-Hip→Leu→Pro→MeOmTyr- D-7-L-6

DIDD pGlu→Gln→Gln→Gln→L-Lac→MeDLeu→Thr→Ist-Hip→Leu→Pro→MeOmTyr- D-12-L-6

DIDE pGlu→Gln→Gln→L-Lac→MeDLeu→Thr→Ist-Hip→Leu→Pro→MeOmTyr- D-11-L-6

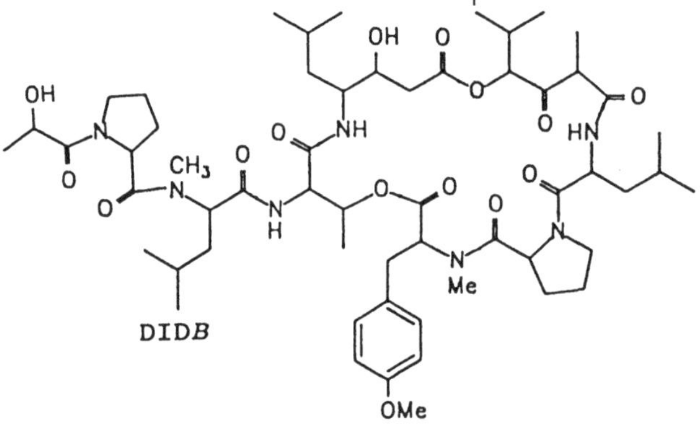

DIDB

Type 7-cycles
subtype cycloheptapeptide C-7
acylcycloheptapeptide R-C-7
branched cycloheptapeptide R-n-P-C-7
lactone L-7
acyllactone R-L-7
branched lactone R-n-P-L-7

C-7

Ilamycin	*Streptomyces islandicus* *S.insubtus*	My	FEBS Lett.17, 145(71)
Phallotoxins	*Amanita sp.*	Mycotoxin actin binding	Crit.Rev.Biochem.5,185(78) Toxicol.Lett. 35,169(85)
Viroisin		Mycotoxin	
Rhizoins	*Rhizopus microsporus*	Mycotoxin	JCS CC 1983, 47
Ulicyclamide	*Lissoclinum patella*	cytostatic	JACS 102,5688 (80),Ang.24 569(85)
Cyanoginosin	*Microcystis aeroginosa*	toxin	JCS Perkin I 1984,2311 1985,2747

ILA cMeLeu→3'NO_2Tyr→Ala→MeLeu→Leu→4,5*d*Ahx→Trp(Nip)→
PHA cAla→Trp→4,5hLeu→Ala→Thr→Cys→4hPro→
(S bridge Trp–Cys)
VIR cAla→Trp(SX)→4,5hLeu→Ala→Thr→Ser→4hPro→
RHI cVal→Val→Val→MeFua→Leu→MeAla→MeFua→
RHI*2* cIle→Val→Val→MeFua→Leu→MeAla→MeFua→
ULI cPro→Ala→Cys→Phe→Ile→Cys→Thr→
CYA*LA* cDAla→Leu→erythroD2miAsp→Ala→Adda→DiGlu→MdAla→
LR cDAla→Leu→erythroD2miAsp→Arg→Adda→DiGlu→MdAla→
YR cDAla→Tyr→erythroD2miAsp→Arg→Adda→DiGlu→MdAla→
YA cDAla→Tyr→erythroD2miAsp→Ala→Adda→DiGlu→MdAla→
YM cDAla→Tyr→erythroD2miAsp→4SmAbu→Adda→DiGlu→MdAla→
RR cDAla→Arg→erythroD2miAsp→Arg→Adda→DiGlu→MdAla→

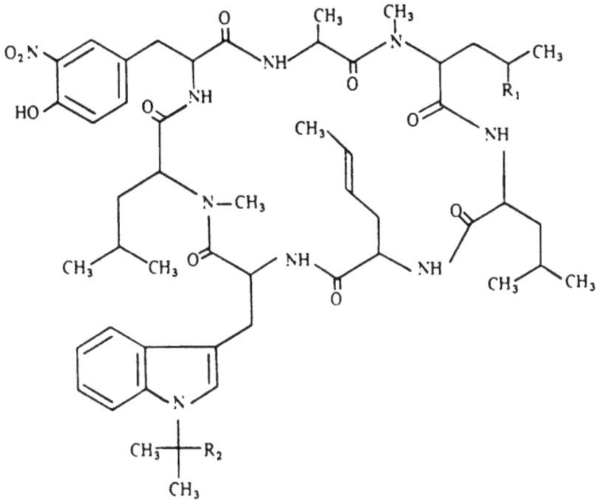

	R1	R
ILA*A*	CHO	$CH(O)CH_2$
ILA*B1*	Me	$CH=CH_2$
ILA*B2*	Me	$CH(O)CH_2$
ILA*C1*	CHO	CH_2CH_2OH

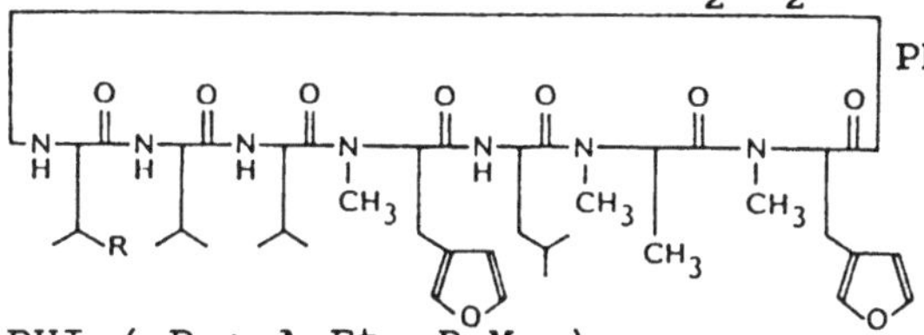

RHI (R : A-Et, B-Me)

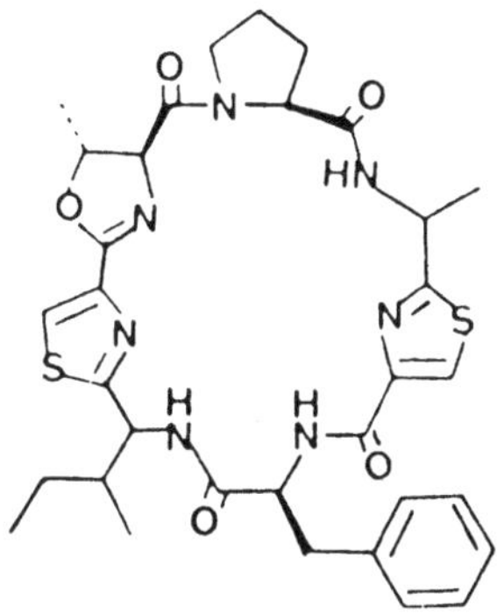

ULI

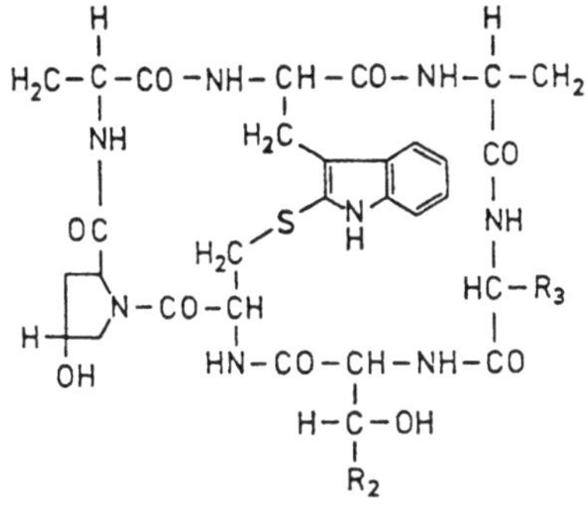

	R1	R2	R3
PHALLOIDIN	$C(OH)CH_2OH$ / CH_3	Me	Me
PHALLOIN	$C(OH)-CH_3$ / CH_3	Me	Me
PHALLISIN	$C(OH)-CH_2OH$ / CH_2OH	Me	Me
PHALLACIDIN	$C(OH)-CH_2OH$ / CH3	CO_2H	iPr

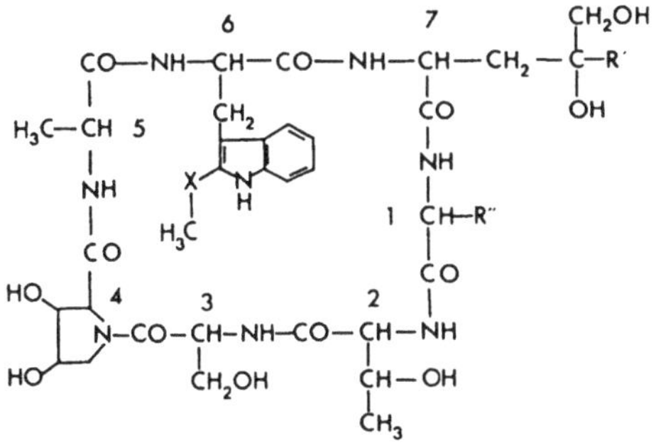

	X	R'	R''
VIROISINE	SO_2	CH_3	CH_3
^{1}A-VIROISINE	SO_2	CH_2OH	$CH(CH_3)_2$
dVIROISINE	SO	CH_2OH	$CH(CH_3)_2$
VIROIDINE	SO_2	CH_3	$CH(CH_3)_2$
dVIROIDINE	SO	CH_3	$CH(CH_3)_2$

n-P-C-7
R-8-P-C-7

Octapeptin	*Bacillus circulans*	G+,G-,My,AF antiprotozoal feed suppl.	JA 30,756(77) 36,625,634(83) US Pat. 4,341,768 Ger.Pat. 2,219,993 Neth.Pat. 72,5751
EM 49	*Bacillus circulans*	antibac.	JA 28,379(75)

R-10-P-C-7

Polymyxin Colistin Circulin	*Bacillus polymyxa, B.colistinus B.polymyxacolistinus*	G+,G-, membrane active	Annu.Rev.Biochem.46,753(77) J.Sci.Ind.Res. 38,695(79) J.Chr.218,653 (81),Br.Pat. 646,258;658,766 742,589,Jpn.Pat 72-27,038

12-P-C-7

Bacitracin	*Bacillus licheniformis, B.subtilis*	inh.cell wall formation, feed additive	Pharmacol.Ther. 16,199(78) Froyshov in Vandamme

OCT*A* $ahC_{11}/ihC_{10}/hC_{10}$→DDab→Dab→Dab→DLeu→Leu→Dab→Dab→Leu→ (↑ at Dab 2)

OCT*B* $hC_{10}/ihC_{10}/ahC_{11}$→DDab→Dab→Dab→DLeu→Phe→Dab→Dab→Leu→ (↑ at Dab 2)

OCT*C* ahC_9→DDab→Dab→Dab→DPhe→Leu→Dab→Dab→Leu→ (↑ at Dab 2)

EM4 $i2hC_{10}/2hC_{10}/a2hC_{11}$→DDab→Dab→Dab→DLeu→Leu→Dab→Dab→Leu→ (↑ at Dab 2)

EM4*2* $i2hC_{10}/2hC_{10}/a2hC_{11}$→DDab→Dab→Dab→DLeu→Phe→Dab→Dab→Leu→ (↑ at Dab 2)

POL*A* aC_9/iC_9→Dab→Thr→Dab→Dab→Dab→DLeu→Thr→Dab→Dab→Thr (↑ at Dab 4)

POL*B* $aC_8/iC_8/C_8$→Dab→Thr→Dab→Dab→Dab→Thr→DPhe→Dab→Dab→Thr→ (↑ at Dab 4)

POL*C* $aC_{9}/_{i}C_8$→Dab→Thr→Dab→Dab→Dab→Phe→Thr→Dab→Dab→Thr→ (↑ at Dab 4)

POL*D* aC_9/iC_8→Dab→Thr→DSer→Dab→Dab→DLeu→Thr→Dab→Dab→Thr→ (↑ at Dab 4)

POL*S* aC_9→Dab→Thr→DSer→Dab→Dab→DPhe→Thr→Dab→Dab→Thr→ (↑ at Dab 4)

POL*T* aC_9→Dab→Thr→Dab→Dab→Dab→DPhe→Leu→Dab→Dab→Leu→ (↑ at Dab 4)

BAC*A* Ile→Cys→Leu→DGlu→Ile→Lys→DOrn→Ile→DPhe→His→DAsp→Asn→ (Ile–Cys bridged; ↑ at Lys)

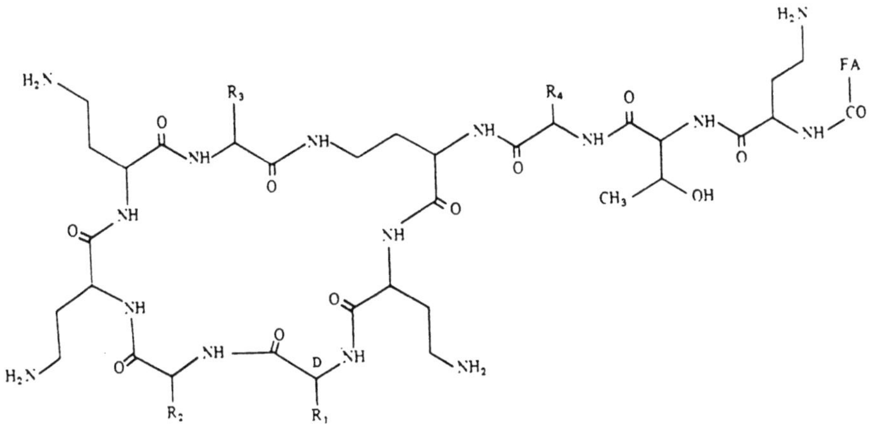

COL/CIR

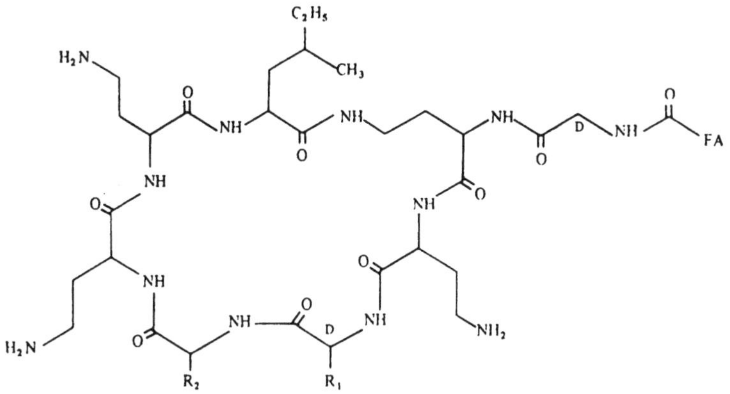

OCT

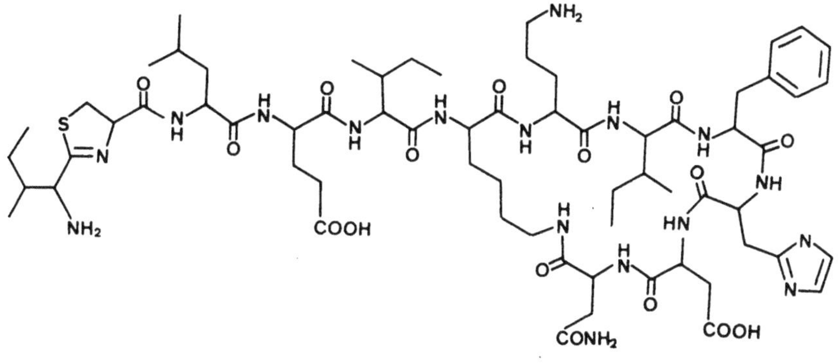

BAC

L-7

A43F	*Fungus ATCC 20529*	AF, phyto-pathogenic fungi	US Pat. 4,254,224

A43 4hC_{14}→Thr→Ala→Pro→Gln→Tyr→Val-

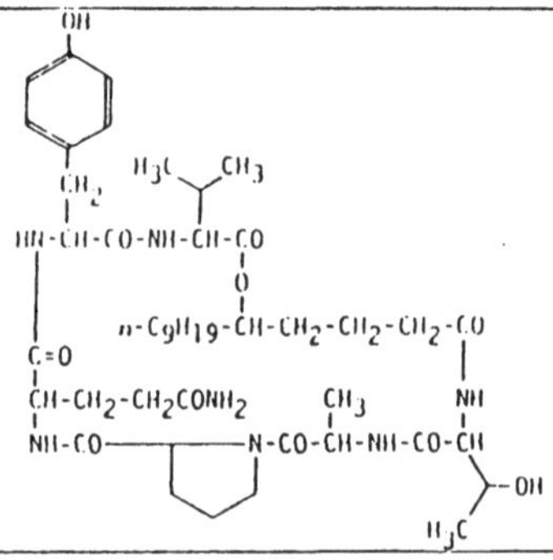

R-L-7

Etamycin	*Streptomyces griseus, S. griseoviridus, S. lavendulus, S. sp.*	G+, My, membrane active, ion carrier broad spectrum	JACS 95,875 (73), Ger.Pat. 1,072,773 Belg.Pat. 855,591
Neoviridogrisein	*Streptomyces griseoviridus S. Sp. P8648*		Okumura, this vol., p.365 ABC 47,1087(83) 48,548(84)

ETA HyPic→Thr→DLeu→Da4hPro→Sar→N2mLeu→Ala→MePhg-

NEOI HyPic→Thr→DLeu→DPro→Sar→N2mLeu→Abu→MePhg-

NEOII HyPic→Thr→DLeu→DPro→Sar→N2mLeu→Ala→MePhg-

NEOIII HyPic→Thr→DLeu→Da4hPro→Sar→N2mLeu→Abu→MePhg-

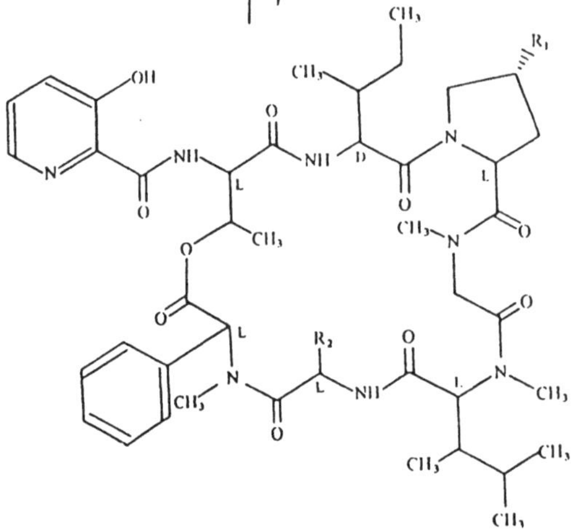

	R1	R2
ETA(NEOIV)	OH	Me
NEOI	H	Et
NEOII	H	Me
NEOIII	OH	Et

R-n-P-L-7
R-9-P-L-7

Viscosin	*Pseudomonas viscosa*	My,AV	TL 1970,1087 Chem.Pharm.Bull. 20,2064(72)

R-14-P-L-7

Stendomycin	*Streptomyces endus*	G+,G-,AF	B.11,4132(72)

VIS $3hC_{10}$→Leu→DGlu→DaThr→DVal→Leu→DSer→Leu→DSer→^{9}Ile-

STE iC_{14}→Pro→MeThr→Gly→DVal→DaIle→DAla→dAbu→DaThr→DVal→^{10}Val→
→^{11}DaThr→Ser→DaIle→Ste-

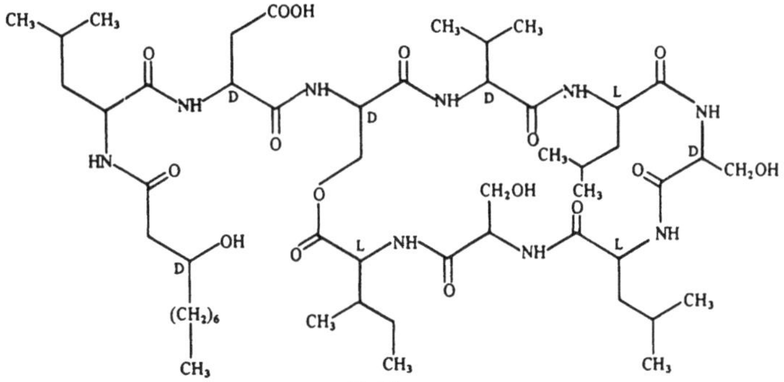

VIS

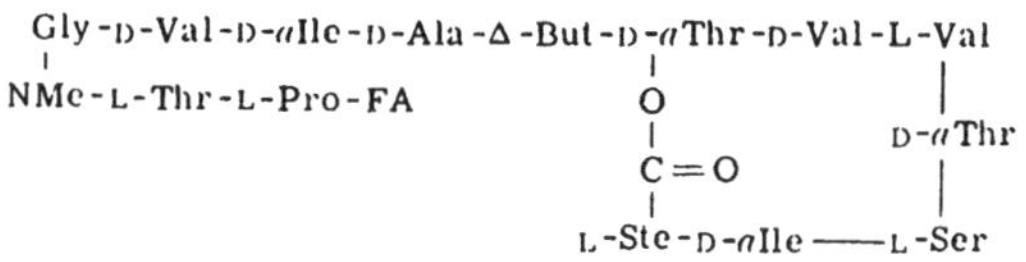

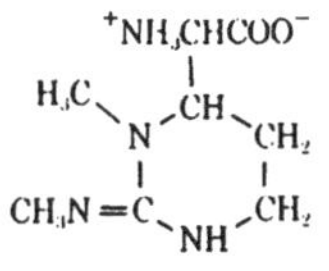

Ste

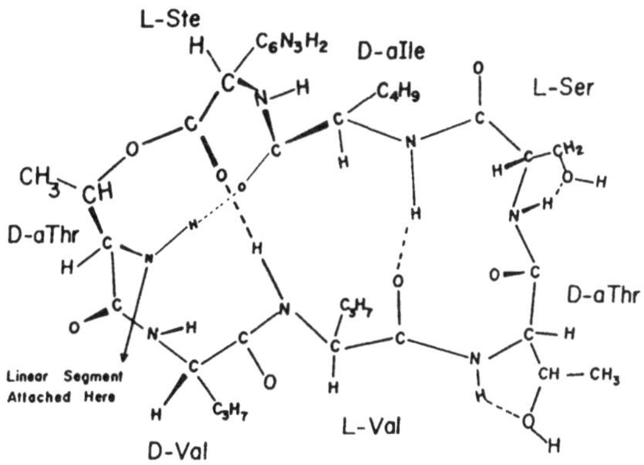

STE

Type 8-cycles

subtype cyclooctapeptide C-8
cyclooctapeptide,symmetrical $c(P-4)_2$
acyl-cyclooctapeptide R-C-8
acyl-cyclooctapeptide,symmetrical $c(R-P-4)_2$
cyclooctapeptide(β-linkage) C-8(β)
cyclooctapeptide,modified C-8-M
branched cyclooctapeptide R-n-P-C-8
lactone L-8
acyl-lactone R-L-8
acyl-lactone, symmetrical $(R-P-4)_2$-L

C-8, $c(4-P)_2$, C-8-M

Asciciacyclamide	*unidentified Ascidian*	cytotoxic	JCS CC 1983, 323, TL 26, 3223(85)
Patellamides	*Lissoclinum patella*		JA 27,267(74)
Ulithiacylamid	*Lissoclinum patella*		JACS 102,5688
Amatoxins	*Amanita sp.*	toxins	Crit.Rev.Biochem.5,185(78)

ASC c(Ile→Thr→Val→Cys→)

PAT*A* cVal→Cys→Thr→Ile→Val→Cys→Ser→Ile→

PAT*B* cPhe→Cys→Thr→Ile→Ala→Cys→Thr→Leu→

PAT*C* cPhe→Cys→Thr→Ile→Ala→Cys→Thr→Val→

ULI $c(Leu→Cys→Cys→Thr→)_2$
-S

AMAα cAsn→4hPro→4,5hIle→6'hTrp→Gly→Ile→Gly→Cys→
O-S

AMAβ cAsp→4hPro→4,5hIle→6'hTrp→Gly→Ile→Gly→Cys*→ β-amanitin

AMAγ cAsn→4hPro→4hIle→6'hTrp→Gly→Ile→Gly→Cys*→ γ-amanitin

AMAε cAsp→4hPro→4hIle→6'hTrp→Gly→Ile→Gly→Cys*→ ε-amanitin
AMA*i* cAsp→4hPro→4,5hIle→Trp→Gly→Ile→Gly→Cys*→ amanin
AMA*a* cAsn→4hPro→4,5hIle→Trp→Gly→Ile→Gly→Cys*→ amaninamide
AMA*u* cAsn→4hPro→Ile→Trp→Gly→Ile→Gly→Cys*→ amanulin
AMA*A* cAsp→4hPro→Ile→Trp→Gly→Ile→Gly→Cys*→ amanullic acid
AMA*P* cAsp→Pro→Ile→Trp→Gly→Ile→Gly→Cys*→ proamanullin

(all Cys* with sulfoxide crosslink to Trp)

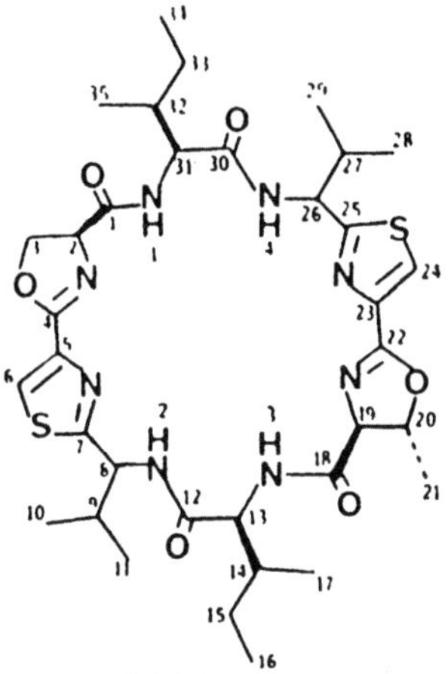

PATA

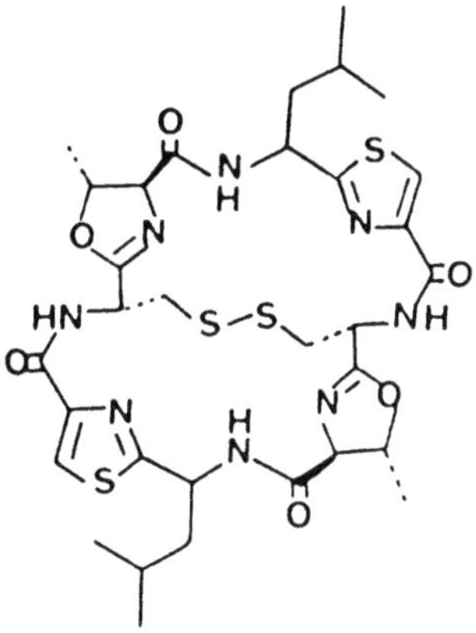

ULT

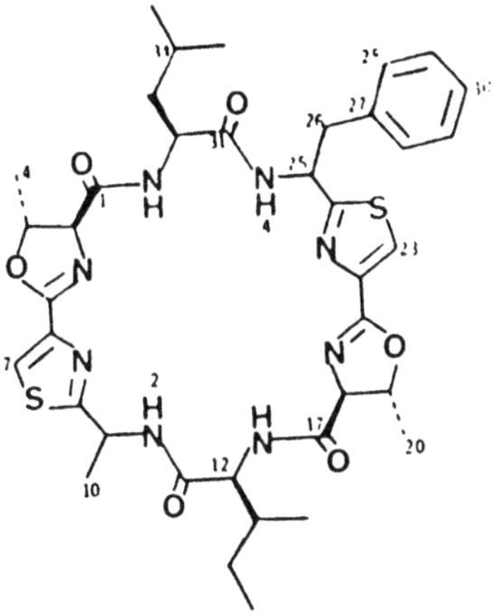

PAT*B*

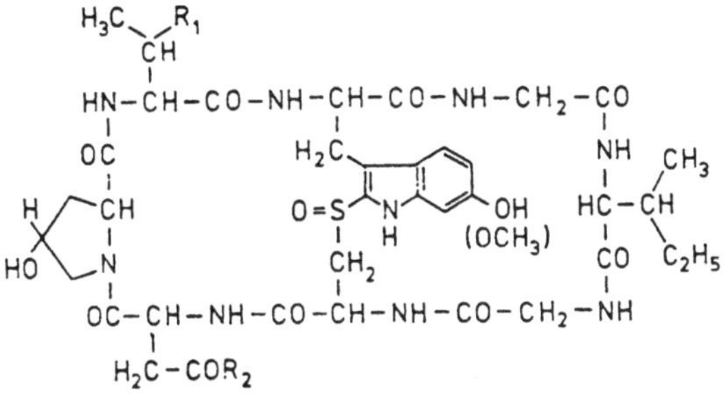

	R1	R2
AMAα	CH(OH)CH_2OH	NH_2
AMAβ	CH(OH)CH_2OH	OH
AMAτ	CH(OH)CH_3	NH_2
AMAu	CH_2CH_3	NH_2

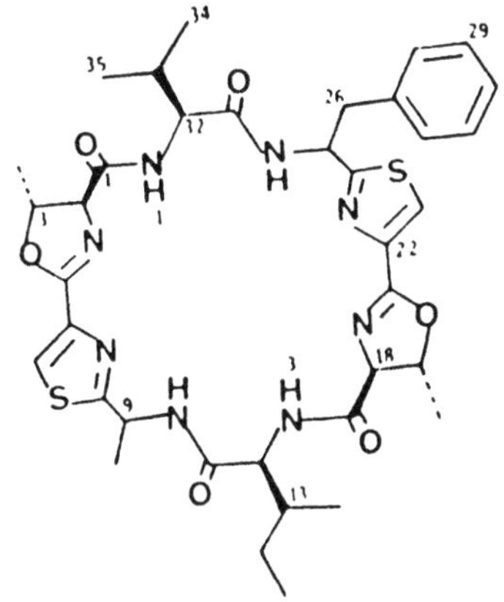

PAT*C*

C-8(β),

Bacillomycin Iturin	***Bacillus subtilis***	AF	JA 35,306(82) 37,1600(84), 40,221(87) 41,282(88)
Mycosubtilin	***B. subtilis, B. subtilis-niger***	AF	JA 39,636(86)

BAI*A* c3aC$_{14/16}$/i3aC$_{14-17}$/a3aC$_{15-17}$→Asn→DTyr→DAsn→Gln→Pro→DAsn→Ser→
BAI*D* → Asn→DTyr→DAsn→Pro→Glu→DSer→Thr→
BAI*L* → Asp→DTyr→DAsn→Ser→Gln→DSer→Thr→
BAI*F* → Asn→DTyr→DAsn→Ser→Pro→DAsn→Thr→
BAI*C* → Asp→DTyr→DAsn→Gln→Pro→DAsn→Ser→
BAI*F** i3aC$_{15-17}$/a3aC$_{15/17}$→ Asn→DTyr→DAsn→Glu→Pro→DAsn→Thr→
MYC → Asn→DTyr→DAsn→Gln→Pro→DSer→Asn→
ITU*A* → Asn→DTyr→DAsn→Gln→Pro→DAsn→Ser→
ITU*C* → Asp→DTyr→DAsn→Gln→Pro→DAsn→Ser→

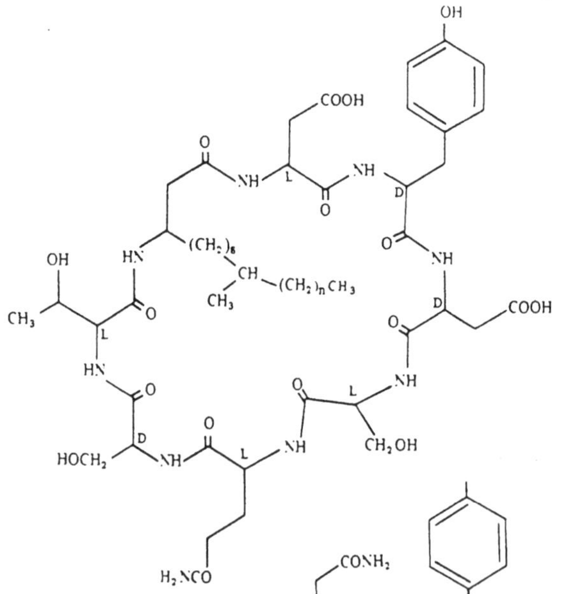

BAI (n=0,1)

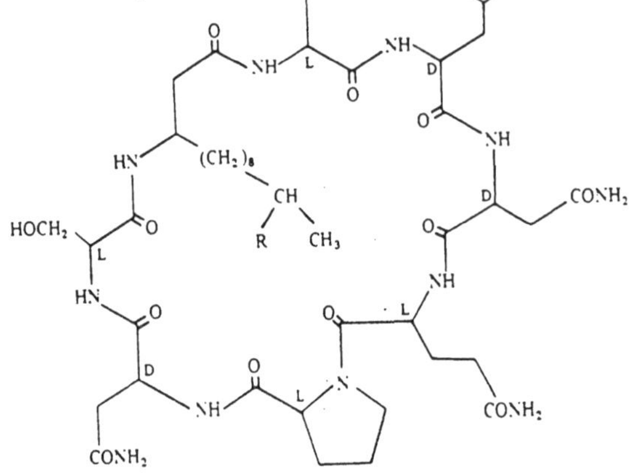

ITU (R= iPr)

L-8, R-L-8

Surfactin **APD**	*B.subtilis* *B.natto*	AF, cAMP-diesterase inhibitor	JA 36,667,674, 679,194(83)
Neopeptin	*Streptomyces sp. K-710*	phyt.bact. cell wall formation	ABC 45,895(81) 46,2621(82) TL 25,423(84)

SUR $i3hC_{14}$→Glu→Leu→DLeu→Val→Asp→DLeu→Leu→

APD*I* $3hC_{10}/i3hC_{13,15}/ahC_{13/15}$→Glu→Leu→DLeu→Val→Asp→Leu→Leu-

NOP
$aC_{15}/iC_{14}/iC_{13}$→Ser→Ser→Aoc→Asp→MePhe→MeAsp→threohGln→*threo*mAsn-

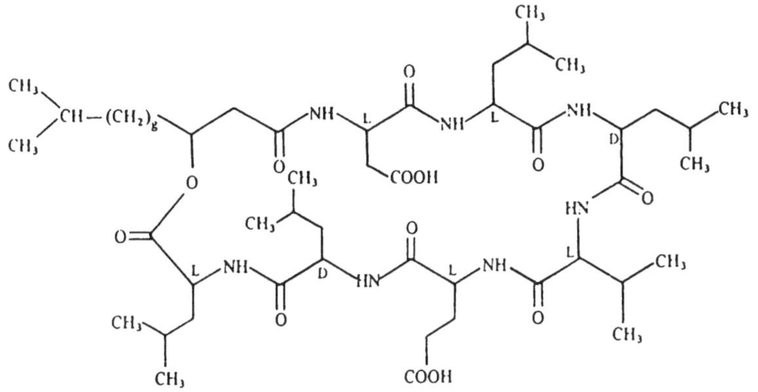

SUR

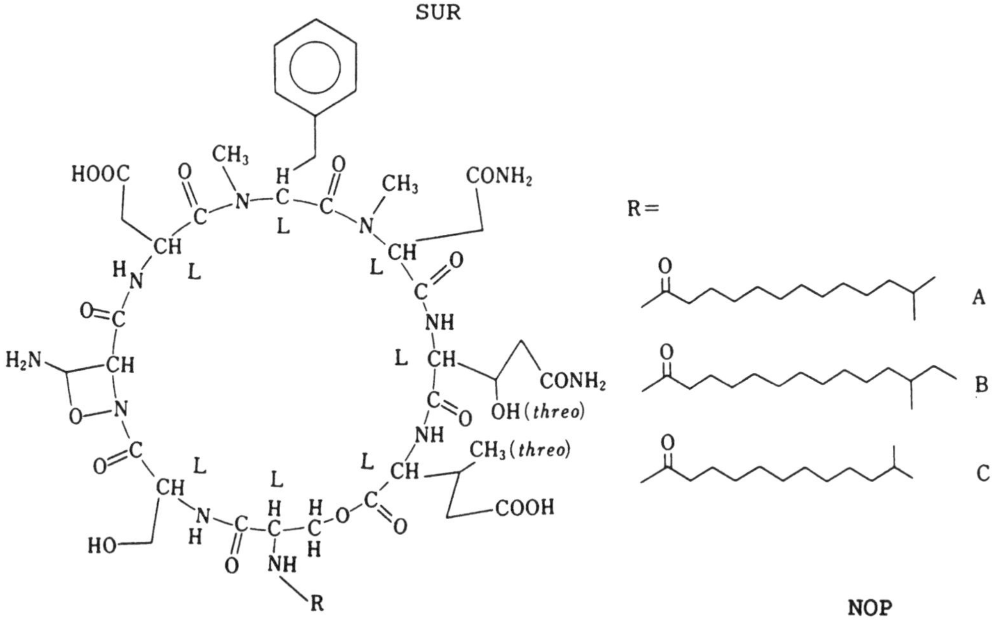

NOP

$(R-P-4)_2$-L

Echinomycin Quinomycin	*Streptomyces echinatus, S. lavendulae,* *S.griseolus* *S.aureus,S.flaveolus,S.Flavochromogenes,S.sp.*	DNA-binding	Chemistry and Biochemistry of Amino Acids, Peptides, and Proteins,Vol.7, (B.Weinstein, ed.)pp.1-33, Dekker,N.Y. JA 28,332(75) JACS 97,2497 (75)
Triostin	*S.sp.* *S.aureus*		TL 1978,1613 Bull.Chem.Soc. Jpn.51,1501(78)
BBM 928A-D	*Actinomyces G455-101* *Actinomadura luzonensis*	DNA-binding	JA 33,1087(80) JACS 103,1241 (80, US Pat.) 4,360,458

QUI*A* c(Qxa/Qaa/Qoa→DSer→Ala→Cys→MeVal-)$_2$ (Ser | ; Cys -S┘)

QUI*C* c(Qxa/Qaa/Qoa→DSer→Ala→Cys→Me4maIle-)$_2$ (Ser | ; Cys -S┘)

QUI*B* c(Qxa/Qaa/Qoa→DSer→Ala→Cys→MeaIle-)$_2$ (Ser | ; Cys -S┘)

TRI c(Qxa/Qaa/Qoa→DSer→Ala→Cys→MeVal)$_2$ (Ser | ; Cys -S┘) and MeS—Cys

BBM c(Qoh→DSer→Thpc→Gly→Me2hVal-)$_2$

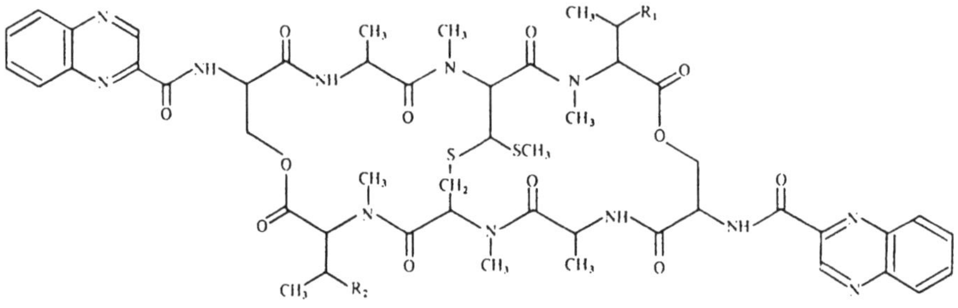

QUI (R: A-CH_3, B-C_2H_5, C-$CH(CH_3)$)

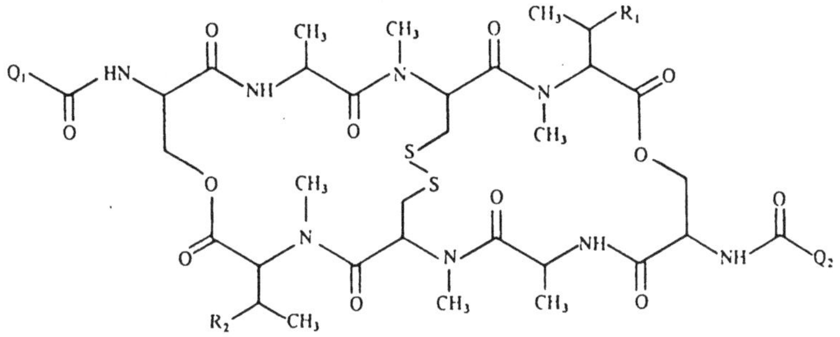

TRI (R : A,B,C wie QUI)

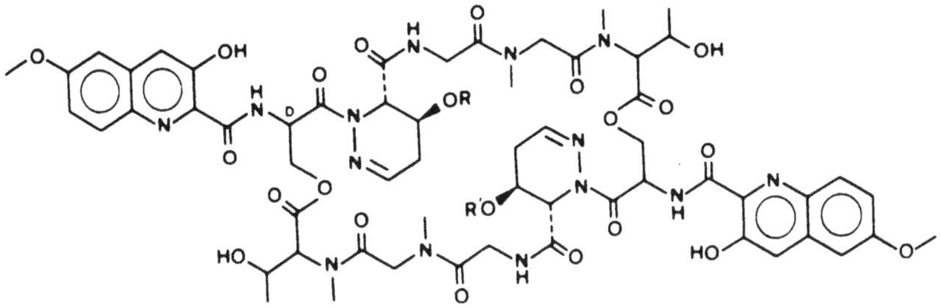

BBM

D-8

Bassianolide	*Beauveria bassiana, Verticillium lecanii*	insecticidal	TL 1977,2167, 4049, ABC 42, 629(78)

BAS c(MeLeu-DHiv→)$_4$

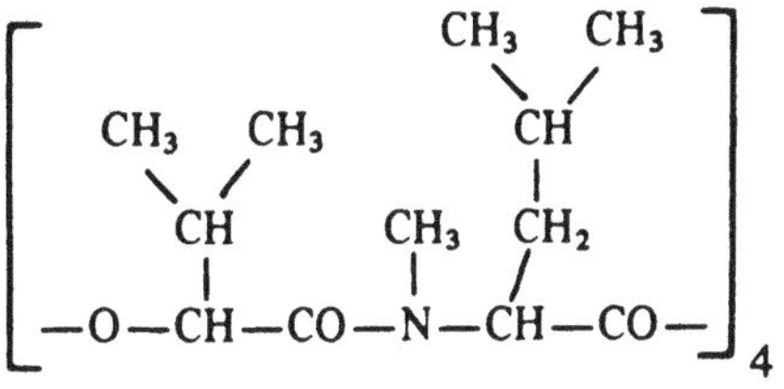

BAS

R-9-P-L-8, R-10-P-L-8

Grisellimycin	*Streptomyces griseus, S. coelicus*	G+,My	Bull.Chem Soc. Fr. 1971,2363 Ger.Pat. 1,805,280 1,180,893
Mycoplanecin	*Actinoplanes sp.No 41042*	My	JA 36,957,961 (83),Jpn.Kokai 81-140,957 81-142,250 82-176,942
Plipastatin	*Bacillus cereus*		JA 39,755,786, 860(86)
Sten**o**thricin			Ann. 1984,1672

GRI 2mBu→MeVal→3EtPro→MeThr→Ile→4mPro→Ile→MeVal→Pro→Ahx-Gly-

GRI*A* AcMeVal→4mPro→MeThr→Leu→4mPro→Leu→MeVal→4mPro→DMeLeu→Gly-

GRI*B* AcMeVal→4mPro→MeThr→Leu→4mPro→Leu→MeVal→Pro→DMeLeu→Gly-

GRI*C* AcMeVal→Pro→MeThr→Leu→4mPro→Leu→MeVal→Pro→DMeLeu→Gly-

GRI*D* AcMeVal→Pro→MeThr→Leu→Pro→Leu→MeVal→Pro→DMeLeu→Gly-

MYC 2oBu→MeVal→4EtPro→MeThr→Val→4mPro→Ile→Val→Pro→MeIle→Gly-

MYC*D* 2oBu→MeVal→4PrPro→MeThr→Leu→4mPro→Hle→MeVal→Pro→MeLeu→Ser-

PLI*1* $2hC_{16}$/$a2hC_{17}$→Glu→DOrn→Tyr→DaThr→Glu→DAla→Pro→Gln→DTyr→Ile-

PLI*2* $2hC_{16}$/$a2hC_{17}$→Glu→DOrn→Tyr→DaThr→Glu→DVal→Pro→Gln→DTyr→Ile-

STO $2oC_n$→DCya→DaThr→Ser→Dpr→dAbu→Val→Ser→Lys→Sar-

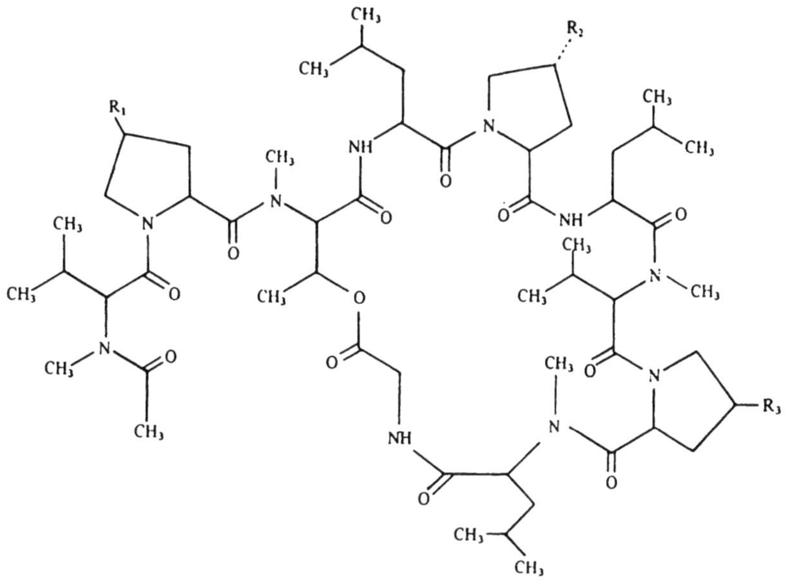
R1
R2
R3
CH3
NH
N
O
CH3
NH

GRI

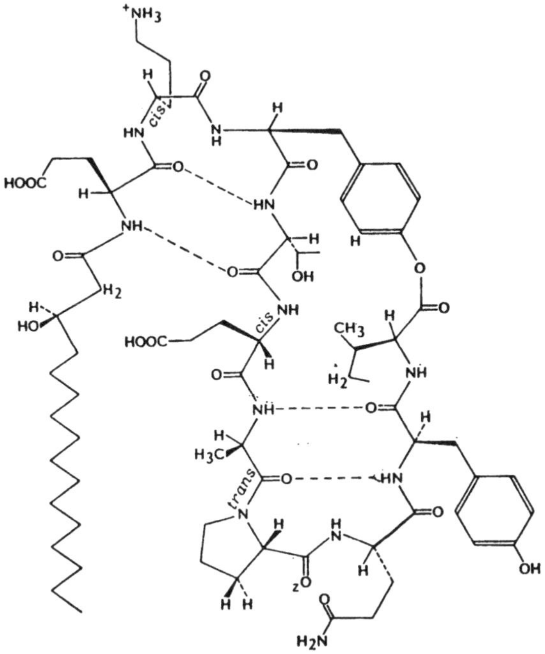
+NH3
cis
cis
trans
HOOC
HOOC
HN
NH
OH
H2N
CH3
H3C

PLI

Type 9-cycles
subtype cyclononalactone L-9
branched lactone R-n-P-L-9

L-9

Empedopeptin	*Empedobacter halobium*		JA 37,949,958 (84),US Pat. 4,409,210, Jpn.Kokai 83-140,053
Majusculamide	*Lyngbya majuscula*	AF	JOC 49,236(84)

EMP 2hC16→DPro→DSer→Pro→Arg→DhAsp→DSer→3hPro→hAsp-

MAJ Hib→Gly→MeIle→Gly→MeVal→MeOmTyr→Aod→Ala→Amp-

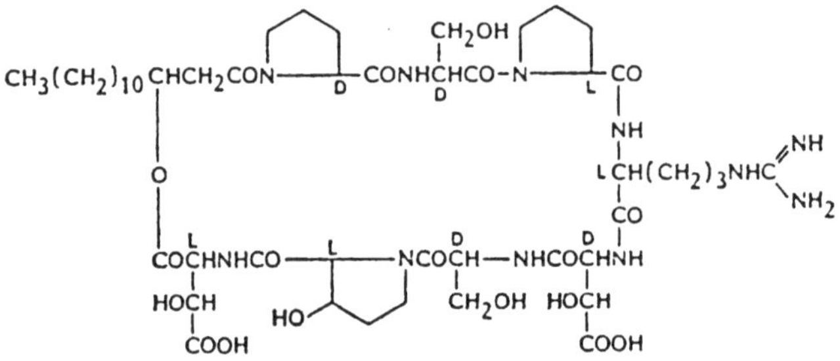

EMP

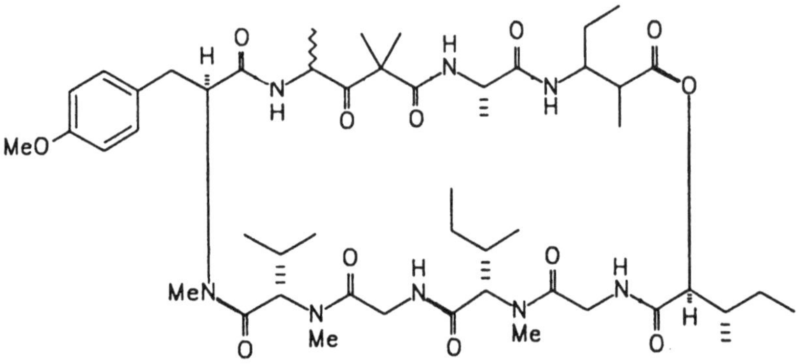

MAJ

10-P-L-9, 11-P-L-9

LL-AO-341	*Streptomyces canidus*	G+	AAC 1966,587, 591, US Pat. 3,377,244
Herbicolin	*Erwinia herbicola*	AF	Ann. 1985,2285
Telomycin **Ne**o**t**elomycin	*Streptomyces canus, S.sp.*	G+	B.12,3811(73) Bioorg.Khim. 3,422(77)
Lysobactin	*Lysobacter sp.*	G+	Jpn.Kokai 86-227,788
Katanosin	*Cytophaga*	G+	JA 41,713(88)

LLA DSer→Thr→Ala→Gly→3hPro→Trp→MeTrp→3hLeu→Pro-

LLA2 DSer→Thr→Ala→Gly→Pro→Trp→MeTrp→3hLeu→Pro-

HER $3hC_{14}$→dAbu→Thr→DaThr→DLeu→Gly→DGln→Gly→MeaThr→Arg-

TEL Asp→DSer→Thr→erythroDThr→Ala→Gly→3hPro→3hLeu→Trp→Trp→3hPro-

NET Asp→DSer→Thr→erythroDThr→Ala→Gly→Pro→3hLeu→Trp→Trp→3hPro-

LYS DLeu→Leu→PhSer→3hLeu→Leu→Arg→Val→aThr→Gly→3hAsn→Ser-

KAT*A* DLeu→Leu→PhSer→3hLeu→Leu→DArg→Val→aThr→Gly→3hAsn→Ser-

KAT*B* DLeu→Leu→PhSer→3hLeu→Leu→DArg→Ile→aThr→Gly→3hAsn→Ser-

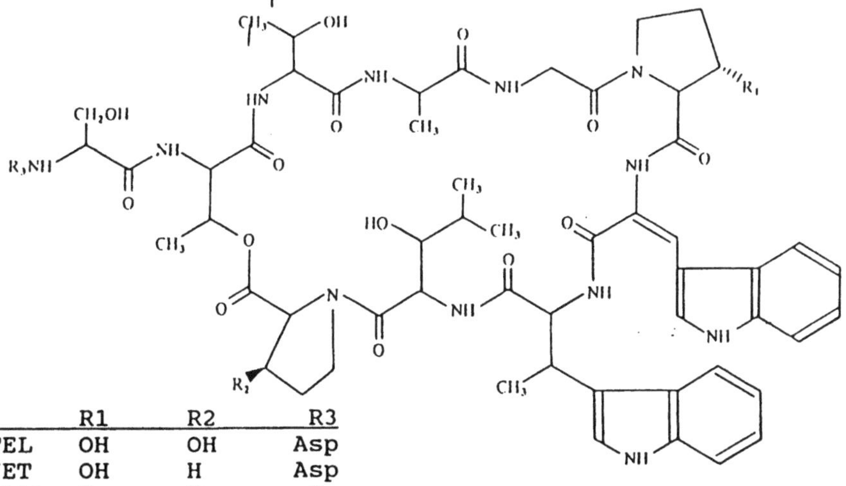

	R1	R2	R3
TEL	OH	OH	Asp
NET	OH	H	Asp
LLAA	H	H	H

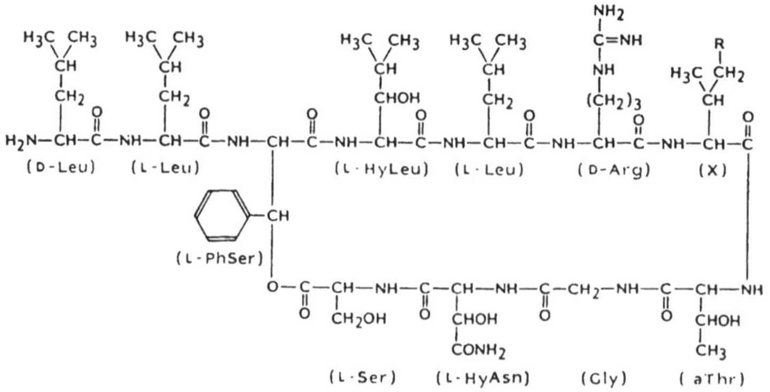

A (1) R = H (X = L-Val)
B (2) R = CH_3 (X = L-Ile)

KAT

Type 10-cycles
subtype cyclodecapeptide C-10, C-10(β)
symmetrical cyclodecapeptide c(P-5)$_2$
cyclodecapeptide lactone L-10
branched cyclodecapeptide lactone R-n-P-L-10

C-10, C-10(β), c(P-5)$_2$

Tyrocidine	*Bacillus brevis ATCC 8185, 10068*	G+, DNA-binding, transcriptional inh.	Exp.26,476,587 (70), this vol., p.6,16
Gramicidin S	*Bacillus brevis ATCC 9999, B. brevis Nagano*	G+, antinematodal, ATP-complexer	Topics in Enzyme Fermentation and Biotechnology (ed. A. Wiseman) p. 187(81), this vol., pp.33-55 JA 37,689(84)
Cortinarins	*Cortinarius speciosissimus C. orellanus C. orellanoides*	vasoconstrictive nephrotoxic toxin	Exp.40,441 (84)

TYA cDPhe→Pro→Phe→DPhe→Asn→Gln→Tyr→Val→Orn→Leu→
TYB cDPhe→Pro→Trp→DPhe→Asn→Gln→Tyr→Val→Orn→Leu→
TYC cDPhe→Pro→Trp→DTrp→Asn→Gln→Tyr→Val→Orn→Leu→
TYD cDPhe→Pro→Trp→DTrp→Asn→Gln→Trp→Val→Orn→Leu→
TYE cDPhe→Pro→Phe→DPhe→Asn→Gln→Phe→Val→Orn→Leu→
GSS1 c(DPhe→Pro→Val→Orn→Leu)$_2$
GSS2 cDPhe→Pro→Val→Orn→Leu→DPhe→Pro→Aba→Orn→Leu→
GSS3 c(Dhe→Pro→Aba→Orn→Leu)$_2$
COR*B* cPhe→hTrp→Val→Orn→Leu→Ile→Cys→DThr→Gly→Lys→
COR*A* cPhe→OmTrp→Val→Orn→Leu→Ile→Cys→DThr→Gly→Lys→
COR*C* cPhe→OmTrp→Val→Orn→Leu→Ile→Ala→DThr→Gly→Lys→

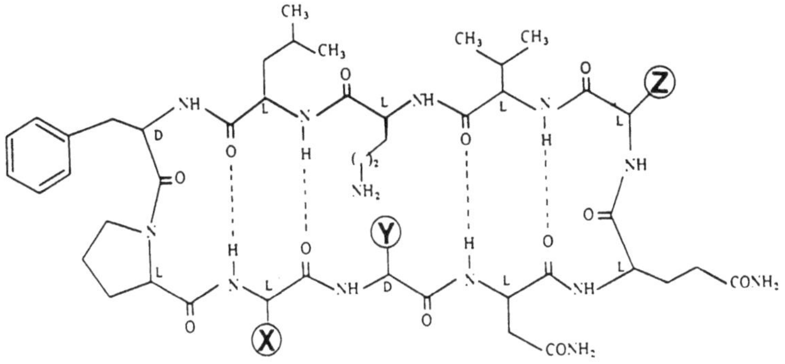

	X	Y	Z
TYA	Phe	Phe	Tyr
TYB	Trp	Phe	Tyr
TYC	Trp	Trp	Tyr
TYD	Trp	Trp	Trp
TYE	Phe	Phe	Phe

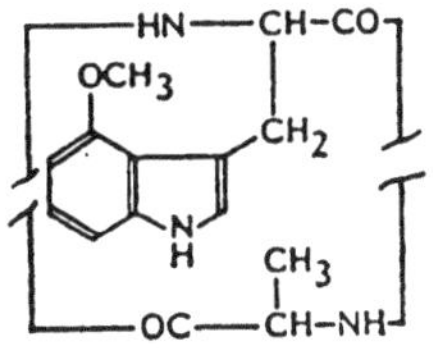

COR*C*

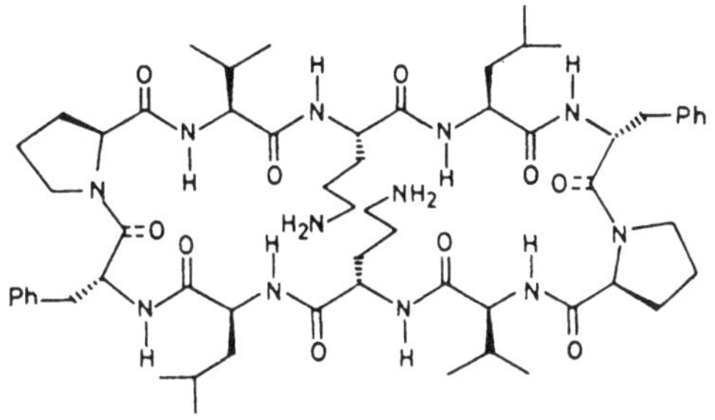

GS

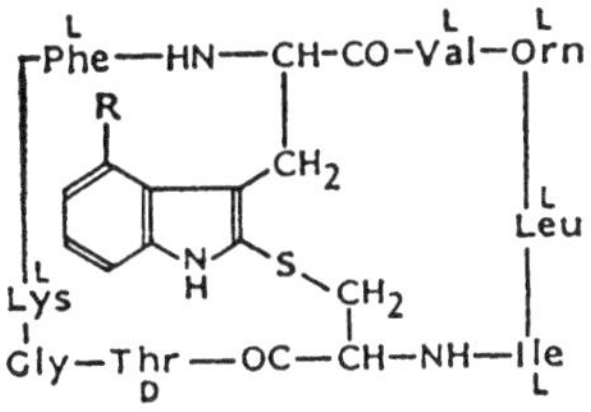

COR (R : A-OMe, B-OH)

L-10(β)

Polypeptin Permetin BMY-28,160	***Bacillus circulans, B. krzemienniewskii***	G+,G-,anaerob. AF	J.Med.Chem. 19,1228(76) JA 37,1257 (84),38,1610 (85)
Puwainaphycin	***Anabaena sp.***	cardiotonic	J.M.Gregson, M.sc.Thesis Univ.Hawaii (86)

PERA ahC_7→Dab→Ile→Dab→DPhe→Leu→Dab→DVal→Leu→Ser-

POYA ahC_7→Dab→Ile→Dab→DPhe→Leu→Dab→DVal→Leu→Thr-

BMY ahC_7→Dab→Val→Dab→DPhe→Leu→Dab→DVal→Leu→Ser-

PUW $2h3a4m14oC_{18}$→Val→dAbu→Thr→Val→Gln→Gly→OmThr→MeAsn→Pro-

PUW*2* $2h3a4m14ClC_{16}$→Val→dAbu→Thr→Ser→Gln→Gly→OmThr→MeAsn→Pro-

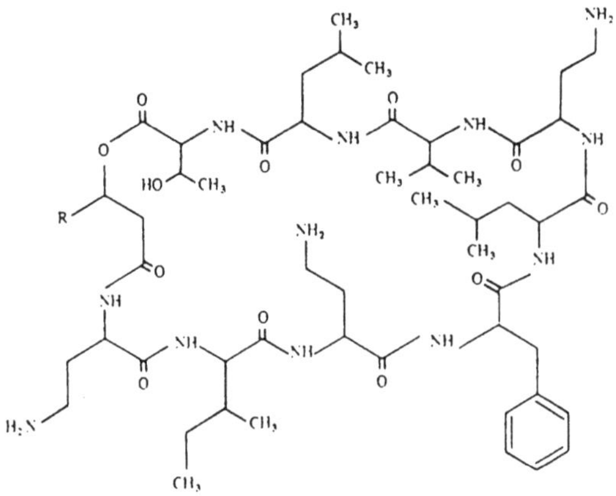

POY

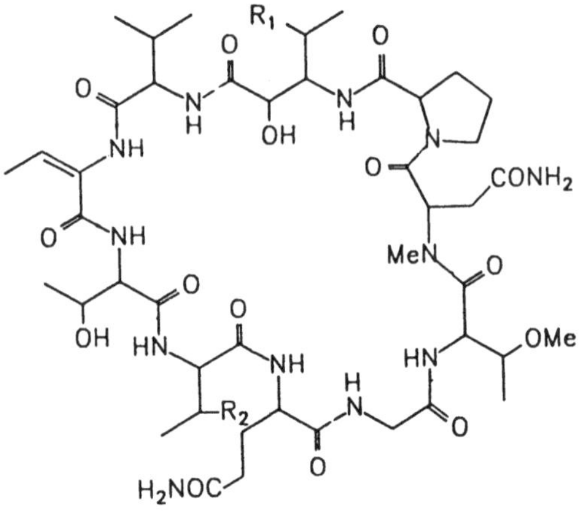

	R1	R2
PUW*A*	$10oC_{14}$	OH
B	$10oC_{14}$	Me
C	$10ClC_{12}$	OH
D	$10ClC_{12}$	Me

R-13-P-L-10

A21978C	*Streptomyces roseosporus NRRL 11379*	G+	JA 40,761 (87)

A21 $aC_{11/13}/iC_{12}$→Trp→Asn→Asp→Thr→Gly→Orn→Asp→DAla→Asp→[10]Gly→[11]DSer→threo3mGlu→Kyn-

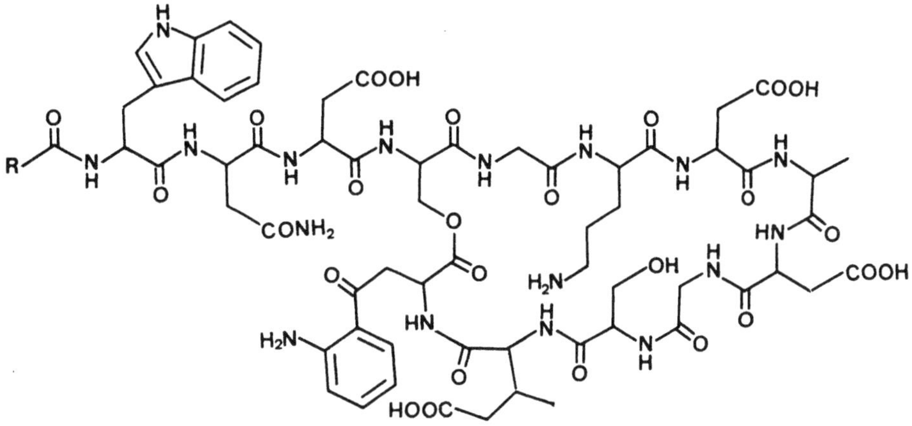

A21

Type higher cycles
subtype undeca-cyclopeptide C-11
acyl-cyclo-undecalactone R-L-11
branched 11-lactone R-13-P-L-11
symmetrical dodecapeptide c(P-6)$_2$
branched dodecapeptide 13-P-C-12-M
dodeca-depsipeptide c(D-4)$_3$
Tridecapeptide C-13
branched 16-lactone R-17-P-L-16

C-11

Epidermidin	*Staphylococcus epidermidis*	G+,G-	Can.J.Microbiol. 18,121(72)
Cyclosporin (CIC)	*Beauveria nivea Trichoderma polysporum Cylindrocarpon lucidum*	AF,immuno-suppressor	this volume pp. ,HCA 60,1568(77), 65,1655(82)

EPD cLys→Ala→Asp→Glu→Ser→Leu→Thr→Gly→Val→Gly→Arg→
CIC*A* cDAla→MeLeu→MeLeu→MeVal→MeBmt→Abu→Sar→MeLeu→Val→MeLeu→Ala→
CIC*B* cDAla→MeLeu→MeLeu→MeVal→MeBmt→Ala→Sar→MeLeu→Val→MeLeu→Ala→
CIC*C* cDAla→MeLeu→MeLeu→MeVal→MeBmt→Thr→Sar→MeLeu→Val→MeLeu→Ala→
CIC*D* cDala→MeLeu→MeLeu→MeVal→MeBmt→Val→Sar→MeLeu→Val→MeLeu→Ala→
CIC*F*
cDAla→MeLeu→MeLeu→MeVal→Me3'dBmt→Abu→Sar→MeLeu→Val→MeLeu→Ala→
CICG cDAla→MeLeu→MeLeu→MeVal→MeBmt→Avl→Sar→MeLeu→Val→MeLeu→Ala→
CIC*K*
cDAla→MeLeu→MeLeu→MeVal→Me3'dBmt→Val→Sar→MeLeu→Val→MeLeu→Ala→
CIC*O* cDAla→MeLeu→MeLeu→MeVal→MeLeu→Abu→Sar→MeLeu→Val→MeLeu→Ala→

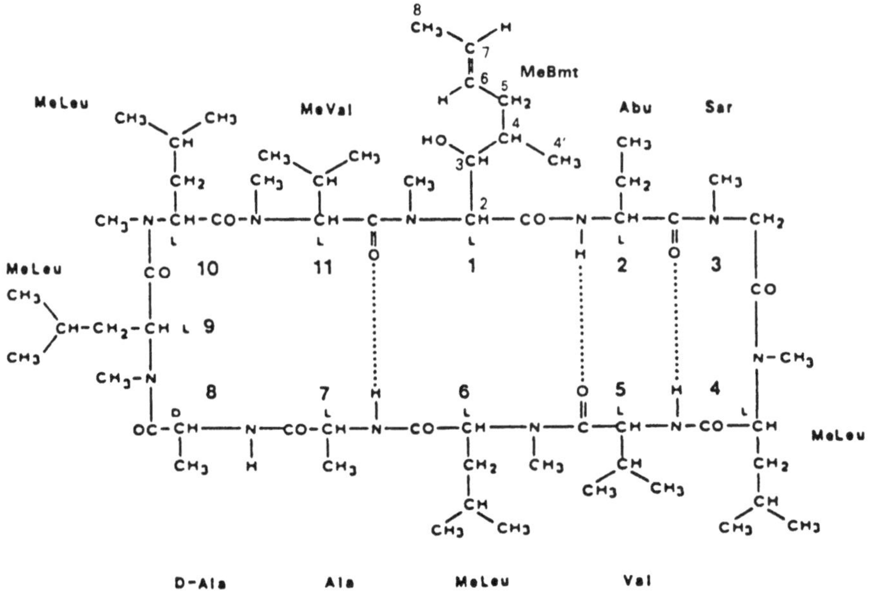

R-L-11

Brevistin	***Bacillus brevis***	G+	JA 29,375,380, (76),Ger.Pat. 2,526,250 Neth.Pat.75-7346

BRE*1* aC$_9$→Thr→DDbu→Glu→Gly→DGln→DGlu→Gly→Trp→Val→Dbu→Phe-

BRE*2* aC$_9$→Thr→DDbu→Glu→Gly→DGln→DGlu→Gly→Trp→Val→Dbu→Phe-

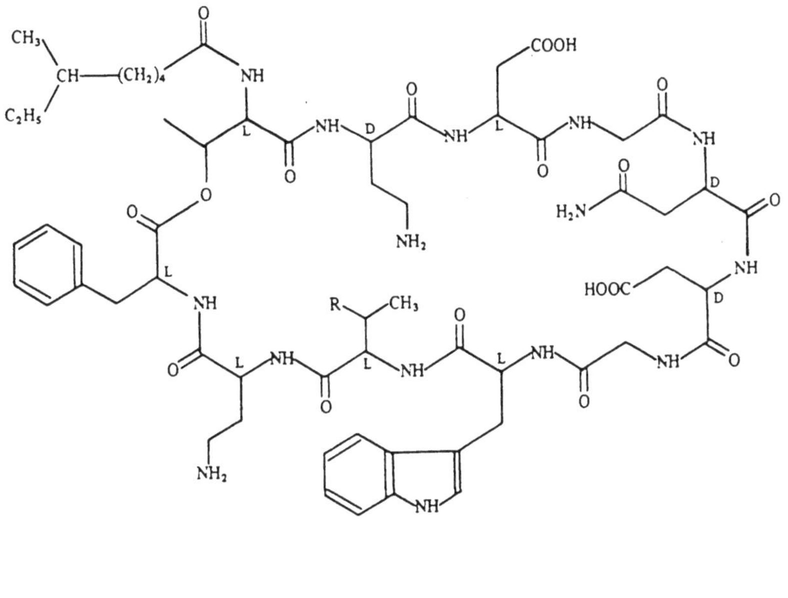

R-13-P-L-11

Theonella- peptolid	***Theonella sp.***	Chem.Pharm.Bull. 35,2129(87), 34,2264(86)

THE

OmAc→[1]Val→DMeLeu→Thr→βAla→DLeu→MeIle→βAla→DaIle→MeVal→[10]MeAla→

[11]βAla→DLeu→DMeaIle-

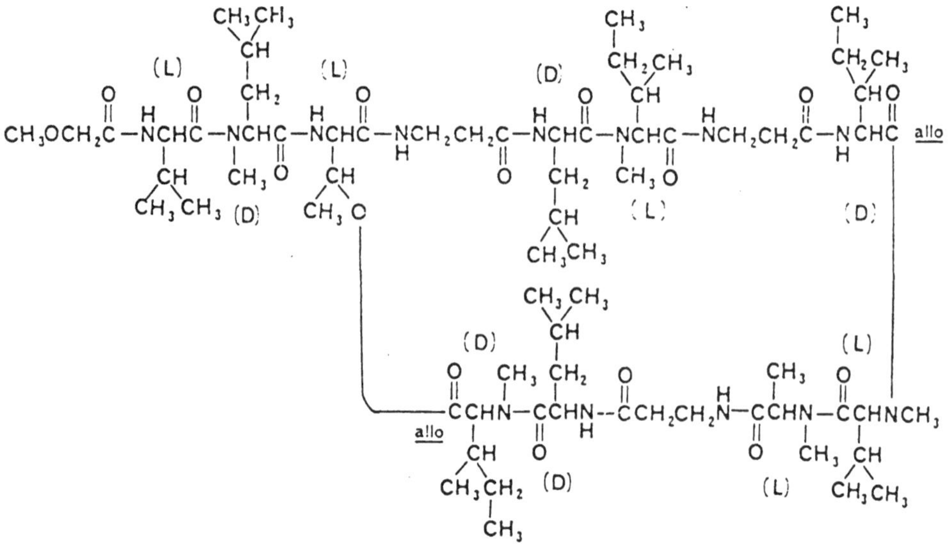

THE

C-12, c(P-6)$_2$

Gratisin	*Bacillus brevis*	G+	JA 34,1227(81) 36,751(83)

GRA c(*Phe→Pro→*Tyr→Val→Orn→Leu)$_2$

D-12, c(D-4)$_3$

Valinomycin	*Streptomyces fulvissimus* *S. tsusimaensis* *S. sp.*	G+,My,AF insecticidal K^+,Rb^+-complexer	JACS 97,7242 (75), US Pat. 3,520,973

VAL c(DHiv→DVal-Lac→Val-)$_3$

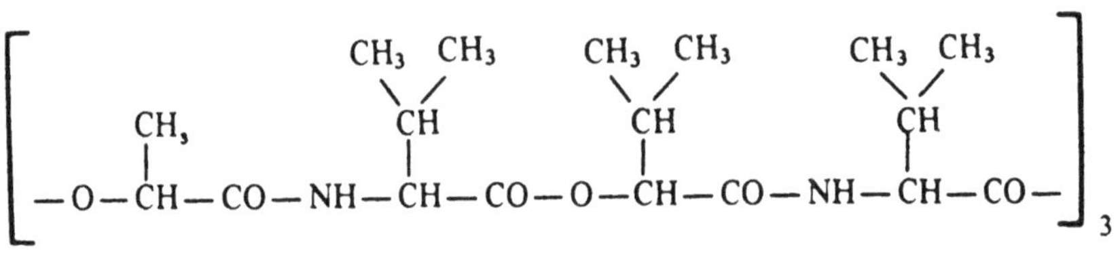

VAL

C-13-(*τ*)

Mycobacillin	*Bacillus subtilis V3*	AF	JA 31,147(78)

MYS cPro→Dap→DGlu-*τ*→Tyr→Asp→Tyr→Ser→Dap→DGlu-*τ*→Dap→Ala→Dap→

R-P-17-L-16

Enduracidine	*Streptomyces fungicidus*	G+,My, murein formation inh. feed add.	JA21,126,138 (68),Chem. Pharm.Bull. 21,1184(73) US Pat. 3,786,142 Jpn.Pat. 70-17,158, Ger.Pat. 1,809,875

END i2,3d4,5dC_{12}/a2,3d4,5dC_{13}→Asp→Thr→D4hPhg→DOrn→DaThr→^{6}4hPhg→

^{7}aThr→DCit→D*allo*End→4hPhg→DSer→3*Cl*4hPhg→Gly→End→DAla→4hPhg-

END*2* i2,3d4,5dC_{12}/a2,3d4,5dC_{13}→Asp→Thr→D4hPhg→DOrn→DaThr→^{6}4hPhg→

^{7}aThr→DCit→D*allo*End→4hPhg→DSer→3,5*Cl*4hPhg→Gly→End→DAla→4hPhg-

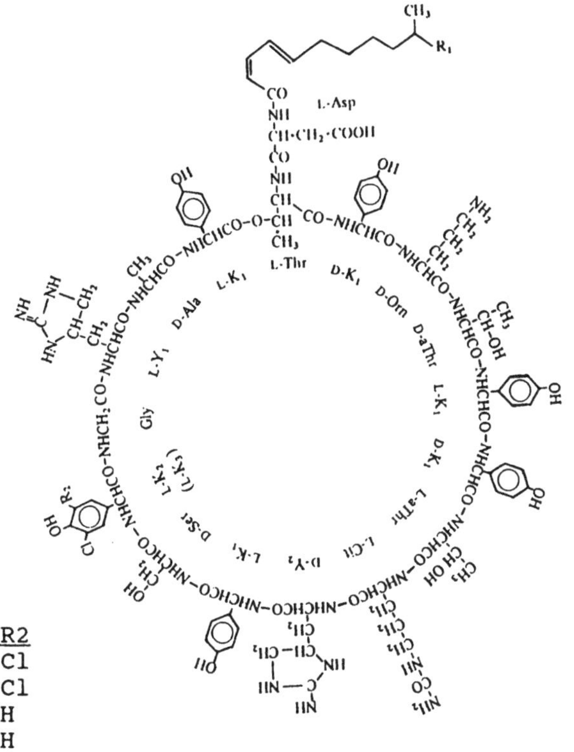

	R1	R2
END*A*	Me	Cl
B	Et	Cl
SA	Me	H
SB	Et	H

type cyclopeptides closed with piperidine-type rings, modified
subtype R-P-13-C-10(PIP)-M
R-P-17-C-10(PIP)-M
P-14-C-13(PIP)-M

R-P-13-C-10(PIP)-M

Nosiheptid Multhiomycin	*Streptomyces antibioticus*	G+	JACS 99,6418 (77),JA 31, 623(78)
Micrococcin	*Micrococcus sp.*	G+	JCS CC 1977 256
Thiocillin	*Bacillus cereus* *B.megaterium* *B.badius*	G+	JA 29,366(76)

NOS [X]→Cys→Thr→dAbu→Cys→3hGlu→Cys→DCys→Cys→[Y]→Cys→*d*Ala($CONH_2$)

MIC [X]→Cys→Thr→dAbu→Cys→Val→Cys→Thr →Cys→[Y]→Cys→Cys→

→dAbu→3h/3oPrA

THI [X]→Cys→Thr→dAbu→Cys→3hVal→Cys →Thr →Cys→[Y]→Cys→Cys→dAbu→EA

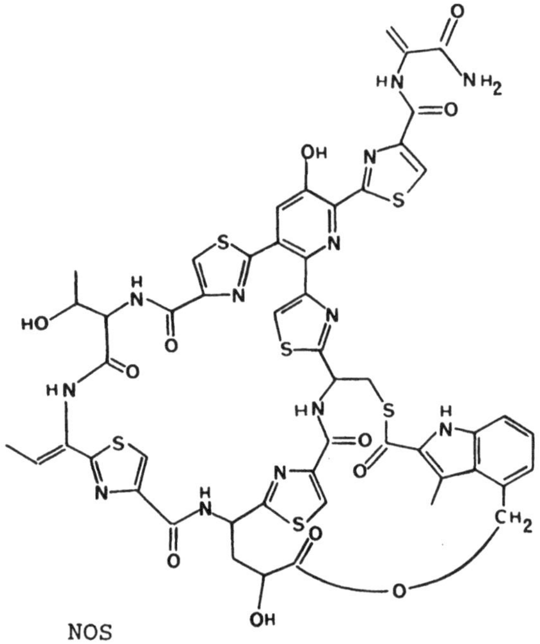

NOS

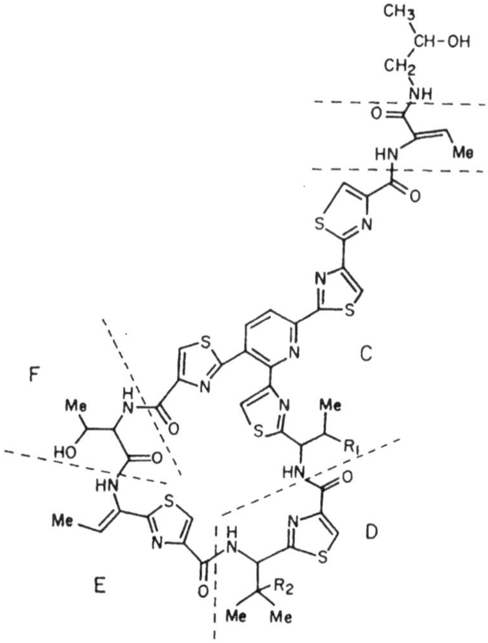

THI/MIC (R_2 OH/H)

R-P-17-C-10(PIP)-M

Thiostrepton Bryamycin Thiactin	*Streptomyces azureus, S. laurentii,S. hawaiiensis*	G+	Nature 225,233 (70),TL 1976, 185
Siomycin Sch 18640 Sporangiomycin Mutabillicin A-59-A Thiopeptin	*Streptomyces sioyaensis* *Planomonospora parontospora* *S.mutabilis* *S.tateyaensis*	G+	JA 33,1563(80) JACS 103,523(81) JA 36,814(83)

THO [Q]-Ile→Ala→dAla→**Ala**→ [X]→Cys→Thr→dAbu→Cys→3,4hIle→ ^{11}Cys-s→ *

→ ^{11}Thr→Cys→[X]→Cys→dAla→ ^{17}dAla(NH_2) *

SIO [Q]-Val→dAla→dAla→Ala→[X]→Cys→Thr→dAbu→Cys→ ^{10}3,4hIle→Cys-s→ *

→ ^{12}Thr→Cys→[X]→Cys→dAla→ ^{17}dAla(NH_2) *

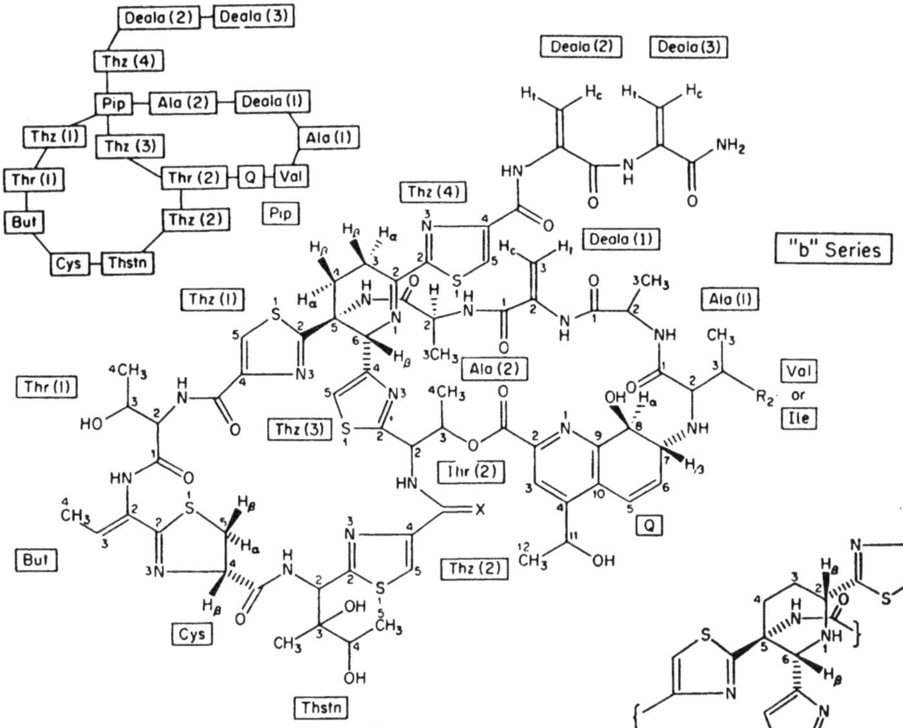

THO

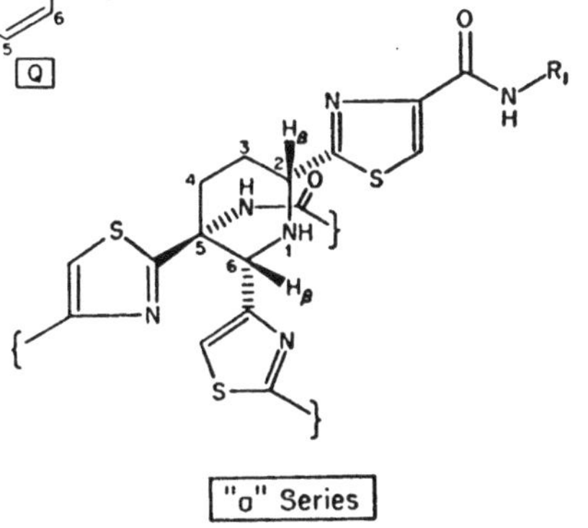

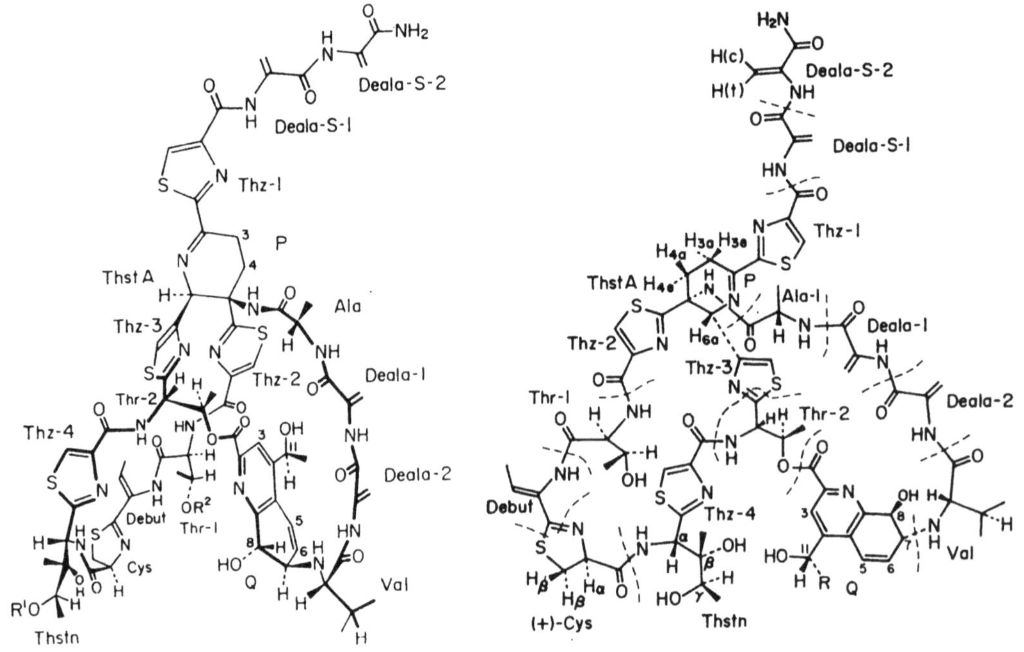

SIO

P-14-C-13(PIP)-M

Berninamycin	*Streptomyces berniensis* *S.verticillus*	G+	TL 1978,2791; 25,2127(84) J.Gen.Microb. 128,875(78) US Pat. 4,001,090

BER

[X]→*d*Ala→Thr→*d*Abu→Thr→*d*Ala→*d*Ala→3hVal→*d*Ala→Thr→*d*Ala→Ser→[Y]→Cys

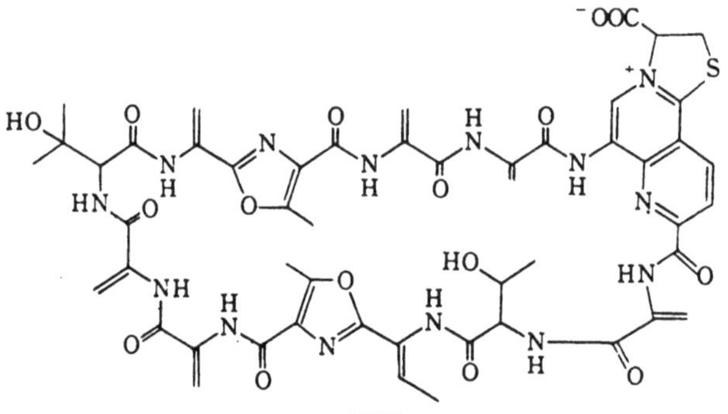

BER

Type **dipeptide P-2**
subtype **dipeptide with bond other than α/α P-2(β,τ)**
modified dipeptide P-2-M
acyldipeptide R-P-2
nucleoside dipeptide N-P-2
glycodipeptide G-P-2

P-2, P-2-M

Bacilysin Tetain Bacillin	*Bacillus subtilis* *Bacillus pumilus*	antibacterial	JA 28,77(75) 34,1608(81)
Alahopcin Nourseimycin	*Streptomyces albulus, S. noursei 804*	G+,G-,AF,Proline antagonist useful against resistant S. aureus	JA 38,302 (85),ABC 47,1599 (83),Jpn. Kokai 84-66,889
LL-BM726	*Streptomycete BM726*	G-	JA 38,541 (85)
Arginyl-glutamine	*Cladophora sp.* *Enteromorpha linza* *Ulva lactuta*		JB 46,63,59 C.A.86,167528 (77)
Citrullyl-arginine	*Grateloupia turutura*		C.A.81,14859u (75)
Citrullyl-ornithine	*G.livida*		49th Jpn.Chem Soc.Conf.(84)
6-**APA**	*Penicillium chrysogenum* *various fungi*		J.Ferment. Technol.53 339(75)
N-(2,6-**Diamino**-6-hydroxymethyl-pimelyl)-**alanine**	*Micromonospora chalcea*		JA 34,374 (81)
Stravidine	*Streptomyces venezuelae*	G-,biotin-antimetabolite	Ann.1985,877
Arphamenine	*Chromobacterium violaceum*	inh. aminopeptidase B	JA 36,1572 (83),37,518 (84)
Bestatin	*Streptomyces olivoreticuli*	inh.proteinase immunomodulator	JA 29,100, 600(76),36, 695(83), Ger.Pat. 2,628,354 Jpn.Pat. 77-116,435
Nitropeptin	*Streptomyces xanthochromogenes*	rice blast(Pyricularia oryzae) Gln-antimetabolite	JA 41,709 (88)
Olivoretin Lyngbyatoxin Teleocidin	*Streptoverticillium olivoreticuli* *Lyngbya majuscula* *Streptomyces*	skin irritant tumor promoter	Chem.Pharm. Bull.32,354, 358,3774(84)

BCI	Ala→Ant
AHO	Ala→4f3(CONHOH)Bu
BAH	βAla→His
LLB	Arg→Cpg
ARG	Arg→Gln
CIA	Cit→Arg
CIO	Cit→Orn
	3 ┌─┐ N
APA	Cys→Val
	S └──┘ 3
DPA	6hmDpm→Ala
STR*1*	Ile→Acm
STR*2*	MeIle→Acm
ARP	[Phe,OAc]→[Arg]
BES	[Phe,OAc]→Leu
STR*3*	MeVal→Acm
NIT	Val→3NO_2Glu
OLI	MeVal→[Trp]OMe

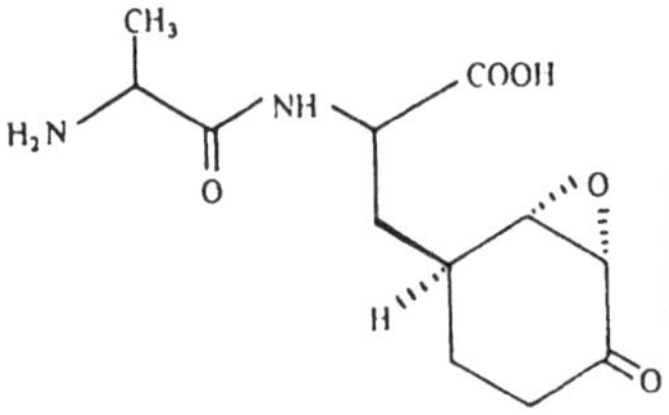

BCI

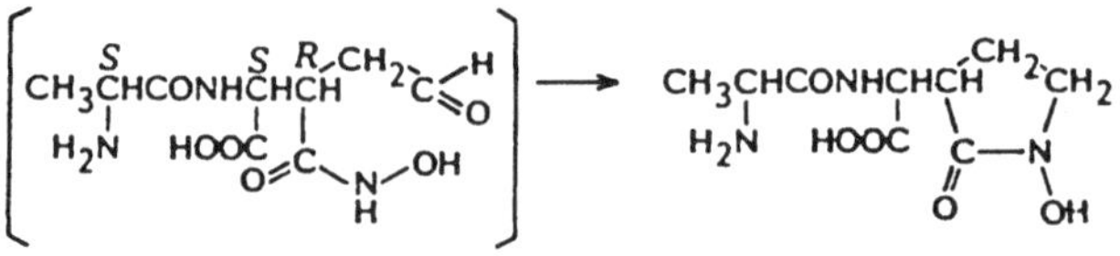

AHO

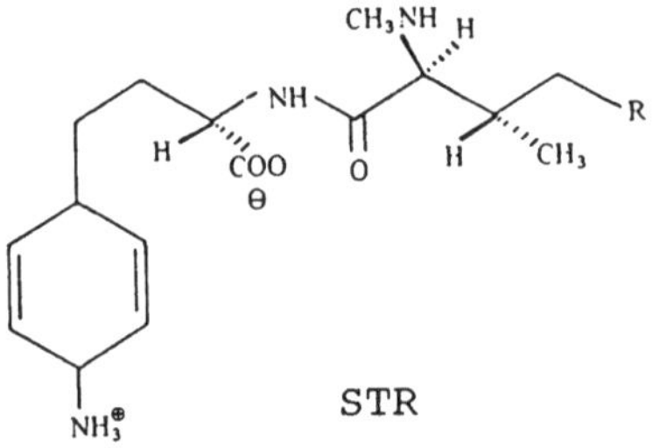

STR

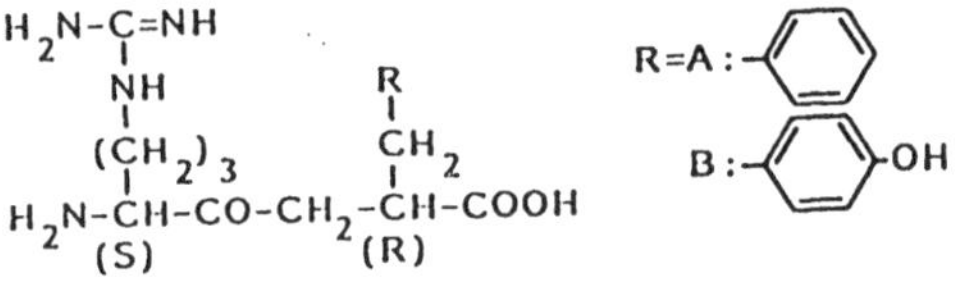

ARP

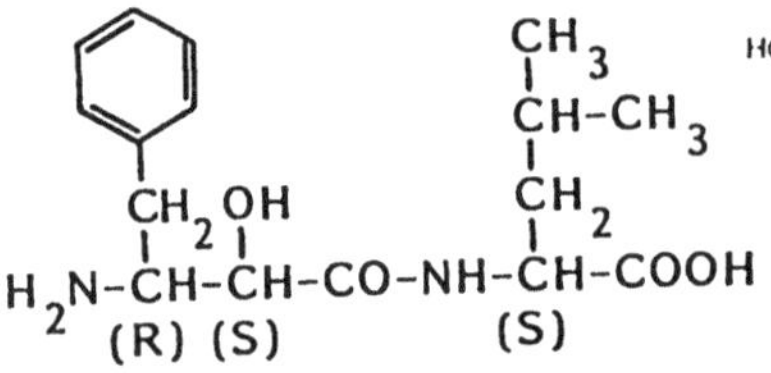

BES

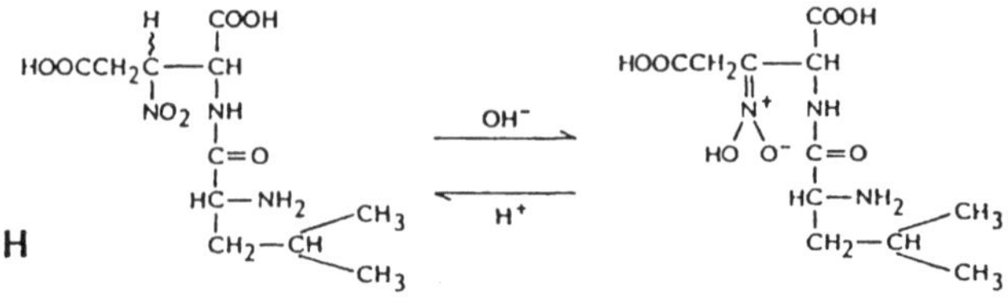

NIT

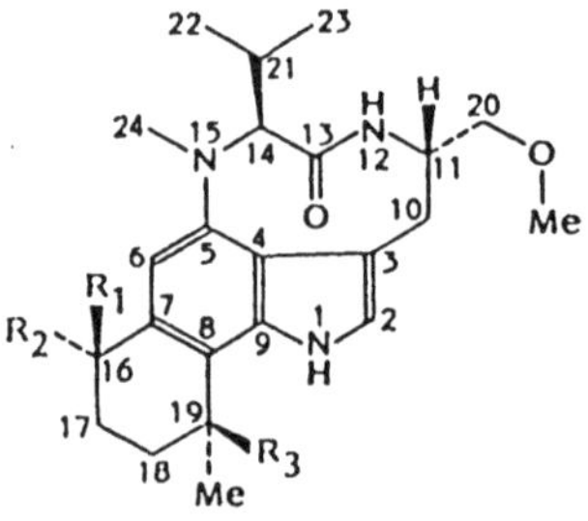

OLI

	R1	R2	R3
A	Me	iPr	vin
B	iPr	Me	vin
C	Me	vin	iPr

P-2-(β)

Carnosin β-Alanyl-histidine	*vertebrates* *Acanthophora dendroides*	JBC 254,6402 (79)
β-Alanyl-ornithine	*vertebrates*	JBC 254,6402 (79)
β-Aspartyl-glycine	*Ceramium rubrum*	Phytochem.17 1659 (78)
γ-Aminobutyryl-histidine	*vertebrates*	JBC 236,409 (61)
γ-Amniobutyryl-ornithine	*vertebrates*	JBC 254,409 (79)
γ-Glutamyl-2-amino-3(Z),5-hexadienoic acid	*Leptinotarsa decemlineata*	Sci.233,221 (86)
(S,S)-Valyl-amidocyclobutane-1-acetic acid	*Streptomyces sp. X-1092*	JA 27,754 (74)

BAH βAla—β→His
BAO βAla—β→Orn
GAH γAbu—γ→His
GAO γAbu—γ→Orn
GAH Glu—γ→Ahd
VAB Val→Acba

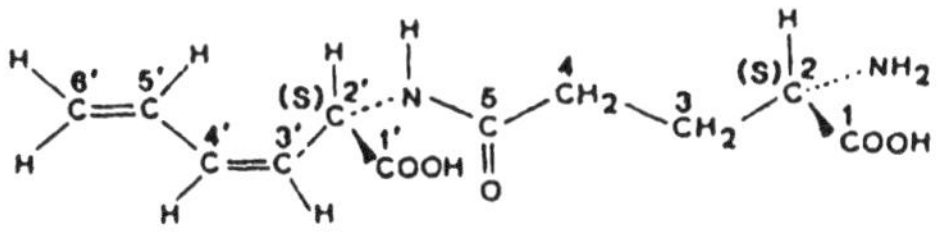

GAD

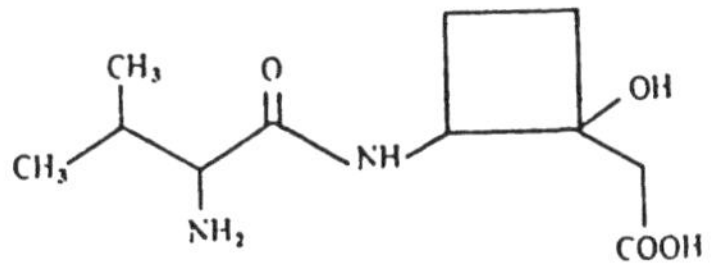

VAB

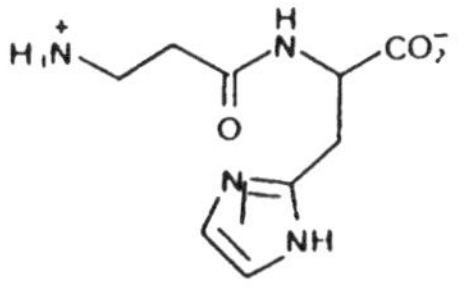

BAH

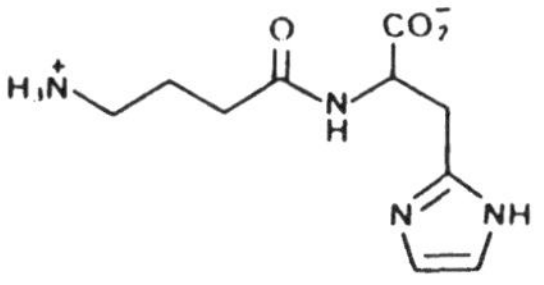

GAH

R-P-2

Bu-2743E	*Bacillus circulans J725-093*		JA 36,1396 (83)
WF-1029	*Doraromyces putredinis*	ACE inh.	JA 41,468 (88)
Propioxatin	*Kitasatospora setae*	inh.enkephalinase	JA 38,1378 (86)
MG245-CF2-A Vanoxinine	*Saccharopolyspora hirsuta*	inh.thymidylate synthetase	Jpn.Pat. 58-134,064 CA 100,4786 JA 36,656 (83),Jpn. Kokai 83-134,064
Majusculamide	*Lyngbya majuscula*	nontoxic	
Indolactam	*Actinomycetes NA 34-17*	induces early antigen EBV	ABC 48,1269 (84),Chem. Pharm.Bull. 30,3457(82)
Penicillin G (V,F,K),	*Penicillium chrysogenum*	G+	J.Ferment. Technol.53 339(75)
KPN(**PEN***)	*Paecilomyces carneus,P.sp,*	G+	J.Ferment. Technol.54 705(76)
RIT-D-2214(**PEN***)	*Acremonium chrysogenum*	G+,G-	JA 29,1258 (76)

BU2 Dhb→Ala→Thr
WF1 $7h4oC_8$→Ala→Tyr
VNO Dhb→Thr→N^6OAchOrn
IND Ac-Val→Trpol (cyclic: Ac–Trpol)

PRP 2pr/iBu(HNOH)suc→Pro→Val
MAU $2m3oC_{10}$→MeDOmTyr→ValNH$_2$

PEN* R→Cys→Val (3 ⌐ N; S ⌐ 3), R: G-Bz, V-PH-O-CH_2CO, F-3,4dC_6, K-C_8, KPN-5hC_5, RIT- Et-S-CH_2CO

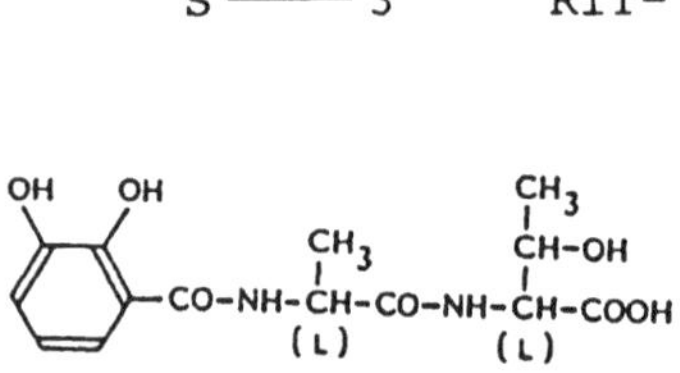

BU2

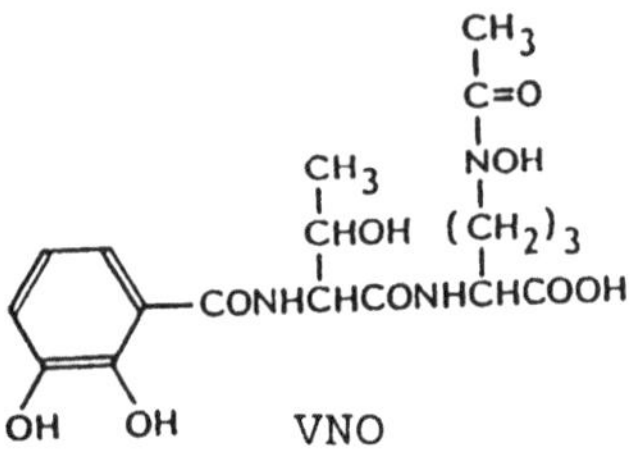

VNO

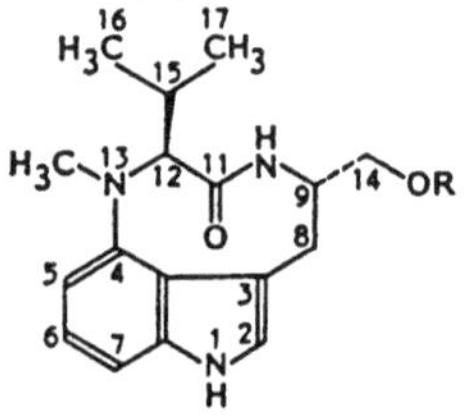

IND

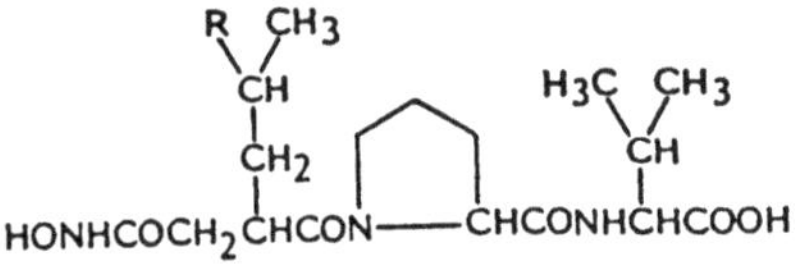

PRP

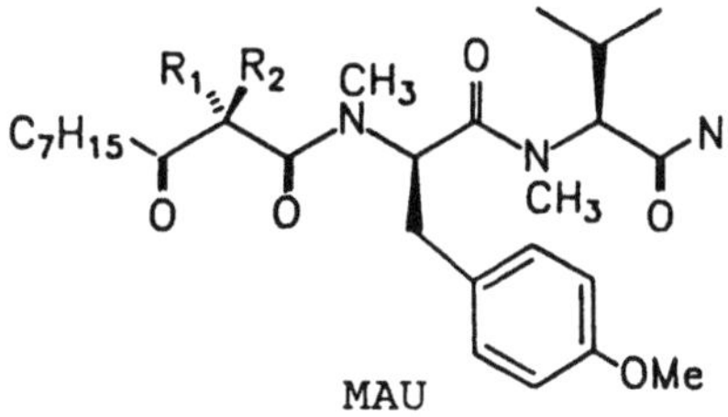

MAU

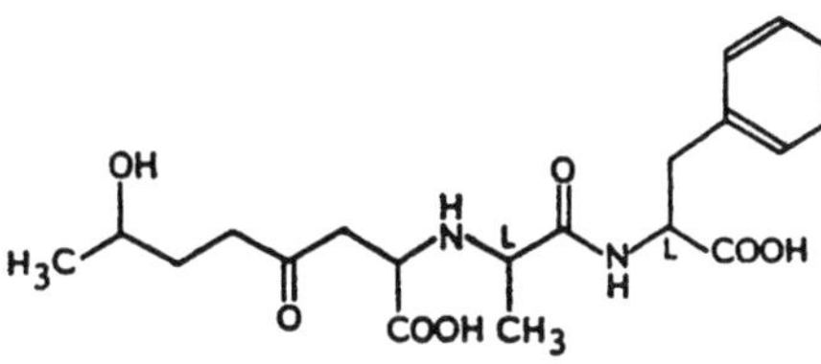

WF-1

N-P-2

Bagougeramine Gougerotin	*Bacillus circulans* *TB-2125*	JA 39,1041 (86)

BAGA MeGly→DGua→N
GOU MeGly→DSer→N

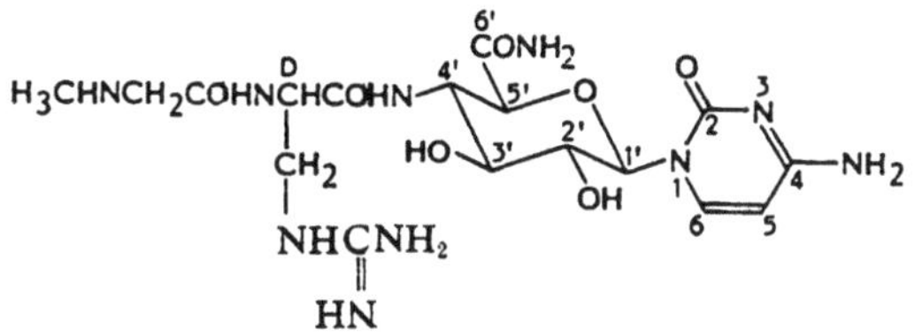

BAGA

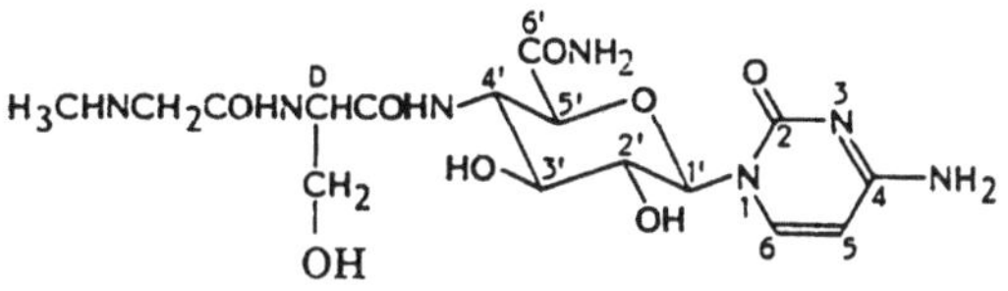

GOU

type linear tripeptides
subtype linear tripeptide P-3
modified tripeptides P-3-M
linear tripeptides with bonds other than α,α P-3-(β,τ)
modified glycotripeptides with β,τ-bonds
acyl-tridepsipeptides R-P-3-(E)

P-3 and P-3-M

Phosphinotricyl-alanylalanine Bialaphos SF-1293	*Streptomyces viridochromogenes* *S.hygroscopicus SF-1293*	herbicide	JA 36,96, 1040(83), Neth.Pat. 72-10,308 JA 38,687 (85)
Phosalacine	*Kitasatospora phosalacinea*	G+,G-, inh. glutaminsynthetase	JA 37,829, 939(84),Jpn. Kokai 85-34,199; Eur. Pat.134,113
Plumbemycin	*Streptomyces plumbeus*	G-	ABC 41,573 (77)
Arginyl-D-allo-threonyl-phenyl-alanine	*Keratinophyton terreum* *Trichophyton indicum*	AF	Ber.106,816 (73)
Aspartyl-N-hydroxy-aspartyl-D-cyclo-serine	*Corynebacterium kutscherii*		J.Bac.137, 243(79)
Fastigiatine	*Pelveta fastigiata*		
Eisenin	*Eisenia bicyclis* *Ecklonia cava*		
Rhizocticin			
Diprotin	*Bacillus cereus*	inh. dipeptidylaminopeptidase	JA 37,422 (84), Jpn. Kokai 84-25,366
K26	*Actinomadura sp.* *Actinomycetes K26*	ACE inh., hypotensive	Jpn.Kokai 82-156,498 83-31,992
K4	*Actinomadura spiculosospora nov.*	ACE inh. hypotensive	JA 37,965 (84)
I5B2	*Actinomadura sp.*	ACE inh.	JA 37,965 (84)
Leucyl-3-epi-deoxy-negamycin	*Streptomyces goshikiensis*	nontoxic	JA 30,1137 (77)
Leucyl-negamycin	*Streptomyces purpeofuscus*	G-	JA 24,732 (71)
L-N^5-Phosphono methionine-S-sulfoxyminyl-alanylalanine	*Streptomyces sp.*	G+,G-	JA 26,261 (73)
OF 4949	*Penicillium rugulosum*	inh. aminopeptidase B	JA 39,1685 1697(86),40 512(87)
Biphenomycin	*Streptomyces griseorubiginosus*	G+	JA 38,1462 (85)
WS-43708A	*Streptomyces griseorubiginosus*	antibacterial	JOC 50,1341 (85)
K-13	*Micromonospora halophytica, subsp. exilisia*		JA 40,455 (87)

PHI	4(mp)Abu→Ala→Ala
PHA	pAbu→Ala→Leu
PLUA	Ala→Asp→5p*cis*3,4*d*DAlv
PLU*B*	Ala→Asn→5p*cis*3,4i*d*DAlv
ATF	Arg→DaThr→Phe
AHC	Asp→NhAsn→DCse
FAS	<Glu→Glu→Gln
EIS	<Glu→Glu→Ala
RHI*C*	Ile→Arg→5pcis3,4dAlv
DIP*A*	Ile→Pro→Ile
K26	Ile→Tyr→Typ
RHI*D*	Leu→Arg→5p*cis*3,4*d*Alv
LEN	Leu-*β*→*β*Lys→MeHaa
LNE	Leu-β→5h*β*Lys→MeHaa
PHM	(Np)Sox,imMet→Ala→Ala
OF*1*	OmTyr-hAsn→Tyr (Tyr–O–Tyr)
OF*2*	Tyr→hAsn→Tyr (Tyr–O–Tyr)
OF*3*	OmTyr→Asn→Tyr (Tyr–O–Tyr)
OF*4*	Tyr→Asn→Tyr (Tyr–O–Tyr)
OF*5*	OmTyr→hAsp→Tyr (Tyr–O–Tyr)
OF*6*	Tyr→hAsp→Tyr (Tyr–O–Tyr)
BIP	2horthoTyr→4hOrn→2hTyr
WS4	orthoTyr→4hOrn→2horthoTyr (Tyr–Tyr)
K13	AcTyr→Tyr→Tyr (Tyr–O–Tyr)
K4	MeVal→Phe→Typ
DIP*B*	Val→Pro→Leu
I5B*1*	MeVal→Phe→Typ
I5B*2*	MeVal→Tyr→Typ

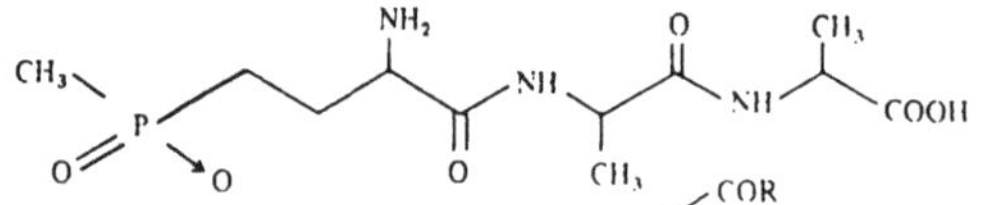

PHI

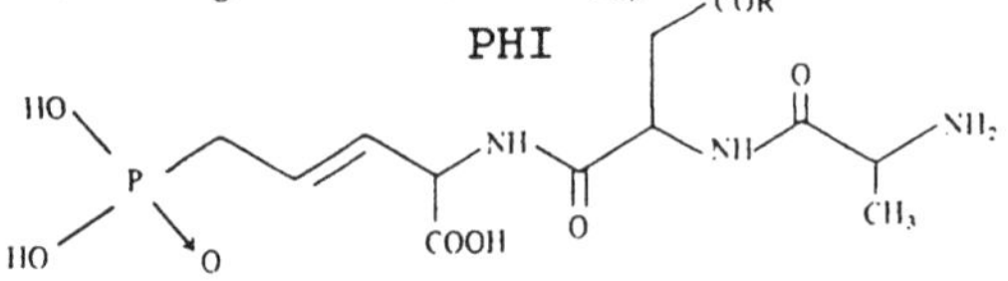

PLU

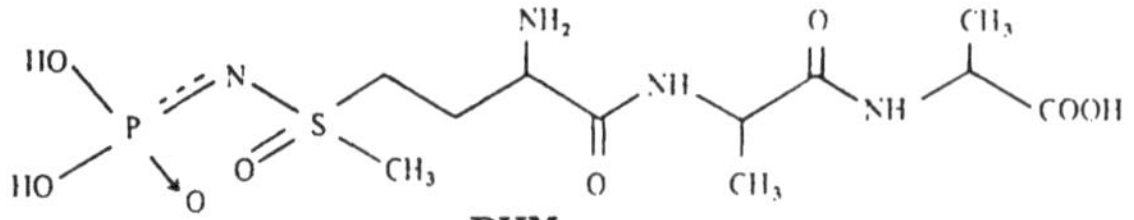

PHM

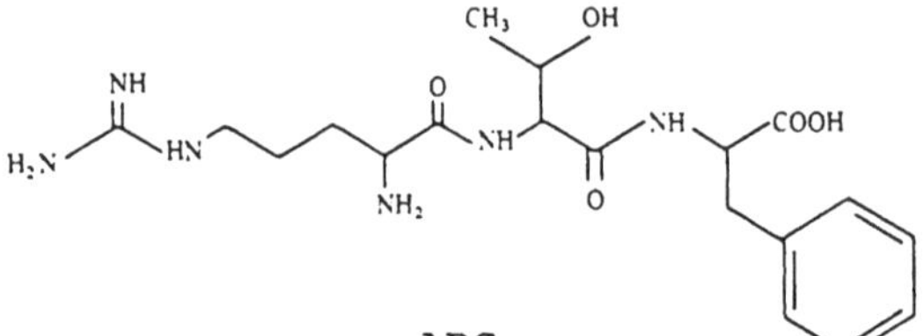

ARG

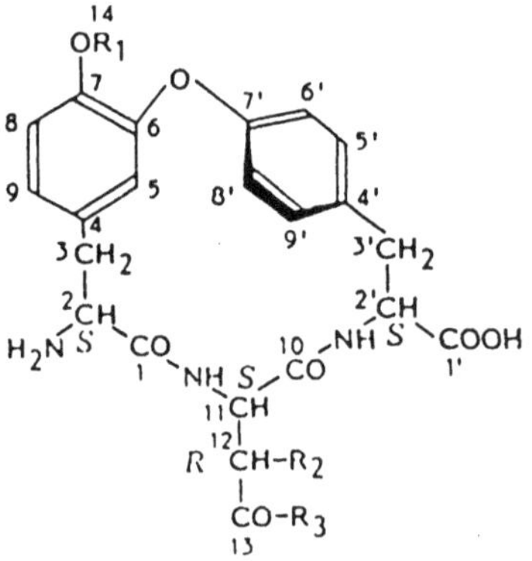

	R1	R2	R3
OFI	CH_3	OH	NH_2
OFII	H	OH	NH_2
OFIII	CH_3	H	NH_2
OFIV	H	H	NH_2
OFD	CH_3	OH	OH
OFF	H	OH	OH

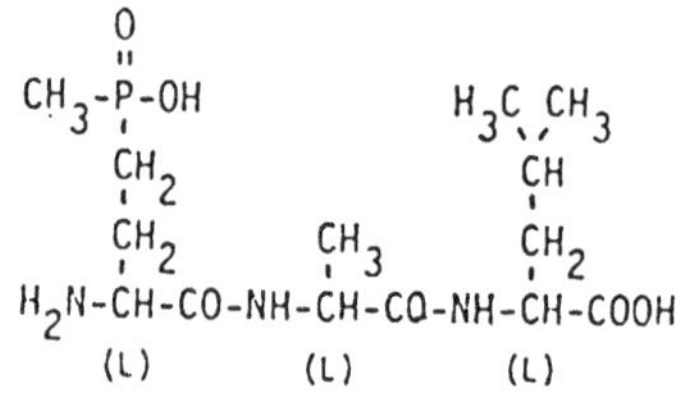

PHA

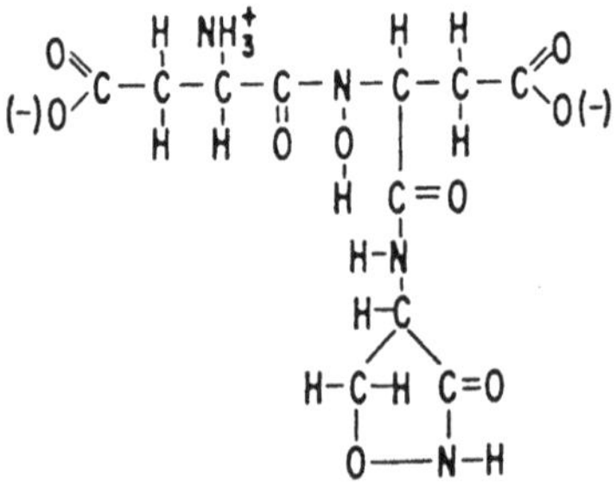

ASP

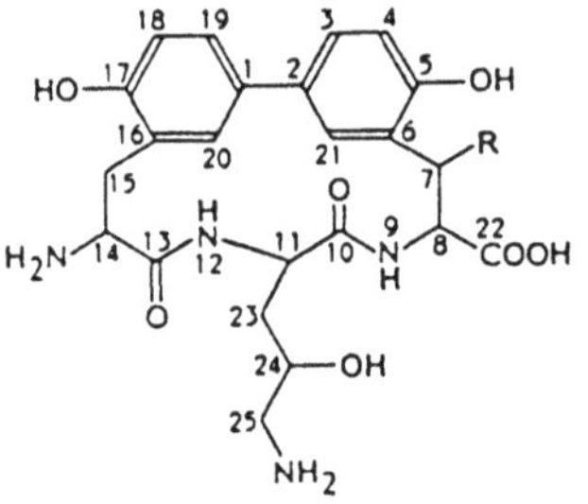

BIP (R : 1-OH,2-H)

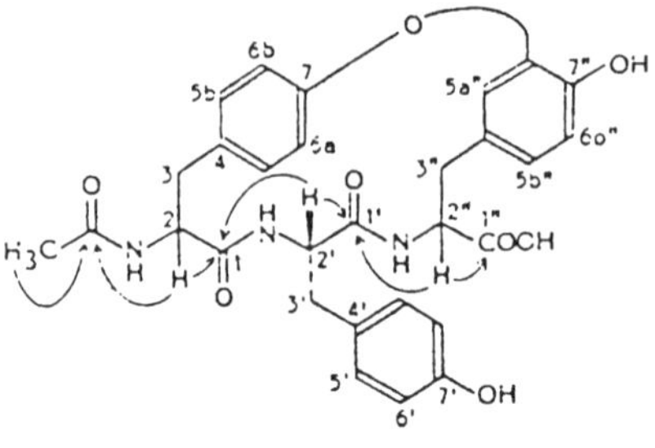

K-13

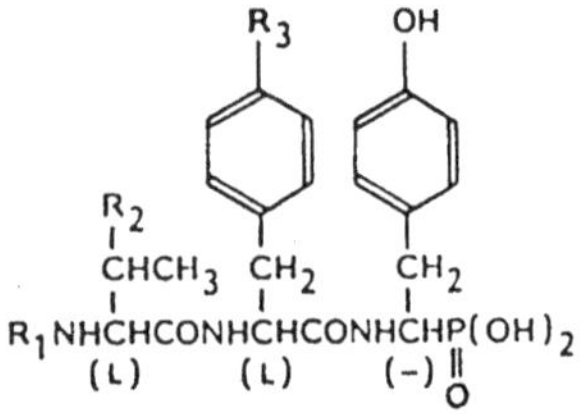

K-4($R1=R2=CH_3$, R3=H)
K-26($R1=COCH_3$, $R2=C_2H_5$, R3=OH)

P-3-(β,τ), P-3-P-(β,τ)-M, R-P-3-(β,τ)-M

ACV_2	*Cephalosporium acremonium N2*		Biotechnology (Rehm and Reed, eds.) Vol.4,
Penicillin N Isopenicillin N	*Penicillium chrysogenum* *various fungi*		
Desacetoxycephalosporin Cephalosporins	*Cephalosporium acremoium, various Streptomycetes*		Vandamme p.45-140
C-1778	*Cephalosporium acremonium*		
Cephabacin Chitinovorin SQ 28,516/7	*Lysobacter lactamgenus, Xanthomonas lactamggenus, Flavobacterium sp. SC12154*		JA 37,773, 1528,1536, 1555,1486 Jpn.Kokai 85-66,987
Janulosamide	*Janolus cristatus*	toxin	TL 27,2505 (86)
Glutathion(**GSH**)	*various organisms*		
4'-**Phosphopantetheine**	*various organisms*		
Nocardicin Formadicin	*Nocardia uniformis* *Streptomyces alcalophilus*	G-(weak)	JA 29,492 (76)
Sulfazecin SQ 26,445	*Pseudomonas* *Gluconobacter* *Acetobacter*	G-(weak)	
PB-5266			

```
                      ┌──────────S-S──────────┐
ACV     Aad-δ→Cys→Val   Aad-δ→Cys→Val
              3┌─┐N                               3┌─┐N
PEN     Aad-δ→Cys→Val            [APA] = →Cys→Val
              S└──┘3                               S└──┘3
IPN     DAad→[APA]
             3┌─┐N                                3┌─┐N
DAC     DAad→Cys→Val             [ACA] = →Cys→Val
             S└──┘4                                S└──┘4
C17     Gla→[ACA]
CEP     DAad→[ACA](7αfNH2)-O-[Arg,Ac]-Ala
     or DAad→[ACA]-O-                -Ala-Ala
                                     -Ala-Ala-Ala
                                     -Ser
                                     -Ser-Ser
                                     -Ser-Ser-Ala
                                     -Val-Orn-Ser-Ala
                                     -Val-Orn-Val-Orn-Ser-Ala
        NAcDAad→[ACA](7αfNH2)-O-[Arg,Ac]-Ala-Ala-Ala
        NAc(CONH2)DAad→[ACA](7αfNH2)-O-[Arg,Ac]-Ala-Ala-Ala
JAN     MeAla→4a3h2mV→Pyd      (MeAla→[Ala,Pr]→[Val,iBu]
GSH     Glu-τ→Cys-Gly
PPA     4'-P-Paa→βAla→Cys(-CO2)
```

NOC Hsr-hPhg(Nh)→[Ser]→Tyr

NOC*C* Hsr-hPhg→[Ser]→Tyr

NOC*G* 4'hPhg→[Ser]→Tyr

SUL Glu-τ→Ala→Ser[N-SO_3]

***iso*SUL** Glu-τ→DAla→Ser[N-SO_3]

PB5*1* 2,3hOPr→Ser→dAsn→Ser[N-SO_3]

PB5*2* 2,3hOPr→Ala→dAsn→Ser[N-SO_3]

PB5*3* 2,3hOPr→Gly→dAsn→Ser[N-SO_3]

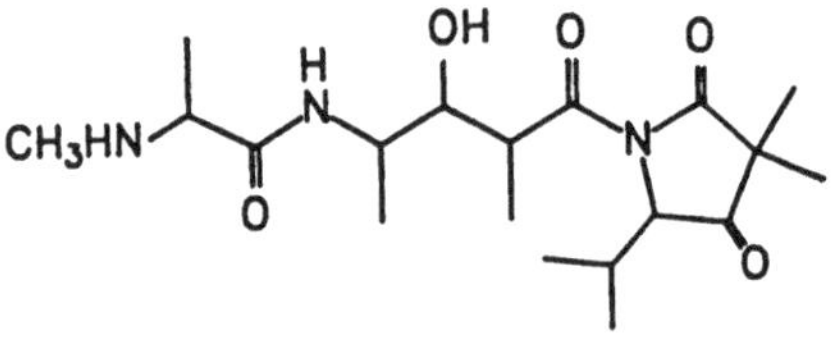

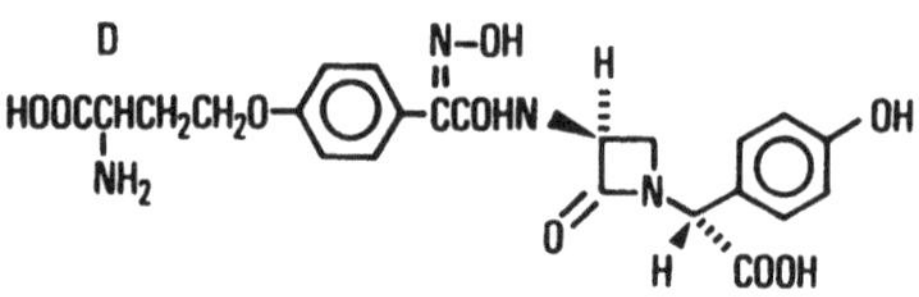

G-P-3-(β,τ)

Muracein A	***Nocardia orientalis***	ACE inh.	JA 37,330,336 (84)

MUR NAcMur→Ala→DGlu-τ→**(L)Dpm(D)NH**$_2$

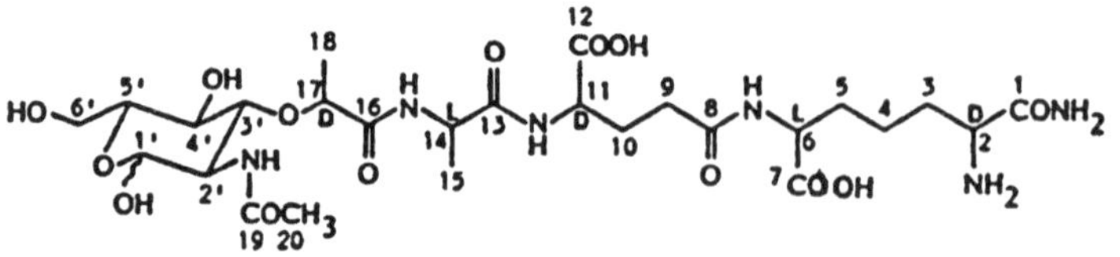

MUR*A*

R-3-P-E

WB-3559	***Flavobacterium sp.*** No.3559	Fibrinolytic	JA 38,1476 (85)

WB3 i4,5dC_{15}/iC_{15}-i3h$C_{16/17}$→Gly→Ser (3h=R)

	R_1	R_2
A	$(CH_3)_2CH(CH_2)_{10}$	$(CH_3)_2CH(CH_2)_7CH=CHCH_2CH_2$
B	$(CH_3)_2CH(CH_2)_{11}$	$(CH_3)_2CH(CH_2)_7CH=CHCH_2CH_2$
C	$(CH_3)_2CH(CH_2)_{10}$	$(CH_3)_2CH(CH_2)_{11}$
D	$(CH_3)_2CH(CH_2)_{11}$	$(CH_3)_2CH(CH_2)_{11}$

WB3

type linear tetrapeptides
subtype tetrapeptide P-4
modified tetrapeptide P-4-M
modified acyltetrapeptide R-P-4-M
modified acyltetrapeptide with β- or τ-bonds R-P-4(β,τ)-M
modified glycotetrapeptide G-P-4-M
linear depsipeptide P-4-(E)
linear peptide with isourea-structure P-4-(U)

P-4

Probestin	*Streptomyces azureus*	proteinase inh.	this volume p. 320ff

PRO 2hHph→Leu→Pro→Pro

P-4-M

Celenamide	*Cliona celata*	JOC 45,3687 (80),Ang.23, 991(84)
Integerrin		
Lasoidin		

CELA Leu→3,5hdTyr→3,5hdTyr→5hdTyr(-CO_2)
CELB aIle→3h5OAcdTyr→3,5hdTyr→5hdTyr(-CO_2)
CELC Ile→3,5hdTyr→BrTrp→5hdTyr(-CO_2)
CELD Ile→3,5hdTyr→BrTrp→dTyr(-CO2)
INT Me_2Val→Phe→Trp→dTyr (3 —O— 4')
LAS $Me_2$3hPhedVal→3hPhe→dTyr(-CO2)

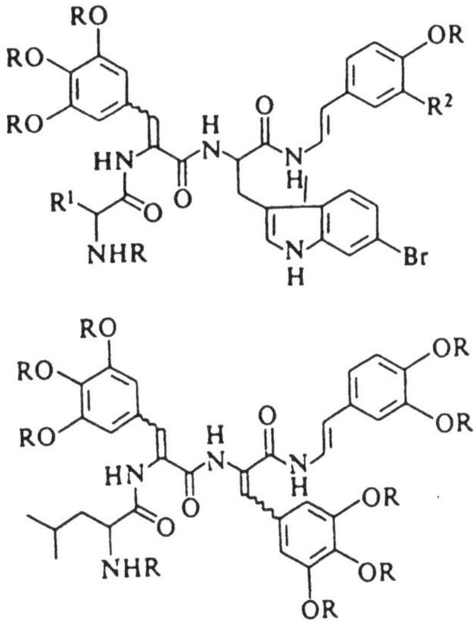

CEL

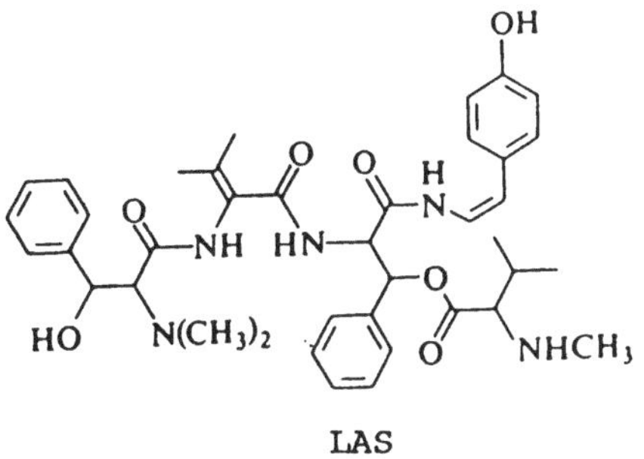

LAS

R-P-4-M

SAB-711	*Streptomyces kaizukaensis*	protease inh.	Jpn.Kokai 83-198,453
Zizyphine	*Zizyphus oenoplia*		TL 1973,2577

SAB OEt→Ile→Leu→Leu→Argal
ZIZ Me_2Ile→Ile→Pro→Pro→β$NH_2$2Om5hSty
3 └──O──┘

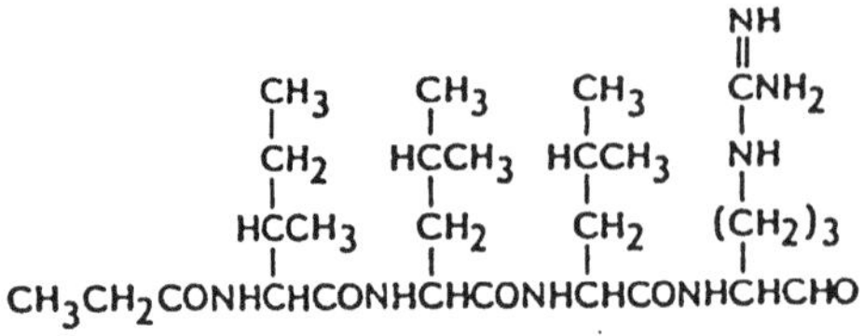

SAB

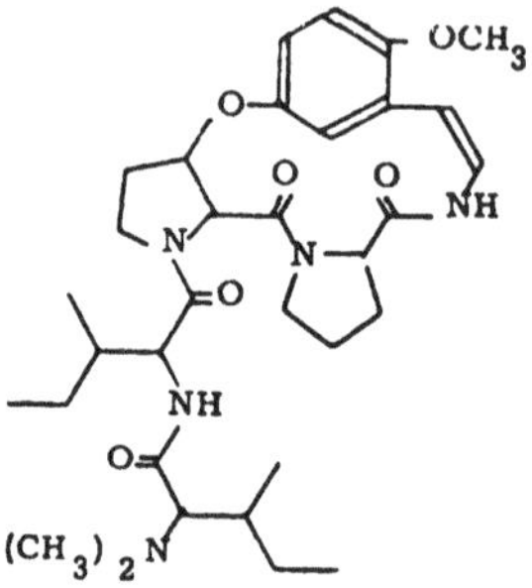

ZIZ

R-P-4-( )

FK-156	*Streptomyces olivaceogriseus* *S.violaceus*	immunostimulatory	JA 35,1280, 1286(35),36, 1045,1051, 1059(83),Chem. Pharm.Bull.32, 2925(84)

FK1 DLac→Ala→DGln-γ→Dpm→Gly

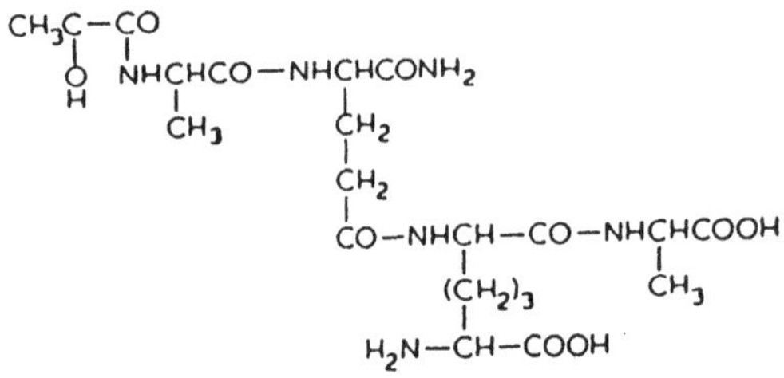

FK1

G-P-4-M

SQ 28,504	***Chromobacterium violaceum***	*β*-lactam potentiator	JOC 51,3942 (86)

SQ2 40(SO_3)NAcGlN-4h5hmPro→Lys→Thr(NACGlN)→Ser

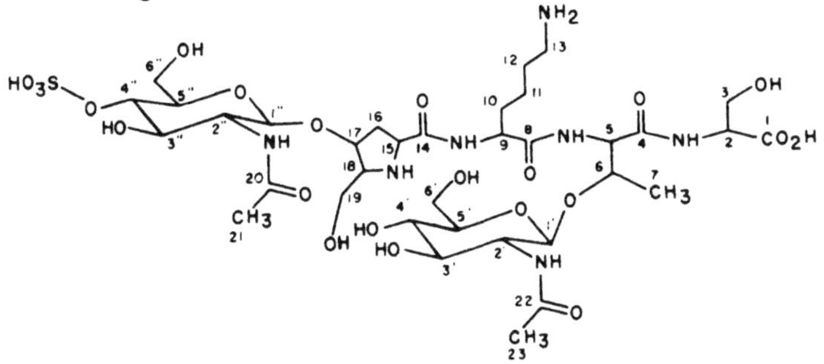

SQ2

P-4-M-(E)

Nocobactin	***Nocardia asteroides***	Fe^{3+}-chelator	BJ 138,407(74)
Mycobactin	***Mycobacterium sp.***	Fe^{3+}-chelator	BJ 111,785(69)

NOC Bz→dThr→N^{7}hOAcLys-3h2mC$_{12}$/C$_{14}$→N^{7}hcLys

MYC*P* 2h5mBz→Ser→N^{7}h,C$_{17}$Lys→3h2mC$_5$→N^{7}cLys

MYC*S* 2hBz→Ser→N^{7}h,C$_{17}$Lys→(S)3hC$_4$→N^{7}cLys

MYC*T* 2hBz→Ser→N^{7}h,C$_{20}$Lys→(R)3hC$_4$→N^{7}cLys

MYC*H* 2h5mBz→Thr→N^{7}h,C$_{17}$Lys→(S)3hC$_4$→N^{7}cLys

MYC*M* 2hBz→Thr→N^{7}hOAcLys→3h2mC$_{18/20}$→N^{7}hcLys

P-4-M-(U)

Antipain	*Streptomyces michganensis* *S.yokosukaensis*	proteinase *inh.*	JA 25,263,267 (72)
Elastatinal	*Streptomyces griseoruber*	proteinase inh.	JA 28,337(75)
Chymostatin	*Streptomyces hygroscopicus*	proteinase inh.	JA 26,625(73)
MAPI	*Streptomyces nigrescens*	proteinase inh.	T 34,223(78),38, 1775(82)
Formestin		proteinase inh.	this vol. p.311-363

ANT Phe-CO→Arg→Val→Argal
ELA Leu-CO→Cap→Gln→Alaal
CHA1 Phe-CO→Cap→Val→Pheal
CHA2 Phe-CO→Cap→Leu→Pheal
CHA3 Phe-CO→Cap→Ile→Pheal
MAP Phe-CO→Arg→Val→Pheal
FORA 2,4,6mC_{10}→Dpr-β→CO-β-Dpr-hAsp-TrpOMe
FORB 2,4,6mC_{10}→Dpr-β→CO-β-Dpr-Asp-TrpOMe

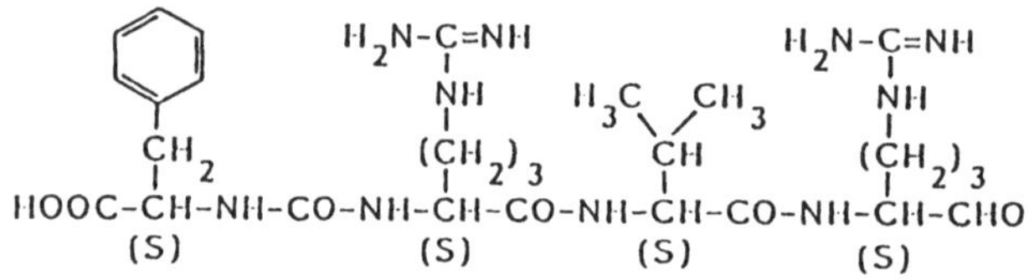

ANT

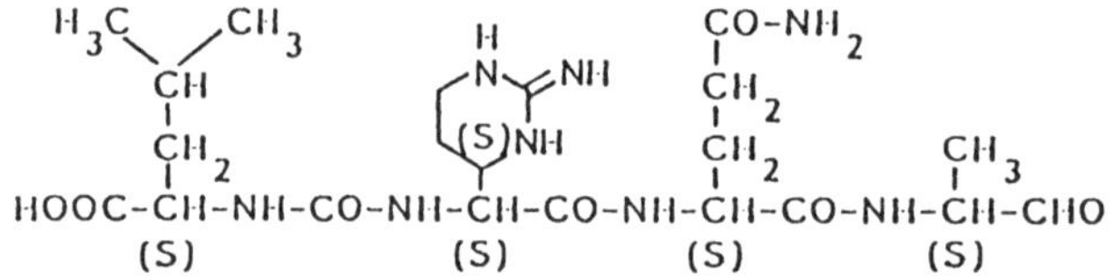

ELA

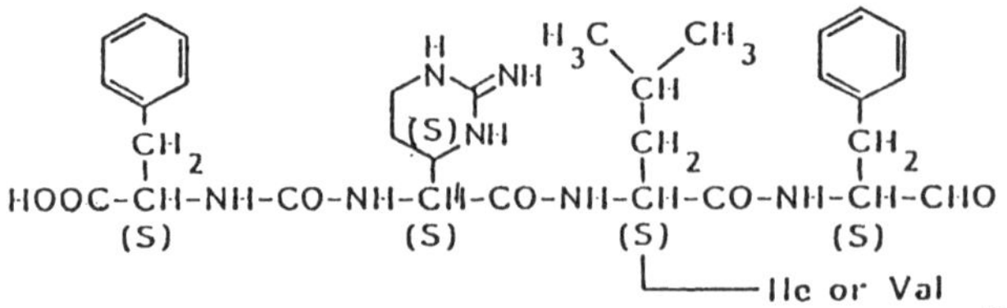

CHA

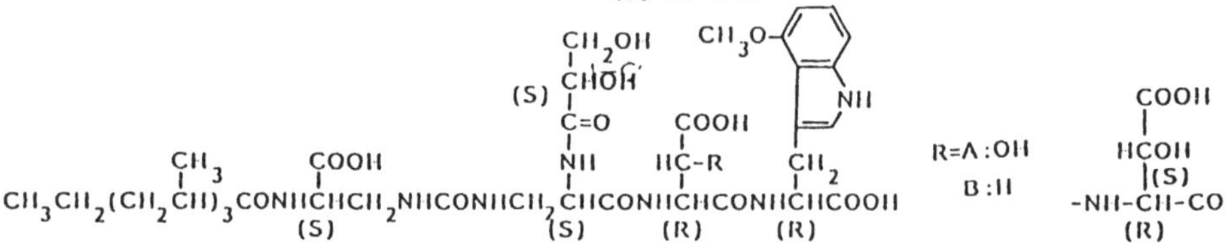

FOR

type linear pentapeptides
subtype pentapeptide P-5
modified pentapeptide P-5-M
modified glycopentpeptide G-P-5-(τ)-M

P-5-M

BMG59-R2	***Bacillus megaterium***	AT, inh. alkal. phosphatase	Jpn. Kokai 83-164,561
Edeine	***Bacillus brevis Vm4***	G+, G-, AF, inh. DNA synth.	Biochem. 9, 1224(70) JA 36,793, 1001(83)
Pepstatine	***Streptomyces testaceus,*** many other strains	proteinase inh.	JA 21,429 (68), this vol., p.
Ahpatinine	***Streptomyces sp. WK-142***	acid protease inh.	JA 39,1079 (86)
Althiomycin	***Streptomyces althioticus*** ***S. matensis*** ***S. bellus*** ***Cystobacter fuscus***	G+, G-, inh. protein synth.	JA 27,897, (74), 35, 635(82)

BMG	Ile→Ile→SerP→Gln→Glu
EDE*A*	βTyr→Ise→Dpr→Dha→Gly→spermidine
EDE*B*	βPhe→Ise→Dpr→Dha→Gly→guanidylspermidine
PEP*A*	iBuO→Val→Val→Ahma→Ala→Ahma
AHP*1*	PrO/iBuO→Val→Val→Ahma→Ala→Ahpp
AHP*2*	iBuO/BuO→Val→Val→Ahma→Ala→Ahma
AHP*F*	iBuO→Phg→Val→Ahma→Ala→Ahma
AHP*G*	iBuO→Val→Val→Ahpp→Ala→AhppOMe
ALT	[Gly]→Cys→Ser→Cys→Gly

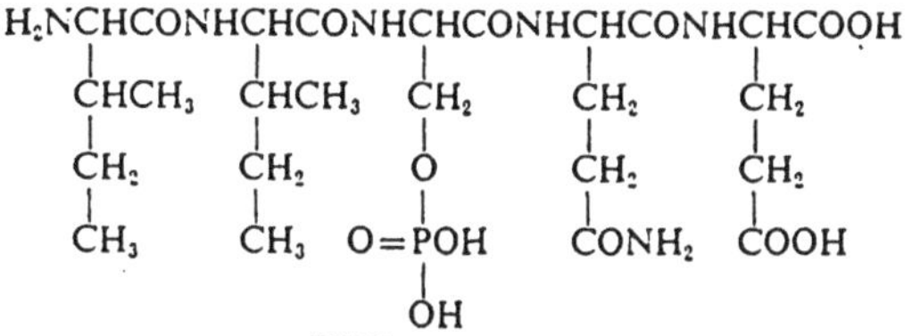

BMG

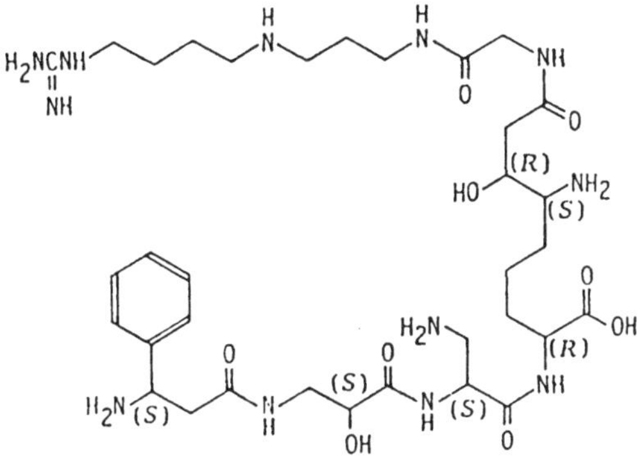

EDEF

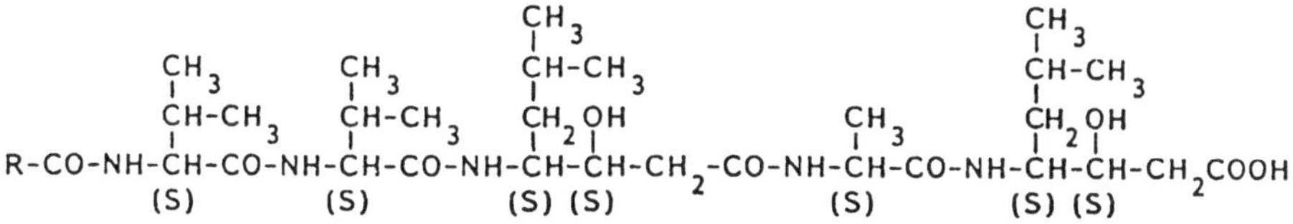

PEP

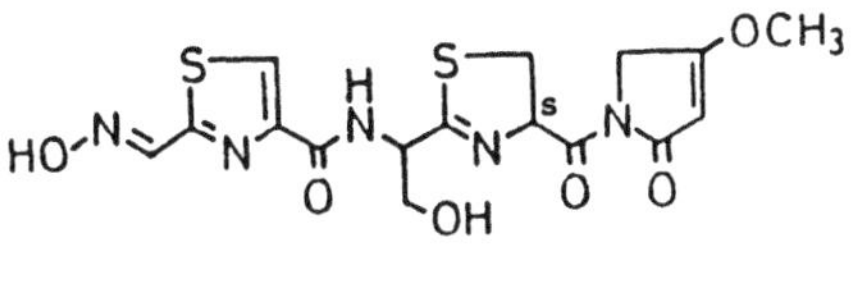

ALT

Muracein B	*Nocardia orientalis*	ACE inh.	JA 37,330,336 (84)

MUB NAcMur→Ala→DGlu-τ→(L)Dpm(D)NH_2→DAla→DAla

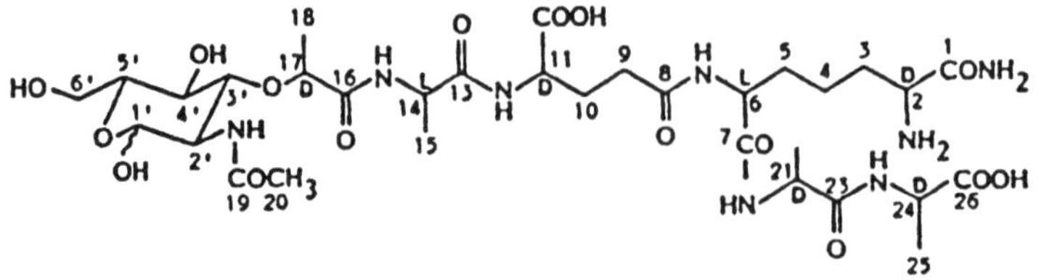

MUB

type linear hexapeptides
subtype modified hexapeptide P-6-M

Lavendomycin	*Streptomyces lavendulae, subsp. brasilicus*	antibacterial	Chem.Pharm. Bull.33,3053 (85)
SQ 28,546	*Chromobacterium violaceum*	β-lactam-potentiator	JOC 51,3942 (86)

LAV 3mArg→Dab→Pip→Pro→dAbu→Ser
SQ8 4O(SO_3)NAcGlN→4h5hmPro→Lys→Thr(NAcGlN)→Ser→Asp→Ser

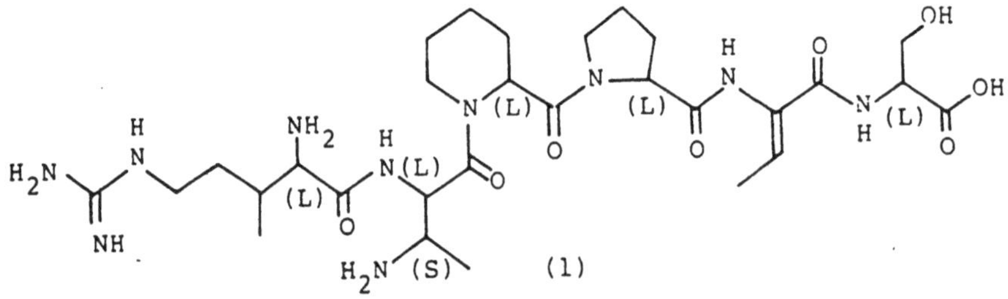

(1)

LAV

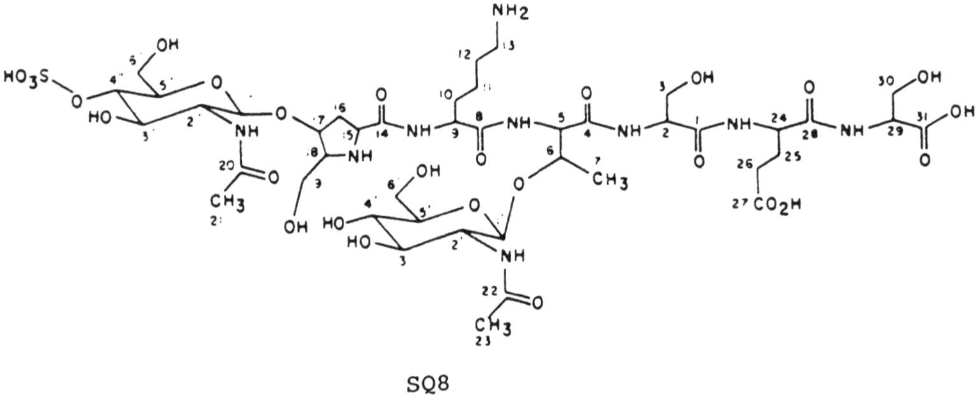

SQ8

type linear heptapeptides
subtype modified heptapeptide P-7-M
modified glycosylated heptapeptide P-7-G-M

P-7-M

Antrimycin Cirratiomycin	*Streptomyces xanthocidicus* *S. cirratus*	narrow range: My, Strepto-coccus, Lacto-bacillus	JA 34,1615 (81), ABC 46, 865,1861,1885 (82), Jpn.Pat. 81-144,306
K-582	*Metarhizium anisopliae*	interferon inducer	JA 33,533(80) 36,335(83) IJPPR 21,546 (83)

ANT 2hmSer→Ala→Dab→Thpc→Ala→dIle→Ser
K58A Arg→threo4hArg→DOrn→Thr→DOrn→Lys→DTyr
B Arg→threo4hArg→DOrn→Thr→DOrn→Lys→DTyr

P-7(τ)

Cadystin	*Schizosaccharomyces pombe*	Cd^{2+}-binding	TL 24,925 (83)

CAD Glu-τ→Cys→Glu-τ→Cys→Cys→Glu-τ→Gly

P-7-G-M

Vancomycin	*Amycolatopsis orientalis**	inh. bact. cell wall formation DAla-DAla-complexer feed add.	Lancini and Cavalleri, this vol. pp. 159-178 Ann.Rev.Microb. 38,339(84) *JACS 110,7896 (88)
A51568			
M43			
PA-42867 Orienticin			
OA-7653	*Streptomyces hygroscopicus subsp. hiwasaensis*		
Eremomycin	*Actinomyces*		
A82846	*Nocardia orientalis*		
Actinoidin	*Proactinomyces actinoides*		
Avoparcin	*Streptomyces candidus*		
Chloropolysporin	*Faenia interjecta*		
Ristocetin	*Nocardia lurida*		
Actaplanin	*Actinoplanes missouriensis*		
A-35512	*Streptomyces candidus*		
A-41030	*Streptomyces virginiae*		
A-47934	*Streptomyces toyocaensis*		
Teicoplanin Teichomycin	*Actinoplanes teichomyceticus*		
Ardacin Aridicin	*Kibdelosporangium aridus*		
A-40926	*Actinomadura sp.*		
Kibdelin	*Kibdelosporangium aridum sp. largum*		
Parvodicin	*Actinomadura parvosata*		

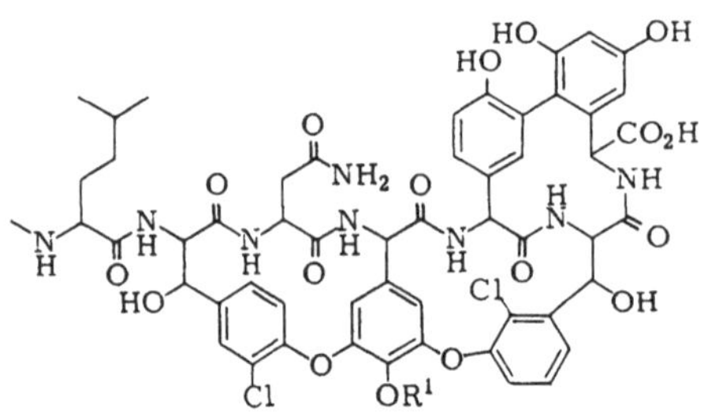

VAN-Type

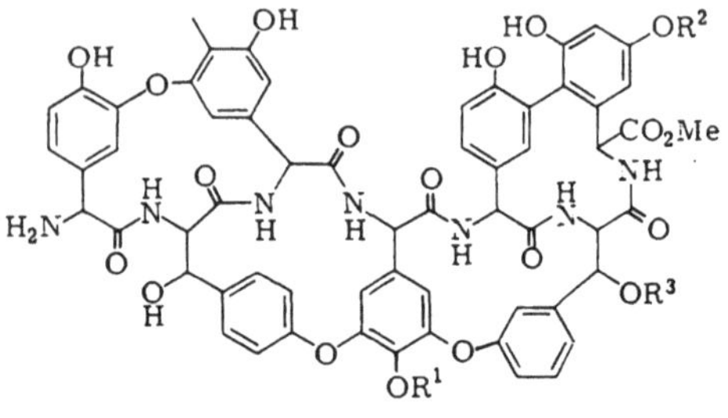

RIS-Type

VAN-Type nMeLeu→2hTyr→Asn→hPhg→hPhg→2h3'ClTyr→3'5'hPhg
(ring links: O from 2hTyr to 3' of hPhg; 5' of hPhg to O of 2h3'ClTyr; 3' of hPhg to 3' of 3'5'hPhg)

VAN MeLeu→2h3'ClTyr→Asn→hPhg(4'glu-van)→hPhg→2h3'ClTyr→3'5'hPhg
A51*A* Leu→2h3'ClTyr→Asn→hPhg(4'glu-van)→hPhg→2h3'ClTyr→3'5'hPhg
A51*B* Leu→2h3'ClTyr→Gln→hPhg(4'glu-van)→hPhg→2h3'ClTyr→3'5'hPhg
M43*A* Me$_3$Leu→2h3'ClTyr→Asn→hPhg(4'glu-van)→hPhg→^{6}2h3'ClTyr→
→^{7}3'5'hPhg
M43*B* Me$_3$Leu→2h3'ClTyr→Asp→hPhg(4'glu-van)→hPhg→^{6}2h3'ClTyr→
→^{7}3'5'hPhg
M43*C* Me$_3$Leu→2h3'ClTyr→Asn→hPhg(4'glu)→hPhg→2h3'ClTyr→3'5'hPhg
M43*D* Me$_2$Leu→2h3'ClTyr→Asn→hPhg(4'glu-van)→hPhg→^{6}2h3'ClTyr→
→^{7}3'5'hPhg
M43*F* MeLeu→2h3'ClTyr→Asp→hPhg(4'glu-van)→hPhg→^{6}2h3'ClTyr→
→^{7}3'5'hPhg

ORI*A* MeLeu→2hTyr→Asn→hPhg(4'glu-4-epi-van)→^{5}hPhg→
→^{6}2h3'ClTyr(2-*4*-epi-van)→3'5'hPhg
ORI*B* MeLeu→2hTyr→Asp→hPhg(4'glu-oli)→^{5}hPhg→
→^{6}2h3'ClTyr(2-*4*-epi-van)→3'5'hPhg
ORI*C* MeLeu→2hTyr→Asn→hPhg(4'glu-4-epi-van)→^{5}hPhg→
^{6}2hTyr(2-*4*-epi-van)→3'5'hPhg
ORI*D* Me$_2$Leu→2hTyrAsn→hPhg(4'glu-4-epi-van)→^{5}hPhg→
→^{6}2h3'ClTyr(2-*4*-epi-van)→3'5'hPhg
OA7
Me$_2$Ala→2h3'ClTyr→Gln→hPhg(4'glu)→hPhg→2h3'ClTyr(2glu)→3'5'hPhg
ERE MeLeu→2h3'ClTyr→Asn→hPhg(4'glu-ere)→hPhg→^{6}2hTyr(2-epi-van)→
→^{7}3'5'hPhg
A82*A* MeLeu→2h3'ClTyr→Asn→hPhg(4'glu-4-epi-van)→^{5}hPhg→
→^{6}2hTyr(4'-*4*-epi-van)→3'5'hPhg
A82*B* MeLeu→2h3'ClTyr→Asn→hPhg(4'glu-4-epi-van)→^{5}hPhg→
→^{6}2h3'ClTyr(4'-*4*-epi-van)→3'5'hPhg
A82*C* MeLeu→2hTyr→Asn→hPhg(4'glu-epi-van)→hPhg→
^{6}2hTyr(4'-*4*-epi-van)→3'5'hPhg

ACT-Type hPhg→2hTyr→Phe/hPhg→hPhg(G)→hPhg→2h3'ClTyr(G)→3'5'hPhg
(ring links: O from 2hTyr to 3' of hPhg(G); 5' of hPhg(G) to O of 2h3'ClTyr(G); 3' of hPhg to 3' of 3'5'hPhg)

ACT*A* hPhg→2hTyr→Phe→hPhg(4'glu-aco)→hPhg→^{6}2h3'ClTyr(2aca)→
→^{7}3'5'hPhg(3Dman)
ACT*B* 2'ClhPhg→2hTyr→Phe→hPhg(4'glu-aco)→hPhg→^{6}2h3'ClTyr(2aca)→
→^{7}3'5'hPhg(3Dman)
AVOα MehPhg(4'rha)→2hTyr(2Dman)→hPhg→hPhg(4'glu-ria)→^{5}hPhg→
→^{6}2h3'ClTyr(2ria)→3'5'hPhg
AVOβ MehPhg(4'rha)→2hTyr(2Dman)→3'ClhPhg→hPhg(4'glu-ria)→^{5}hPhg→
→^{6}2h3'ClTyr(2ria)→3'5'hPhg
AVOε MehPhg(4'rha)→2hTyr→3'ClhPhg→hPhg(4'glu-ria)→^{5}hPhg→
→^{6}2h3'ClTyr(2ria)→3'5'hPhg
CLP*B* MehPhg(4'rha)→2h3'ClTyr→hPhg→hPhg(4'glu)→^{5}hPhg→
→^{6}2h3'ClTyr(2ria)→3'5'hPhg
CLP*C* MehPhg→2h3'ClTyr→hPhg→hPhg(4'glu)→hPhg→^{6}2h3'ClTyr(2ria)→
→^{7}3'5'hPhg
ACT*A2* hPhg→2h3'ClTyr→Phe→hPhg(4'glu-rha)→hPhg→^{6}2h3'ClTyr(2aca)→
→^{7}3'5'hPhg(3Dman)

RIS-Type

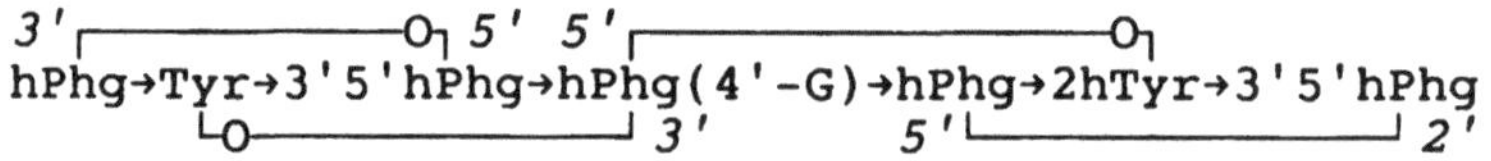

RIS*A* hPhg→2hTyr→3'h4'mPhg→hPhg(4'rha→Dglu-Dman-Dara)→^{5}hPhg→
→^{6}2hTyr(2ria)→3'5'hPhg(5'Dman)

RIS*B* hPhg→2hTyr→3'h4'mPhg→hPhg(4'rha→Dglu)→hPhg→^{6}2hTyr(2ria)→
→^{7}3'5'hPhg(5'Dman)

ACP*A* hPhg→Tyr→3'h4'mPhg(3'man)→hPhg(4'glu-man)→^{5}hPhg
→^{6}2h3'ClTyr(2ria)→3'5'hPhg(5'man)

ACP*B1* hPhg→Tyr→3'h4'mPhg(3'man)→hPhg(4'glu-rha)→^{5}hPhg→
→^{6}2h3'ClTyr(2ria)→3'5'hPhg(5'man)

ACP*B2* hPhg→Tyr→3'h4'mPhg(3'man)→hPhg(4'glu)→^{5}hPhg
→^{6}2h3'ClTyr(2ria)→3'5'hPhg(5'man)

ACP*B3* hPhg→Tyr→3'h4'mPhg→hPhg(4'glu-man)→hPhg→^{6}2h3'ClTyr(2ria)→
→^{7}3'5'hPhg(5'man)

ACP*C1* hPhg→Tyr→3'h4'mPhg→hPhg(4'glu-rha)→hPhg→^{6}2h3'ClTyr(2ria)→
→^{7}3'5'hPhg(5'man)

ACP*C3* hPhg→Tyr→3'h4'mPhg(3'man)→hPhg(4'glu)→^{5}hPhg→
→^{6}2h3'ClTyr(2ria)→3'5'hPhg

ACP*G* hPhg→Tyr→3'h4'mPhg→hPhg(4'glu)→hPhg→^{6}2h3'ClTyr(2ria)→
→^{7}3'5'hPhg(5'man)

A35 hPhg→2hTyr→2'Cl3→hPhg(3'fuc)→hPhg(4'gluha)→^{5}hPhg
→^{6}2hTyr(2-*3*-epi-van)→3'5'hPhg(5'man)

A41*A* hPhg→3'ClTyr→3'5'hPhg→hPhg→3'ClhPhg→2h3'ClTyr→3'5'hPhg

A41*B* hPhg→3'ClTyr→3'5'hPhg→hPhg→3'ClhPhg→2hTyr→3'5'hPhg

A41*C* hPhg(4'gal)→3'ClTyr→3'5'hPhg→hPhg→3'ClhPhg→^{6}2h3'ClTyr
→^{7}3'5'hPhg

A41*E* hPhg→Tyr→3'5'hPhg→hPhg→hPhg→2h3'ClTyr→3'5'hPhg

A41*F* hPhg(4'gal→gal)→3'ClTyr→3'5'hPhg→hPhg→^{5}3'ClhPhg→
→^{6}2h3'ClTyr→3'5'hPhg

A47 hPhg(4'SO_3H)→3'ClTyr→3'5'hPhg→hPhg→3'ClhPhg→^{6}2h3'ClTyr→
→^{7}3'5'hPhg

TEI-Type

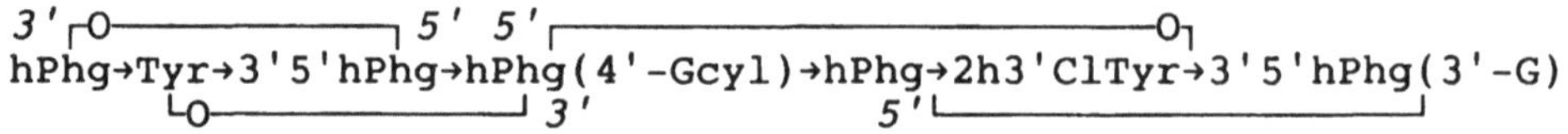

TA2
hPhg→2'ClTyr→3'5'hPhg→^{4}hPhg(4'glA,5dC$_{10}$/iC$_{10}$/C$_{10}$/aC$_{11}$/iC$_{11}$)→
^{5}→hPhg→2h3'ClTyr(2Dman)→3'5'hPhg(3'man)

ARD Me_2hPhg→2h2'ClTyr→2'Cl3'5'hPhg→^{4}hPhg(4'Aga$_{10}$/iC$_{11}$/iC$_{12}$)→
→^{5}hPhg→2h3'5'ClTyr→3'5'hPhg(3'man)

A40 Me_2hPhg→Tyr→2'Cl3'5'hPhg→hPhg(4'Aga$_{11}$/iC$_{12}$)→^{5}hPhg→
^{6}2h3'ClTyr→3'5'hPhg(3'man)

KIB Me_2hPhg→2h2'ClTyr→^{3}2'Cl3'5'hPhg→
^{4}hPhg(4'glA$_{10}$/iC$_{11}$/iC$_{12}$/C$_{12}$/4,5dC$_{10}$)→hPhg→^{5}2h3'5'ClTyr→
^{6}3'5'Phg(3'man)

PAR Me_2hPhg→Tyr→2'Cl3'5'hPhg→^{4}hPhg(4'Aga$_{10}$/iC$_{11}$/C$_{11}$/iC$_{12}$)→
^{5}hPhg→2h3'ClTyr→3'5'hPhg(3'man/manOAc)

Type linear octapeptides
subtype acyloctapeptide,modified R-P-8-M
octapeptide(β,τ),modified P-8-M(β,τ)

R-P-8-M

Lophyrotomin	*Lophyrotoma interrupta*	toxin	JCS CC 1982, 1394
Stenothricin	*Streptomyces sp.*	G+	Ann.1976,2011
Moroidin	*Laportea moroides*	skin irritant upon cutaneous contact	T 42,3333(86)

LOP PhCO→DAla→DPhe→Val→Ile→DAsp→Asp→DGlu→Gln
STT βketoacyl→DCya→Val→DSer→Lys→Sar→DaThr→Ser→Dpr
MOR 5oPro→Val→Leu→Val→Trp→Arg→Gly→His (Leu-3 linked to Trp; Trp linked to His N)

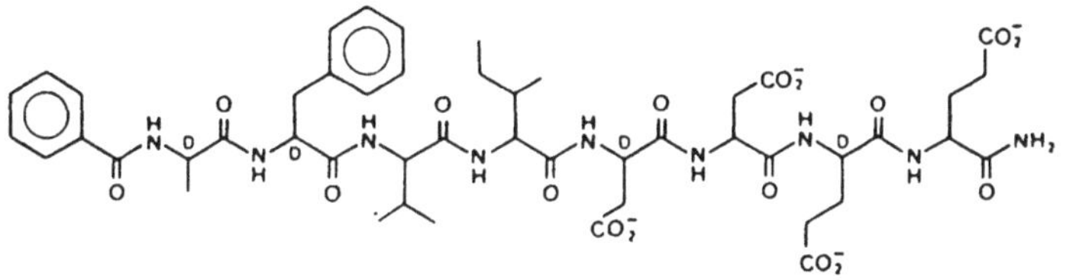

LOP

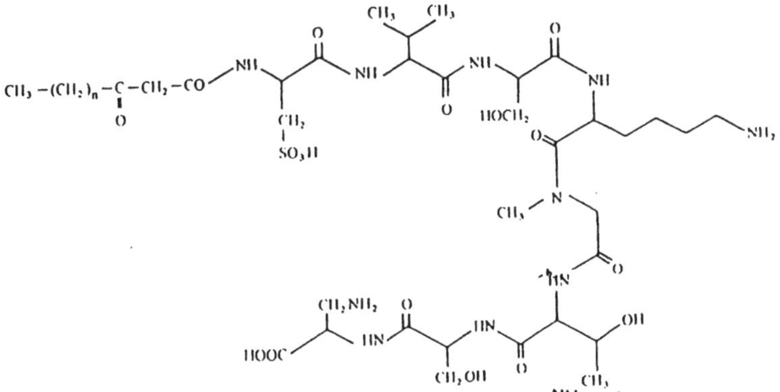

STT

MOR

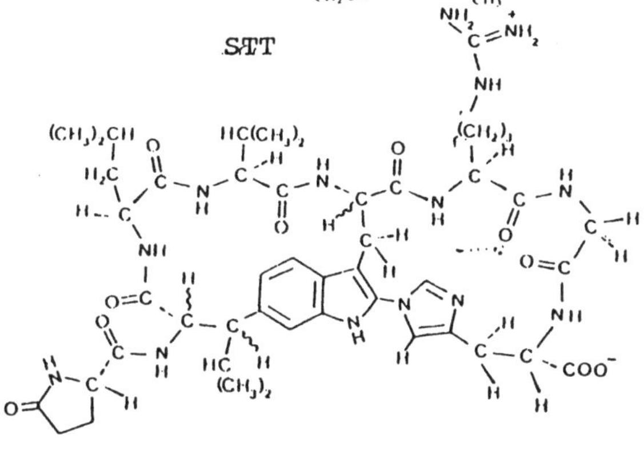

P-8-G-M(β,τ)

Phleomycin	*Streptomyces verticillus*	G+,G-,My,AT	JA 25,752,755 (72)
Bleomycin	*S. verticillus*	G+,G-,My,AF, AT	JA 36,92(83), Takita & Muraoka, this volume, p.289 -309
Tallysomycin	*Streptoalloteichus hindustanus*	G+,G-,AF,AT	JA 30,779,789 (77)
Zorbamycin	*Streptomyces bikiniensis*	G+,G-,My,AF, AT	JA 24,543(71)
YA-56-X			

BLE [Ser]→Asn→Asn→3h(G)His→[Ala,Ac]→Thr→βAla→Cys→Cys→amine

PHL [Ser]→Asn→Asn→3h(G)His→[Ala,Ac]→Thr→βAla→Cys→Cys→amine

TAL [Ser]→Asn→Asn→3h(G)His→[Ala,Ac]→Thr→2,3h3(G)βAla→Cys→Cys-
→βLys→amine

YA5 [Ser]→Asn→Asn→His→[4a3,6h2mC_6]→3hVal→βAla→Cys→Cys→amine

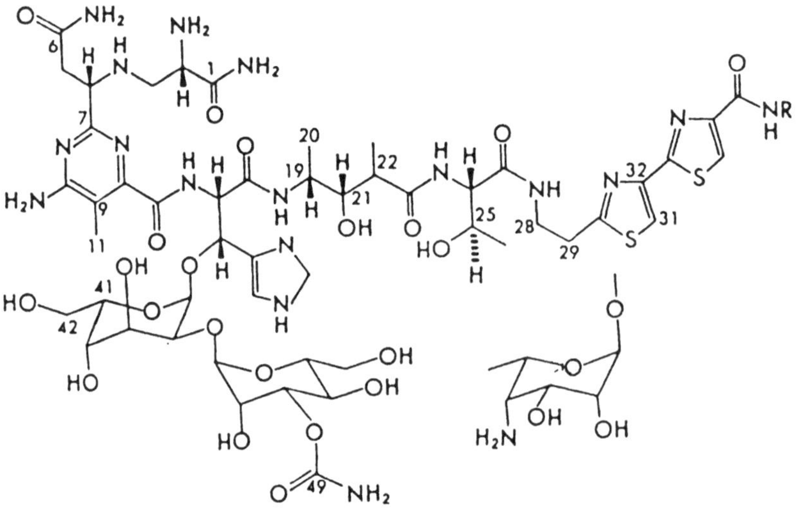

BLE-Type

R represents various amine compounds like $NH(CH_2)_3SOCH_3$, $NH(CH_2)_3NH(CH_2)_4NH_2$. In phleomycin, the C 31-32 double bond is missing. Zorbamycins (YA-56-X) contain an extra CH_2OH at methyl group 20, an extra methyl group at 25, no double bond at 31-32, and no hydroxyl group at 42. In YA-56-Y the hydroxyl group at 21 and the methyl group at 22 are missing, while an methyl is found at 25. In tallysomycin an extra aminosugar is found at 28, a hydroxyl group at 29, and the methyl group at 22 is missing.

Type linear nonapeptides
subtype nonapeptide P-9
acyl-nonapeptide R-P-9

R-P-9

Trichopolin	*Trichoderma polysporum*	G+,AF phyt.fungi	Exp.34,238(78) JCS CC 1981,585

TRIA 2mC$_{10}$→Pro→Ahmod→Ala→Aib→Aib→Ile→Ala→Aib→Aib→mPdaEtOH
TRIB 2mC$_{10}$→Pro→Ahmod→Ala→Aib→Aib→Val→Ala→Aib→Aib→mPdaEtOH

Type linear decapeptides
subtype acyl-decapeptide R-L-10

R-P-10

Leucinostatin 1907	*Paecilomyces lilacinus* *P.marquandi*	G+,AF,ATPase inh.	JA 36,1084, 1606,1613 JCS CC 1982,94
Cerexin	*Bacillus cereus*	G+	JA 29,1281 (76),32,313 (79) US Pat. 3,590,436 Ger.Pat. 2,420,103/4

LECA a2,3dC$_6$→4mPro→Ahmod→2hLeu→Aib→Leu→Leu→Aib→Aib→βAla→Ala
LECB a2,3dC$_6$→4mPro→Ahmod→2hLeu→Aib→Leu→Leu→Aib→Aib→βAla→MeAla
LECD a2,3dC$_6$→4mPro→Leu→2hLeu→Aib→Leu→Leu→Aib→Aib→βAla→MeAla
LECH a2,3dC$_6$→4mPro→Leu→2hLeu→Aib→Leu→Leu→Aib→Aib→βAla→AlaNOMe$_2$
LECK a2,3dC$_6$→4mPro→Ahmod→2hLeu→Aib→Leu→Leu→Aib→Aib→βAla→AlaNOMe$_2$
CERA i3hC$_{11}$→DAsn→Val→DVal→Asn→DAsn→4hLys→aThr→Ser→DTrp→aIle
CERB i3hC$_{10}$→DAsn→Val→DPhe→Asn→DAsn→4hLys→aThr→Gly→DTrp→aIle
CERC i3hC$_{11}$→DAsn→Val→DVal→Asn→DAsn→Lys→aThr→Ser→DTrp→aIle
CERD i3hC$_{10}$→DAsn→Val→DPhe→Asn→DAsn→Lys→aThr→Gly→DTrp→aIle

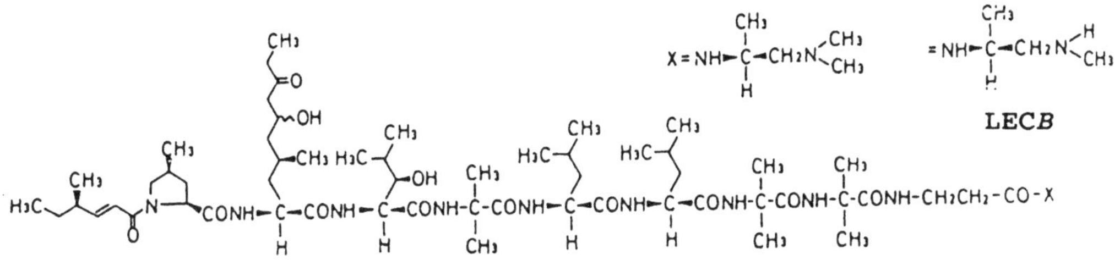

LECB

LECA

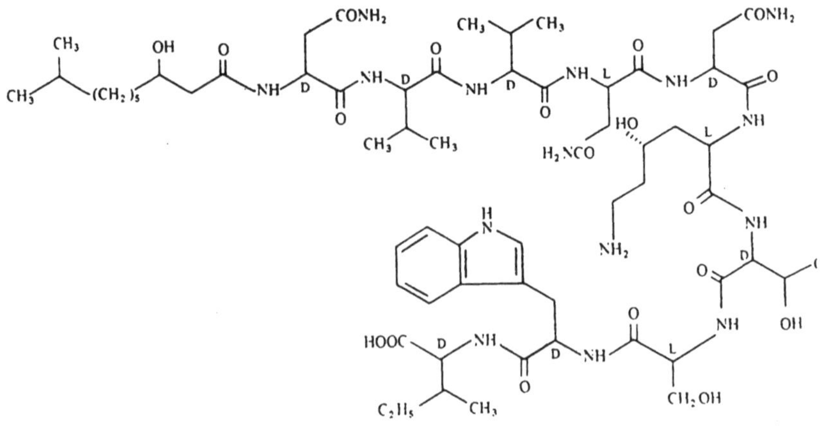

CERA

R-P-13

Tridecaptin	*Bacillus polymyxa*	G+,G-	JA 32,305,313 (79),Jpn.Pat. 76-144,796

TRD ahC9→DVal→DDab→Gly→DSer→DTrp→Ser→Dab→DDab→Phe→10Glu→
→Val→aIle→Ala
ahC9→DVal→DDab→Gly→DSer→DTrp→Ser→Dab→DDab→Phe→10Glu→
→Val→Val→Ala
aC9→Gly→DDab→Gly→DSer→DTrp→Ser→Dab→DDab→Ile→10Glu→
→Val→aIle→Ser
aC9→Gly→DDab→Gly→DSer→DTrp→Ser→Dab→DDab→Ile→10Glu→
→Val→Val→Ser
aC9→Gly→DDab→Gly→DSer→DTrp→Ser→Dab→DDab→Val→10Glu→
→VAL→aIle→Ser
aC9→Gly→DDab→Gly→DSer→DTrp→Ser→Dab→DDab→Val→10Glu→
→Val→Val→Ser
a3hC11→DVal→DDab→Gly→DSer→DTrp→Ser→Dab→DDab→Phe→10Glu→
→Val→Val→Ser
i3hC10→DVal→DDab→Gly→DSer→DTrp→Ser→Dab→DDab→Phe→10Glu→
→Val→Val→Ser
a3hC11→DVal→DDab→Gly→DSer→DTrp→Ser→Dab→DDab→Phe→10Glu→
→Val→aIle→Ser

R-P-15-M

Gramicidin	*Bacillus brevis ATCC 8185*	G+,G-	J.Sci.Ind.Res. 34,249(75) Annu.Rev. Sci.
Emericin	*Emericellopsis microspora*	G+	JACS 99,5205 (77),US Pat. 3,821,367 Ger.Pat. 1,809,875
Discodermin	*Discodermia kiiensis*	G+,G-	TL 26,855(85)

GRA fVal→Gly→Ala→Dleu→Ala→DVal→Val→DVal→Trp→^{10}DLeu→
→Trp→DLeu→Trp→DLeu→Trp→EtOH

GRA fIle→Gly→Ala→DLeu→Ala→DVal→Val→DVal→Trp→^{10}DLeu→
→Trp→DLeu→Trp→DLeu→Trp→EtOH

GRA fVal→Gly→Ala→DLeu→Ala→DVal→Val→DVal→Trp→^{10}DLeu→
→Tyr→DLeu→Trp→DLeu→Trp→EtOH

GRA fIle→Gly→Ala→DLeu→Ala→DVal→Val→DVal→Trp→^{10}DLeu→
→Tyr→DLeu→Trp→DLeu→Trp→EtOH

EME1 Ac→Phe→Aib→Aib→Aib→Val→Gly→Leu→Aib→Aib→^{10}4hPro→
→Gln→Iva→4hPro→Aib→^{15}Pheol

EME2 Ac→Phe→Aib→Aib→Aib→Val→Gly→Leu→Aib→Aib→^{10}4hPro→
→Gln→Iva→4hPro→Ala→^{15}Pheol

DISA fDAla→Phe→Phe→Pro→DtLeu→tLeu→DTrp→Arg→DCya→^{10}Thr→
→MeGln→DLeu→Asn→Thr→^{15}Sar

DISA fDAla→Phe→Phe→Pro→DtLeu→tLeu→DTrp→Arg→DCya→^{10}Thr→
→MeGln→DLeu→Asn→Thr→^{15}Sar

DISB fDAla→Phe→Phe→Pro→DVal→tLeu→DTrp→Arg→DCya→^{10}Thr→
→MeGln→DLeu→Asn→Thr→^{15}Sar

DISC fDAla→Phe→Phe→Pro→DtLeu→Val→DTrp→Arg→DCya→^{10}Thr→
→MeGln→DLeu→Asn→Thr→^{15}Sar

DISD fDAla→Phe→Phe→Pro→DVal→Val→DTrp→Arg→DCya→^{10}Thr→
→MeGln→DLeu→Asn→Thr→^{15}Sar

R-P-16-M

Zervamycin	*Emericellopsis salmosynnemata*	antibacterial pore forming	JA 27,274, 321(74) JACS 103, 6517(81)

ZERIA Ac→Trp→Ile→Glu→Iva→Val→Thr→Aib→Leu→Aib→^{10}4hPro
→Gln→Aib→4hPro→Aib→Pro→^{16}Pheol
ZERIB Ac→Trp→Val→Glu→Iva→Ile→Thr→Aib→Leu→Aib→^{10}4hPro
→Gln→Aib→4hPro→Aib→Pro→^{16}Pheol
ZERIB' Ac→Trp→Ile→Glu→Aib→Ile→Thr→Aib→Leu→Aib→^{10}4hPro
→Gln→Aib→4hPro→Aib→Pro→^{16}Pheol
ZERC Ac→Trp→Ile→Glu→Iva→Ile→Thr→Aib→Leu→Aib→^{10}4hPro
→Gln→Aib→4hPro→Aib→Pro→^{16}Pheol
ZERIIA Ac→Trp→Ile→Gln→Aib→Ile→Thr→Aib→Leu→Aib→^{10}4hPro
→Gln→Aib→4hPro→Aib→Pro→^{16}Pheol
ZERIIB Ac→Trp→Ile→Gln→Iva→Ile→Thr→Aib→Leu→Aib→^{10}4hPro
→Gln→Aib→4hPro→Aib→Pro→^{16}Pheol
ZERII1 Ac→Trp→Ile→Gln→Aib→Val→Thr→Aib→Leu→Aib→^{10}4hPro
→Gln→Aib→4hPro→Aib→Pro→^{16}Pheol
ZERII2 Ac→Trp→Ile→Gln→Aib→Ile→Thr→Aib→Val→Aib→^{10}4hPro
→Gln→Aib→4hPro→Aib→Pro→^{16}Pheol
ZERII3 Ac→Trp→Val→Gln→Aib→Ile→Thr→Aib→Leu→Aib→^{10}4hPro
→Gln→Aib→4hPro→Aib→Pro→^{16}Pheol
ZERII4 Ac→Trp→Ile→Gln→Iva→Val→Thr→Aib→Leu→Aib→^{10}4hPro
→Gln→Aib→4hPro→Aib→Pro→^{16}Pheol
ZERII5 Ac→Trp→Ile→Gln→Iva→Ile→Thr→Aib→Val→Aib→^{10}4hPro
→Gln→Aib→4hPro→Aib→Pro→^{16}Pheol

P-19-M

Lanthiopeptin Ro09-0198	***Streptoverticillium* No.1337-2** *Sv.griseoverticillatum*	AV(HSVI) immuno-potentiator	TL 29,4771(88) HCA 70,726(87)
Ancovenin	***Streptomyces sp.***	angiotensin conv.enzyme inhibitor	TL 26,665(85)
Duramycin			HCA 70,726(87)
Trichorzianine	*Trichoderma harzianum*	phyt.fungi	JACS 107,6011 (85),JMB 170, 795(83)

LAN

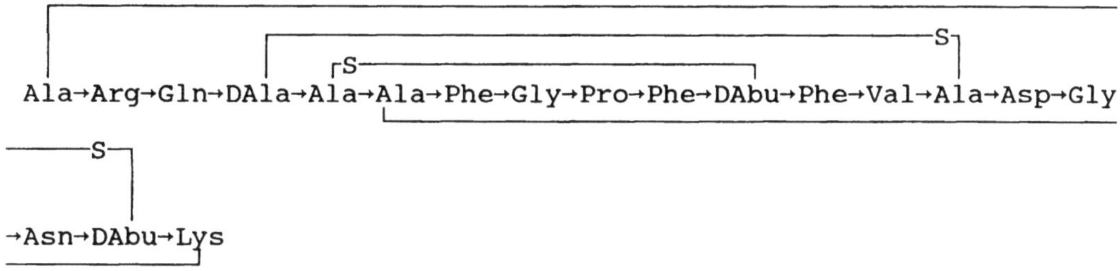

ANC

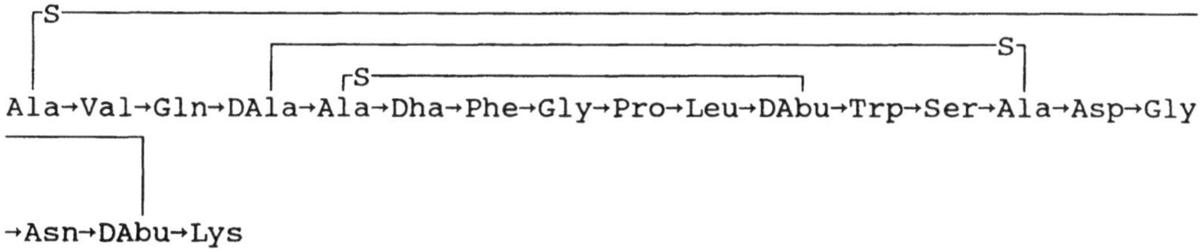

DUR

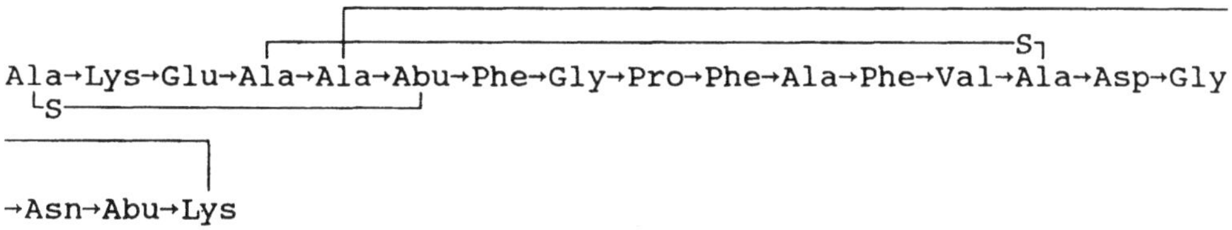

preLAN Cys→Arg→Gln→Ser→Cys→Ser→Phe→Gly→Pro→[10]Phe
→Thr→Phe→Val→Cys→Asp→Gly→Asn→Thr→[19]Lys
preANC Cys→Val→Gln→Ser→Cys→Thr→Phe→Gly→Pro→[10]Leu
→Thr→Trp→Ser→Cys→Asp→Gly→Asn→Thr→[19]Lys
preDUR Cys→Lys→Glu→Ser→Ser→Thr→Phe→Gly→Pro→[10]Phe
→Val→Phe→Val→Cys→Asp→Gly→Asn→Thr→[19]Lys

TRI Ac→Aib→Ala→Ala→Aib→Aib→Gln→Aib→Aib→Aib→[10]Ser→
→Leu→Aib→Pro→Val→Aib→Ile→Gln→Gln→[19]Trpol

R-P-20-M

Alamethicin	*Trichoderma viride*	G+,G-,AT	JACS 99,8469 (77)
Hypelcin	*Hypocrea peltata*	G+,AF	Chem.Pharm.Bull. 32,1822,2870(84) JCS CC 1979,413
Suzukacillin	*Trichoderma viride*	G+,phyt. bact	JACS 99,8469 (77),Ann.1985, 1041
Gliodeliquescin	*Gliocladium deliquescens*	membrane active	Chromatogr.19 188(84)
Paracelsin	*Trichoderma resei*		Exp.40,1189(84)

AME Ac→Aib→Pro→Aib→Ala→Aib→Ala→Gln→Aib→Val→^{10}Aib→
→Gly→Leu→Aib→Pro→Val→Aib→Aib→Glu→Gln→^{20}Pheol

AMEB Ac→Aib→Pro→Aib→Ala→Aib→Aib→Gln→Aib→Val→^{10}Aib→
→Gly→Leu→Aib→Pro→Val→Aib→Aib→Glu→Gln→^{20}Pheol

HYPA1 Ac→Aib→Pro→Aib→Ala→Aib→Aib→Gln→Aib→Leu→^{10}Aib→
→Gly→Aib→Aib→Pro→Val→Aib→Aib→Gln→Gln→^{20}Leuol

HYPA2 Ac→Aib→Pro→Aib→Ala→Aib→Ala→Gln→Aib→Leu→^{10}Aib→
→Gly→Aib→Aib→Pro→Val→Aib→Aib→Gln→Gln→^{20}Leuol

HYPA4 Ac→Aib→Pro→Aib→Ala→Aib→Aib→Gln→Aib→Ile→^{10}Aib→
→Gly→Aib→Aib→Pro→Val→Aib→Aib→Gln→Gln→^{20}Leuol

SUZA1 Ac→Aib→Ala→Aib→Ala→Aib→Ala→Gln→Aib→Aib→^{10}Aib
→Gly→Leu→Aib→Pro→Val→Aib→Aib→Glu→Gln→^{20}Pheol

SUZA2 Ac→Aib→Ala→Aib→Ala→Aib→Aib→Gln→Aib→Aib→^{10}Aib→
→Gly→Leu→Aib→Pro→Val→Aib→Aib→Glu→Gln→^{20}Pheol

SUZA3 Ac→Aib→Ala→Aib→Ala→Aib→Ala→Gln→Aib→Leu→^{10}Aib→
→Gly→Leu→Aib→Pro→Val→Aib→Aib→Glu→Gln→^{20}Pheol

SUZA4 Ac→Aib→Ala→Aib→Ala→Aib→Ala→Gln→Aib→Aib→^{10}Aib→
→Gly→Leu→Aib→Pro→Val→Aib→DIva→Glu→Gln→^{20}Pheol

GLIA Ac→Aib→Ala→Aib→Ala→Aib→Ala→Gln→Aib→Val→^{10}Aib→
→Gly→Leu→Aiob→Pro→Val→Aib→Aib→Gln→^{20}Pheol

PARA Ac→Aib→Ala→Aib→Ala→Aib→Ala→Gln→Aib→Val→^{10}Aib→
→Gly→Aib→Aiob→Pro→Val→Aib→Aib→Gln→^{20}Pheol

PARB Ac→Aib→Ala→Aib→Ala→Aib→Ala→Gln→Aib→Leu→^{10}Aib→
→Gly→Leu→Aiob→Pro→Val→Aib→Aib→Gln→^{20}Pheol

PARC Ac→Aib→Ala→Aib→Ala→Aib→Aib→Gln→Aib→Val→^{10}Aib→
→Gly→Leu→Aiob→Pro→Val→Aib→Aib→Gln→^{20}Pheol

PARD Ac→Aib→Ala→Aib→Ala→Aib→Aib→Gln→Aib→Leu→^{10}Aib→
→Gly→Leu→Aiob→Pro→Val→Aib→Aib→Gln→^{20}Pheol

P-21-M

Epidermin	***Staphylococcus epidermidis***	G+, acne bacteria	EJB 160,9(86)
Gallidermin	***Staphylococcus gallinarium***	propioni-bacteria	EJB 177,53(88)

EPI

Ile→Ala→DAla→Lys→Phe→Ile→Ala→DAbu→Pro→Gly→Ala→Ala→Lys→dAbu→Gly→

→DAla→Phe→Asn→DAla→Tyr→Ala→Ala(-CO_2)

GAL

Ile→Ala→DAla→Lys→Phe→Leu→Ala→DAbu→Pro→Gly→Ala→Ala→Lys→dAbu→Gly→

→DAla→Phe→Asn→DAla→Tyr→Ala→Ala(-CO_2)

P-32-M

Subtilin	***Bacillus subtilis***	G+ food preser-vative	BBRC 50,559(73) JBC 263,9508 (88)

SUB

Trp→Lys→DAla→Glu→Dha→Leu→Ala→DAbu→Pro→Gly→Ala→Val→DAbu→Gly→Ala→Leu

→Gln→Dhb→Ala→Phe→Leu→Gln→DAbu→Leu→DAbu→Ala→Asn→Ala→Lys→Ile→Dha→Lys

preSUB Trp→Lys→Ser→Glu→Ser→Leu→Cys→Thr→Pro→^{10}Gly
→Cys→Val→Thr→Gly→Ala→Leu→Gln→Thr→Cys→^{20}Phe
→Leu→Gln→Thr→Leu→Thr→Cys→Asn→Cys→Lys→^{30}Ile
→Ser→^{32}Lys

P-33-M

Subtilosin	***Bacillus subtilis* *168***	G+	JB 98,585(85)

SUO

```
                                        ┌──────SS──────┐
Gly→Leu→Gly→Leu→Trp→Gly→Asn→Lys→Gly→Cys→Ala→Thr→Cys→Ser→Ile→Gly→
└──────────────────────────────────────────────────────────────┐
Ala→Ala→Cys→Leu→Val→Asp→Gly→Pro→Ile→Pro→Asp→Glx→Ile→Ala→Gly→Ala-X
        └──────────────────Xu──────────────┘
```

preSUO Gly→Leu→Gly→Leu→Trp→Gly→Asn→Lys→Gly→^{10}Cys
→Ala→Thr→Cys→Ser→Ile→Gly→Ala→Ala→Cys→^{20}Leu
→Val→Asp→Gly→Pro→Ile→Pro→Asp→Glx→Ile→^{30}Ala
→Gly→Ala-^{33}X

P-34-M

Nisin	*Streptococcus lactis*	G+,My food preservative	Bull.Chem.Soc. 56,2044(83)
Pep-5	*Staphylococcus* **epidermidis**	G+,membrane disruption	AAC 27,836, 841(85)

NIS

```
          ┌────────────────S┐    ┌──────────S┐      ┌──────────
Ile→Dhb→DAla→Ile→Dha→Leu→Ala→DAbu→Pro→Gly→Ala→Lys→DAbu→Gly→Ala→Leu

──────────S┐                 ┌───────────S┐
→Met→Gly→Ala→Asn→Met→Lys→DAbu→Ala→DAbu→Ala→His→Ala→Ser→Ile→His→Val
                                   └────────S┘
→Dha→Lys
```

PE5

```
                                ┌─────────────────────────S┐
2oC4→Ala→Gly→Pro→Ala→Ile→Arg→*Ala→Ala→Val→Lys→Gln→Ala→Glu→Lys→Dhb→
                              ┌──────────S┐
→Leu→Lys→Ala→Dhb→Arg→Leu→Phe→Abu→Val→Ala→Ala→Lys→Gly→Cys→Asn→Gly→
                                      └──────────────────────────
→Ala→Lys
-S┘
```

preNIS Ile→Thr→Ser→Ile→Ser→Leu→Cys→Thr→Pro→^{10}Gly
→Cys→Lys→Thr→Gly→Ala→Leu→Met→Gly→Cys→^{20}Asn
→Met→Lys→Thr→Ala→Thr→Cys→His→Cys→Ser→^{30}Ile
→His→Val→Ser→^{34}Lys

prePE5 Thr→Ala→Gly→Pro→Ala→Ile→Arg→Ala→Ser→^{10}Val
→Lys→Gln→Cys→Glu→Lys→Thr→Leu→Lys→Ala→^{20}Thr
→Arg→Leu→Phe→Thr→Val→Ser→Cys→Lys→Gly→^{30}Lys
→Asn→Gly→Cys→^{34}Lys

Index of tables

Cyclodipeptides		
C-2	GLI	Gliotoxin
C-2-M	CWT	Tryptophanyldehydrobutyrin
	SPD	Sporidesmolide
	VET	Verticillin
	CAI	Cairomycin
	BIC	Bicyclomycin
	CLF	Cyclo-leucyl-N-methyl-phenylalanine
	CFP	Cyclo-D-phenylalaninyl-proline
	HYA	Hyalodendrin
	ALB	Albonoursin
	ALB*m*	(3Z,6E)-1-N-Methyl-albonoursin
	FUM	Fumitremorgin
	VER	Verruculogen
	BEP	3-Benzyl-6-ethyl-3,6-bis-(methylthio)-piperazine-dione
	CPY	Cyclo-prolyl-tyrosine
	CPV	Cyclo-prolyl-valine
	PIZ	Piperazinedione
	CHA	Chaetocin
	CHA*m*	11,11-dihydroxy-chaetocin
	CHE	Chaetomin
branched cyclodipeptides		
R-4-P-C-2	FOR	Foroxymithine
R-11-P-C-2	AMP	Amphomycin
cyclotripeptides		
R-C-3	ASP	Aspochracin
P-C-E-3	ENT	Enterobactin
R-C-D-3	PYR	Pyridomycin
cyclotetrapeptides		
C-4	FR9	FR900261(WF 3161)
	HC	HC-toxines I,II
	CHL	Chlamydomycin
	CY	Cyl-1, Cyl-2
	TEN	Tentoxine
	DIH	dihydrotentoxine
branched cyclotetrapeptides		
7-P-C-4	BOT	Bottromycin A_2
tetrapeptidolactones		
	ALT	Alternarolide
	BEA	Beauverolide
	JAS	Jasplamide
branched tetrapeptidolactones		
	A-3	A-3302
tetrapeptidolactone		
D-4	ANG	Angolide
	SER	Serratamolide
Cyclopentapeptides		
C-5(β)	CYC	Cyclochlorotine
C-5-M	MA	Malformin

branched pentapeptides		
6-P-C-5(β)-M	CAP	Capreomycin
	LLB	LL-BM-547
	TUB	Tuberactinomycin
	VIO	Viomycin
pentapeptidolactone		
R-L-5	BRV	Brevigellin
R-10-P-L-5	IMA	Imacidine
$R-(5-L)_2$	ACT	Actinomycin
cyclohexapeptides		
C-6	ASP	Asperchrome
	FERc	Ferrichrome
	FERy	Ferrichrysine
	FERc	Ferricrocine
	FERo	Ferrirhodin
	FERu	Ferrirubin
	MAC	Malonichrome
	MAL	
	PHO	Phomopsin
	RA	RA-peptide
	BOU	Bouvardin
R-C-6		
	MUL	Mulundocandin
	ECH	Echinocandin
	S41	S-41062-F
	ACU	Aculeacin
hexapeptidolactones		
L-6	SF	SF-1902-A2
	GLO	Globomycin
	DES	Destruxin
	ISA	Isariin
	MON	Monamycin
	ROS	Roseotoxin
	AZI	Azinothricin
	A83	A83586C
R-L-6-M	DOR	Doricin
	OST	Ostreogrycin
	YAK	Yakusimycin
7-P-L-6	CYH	Cycloheptamycin
8-P-L-6	ESP	Esperin
cyclohexadepsipeptides		
n-D-L-6	DID	Didemnin
D-6	ENN	Enniatin
	BEV	Beauvericin
	NEU	Neurosporin
	SPO	Sporidesmolide
cycloheptapeptides		
C-7	ILA	Ilamycin
	PHA	Phallotoxin
	VIR	Viroisin
	RHI	Rhizoin
	ULI	Ulicyclamide
	CYA	Cyanoginosin
R-C-7	OCT	Octapeptin

branched heptapeptides		
R-8-P-C-7	EM4	EM-49
R-10-P-C-7	POL	Polymyxin
	COL	Colistin
	CIR	Circulin
12-P-C-7	BAC	Bacitracin
heptapeptidolactones		
L-7	A43	A-43-F
R-L-7	ETA	Etamycin
	NEO	Neoviridogrisein
R-9-P-L-7	VIS	Viscosin
R-14-P-L-7	STE	Stendomycin
cyclooctapeptides		
C-8-M	ASC	Ascidiacyclamide
	PAT	Patellamide
	AMA	Amatoxin
$C-(P-4)_2$	ULT	Ulithiacyclamide
C-8(β)	BAI	Bacillomycin
	ITU	Iturin
	MYC	Mycosubtilin
octapeptidolactones		
R-L-8	SUR	Surfactin
	APD	APD
	NOP	Neopeptin
	LIP	Lipopeptin
R-9-P-L-8	GRI	Grisellimycin
R-10-P-L-8	STO	Stenothricin
	MYC	Mycoplanecin
	PLI	Plipastatin
$(R-P-4)_2-L$	QUI	Quinomycin
	TRI	Triostin
	BBM	BBM-928A
octadepsipeptide		
D-8	BAS	Bassianolid
nonapeptidolactones		
L-9	EMP	Empedopeptin
	PER	Permetin
	BMY	BMY-28,117
	MAJ	Majusculamide
10-P-L-9	LLA	LL-AO-341
	HER	Herbicolin
	TEL	Telomycin
	NET	Neotelomycin
	LYS	Lysbactin
	KAT	Katanosin
cyclodecapeptides		
C-10	TY	Tyrocidine
	GSS	Gramicidine S
C-10-M	COR	Cortinarin
C-10(β)	POY	Polypeptin
	PER	Permetin
	BMY	BMY-28,160
	PUW	Puwainaphycin
cycloundecapeptides		
	EPD	Epidermin
	CIC	Cyclosporin
undecalactones		
R-L-11	BRV	Brevistin
R-13-P-L-11	THE	Theonellapeptolide

higher cycles		
C-12 = c-(P-6)$_2$	GRA	Gratisin
D-12 = c-(D-4)$_3$	VAL	Valinomycin
C-13(2γ)	MYB	Mycobacillin
R-P-17-C-16	END	Enduracydine
Cyclopeptid3e closed with piperidine-type ring		
R-P-13-C-10(PIP)-M	NOS	Nosiheptide
	THI	Thiocilline
	MIC	Micrococcin
R-P-17-C-10(PIP)-M	THO	Thiostrepton
	SIO	Siomycin
R-P-14-C-13(PIP)-M	BER	Berninamycin
linear Peptides		
dipeptides		
P-2	BCI	Bacilysin
	AHO	Alahopcin
	LLB	LL-BM-726
	ARG	Arginyl-glutamine
	CIA	Citryl-arginine
	CIO	Citryl-ornithine
	DPA	N-2,6-Diamino-6-hydroxy-methyl-pimelyl-alanine
	STR	Stravidine
	ARP	Arphamenine
	BES	Bestatine
	NIT	Nitropeptin
	OLI	Olivoreticulin
P-2-(β)	BAH	β-Alanyl-histidine (Carnosine)
	BAO	β-Alanyl-ornithine
	BAG	β-Aspartyl-glycine
	GAH	γ-Aminobutyryl-histidine
	GAO	γ-Aminobutyryl-ornithine
P-2-(γ)	GAH	γ-Glutamyl-2-amino-3(Z),5-hexadienoic acid
	VAB	(S,S)-Valyl-amidocyclobu-tane-1-acetic acid
R-P-2	BU2	Bu-2743
	WF1	WF-1029
	VNO	Vanoxinine
	IND	Indolactame
	PEN	Penicillins
glycodipeptides		
G-P-2	PAM	Phosporamidon

tripeptides		
P-3	PHI	Phosphinotricinylalanyl-alanine
P-3-M	PHA	Phosalicine
	PLU	Plumbemycin
	ATF	Arginyl-D-allothreonyl-phenylalanine
	ASP	Aspartyl-N-hydroxy-aspartyl-D-cycloserine
	FAS	Fastigiatine
	EIS	Eisenine
	RHC	Rhizocticine
	DIP	Diprotine
	K26	K-26
	K4	K-4
	I5B	I5B2
	LEN	Leucyl-3-epi-deoxy-negamycin
	LNE	Leucyl-negamycin
	PHM	L-N^5-Phosphono-methionine-S-sulfoximinyl-alanyl-alanine
	OF4	OF-4949
	BIP	Biphenomycin
	WS4	WS-43705-A
	K13	K-13
P-3-(β,γ)	MUR	Muracein
	FEL	Feldamycin
	ACV	$(ACV)_2$
	PEN	Penicillin N
	IPN	Isopenicillin N
	DAC	Deacetoxycephalosporin
	C17	C1778
	CEP	Cephalosporin
	JAN	Janulosamide
	GSH	Glutathione
	PPA	4'-Phosphopantetheine
	NOC	Nocardicine
	SUL	Sulfazecin
	PB5	PB-5266
G-3-P-M	MUR	Murcein A
R-3-P-(E)	DET	Detoxin
	WB3	WB-3559
tetrapeptides		
P-4	AMA	Amastatine
	PRO	Probestin
P-4-M	CEL	celenamide
	INT	Integerrin
	LAS	Lasoidin
R-P-4-M	SAB	SAB-711
	ZIZ	Zizyphine
R-P-4-(γ)	FK1	FK-156
G-4-P-M	SQ2	SQ 28,504
P-4-M-(E)	NOC	Nocobactin
	MYB	Mycobactin

P-4-M-(U)	ANT	Antipain
	ELA	Elastatinal
	CHA	Chymostatin
	MAP	MAPI
	FOR	Formestin
pentapeptides		
P-5-M	BM5	BMG59-R2
	EDE	Edeine
	PEP	Pepstatin
	AHP	Ahpatinine
	ALI	Althiomycin
R-P-5	ACTa→	Actinomycin,acidic
G-P-5-M	MUB	Muracein B
hexapeptides		
P-6-M	LAV	Lavendomycin
	S28	SQ 28,546
heptapeptides		
P-7-M	ANI	Antimycin
	K58	K-582
P-7-(γ)	CAD	Cadystin
G-P-7-M	VAN	Vancomycin
	A51	A-51,568
	M43	M43
	ORI	Orienticin
	OA7	OA-7653
	ERE	Eremomycin
	A82	A-82,846
	ACO	Actinoidin
	AVO	Avoparcin
	CLP	Chloropolysporin
	RIS	Ristocetin
	ACP	Actaplanin
	A35	A-35,512
	A41	A-41,030
	A47	A-47,934
	TA	Teicoplanine
	ARD	Ardacin
	A40	A-40,926
	KIB	Kibdelin
	PAR	Parvodicin
octapeptides		
R-P-8-M	LOP	Lophyrotomin
	STT	Stenothricin
	MOR	Moroidin
P-8-M(β,γ)	PHL	Phleomycin
	BLM	Bleomycin
	ZOR	Zorbamycin
	YA5	YA-56-X
nonapeptides		
P-9	TEP	Teprotide
R-P-9	TRI	Trichopolin
R-P-9-(β,γ)	TAL	Tallysomycin
peptides with 10 or more residues		
R-P-10	LEC	Leucinostatin
	CER	Cerexin
R-P-13	TRD	Tridecaptin
R-P-15-M	GRA	Gramicidine
	EME	Emericin
	DIS	Discodermin

P-19-M	LAN	Lanthiopeptin
	ANC	Ancovenin
	DUR	Duramycin
	TRO	Trichorzianine
R-P-20-M	AME	Alamethicine
	HYP	Hypelcin
	SUZ	Suzukacillin
	GLI	Gliodeliquescin
	PAR	Paracelsin
P-21-M	EPI	Epidermin
	GAL	Gallidermin
P-32-M	SUB	Subtilin
P-33-M	SUO	Subtilosin
P-34-M	NIS	Nisin
	PE5	Pep-5

List of abbreviations

a) Compounds

2,5h3m*d*Ile - 2,5-dihydroxy-3-methyl-dehydro-isoleucine
2'Cl3'5'hPhg - 2'-chloro-3',5'-dihydroxy-L-phenylglycine
2h3'5'ClTyr - 2-hydroxy-3',5'-dichloro-L-tyrosine
2h3'ClTyr - 2-hydroxy-3'-chloro-L-tyrosine(β-hydroxy-3-chloro-)
2h3'ClTyr(2Dman) - 2-mannosyl-3'chloro-L-tyrosine
2h3hHty - 2,3-dihydroxy-L-homotyrosine
2hBu - 2-hydroxy-butyric acid
2hHph - 2-hydroxy-homophenylalanine
2hHty - 2-hydroxy-L-homotyrosine
2hmSer - 2-hydroxy-2-methyl-L-serine
2hTrp - 2-hydroxy-L-tryptophane
2hTyr - 2-hydroxy-L-tyrosine (β-hydroxy-L-tyrosine)
2mBu - 2-methyl-butyric acid
2miAsp - β-methyl-iso-aspartic acid
2oBu - 2-oxo-butyric acid
3'5'h4'mPhg - 3',5'-dihydroxy-4'-methyl-L-phenylglycine
3'5'hPhg - 3',5'-dihydroxy-L-phenylglycine
3'5'hPhg(-man) - mannosyl-3',5'-hydroxy-L-phenylglycine
3'5'Phg(3'fuc) - 3'-fucsyl-5'-hydroxy-L-phenylglycine
3NO$_2$Glu - 3-nitro-glutamic acid
3'NO$_2$Tyr - 3'-nitro-L-tyrosine
3,4ClPro - 3,4-dichloro-L-proline
3,4*d*Pro - 3,4-dehydro-L-proline
3EtPro - 3-ethyl-L-proline
3h4mPro - 3-hydroxy-4-methyl-L-proline
3h4o5mPro - 3-hydroxy-4-oxo-5-methyl-L-proline
3h5mPro - 3-hydroxy-5-methyl-L-proline
3hAvl - 3-hydroxy-norvaline
3hGlu - 3-hydroxy-glutamic acid
3hPic - 3-hydroxy-picolinic acid
3mArg - 3-methyl-L-arginine
3mPhe - 3-methyl-L-phenylalanine
3mPro - 3-methyl-L-proline
4,5*d*Ahx - 4,5-dehydro-norleucine(=2-amino-hexanoic acid)
4,5hIle - (2S,3R,4R)-2-amino-3-methyl-4-dihydroxy-valeric acid
4,5hIle - 4,5-dihydroxy-L-isoleucine
4,5hLeu - 4,5-dihydroxy-L-leucine
4ClPro - 4-chloro-L-proline
4f3(CONHOH)Bu - 4-formyl-3-N-hydroxy-carboxamido-butyric acid
4h5hOrn - 4,5-dihydroxy-L-ornithine
4hβLys - 4-hydroxy-β-lysine
4hIle - 4-hydroxy-L-isoleucine
4hPia - 4-hydroxy-L-piperazic acid
4hThr - 4-hydroxy-L-threonine
4m3hA - 4-methyl-3-hydroxyanthranilic acid
4m3hB - 4-methyl-3-hydroxybenzoic acid
4m*d*Hiv - 4-methyl-2,3-dehydro-hydroxy-isovaleric acid
4m*d*Hiv -4-methyl-dehydro-hydroxyisovaleric acid
4mHiv - 4-methyl-hydroxyisovaleric acid
4o5hPip - 4-oxo-5-hydroxy-pipecolic acid
4o5mPro - 4-oxo-5-methyl-L-proline

4oPip - 4-oxo-L-pipecolic acid
4oPro - 4-oxo-L-proline
4SmAbu - 4-sulfomethyl-aminobutyric acid
5ClPig - 5-chloro-piperidyl-2-glycine
5hCap - 5-hydroxy-capreomycidine
5h*d*Tyr - 5'-hydroxy-2,3-dehydrotyrosine
5mPro - 5-methyl-proline
5OmTrp - 5-O-methyl-L-tryptophane
6'hTrp - 6'-hydroxy-L-tryptophane
6hmDpm - 6-hydroxy-6-methyl-diaminopimelic acid
Aad - aminoadipic acid
βAbu - β-amino-butyric acid
γAbu - γ-amino-butyric acid
Abu - aminobutyric acid
ACA - 7-β-aminocephalosporanic acid (7-ACA)
Acba - 1(S)-hydroxy-2(S)-aminocyclobutane-1-acetic acid
Acm - 2-amino-4(cyclohexa-2,5-dienylaminyl)-butyric acid
AcMeVal - N-acetyl-N-methyl-L-valin
Act - 2-amino-4,6-dimethylphenoxazinone-(3)-1,9-dicarboxylic acid
Adda.- 3-amino-9-methoxy-2,6,8-trimethyl-10-phenyldeca-4,6-dienoic
Ahmod - 2-amino-6-hydroxy-4-methyl-8-oxo-decanoic acid
Ahmpp - 4-amino-3-hydroxy-2-methyl-5-(3-pyridyl)-pentanoic acid
Ahp - 2-aminopeptanoic acid
Ahx - 2-aminohexanoic acid (norleucine, Nle)
Aib - α-amino-isobutyric acid
βAla - β-alanine
$AlaNMe_2$ - 3-Dimethylamino-L-alanine
*allo(cis)*4hPro - *allo(cis)*-4-hydroxy-L-proline
*allo(cis)*4mPro - *allo(cis)*-4-methyl-proline
Amp - 3-amino-2-methyl-pentanoic acid
Aoc - 3-amino-2-oxazetidine-4-carboxylic acid
Aod - 4-amino-3-oxo-2,2-dimethyl-pentanoic acid
Aoe - 2-amino-8-oxo-9,10-epoxidecanoic acid
APA - 6-amino-penicillanic acid (6-APA)
App - 2-amino-5-phenyl-pentancarboxylic acid
aThr - *allo*-L-threonine
Avl - 2-aminovaleric acid (norvaline, Nva)
Aze - azetidine-2-carboxylic acid
Bz- - benzyl
c - cyclo
Cap - capreomycidine, 2-(2-iminohexahydro-(4S)-pyrimidyl)-(S)-
*cis*5hPro - *cis*-5-hydroxy-L-proline
Cit - L-citrulline
$(CONH_2)$ - amido
Cpp - 3-(3-carboxypropyl)-2,3-dehydro-pyrrolin
Cya - cysteic acid
Cys - L-cysteine
Cys^* - L-cysteine with sulfoxide (only amatoxins)
$Cys(-CO_2)$ - cysteamine
D4ClPia - 4-chloro-D-piperazic acid
Da4hPro - D-allo-4-hydroxy-L-proline
DAad - D-aminoadipic acid
Dab - diaminobutyric acid
*d*Abu - 2,3-dehydro-aminobutyric acid

DAbu - D-aminobutyric acid
DaIle - D-*allo*-isoleucine
DaIle - D-*allo*-isoleucine
*d*Ala - 2,3-dehydro-alanine
DAla - D-alanine
D*allo*End.- α(R)-amino-β-4(R)-(2-iminoimidazolidinyl)-propionic
*d*Asn - 2,3-dehydro-L-asparagine
DAsn - D-asparagin
*d*Asp - 2,3-dehydro-L-aspartic acid
D*a*Thr - D-*allo*-threonine
DCya - D-cysteic acid
DDbu - D-diaminobutyric acid
*d*Dpr - 2,3-dehydro-diaminopropionic acid
*d*Dpr($CONH_2$) - 2,3-dehydro-diaminopropionic acid amide
D*erythro*Dbu - D-*erythro*-diaminobutyric acid
DGlu - G-glutamic acid
Dhb - dihydroxybenzoic acid
DHiv - D-hydroxyisovaleric acid
DiGlu - D-iso-glutamic acid
DiGlu - D-iso-glutamic acid(β-carboxyl involved in binding)
*d*Ile - 2,3-dehydro-isoleucine
*d*Ile - 2,3-dehydro-isoleucine
DIle - D-isoleucine
DLeu - D-leucine
DOmTrp - D-O-methyl-tryptophane
DOmTyr - D-O-methyl-tyrosine
DOrn - D-ornithine
*d*Phe - dehydro-phenylalanine
DPip - D-pipecolic acid
Dpm - diaminopimelic acid
Dpr - diaminopropionic acid
DPro - D-proline
DSer - D-serine
DthreoDbu - D-*threo*-diaminobutyric acid
DVal - D-valine
End - α(S)-amino-β-4(R)-(2-iminoimidazolidinyl)-propionic acid
*erythro*DIle - *erythro*-D-isoleucine
*erythro*D2miAsp - *erythro*-D-2-methyl-iso-aspartic acid
f - formyl
Fua - furanylalanine
fuc - fucose
fVal - formyl-L-valine
gal - galactose
Gla - glutaric acid
glu - glucose
Gly - glycine
hAsn - hydroxy-asparagine
Hib - hydroxy-isobutyric acid
Hip - hydroxyiosvalerylpropionate
His - L-histidine
Hiv - hydroxyisovaleric acid
Hle - homoleucine
hPhg - p-hydroxy-phenylglycine
hPhg(4'glu-4-*epi*-van) - 4'-glucosyl-4-epi-vancosamine)-L-phenyl-

hPhg(4'glu-aco) - 4'-glucosyl-acosamine-L-phenylglycine
hPhg(4'glu-oli) - 4'-glucosyl-olivose-L-phenylglycine
hPhg(4'glu-rha) - 4'-glucosyl-rhamnose-L-phenylglycine
hPhg(4'glu-ria) - 4'-glucosyl-ristosamine-L-phenylglycine
hPhg(4'glu-van) - 4'-glucosyl-vancosamine)-phenylglycine
hPhg(4'rha-Dglu) - 4'-rhamnosyl-D-glucose-L-phenylglycine
hPhg(Nh) - N-hydroxy-phenylglycine
HyPic - 3-hydroxypicolinic acid
Ile - L-isoleucine
Ist - Isostatine
Leu - L-leucine
βLys - β-lysine
Lys - L-lysine
man - mannose
M*d*Ala - N-methyl-2,3-dehydro-alanine
M*d*Ala - N-methyl-dehydro-alanine
Me2hOmTyr - N,O -dimethyl-2-hydroxy-L-tyrosine
Me3Cl5hTyr - N-methyl-3-chloro-5-hydroxy-L-tyrosine
Me3hTyr - N-methyl-3-hydroxy-L-tyrosine
Me_3Leu - N,N,N-trimethyl-L-leucine
MeβArg - N-methyl-β-arginine
MeAla - N-methyl-L-alanine
MeBrTrp - methyl-bromo-L-tryptophane
Me*d*Ala - N-methyl-dehydro-alanine
MeDleu - N-methyl-D-leucine
Me*d*Phe - N-methyl-2,3-dehydro-phenylalanine
MeDPhe - N-methyl-D-phenylalanine
MeDSer - N-methyl-D-serine
MeDVal - N-methyl-D-valine
MeFua - N-methyl-furanoyl-L-alanine
MeOmTyr - N,O-dimethyl-L-tyrosine
MePhe - N-methyl-L-phenylalanine
MePhe(NHme) - N-methyl-p-amino-methyl-L-phenylalanine
MePhe($pNMe_2$) - N-methyl-p-amino-dimethyl-L-phenylalanine
MeSer - N-methyl-L-serine
MeThr - N-methyl-L-threonine
MeTrp - N-methyl-L-tryptophane
MeTyr - N-methyl-L-tyrosine
MeVal - N-methyl-L-valine
mPda - (2S)-N^1-methyl-propane-1,2-diamine
N2mLeu - N-dimethyl-L-leucine
N^6h*cis*hipOrn - N^6-hydroxy-N^6-cis-hydroxyisoprenyl-L-ornthine
N^6hipOrn - N^6-hydroxyisoprenyl-L-ornithine
N^6hmalOrn - N^6hydroxy-N^6-malonyl-L-ornithine
N^6hOAcOrn - N^6-hydroxy-N^6-acetyl-L-ornithine
N^6h*trans*hipOrn - N^6 -hydroxy-N^6-trans-hydroxyisoprenyl-L-ornithine
NAcMur - N-acetyl-muraminic acid
NfhOrn - N^6-formyl-N^6-hydroxy-L-ornithine
NhAla - N-hydroxy-L-alanine
NhOrn - N^6-hydroxy-L-ornithine
NhOmSer - N-hydroxy-O-methyl-L-serine
OAc - acetyl
oli - olivose
OPr - propyl

Orn - L-ornithine
Paa - pantoic acid
Pda - propane-1,2-diamine
βPhe - β-phenylalanine
Phe - L-phenylalanine
Phg - L-phenylglycine
Pia - piperazic acid = hexahydropyridazine-3-carboxylic acid
Pip - L-pipecolic acid
Qaa - quinaldinic acid
Qoa - quinazol-4-one-3-acetic acid
Qoh - 6-O-methyl-3-hydroxy-quinoline-2-carboxylic acid
Qxa - quinoxaline-2-carboxylic acid
rha - rhamnose
ria - ristosamine
Sar - sarcosine, N-methyl-glycine
Ser - L-serine
Ste - stendomycidine: 2-imino-hexahydro-4-pyrimidyl-glycine
Thpc - 2,3,4,5-tetrahydro-pyridazine-3-carboxylic acid
Thpc - 2,3,4,5-tetrahydropyridazine-3-carboxylic acid
Thr - L-threonine
*threo*mAsn - *threo*-methyl-L-asparagine
*threo*3mAsp - *threo*-3-methyl-L-aspartic acid
*threo*hGln - *threo*-hydroxy-L-glutamine
*threo*hGlu - *threo*-hydroxy-L-glutamic acid
*threo*mAsn - *threo*-methyl-L-asparagin
tLeu - L-*tert*-leucine
*trans*3hPro - *trans*-3-hydroxy-l-proline
*trans*4hPip - *trans*-4-hydroxy-L-pipecolic acid
*trans*4hPro - *trans*-4-hydroxy-L-proline
*trans*4mPro - *trans*-4-methyl-L-proline
Trp - L-tryptophane
Trp(Nip)- L-tryptophane(N-isopentenyl)
Trpol - L-tryptophanol
βTyr - β-tyrosine
van - vancosamine

b) Sources (Common short versions deviating from CABS-standard)

AAC : Antimicrob. Agents Chemother.
ABC : Agr. Biol. Chem.
Ann. : Liebigs Ann.
Ant. : Antibiotiki
B. : Biochem.
BBRC : Biochem. Biophys. Res. Comm.
BJ : Biochem. J.
Chromatogr. : Chromtographia
EJB : Eur. J. Biochem.
Exp. : Experientia
HCA : Helv. Chem. Acta
IJPPR : Int. J. Pept. Prot. Res.
J. Bac. : J. Bacteriol.
JA : J. Antibiot.
JACS : J. Am. Chem. Soc.
JB : J. Biochem.(Tokyo)JBC : J. Biol. Chem.
JCS CC : J. Chem. Soc. Chem. Comm.
JCS Perkin I : J. Chem. Soc. Perkin Trans. I
JOC : J. Org. Chem.
Phyt. : Phytochem.
Sci : Science
T : Tetrahedron
TL : Tetrahedron Lett.

Books:
Biotechnology: eds. H.-J. Rehm and G.Reed, Vol.4: Microbial Products II (Vol.eds. H.Pape and H.-J. Rehm), VCH Verlagsgesellschaft, Weinheim 1986

Vandamme: Biotechnology of Industrial Antibiotics (ed. E.J.Vandamme), Marcel Dekker, New York and Basel 1984

List of Contributors

Takaaki Aoyagi, Ph. D.
Institute of Microbial Chemistry
3-14-23 Kamiosaki
Shinagawa-ku
Tokyo 141
Japan

Dr. Andreas Billich
Institut für Biochemie und Molekulare Biologie
Technische Universität Berlin
Franklinstr. 29
D-1000 Berlin 10
F.R.G.

Prof. Dr. Volkmar Braun
Mikrobiologie II
Universität Tübingen
Auf der Morgenstelle 28
D-7400 Tübingen
F.R.G.

Dr. Bruno Cavalleri
Merrel Dow Research Institute
Lepetit Research Center
Via R. Lepetit 34
I-21040 Gerenzano (Varese)
Italy

Hans G. Fliri, Ph. D.
Preclinical Research
Sandoz AG
CH-4002 Basel
Switzerland

Setsuo Harada, Ph. D.
Central Research Division
Takeda Chemical Industries, Ltd.
17-85 Jusohonmachi 2
Yodogawa-ku
Osaka 532
Japan

Prof. Nobuo Izumiya, Ph. D.
Laboratory of Organic Synthetic Chemistry
Department of Industrial Chemistry
Faculty of Engineering
Kyushu Sangyo University
Higashi-ku
Fukuoka 813
Japan

Prof. Dr. Volker Kasche
AB Biotechnologie II
Technische Universität Hamburg-Harburg
Postfach 90 14 03
D-2100 Hamburg 90
F.R.G.

Prof. Dr. Horst Kleinkauf
Institut für Biochemie und Molekulare Biologie
Technische Universität Berlin
Franklinstr. 29
D-1000 Berlin 10
F.R.G.

Dr. Günther Michaelis
AB Biotechnologie II
Technische Universität Hamburg-Harburg
Postfach 90 14 03
D-2100 Hamburg 90
F.R.G.

Yasuhiko Muraoka, Ph. D.
Microbial Chemistry Research Foundation
Institute of Microbial Chemistry
14-23 Kamiosaki 3-chome
Shinagawa-ku
Tokyo 141
Japan

Dr. Giancarlo C. Lancini
Merrel Dow Research Institute
Lepetit Research Center
Via R. Lepetit 34
I-21040 Gerenzano (Varese)
Italy

Yasushi Okumura, Ph. D.
Laboratory of Applied Microbiology
Central Research Laboratories
Asahi Breweries, Ltd.
2-13-1 Omori-Kita, Ota-Ku
Tokyo 143
Japan

Hideo Ono, Ph. D.
Central Research Division
Takeda Chemical Industries, Ltd.
17-85 Jusohonmachi 2
Yodogawa-ku
Osaka 532
Japan

Tomohisa Takita, Ph. D.
Microbial Chemistry Research Foundation
Institute of Microbial Chemistry
14-23 Kamiosaki 3-chome
Shinagawa-ku
Tokyo 141
Japan

Prof. Dr. Ir. Erick J. Vandamme
Laboratory of General and Industrial Microbiology
Faculty of Agricultural Services
State University of Gent
Gent
Belgium

Dr. Joachim Vater
Institut für Biochemie und Molekulare Biologie
Technische Universität Berlin
Franklinstr. 29
D-1000 Berlin 10
F.R.G.

Dr. Hans von Döhren
Institut für Biochemie und Molekulare Biologie
Technische Universität Berlin
Franklinstr. 29
D-1000 Berlin 10
F.R.G.

Prof. Michinori Waki, Ph. D.
Laboratory of Biochemistry
Department of Chemistry
Faculty of Science
Kyushu University
Higashi-ku
Fukuoka 812
Japan

Prof. Jonathan D. Walton, Ph. D.
Department of Energy Plant Research Laboratory
Michigan State University
East Lansing
Michigan 48824
U.S.A.

Dr. Roland M. Wenger
Preclinical Research
Sandoz AG
CH-4002 Basel
Switzerland

Dr. Rainer Zocher
Institut für Biochemie und Molekulare Biologie
Technische Universität Berlin
Franklinstr. 29
D-1000 Berlin 10
F.R.G.

Index

(3Z,6E)-1-N-methyl-albonoursin 58, 415
11,11-Dihydroxy-chaetocin 416
1907 487
3-AMA 404
3-amino-2-piperidone 38, 45
3-benzyl-6-ethyl-3,6-bis(methylthio)piperazinedione 416
3-hydroxy-4-oxopipecolic acid 367
3-methyl-lanthionine 4
3-methyl-proline 368
4′-phospho-pantetheine 9 f., 473
4-chloro-proline 368
4-hydroxy-D-proline 367
4-hydroxypipecolic acid 367
4-oxopipecolic acid 367
5-methyl-proline 368
593A 416
6-amino penicillanic acid 88
6-APA 88, 392 ff.
7-ACA 397 ff.
7-ADCA 400 f.
7-formylamino-DCPC 139
7-ACA, production 397 ff.
A21978C 457
A22, 765 123
A30912 430
A3302-A 422
A35512 160, 164, 171, 482
A40926 160, 165, 171, 482
A41030 160, 164, 171, 482
A42867 160, 172
A43F 442
A47934 160, 164, 171, 482
A51568A 160 f., 482
A59-A 463
A82846 160, 162, 482
A83586C 434
AB-65 160
abr-B-protein 17
Acanthophora dendroides 467

Acetobacter 437
Acetobacter pasteurianus 132 f.
Acetobacter turbidans 132 f.
acetyl-transferase, N^6-hydroxy-lysine:acetyl-CoA 107 ff.
acetylation, of aerobactin 111
acetylesterase, cephalosporin 400
Acrostalagmus cinnabarinus 416
actaplanin 160, 164, 171, 482
actinoidin 160, 163, 482
actinoidin-type 166
Actinomadura luzonensis 58, 448
Actinomadura parvosata 160, 482
Actinomadura sp. 160, 470, 482
Actinomadura spiculosospora nov. 470
Actinomadura verrocosospora 2
Actinomyces 160, 482
Actinomyces albus 415
Actinomyces G455–101 448
Actinomycetales 161
Actinomycetes 329
Actinomycetes K26 470
actinomycin 9 f., 13 f., 58, 426
actinomycin synthetase 75 f.
actinonin 320 ff., 357
Actinoplanes missouriensis 160, 482
Actinoplanes sp. 160
Actinoplanes sp. No 41042 450
Actinoplanes teichomyceticus 160, 482
actinoxanthin 2 f.
activation 8 ff.
activation, of amino acids 40 f.
activation, specificity of 41 f.
ACV 7, 9 f., 12, 387, 473
ACV synthetase 7, 12, 17 ff.
acyl-agmatine hydrolase 295
acyl-cyclo-octapeptide 444
acyl-cyclo-undecalactone 458
acyl-decapeptide 487
acyl-enzyme 83 ff.
acyl-nonapeptide 487
acyl-octapeptide, modified 485
acyl-phophate 10 f.
acyl-transfer 81 ff.
acyl-tridepsipeptide 469 ff.
acyladenylates 9
acylase, penicillin 392 ff.
acylases, extracellular 401
acyldipeptide 465
adenylate formation 38 ff.
Aerobacter aerogenes 10, 104 ff., 113, 419
aerobactin 14, 103 ff., 111 ff.
aerobactin, biosynthesis of 111 ff.
affinity chromatography, gramicidin S synthetase 34 f.
Agrobacterium 402
Agrobacterium radiobacter 132 f., 403
AHMHA 316 f.
ahpatinine 479
alahopcin 465
alamethicin 7, 9 f., 492
albomycin 122 f.
albonoursin 415

alkaline phosphatase, inhibitor 329
α,α'-dipyridyl 372 ff.
α-amino-n-butyric acid 369
Alternaria 181
Alternaria alternata 10, 181, 194, 415, 420
Alternaria brassicae 183 f.
Alternaria citri 182
Alternaria fragariae 182
Alternaria kikuchiana 181
Alternaria longipes 182
Alternaria mali 59, 182, 420, 421
Alternaria tenuis 194, 420
alternarolide 182, 421
althiomycin 20, 479
AM-toxins 182 f., 421
AM374 160
amanin 444
amaninamide 444
Amanita sp. 438, 444
amanitine 246 f.
amanullin 444
amanullic acid 444
amastatin 320 ff., 331, 356
amatoxins 444
amidase, penicillin 94 ff.
amikacin 405
amino acids, effect on peptide production 368 ff.
amino acyl-tRNA 11
aminoacyl tRNA synthetase 83
aminoacyladenylates 9
aminoacylarginine 320 ff.
aminopeptidases 313 ff.
amoxycillin 392
amoxycillin, synthesis 396
amphomycin 418
ampicillin 392
ampicillin, acylase 396 f.
ampicillin, semisynthesis of 94 ff.
Amycolatopsis orientalis 172
AN-peptides 2
Anabaena cylindrica 10
Anabaena sp. 456
ancovenin 4, 491
angiotensin converting enzyme 313, 321, 349 ff.
angolide 422
antiamoebin 7
antibiotic genes, cloning 406
antibiotic, term 1
antibiotics, enzymatic synthesis 381 ff.
antibiotics, production 379 ff.
antibiotics, uses of 380
antimetabolite, glutamine 465
antipain 313 ff., 478
antrimycin 481
AOE 189
APD 447
ardacin 160, 165, 171, 482
arginyl-D-*allo*-threonyl-phenylalanine 470
arginyl-glutamine 465
aridicin 482
arphamenin 13, 320 ff., 326, 357, 465
Arthrobacter 398
ascidiacyclamide 444
asparenomycin 403
aspartic proteinases 312 ff.
aspartyl-N-hydroxy-aspartyl-D-cycloserine 470
asperchrome 428
Aspergillus 197, 428
Aspergillus 42355 430
Aspergillus caespitosus 415
Aspergillus fumigatum 415
Aspergillus nidulans 10, 17 ff., 430
Aspergillus nidulans-echinolatus 430
Aspergillus niger 423
Aspergillus NRRL 8113 430
Aspergillus ochraceus 59, 419, 428, 432
Aspergillus quadricinctus 10, 15, 197
Aspergillus rugulosis 430
Aspergillus rugulosus, var. roseus 430
Aspergillus sydowi 430
echinocandin 430
aspochracin 59, 419
ATP consumption 12
ATP, analogs of in amino acid activation 41
ATP-PP_i-exchange reaction 40 f.
attacins 3 f.
avoparcin 160, 163, 171, 482
azetidine-2-carboxylic acid 368, 370
azinothricine 58, 434
Azospirillum sp. 132 f.

B-cells 345 f.
B-cells, cyclosporine 277 ff.
bacillin 465
bacillomycin 446
Bacillus 180, 402
Bacillus badius 462
Bacillus brevis, gramicidin S negative mutants 49 f.
Bacillus brevis 8, 10, 13, 33, 206 f., 385, 415, 459, 460
Bacillus brevis ATCC 8185 13, 17, 454, 489
Bacillus brevis ATCC 9999 13, 384 f., 454
Bacillus brevis Nagano 49 f., 454
Bacillus brevis Vm4 479
Bacillus brevis Y-33 208
Bacillus cereus 450, 462, 470, 487
Bacillus circulans 396, 405 f., 440, 456
Bacillus circulans J725-093 468
Bacillus circulans TB-2125 469
Bacillus colistinus 440

Bacillus krzemieniewskii 456
Bacillus licheniformis 10, 440
Bacillus megaterium 396, 400 f., 462, 479
Bacillus mesentericus 436
Bacillus natto 16, 447
Bacillus polymyxa 440, 448
Bacillus polymyxacolistinus 440
Bacillus pumilus 465
Bacillus sphaericus 396
Bacillus subtilis 4 f., 10, 11, 21, 230 ff., 422, 440, 446, 447, 465, 493
Bacillus subtilis 168 4 f., 16 f., 21, 494
Bacillus subtilis ATCC 6633 4 f., 21
Bacillus subtilis B3 11
Bacillus subtilis F-29-3 21
Bacillus subtilis V3 460
Bacillus subtilis-niger 446
bacilysin 21, 465
bacitracin 8 f., 16, 440
bacitracin synthetase 381
bactenicin 3
Bacterium prodigiosum 422
bagougeramine 469
bassianolide 59, 449
BBM 928A-D 448
Beauveria bassiana 10, 59, 421, 437, 449
Beauveria nivea 8, 10, 458
beauvericin 9 f., 14, 59 ff., 437
beauvericin synthetase 72 ff.
beauvericin synthetase, methyltransferase 72 f.
beauverolide 421
benzylmalic acid 320 ff., 357
berninamycin 464
bestatin 320 ff., 331 ff., 356, 465
bestatin, analogs 323 ff.
bestatin, oscillations 346 ff.
bestatin, stereoisomers 346 ff.
β-alanyl-histidine 467
β-aspartyl-glycine 467
β-lactam antibiotics, acyl transfer 82
β-lactam precursor 7
β-lactam producer 13
β-lactamase, cephabacins 151
β-lactamase, induction of 134
β-lactamase, inhibition 134
β-lactamase, lactivicin 152
β-lactams, bacterial 131
β-lactams, bioconversions 392 ff.
β-lactams, distribution 21
β-lactams, semisynthesis of 94 ff.
β-lactams, structure determination 142 ff.
β-lactams, spectral characteristics 140 ff.
β-N-dimethyl-leucine 368
bialaphos 470
bioactive, term 1
biocatalysts, immobilized 379 ff.
bioconversion 381, 392 ff.
biological response modifier 325
biosynthesis, antibiotic genes 406
biosynthesis, arphamenins 326
biosynthesis, directed 13, 365 ff.
biosynthesis, general mechanism 411 ff.
biosynthesis, glycopeptides 167 ff.
biosynthesis, hybrid 380
biosynthesis, imino acid 375
biosynthesis, of enterochelin 113 ff.
biosynthesis, of HC-toxin 191
biosynthesis, of neoviridogrisein 365 ff.
biosynthesis, of penicillin G 386
biosynthesis, of siderophores 103 ff.
biosynthesis, of victorin 187
biosynthesis, siderophores 197 f.
biosynthesis, vancomycin group 167 ff.
Biphenomycin 470
Bipolaris 181
bleomycin 13, 246, 248, 289 ff., 486
bleomycin, analogs 293 ff., 305 f.
bleomycin, biosynthesis 292 ff., 300 ff.
bleomycin, chemical synthesis 297 ff.
bleomycin, discovery 289 ff.
bleomycin, semisynthesis 294 f.
bleomycin, synthetic analogs 306
bleomycinic acid 295
BMG59-R2 479
BMY-28,117 456
BMY-28,160 456
bottromycin A2 421
Bouvardia terniflora 428
bouvardin 428
brevigellin 425
brevistin 459
bromelin 312 ff.
bryamycin 463
Bu-2743E 468
butirosin 404 f.

C-1778 473
C-19393 139
cadystin 481
caeruleins 3 f.
cairomycin 415
capreomycin 424
carboxyl-activation 10 f.
carboxypeptidase II 90
carboxypeptidase Y 90, 93
carboxypeptidases 313 ff.
carnosin 10
cathepsins 312 ff.
cecropins 3 f.
celenamide 475
cell membrane, changes 318 ff.
cell sociology 318 ff.
cells, immobilized 390 ff.
central dogma 6

cephabacin 22, 132, 137, 138 ff., 144 ff., 403, 423, 473
cephabacin, biological activity 147 ff.
cephalexin 401 ff.
cephalexin, semisynthesis of 94 ff.
cephalosporin, bioconversions 397 ff.
cephalosporin, Dhb substitution 124
cephalosporin 21, 139, 473
cephalosporins, production 397 ff.
Cephalosporium 13, 17 ff.
Cephalosporium acremonium 17 ff., 388, 397, 473

Cephalosporium acremonium N2 473
Ceramium rubrum 467
cerexin 487
chaetocin 416
chaetomin 416
Chaetomium cochloides 416
Chaetomium minutum 416
Chaetomium uniporum 59 f.
chain length 413
chitinovorin 132 f., 403, 473
chlamydocin 188 ff., 246 f., 420
chloromethyl ketones 43 f.
chloropolysporin 160, 163, 482
Chromobacterium violaceum 13, 132, 465, 477, 480
chromophores 2
chymosin 312 ff.
chymostatin 313 ff., 320 ff., 357, 478
chymotrypsin 312 ff.
chymotrypsin, bovine 90 f., 93
cinnamycin 6
circulin 440
cirratiomycin 481
citrate, Fe^{3+}-transport 115
citrullyl-arginine 465
citrullyl-ornithine 465
Claviceps purpurea 13
clavulanic acid 139 f., 403
Cliona celata 475
clostripain 92
Cochliobolus carbonum 180, 187 ff, 195
Cochliobolus heterostrophus 186
Cochliobolus sacchari 180 ff.
Cochliobolus victoriae 180, 184 ff.
Col V-plasmid 105 ff.
colistin 440
collagenase 313 ff.
Comamonas sp. 398 f.
complementation, of mutant enzymes 50
condensation 83 ff.
condensation product, yield 87
conformations, gramicidin S 221 ff.
conformations, of cyclosporins 270 ff.
conformations, tyrocidine 221 ff.
coprogen 113
cortinarins 454
Cortinarius orellanoides 454
Cortinarius orellanus 454
Cortinarius speciosissimus 454
Corynebacterium kutscherii 470
Cryptosporiopsis ATCC 20594 430
cyanoginosin 23 ff., 438
cyclo-D-phenylalanyl-proline 415
cyclo-leucyl-N-phenylalanine 415
cyclo-leucyl-proline 415
cyclo-prolyl-tyrosine 416
cyclo-prolyl-valine 416
cyclochlorotine 423
cyclodecapeptide 454
cyclodecapeptide lactone 454
cycloheptamycin 58, 415, 436
cycloheptapeptide 438
cycloheptapeptide lactone 438
cyclohexapeptide 428
cyclohexapeptide lactone 428
cycloheximide 187, 193
cyclononalactone 452
cyclooctapeptide 444
cyclooctapeptide lactone 444
cyclopeptin 10
cyclopeptides 59
cyclosporine 8 f., 12 ff., 59, 458
cyclosporine synthetase 8, 12, 73 f.
cyclosporine, analogs 252 ff.
cyclosporine, biosynthesis 274 ff.
cyclosporine, chemical modification 266 ff.
cyclosporine, conformations 270 ff.
cyclosporine, isolation 250
cyclosporine, structure determination 250 ff.
cyclosporine, structure-activity 245 ff., 267 ff.
cyclosporine, synthesis 245 ff., 253 ff.
cyclosporine, mechanism of action 276 ff.
cyl-1 420
cyl-2 188 ff., 420
Cylindrocarpon lucidum 59, 458
Cylindrocladium scoparium 188, 420
cysteine proteinase 92, 312 ff.
Cystobacter fuscus 479
cytochrome c 20
Cytophaga 21 f., 453
Cytophaga group 20 ff.
Cytophaga johnsonae 132 f.
Cytophaga phytophila 22, 133 f.

D-alanine carboxypeptidase 134
D-alanine, in ribosomal peptide 3
D-alanyl-D-alanine peptides 161, 170
D-carbamoylase 403
D-hydantoinase 403
Dactylosporangium variesporum 424
danomycin 123

DCPC 139, 146
deacetoxy-cephalosporin synthase 17 ff.
deacetoxycephalosporin C 387 ff.
deacetoxycephalosporin C synthase 387 ff.
deacetylcephaosporin C 387 ff.
deacylation 83 ff.
decapeptides, linear 487
defensins 3 f.
deglyco-bleomycin 300 ff.
dehydroamino acids 4 ff.
demethyl-enniatin 64 f.
demyelination diseases 359
dermorphin 3
δ-(α-aminoadipyl)-cysteinyl-D-valine 7
desacetoxycephalosporin 473
design, of peptides 224 ff.
destruxin 10, 59, 183 f., 432
didemnin 25, 59
Diheterospora chlamydospora 188, 420
dihydro-tentoxine 420
diketopiperazines 59
dipeptide 465
dipeptidylaminopeptidases 313 ff.
diprotin 320 ff., 357, 470
directed biosynthesis 13
Discodermia 23
Discodermia kiinensis 59, 489
discodermin 23, 25, 59, 489
DNA strand breaks 2
DNA synthesis 318 ff.
dodeca-depsipeptide 458
dodecapeptide 458
Dolabella auricularia 23, 25
dolastatin 23, 25, 246, 248
domain structure 46
Doraromyces putredinis 468
doricin 434
DPhe-Pro-piperazinedione 38
duramycin 491
dystrophy 330 ff., 352 ff.
dystrophy, Duchenne 335 ff.
dystrophy, muscular 253 ff.

ebelactones 329, 356
echinomycin 13, 448
Ecklonia cava 470
ecological functions 180 f.
edeine 9 f., 479
Eisenia bicyclis 470
ektobiology 318 ff.
elastase 90, 312 ff.
elastatinal 313 ff., 478
electron microscopy, of gramicidin S
  synthetase 47 f.
EM 49 440
Emericellopsis microspora 7, 489
Emericellopsis poonensis 7
Emericellopsis salmosynnemata 7, 490
Emericellopsis synnematicola 7
emericin 7, 489
Empedobacter halobium 452
Empedobacter lactamgenus 22, 133 f.
empedopeptin 452
endopeptidases, dystrophy 330 ff.
enduracidine 461
enniatin 9 f., 12 ff., 59 ff., 196, 385, 437
enniatin synthetase 12, 385 f.
enniatin synthetase, methyltransferase 62 ff.
enniatin synthetase, monoclonal
  antibodies 70 ff.
enniatin synthetase, photoaffinity
  labelling 68 ff.
Enterobacter 104 ff.
enterobactin 14, 419
enterochelin 10, 14 f.,103 ff., 113 ff.
Enteromorpha linza 465
enzyme networks 330 ff.
enzymic oscillations 346 ff.
enzymic synthesis 12 ff.
epidermidin 458
epidermin 4 ff., 493
epithienamycin 139
eremomycin 160, 162, 482
ergot-peptides 13
Erwinia aroideae 396, 401
Erwinia carotovora 132 f.
Erwinia herbicola 453
Erwinia herbicolata 132 f.
Escherichia coli 10, 14, 113, 120, 123, 171,
  228 ff., 419
Escherichia coli K-12 104, 112, 115
Escherichia coli PG8 135
esperamicidin 2
esperin 436
esterase 87
esterase, cell surface 319
esterastin 328 ff., 331, 356 f.
etamycin 442
evolution 17 ff.
exopeptidase, inhibitors 320 ff.
exopeptidases 313 ff.
expandase 381, 387 ff.

factor A 381
Faena interjecta 160, 482
fastigiatine 470
fengymycin 21
fermentation, antibiotic 390 ff.
fermentation, of *B.brevis* 385
ferrichrome 10, 15, 111, 113, 123, 197, 428
ferrichrysine 428
ferricrocin 123, 428
ferrimycine 123
ferrioxamine 123

ferrirhodin 428
ferrirubin 428
ficin 312 ff.
FK-156 476
Flavobacterium 13, 17 ff., 21 f.
Flavobacterium chitinovorum 22, 132 f.
Flavobacterium sp. SC 12154 473
Flavobacterium sp. 132 f.
Flavobacterium sp. No 3559 474
Flexibacter 21 f., 404
Flexibacter alginoliquifaciens 22 f., 133 f.
Flexibacter sp. 132 f.
formadicin 22, 133, 136, 138 ff., 146, 403, 473
formadicin, biological activity 147 ff.
formestins 320 ff., 357, 478
formylamino group, configuration 146
foroxymithine 320 ff., 349 ff., 418
forphenicin 328 ff., 332 ff., 357
forphenicinol 332 ff., 357
FR 900261 420
fumitremorgin B 415
fungus ATCC 20529 442
Fusarium 59
Fusarium anguioides 295
Fusarium latentium 437
Fusarium lateritium 196
Fusarium orthoceras 437
Fusarium oxysporum 10, 59 ff., 196, 385, 437
Fusarium roseum 428
Fusarium sambucinum 437

gallidermin 6, 493
γ-DGlu-capsule 10
γ-Glu-Cys synthetase 7, 13
γ-Glutamyl-2-amino-3(Z),5-hexadienoic acid 467
γ-lactame 146 f.
γ-lactone 147
gastricisin 312 ff.
gene expression, iron transport 103 ff.
gene transfer 17 ff.
genes, aerobactin 108
genes, cloning 406
genes, enterobactin 115
genes, Fe^{2+} dicitrate transport 115
genes, β-lactam biosynthesis 17 ff.
genetic instability 21
Gliocladium deliquescens 416, 492
Gliocladium fimbriatum 415
gliodeliquescin 492
gliotoxin 13, 415
globomycin 58, 432
Gluconobacter sp. 132 f., 404, 473
glucuronidase 146
glutamine synthetase 470
glutaminyl cyclase 3
glutaryl-7-ACA acylase 381
glutathion 7, 10, 13, 34, 473
glutathion production 13
glycodipeptide 465
glycopeptide antibiotics 159 ff.
glycopeptide, synthesis 303 ff.
gougerotin 469
gramicidin 9 f., 489
gramicidin S 8 f., 12 ff., 16, 33 ff., 382 ff., 454
gramicidin S synthetase 33 ff., 381 ff.
gramicidin S synthetase, electron microscopy of 47 f.
gramicidin S synthetase, reaction mechanism 37 ff.
gramicidin S, analogs 212 ff.
gramicidin S, biological activity 205 ff.
gramicidin S, biosynthesis 37 ff.
gramicidin S, chemical synthesis 205 ff.
gramicidin S, congeners of 206 f.
gramicidin S, discovery 205 f.
gramicidin S, semi- 211
gramicidin S, structure-activity 212 ff.
Grateloupia livida 465
Grateloupia turutura 465
gratisin 208, 460
gratisin analogs 221 ff.
gratisin, structure-activity 212 ff.
grisein 122
grisellimycin 58, 450
GSH synthetase 7, 133

HABA acylase 381
HC-toxin 10, 188 ff., 195
HC-toxin I 420
HC-toxin II 420
Helminthosporium 180 ff.
Helminthosporium carbonum 10, 420
heptapeptide
heptapeptide, glycosylated 481 ff.
herbicolin 453
Herpes simplex 6
hexapeptide 480
histargin 320 ff., 357
Hitomyces sacchari 422
HMT-toxin 186
homeostasis, modification 348 ff.
hormones 3 f.
Humicola ATCC 20, 620 405 f.
hyalodendrin 415
Hyalodendron sp. 415
hydantoinase 393, 402 f.
hydrolase 82 ff.
hydrolysis rate 85 ff.
hydrolytic enzymes, diseases 330 ff.
hydroxamate siderophores 113
hydroxamate, biosynthesis 105 ff.
hydroxylase, prolyl 369
hydroxylation, proline 369 ff.

hydroxypepstatine 316
hypelcin 7, 492
hypertension 337 ff.
Hypocrea peltata 7, 492
Hypomycetes 59 f.

I5B2 470
Ilamycins 58, 438
Imacidine 425
immobilized cells 390 ff.
immune system, bestatine 324 ff.
immunological diseases 342 ff.
immunopotentiation 357
immunostimulator 6
infection, viral 318 ff.
information, flow of 6
inhibition, hydroxylase 371 ff.
inhibition, of N-methyltransferase 64 ff.
inhibitor, alkaline phosphatase 329
inhibitors, aminopeptidase 326
inhibitors, carboxypeptidase 327
inhibitors, cell surface enzymes 320 ff.
inhibitors, endopeptidases 313 ff.
inhibitors, membrane enzymes 328 ff.
inhibitors, protease 311 ff., 327
insulin, semisynthesis of 94
integerrin 475
iron proteins 104
iron transport 103 ff.
Isaria cretacea 432
isariin 432
isolation, of β-lactams 135 ff.
isolsulfazecin 143
isopenicillin N 387, 473
isopenicillin N epimerase 17 ff.
isopenicillin N synthase 13, 17 ff., 381, 387 f.
isosulfazecin 131 ff., 136, 138 ff.
isotentoxin 59
iturin 446
izupeptin 160

Janolus cristatus 470, 473
janulosamide 470, 473
jaspamide 421
Jaspis sp. 421
jasplakinolide 421

K4 470
K26 470
K-13 470
K-288 160
K-582 481
kallikrain 312 ff.
kanamycin A 404 f.
katanosin 21 f., 453
Keratinophyton terreum 470
kibdelin 160, 165, 482
Kibdelosporangium aridum 160, 168, 482
Kitasatoa 421
Kitasatospora phosalacinea 470
Klebsiella pneumoniae 113

L-N^5-phosphono-methionine-S-sulfoxyminyl-alanylalanine 470
lactivicin 22, 133, 138 ff., 146
lactivicin, biological activity 147 ff.
lanthionine 4
lanthiopeptin 6, 491
lantibiotics 4 ff.
Laportea moroides 485
lasoidin 475
lavendomycin 480
Leptinotarsa decemlineata 467
leucinostatin 7, 487
leucyl-3-epi-deoxy-negamycin 470
leucyl-negamycin 470
leucyl-tryptophan-N-phosphate 315
leupeptin 10, 313 ff., 332 ff., 356 f.
liblomycin 294
limited proteolysis, gramicidin S synthetase 46 f.
Lissoclinum patella 438, 444
LL-AO-341 453
LL-BM-547 424
LL-BM726 465
Lophyrotoma interrupta 485
lophyrotomin 485
luzopeptin 58
Lyngbya majuscula 23, 25, 59, 452, 465
lyngbyatoxin A 23, 25, 465
lysine N^6-oxidase 107 ff.
lysine N^6-oxidase, overproduction of 110
Lysobacter albus 22, 133 f.
Lysobacter lactamgenus 22, 133 f., 473
Lysobacter sp. 21 f., 453
lysobactin 9, 453

M43 160, 162, 482
macromomycin 2 f.
magainins 3 f.
majusculamide 59, 452
malformin 423
malonichrome 428
MAPI 478
marine organisms 21 ff.
melinacidin 416
membranes, activities 317 ff.
membranes, bacterial 230 ff.
membranes, model 230 ff.
metal proteinases 312 ff.
Metarhizium anisopliae 10, 59, 183 f., 432, 481
micrococcin 462
Micrococcus sp. 462
Microcystis aeruginosa 23, 25, 438

Micromonospora chalcea 465
Micromonospora echinospora 2
Micromonospora floridensis 58
Micromonospora halophytica, subsp. exilisia 470
mikamycin 434
mitochondrial respiration 20
modifications, side chain 413
monamycin 58, 432
monobactams 21, 131
Monocillium ATCC 20, 621 405 f.
moroidin 485
multhiomycin 462
multi-Arg-poly-(Asp) 10
multienzyme 6 ff., 14, 33, 46
multienzyme system 6 ff.
multienzyme, structure 46 ff.
multigene 6, 14
mulundocandin 430
muracein A 474
muracein B 480
murein 10
murein pentapeptide 21
murine muscular dystrophy 330 ff.
mutabilicin 463
mutants, aerobactin synthesis 105 ff.
mutasynthesis 380
mycobacillin 8 ff., 34, 460
Mycobacteria 329
Mycobacterium sp. 477
mycobactin 477
mycolic acid 329
mycoplanecins 58, 450
Mycoplasma 134
mycosubtilin 446
Myxobacteria 20 f.
myxothiazol 20
myxovalargin 20

N-(2,6-Diamino-6-hydroxymethyl-pimelyl)-alanine 465
N-acetylase 107 ff.
N-hydroxy-lysine 107 ff.
N-hydroxy-ornithine 107 ff.
N-methyl amino acids 60 ff.
N-methyl-transferase 61 ff.
N-methyl-valine 63
N-methylated peptide bonds 57 ff.
N-methyltransferase, in beauvericin synthetase 72 f.
N-methyltransferase, in cyclosporine synthetase 73 f.
N-methyltransferase, in enniatin synthetase 70 ff.
NCS-chrom 2
Neisseria gonorrhoeae 171
neocarzinostatin 2 f.
neopeptin 58, 447
neotelomycin 453
neoviridogrisein 14, 365 ff., 442
Neovossia indica 16
neuropeptides 3 ff., 81
Neurospora 428
Neurospora crassa 437
neurosporin 437
nisin 4 f., 494
nitropeptin 465
Nocardia asteroides 477
Nocardia lurida 160, 482
Nocardia orientalis 160, 474, 480, 482
Nocardia sp. 424
Nocardia uniformis 473
nocardicin 473
nocardicin, bioconversion 403
nocobactin 477
nonapeptide 487
nosiheptid 462
nucleoside dipeptide 465

OA-7653 160, 162, 482
octapeptide 485
octapeptin 440
OF 4949 470
oligonucleotides 83
oligosaccharides 83, 87
olivanic acid 403
olivoretin 465
Oospora destructer 432
orienticin 160, 162, 482
ostreogrycin 367, 434
ostreogrycin B 434

PA-42867 482
Paecilomyces fumosoroseus 59, 437
Paecilomyces lilacinus 7, 487
Paecilomyces marquandi 487
pantetheine 9
papain 92, 312 ff.
paracelsin 492
Paracoccus denitrificans 20
parvodicin 160, 165, 482
patellamide 25, 246, 248, 444
patricin 367
PB-5266 132, 144, 473
PB-5582-A 132
PC-toxin 192 f.
Pelveta fastigiata 470
eisenin 470
penicillin amidase 85, 94 ff.
penicillin binding proteins 153
penicillin G 392 f.
penicillin G acylase 381, 392 ff., 404
penicillin N 139 f., 473
penicillin V 392 ff.

penicillin V acylase 381, 392 ff., 401
Penicillium 17 ff., 181, 197, 428
Penicillium brevicompactum 425
Penicillium chrysogenum 17 ff., 386, 388 f., 397, 473
Penicillium cyclopoium 10
Penicillium islandicum 423
Penicillium lanosum 415
Penicillium rugulosum 470
Penicillium sp. 415
Penicillium verruculosum 415
pentapeptide 478 ff.
Pep-5 494
peplomycin 294
pepsin 312 ff.
pepstanone 316 ff.
pepstatin 315 ff., 331, 357, 479
peptide bond 9 f.
peptide bond, N-methylation 57 ff.
peptide bond, synthesis of 81 ff., 89 ff.
peptide bond, types 412
peptide hormone 3 f., 81
peptide synthesis, carboxyl group protection 94 f.
peptide synthesis, enzymatic, ionic strength 97
peptide synthesis, enzymatic, pH 95 f.
peptide synthesis, enzymatic, solvent composition 97
peptide synthesis, enzymatic, stereospecificity 95
peptide synthesis, enzymatic, temperature 96 f.
peptide synthesis, ribosomal 81
peptides, linear 465
peptidoglycan synthesis 169 ff.
peptidyl transfer 81 ff.
peptolides 57 ff.
Periconia circinata 192
permetin 456
Petriella guttulata 188, 420
phallotoxins 438
phaseolotoxin 179
phenylacetyl-cysteinyl-D-valine 389
phleomycin 486
phomopsin 428
Phomopsis leptostromiformis 428
phosalacine 470
phosphatase, alkaline 319
phosphinotricyl-alanylalanine 470
phosphopantetheine 9
phosphoramidon 315 ff., 357
phytotoxins 179 ff.
pipecolic acid 370
piperazinedione 38, 415 f.
Pithomyces chartarum 59
Pithomyces dartum 415, 437
Planomonospora parontospora 463
plasmin 312 ff.
plipastatin 328 ff., 450
plumbemycin 470
polyenzyme 33
polyketides 20
polymyxin 9 f., 16, 440
polypeptide chain, identification of 46
polypeptin 456
Polysporus sulphureus 59 f., 437
porin 171
pristinamycin 434
pro-peptides 4 ff.
Proactinomyces actinoides 160, 482
Proactinomyces fructiferi 160
proamanullin 444
probestin 320 ff., 475
processing, posttranslational 81
production, 6-APA 394 ff.
proliferation 318 ff.
proline, analogs 368 ff.
prolyl hydroxylase 369
Propionebacterium acne 4
prostatin 320 ff.
protease inhibitors 311 ff.
proteases, cellular functions 312
proteases, roles of 311 ff.
protection, carboxyl group 94
protein antibiotics 2
proteinase 87, 89 ff.
proteinase inhibitors 413
proteolysis, limited, of gramicidin S synthetase 46 f.
Proteus rettgeri 401
Proteus vulgaris 171
pseudobactin 15
Pseudomonas 15, 131 f., 402 ff., 473
Pseudomonas acdiophila 132 f.
Pseudomonas aeruginosa 15, 135, 171
Pseudomonas melanogenum 396
Pseudomonas putida 15, 398 f.
Pseudomonas schuylkillensis 403
Pseudomonas sp. 398 f.
Pseudomonas syringae 15, 179 f.
Pseudomonas viscosa 443
purification, of gramicidin S synthetase 34 f.
puwainaphycin 456
pyoverdine 15
pyridomycin 419
pyroglutamyl 2 f.

quinomycin 448
quinoxalines 58

RA 428
rate, of hydrolysis 85 ff.
regulation, blood pressure 337 ff.
regulation, Fe^{2+}-dicitrate transport 117 f.
regulation, iron transport 118 ff.

regulation, of gene expression 118 ff.
regulation, of receptor activity 120 ff.
renin 312 ff.
repressor protein, Fur 119
rhizocticin 21, 470
rhizoins 438
Rhizopus microsporus 438
Rhodotorula 197
Rhodotorula gracilis 398
rhodotorulic acid 113
rifamycin 405
ristocetin 160, 164, 171, 482
Ro09-0198 491
roseotoxin 432
Rubiae radix 428

S-41062-F 430
S-adenosyl-L-homocysteine 65 ff.
S-adenosyl-methionine 60 ff.
SAB-711 476
Saccharomonospora viride 160
Salmonella typhimurium 104 f., 419
Salmonella typhosa 228
Sarcophaga peregrina 10
sarcophagine 10
Sch 18640 463
Schizosaccharomyces pombe 481
screening, of β-lactams 134
secondary metabolites 16
semi-gramicidin S 211
semisynthesis, of ampicillin 94 ff.
semisynthesis, of cephalexin 94 ff.
seminisynthesis, of insulin 94
serine proteinases 312 ff.
serine proteinases, specificity 90 f., 93
serratamolide 422
Serratia marcescens 422
Serratia sp. 132 f.
SF-1902-A2 432
Shigella 105
Shigella flexneri 113
Shigella sonnei 113
sideromycins 103 ff., 196 ff.
siderophores 103 ff., 196 ff.
siderophores, synthetic 123 f.
sinefungin 65 ff.
siomycin 463
sporangiomycin 463
sporidesmolide 59, 415, 437
SQ 26, 180 132, 143
SQ 26, 445 132, 473
SQ 26, 700 132, 143
SQ 26, 812 143
SQ 26, 823 143
SQ 26, 875 143
SQ 26, 970 143
SQ 27, 860 132, 144
SQ 28, 332 132, 138 ff, 403
SQ 28, 504 477
SQ 28, 516 132, 473
SQ 28, 546 480
stability, of β-lactams 139
Staphylococcus aureus 123 f., 228, 231 f., 234
Staphylococcus epidermidis 6, 458, 493, 494
Staphylococcus gallinarium 493
staphylomycin 434
stendomycin 58, 443
stenothricin 58, 450, 485
stereospecificity, enzymatic peptide
  synthesis 95
stravidine 465
Streptoalloteichus hindustanus 486
Streptococcus lactis 4 f., 494
Streptomyces 2, 6, 17 ff., 20, 23, 58, 180, 465
Streptomyces abikoensis 424
Streptomyces albidofuscus 419
Streptomyces albogriseolus 418
Streptomyces albulus 465
Streptomyces albus 59
Streptomyces alcalophilus 473
Streptomyces althioticus 479
Streptomyces antibioticus 426, 434, 462
Streptomyces aureus 448
Streptomyces azureus 320, 463
Streptomyces bellus 479
Streptomyces berniensis 464
Streptomyces bikiniensis 486
Streptomyces bottropensins 421
Streptomyces californicus 424
Streptomyces candidus 160, 453, 482
Streptomyces canus 418, 453
Streptomyces capreolys 424
Streptomyces cattleya 21
Streptomyces chrysomallus 10, 426
Streptomyces cirratus 481
Streptomyces clavuligerus 18, 388, 391
Streptomyces coelicus 450
Streptomyces eburosporeus 160
Streptomyces echinatus 13, 448
Streptomyces endus 59, 443
Streptomyces flaveolus 448
Streptomyces flavochromogenes 448
Streptomyces floridae 424
Streptomyces fradiae 404
Streptomyces fulvissimus 2 460
Streptomyces fungicidus 461
Streptomyces gancidicus 415
Streptomyces gelaticus 416
Streptomyces goshikiensis 470
Streptomyces graminofaciens 434
Streptomyces griseolus 448
Streptomyces griseoluteus 416
Streptomyces griseoruber 478
Streptomyces griseorubiginosus 470

Streptomyces griseoverticillatus-tuberacticus S. 424
Streptomyces griseoviridus 365 ff., 442
Streptomyces griseoviridus P8648 365, 369
Streptomyces griseus 13, 442, 450
Streptomyces griseus-purpureus 424
Streptomyces halstedii 432
Streptomyces haranomachiensis 160
Streptomyces hawaiiensis 463
Streptomyces hygroscopicus 59, 160, 432, 478
Streptomyces hygroscopicus SF-1293 470
Streptomyces hygroscopicus, subsp. hiwasaensis 482
Streptomyces insubtus 59, 438
Streptomyces islandicus 59, 438
Streptomyces jamaicensis 59, 432
Streptomyces jumojinensis 18
Streptomyces kaizukaensis 476
Streptomyces kanamyceticus 404 f.
Streptomyces karantakensis 434
Streptomyces lactamdurans 388 f.
Streptomyces laurentii 463
Streptomyces lavendulae 448
Streptomyces lavendulae, subsp. brasilicus 480
Streptomyces lavendulus 442
Streptomyces lipmanii 18
Streptomyces loidensis 369, 434
Streptomyces matensis 479
Streptomyces mediocidicus 23
Streptomyces michiganensis 478
Streptomyces mitakaensis 434
Streptomyces mutabilis 463
Streptomyces nigrescens 478
Streptomyces nitrosporeus 418
Streptomyces noursei 415
Streptomyces olivaceogriseus 476
Streptomyces olivaceus 425, 434
Streptomyces olivoreticuli 424, 465
Streptomyces orientalis 159 ff., 482
Streptomyces ostreogriseus 434, 369
Streptomyces parvulus 426
Streptomyces pimprina 7
Streptomyces piralis 434
Streptomyces plumbeus 470
Streptomyces pristinae 434
Streptomyces puniceus 424
Streptomyces purpeofuscus 470
Streptomyces pyridomyceticus 419
Streptomyces roseosporus NRRL 11379 457
Streptomyces roseus 10
Streptomyces sioyaensis 463
Streptomyces sp. X1092 467
Streptomyces sp. 424, 434, 436, 442, 448, 460, 467, 470, 485, 491
Streptomyces sp. K-710 447
Streptomyces sp. P8648 442
Streptomyces sp. WK-142 479
Streptomyces spectabilis 415
Streptomyces tanashiensis 317, 416
Streptomyces tateyaensis 463
Streptomyces testaceus 317, 479
Streptomyces thermotoleranz 404
Streptomyces toyocaensis 160, 168, 482
Streptomyces triostinicus 13
Streptomyces tsusimaensis 16, 460
Streptomyces venezuelae 465
Streptomyces verticillus 13, 464, 289 ff., 486
Streptomyces vinaceus 424
Streptomyces violaceus 418, 476
Streptomyces virginiae 160, 369, 434, 482
Streptomyces viridochromogenes 470
Streptomyces xanthochromogenes 465
Streptomyces xanthocidicus 481
Streptomyces sp. AS-C-19 415
Streptomyces BM726 465
Streptomycetes 473
Streptoverticillium cinnamoneum 432
Streptoverticillium griseoverticillatum 6, 491
Streptoverticillium No. 1337-2 491
Streptoverticillium olivoreticuli 465
subtilin 4 ff., 21, 493
subtilosin 4 f., 21, 494
sulfazecin 131 ff., 136 ff., 143, 403 f., 473
sulfhydryl groups, determination 36, 44 f.
sulfide rings 4 ff.
surfactin 11, 447
suzukacillin 7, 492
synmonicin 160, 172
Synnemomyces mamnoori 160
synthesis, equilibrium controlled 82 ff.
synthesis, kinetically controlled 82 ff.
synthesis, of gramicidin S 208 ff.

T-cells 345 f.
T-cells, cyclosporine 277 ff.
tabtoxin 179
tallysomycin 486
TAN-458 132, 144
TAN-542 132
TAN-643 183 ff., 143
TAN-667 22, 133, 140, 146
TAN-850 132, 137, 138 ff., 144
teichomycin 482
teicoplanin 160, 165, 171, 482
teicoplanin aglycone 167
teicoplanin group 166
teleocidin 23, 25, 465
telomycin 453
tentoxin 10, 194 ff., 420
tetain 465
tetraglycylferrichrome 16
Theonella sp. 59, 459
theonellamine 59
theonellapeptolid 459

thermolysin 313 ff.
thiactin 463
thienamycin 403
thiocillin 462
thioesters, stability 38 ff.
thiol groups, gramicidin S synthetase 36
thiol groups, inhibitor design 43 f.
thiol groups, labelling of 43 ff.
thiopeptin 463
thioproline 368
thiostrepton 463
thiotemplate mechanism 8 f., 34, 60 f.
thrombin 312 ff.
TL-119 422
Tolypocladium inflatum 8, 10, 13, 59
Ton B box 121
transfer, of genes 17 ff.
transferase 82 ff.
transpeptidation 37 ff.
transport of intermediates 9 ff.
transthiolation 37 ff.
TRH 7
Trichoderma harzianum 491
Trichoderma polysporum 7, 458, 487
Trichoderma reesei 7, 492
Trichoderma viride 10, 13, 415, 492
Trichophyton indicum 470
trichopolyn 7, 487
trichorzianine 491
Trichothecium roseum 432
trichotoxin 7
tridecapeptide 458
tridecaptin 488
Trididerimnum sp. 59
Trigonopsis variabilis 398
triostin 13, 448
tripeptide 469 ff.
trypsin 90 f., 312 ff.
tryptophane dehydrobutyrine
piperazinedione 415
reverse transcriptase 415
tryptophane-dehydrobutyrine
diketopiperazine 58
tuberactinomycin 424
tylosin 404 f.
tyrocidine 8 f., 16, 287, 454
tyrocidine synthetase 16 f., 381
tyrocidine, analogs 219 ff.
tyrocidine, structure-activity 212 ff.

UDP-MurNAc-pentapeptide,
accumulation 169 ff.
ulicyclamide 438
ulithiacyclamid 444
Ulva lactuta 465
undeca-cyclopeptide 458
Ustilago 428
Ustilago maydis 197
Ustilago spaerogena 197

valinomycin 16, 468
(S,S)-Valyl-amidocyclobutane-1-acetic
acid 467
vancomycin 21, 246, 249, 482
vancomycin group 159 ff.
vancomycin, complex formation 170 f.
vancomycin, structure determination 161 ff.
vernamycin 367, 434
verruculogen 415
verticillin 246, 249, 415
Verticillium lecanii 59, 449
Verticillium sp. 415
Verticillium tererum 416
victorin 184 ff., 190
viomycin 424
virginiamycin 58, 367, 434
virginiamycin group 365
viridogrisein 13, 365 ff.
viroisin 438
viscosin 443

WB-3559 474
WF-3161 188 ff., 420
WF-1029 468
WS-43708A 470

Xanthomonas 21 f.
Xanthomonas citri 400
Xanthomonas lactamgena 22, 133 f., 473
Xenopus laevis 81

YA-56-X 486
yakusimycin 434
Yersinia 105
yield, of condensation product 87

zervamycin 7, 490
zizyphine 476
Zizyphus oenoplia 476
zorbamycin 486